D1677485

Malts and Malting

Malts and Malting

Dennis E. Briggs

Department of Biochemistry
University of Birmingham
Birmingham
UK

BLACKIE ACADEMIC & PROFESSIONAL

An Imprint of Chapman & Hall

London · Weinheim · New York · Tokyo · Melbourne · Madras

Published by Blackie Academic & Professional, an imprint of Thomson Science, 2–6 Boundary Row, London SE1 8HN, UK

Thomson Science, 2–6 Boundary Row, London SE1 8HN, UK

Thomson Science, 115 Fifth Avenue, New York, NY 10003, USA

Thomson Science, Suite 750, 400 Market Street, Philadelphia, PA 19106, USA

Thomson Science, Pappelallee 3, 69469 Weinheim, Germany

First edition 1998

Thomson Science is a division of International Thomson Publishing I(T)P®

Typeset in 10/12pt Times by Type Study, Scarborough

Printed in Great Britain by T. J. International Ltd, Padstow, Cornwall

ISBN 0 412 29800 7

A catalogue record for this book is available from the British Library

Library of Congress Catalog Card Number 97-73406

Printed on permanent acid-free text paper, manufactured in accordance with ANSI/NISO Z39.48–1992 and ANSI/NISO Z39.48–1984 (Permanence of Paper).

Contents

Preface

About 15 years have passed since the appearance of the 2nd edition of *Malting and Brewing Science*. Many changes have occurred in the intervening years, not least the early death of James S. Hough, the retirement of Roger Stevens and the closure of the Birmingham-based British School of Malting and Brewing. For a time it seemed impossible to undertake another revision of *Malting and Brewing Science*. Furthermore a need for a more extensive treatment of malts and the malting process was recognized. In consequence, this volume was prepared. It is the author's intention that the 'brewing science' will be covered by another book in the future.

Malting, one of the earliest 'biotechnologies', is a fascinating blend of pure and applied sciences (including plant and microbial biochemistry and physiology, chemistry, physics and engineering) with personal skills and judgements, all being used with industrial and commercial restraints and requirements in view. In a little more than one century, malting has evolved from an 'art' to a (nearly) routine production process. The differences between grain samples from different areas, grown in different years, and varietal differences as well as variations in users' requirements will never permit the production of malt to become a 'merely routine' process. The present book, which emphasizes principles rather than the details of malting, outlines the uses to which malt is put. Then the intention has been to describe the various aspects of malts and malting to give a sound technical introduction to understanding the various topics involved. A complete account of the history of malting has not been attempted, but some historical information is given both for its intrinsic interest and because the rich records of malting trials and practice cannot be understood without it. Deciding how much information to include has been difficult. For example, while the text includes some discussion of analytical methods, the details of these methods will have to be sought in the official, definitive publications, which are cited. As far as possible, measurements are given in metric units, with equivalents in imperial, American and traditional units, when this is appropriate. The equivalence – and lack of true equivalence – between some units is explained. The Appendix contains tabulations of units and some other useful data. The, apparently antiquarian, description of some older units is strictly necessary since, unless they are understood, maltsters cannot follow the older reports and literature. This constitutes an enormous store of useful data and observations. It is hoped that the limited number of references cited will be sufficient to give an entry into the more specialized literature. Inevitably many key contributions have not been individually acknowledged. Those who need to delve further should consult the collective indices of the major journals and the relevant computer-based data banks.

The inadequacy of the acknowledgements to individuals for help received is a source of embarrassment. I hope it will be forgiven when I explain that in the last 36 years I have always received friendly and generous help from any maltster I have approached with a reasonable request for assistance, and help was also nearly always given when the requests were less than reasonable. These helpful friends numbered over 100 when I gave up counting and they have been or are connected with most of the malting companies in the UK and with several overseas. I thank them all for making the preparation of this book an interesting and enjoyable activity. I hasten to add that any faults that may be present are mine.

I would like to thank four people without whose help the exercise would have failed: Mrs S. Maltby for her typing from my imperfect manuscripts; Mrs P. Hill for drawing and redrawing diagrams and Mrs M. Pass whose efforts in supplying information have been unstinting.

Above all I must thank Rosemary, my wife, for unfailing support and good humoured resignation as yet another morning was spent writing. I also thank those who have provided data or have permitted me to use, in original or modified forms, various diagrams and tables. These are acknowledged in the text.

Abbreviations

°G	SG-1000
°L	degrees Lintner (diastatic power; DP)
°W-K	degrees Windisch-Kolbach (diastatic power; DP)
70 °C viscosity	viscosity of wort from a small analytic mash, made at 70 °C
ASBC	American Society of Brewing Chemists
BOD	biological oxygen demand
brl	barrel (volume; see Appendix)
c. conc. mash	coarse ground grist in a concentrated analytical mash
c.g.	coarse grind
CIP	cleaning in place
coag.N	coagulable nitrogen
COD	chemical oxygen demand
cP	centipoise (unit of viscosity)
CWE	cold water extract
d.b.	dry basis
d.wt.	dry weight
DE	dextrose equivalent
DFP	diisopropyl fluorophosphate
d.m.	dry matter
DMS	dimethyl sulphide
DMS-P	dimethylsulphide precursor (SMM)
DMSO	dimethylsulphoxide
DP	diastatic power
DU	dextrinizing units (α-amylase)
EBC	European Brewery Convention
EBC unit	unit of colour
EDTA	ethylenediamine tetraacetic acid
F(app.%)	apparent percentage fermentability (attenuation)
F(real, %)	real percentage fermentability
f.-c. extract difference	difference in hot water extract yields obtained with finely and coarsely ground samples of malt
f.g.	fine-grind
f.wt.	fresh weight
FAN	free amino nitrogen

FE	fermentable extract
formol N	nitrogen fraction reacting with formaldehyde
GC	germinative capacity
GE	germinative energy
GKVs	germination and kilning vessels
GLC-FID	gas-liquid chromatography using a flame ionization detector
GLC-TEA	gas-liquid chromatography using a thermal energy analyser
h	h
hl	hectolitre
HMF	hydroxymethylfurfural
HMW-N	high-molecular-weight nitrogen
HPLC	high performance liquid chromatography
HWE	hot water extract
IoB	Institute of Brewing
KI	Kolbach index
lb	pounds weight or brewers' pound of extract
LOX	lipoxygenase
m.c.	moisture content
m/m	mass/mass
MAFF	Ministry of Agriculture, Fisheries and Food
MC	measurable cyanide
min	minute(s)
ML	malting loss
mPa/s	millipascal/s (units of viscosity)
MRL	maximum residue limit
MY	malt yield
N	nitrogen
NDMA	*N*-nitroso-dimethylamine
NIAB	National Institute of Agricultural Botany
NIR	near infrared spectroscopy
NMR	nuclear magnetic resonance
NPN	non-protein nitrogen
OG	original gravity (water = 1000)
p.p.b.	parts per billion
p.p.m.	parts per million (usually mg/l or mg/kg)
PG	present gravity (water = 1000)
PSN	permanently soluble nitrogen
PSY	predicted spirit (alcohol, ethanol) yield
PV	permanganate values for COD
PVPP	polyvinylpolypyrrolidone
Qr	quarter (various values; see Appendix)
QV	water-absorbing power (das Quellvermögen)

r(95)	repeatability value
R(95)	reproducibility value
RH	relative humidity
RQ	respiratory quotient
s	second(s)
SASPL	saturated ammonium sulphate precipitation limit
SG	specific gravity (water = 1000)
SGKVs	steeping, germination and kilning vessels
SGVs	steeping and germination vessels
SMM	*S*-methylmethionine
SNR	soluble nitrogen ratio
SOD	superoxide dismutase
t	tonne (1000 kg)
TCW	thousand corn weight
TN	total nitrogen
TSN	total soluble nitrogen
v/v	volume/volume
WS	water sensitivity

1 An introduction to malts and their uses

1.1 Introduction

Malting is the limited germination of cereal grains or, occasionally, the seeds of pulses (peas and beans), under controlled conditions. Sometimes malt is used 'green' (undried), but it is usually used after drying in the sun or in a current of warm air. Malts are used around the world in making foods and drinks. In the industrialized regions of the world, malt made from barley is by far the most important, but malts are, or have been, made from wheat, rye, oats, triticale, maize, sorghum, various millets and even rice. In Asia, pulses are widely germinated for use in foods. The word 'malt' is derived from Anglo-Saxon mealt or malt and perhaps has the same root as melt, referring to the softening that occurs during germination (Anglo-Saxon *metan*, to melt, to burn up, to dissolve), or perhaps to the word *malled* (mauled: broken or ground), as malts are milled before being used in brewing (cf. maul, mallet).

In this book, most attention is paid to barley malt. First, the malting processes are outlined to orientate the reader, then the main ways in which malts are used will be noted, with particular emphasis on making clear, European-style beers. The reason for this presentation is that brewers have been seeking high-quality malts, using scientific methods, for about 200 years; consequently most studies have been carried out with brewing quality in view. Frequently other industrial users have adopted brewers' methods for evaluating malts.

Malting is perhaps the oldest biotechnology. The cultivation of barley and wheat was probably beginning in the near-eastern fertile crescent about 10 000 BC (Briggs 1978). Wild grain must have been collected earlier still. The improvements in texture and flavour of foods prepared from grains following their accidental germination would soon have been noted and followed by deliberately sprouting grain (Singer *et al.*, 1954; Briggs, 1978). The method of making beer from bread must have been discovered soon afterwards. Malting and brewing are believed to have been practised for at least 6000 years. They are mentioned in some of the earliest written records (Singer *et al.*, 1954). In one such reference to malting, in a 'city' in Sumer, the Goddess Ninkasi was glorified as brewster to the Gods, who 'bakes the bappir-malt in the lofty kiln' (*ca.* 2500 BC: Kramer, 1963). Beer-making is often illustrated in ancient Egyptian tomb-paintings (Singer *et al.*, 1954; Wild, 1966). The Egyptians were still making and exporting beers (which were taxed) in Roman times. The Celts were brewing within the Empire, as were the Germanic tribes without (White, 1860; Arnold, 1911; Singer *et al.*, 1954). Beer, or more correctly ale, was well known to the peoples of the migration time, the Anglo-Saxons, Scandinavians, Teutons, Franks and so on. Indeed, its manufacture has never been discontinued. The old writings contain many references to ale, not all of a cheerful nature. In a mistranslation of the Icelandic original one reads that when the Scandinavian Viking king Ragnar Lodbrok was captured and thrown into a pit with poisonous snakes, his dying ode included, ‘soon, soon in the splendid abode of Odin shall we drink beer out of the skulls of our enemies’. (The original refers, less grimly, to horns and not skulls; B. Benedicz, private communication.) Unfortunately, archaeological remains of maltings of this period have not been recognized. No doubt they would be hard to distinguish from other farm buildings with associated corn-drying kilns (Scott, 1951; Nordland, 1969). It is questionable if the ancient ales would be readily appreciated now. Distilled alcoholic products were probably not known before about 1300 AD. One may suppose that the ancient Nubians, as neighbours of the Egyptians, knew how to make beer, and so the traditional opaque beers of the black tribes of Africa probably have very ancient origins also.

The making and selling of malts was often controlled. Thus, in Nürnberg in 1290 only barley was allowed to be malted, while in Augsberg between 1433 and 1550 beer was only to be made from malted oats (Smith, 1905). In England, malt carried a tax for many years (until 1880) and the consequent rigid controls and legalized harrying of maltsters caused a fossilizing of industrial practices. Maltsters devoted much effort to circumventing or breaking the laws (White, 1860; Stopes, 1885; Mathias, 1959; Brown 1983). A consequence was that British and mainland European practices diverged in malting and brewing.

British laws effectively compelled brewers to use only well-modified malts for mashing. If grain sprouted in the stack before it was threshed, this ‘providence malt’ escaped tax. No doubt many stacks were well watered in consequence. Malt sales were regulated very early on. In an act of Richard II, 1393, it was laid down that if malt was sold insufficiently cleaned the maltster had to make redress. If he refused he might be imprisoned or have his ear slit and be put in a pillory (White, 1860). This law no longer operates!

Malt is an intermediate and is used in making other products such as malt coffee, sweets, biscuits, malted bakery products, syrupy malt extracts, breakfast cereals, diastase, beers, whiskeys and malt vinegars. Consequently, there has been an extensive trade in malt over a long period and the qualities and values of malts have always been of interest. The earliest methods for assessing quality (smell, flavour, texture when chewed, appearance when cut or broken, and the ability, or not, to make a white mark on slate) were of limited value, as were determinations of bulk density. For making beer, and for many other purposes, the most important characteristic of malt is its ability to yield ‘extract’, of the correct quality and in the greatest possible amount. Extract is the solid material that dissolves when ground malt is mashed; that is, it is mixed with a certain proportion of warm water and is held for a specified time at closely regulated temperatures. The liquid containing the dissolved extract, 'wort' to brewers, has a greater density than water. The hydrometer, or saccharometer, was introduced into brewing to determine wort strengths centuries ago and was soon being used to evaluate malts. Thrale was testing extracts from malts with Baverstock's hydrometer in 1770 (Mathias, 1959). By 1806, a Royal Commission had established that, in general, malts from two-rowed barleys yielded more extract than those made from Scottish six-rowed barley, called bere or bigg (elsewhere called bear, beare, biggle; cf. modern Danish *byg* for barley; Anglo-Saxon *bere* and *berecorn* (barley corn) (Thomson *et al.*, 1806)). In 1898, at the Third International Congress of Applied Chemistry, convened in Vienna, agreement was reached on the 'Congress' analyses to be used in Germany and Austria. Standardized methods of analysis were introduced in North America in 1902–3 and, after much controversy, in the UK in 1906 (Hudson, 1960). At present there are too many 'agreed' systems of malt analysis, and some 'unofficial' methods are also used (Chapter 13).

Originally malting was carried out on a small scale. Within the 19th century, on some Scandinavian farms, grain was steeped by alternately immersing partly filled sacks in streams and draining, or by soaking in small wooden troughs. The grain was allowed to sprout, in a sack or in a trough, and was finally kiln-dried over a stove or in a 'dry' bath-house (Nordland, 1969). It is believed that malting was carried out on this scale in England for many years, but the need to avoid paying malt tax made these activities clandestine. Floor-malting was the method used for making larger quantities of malt and was regarded as traditional in the 17th century (Markham, Gervase, 1615). This method was in use all over Europe and was being used by the monks of St Gall, Switzerland in about 830 AD. At least by 1588, European settlers in North America were trying to make beer from malted maize (corn, Indian corn) (Brown, 1895). It was stated by a Mr Winthorp (1678) and frequently confirmed (e.g. Thomson, 1849) that those used to making malted barley made bad malt from maize because in the latter case malting was not complete until large roots and leaves had grown. Maize was 'malted' by growing it under 2 or 3 in (2.7–7.6 cm) of soil for 10–14 days. When the green shoots showed, the mat of tangled roots, shoots and grains was lifted, washed and sun-dried or kilned. The mellow, sweet and floury grain reportedly produced good beer, but 'beer made from bread was more usual' (see below).

The development of large breweries, particularly in the 19th century, led to the industrialization of malting and a progressive increase in the size of production units and of the batches of grain processed. This trend was strongly apparent in the UK by about 1880 and has continued to the present day. More recently, increased mechanization has dramatically increased the amount of malt a person can produce and has increased our understanding of grain physiology and biochemistry; in addition, improvements in technology have reduced the time it takes to convert barley into malt. Therefore, in 1880 the average quantity of malt produced, per person employed, probably just exceeded 100 tonnes in a malting season, which ran from about November to April. In 1985, the average value was between 500 and 600 t (tonnes), while in the most modern maltings the value is about 6500 t. Malt is now made the whole year round (Hudson, 1986a). In the 1830s, it was usual to steep 15–30 Qr (about 3–6.1 t) of barley in each batch, but smaller quantities were processed and some were as large as 50 Qr (7.6 t) (Brown, 1983) (Qr, quarter; see Appendix). By 1900 several UK maltings had batch sizes of 150–250 Qr (30.5–50.8 t), and buildings designed to process batches of 70–100 Qr (14.2–20.2 t) were built. In traditional floor-maltings each quarter of barley (0.2024 t) steeped needed a floor area of about 200 ft^2 (18.6 m^2). In 1850, malting floors were 'typically' 80–90 ft × 20–30 ft (24.4–27.4 m × 6.1–9.1 m). In 1882, a particular malthouse had floors 115 ft × 65 ft (35.1 m × 19.8 m) and at a house built in Ware in 1907 the floors were 137 ft × 47 ft (41.8 m × 14.3 m) (Brown, 1983). In a modernized and mechanized floor-malting in 1983, about 4000 Qr (608 t) were produced for each person employed, while in a mechanized box- or compartment-malting the value was 25 000 Qr (3800 t). Of course, not all the people in a maltings are involved in the actual production process (Brown, 1983). The rapid disappearance of old floor-malting buildings, leaving few records of their construction details, has triggered attempts to record and classify the types of building once used (Patrick, 1996).

In the 1950s in the UK, many boxes and drums were installed having capacities of 75–100 Qr. In the 1960s boxes of 375 Qr were built, while in the 1970s capacities reached 1000–1500 Qr (Brown, 1983). In the 1980s tower-maltings of about 200 t capacity, and the largest combined steeping–germinating–kilning vessel of 520 t capacity was completed (Chapters 9 and 10). In old floor-maltings, which lacked any means of controlling the air or water temperatures, it might take from 15 to 32 days to make a batch of malt from putting the grain into the steep to taking it from the kiln. In most modern maltings, the equivalent period is 5.8–7.5 days (Briggs, 1987a) (Table 15.1). At the present time (1997) there are still a few floor-maltings in the UK.

1.2 Malting in outline

Maltsters may be divided into sales-maltsters, contract-maltsters and user- (i.e. brewer- or distiller-) maltsters. These divisions are not necessarily firm; for example, a sales-maltster may contract to malt a particular batch of grain, and a user-maltster may sell malt. User-maltsters target their production for only one, or a few, customers and so frequently make only a relatively few types of malt. This contrasts with sales-maltsters who undertake to make many different types of product.

The stages of malting are usually given as steeping, germination and kilning. In reality, malting involves more than this and the divisions between these 'classical' stages are no longer clear-cut. Before the production process can begin, the maltster must obtain his barley, directly or indirectly, from the farmer and he must clean and store it ready for use. Even before this, the farmer must have been persuaded to grow acceptable varieties of grain, using conditions and techniques able to give samples that meet malting quality standards. In turn, plant breeders must have produced these varieties. Then, after the kilning process, malt must be 'dressed', or cleaned, and stored before being blended and dispatched to the user. The by-products of the process are collected and sold (Fig. 1.1)

Older barley varieties were inferior to those presently available. Breeders work to produce plants that yield more grain, have superior agronomic characteristics (such as improved resistance to diseases and to lodging) and produce grain of superior quality. It is simpler to breed for yield alone, rather than yield and malting quality. Maltsters cooperate with breeders in testing new varieties of barley for malting quality.

Farmers choose which variety of grain to grow with a view to maximizing their profits. UK malting barleys often yield less grain (tonnes per hectare) than 'feed' barleys, so more is paid for them (i.e. there is a price 'premium') to ensure that they are still grown. However, adverse weather or diseases may reduce the quality of a crop to such an extent that a batch of grain is not acceptable for malting. To produce barley of malting quality more care is needed than when it is to be used for animal-feed or for seed. Therefore, it must be an accepted malting variety, not appreciably 'contaminated' by admixture with other varieties and grown with the use of only permitted insecticides, herbicides and fungicides and with limited amounts of nitrogenous fertilizer. The harvested grain must have received minimal physical damage and be fully viable.

Maltsters may receive their grain directly from farms or from grain merchants who act as intermediaries. At every stage and during the delivery of the grain, it is tested in various ways. The grain is usually delivered to the maltings in bulk and is roughly pre-cleaned on its passage to the green grain (that is undried grain; it is *not* coloured green) store. As soon as possible, the grain is dried in a current of warm air, is thoroughly cleaned and graded and is carefully stored. Storage conditions are arranged so that grain deterioration is minimized and the grain undergoes 'post-harvest maturation'. In warmer and drier climates, south Australia for example, the grain leaves the fields dry, but it may need to be cooled during storage. Dust, thin and broken grains recovered in the cleaning process, together with rootlets

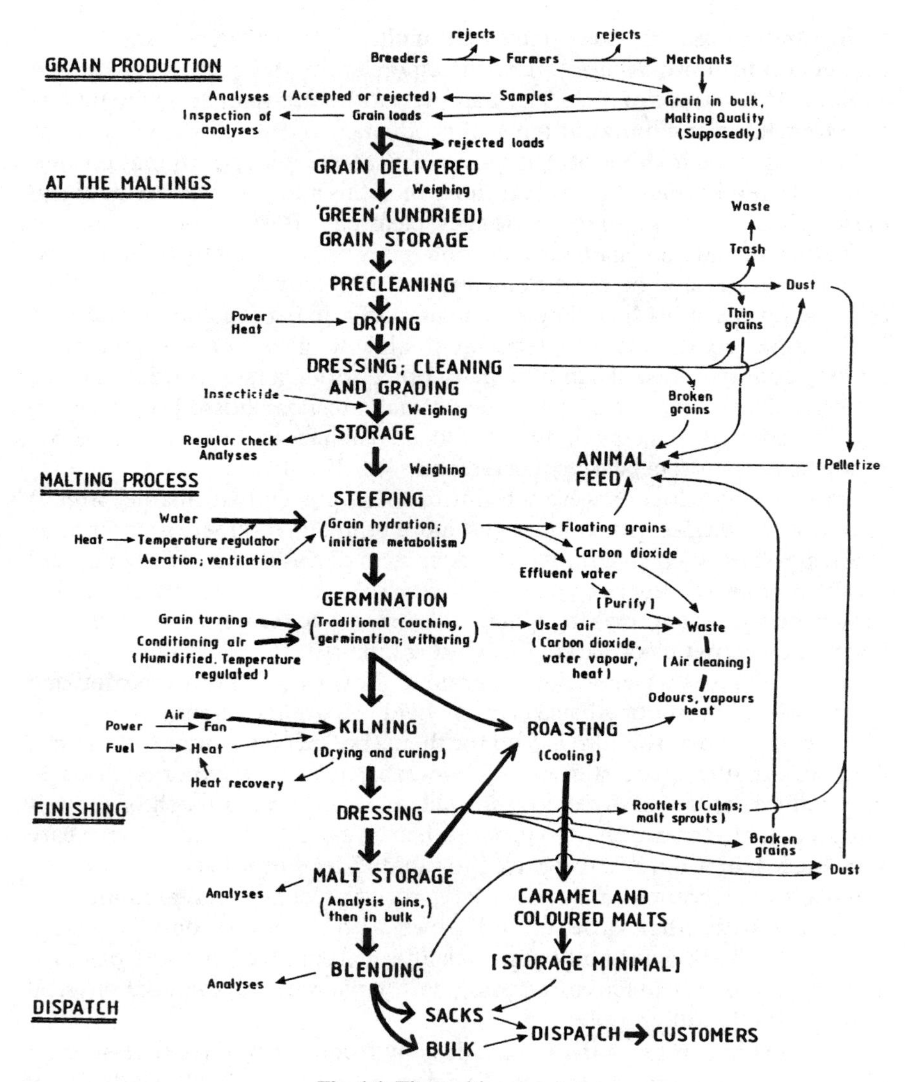

Fig. 1.1 The malting process.

and other by-products of the malting operations, are used in animal feedstuffs. Often these are compressed into pellets for ease of handling.

When the grain is sufficiently mature, that is dormancy has declined sufficiently for adequate germination to occur, it may be used. Because there are very many types of malting plant in use, the generalized account given here is not universally correct. Because malting has been carried out for so many centuries, many of the English terms and phrases that are used are ancient. Originally they were used in floor-maltings, but they are applied equally in the most modern maltings.

In steeping, grain is allowed to imbibe water to a controlled extent, e.g. 42–48%, usually by immersing it in water, but sometimes by spraying water over it. The temperature of the water may be controlled. The steep water becomes dirty and is changed at least once, to keep the grain fresh and 'sweet'. Each batch of grain being steeped is chosen to be of the quality appropriate for the kind of malt that is being made. During steeping, the grain swells and softens, while the living tissues resume their metabolism, which ceased as the grain ripened and dried. Sometimes air is blown through the grain–water mixture, aeration, or the grain may be 'air-rested', that is the water is drained away and air is sucked downwards through the grain. When the grain has reached the correct moisture content, the water is removed. Usually the grain is cast, that is it is transferred to a germination vessel, but in some plants steeping and germination and, less usually, kilning take place in one container. In traditional malting, the first visible signs of germination do not appear until after 'steep-out', but in most modern maltings the grain is 'further forward' and is chitting at steep-out. Sometimes other treatments, such as the application of a solution of gibberellic acid, may take place at this stage.

In malting, each batch of grain being processed is a 'piece'. In a floor-malting, each piece may be couched after steeping, that is it is heaped up to warm and begin germinating. Afterwards it is 'floored' by being spread more thinly. It grows and is regularly turned (disturbed and mixed) to prevent the rootlets matting and tangling together and to control the temperature in the grain bed. Sometimes germinating pieces may be sprinkled (water is sprayed onto them) but this is not a universal practice. In mechanical 'pneumatic' malting plants, the temperature of the grain is controlled by forcing a stream of damp air, at a selected temperature, through the bed of grain. In these plants, the deep layers of grain, often 3 to 8 ft (0.91–2.4 m) deep, are turned mechanically. During germination the grain undergoes 'modification'. Modification is an imprecise term that signifies all the desirable changes that occur when grain is converted into malt. Modification continues during the initial stages of kilning. Three aspects of modification are: (i) the accumulation of hydrolytic enzymes; (ii) the variety of chemical changes that occur in the grains; and (iii) the physical changes, which appear as a weakening and softening of the grains. Visible signs of germination include the initial appearance of a white 'chit' at the end of each grain, followed by a tuft of rootlets or 'culms'. At the same time the 'acrospire' (coleoptile or shoot) grows. It is covered by the husk in barley, but it grows freely in many other grains. After a sufficient period of growth, barley used to be allowed to dry partly, but this 'withering' stage is now omitted and the germinated grain is transferred to the kiln while it is still fresh (green or undried; not green-coloured).

In villages in Africa and India, malting grains are dried in the sun instead of being kilned; historically, in Europe, 'wind-malts' were prepared by drying malts in lofts. At present limited amounts of green malt are used in some alcohol, whiskey and vodka distilleries to avoid the cost of kilning and to take advantage of the full range of enzymes that are present. However, the overwhelming majority of malts are kilned, that is they are dried and partly cooked, or cured, in a forced flow of hot air. The exact conditions used depend on the type of malt being made. Great care must be taken to achieve the correct colour and level of enzyme survival. In addition, small amounts of special 'coloured' malts are made by cooking green or kilned malts in roasting drums. Kilning dries the malt, stopping germination and giving a brittle and 'friable' product (compared with barley, which is tough), that is easily crushed when milled. The dry rootlets are brittle and are separated and sold for animal feed, often after being mixed with grain-dust and being made into pellets. Unlike green malt, kilned malt is comparatively stable and, if kept dry, can be stored for extended periods. Kilning also reduces the enzyme complement of the malt but adds to its 'character', its colour, flavour and aroma. Malts flavoured with woodsmoke are no longer made in Britain, although some are made in Germany, but malt for Scotch whisky is 'peated'; that is it exposed to peat smoke to obtain its characteristic flavour.

Each batch of de-culmed (de-rooted) malt is held separately until it has been analysed, then it may be added to and mixed (blended) with a similar bulk. After more cleaning it will be sent to a user, usually in bulk.

The malting process uses water and produces dirty water effluent, which must be disposed of in an acceptable way. Power, now usually electrical, is used to drive refrigeration equipment, grain conveyors, turners, pumps, grain dressers and so on. Fuel (typically coal, gas or oil) is used to provide heat for drying, kilning and occasionally roasting. Human resources are required to run and maintain the business, and the grain stocks, new equipment and premises must be financed. In addition, there are weight losses when barley is converted into malt. All these items add to the production costs.

The various topics summarized here are amplified in subsequent chapters. Some of the uses to which malt is put will now be outlined in sufficient detail to permit the need for malts of different qualities and characteristics to be appreciated.

1.3 Coffee substitutes

Malt coffee (malzkaffee) is a drink prepared by infusing ground, roasted malt with hot water. Often it is consumed after adding milk and/or sugar. Also hot water infusions of roasted malt and roasted cereal grains (often barley, wheat and rye) are dried and the soluble powders, with or without added milk powder, are sold as ‘instant’ hot beverages.

Originally coffee malts were made by heating malts in special kilns equipped with metal decks. Now the cooking is carried out in roasting cylinders. The Japanese summer drink, mugi-cha or barley-tea, is prepared by roasting raw barley at 160 °C for 20 min. Roasting inactivates enzymes and denatures proteins, causes the development of dark colours and flavour and increases the acidity and solubility of grain components. Thermal degradation of polysaccharides occurs, sugars are caramelized and melanoidins are formed (Briggs, 1978).

1.4 Malt flours and malted wheat used in baking

For baking purposes, the level of α-amylase present in malt flour is critical. For example, when making white bread the level of enzyme must be adjusted to suit the baking system. Therefore, small additions of enzyme may be made to a dough-mix to increase the loaf volume. The enzyme may be provided by additions of barley- or, less usually, wheat-malt flour, particularly when the traditional, extended bulk-dough fermentation methods are used. Between 0.09 and 0.36 kg of high diastatic malt flour may be added to 100 kg white flour. However, crumb firmness is decreased and an excess of enzyme causes the loaf to collapse, while the interior becomes so sticky that it is impossible to cut slices cleanly.

During baking the malt α-amylase attacks the damaged starch granules and, together with the β-amylase, produces a mixture of sugars. These are metabolized by the yeast, producing carbon dioxide gas, which causes the bread to rise during the extended fermentation periods. In newer methods of baking, involving the rapid mechanical mixing of doughs and short fermentation periods, additions of fungal α-amylase are preferred, partly because the more rapid heat inactivation of this enzyme during cooking usefully limits its activity.

Sometimes enzyme-free malt flours are used. These are made from highly coloured roasted or crystal barley malts and they are used to give particular colours, flavours and aromas to the products and to modify their textures. Malt extracts, which are syrupy, are also used in baking.

‘Granary type’ breads are usually made with wholemeal wheat flours to which are added preparations of malted wheat that have been slightly crushed and then cooked (torrefied or micronized). The fragments of malt impart a characteristic texture and flavour.

Malt fruit loaves have become specialist products. The dark brown, luscious, rich-smelling and sticky loaves are distinctive. Most recipes are secret, but generally white and/or wholemeal flour may be supplemented with low-diastatic malt flour, dried or syrupy malt extract, possibly treacle, sultanas, water, yeast, salt and, on occasions, sugar, skimmed milk powder, caramel and soya flour (Brown, 1982). These loaves are cooked and then are briefly cooled in a vacuum to reduce their moisture content; this prevents them collapsing when they are turned out of their metal moulds. Non-sticky malt breads contain substantially less malt flour, relative to the wheat flour.

Malt flours are usually made by dry-milling appropriate grades of barley malt and separating the fractions by sieving. They are extremely hygroscopic, so are packaged in waterproof, multi-wall sacks.

Grape-nuts™, one of the earliest breakfast cereals, was made by making a dough of wheat flour mixed with barley malt flour plus water, salt and yeast (Jacobs, 1944). After tempering (when much starch liquefaction occurred) and allowing the mixture to rise for 4.5–5.0 h, the risen dough was moulded into loaves and baked. The loaves were later crumbled and rebaked or flaked to give breakfast cereals. The process was dependent on the proper control of the amylase activity of the flour. About 45% of the final product was soluble in water.

Recently there has been disquiet over the use of some caramels to colour foods. Extracts and flours made from roasted and highly coloured malts and from paler 'white' malts can provide a wide range of enzyme supplements, colours, flavours, aromas and degrees of sweetness that have appreciable nutritional value. These characteristics are being exploited in foodstuffs, for example breakfast cereals.

1.5 Ancient Middle Eastern methods of brewing

Ancient written and pictorial records of brewing and, less commonly, malting have been found in Egypt and Mesopotamia. It is probable that these practices were known even earlier, in the neolithic period (White, 1860; Arnold, 1911; Lutz, 1922; Hartmann and Oppenheim, 1950; Singer *et al.*, 1954; Lucas, 1962; Wild, 1966; Darby *et al.*, 1977; Samuel and Bolt, 1995; Samuel 1996). Even so, details of the processes used are far from clear. Consequently the account that follows may contain substantial errors. It is often presumed, without proof, that the ancient brewers' techniques are largely reproduced by the modern makers of bouza (section 1.6).

Perhaps 'beer' was discovered when gruels, prepared from grain or bread, accidentally underwent alcoholic and/or lactic fermentation (Darby *et al.*, 1977). Brewing and baking were closely related practices. Grain, variously barley or emmer wheat, was malted. It was wetted and sprouted in jars, which were held at an angle to the vertical. When germination was sufficiently advanced the green malt was dried in the sun. The roots may have been separated from the dried product. The malt was ground, possibly it was mixed with some flour from raw grain (or, indeed, raw-grain flour may have predominated) and mixed into a yeast-rich dough. Whether or not spices were also mixed into the dough is a matter of dispute (Lucas, 1962; Darby *et al.*, 1977). The dough pieces were formed into traditional shapes and were lightly baked to give beer-bread.

The beer-bread was broken into small pieces, which were mixed with water in jars. Models exist that show people standing in the jars, which reach their waists, apparently mixing mashes by trampling them. Residual enzymes from the bread, and adventitious microbes, together with an inoculum carried over on the jar would have ensured that a mixed fermentation of the sugars occurred. 'Lees' from a successful fermentation might be added to a fresh brew, so inoculating it with a favourable mixture of microbes and indeed propagating them. After a period of fermentation the liquid was strained into a second vessel, which was sealed to exclude the air. The residual solids, separated by straining, were probably eaten or fed to animals. Sometimes fuller's earth was added to the fermented liquid to assist the precipitation of unwanted, suspended solids. Sometimes the beers were diluted with water. At other times they may have been flavoured or strengthened by additions of dates or honey, but there is no evidence for this. The majority view is that herbs and spices were eaten at beer-drinking sessions, rather than being used in the brewing process (Lucas, 1962; Darby *et al.*, 1977). The beers were drunk through tubes, probably hollow reeds. A considerable number of different beers was available, so doubtless a variety of brewing practices was employed. Sumerian texts refer to eight barley beers, eight emmer beers and three mixed beers. The Egyptians probably had even more varieties (Singer *et al.*, 1954). These beers were probably acidic, containing lactic and acetic acids. Tomb paintings of drunken parties indicate their intoxicating powers, but the estimate that the beers contained as much as 12% alcohol seems improbably high. The Egyptians believed that 'mankind' owed its survival to the red-coloured beer made by the God Ra, who used it to intoxicate the Goddess Hathor, so diverting her from completing mankind's destruction (Darby *et al.*, 1977).

The manufacture of beer was an important occupation. It is estimated that, at one time, about 40% of all the cereals produced in Mesopotamia were used in brewing (Singer *et al.*, 1954). Despite a large local production, the Egyptians imported beer from 'Kedi' (Darby *et al.*, 1977). Egyptian barley-beer, wrote Diodorus, was 'for smell and sweetness not much inferior to wine' (Lucas, 1962).

1.6 Bouza, merissa and busaa

Bouza (bouzah, bowza, booza (cf. English slang, a boozer) is made by Nubians resident in Egypt and the Sudan. It is a bread-beer, and its methods of manufacture are said to resemble those used by ancient Egyptian brewers (Arnold, 1911; Lucas, 1962; Darby *et al.*, 1977). Several methods for preparing this drink have been described. All agree that no dates or other flavouring substances are used.

Cereal grain, wheat, barley or millet, is picked over to remove rubbish. About 0.75 of it is coarsely ground and kneaded with water and leaven. Alternatively, the dough is made with a mixture of malted and unmalted grains, with yeast added as sourdough. After a period the dough is formed into loaves, which are lightly baked. Presumably some of the yeast and grain enzymes survive. The remaining grain is wetted and is exposed to the air to let it malt; alternatively, limited germination in the soil may be arranged. Wetted wheat may germinate for 3.5–4.0 days in the shade, where the temperature may reach 45 °C (112 °F). The matted roots may be removed, after the malt has been dried, by hand rubbing to break them up and then by sifting to separate the fragments. The green malt may be used directly, or after drying in the sun. It is crushed and mixed with broken lumps of the bread and with water. The mixture begins to ferment, sometimes after some old bouza has been added to provide an inoculum of microbes. After a period the mixture is roughly filtered, for example through a hair sieve, to remove coarse solids. This process mixes in air, which checks the fermentation, but this soon resumes. The beverage is thick and yeasty and contains suspended starch and other materials from the grain. It is consumed while it is frothy and fermenting. It may contain 6–8% alcohol, by volume, and 11.8–13.1% crude protein (Hesseltine, 1979). It does not keep, as it becomes acetous on storage. During the preparation of the Sudanese drink called merissa, the women chew some of the grain and spit the pulp into the mash, so grinding it and supplementing it with salivary α-amylase, which may help surviving grain enzymes to 'convert' or saccharify starch damaged or gelatinized during the cooking of the loaves. Arnold (1911) noted that pastes of flours made from Egyptian cereals carry a considerable microbiological load, since they readily undergo rapid, spontaneous fermentations.

Busaa is a traditional, home-brewed, Kenyan opaque maize beer (Nout, 1980; Kretschmer, 1981), which contains 0.5–1.0% lactic acid and 2–4% alcohol. It is consumed lukewarm, while still fermenting. Experimentally, a pasteurized and comparatively stable product has been made. Brewing begins with two process streams. In the first, maize grits are mixed with water and the mixture is allowed to 'sour' at 25 °C (77 °F) for 2–3 days. This occurs spontaneously. The microbes involved are a complex mixture that includes yeasts and bacteria, among which are *Lactobacilli* and acid-producing *Pediococci*. The soured 'mash' is lightly roasted on steel sheets for 3 h to develop colour and flavour. The loaves are broken up and the 'soured maize crumb' is mixed with water and is boiled for *ca.* 10 min. The paste is then cooled. In the second, malt process stream, finger-millet grain is steeped for 12–24 h at about 25 °C (77 °F) and is germinated in jute bags for 2–3 days, also at *ca.* 25 °C. The green malt is dried in the sun for 2–3 days. The product, which has a substantial population of microbes, is coarsely ground and 0.1 part is mixed with the cooled (50 °C; 122 °F) paste from the first process. This 'mixed mash' acidifies and ferments spontaneously. After 2–4 days, the product is strained to remove coarse particles. It is consumed before it is spoiled by the continuing acidification, which causes the suspended particles to separate. In experiments designed to lead to the small-scale commercial brewing of busaa, the various stages were carried out using controlled temperatures and pure cultures of a yeast and a *Lactobacillus* sp. The bottled product was pasteurized and was acceptable. It contained 0.8% of lactic acid and 3.25% alcohol, as well as residual starch (Nout, 1980).

1.7 Other opaque beers, fermented gruels and porridges

In other parts of Africa, thick, opaque beers and, less usually, clear beers are made without using beer-breads. In some areas (South Africa, Zimbabwe) home-produced products are being displaced by beers that are made commercially in large, modern factories. The transition in South Africa is well documented (Chapman and Baker, 1907; Oxford, 1926; Fox, 1938; Fox and Stone, 1938; Crone, 1941; Schwarz, 1956; van der Walt, 1956; Novellie, 1968, 1977, 1981; Novellie and de Schaepdrijver, 1986). Elsewhere, considerable 'home production' continues (Young, 1949; Perissé *et al.*, 1959; Platt, 1964; Peterson and Tressler, 1965; Faparusi, 1970; Faparusi *et al.*, 1973; Chevassus-Agnes *et al.*, 1976; Muller, 1976; Hesseltine, 1979; Novellie and de Schaepdrijver, 1986). Numerous different products are made using malted sorghum and millets, often supplemented with raw cereals or non-cereal starchy materials, such as cassava or bananas. Some of the lactic-soured porridges are made without using malt. Confusion is compounded because of the different names given to similar products either because they differ in quality or because of the many different languages in use. Nigerian beers include otika, burukutu, sekele and pito, which are normally made from sorghum and maize. Merissa is made in the Sudan and pombe in east Africa. Amgba and affouk occur in the Cameroons. In South Africa, the kaffir or bantu beers have numerous tribal names, including joala (Basuto) and utshwela (Zulu).

In the traditional production of bantu beers, the stages are malting, mashing, souring, boiling, conversion, straining and alcoholic fermentation. However, the lack of microbiological control makes any division between souring and fermentation irregular and incomplete. Many recipes have been recorded (e.g. Schwarz, 1956). The following account is of an elaborate method given by Fox (1938). At that time the preferred malts were made of sorghum, but maize and millet malts were also used. Grain sewn into sacks or pockets was steeped by soaking in streams (1–2 days for sorghum, up to 4 days for maize). The grains were sprouted for 2–3 days for sorghum (when the shoots reached 1.9 cm (0.75 in) or 2–5 days for maize (when the shoots reached 1.3 cm (0.5 in). Some dried the malt in the sun, others in a hut, while yet others used it undried. In the first stage of brewing, unsprouted grain or, better, a mixture of grain and malt was soaked for a day and was finally ground while still wet. The paste was moulded into lumps of dough in the morning; in the afternoon it was put in a pot and covered with boiling water. Cold water was added to adjust the temperature, giving conditions favourable for the thermophilic microbes that make lactic acid. Next day the water above the dough was collected and boiled with extra water. The soaked dough was mixed with more water to give a thin porridge. This was added to the boiling water, terminating the spontaneous, largely lactic acid fermentation that was occurring in the dough and had soured both it and the water. After a 20–40 min boil, the acidified 'mash' had the consistency of thick porridge, because the starch had gelatinized. The pots were allowed to cool in a breezy spot. Meanwhile a small amount of the 'porridge' was quickly cooled and mixed with dry, ground malt. Starch conversion to sugars and a spontaneous fermentation began, giving isiXubo. When the main mashes were cool enough, more ground malt was mixed in and the isiXubo was added, precipitating the onset of fermentation. When fermentation was rapid, coarse solids were removed by filtration, often using a woven grass strainer. The solids were retained for making other 'small beer'. The liquid was allowed to stand. As the dissolved oxygen from the entrained air was used up, fermentation restarted and in 1–2 days the fermenting liquid was ready for consumption. If this was delayed, the beer became too acid and, particularly if there was an accumulation of acetic acid, it was spoilt. Such beers contain living yeasts, as well as other microbes. The streams of carbon dioxide generated keep the solids, which include starch grains, in suspension. Such products are pale buff to pink–brown in colour and, because of the suspended solids, they are opaque. Analyses of some home-brewed beers gave total solids contents of 5–13%, alcohol levels of 0.5–8.0% by volume (4% when fresh), crude protein contents of 0.7–1.1% and mineral salt contents of 0.18–0.36% (Fox, 1938). Young (1949) noted solids contents of up to 20%. All such beers contain lactic acid, which gives them their desired sour flavours. In the past such beers were an important part of the villagers' diet, and men were said to exist on them exclusively for extended periods. Their calorific values were considerable; they supplied some B vitamins and supposedly some vitamin C, lysine and available iron (Fox and Stone 1938; Platt, 1964; Derman *et al.*, 1980). When these beers were pasteurized, the nutritional availability of some B vitamins was greatly increased (Van Heerden and Glennie, 1987). Unfortunately, the trend towards more 'refined' cereal products, the use of maize grits instead of sorghum adjuncts and foods manufactured by various more rapid processes is seriously reducing the food values of many foods and opaque beers (Novellie, 1968; Van Heerden, 1989).

Acidified gruels and porridges are made by boiling cereal mashes after an acidification or souring stage. They contain negligible amounts of alcohol. The acidity and boiling make these foods comparatively safe to eat as both are destructive to pathogens. Mycotoxins from grain can be carried through into opaque beers (e.g. burukutu) but this is not known to be a serious problem.

Many of the tribal brewing practices, no doubt arrived at by trial and error, have elements that appear to be nearly optimal. For example, boiling mashes sterilizes them and gelatinizes some of the starch, so making it more readily degraded to sugars by enzymes from added malt. The Moba tribeswomen of Togo grind their sorghum malt, mix it with water at about 29 °C (84 °F) then allow the mixture to settle. The supernatant, which contains extracted enzymes, is decanted and reserved while the solids are boiled with more water, thus gelatinizing and dispersing the starch. The original extract and the boiled solids are recombined, achieving a temperature of about 70 °C (158 °F) (Perissé *et al.*, 1959). The enzymes from the supernatant quickly dextrinize and sacharrify the starch under these conditions. To a brewer used to using barley malt, sorghum malt appears deficient in amylases and its starch has an inconveniently high gelatinization temperature. In principle, the Moba process minimizes these defects and is nearly identical to a single, limited-decoction process developed for processing poor-quality barley malts in lager brewing (Sykes, 1897; Hind, 1940). Novellie and de Schaepdrijver (1986) compare various African brewing practices.

In South Africa, the brewing of Bantu beer has become industrialized, being the first and only industry based on an African tribal art (Novellie, 1968, 1977, 1981). This development has been supported by a substantial body of research, including work on making sorghum malts (see Chapter 15). Brewing is now carried out in stainless steel equipment. A 10% slurry of sorghum malt in water is inoculated with a pure culture of thermophilic *Lactobacillus leichmanii* and is held at 50 °C (122 °F) for 12–16 h, until the pH is about 3.3 and the lactic acid content is 0.8–1.0%. Unmalted cereal, usually maize grits, and water are added and the mixture is cooked, under pressure. After cooling to about 60 °C (140 °F), more ground sorghum malt is added and the mixture is held for 45–90 min, until sufficient fermentable sugars have been generated. At this stage, the acid wort has a pH of 3.9, 0.16% lactic acid, a specific gravity (s.g.) of 1.037 and a high total solids content. It is centrifuged to remove coarse particles (above 0.25 mm) and is inoculated with a selected strain of yeast, *Saccharomyces cerevisiae*. Fermentation proceeds for about 24 h at 25–30 °C (77–86 °F). The product is consumed promptly, before it deteriorates, while fermentation is active and before the alcohol content rises above the legal limit. Deterioration is marked by the accumulation of acetic acid. All Bantu beers are rich in fusel oils. Other analyses are: pH 3.2–3.9, lactic acid 0.16–0.25%, volatile acids (as acetic acid) 0.012–0.019%, total solids 2.6–7.2%, insoluble solids 1.6–4.3%, alcohol 2.4–4.0% (by weight) and nitrogen 0.065–0.115% (Novellie, 1968). As these beers are distributed while still fermenting, they are held in vented containers. They are drunk at room temperature and are viscous with a fruity odour.

1.8 Clear beers and stouts

Probably medieval and earlier European beers were also acidic and turbid. The most 'primitive beers' still made in Europe (Belgium) are Lambic and Gueuze, lactic beers made using a complex spontaneous fermentation in which a large number of microorganisms take part (Van Oevelen *et al.*, 1976; Van Oevelen, 1978). Mashes consist of 30–40% unmalted wheat and 60–70% under-modified barley malt. Fermentation and conditioning may go on for 2–3 years. Beers of different ages are often blended.

European-style hopped beers and stouts are now made on the industrial scale in most regions of the world. Home brewers may make batches as small as 25 l (around 5 UK gal), but commercial breweries with annual outputs of more than 50 000 hl (about 30 000 UK gal) are common. The types of brewing equipment, raw material and product and the degrees of technical and scientific expertise used vary very widely indeed. Consequently only a generalized account of brewing is given here, taking note of only the most usual practices and putting emphasis on the mashing operations because an understanding of mashing is indispensible for understanding the qualities of malts. This is true of malts used in making beers, whiskeys, malt vinegar, malt extracts or diastase. For this reason, Chapter 5 covers the biochemistry of mashing. There are many accounts of brewing available (Hind, 1940; de Clerck, 1952; Vermeylen, 1962; Ohlmeyer and Matz, 1970; Hough *et al.*, 1975, 1982; Narziss, 1976, 1992; Broderick, 1977; Briggs *et al.*, 1981; Pollock, 1987 (and earlier references); Kunze, 1994; Lewis and Young, 1995).

The indispensible raw materials for brewing are water, malt (or malt extracts), hops (or hop-based preparations) and yeast. In most parts of the world 'starchy adjuncts' and/or sugars and syrups may be used, as are microbial enzymes, chiefly as ways of reducing malt usage. In addition, process aids such as copper finings, yeast finings, caramels, selective adsorbents and filter aids may be used.

Water is a major raw material. Besides its use in the brewing process, it is used for cleaning and in raising steam. Water used in brewing must have a number of characteristics: it must be microbiologically pure, that is free of pathogens or organisms that could cause beer spoilage; it must be free from undesirable flavours, odours and taints; it must be potable and meet all the quality criteria used for drinking water; and, finally, it must have a suitable ionic composition (permanently hard water for pale ales, carbonate water for mild ales, very soft water for Pilsen-type lagers, and so on). Often brewers adjust the composition of their 'brewing liquors'.

The malts used in brewing range from very pale, white malts, such as pale ale and Pilsener-lager malts, through malts of progressively darker colours to 'special' or coloured malts, which are partly caramelized or roasted. The great majority of brewing malts is made from barley, but in Europe wheat malts are used, either as a small proportion of the 'grist' to give a beer some special characteristics or exclusively to make, for example, German weissbier (weizenbier). Rarely, European brewers use malted oats and malted rye; in Africa, malted sorghum is used to some extent. Pale malts contain the enzymes needed in mashing. They provide most of the 'extract' used in beers made exclusively from malts, and their enzymes convert the starch of unmalted adjuncts and coloured malts in other brews. 'Extract', the dissolved solids that are derived from the malt, the adjuncts and any sugars or syrups that may be used, is a complex mixture of carbohydrates, amino acids, peptides, minerals, polyphenolic materials, vitamins and many other substances. Extract includes the nutrients that support yeast growth and, most obviously, are converted into alcohol and carbon dioxide during fermentation. However, many substances that add to beer's flavour and aroma are produced in small amounts. Materials from malt are chiefly responsible for beer's head, or foaminess, and its 'body', or mouth-feel. Small proportions of dark or special malts are used to give darker colours, aromas, characteristic flavours and better flavour and haze stabilities.

Adjuncts, microbial enzymes and some plant enzymes may be used in brewing. Unmalted adjuncts are nearly always made from cereals and are used as sources of extract and to adjust beer characteristics. These preparations include maize, rice and sorghum grits, which all need to be cooked before use, and flaked maize and flaked rice grits, wheat flour and cooked (torrefied or micronized) whole wheat grains, which may be used in moderate amounts in mashing and which do not need to be cooked in the brewery. Less commonly, cassava and potato starches have been used. Other sources of extract include sucrose (cane or beet sugar) and sugars and syrups obtained by the controlled hydrolysis of purified starch (wheat or maize) or whole cereal grains.

Often it is helpful, or even essential with some grists, to supplement mashes with non-malt enzymes. These are derived almost exclusively from bacteria and fungi and are most likely to be used to assist starch conversion when the 'mash bill' contains a large proportion of unmalted adjuncts. In addition, enzymes may be added to worts to obtain maximal carbohydrate degradation and utilization by yeast, to give 'low-calorie' beers. Proteolytic enzymes, typically papain, may be added to beer to degrade residual large polypeptides and so reduce the chances of haze formation during storage.

Apparently worts have always been boiled with herbs to give beers particular flavours. In early medieval times, mixtures of herbs (gruit) were used to flavour the 'ales'. The only extant echo of this practice may be the use of spruce twigs as combined filter-aids and flavouring agents in small-scale Norwegian farmhouse brewing. In the later Middle Ages, gruit was displaced by hop cones. At first, the competition between the ale-brewers and the beer-brewers was intense. Hops (*Humulus lupulus*) are climbing plants of the family Cannabinaceae (Neve, 1991). They are cultivated exclusively for use in brewing. The cones are the female inflorescences and made up of a bundle of bracts attached to a stalk. They may be used with seeds or seedless. Each green, leafy bract is powdered with tiny lupulin glands that contain the substances which give the hop character to the beer, i.e. the resins, which are partly extracted and isomerized during the boil to give bitter flavours, and the hop oils, which contribute to the aroma. Hops are also rich in tannins, which encourage 'trub' (precipitate) formation during wort boiling. Hops vary widely in their characteristics (Hough *et al.*, 1982; Neve, 1991). Whole hop cones, ground hops or hop pellets may be used in the copper (kettle) boil, as may various kinds of prepared extract. Iso-α-acid preparations, made by isomerizing the α-acid fraction of hop resins, may be added directly to beer to adjust the bitterness level. Specially selected hop cones may be added to beer ('dry hopping') to give the product a particular aroma.

With the exception of rare 'special beers', mixed acidic and alcoholic fermentations are prevented. Because they are boiled, brewers' worts are initially sterile. After cooling they are pitched or inoculated with selected strains of yeast (*Saccharomyces cerevisiae*; 'sugar-fungi of beer'); this is a single-celled, ovoid (10–15 μm diameter) microfungus. Generally a single strain is used, but some brewers use mixtures and occasionally a tiny population of bacteria is tolerated. Yeast strains vary, for example in their preferred fermentation temperatures, the sugars they will utilize and the flavours they give to beers. Practically, yeasts are divisible into top, powdery and bottom strains according to whether, as fermentation progresses, the yeast rises to the top to form a frothy layer (the head), it stays in suspension or it settles to the bottom. Top-yeasts used to be a characteristic of British-style ale breweries, while bottom-yeasts (which were often classified as *Saccharomyces carlsbergensis*) were typical of lager breweries. However, newer methods of carrying on fermentations and separating yeasts at the end of this process have now blurred these distinctions.

The stages of brewing are summarized in Fig. 1.2. Some breweries have their own maltings, and sometimes their own hop farms. All materials are checked on delivery. When brewing, malt is drawn from store and is cleaned to remove dust and fragments of metal and stone, then it, and any adjuncts that are used, are weighed. Torrified and roasted cereals are mixed with the malt, ready to be milled. This mixture is milled to give the grist.

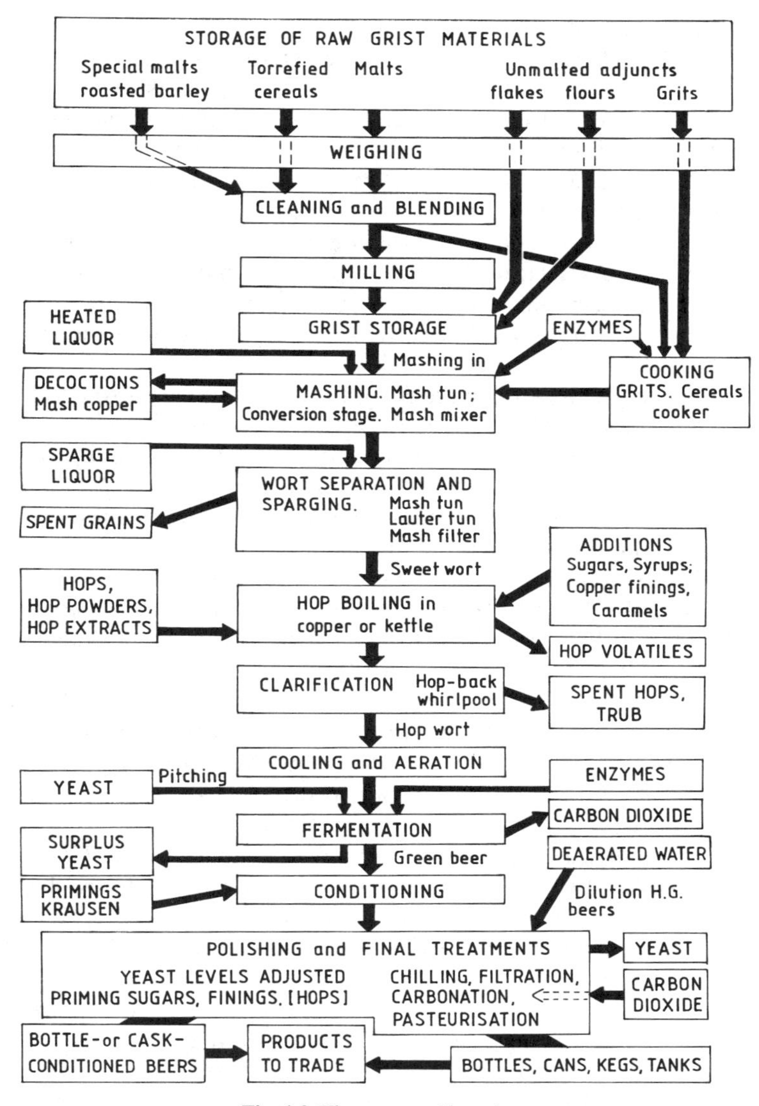

Fig. 1.2 The stages of brewing.

Raw barley, if used, may be crushed at this stage, if a wet milling system is being used.

Roller mills are usually used because, although many other types of mill are available, crushing malt gives a grist with the range of particle sizes needed in brewing. Mash tuns need the coarsest grists, lauter tuns use finer grists, while the finest are used in mash filters. Some mash filters can handle flour-like grists made of hammer-milled malt. Husk fragments 'open-out' the mash bed and help wort drainage at the end of mashing. Various milling systems are in use. Dry milling may involve successive passage between two pairs of rollers, or it may involve three pairs of rolls with the grist being separated by sieving into different streams, which are directed either to the 'grist case', or to pairs of rolls for further crushing. Dry husks are brittle, and larger fragments are obtained if the malt is 'conditioned' by slightly dampening the malt's surface, using steam or fine sprays of water, immediately before it enters the mill. The extreme approach is wet milling, in which the malt is steeped by immersion in warm water before it is roller-milled and mixed with the mashing liquor. With this system, milling and mashing-in (doughing-in) are combined.

When mashing-in dry-milled grists, a chosen quantity is thoroughly mixed with the correct volume of brewing liquor at the right temperature, to achieve a mash of the correct temperature and consistency. After a stand, during which conversion (starch hydrolysis) occurs and about 80% of the malt is dissolved, the wort or liquid, which contains the dissolved extract, is separated leaving the wet spent grains or 'draff'. The draff, which represents about 20% (dry basis) of the grist, is used for animal feed. Experimentally, other uses have been found for the draff, such as in making fibre-rich biscuits, breads and cookies. Biochemically, the mashing process is complex, involving the solution of pre-formed soluble materials, the alteration of these and of some of the insoluble grist materials, notably starch (which is converted to soluble dextrins and sugars), by enzymes acting at temperatures at which they are unstable and by which they are progressively inactivated (Chapter 5). Elevated temperatures cause the starch to swell and gelatinize in water, rendering it susceptible to enzymic attack. Cereal grits (maize, sorghum or rice) must be cooked at 80–100 °C (170–212 °F) before mixing with the malt mash, to disrupt them and to ensure that their starches are gelatinized. The slurried grits are heated with small proportions of highly enzymic malt or some heat-resistant bacterial α-amylase to liquefy the starch as it gelatinizes.

Mashing programmes are conveniently, if artificially, grouped under four headings: infusion mashing, decoction mashing, double-mashing and temperature-programmed infusion mashing. Infusion mashing used to be a characteristic feature of British breweries, decoction mashing of lager breweries and double-mashing of North American breweries. These distinctions are now blurred and temperature-programmed mashing has been widely adopted. Traditional British ales were (some still are) made using an infusion mashing system, carried out in a mash tun, which contains the mash during the conversion phase and allows wort separation at the end of the mash. These large, well-insulated vessels of circular cross-section are equipped with a false bottom of slotted plates, which retains the residual grist solids when the wort is being run off. At the start of mashing, the plates are covered with warm water, or 'liquor', and the grist and warm water are mixed in a mashing machine that directs the slurry into the tun. Eventually mash depths of 0.9–1.2 m (3–4 ft) are reached. Because air is entrained, the mash tends to float. At the end of the stand period wort is drawn through tubes in the base of the tun, and the suspended draff sinks onto the plates of the false bottom. Initially the wort is sparged (sprayed) back onto the top of the mash and this recirculation is continued until it 'runs bright', that is it is absolutely clear. The flow of wort is then diverted to the copper (hop kettle). When the liquid level has fallen enough, sparging begins. Hot water, the sparge liquor, is sprayed onto the surface of the mash at a rate that matches wort withdrawal. As the liquor percolates downwards it leaches the residual soluble materials from the grist and so most of the extract is recovered. The spent grains are drained and are discharged mechanically by a rotating arm. Traditional ale grists are of well-modified malts, often with a small proportion of coloured malt and perhaps some flaked rice or maize. Mashes are thick, with a liquor:grist ratio of about 2.7 hl/100 kg (16.7 brl/t) (brl, barrel; see Appendix). Older mash tuns were fitted with rakes or mixing machinery, so by mixing the mash and by adding hot water beneath the plates – underletting – the mash could be diluted and the temperature raised. For example, for a mild ale the initial temperature of 64 °C (147 °F) might be increased to 67 °C (153 °F) and finished with a sparge at 77 °C (170 °F) (Hind, 1940). Old mashing programmes might last for 6.5 h. Modern mash tuns lack internal mixing machinery and so the stands are isothermal, often in the range 63–68 °C (145–155 °F). Filling time is about 30 min, stand times are in the range 0.25–2.0 h and the sparge and drain period lasts 2–4 h.

In decoction brewing, three vessels are used: a mash-mixing vessel, a smaller mash copper (kettle) and a lauter tun or, alternatively, a mash filter. Traditionally this system was used with malts that were, by British standards, severely undermodified. This is no longer true. Mashes are made comparatively thin (3.5–5.4 hl/100 kg; 20–33 brl/ton) to allow them to be pumped between the vessels. Initially the grist and liquor are mixed and the mix is discharged into the mash-mixing vessel, which is stirred. In the old, and long three-decoction process, three samples of the mash, each of about 0.3 the total volume, are successively transferred to the stirred mash copper and are heated, at a controlled rate, to boiling. The hot sample is then pumped back to the main mash and is mixed with it. Thus, over a 6 h period, the main mash is held at several temperatures, initially at about 40 °C (104 °F), then after successive decoctions, at about 52 °C (125.6 °F), 65 °C (149 °F), and 76 °C (168.9 °F). Then the mash is transferred to a lauter tun, which usually resembles a mash tun but with a shallow bed depth and internal raking machinery, or to a plate and frame filter in which the mash is held between filter-cloths. Worts are recirculated until they are bright, then they are collected and the grains are sparged. Double- and single-decoction programmes are also used, but these need better modified malts.

In some breweries the mash copper is also used to cook raw cereal grits. The cooked grits are then mixed with the main mash.

The double-mashing system was developed in North America, to deal with grists containing large proportions of rice or maize grits (as much as 60% of the mash bill) and to utilize the nitrogen- and enzyme-rich malts that were available. The adjunct mash, containing the grits and a small proportion of enzyme-rich malt or bacterial α-amylase, is doughed-in at about 35 °F (95 °F) in the cereal cooker. The stirred mash is heated to about 70 °C (158 °F) and after remaining at this temperature for, say 20 min is brought to 85–100 °C (185–212 °F). It is held at this temperature for 45–60 min to ensure that any starch that has not been liquefied is gelatinized. Meanwhile, the malt mash has been doughed-in at 35 °C (95 °F). After a stand of about an hour the adjunct mash is pumped in, with mixing, so that the final temperature of the combined mash is around 65 °C (154.4 °F). The whole process may take 3.5 h. Finally the mash is transferred to a lauter tun and the wort is collected.

The characteristic vessels in temperature-programmed infusion mashing are the mash-mixing unit and a wort-separating device (a lauter tun or a mash filter). However, if grits are to be used then a cereal cooker is also needed. The mash mixer is equipped to stir the mash and to raise its temperature according to a pre-chosen programme. This equipment, which can be used to efficiently convert many types of grist, is gaining favour with ale- and lager-brewers. The temperature programme, which typically lasts for 2–3 h, is chosen to suit the type of grist being used and the product being made. As an example, the grist may be mashed in at 45–55 °C (113–131 °F), then the temperature will be raised to, say, 63–66 °C (145.5–150.8 °F). After a stand at this temperature (for example, for 30–60 min) it will be heated again to 75 °C (167 °F) and transferred to the wort-separation unit. A modern brewery with a mash mixer and lauter tun may produce wort every 2 h.

Hops, or hop preparations, are added to the freshly collected wort, which is boiled, typically for 60–90 min. During this process, the wort is concentrated and sterilized, residual enzymes from the mash are inactivated, there is an increase in colour, proteins are denatured and separated, in association with some of the polyphenolic tannins from the malt and the hops, as flocculent aggregates of 'hot break' or trub. Meanwhile, some unwanted hop oils and volatile grain flavouring substances evaporate and α-acids are extracted from the hops and are isomerized to the bitter iso-α-acids. If brewing sugars or syrups are used, they are mixed with the wort at this stage. Various devices are used to remove the trub and residual hop materials from the wort. After cooling and aeration or oxygenation, the wort is transferred into a fermenter.

At this stage the wort is pitched or inoculated with the appropriate yeast (say 0.2 kg/hl). Enzymes may also be added to hydrolyse dextrins to fermentable sugars if a 'low-carbohydrate' beer is being made. During the fermentation, temperatures rise and fall, often in the ranges 8–15 °C (46.4–59 °F) with lagers and 15–20 °C (59–68 °F) with ales. Cooling is used during the fermentation and at the end, temperatures are reduced to around 15 °C (59 °F) for ales and 4 °C (39.2 °F) for lagers to assist yeast separation. Nearly all beers are now made in batch fermenters of which many patterns are in use. The favourites are Nathan or 'cylinder-conical' fermenters: free-standing, well-insulated, vertically mounted, stainless steel cylinders that are equipped with cooling jackets and conical bottoms which allow the yeast to be collected after it has sedimented (Hough *et al.*, 1982; McDonald *et al.*, 1984). The carbon dioxide that is evolved during the fermentation may be collected and, after purification, be used at a later stage in the process.

After the primary fermentation, the green beer is 'conditioned' to improve its flavour and to reduce its tendency to become hazy. The beer, which contains a residue of yeast, is normally supplemented with a small quantity of priming sugar, or fermenting wort (krausen). The secondary fermentation that occurs removes unwanted flavours and charges the beer with carbon dioxide. In naturally conditioned beers, the secondary fermentation occurs in bottle or in cask and the beer, to which finings have been added, is dispensed from above a layer of living yeast. Hops may be added to casks of beer to give a characteristic 'dry hopped' flavour. In other beers, the secondary fermentation occurs in fermenters or specialized lager tanks. In the latter case, the beer is held at about 13–16 °C (55.4–60.8 °F) for an extended period. Other beers are chilled and filtered to remove residual yeast and suspended solids; the levels of proteins and tannins may be reduced by adsorption onto solid silica hydrogel or polyvinylpolypyrrolidone (PVPP). The pasteurized beers are packaged under a top pressure of carbon dioxide. They may be supplied in bottles, cans or kegs.

Beers differ in their colours (pale straw, through shades of brown to turbid black), flavours, degrees of bitterness and alcohol contents (% v/v), low (0.5–2.0%) through normal strengths (4–5%) up to barley wines (*ca.* 10–11%). Alcohol-free beers (less than 0.1%) are made by removing alcohol after fermentation. They are still classified as pale ales, mild, stouts (dark beers with characteristic flavours), lagers (which traditionally received a cold storage period lasting months). Pilsener pale lagers, Munich lagers (which are dark and aromatic) and so on. However, as brewing practices continue to evolve, the traditional distinctions are disappearing.

In addition to water, alcohol, carbon dioxide, hop oils and hop bitter substances, beer contains hundreds – possibly thousands – of other compounds that add to its flavour, aroma, the ability to form a foamy head and give it the desired mouth-feel. On aging, beers deteriorate, even if they remain sterile. Flavour deterioration is followed by increasing turbidity caused by the formation of protein–tannin haze particles.

1.9 Malt extracts

Malt extracts, which may have a syrupy consistency or may be powders, are made by mashing ground malt, usually barley malt, in conventional brewery equipment, collecting the wort and concentrating it or drying it. A very wide range of product types is produced, by varying the type(s) of malt used, the mashing programme and the ways in which the wort is subsequently handled (Weichherz, 1928; Hind, 1940; Burbridge and Hough, 1970; Briggs, 1978; Briggs *et al.*, 1981). If materials other than malt are used in the mash, the product may be classed as a cereal syrup. In the past extracts, syrups and powders were made by concentrating worts in vacuum pans. Now they are generally prepared in single-, double- or triple-effect climbing-film evaporators. To achieve high concentrations of solids in the initial worts and so reduce the costs of evaporation, the first worts are directed to the evaporator, while the weak 'last runnings' are reserved and are used to make the next mash. Powdered products are made by spray-drying worts, or by drying them in moving-belt vacuum ovens.

Enzyme-rich, diastase-rich extracts are made from nitrogen-rich barleys, 1.7–2.2% nitrogen in the UK, chosen for their ability to generate high levels of enzymes; they are malted to maximize enzyme production. Care is taken to kiln at low temperatures to minimize enzyme inactivation. The malt is ground and maybe mashed twice, first at a low temperature (less than 49 °C; 120.2 °F) to extract heat-labile enzymes, then at about 68 °C (154.4 °F) to recover the rest of the extract. The cooled worts are combined and are evaporated at temperatures that rise from 32 °C (89.6 °F) to 46 °C (114.8 °F) as concentration proceeds.

In contrast, for some uses, extracts must be entirely free from enzyme activity. To achieve this, worts may, if the customer permits, be slightly acidified and be heated to 80–90 °C (176–194 °F) before concentration.

Malt extracts are expensive to make both because of the cost of the malt and because the concentration and drying stages are expensive. Cheaper syrups can be made by replacing some of the malt with unmalted cereal adjuncts, or by converting raw cereals in mashes made with microbial enzymes (Chapter 5). Another method was to make extracts from green, unkilned malt. Technical problems were encountered. The green malt was ground with special mills, and the wort was separated from the mash using bowl centrifuges. The costs of kilning were avoided, full use was made of the enzymic potential of the green malt and the product, sometimes called 'green malt', was low in anthocyanogens (Chapter 4). The unpleasant flavour that is a characteristic of worts made from unkilned malts was removed during the evaporation process.

Dried malt extracts, which contain 95–98% solids, are very hygroscopic and must be stored in sealed containers. They vary in colour, flavour and enzyme content. Textures range from fine, soft powders to being coarse, gritty and semi-crystalline. Malt extracts are highly viscous and must be warmed to about 49 °C (120.2 °F) to permit them to flow. However, warm periods are kept short to minimize darkening and flavour changes. Dark products are obtained by using dark malts and sometimes by heating worts. Colours range from pale straw to very dark chestnut brown (3–500 EBC units, 3–50 EBC units for diastatic extracts (EBC units of colour)). Diastase levels range from 0 to 400 °L. Diastase is a collective name for the malt enzymes involved in the breakdown of starch in mashing. It is measured as diastatic power (DP) in °Lintner (°L). Proteases, lipases and other enzymes are present besides the amylases. Flavours vary widely, pale extracts being sweet and comparatively bland, while darker extracts used in confectionery are rich in melanoidins and are aromatic, sweet and richly luscious in flavour. Malt extract syrups contain 75–82% solids, with specific gravities of 1.40–1.45. The high concentrations of solids check microbial growth. The higher concentrations are used in the tropics, where high temperatures and humidities favour the growth of spoilage organisms. At ambient temperatures, syrups are stable, unless they are diluted, for example by condensate from the top of the container dripping onto the surface of the syrup. To prevent this, small containers are filled completely while the spaces in large containers may be ventilated with warm, filtered air. Extracts may be distributed in drums or in bulk tankers. Specifications normally limit the levels of microbiological contamination. Syrups are a compact form of 'extract', in that 3.76 volumes are equivalent to about 10 volumes of malt. Values of around 302 l°/kg (20 °C) are normal. (IoB units of hot water extract (HWE) are l°/kg.) Other specification values may include pH, buffering capacity, ranges of sugar composition or fermentability, ash content and the protein content (nitrogen × 6.25; commonly 4–9%). Malt extracts may be made from 'naturally acidified' worts that have been inoculated with lactic acid bacteria, such as *Lactobacillus delbrückii*. These preparations are used when 'chemically prepared' acids are unacceptable in foods or beverages. In some countries, acidified worts are used to adjust the pH of brewery mashes. Pale, enzyme-rich extracts are highly fermentable and are rich in soluble nitrogen, much of which is present as amino acids and other low-molecular-weight substances (determined as 'formol-nitrogen'). Extracts may have fermentabilities in the range 58–96% and glucose equivalents in the range 65–93%.

Substantial quantities of malt extract are made (*ca.* 70 000 t per annum in 1970 in the UK, Burbridge and Hough, 1970). The major uses have changed with the years. Therefore, the use of high-diastase extracts (or diastase, prepared by alcohol precipitation of the enzymes in the extract) for desizing textiles (dissolving protective starch coatings) has been discontinued and has been replaced by bacterial α-amylase.

To a limited extent, malt extracts are used as carriers or emulsifiers in pharmaceutical preparations, as their rich, pleasant flavours mask other bitter or unpleasant flavours and they are easily digestible. A well-known use is as carriers of as much as 10% of fish-liver oils (cod, halibut or shark), which are rich in oil-soluble vitamins. One standard (Anon., 1973) specifies that the extract must be made with barley malt and not more than 33% of malted wheat. It should contain more than 4% 'protein', have a density of 1.38–1.42 g/ml (20 °C) and a refractive index of 1.489–1.498 (Reynolds and Prasad, 1982). Elsewhere, 10% of glycerol may be incorporated into these extracts.

Enzyme-rich extracts, or diastase, may be used to prepare pre-digested foods for invalids or infants. Tentative standards have been proposed for Indian sorghum (cholam) malt extracts (Siddappa, 1954). Because malted sorghum contains little diastase (DP 2–4°L) the values in the extracts are always low (max 2°L).

At present malt extracts are little used in large-scale brewing, but they are used extensively by home brewers and in some small or isolated production units, such as 'pub breweries'. Non-diastatic malt extracts, which may or may not have been bittered with hops, are dissolved in boiling water to produce sweet or hopped wort. Home brewers usually purchase their extracts in cans. In the Con-brew System hopped wort concentrates prepared in Newfoundland were shipped in 19 hl (500 US gal) collapsible rubber containers to the Bahamas. On arrival, the syrup was diluted and fermented. Acidified extracts may be added to mashes to adjust the pH. Diastatic malt extracts used to be added to the mash (e.g. at the rate of 1 part per 36 parts (w/w)) when extra enzymes and nitrogenous yeast nutrients were needed. At present enzyme-rich extracts may be added to the cooled, hopped wort (despite the microbiological risks entailed) to permit the enzyme-catalysed hydrolysis of the dextrins, so giving a highly fermentable wort and a low-carbohydrate' beer. A minor use of dried malt extracts is as a component of solid or liquid nutrient media used by microbiologists in culturing yeasts and other microorganisms (e.g. MYGP: malt, yeast-extract, glucose and peptone medium).

Malt extracts are used in making some breads, biscuits and other bakery products, in confectionery and in a range of other foodstuffs, including some ready-to-eat, pre-cooked breakfast cereals, desserts, ice-creams, meat products, pickles, sauces, soft drinks, stock cubes and wines (Turner, 1986). Coloured malt extracts may displace caramels in some manufactured foodstuffs, as some of the latter have come under suspicion as health risks. Ironically, some of the cheaper malt extracts used to be coloured with caramels.

Malt extracts are used in making, or constitute part of, malted beverages (*sic*). Malted milk, malted chocolate milk and malted cereal grain preparations are commonly available. They are usually spray-dried powders that may be reconstituted with hot water or milk. Other cereal components sometimes used include roasted wheat, rye and barley. Milk or milk powders may be used in the manufacturing process. For instance, malted milk may be made by mixing wort, from a temperature-programmed barley malt and wheat malt mash, with milk to give a mixture containing more than 7.5% milk solids (dry basis). Salt and sodium bicarbonate are added to adjust the pH and flavour. The mixture is pre-warmed to 65.6 °C (150 °F) and is concentrated to 68–70% solids. Lecithin may be added to increase dispersibility, then the syrup is dried to about 98% solids in vacuum pans (Marth *et al.*, 1967).

1.10 Diastase

Diastase is the name given to the mixture of enzymes from grains that degrade starch. Preparations of diastase are not uniform in composition, The original preparation, described in 1833 by Payen and Persoz, was made by precipitating the diastase from a malt extract by the addition of alcohol. This type of preparation may still be used in the treatment of dyspepsia and in the preparation of pre-digested starchy foods. It is a yellow-white, amorphous and tasteless powder (Wade and Reynolds, 1977). In addition to the α- and β-amylases, diastase contains a complex mixture of other enzymes. Other preparations of diastase are enzyme-rich malt extracts, either syrups or powders.

1.11 Breakfast cereals and pre-digested foods

Breakfast cereals may be divided into those that require cooking (e.g. porridge made from rolled oats) and those that are pre-cooked and ready for eating, such as corn (maize) flakes, wheat flakes, pretzels and puffed or shredded preparations. These may be flavoured with malt extracts (Matz, 1970; Kent, 1983; Hoseney, 1986). Granular, pre-cooked cereals may be made from malt flour, for example the earliest cereal was Grape-nuts™ introduced in 1896 (Jacobs, 1944) (Section 1.4).

Many commercial infant and invalid foods are made with cooked cereals, pulses and vegetables. Sometimes α-amylase (supplied as malt diastase, malt flour, or bacterial or fungal preparations) is used to liquefy starch as it cooks to reduce the viscosity of the final products. In India and the Far East, domestic techniques are being adapted to the preparation of high-quality weaning foods from locally available crops (Ramakrishnan, 1979; Desikachar, 1980, 1982; Gopaldas *et al.*, 1982; Malleshi *et al.*, 1986b; Chavan and Kadam, 1989). Malting or sprouting cereals and pulses are common practices; indeed sprouted mung-beans (moong-beans) as Chinese vegetables are now commonly available. Both cereals and pulses often become more attractive, palatable and digestible when they are sprouted, through alterations in flavour and texture. Often there are declines in stress and anti-nutritional factors, such as phytate (which interferes with the absorption of metal ions), oxalate, tannins (which inactivate many enzymes and reduce the digestibilities of proteins), trypsin inhibitors and possibly other enzyme inhibitors and flatus (flatulence) factors. At the same time there are often increases in digestibility and in the levels of sugars and free amino acids, including available lysine, and sometimes various vitamins, including B_1, B_2, B_6, niacin, folic acid, tryptophan, biotin and vitamin C (ascorbic acid) (Wang and Fields, 1978; Wu and Wall, 1980; Teutonico and Knorr, 1985). Sprouted pulses are poor in amylase content but are rich in high-quality proteins (19–29%). Cereals are comparatively poor in proteins, which are of lower nutritional value. Therefore balanced infant foods should contain both pulses and malted cereals. Often supplementation with milk or milk powder is desirable but may not be feasible (Dean, 1953). Pulses that have been tested include mung beans, green gram, Bengal gram, groundnuts and cow peas. The malted cereals used include wheat, barley, bajra (*Pennisetum typhoides*) ragi (*Eleusine corocana*), jowar (sorghum) and navane (*Setaria italica*) (Finney, 1982; Chavan and Kadam, 1989). In central and south India, ragi and green gram (*Phaseolus radiatus*) are grown widely. Raw ragi is poorly digested and it is rich in phytate. However, it malts well and produces substantial quantities of amylases. Various trials have been carried out to optimize the use of these materials in making weaning foods. Ragi and green gram were steeped for 16 h and germinated for 48 and 24 h, respectively. After drying, the malts were milled and the flours were mixed in the ratio 70:30 (ragi:green gram). When this mixture was made up with warm water and slowly heated, substantial amylolysis occurred and thinned the paste. A final concentration of solids, acceptable to infants, was achieved that was three times that of the unmalted raw materials, with consequent increases in the nutrient and calorie densities. Since infants will only eat limited volumes of food, the high 'nutritional density' of the malted foods is of great value (Brandtzaeg *et al.*, 1981; Desikachar, 1982; Gopaldas *et al.*, 1982; Malleshi and Desikachar, 1982; Mosha and Svanberg, 1983).

1.12 Distilled products

For many purposes ethanol is made from petrochemicals. However, in some countries ethanol for use as a solvent or fuel is made by fermentation and distilled alcoholic beverages (legally prepared) are always based on fermentations. Fuel alcohol or 'neutral spirits', for making gins, aqua vitae or vodkas for example, are distilled from fermented mixtures based on sucrose (cane or beet sugar) or starch from any convenient source, converted with acids, microbial enzymes or malt enzymes (Briggs, 1978; Margiloff *et al.*, 1981; Lyons, 1983; cf. Chapter 5). In some cases, the source of the fermentable carbohydrate is immaterial because the alcohol is highly purified by fractional distillation. In the preparation of some vodkas the last traces of 'impurities' are removed by percolating the spirit through beds of activated charcoal. Malt is still used in the preparation of some gins (Simpson, 1966; Laatsch and Sattelberg, 1968). In gins, as understood in the UK, flavour is imparted by distilling or redistilling rectified spirit with 'botanicals', spices such as juniper berries, coriander, angelica root, cinnamon bark, and orange or lemon peel. Alternatively, extracts, or essences, of the botanicals may be used. The characteristic flavours are imparted by the essential oils of the botanicals.

Whiskeys (Irish; US: whiskeys; Scotch whisky; Gaelic: *uisge beathe*, water of life) are distilled from fermented mixtures, either fermented worts or thin fermented mashes, based on malts and, in the cases of grain whiskeys, using some high proportion of unmalted cereal adjuncts. The 'beers' or fermented mashes contrast with true beers in that they are not flavoured with hops. In addition, because they are not boiled before fermentation, they are not sterile when they are pitched with yeast and enzymes surviving from the malt continue to degrade dextrins; as a result more carbohydrate is fermented and higher alcohol yields are achieved. Many different whiskeys are made. They differ in the raw materials used, and the ways in which the mashing, fermentation and distillation processes are performed (Hough *et al.*, 1975; Dolan, 1976; Briggs, 1978; Kahn, 1979; Nykänen and Lehtonen, 1984; Piggott *et al.*, 1989).

Scotch malt whisky is made exclusively from an all barley malt grist. The malt, often with a DP of 67–75°L and chosen to give a good yield of spirit, is lightly kilned and is peated (exposed to smoke from smouldering peat) to various extents to give products with characteristic flavours. Malts for Islay whiskys, for example, are highly peated. The grist may be mashed three times at successively higher temperatures, e.g. 63 °C (145.4 °F), 75 °C (167 °F) and 85 °C (185 °F). The first and second worts are pooled and after cooling are inoculated with yeast. The third wort is used to mash the next grist, so maximizing extract recovery without further diluting the extract. Commonly the yeast used to ferment the hazy wort is a mixture of brewer's yeast (*Saccharomyces cerevisiae*) and a distiller's yeast, which may be *Saccharomyces diastaticus*, a species that is able to ferment dextrins. Fermentation is vigorous, with temperatures reaching 32 °C (90 °F). Bacteria, which are inevitably present, may cause appreciable reductions in the yield of alcohol (Dolan, 1976, 1979). The fermented 'beer', which may contain 7–10% alcohol, is distilled in pot stills, which were traditionally made of copper but which are now frequently made of stainless steel. Each still has a bulbous, onion-shaped body, a rising swan neck, a horizontal lyne arm and a condenser. The degree to which the vapours are rectified before they are condensed regulates the composition and hence the flavour of the spirit. Directly heated stills generate pyrolysis products, which give flavours that are absent from spirits rectified in indirectly heated stills. The first and last runnings of each distillation, respectively the foreshots and the feints, are collected separately and are returned to be redistilled. The first distillation yields the low wines, containing 25–30% (v/v) alcohol. These are redistilled in a spirit still to give raw whisky, which at this stage contains 65–70% (v/v) alcohol, is colourless and tastes raw or fiery. It is matured by storing in used wooden sherry casks for 3–12 years. Complex changes occur, some alcohol is lost, materials are leached out from the wood and the flavour improves. Initially the product, which may or may not be blended, is diluted to strength, caramel is added to adjust the colour and it is bottled. Unblended Scottish whiskeys are sold as straight malts. The characteristic flavours of whisky result from the presence of many hundreds of minor substances called congeners. These have been extensively investigated (Briggs, 1978; Kahn, 1979; Nykänen and Lehtonen, 1984). Besides water and major amounts of ethanol, all whiskeys, regardless of origin, contain a range of other alcohols, acids, esters, carbonyl compounds, *N*-heterocyclic compounds and Maillard products. Polysulphides, such as dimethyl trisulphide, and thiophenes have been detected. In Scotch whiskeys, phenols, derived from the peat smoke, are also found. Phenol, *o*-, *m*- and *p*-cresols, guaiacol, *p*-ethyl phenol, *p*-ethyl guaiacol, eugenol and 2,4-dimethylphenol have all been detected at levels exceeding their flavour thresholds, indicating that they contribute to the product's flavour.

Irish whiskey is also prepared by distillation in pot stills. However, the product is redistilled three times, the alcohol content is higher in the later stages than is usual in Scotland, and the grist consists of 40–60% of finely ground, 'low dried' barley (4–5% moisture) with most of the remainder being unpeated barley malt together with a small percentage of wheat or oats. Although other whiskey distillers use unmalted adjuncts in their mashes, the Irish are unique in not pre-cooking the raw barley before it is mixed with the malt.

Scottish grain whisky is made rather differently. The products are used almost wholly in making blends with other whiskeys. The grist consists of 55–98% maize grits, or wheat. These are mixed with the 'fourth water' (the weakest wort) from the previous mash and are cooked with a little enzyme-rich, highly nitrogenous malt, which provides the α-amylase needed to liquefy the starch. After cooking, the mash is cooled, mixed with more enzyme-rich malt and is held at 65 °C (149 °F) to complete the starch conversion and supply nitrogenous and other yeast nutrients. Traditionally a cloudy wort, or wash, is collected, then the grist is remashed and the second wort is added to the first. Two more successive remashes give rise to the third and fourth worts, which are used in the first remash and in the grits cooker, respectively, in the next mash sequence. The unboiled worts are then fermented. Some distillers have adopted the practice of fermenting the whole mash, without first separating the grist solids. The fermented wort, or mash, is distilled in a continuous patent still of a type first devised by Aeneas Coffey. The fermented, pre-heated liquid, with the yeast still present, is combined with wash returned from the next stage and works its way down the first, or analyser column, where a rising stream of live steam strips it of volatiles, including alcohol. The vapour is led to the second, or rectifying, column, where the volatiles are fractionally condensed. The latent heat of condensation is used to pre-heat the incoming stream of wash on its way to the analyser column (Hough *et al.*, 1975). By collecting spirit from selected points on the rectification column, it is possible to obtain liquid containing as much as 95% alcohol and very low levels of congeners.

Whiskeys made in North America differ again in their methods of preparation. They are named with reference to the grist materials. Thus Bourbon whiskey grist contains at least 51% maize. A typical mash bill might be 70% maize, 15% rye and 15% highly diastatic barley malt. Grists for rye whiskeys contain at least 51% rye, as well as wheat and rye malts. The raw grains, mixed with a little enzyme-rich malt, are cooked batchwise or continuously to gelatinize and liquefy the starch. The cooled mash is doughed-in (mixed) with the rest of the malt. After a period of conversion and cooling this sweet mash may be pitched directly with yeast, but if a sour mash is used the mash will be inoculated with de-alcoholized waste from the still. After fermentation, the entire, alcoholic mash is distilled, often in a continuous still. The collected spirit is matured in oak casks, which, in the case of American Bourbon whiskey, are made of new, white oak and are charred on the inside before use.

1.13 Malt vinegar

Vinegars (French: *vin aigré*, soured wine) are prepared by double-fermentation processes. They were being made in Babylonia as early as 4000 BC (Conner and Allgeir, 1976). Malt vinegar was once called alegar, indicating that it was made from ale or unhopped beer. In the first fermentations, a sugary solution derived from malt, grape juice, other fruit juices or honey is fermented anaerobically by yeasts to produce an alcoholic solution. The yeast is removed, and the ale, cider, wine or mead is acetified by bacteria, which, under the aerobic conditions used, oxidize the alcohol to acetic acid (ethanoic acid), the characteristic component of vinegars. A complex mixture of minor components derived from the raw materials and from the metabolic activities of the microbes and changes that occur during maturation combine to give vinegars their individual characteristics. These minor components include ethanol and other alcohols, esters, acetoin, diacetyl, residual carbohydrates, amino acids, polyphenols, proteins, gums, mineral ions, vitamins, glycerol and lactic acid (Prescott and Dunn, 1959; White, 1966, 1970, 1971; Bunker, 1972; Hough *et al.*, 1975; Conner and

Allgeir, 1976; Briggs, 1978). Spirit vinegars are made from rectified spirits (alcohol) supplemented with bacterial nutrients. Other vinegars are themselves distilled before use. These white vinegars are particularly suitable for pickling foods that must not be discoloured. Today, vinegars are used as pickling agents and condiments and in sauces, but in the past they were also used as beverages, as hair rinses and as antiseptics for dressing wounds. World production probably runs in millions of hectolitres.

Typically, the malt vinegar brewer makes his wort from pale, lightly kilned malts (rich in nitrogen and enzymes) mixed with unmalted adjuncts. Usually his milling and mashing equipment resembles that used by beer brewers, but in older types of plant wort might be collected from below, from the sides and from the top of the mash. Generally an infusion mash is prepared at 63–65 °C (145.4–149 °F) or a lower initial temperature may be used to ensure the survival of enzymes that, in turn, will ensure that a highly fermentable wort will be obtained. The mashing liquor is the weakest wort from the previous mash. The first wort is collected as an SG of 1.040–1.080. The spent grains may be sparged with sprays of hot water to recover the residual soluble extract, or they may be remashed once or twice. Stronger worts are mixed and fermented, the weak, last worts being used to make the next mash. The cooled, unboiled worts are heavily inoculated with yeast (*ca.* 5.5 g/l, approximately 2 lb/brl). Yeasts used are *Saccharomyces cerevisiae* or *S. ellipsoideus*, sometimes mixed with highly attenuative *S. diastaticus*. Maximum alcohol production, and hence maximum carbohydrate utilization, is desired and this is favoured by the choice of yeasts, the survival of malt enzymes in the unboiled wort and, sometimes, by the addition of fungal amyloglucosidase. Vigorous fermentation occurs at 24–30 °C (75.2–86 °F) and is complete in 2–4 days, when the alcohol content may reach 8% and the pH is 3.4 or less. The yeast is removed and the beer (liquor; wash; mash) is either directed immediately to the acetifiers, or it may be stored after the addition of preservatives, such as 1% sodium chloride or 70 p.p.m. sulphur dioxide.

For many years, acetification was carried out in partly filled barrels having air holes. A bacterial slime or zoogleal mat, the so-called mother of vinegar, floated on the surface partly supported by a floating wooden grating. In the field process the barrels were in the open, while in the Orleans process they were stored in heated cellars. This process was slow. For acetification to proceed faster, both mixing and better aeration were needed. Various acetifiers are now in use (Prescott and Dunn, 1959; White, 1966, 1970, 1971; Conner and Allgeier, 1976). In the older quick acetifiers, the sump of a wooden vat (up to 160 hl (3500 gal) capacity) is charged with beer. This is pumped to the top of the vessel and is distributed over the packing, which is often of beechwood shavings or birch twigs and is supported on a false bottom. The packing is covered with bacterial slime. As the alcoholic liquor drains over the packing and back to the sump, the bacteria oxidize a proportion of the alcohol to acetic acid. Heat is generated and, by using the chimney effect, is used to draw the required, large volumes of air into the base of the vessel, through downwardly sloping pipes near the base but above the liquid level in the sump. The air escapes from the top of the vessel. In other, otherwise similar vessels, generators, air is pumped in at the bottom rather than being drawn in by convection. The temperature of the beer may be reduced by passing it through a cooler as it is recirculated. Typically acetification is completed in 4–5 days. Newer acetifiers use the technology appropriate to submerged bacterial cultures and may be operated batchwise or continuously. Usually they are large, stainless steel vessels of circular cross-section fitted with a variety of devices to ensure vigorous mixing and aeration. They may be highly automated. Usually they have internal cooling coils and are thermostated to maintain the optimal temperature for the strain of bacteria. This may be as high as 40 °C (104 °F). Large volumes of air are pumped into the base of each vessel through dispersing devices. A 15 s failure of aeration can result in the death of the bacteria, probably by acetaldehyde poisoning. Consequently emergency back-up aeration pumps, with automatic switching designed to cut in within seconds of a failure, are usually installed.

Old, quick acetifiers achieved about 65% of the theoretical yield of acetic acid. Batch-operated, deep-culture acetifiers achieve 90–95% recoveries, while values of 98% are claimed for continuously operated plant. Losses from old-fashioned plant are the result of evaporation of alcohol and acetic acid, the formation of bacterial mass at the expense of the organic substances and the 'over-oxidation' of acetic acid to carbon dioxide and water. Normally pure cultures of acetic acid bacteria are used, strains being chosen for their particular characteristics, such as the taste they give to the product. Acetic acid bacteria are difficult to classify because of their variable morphologies and mutabilities. Originally called *Mycoderma aceti*, they are now classified in the genus *Acetobacter*, e.g. *A. aceti*; *A. rancens*. When the alcohol content of the beer falls below 0.1%, acetification is essentially complete and the bacteria are removed. The clarified product is matured by storage in vats containing chips of beechwood, or other packing, for up to 6 weeks. Air is excluded. Hazy materials separate and minor chemical changes, such as esterification, occur with consequent improvements in flavour. Finally the product may be treated with adsorbents such as PVPP to reduce the haze potential, and it may be centrifuged, filtered or fined. Colour may be adjusted by additions of a caramel, and the product is pasteurized. Malt vinegars contain 4% (minimum) to 8% acetic acid and have pH values of 2.8–3.2 (at 5% total acid). Pasteurized malt vinegar is reasonably stable, but it may become cloudy if infected with microbes that can oxidize acetic acid, or if it 'throws' a protein–tannin haze. It may also become infected with tiny nematode worms, which are indistinguishable from potato eelworms. These vinegar eels (*Anguillula aceti*) swim freely in the vinegar, living on the acetic acid. If they are numerous the vinegar appears to shimmer.

1.14 Other uses of malt

In principle, extracts of malt or malt rootlets might be used in media for growing many microbes and producing very many different products (Prescott and Dunn, 1959). However the bulk chemicals that might be produced are now made almost exclusively from petrochemicals, while some of the fine chemicals such as antibiotics appear to be made using media from which the products can be separated more simply or which have other advantages.

2 Grains and pulses

2.1 The cereal grains, peas and beans

Cereal grains are the fruits of monocotyledonous grasses of the family Gramineae, while peas and beans (the pulses) are the dicotyledonous seeds of members of the family Leguminosae. Barley is widely malted. Smaller amounts of malt are made from wheat, rye, triticale, sorghum, millets, oats, maize and pulses (Chapter 1). Attempts to malt rice have had only limited success (Chapter 15). Unmalted preparations of cereals and sugar (cane or beet) provide the 'non-malt' extract used in brewing and distilling (Chapter 12). The cereals are herbaceous annuals, superficially similar in appearance but differing significantly in details of structure, physiology and biochemistry, even in details of photosynthetic carbon dioxide fixation. They occur in widely separated sub-divisions of the Gramineae (Bowden, 1965; Soderstrom *et al.*, 1988). Consequently, findings cannot safely be extrapolated from species to species. The broad-leaved, tropical species maize, sorghum and the millets are quite distinct from rice. The members of the tribe Festucoideae differ from all of these; oats are clearly different to the more closely related wheats, rye, triticales and barley.

The pulses or legumes are an entirely distinct group. They belong to a large and complex family (Martin, 1984). They are not malted commercially but sprouted seeds of some species, such as mung beans, green gram and black gram, are important foods.

2.2 The barley plant

The barley plant has been described in many publications (Cook, 1962; Briggs, 1978; Gill *et al.*, 1980; Rasmusson, 1985; MacGregor and Bhatty, 1993; NIAB, 1996). The stages of growth of the barley plant are defined by the Large–Feekes scale, or its decimal equivalent (Briggs, 1978; Tottman *et al.*, 1979). They are important in deciding when, for example, fertilizers and fungicides should be applied. The first visible sign of germination is the appearance of the small, white chit (root-sheath, coleorhiza, bud) at one end of the grain. This splits and a variable number of seminal roots (one to ten, often five) emerge and grow into the soil. Meanwhile the coleoptile (leaf-sheath, acrospire) emerges from beneath the husk. When it reaches the soil surface, the first leaf emerges from a pore in the tip. Other leaves appear

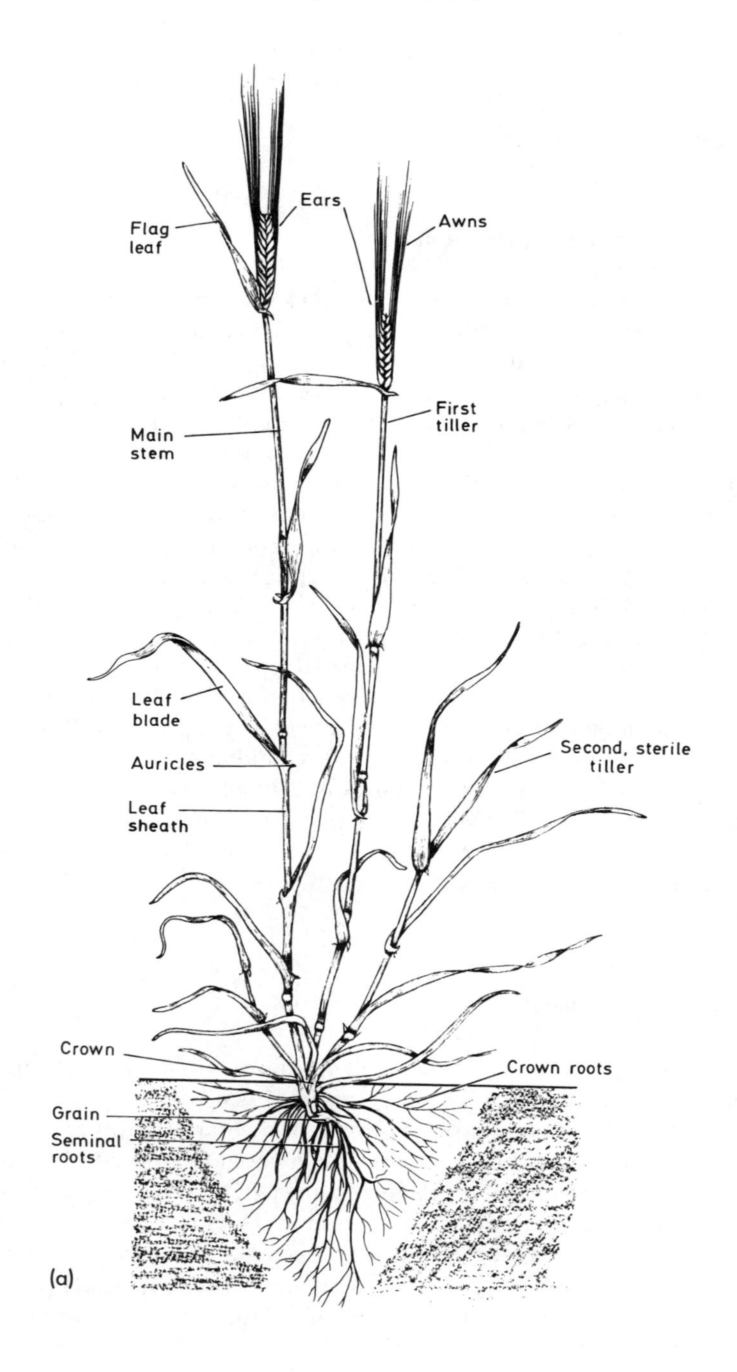
Ears
Awns
Flag leaf
First tiller
Main stem
Leaf blade
Second, sterile tiller
Auricles
Leaf sheath
Crown
Crown roots
Grain
Seminal roots
(a)

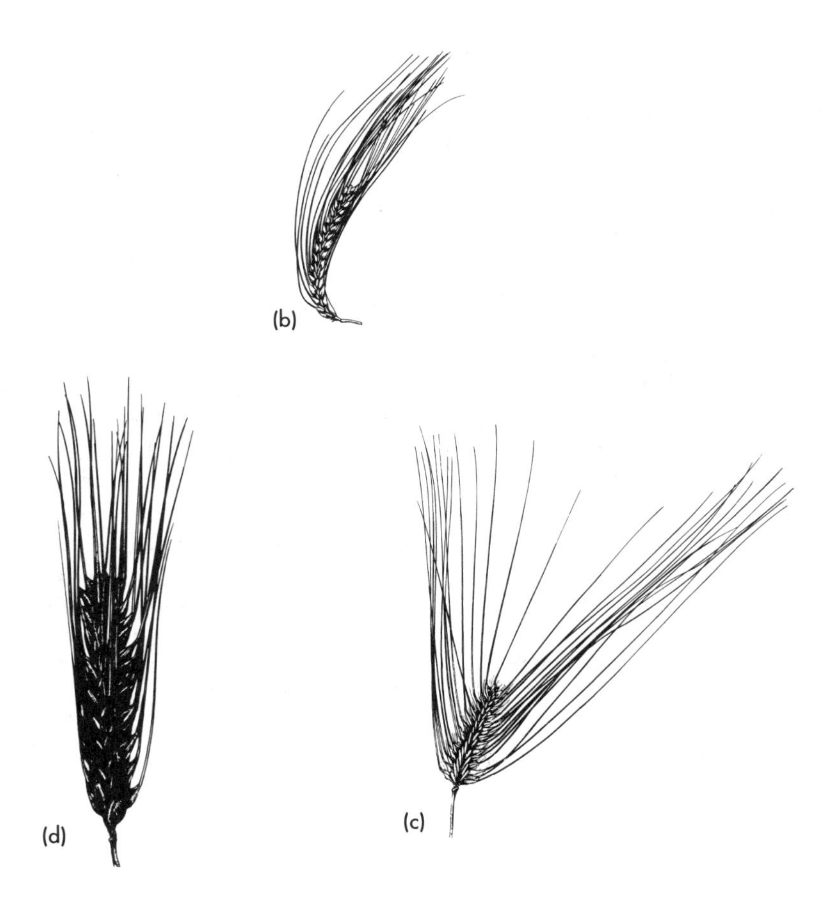

Fig. 2.1(a) A general view of a fully grown but unripe barley plant, having a main stem and two tillers, one of which is sterile, having no ear. The ears are erect (var. *erectum*), and two rowed. (b) A nodding, lax ear of a two-rowed barley (var. *mutans*) (after Briggs *et al.*, 1981); (c) a dense two-rowed ear (var. *zeocrithon*), a type formerly called Spratt, Fan, Battledore or Peacock (after Briggs, 1978); (d) a compact, six-rowed ear of Trebi barley (after Briggs, 1978).

and the plant assumes a rosette form. At a particular stage, the main stem and secondary stems (tillers) elongate. Each elongated leaf consists of a strap-like part extending outwards and a sheath that is rolled around the stem. Each originates at a swollen node or joint on the stem. Where the blade joins the sheath it is extended sideways as two small auricles and upwards as a small, papery ligule, which acts as a sliding seal around the stem. The bases of the main stem and tillers swell at the soil surface and form a crown from which secondary, adventitious or coronal roots emerge, creating a second root system (Fig. 2.1a). The sizes and relative importance of the root systems depends on many factors, but in well-drained soil they may reach to a depth of 1.8 m (6 ft). Potentially many tillers may form, but in practice numbers are controlled by the competition between the plants. One or two tillers surviving to harvest would be usual. As flowering time approaches the bases of the upper, flag-leaves of the main stem and fertile tillers swell as the heads, or ears, the inflorescences (technically spikes), enlarge. Final stem heights depend on cultural conditions and the variety. Older barleys often reached 150 cm (*ca.* 60 in) in height, but now shorter stems of 60–90 cm (*ca.* 24–36 in) are more common. Shorter, stronger stems are preferred because they are less likely to bend, break or collapse (lodge), so the grain is easier to harvest. At first, the ear is confined within the sheath of a flag leaf, known as the boot. Usually it emerges at flowering time. The axis of the ear is the rachis, which is a modified extension of the stem; in cultivated barleys this must be tough and firmly attached to the stem to minimize grain losses. A triad or group of three spikelets (each containing a single, simple flower) is situated at each node. Triads occur on alternate sides of the rachis. In varieties in which all the spikelets are fertile, giving rise to grains, the ear carries six rows of grains and is termed 'six-rowed' (Figs 2.1d, 2.2a and 2.3). The central spikelets produce symmetrical grains; the grains of the lateral spikelets are skewed, but in opposite senses (Fig. 2.4a). In many cultivars the lateral spikelets are small and sterile. Grain forms exclusively in the central florets and tends to be plump and symmetrical. The ears carry two rows of the grain and are termed two-rowed (Figs

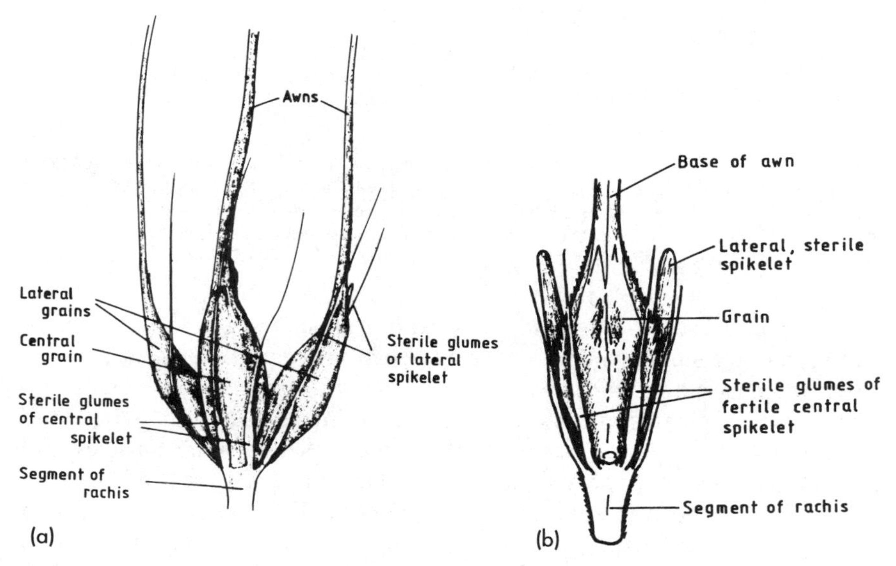

Fig. 2.2 Triads, or triplets, of spikelets of (a) a six-rowed barley (after Wiebe and Reid, 1961) and (b) a two-rowed barley (after NIAB, 1996)

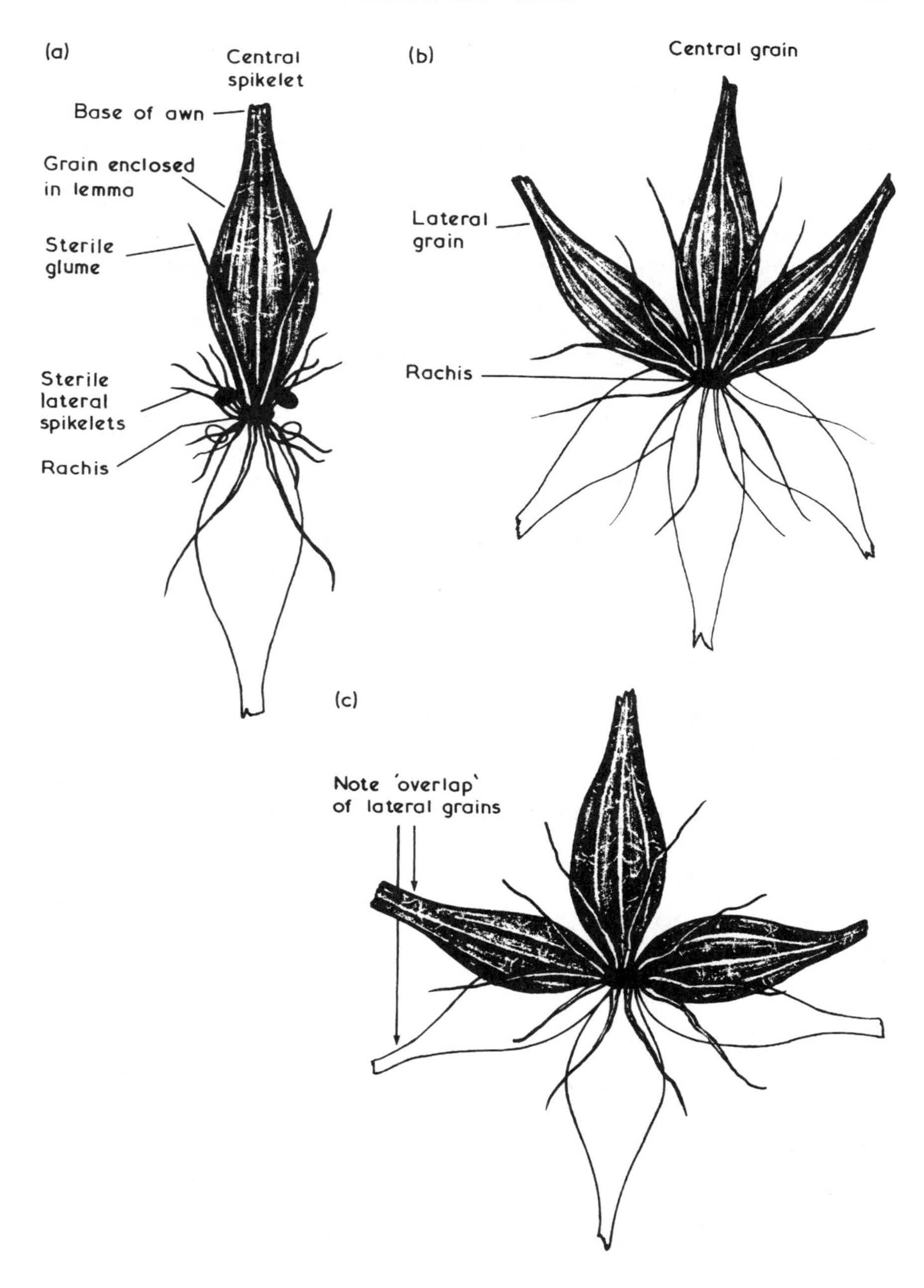

Fig 2.3 The general arrangements of grains in barley ears, seen from below. The grains shown in most detail arise at one node. The grains shown in outline might be of the node above or the node below (after Briggs *et al.*, 1981). (a) A two-rowed ear; (b) Six-rowed, moderately dense ear; (c) Lax, six-rowed ear; a so-called four-rowed form.

2.1a–c, 2.2b and 2.3). When the rachis internodes are short, the grains are closely packed and are pressed together, giving a dense ear (Fig. 2.1a). When the internodes are longer the grains have more space and the ears are lax (Fig. 2.1b). In two-rowed barleys, dense ears tend to be upright (*erectum* forms) and in some cases the grains and awns may be spread to give fan-type ears (*zeocrithon* type) (Fig. 2.1c), while lax ears tend to bend over and nod (*nutans* forms) (Fig. 2.1b). When six-rowed barleys have lax ears, the lateral rows of grains tend to overlap, giving rise to the misnamed four-rowed barleys (Fig. 2.3c) (Briggs, 1978). The traditional malting barleys of Europe are two-rowed, and these are grown in many other regions. However, worldwide, six-rowed barleys are more widely grown and many are malted.

At flowering, each fertile spikelet consists of an inconspicuous pedicel, or short stalk, attached to the rachis and carrying two small, green sterile glumes and two larger bracts, the flowering glumes. The glumes are (i) the lemma (palea inferior), which is furthest from the rachis and in most varieties is extended as a long, thin awn (Figs 2.1 and 2.2) or, rarely, as an appendage called a hood or with no extension and (ii) the palea (palea superior), which is nearest to the rachis. The lemma and palea form the husk of the mature grain. The awns give barley ears their bearded or brush-like appearance. Within the flowering glumes, the simple flower consists of a unilocular ovary with two feathery stigmas, three anthers and two small, hairy scales, the lodicules. The inflorescence is designed for wind pollination. A little cross-pollination does occur, but usually about 99% self-pollination occurs, favouring the maintenance of varietal purity (Briggs, 1978). Two fertilizations are necessary for grain formation, one between a

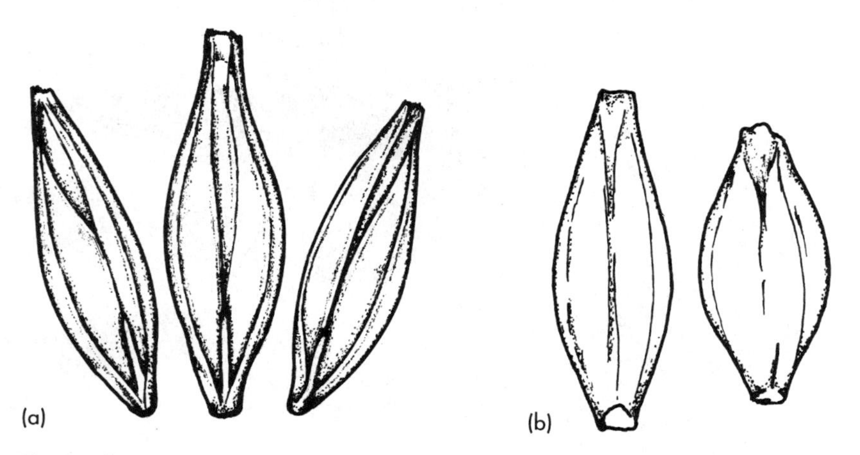

Fig. 2.4 (a) Three grains of a six-rowed barley, viewed from the ventral (furrowed) side, showing a symmetrical central grain and two asymmetrical lateral grains. (b) Symmetrical grains of two, two-rowed barleys. One is elongated and comparatively slim; the other is short and plump.

haploid pollen nucleus and the egg cell to give a diploid embryo ($2n = 14$), and another between a second pollen nucleus and two polar nuclei, giving rise to the triploid tissues of the aleurone layer and the starchy endosperm ($3n = 21$). As these fertilized cells grow, divide and differentiate into tissues, they crush and displace the surrounding nucellar tissue, of which traces can be found in the mature grain as the hyaline layer, in the sheaf cells and possibly in the crushed cell layer. These tissues are invested by the testa (spermoderm, tegmen, seed-coat), which is formed from the integument of the nucellus and the inner integument of the ovary and is the limit of the true seed. In turn the testa is enclosed by and is fused to the pericarp (fruit coat), a maternal tissue derived from the ovary wall. Grains are fruits (caryopses) because the seeds are enclosed within pericarps. Some barleys have 'naked grains' in which the pericarp is the outer layer, as is the case in wheat and rye. In the great majority of cases, and in all malting barleys, the grains are husked. During grain development, the palea and lemma become glued to the pericarp and become the husk. Therefore, to call barley grains (corns, or kernels) seeds is incorrect.

2.3 The quiescent barley grain

There is considerable variation in quiescent barley grains (Cook, 1962; Rohrlich and Brückner, 1966; Briggs, 1978). Even within grains of one variety there are differences in size and in the relative proportions of the parts. Lengths are commonly 6–12 mm, widths 2.7–5.0 mm and thicknesses 1.8–4.5 m. Varieties are known with thousand corn dry weights (TCW)

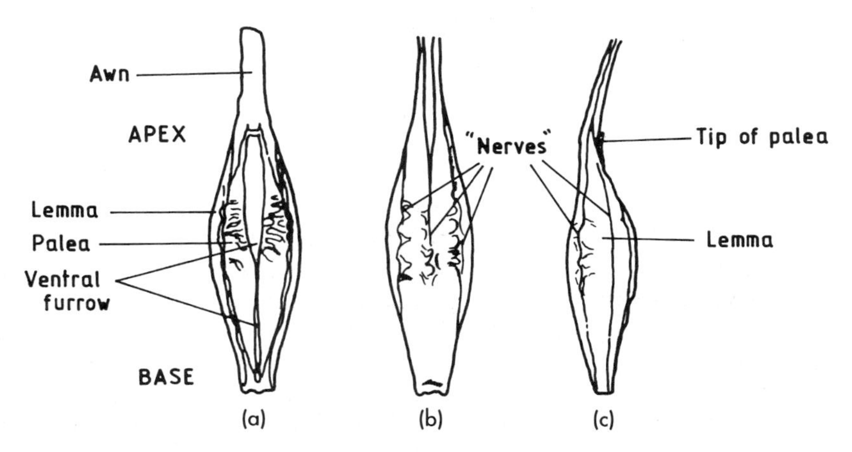

Fig. 2.5 Three views of a barley grain, with the base of the awn still attached. (a) Seen from the ventral, furrowed side; (b) seen from the dorsal, rounded side; (c) a lateral view (after Briggs *et al.*, 1981).

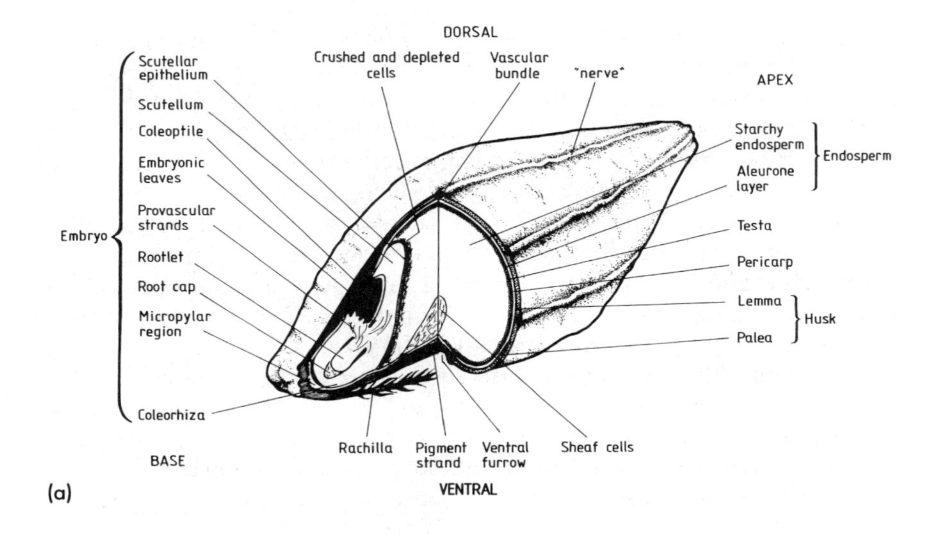

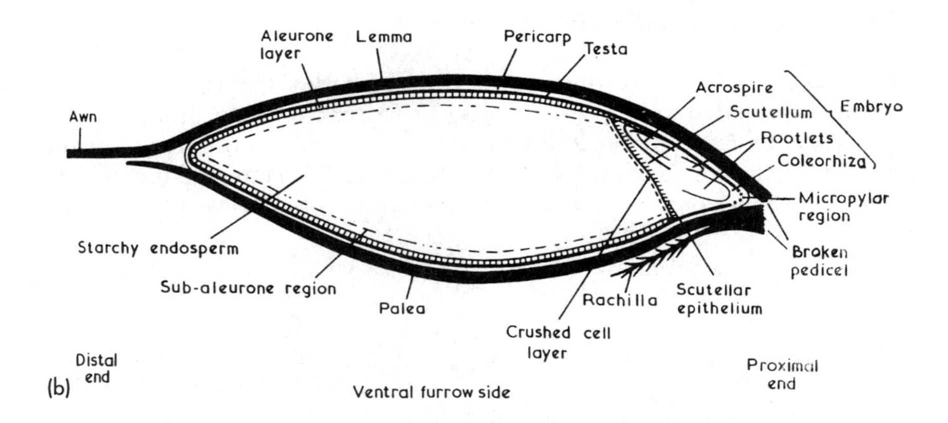

ranging between 5 and 80 g. Barley hectolitre weights range between 50 and 76 kg. Malting barleys are commonly screened over 2.2, 2.3 or 2.5 mm slotted sieves to retain the plump grains. TCW values of 32–44 g and 30 g are usual for two- and six-rowed varieties. Grains appear elongated, roughly cigar shaped and slightly compressed (Figs 2.4, 2.5 and 2.6). The proximal, basal or embryo, end may still have the pedicel in place, or this may have been roughly broken away during threshing. The apical end may carry the base or stub of the awn but this too may have been removed by rough handling. The husk completely encloses the grain, the lemma on the dorsal rounded side and the palea on the ventral furrowed side. The lemma

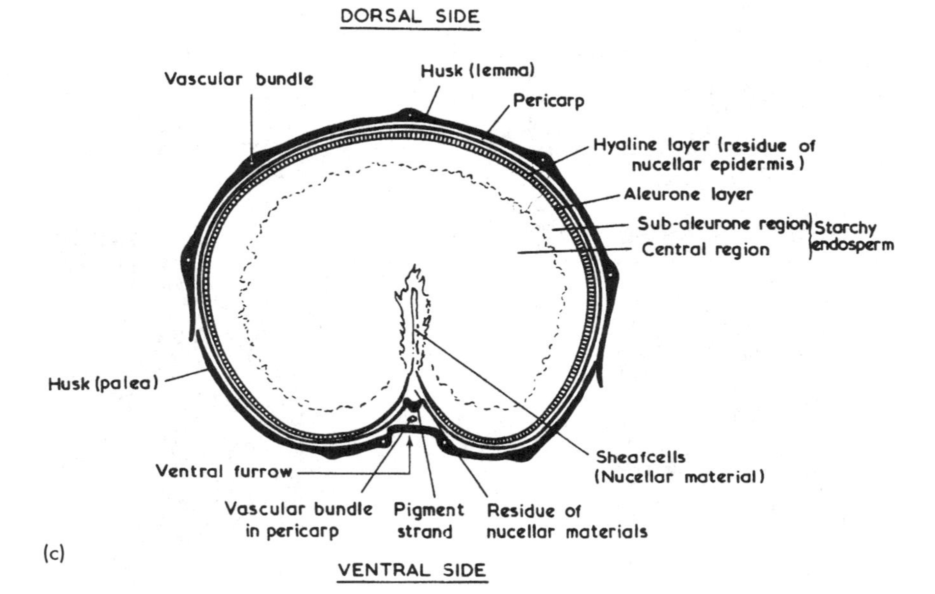

Fig. 2.6 The barley grain. (a) A grain, with a sector removed to show the disposition of the tissues; (b) a longitudinal section, to one side of the furrow and the central structures; (c) a transverse section through the broadest part of a plump barley grain (after Briggs, 1973; Briggs *et al.*, 1981).

overlaps the palea along the edges. Sound, ripe barleys are usually pale, golden brown with darker or even reddish veins. However, varieties are known in which grains are various shades of black, blue, green or red. At present the only coloured barleys known to be grown commercially are greenish in appearance. The colour results from their blue aleurone layers showing through the yellow husk. The rachilla, or basal bristle, can often be found in the furrow, even in threshed grains. Each palea has two vascular bundles running along its length, while each lemma has five. Vascular bundles are detectable externally as small ridges known as veins or nerves, which may be pigmented and may carry small 'teeth'. The husk contains about four types of cell, including a layer of fibres (Fig. 2.7). It has cuticles on the outer and inner epidermal layers. It is the only grain tissue that is lignified and its external epidermis is silicified. This makes barley grain very abrasive and causes wear on conveying machinery. The husk, the pericarp and the testa are dead tissues in mature grain. Even in healthy grains, microbes occur on the husk, possibly in the husk tissues, between the husk and the pericarp and even within the pericarp tissue. In diseased grains, the microbes may even have penetrated the living tissues. Here and there air gaps occur between the husk and pericarp (Fig. 2.8). The lodicules are attached to the pedicel and fit against the embryo between the husk and the

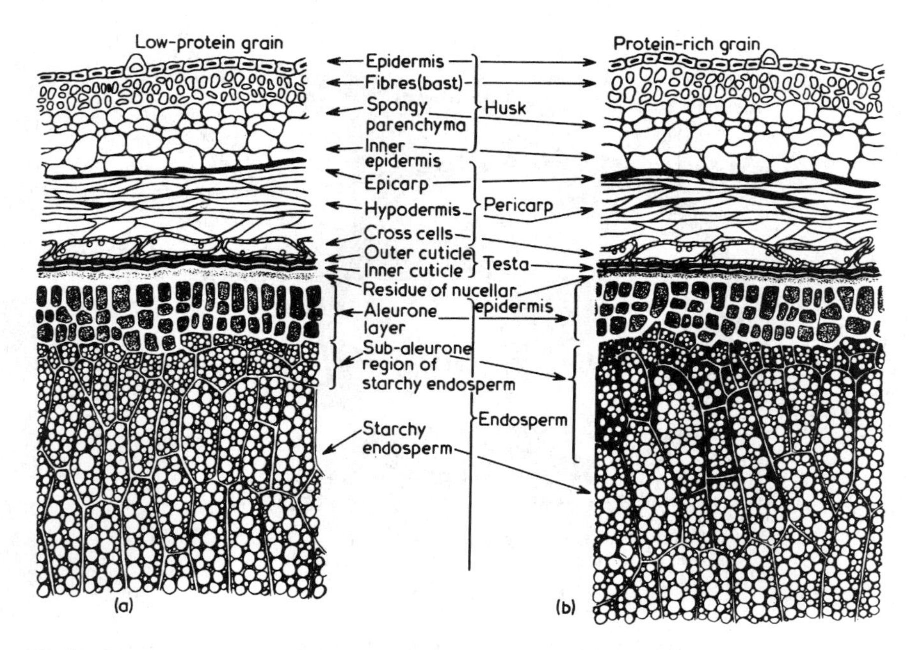

Fig. 2.7 Transverse sections on the dorsal sides of grains. (a) A protein-poor (low nitrogen) grain; (b) a protein-rich (high nitrogen) barley grain. (After Briggs, 1978; copyright Academic Press.)

pericarp (Fig. 2.9a–c). Together the husk, pericarp and lodicules make up about 9–14% dry weight (d.wt) of malting barleys (Table 2.1).

The pericarp consists of several types of cell. Its outermost layer is a cuticle, which adheres to the inner face of the husk. In contrast to the wheat pericarp, this tissue shows no signs of separating into layers in the mature grain (Fig. 2.7). The inner face of the pericarp is fused to the testa or the pigment strand (Fig. 2.6a). By the pigment strand, in the furrow region, the pericarp is thicker and contains a single vascular bundle (Fig. 2.bc).

The testa makes up perhaps 1–3% of the grain's weight. This thin, dead and rather featureless tissue is made of two waxy cuticles separated by the remains of cell walls. The outer cuticle is the thicker. Polyphenols (proanthocyanins) occur between the cuticles in most varieties. The testa is thickest over the flanks of the furrow and at the apex of the grain and is thinnest over the embryo. In the micropylar region it has a more porous structure (Fig. 2.8). At the furrow, the edges of the testa do not meet, the gap being closed by the pigment strand; this is a tissue made of dead cells in which the thick brown cell walls are suberized and enclose dark, waxy material (Fig. 2.6ac). The husk and pericarp protect the grain mechanically and are barriers to the diffusion of gases, water and solutes. In many respects it is the testa and pigment strand that contribute most to the

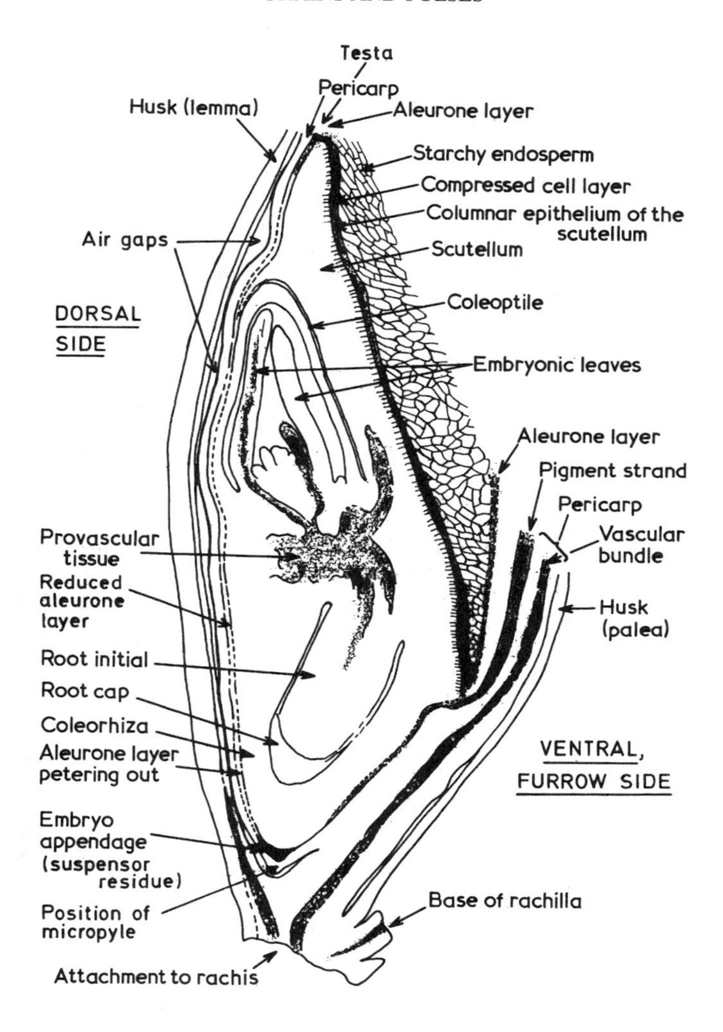

Fig. 2.8 A longitudinal section of the embryo or basal end of a barley grain (after Lermer and Holzner, 1888; Briggs, 1978).

separation of the living tissues within the grain from the outside environment. Grains may be decorticated with 50% sulphuric acid, when the husk, pericarp and lodicules break up and separate leaving the remainder of the grain intact, a striking demonstration of the resistance and impermeability of the testa to strong chemical agents.

Residues of the nucellus occur as the thin, hyaline layer, which is situated between, and fused to, the testa and the aleurone layer as the crest of sheaf cells; this extends, to various extents, into the starchy endosperm above the

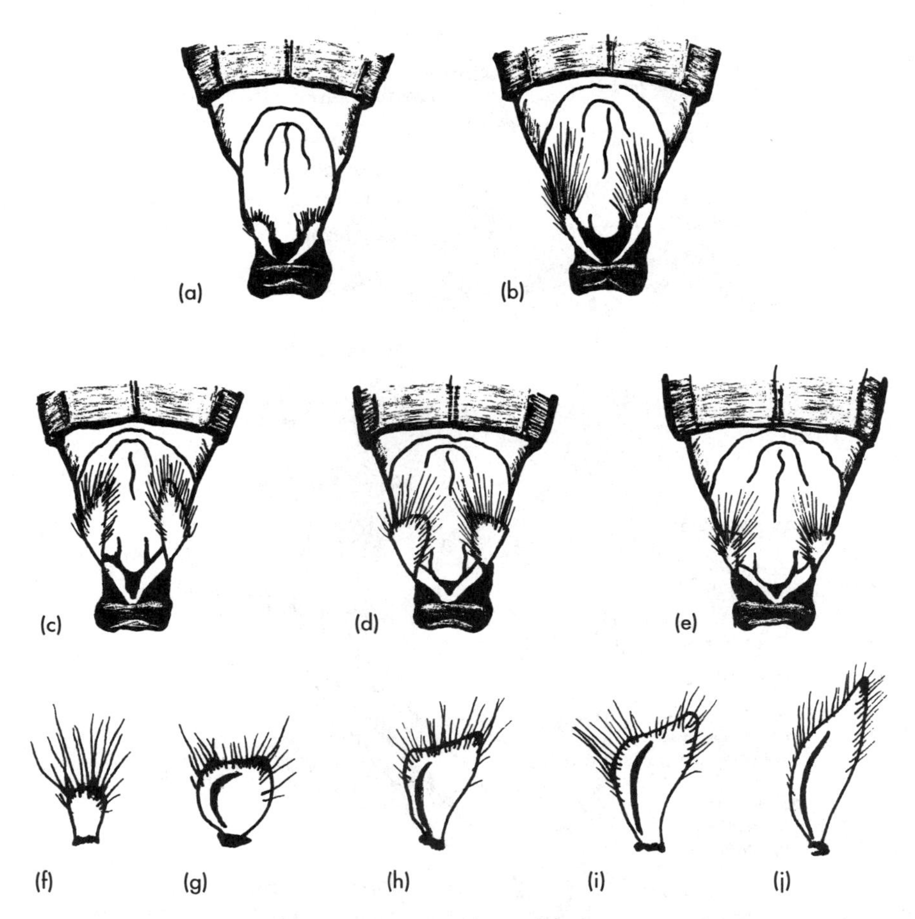

Fig. 2.9 The positions and shapes of lodicules. (a–e) Lodicules in place, next to the embryo, but with the lemmae removed (after Renfrew, 1973); (f–j) lodicules of different shapes and sizes (after EBC, 1978). In (a) and (b) bib-type lodicules with short and long hairs, respectively, are shown; collar-type lodicules are shown in (c) to (e) with long, medium and short lodicules, respectively.

pigment strand (Figs 2.6a and 2.10) and, possibly, between the embryo and the starchy endosperm as the 'crushed' or depleted cell layer (Fig. 2.8; Briggs and MacDonald, 1983b). In grain sections, the hyaline layer appears bright and featureless unless it is swollen with caustic alkali, when cell walls can be detected. The sheaf cells lack contents and consist of cell walls closely pressed together. In some grains these may tend to pull apart as the grain swells during water uptake. The cell walls of the depleted layer, between the embryo and the starchy endosperm, are crushed together and the cells lack contents. Unlike the cell walls of the starchy endosperm, the walls of the embryo, the aleurone layer, the crushed cell layer, the hyaline layer and

Table 2.1 The approximate proportions of the different parts (% d.wt.) in cereal grains. The exact proportions will vary with grain size, variety, etc.

	Husk or hull	Pericarp plus testa	Aleurone layer	Starchy endosperm	Embryo or germ
Barley	10–13 (+ pericarp)	3 (testa)	5	76–82	2.3–3.5
Wheat	–	7.5–9.5	6.5–7.0	81–84	3
Rye	–	10[a]		86.5	3.5
Oats					
Entire	25	9[a]		63	2.8
Kernel	–	12[a]		84	3.7
Rice					
Entire	20	4.8[a]		73	2.2
Kernel	–	5–7[a]		90–92	3.3
Maize	–	5.0–6.5	2–3	76–80	10–15
Sorghum	–	6–8[a]		82–84	10
Millets					
Proso	16	3	6	70	5
Milo	–	5.5[a]		84	11
Feterita	–	6.6[a]		86	7.3

After Rohrlich and Brückner (1966); Briggs (1978); Hulse *et al.* (1980); Kent (1983); Watson (1987).
[a] Combined figure for pericarp, testa and aleurone layer.

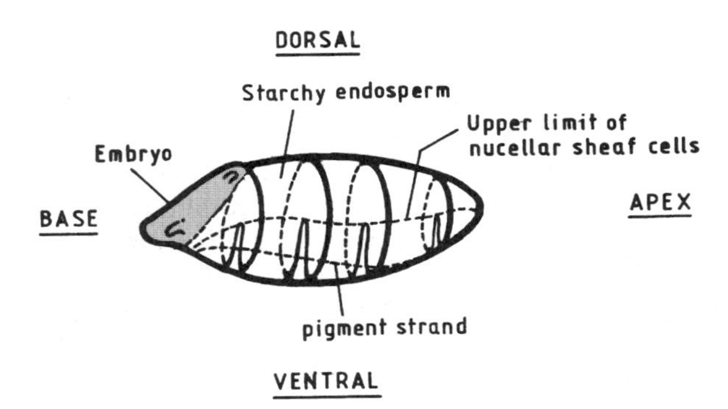

Fig. 2.10 The positions of the pigment strand and the innermost extension of the sheaf cells, which are residual nucellar tissue (after Collins, 1918; Briggs and MacDonald, 1983a).

the sheaf cells all fluoresce strongly in ultraviolet light and are comparatively resistant to enzymic degradation (Briggs, 1987b).

The aleurone layer is a living tissue that respires (reducing tetrazolium salts to coloured formazans) and metabolizes, but its cells do not grow or divide during grain germination. The hyaline layers and the aleurone layer together make up 8–15% of the dry grain (Figs 2.6, 2.7 and 2.8). Botanically this tissue is part of the endosperm but its properties are completely different to those of the starchy endosperm. The triploid aleurone cells are cuboidal, arranged in a layer that is usually about three cells thick (50–110 μm). One side of the aleurone layer is fused to the hyaline layer. The other is attached to the starchy endosperm, which it invests except at the sheaf cells, over the furrow and at the interface with the embryo. This tissue thins to a single layer of smaller, flattened cells, which extend partly over the surface of the embryo (Fig. 2.8). The walls of the aleurone cells are thick (3–5 μm) and are made of the two distinct layers that differ in their resistance to enzymolysis. They are crossed by plasmodesmata. The cells have dense cytoplasm and prominent nuclei and they stain with tetrazolium salts. Starch is absent, but the tissue is rich in oil, held in spherosomes, and mineral, protein and possibly carbohydrate reserves deposited in aleurone grains. Mitochondria, Golgi bodies, microbodies, rough endoplasmic reticulum and proplastids have also been detected. The aleurone layer is a major reserve of phosphate and other mineral ions (ash). It is also the source of many of the enzymes that cause modification.

The diploid embryo is alive: it respires, it stains with tetrazolium salts, and its axis grows and gives rise to the mature plant. The quiescent embryo is 3–5% of the dry weight of the grain. It consists of an elongated axis joined at the primary node to the expanded and flattened scutellum (Latin: shield; Figs 2.6 and 2.8). The axis consists of the shoot (the coleoptile surrounding three to four leaf initials) pointing towards the apex of the grain and the root initials, with their caps, surrounded by the coleorhiza, or root sheath, pointing towards the base. The number of rootlets varies (one to ten, often five). Between the tip of the coleorhiza and the micropyle there is a pad of dead cells, the 'embryonic appendage', which may be the residue of the suspensor. Its location by the micropyle suggests that it may contribute to the selective permeability of the surface layers of the grain (Chapter 3). During germination the axis grows, but the barley scutellum (unlike that of oat) does not. Provascular traces extend between the scutellum and the axis. The flattened side of the scutellum, facing towards the endosperm, is surfaced with a single layer of columnar epithelial cells. In the dry grain this adheres to the crushed cell layer, but in the hydrated grain the adhesive layer is weakened and allows the embryo to be removed. Most of the embryo's cells are parenchymatous, with thin walls crossed by plasmodesmata. Generally starch is absent, but the cytoplasm contains nuclei, mitochondria and other sub-cellular organelles.

The largest tissue in the grain, its major foodstore and the major source of malt's extract, is the triploid starchy endosperm or, less exactly, the endosperm (Figs 2.6, 2.7 and 2.8). This dead tissue, together with the sheaf cells and the depleted cell layer, makes up some 75–80% of a grain's dry weight. In transverse section, the thin walls (*ca.* 2 μm) of the polygonal cells are seen to radiate from the sheaf cells. Next to the aleurone layer, in the sub-aleurone region, the endosperm cells are smaller and more cuboidal. The cells are packed with starch granules embedded in a protein matrix that also contains some small protein granules. The starch granules are concentrated into two main size ranges, 1.7–2.5 μm and 22.5–47.5 μm; there are differences between barley varieties (Briggs, 1978). In the sub-aleurone region, particularly in protein-rich (high-nitrogen) grains, starch granules are less closely packed, there is a higher proportion of small granules, the protein matrix is more apparent and, indeed, the concentration of β-amylase is elevated (Fig. 2.7; Chapter 4). Small protein bodies occur, embedded in the matrix. No other organelles have been detected in mature endosperms, but in immature, incompletely ripened grains the remains of nuclei may be distinguished. Their presence has been used as a criterion of unripeness.

When plump, well-filled grains are cut across they have a rounded, kidney shape, while poorly filled thin grains have an angular appearance (Fig. 2.11). The endosperm of a good malting barley should have an opaque, white, floury or 'mealy' appearance owing to the numerous small cracks and air spaces in the matrix around the starch granules. In contrast, the endosperms of poor malting-quality grains, which are often immature and/or rich in protein, are more or less grey and translucent: known as hard, horny, vitreous, glassy or steely. The specific gravity of some steely endosperms was reported to be 1.345 as against 1.305 for mealy endosperms. Grains that show 'permanent steeliness' tend to malt badly, while mealy grains and grains in which steeliness disappears on wetting and re-drying malt better. Steely endosperms are under tension, since attempts to cut thin sections causes them to become mealy. In steely grains, the endosperms are more tough and dense than those of mealy grains; protein adheres more strongly

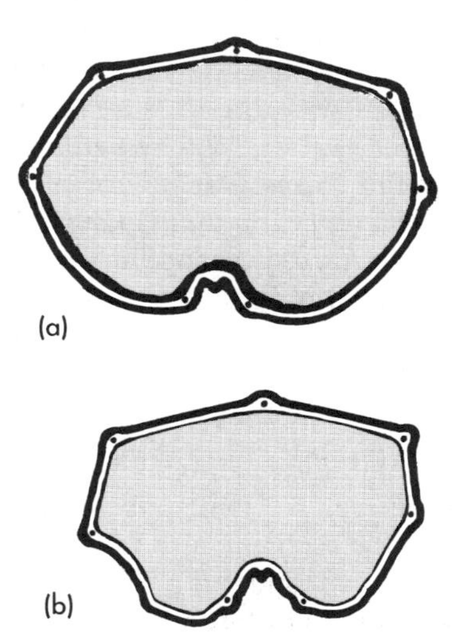

Fig. 2.11 Transverse sections through the broadest parts of (a) a plump and well-filled barley grain; (b) a thin, poorly filled grain (after Percival, 1902).

to the starch granules in steely grains and water is taken up more slowly by steely grains during steeping, probably because their tough structure resists swelling.

2.4 Barley classification and varietal identification

All cultivated barleys and the weed grasses *Hordeum spontaneum*, *H. agriocrithon*, *H. paradoxon* and *H. lagunculiforme* belong to one species and should, therefore, have one specific name. In practice, for cultivated barleys, *H. vulgare* (Linn) and *H. sativum* (Jessen) are used most often. Many thousands of types and strains (varieties; cultivars) of barley are available (Carson and Horne, 1962; BC, 1976; Briggs, 1978; Anon., 1996; NIAB, 1996). Incorrectly these have often received Latin binomial, 'specific' names. Maltsters have to distinguish between grains of different varieties. To help in this, 'keys' are available. Distinctions are often based on quite small morphological differences. There are four important groups of barleys, known by their (incorrect) Latin names.

- *Hordeum vulgare*: six-rowed barleys, with all grains of about the same size. The lemmae are awned or hooded. Many malting barleys occur in this group but are rarely used in the UK (Fig. 2.1d). Sometimes hooded barleys with naked grains are called *H. trifurcatum*.
- *Hordeum intermedium*: six-rowed barleys in which the lateral grains are small and have lemmae with no awns or hoods. The wide spread of grain size that occurs makes these barleys unsuitable for malting.
- *Hordeum distichon* (syn. *H. distichum*): two-rowed barleys with reduced lateral florets having rudimentary sexual organs (Fig. 2.1a–c). Subdivisions include cultivars with broad, spreading, fan-, or battledore-shaped ears (*H. zeocriton* or *H. zeocrithon*; Fig. 2.1c); broad, parallel-sided, erect-eared forms (var. *erectum*) and narrow-eared, lax forms that bend over or nod (var. *nutans*) (Fig. 2.1b). Traditional European malting barleys fall into this group, in which the grains are plump and comparatively even in size.
- *Hordeum deficiens* (syn. *H. decipiens*): two-rowed barleys in which the lateral florets are so reduced that no traces of sexual organs remain. These varieties seem of little practical importance.

Not all barleys are suitable for malting and those that are vary in their malting characteristics. Therefore, maltsters must evaluate varieties for their maltabilities and be able to distinguish between the grains of different cultivars. They must be able to detect if a sample of grain is pure, or if it contains a mixture of varieties. This is an important and difficult task, and operators require practice and training to carry it out well. Unlike breeders, who can use all the characteristics of a plant from seed to maturity for recognition purposes, a grain buyer must use the, often minor, differences in grain morphology.

Only husked barleys are used for malting. Naked barleys are found to be unsuitable although they may find a place in foodstuffs. Awns are broken and removed in threshing. In heavily threshed grains even the bases of the husks may be broken, reducing further the number of characters available for grain recognition. Grain should be inspected against a black background under a northern light or a carefully chosen artificial equivalent. A lens or low-power binocular microscope, a handbook and a reference collection of barleys should be available. The sample must be free from significant admixture and should have the appearance characteristic for a healthy sample of that variety. It must also be uniform, in the sense that all the grains in the bulk that the sample being inspected represents are of one quality. Good-quality grain must not be diluted with inferior (thinner, damaged or diseased) grain, even of the same variety.

Two-rowed and six-rowed grains can be distinguished by differences in grain symmetry (Figs 2.3 and 2.4). Grain shapes vary within a sample and between samples of one variety grown under different conditions, but they

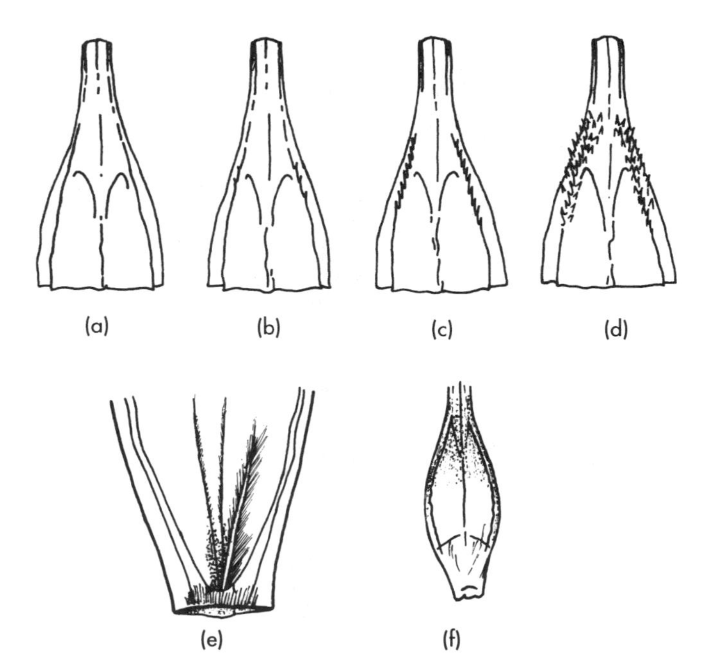

Fig. 2.12 (a–d) Increasing amounts of spicule developments on the inner nerves of four barleys (after Renfrew, 1973). Spicules may also occur on the outer nerves. (e) Minute hairs on the inner edges of the furrow, or the palea, beneath the rachilla (after NIAB, 1996); (f) rough 'sharkskin', an area of the lemma covered with minute teeth (after NIAB, 1996).

also vary between varieties. Some are short and plump, while others are thin and elongated; some have a depression and are 'dished' over the embryo. There are variations in how far along the grain the widest part occurs. Grains can vary widely in colour, but of those grown commercially the important distinctions are between grains with or without blue aleurone layers and between grains lacking or with varying degrees of red or purple pigmentation in the husk, usually along the nerves. Husks may appear coarse or fine and delicately wrinkled or even smooth, but these characters are strongly influenced by differences in growing conditions. In some varieties, the lemma does not fully overlap the palea along the sides of the grains, so the pericarp can be seen in the 'gape'. Husks also differ in the presence or absence of spicules (barbs) on the inner or outer nerves of the lemmae (Fig. 2.12a–d), and in the presence and distribution of hairy or rough patches (Fig. 2.12e) or 'sharkskin', caused by small barbs between the nerves (Fig. 2.12f). Rachillae, present in the base of the ventral furrow, are classified as long, medium or short and as having long or short hairs (Fig. 2.13). Rachillae may be lost in threshing, but rachilla characteristics are stable, unlike some others that may vary between seasons or even, in extreme cases, between grains on one ear. Ventral furrows vary in shape

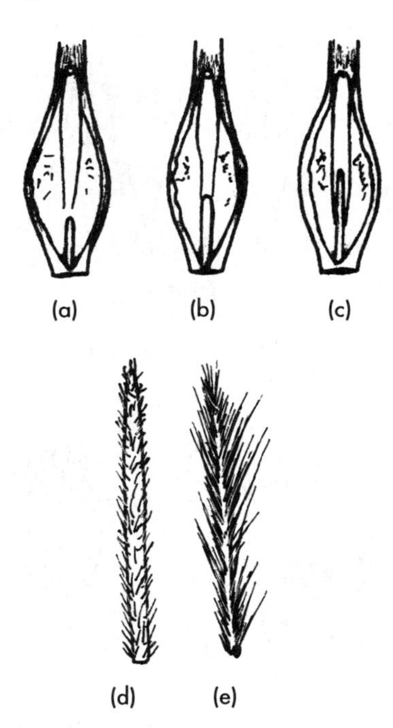

Fig. 2.13 (a–c) The ventral sides of three grains, showing the relative lengths of short, medium and long rachillae (after NIAB, 1996); (d, e) rachillae with short and woolly hairs and with long, straight hairs, respectively.

from wide, open and V-shaped to narrow and almost closed. The flanks may carry hairs in various patterns or they may be hairless or 'glabrous'. The tips of the paleae vary in shape; they may be rounded, square-shaped or pointed (Fig. 2.14a–c).

If unbroken, the shape of the base of the lemma is often a useful characteristic in identifying grains, although it is variable in some varieties (Fig. 2.15a–i). The major types are those in which the base of the lemma is (i) plain (*spurium* forms), (ii) bears an angular depression, (iii) has a centre that is raised, (iv) is bevelled (*falsum* form), and (v) carries a complete transverse nick (*verum* form).

The lodicules are revealed by pealing the lemma from over the embryo. These vary widely in size, shape, hairiness, position and the ability to retain stains. Consequently they are very useful in varietal identification and complex classification systems have been based on them. The main divisions are between small, 'parvisquamose' bib types, of varying degrees of hairiness, and large, 'latisquamose', collar types, which also vary in hairiness and lie curved partly around the embryo (Fig. 2.9).

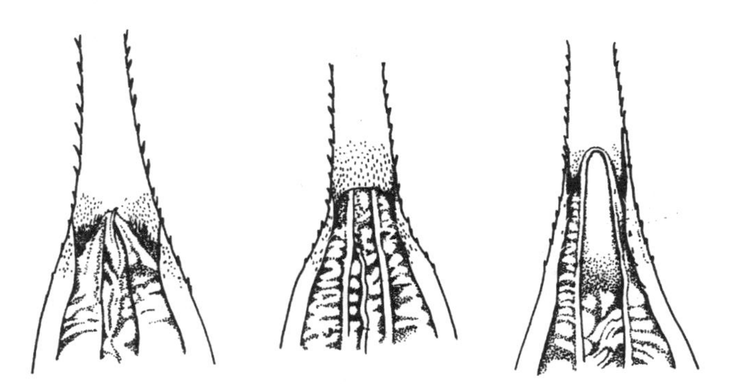

Fig. 2.14 Three types of palea apex. Note how the lemmae overlap the paleae at the sides, the awns (all with teeth at the sides) projecting at the top and the varying degrees of hairiness within the lemma, at the base of the awn. (After Bergal and Friedberg, 1940).

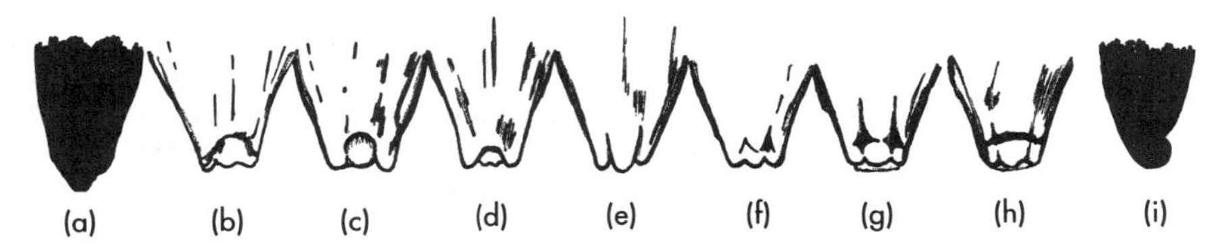

Fig. 2.15 Types of lemma bases of barley grains. (a, b) Vertical section (black) and dorsal view of base of a grain with a bevelled base (*falsum* form); (c, d) two forms with more deeply bevelled bases; (e) base with raised centre; (f) plain base (*spurium* form); (g) base with angular depressions; (h, i) surface view and section (black) of the base of a grain having a 'nick' (*verum* form). (Mostly after NIAB, 1996.)

Disputes regarding the purity of grain samples readily arise. An additional test increasingly used in these cases is based on the characteristics of the hordeins: alcohol-soluble proteins that are easily extracted from grains. These vary in their detailed characteristics and so give rise to different banding (or peak) patterns when separated by electrophoresis or, less usually, high performance liquid chromatography (HPLC). The patterns are characteristic for groups of varieties. Probably, in the future, it will be possible to distinguish the variety of an individual grain by the electrophoretic patterns given by other proteins, including enzymes, and ultimately by using DNA fingerprinting.

2.5 Other cereal grains

The varieties of the other cereal grains vary as much, in detail, in their characteristics as those of barley, the structure of which they generally resemble. (Percival, 1921; Hector, 1936; Rohrlich and Brückner, 1966; MacMasters *et al.*, 1971; Bushuk, 1976; Baum, 1977; Gill *et al.*, 1980; Hulse *et al.*, 1980; Kent, 1983; Watson and Ramstad, 1987; Doggett, 1988; Welch, 1995). The grains mentioned here are either malted (Chapter 15) or they are used to make adjuncts (Chapter 12). The grains of most of the cereals are naked (thresh free, i.e. they do not have adherent husks). The embryos of the 'tropical cereals' make up relatively large proportions of the grains (Table 2.1).

2.5.1 Wheats

The wheats (*Triticum* spp.) include diploid ($2n = 14$), tetraploid ($2n = 28$) and hexaploid ($2n = 42$) species. It seems that only hexaploid bread wheats (*Triticum aestivum* L.) are now malted. Wheat malts are used in brewing beer (especially German-style weissbier/weizenbier) and in baking. Wheat grains are naked, ovoid, generally rounded and have deep furrows or creases (Fig. 2.16; see also Fig. 2.22a, below). Without husks, the grains lack rigidity, which can cause compaction and handling problems during malting when the grain is wet. The exposed acrospires (coleoptiles, shoots) are easily damaged by turning during malting. The pericarp has a weak plane, which can allow the outer part to separate and come away as 'beeswing'. In red wheats the colour is situated in the testa. The embryo resembles that of barley, except for the presence of a small scale (the epiblast) on the outer surface, which may be the evolutionary relic of a second cotyledon or scutellum. The aleurone layer is only one cell thick and the endosperm varies in texture in different varieties. TCW values of commercial varieties vary widely: 15–60 g with about 40 g being usual. Hectolitre weights are in the range 65–84 kg and dimensions are length 5–8.5 mm, width 1.6–4.7 mm and thickness 1.5–3.5 mm.

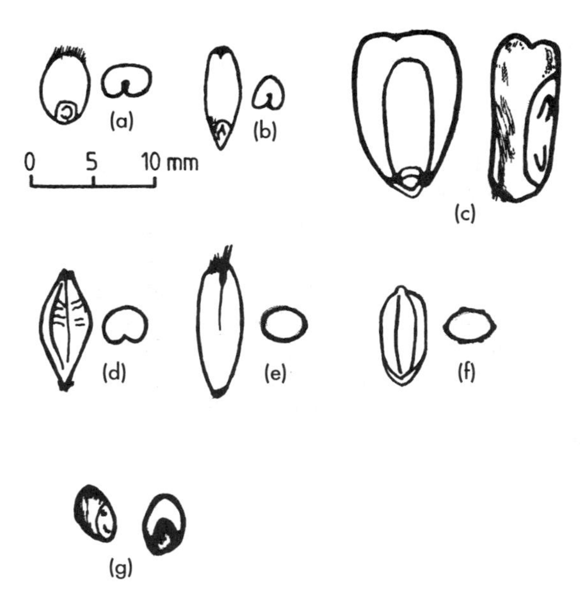

Fig. 2.16 Some cereal grains. (a) Wheat and (b) rye in surface view and cross-section; (c) maize in plan view and longitudinal section; (d) barley, (e) oats and (f) rice in surface view and transverse section; (g) sorghum (plan view and longitudinal section). Oats and rice are husked; the others are naked. Note the relatively large embryos in maize and sorghum. (After Kent, 1983.)

2.5.2 Rye

Malted rye (*Secale cereale* L) is used in making rye whiskey and possibly some other alcoholic beverages. Rye is a diploid species ($2n = 14$). The naked caryopsis resembles wheat in structure but in most varieties is more slender, with a pointed basal embryo end and a more angular, triangular cross-section (Fig. 2.16). Newer varieties are plumper and resemble wheat more closely. This cereal is cross-pollinated and will grow on poor soils. Typical characteristics are: TCW, 13–42 g; hectolitre weights, 60–80 kg; dimensions, length 4.5–10.0 mm, width 1.5–3.5 mm and thickness 1.5–3.0 mm.

2.5.3 Triticale

Triticale cereals (*Triticosecale* spp.) are fertile, amphiploid hybrids from crosses between wheats and rye. They retain the full chromosome complements of both parents. Grains are naked and in early varieties were often narrow and shrivelled, but newer forms are plump and look like good wheat grains. In some trials new varieties have substantially outyielded wheats. At present triticales are probably not being malted. Typically the TCW range is 28–45 g.

2.5.4 *Oats*

Various oats (*Avena* spp.) such as the common oat (*Avena sativa* L), are cultivated. Oat malts were used in the UK for making special beers such as oat stouts. Probably oat malts are now only being used regularly for making special breads. The grains are husked and are long and slender (Fig. 2.16). Typical characteristics are TCW 16–36 g, hectolitre weight 33–60 kg and dimensions are length 6–17 mm, width 1.5–4.5 mm and thickness 1.6–2.8 mm. The notably high husk content (Table 2.1) is reflected in the high fibre content: 10–11%, compared with 4–5% in barley and 2% in wheat. Brewers used to utilize the husky character of oat malts, or even oat husks, to give a more open texture to their mashes and so accelerate wort separation (Briggs *et al.*, 1981).

2.5.5 *Rice*

Rice (*Oryza sativa*) has not been malted successfully commercially despite various attempts to do so (Chapter 15). However, unmalted rice is an important source of extract for brewers (Chapter 12). In the tropics sprouted rice is used as a food. Rice grains are hulled and tough (Table 2.1; Fig. 2.16). They are flattened laterally and have no furrow. Rice is only distantly related to the other cereals. Its grains are unusual in being able to germinate under water. The dimensions of paddy rice are length 5–10 mm and width 1.5–5 mm. The TCW averages about 27 g.

2.5.6 *Maize*

Maize (*Zea mays* L) is also called Indian corn and corn (North America). The botany of the maize plant is unique in that the male inflorescence is a staminate tassel and the separate female inflorescence eventually turns into a highly specialized structure, the cob, which carries as many as 800 exceptionally large grains. Maize was malted by the earliest settlers in North America (Chapter 1) and is malted in African villages (Chapter 15). Preparations made from the raw grain are major sources of extract for brewers and distillers (Chapter 12). Many varieties of maize are grown and grains differ substantially in their sizes, colours, textures and other characteristics. TCW values are 100–600 g and dimensions are length 8–17 mm and width 5–15 mm. For an average Dent corn grain, TCW is 150–300 g, length is 12 mm, width 8 mm and thickness 4 mm. The flattened, naked grain has a narrow base and a broad apex; there is no furrow (Fig. 2.16). The pericarp and testa are fused as the tough hull, which may have various colours but is frequently yellow. The embryo is comparatively large, often 10–12% of the grain, and is placed to one side, at the base. The endosperm contains both flinty and floury regions. The shoot grows 'free' as the grain germinates (see Fig. 2.22b, below).

2.5.7 *Sorghum*

Sorghum (*Sorghum vulgare Pers.*) is known by a range of names: *S. vulgare* Pers., *S. bicolour* Linn., Moench., *Andropogon sorghum* (Brot.), Kaffir corn, Milo, Milo maize, Durra, the Great Millet, Feteritas, Shallu, Jowar, Cholam, Indian millet, Intama, Guinea corn and Koaliang. The list of synonyms (incomplete) indicates how widely sorghum is grown in warmer regions. This cereal is hardy and will grow under comparatively arid conditions. Sorghums are very variable in form and colour. Malted and unmalted grain is used in making African opaque beers, gruels and porridges (Chapter 1). Raw grain is used to make brewers' adjuncts (Chapter 12). In Africa, clear beers are being made from sorghum (raw and malted) to economize in the use of imported barley malt. The grains are carried on panicles of very varied shapes. About 50% cross-pollination is usual. The naked grains are rounded and, although they are generally smaller than barley corns (Fig. 2.16), they vary in size and shape. TCW values of 7–61 g are known, but values of 10–38 g (average 28 g) are usual. Dimensions are commonly length 3–5 mm and width 2–5 mm. Colours range from pale cream through reddish-brown to dark purplish-brown. Yellow colour may occur throughout the grain, but the darker pigments are confined to the outer layers, particularly the testa. Birdproof sorghums, which are intended to be unpalatable to birds, are dark in colour and are rich in polyphenolic tannins in the outer layers. These give problems when malted and used for brewing, and birdproof grains need special processing.

2.5.8 *The millets*

The millets are a mixture of small-grained annual grass species (tribe Paniceae) that grow in warmer regions. Despite the small size of the grains, they are often malted (Chapter 15). Only the more important species are indicated.

Pearl millet (*Pennisetum americanum* (L.) Leeke, *P. typhoides*, bullrush millet, cattail, candle millet, bajra, sajja, cumbu), is a widely grown and hardy millet that has a wide variety of different forms. The grains are ovoid and, with lengths of 3–4 mm (TCW 5–10 g); they are the largest of the millets (Table 2.1). Proso or common millet (*Panicum miliaceum* L., hog millet, broom corn, French millet) has rounded, unfurrowed grains, about 3 mm long (TCW 6 g). Each grain is invested with a thick, shiny palea and a lemma, which may be white, yellow, grey, brown or black. Foxtail or Italian millet (*Setaria italica* L.) has grains that are smaller than those of common millet and flattened on one side. Finger millet (*Eleusine coracana* Gaertn., ragi, birdsfoot millet, marica) is an important cereal in India and parts of Africa.

Other millets include the little millet (*Panicum miliare* Lam.), Japanese barnyard millet (*Echinochloa frumentacea*), kodo (*Paspalum scrobiculatum*, varagu, ditch millet), fonio or acha (*Digitaria exilis* (TCW 0.65 g) (Nzelibe and Nwasike, 1995)), Job's tears (*Coix lachryma-jobi*) and teff (*Eragrostis tef, Amaranthus paniculatus*). Teff, which is confined to the Ethiopian highlands, has the smallest grain of all (TCW 0.14–0.2 g).

2.6 Pulses or legumes

Some peas and beans, which are seeds of plants belonging to a dicotyledenous family, the Leguminosae, are sometimes malted or sprouted (Chapter 1). This family is large and complex (Gill *et al.*, 1980; Martin, 1984). The seeds form in pods; although they are sometimes called grains, this is incorrect. In the seed the embryo plant is positioned between the edges of two swollen organs, the cotyledons, which are rich in reserve substances. All these tissues are enclosed in a tough seed coat, the testa (Fig. 2.17). The seeds of different species germinate in one of two ways, either with the cotyledons being retained within the testa while the seedling plant grows or the cotyledons may be pulled from within the investing testa by the growing plant. They may turn green in light. The best known pulses that are sometimes sprouted are the mung bean (*Vigna radiata* (L.) Wilczek, syn. *Phaseolus aureus* (Roxb.), *P. radiatus* (L.), green gram and golden gram, with a thousand seed weight of about 50 g, and black gram or urd (*Vigna mungo* (L.) Heffer, *Phaseolus mungo* (L.)), a closely related species.

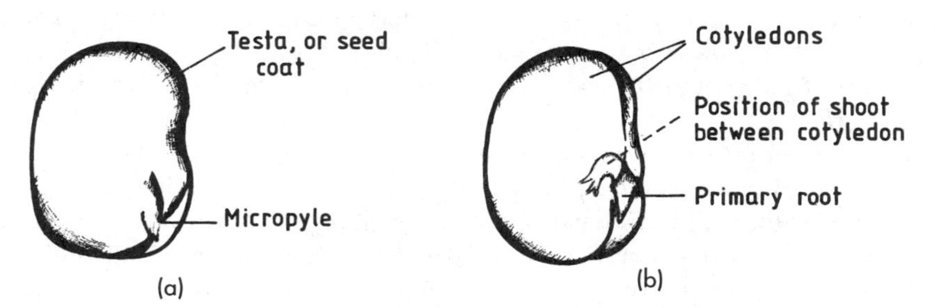

Fig. 2.17 A generalized legume seed: (a) intact; (b) with the testa removed, indicating the position of the embryonic axis between the cotyledons.

2.7 Physical changes occurring in malting barley

During steeping, the surface layers of the barley grain are rapidly wetted, but the interior of the grain is not thoroughly hydrated for 2 or 3 days (Chapter 3). During imbibition the grain swells, its bulk volume increasing by about a quarter. Allowance is made for this 'swell' when steep tanks are filled. The widths and depths of the corns increase; the lengths do not alter. As the embryo hydrates it swells, becomes turgid and loses its scale-like appearance (Fig. 2.18). The junction between the face of the scutellum and the crushed cell layer weakens and becomes slippery, so that they are easily separated. The embryo hydrates before the endosperm (Chapter 3).

The first visible sign of germination is the appearance of the chit (pip, bud), the yellowish coleorhiza or root sheath, which breaks through the testa and pericarp and protrudes between the valves of the husk at the base of the grain (Fig. 2.19). The chit splits and the rootlets (seminal roots, culms, coombes, cummins; incorrectly sprouts) grow and form a tuft at the end of the grain. In humid conditions, they are coated with a fine pile of root hairs. At the same time the coleoptile (the maltster's acrospire, spire or blade) grows along the dorsal side or back of the grain, usually between the testa and the pericarp. This is most readily perceived in naked grains. In husked barleys the growing acrospire presses a groove into the tissues below. It usually emerges from beneath the lemma near the apex of the grain, but it may emerge at one side particularly if the grain has been physically

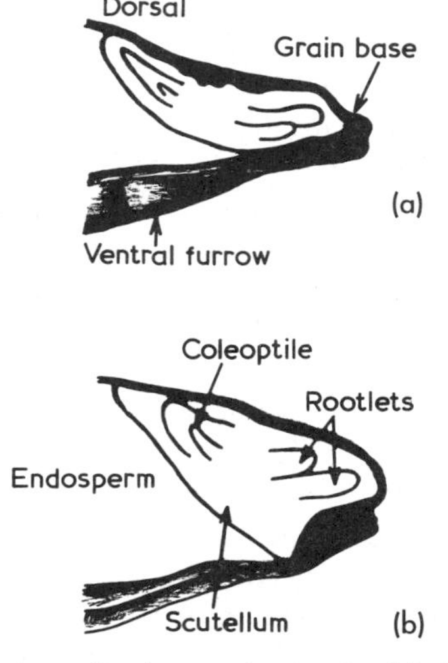

Fig. 2.18 Longitudinal sections of embryo ends of grains (a) air dried; (b) after steeping. The embryo swells markedly during imbibition and presses the husk up, giving it a smoother appearance. (After Briggs, 1978.)

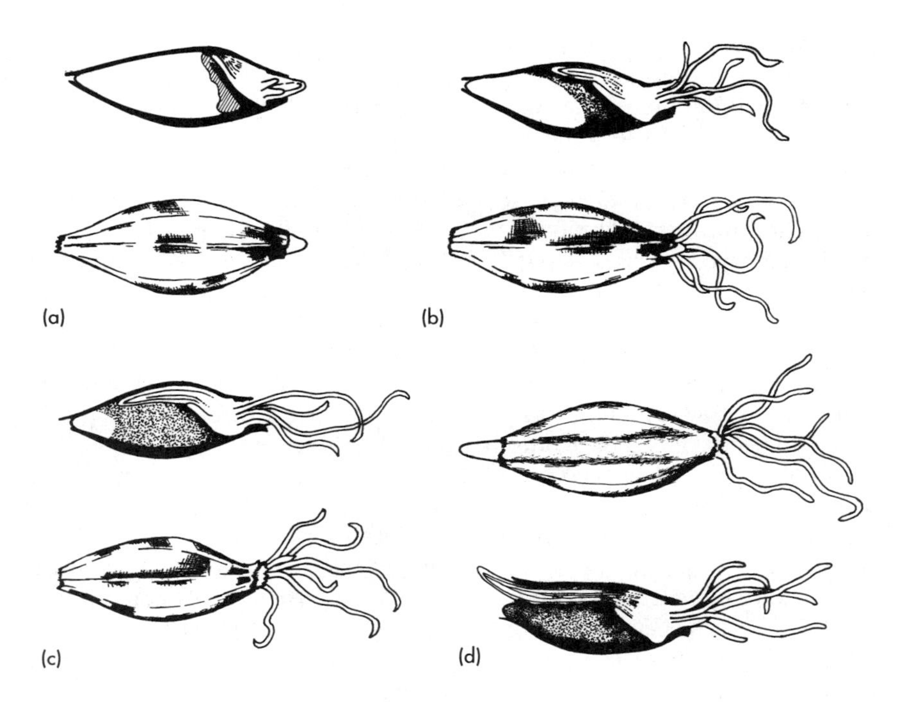

Fig. 2.19 Four successive stages of germination in the malting barley grains (a–d). Each pair of diagrams shows a vertical longitudinal section of the grain to one side of the sheaf cells, and the approximate extent of modification of the starchy endosperm, and a view of the exterior of the grain from the dorsal side. In (c) the grain is slightly undermodified while in (d) it is overgrown or 'overshot' and is known as a huzzar or a bolter. (After Briggs, 1978.)

damaged. The extents of acrospire growth and, less usually, rootlet growth are traditional guides to the progress of malting. Lengths of acrospires are noted, each as a fraction of the grain's length. Length may be judged by eye, or with the aid of grids mounted on glass or on a magnifying lens. Thus, when the tip of the acrospire is 50% of the way 'up the back' its length is 1/2; when it reaches the apex it is 1; if it exceeds the length of the grain it is 1+ (overshot, overgrown, bolter, huzzar). Traditionally, in the UK, malt was kilned when the average length of the acrospires was between 3/4 and 7/8. To make the acrospires more easily seen, the grains may be cut or peeled or they may be boiled in a solution of copper sulphate (Chapter 13).

After imbibition, starch appears in the embryo (Brown and Morris, 1890). It is formed from the embryo's own reserves, probably the sugars, because it occurs before substances are taken up from the endosperm and because it occurs in isolated embryos incubated on water. The appearance of starch in embryos incubated on water has been used as a test of vitality (French, 1959). As germination proceeds, the embryo relies increasingly on nutrients that diffuse to it from the endosperm and the aleurone layer. The vascular system develops and extends through the axis and the scutellum (Briggs, 1987b). Changes occur in the cytoplasm of the scutellar cells and, at least under some conditions, the cells of the scutellar epithelium separate along their lengths and elongate (by two to four times), forming a fine pile directed towards the endosperm (Briggs, 1973, 1978; Gram, 1982). The increased surface area favours the epithelium's functions, the release of hydrolytic enzymes and gibberellin hormones and the uptake of low-molecular-weight nutrient substances.

After 1–3 days of germination, depending on the malting conditions, the aleurone layer responds to gibberellin hormone(s) from the embryo and releases mineral nutrients and hydrolytic enzymes. This process begins near the embryo and progressively extends towards the apex of the grain (Briggs, 1973, 1978, 1987b; Gibbons, 1980; Briggs and MacDonald, 1983a; Oh and Briggs, 1989). Adding gibberellic acid to the grain enhances enzyme production and release. As the aleurone tissue hydrates, its cytoplasm alters substantially. The aleurone grains swell, the mitochondria develop more cristae and the respiration of the tissue increases. Subsequently, when the gibberellin arrives, other alterations occur. The respiration rate increases further and the Golgi bodies and the rough endoplasmic reticulum increase in amount, changes associated with enhanced rates of protein synthesis. The lipids in the sphaeroplasts and the materials in the aleurone grains are gradually utilized and these organelles are replaced by vacuoles. Meanwhile the cell walls are partly eaten away, particularly on the side nearest to the starchy endosperm, and the channels through the walls around the plasmodesmata are enlarged. These changes are brought about by hydrolytic enzymes, which are, therefore, directionally secreted towards the endosperm and later assist in its modification.

Modification is the collective term for the physical and chemical changes that occur to grains during malting, in particular the partial degradation of the starchy endosperm and the associated biochemical changes (Briggs, 1987b,c, 1992). The first sign of cytolysis in the endosperm, caused initially by hydrolytic enzymes from the scutellum and subsequently the aleurone layer, is the degradation of the cell walls. Subsequently the matrix proteins are attacked, freeing the starch granules. Slightly later the starch granules are seen to have been attacked as they become etched and pitted and the layers tend to separate. The endosperm degradation which occurs during malting is less extreme and is, therefore, less easy to detect than that which occurs in grains germinated in contact with water (Briggs, 1987b; 1992). Maltsters traditionally detect modification in germinating grains by splitting corns with the thumb nail and rubbing out the endosperm between fingers and thumb. The endosperm of steeped barley is tough, but as it modifies it becomes soft and doughy. Dry barley is tough and hard to bite or crush, but well-made, kiln-dried malt is friable, that is it is brittle and readily crushed and the endosperm breaks up as a flour. When malt corns are bitten, very few should have unmodified parts or hard ends that resist crushing. Some investigative techniques allow the extent of modification to be mapped with some precision (Chapter 13). For example, thin (10 μm) sections may be made from frozen grains and may be inspected by transmitted light in the microscope either with or without prior staining. The stains Trypan Blue and Congo Red selectively colour undegraded cell walls (Fig. 2.20).

The patterns of modification that occur in malting grains have been the

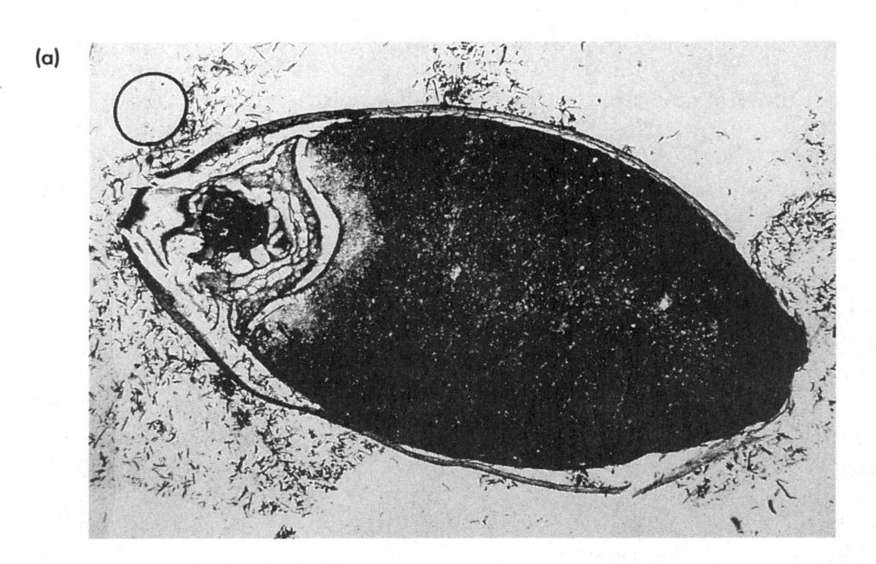

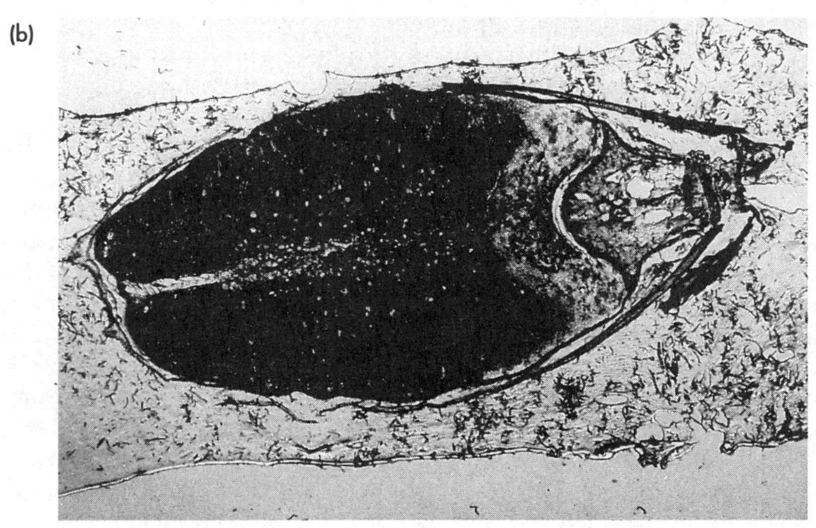

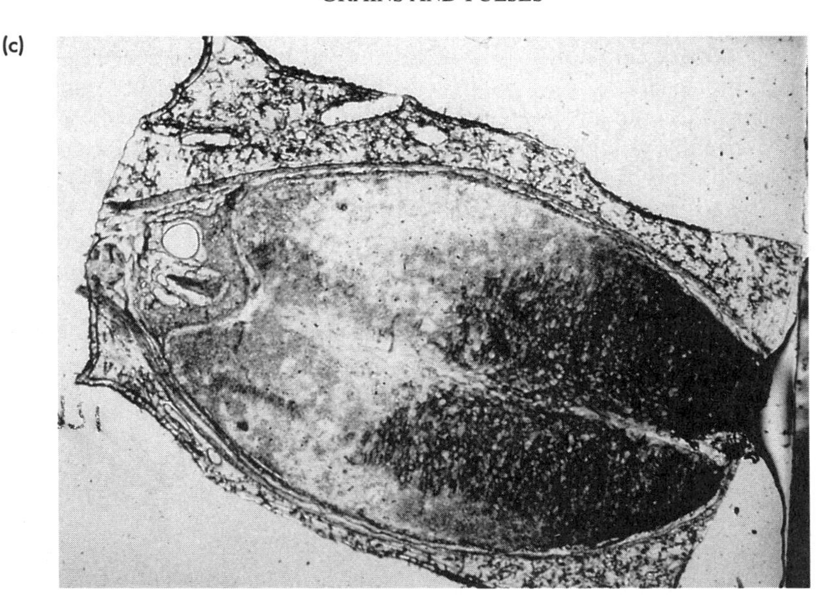

Fig. 2.20 Photographs of horizontal, longitudinal sections of grains at three stages of malting. The undegraded cell walls are stained with Trypan Blue. (a) Initial endosperm breakdown 'modification' beneath the scutellum; (b) modification extended further beneath the scutellum and just beginning beneath the aleurone layer; (c) endosperm modification slightly more than half complete, showing most advance at the sides, adjacent to the aleurone layer. This grain had germinated for 4 days. (D. E. Briggs, unpublished data.)

subject of dispute. However, discordant results have usually been reported in publications in which invalid experimental or undefined techniques were used (Briggs, 1987b). Within a batch of malting barley, grains modify at various rates and different patterns of modification occur. Modification begins next to the scutellum, which is the first tissue to release hydrolytic enzymes, and spreads into the endosperm at a uniform rate in some corns but more rapidly in the mid-region on the ventral side around the sheaf cells in others (Fig. 2.21), probably because in these latter grains there is a tract of tissue that is more pervious to hydrolytic enzymes. The walls of the crushed cell layer are degraded more slowly than the walls of the adjacent endosperm and the walls of the sheaf cells and the aleurone cells are not greatly degraded in malting. After 1–3 days of germination, and as a consequence of the release of enzymes from the aleurone layer, which join those from the scutellum, modification starts to advance more quickly immediately beneath the aleurone layer (Fig. 2.21). Sub-aleurone modification is particularly pronounced in grains dosed with gibberellic acid (Briggs, 1987b). The scutellum is at an angle to the long axis of the grain (Figs 2.6 and 2.21) and, as modification advances roughly in parallel to the scutellar face, it tends to be more advanced on the dorsal side of the grain.

Despite contrary statements, it does not move faster on the dorsal side. However, in a proportion of grains, it advances faster in the ventral mid-region near to the sheaf cells (Fig. 2.21). Ideally all grains in a batch of malt would be equally modified. In fact, a range of degrees of modification is always found: one of the maltster's objectives is to minimize this heterogeneity. Dead and dormant grains will be wholly unmodified, a few exceptionally vigorous grains may be overgrown. Most will be well modified but a small proportion of these will contain unmodified regions, which in some cases are confined to grain apices but which also occur in more central locations in the starchy endosperm. The occurrence of heterogeneity in

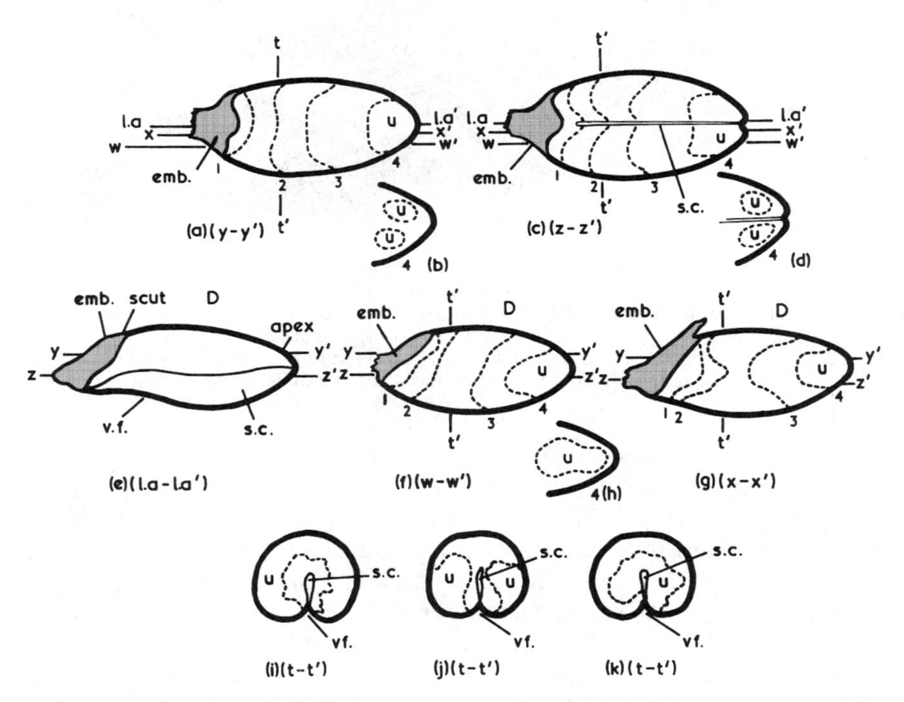

Fig. 2.21 Idealized schemes of the most common patterns of modification encountered in untreated Maris Otter barley micromalted at 16 °C, without additives, after an air-rest steeping sequence. The dashed lines indicate boundaries between the modified and unmodified (u) regions on germination days 1–4, respectively. The planes in which various sections are taken are indicated. (a) and (b) Plane y–y′; horizontal, longitudinal sections, above the sheaf cells, showing alternative patterns of modification; (c) and (d) plane z–z′; horizontal longitudinal sections closer to the ventral side than (a) and (b), which intersect the crest of sheaf cells; (e) vertical, longitudinal section in the mid-line (plane 1.a–1.a.′) of an ungerminated grain showing the disposition of the sheaf cells; (f) vertical longitudinal sections cut to one side of the sheaf cells (plane w–w′; (g) cut close to the sheaf cells in plane x–x′; (h) as (g) but an alternative pattern of modification that sometimes occurs; (i), (j) and (k) vertical, transverse sections, plane t–t′, showing various patterns of modification encountered in grains that had been germinated for 2 days. D, Dorsal side; emb., embryo; scut, scutellum; s.c., sheaf cells; v.f., ventral furrow; u, unmodified region. (After Briggs and MacDonald, 1983a.)

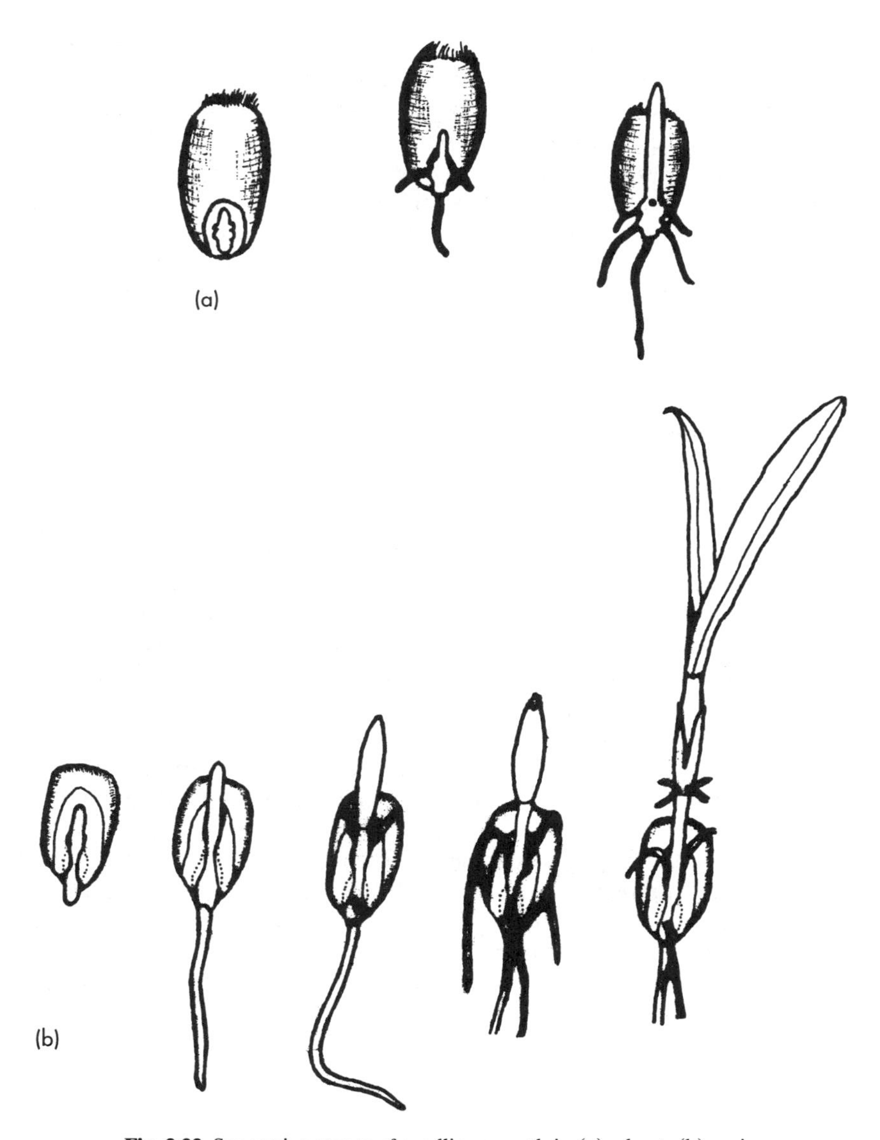

Fig. 2.22 Successive stages of seedling growth in (a) wheat; (b) maize.

malt, and especially of substantial numbers of unmodified grains, is undesirable. It may be monitored using several 'sanded block' techniques or with a machine such as the Friabilimeter (Chapter 13). As endosperm modification progresses, so the acrospire and the rootlets are growing. The relative rates of these processes are roughly related to each other, but the relationships may differ when different barleys are processed or when different malting systems are used. If the husk splits or is otherwise damaged, the acrospire may grow away from the grain. Under these circumstances, modification proceeds normally and so is not dependent on the location of the acrospire. Physical modification is accompanied by massive accumulations of some enzymes, increases in the levels of soluble substances, such as amino acids and peptides (soluble nitrogen) and sugars (which may be determined as the 'preformed soluble materials' or cold water extract (CWE) of the grain) (Chapters 4, 13). Ease of recovery of extract when malting grain is mashed also increases with time, up to a maximum value (Chapter 14).

2.8 Germination and modification in other species

Germination in wheat, triticale and rye resembles that of barley, except that the coleoptile breaks through the pericarp and grows away from the grain (Fig. 2.22a). In malting maize and sorghum seedling, growth is quite extensive (Fig. 2.22b) causing substantial malting losses (Chapter 15).

From the limited data available, the patterns of endosperm modification that occur in most small grains seem to resemble those of barley. Oats is an exception, because in this species the scutellum separates from the other tissues and grows into the endosperm along the dorsal side, releasing enzymes as it goes (Briggs, 1987b). In the tropical cereals maize and sorghum, the enzymes that catalyse modification appear to originate exclusively from the scutellum (Aisien, 1982a; Aisien and Palmer, 1983; Glennie *et al.*, 1983).

In the pulses, the changes that occur seem to have been little studied under malting conditions. In contrast to the endosperms of cereals, the reserves of legume seeds are mainly located in the pair of cotyledons, which are living organs. The mechanisms by which reserve starch and protein are mobilized are clearly different to those that occur in the dead starchy endosperms of cereals.

2.9 Barley cultivation

Barley is grown from sub-arctic Scandinavia to near the equator; in the mountains of Ethiopia and in South America; from below sea level near the Dead Sea to great altitudes in the Andes and Himalayas; from humid, temperate regions, like western Europe to dryland areas in parts of North America, to irrigated areas in deserts, such as the Sahara (Hunter, 1962; Briggs, 1978; Rasmusson, 1985). Production areas occur worldwide, outside the humid tropics, and grain is traded widely. The British crop has been around 10 million tonnes per annum, of which about 20% is used in malting. The remainder is used chiefly for animal-feed, in which a high nitrogen content is desirable. This, combined with the need to use nitrogenous fertilizers to maximize yields, has resulted over recent decades in a rise in the nitrogen content of the crop as a whole, including that part used in malting. Many feed-grade barleys tend to outyield the genetically more refined and more difficult to breed malting barleys. Because of this, the price of malting-quality barley is above that of feed grain. Because of climatic vagaries, there is never any guarantee that a crop of barley of a malting variety, even grown with limited applications of nitrogenous fertilizer, will be able to meet the maltsters' stringent quality requirements. For preparing most British malts, moderate levels of grain nitrogen (or 'protein') are preferred (e.g. 1.5–1.65% N; 9.4–10.3% crude protein). However, the maltster can only use the grain available, and substantial differences may occur between seasons. For making highly enzymic malts, barleys having high nitrogen contents (2.0–2.2% N; 12.5–13.8% crude protein) are required.

In W. Europe and some other areas of the world, barley for malting is grown from clean, disease-free seed, and of a single variety that maltsters will accept. Seed-dressings may be applied to confer resistance to soil-borne pests and diseases and a systemic fungicide may be used to combat mildew. The chemicals used must have been cleared as 'non-hazardous' for use on malting barley. The use of organomercuric seed dressings is now banned. Treated seed **must not** be diverted and used for malting. Some varieties are sown in the autumn (fall), so called winter barleys, while others are sown in the spring. The longer growing period of winter barleys (about 9 months) compared with spring barleys (90–100 days in Scandinavia, approximately 120 days in Continental Europe, approximately 140 days in the UK and 92–98 days in Canada) gives them higher yield potentials; higher yields are usually obtained when sufficiently hardy strains are available to prevent extensive seedling killing by frost. However, the extended growing season gives greater opportunities for disease development within the crop, and so the use of fungicides, for example, may be high. Barleys that are hardy in the UK often cannot survive colder winters, those of central Europe for example. Severe winter killing may necessitate resowing fields in the spring, with a different variety. The greater yields explains the increased popularity of winter barleys, despite the risk of winter killing, and the difficulties of getting fields clean and weed free before autumn sowing. It is proving difficult to breed very hardy winter barleys having good malting qualities for use in central Europe, although in England Maris Otter and subsequently several other varieties (such as Puffin, Halcyon, Pipkin, Melanie and Angora) have been extremely successful and have excellent malting qualities (Schildbach and Burbridge, 1996). It is best to sow spring barleys as early as the season (wet and frost) permits, as the longer growing period gives plumper, better-quality grains and higher yields (Table 2.2).

Table 2.2 The effect of variety and time of sowing on the yield, TCW and nitrogen content of harvested barley

Date of sowing	Carlsberg	Proctor	Provost	Spratt-Archer
Yield of grain (kg/ha)[a]				
16.3.55	4374	4161	3746	3658
5.4.55	3620	3482	3256	3143
27.4.55	1911	2124	1571	1094
Nitrogen content of grain (%)				
16.3.55	1.46	1.64	1.83	1.63
5.4.55	1.61	1.68	1.78	1.69
27.4.55	1.98	2.26	2.08	2.25
TCW(g)				
16.3.55	42.5	35.7	31.9	38.2
5.4.55	37.7	32.9	31.4	34.6
27.4.55	35.9	32.5	34.6	33.3

[a] To convert yield to cwt/acre, divide by 125.7.
From Bell and Kirby (1966).

Each barley variety must meet the farmers' requirements. It must give a good yield of grain, it is likely to have short and stiff straw to increase its resistance to lodging (stem collapse, which is likely to occur in strong winds and rain, especially if nitrogen fertilizer has been applied) and make it suitable for combine-harvesting. In addition, it should have tolerance or resistance to the locally important pests and diseases. In many countries, lists of recommended varieties are available, for example those produced by the National Institute of Agricultural Botany (NIAB) and the Scottish Agricultural Colleges in the UK. These lists are drawn up annually with reference to the results of comparative field trials. Often barleys perform quite differently in different areas. For example, Golden Promise performed well in Scotland for many years, while its performance in central and southern England was disappointing. Farmers often grow several varieties of barley, partly to take advantage of the slowing effect this has on the spread of diseases between fields when the cultivars are chosen for diversity of disease-resistance characteristics, and partly to take advantage of different ripening times, which allows the work of harvesting to be spread over a long period. Usually winter barleys ripen before spring barleys.

The traditional method of seedbed preparation is to plough the soil once, or more, with mouldboard ploughs, then to reduce it to a fine tilth by harrowing. The seed may be drilled in (with or without fertilizer) or be broadcast and harrowed in. Other methods are also used, to save cultivation costs, to reduce soil preparation times or to meet local conditions, for example to conserve soil moisture and minimize erosion. These methods necessitate the use of selective herbicides to control weeds. The ground may be broken with chisel- or disc-ploughs or, in dryland areas, the soil may be disturbed by undercutting with a duck-foot blade, leaving the stubble of the previous crop in place to act as a mulch. On some soils it is feasible to drill seed directly into the stubble of a previous crop, without working the soil. It is probably 'best practice' to rotate crops, but when the alternative crops are not as profitable as cereals, the cereals may be grown continuously on the same soil, in successive years.

In wet areas, fields are drained to prevent water-logging, which checks seed germination, limits root penetration and stunts growing plants. In dry areas, barley may be irrigated, or soil may be allowed to stand fallow for 1–2 years to allow soil water levels to build up enough to support a crop.

Seeding rates vary widely: too high and the seedlings are crowded, resulting in waste; too low and ground is wasted and weeds readily grow between the barley plants, the widely spaced barley plants tiller profusely and produce ears over an extended period, so that many will not be ripe at harvest. However, in dryland areas, thin planting may be desirable, to allow sufficient soil moisture for each plant. At present in the UK, the average seed rate is probably about 95 kg/ha (85 lb/acre) while in North America rates vary between 40 and 160 kg/ha (36–145 lb/acre). In well-favoured areas, seed is normally drilled at a depth of 2–6 cm (1–2.4 in) in rows about 18 cm (7 in) apart. In dry areas, these values may be as much as 7.6 cm (3 in) deep and 25–36 cm (10–14 in) apart. Barleys will grow on a wide variety of soils, but English-type, low-nitrogen malting barleys are best grown on light soils, with pH values about 6.5. Chalky sub-soils are suitable. Usually fertilizers are used to supply major quantities of nitrogen, potassium and phosphorus. However, soil deficiencies such as a deficiency of sulphur may need to be remedied. Usually applications of nitrogenous fertilizers have the most dramatic effects on grain yield and quality. It is now usual to split applications of fertilizers, drilling some in with, or next to, the rows of seed grains, and broadcasting one, or more, later applications. It is important, especially when growing grain for malting, that applications should be made at the most beneficial time, and these are normally recognized by the growth stage of the plants (Briggs, 1978; Tottman *et al.*, 1979). Within limits, increasing doses of nitrogenous fertilizer result in larger plants (which may, however, be weak-strawed) with increased tillering, increased ear and tiller survival to harvest and increased yields of grain. However, ripening may be delayed and be a little uneven, and grain nitrogen contents may be increased. Early applications of nitrogen, up to the first tiller or erection of tiller stage, increase grain yields and may produce more plump grains with desirably low nitrogen contents (Table 2.3). Late applications may increase grain nitrogen levels to undesirable extents, with little or no increase in yield (Table 2.3). Late applications may also allow the survival of late tillers, which, at harvest, contribute thin, unripe grain, seriously reducing the malting quality of the sample. Organic fertilizers, such as farmyard manure, or ploughed-in legumes, release assimilable nitrogen throughout the growing season so are avoided when malting quality grain is being produced. Sometimes straw-stiffening and straw-shortening chemicals are

Table 2.3 The influence of applications of nitrogenous fertilizer (sodium nitrate) made at different dates to Spratt-Archer barley. Seeds were drilled on 16 April, and the grain was harvested on 18 August. Results are the means of seven, replicated plots

Sodium nitrate applied (125 kg/ha; 1 cwt/acre)	Grain yield (g/plot)	Tillers/ plant at harvest	Grain N (%)	TCW (g)	Grain/ plant (g)	Plants/ plot	Ears/ plot
None	137	0.95	1.66	39.2	1.41	98.1	190
At sowing (16/4)	207	1.50	1.61	39.8	2.14	97.1	243
Appearance of 1st tiller (18/5)	218	1.60	1.62	41.1	2.17	101.6	264
At erection of tillers (4/6)	211	1.73	1.60	40.6	2.07	101.9	278
At flowering (3/7)	146	0.91	2.04	41.6	1.49	97.6	187
Three weeks after flowering (23/7)	135	0.81	2.01	40.2	1.37	98.6	179
At sowing (16/4) and 3 weeks after flowering (23/7)	214	1.51	1.91	41.8	2.16	98.7	247

Data of Hunter and Hartley (1938).

sprayed onto plants to offset the effects of nitrogenous fertilizers and reduce the risk of lodging.

Farmers also use chemicals to control weeds, pests and diseases. Weeds reduce yields, interfere with ripening and harvesting and their seeds or propagules may be undesirable contaminants in grain. For example, seeds of darnel (*Lolium temulentum*) are poisonous, and bulbils of wild garlic taint grain. In the UK, the practices used to combat weeds, since about 1940, have been so successful that grain is much less contaminated with weed seeds than was once the case; in consequence, maltsters now use less complex grain-cleaning machinery than they once did (Chapter 8).

Pests that attack barley include birds, grazing animals, rodents, insects, slugs, snails and nematodes. Degrees of control are variable but, for example, rabbit burrows may be gassed, poisoned bait may be used to combat slugs and snails and the crop may be sprayed to control insects. Some insects act as vectors and spread diseases so causing harm in addition to the direct damage they do by feeding on the plants.

Like other cereals, barley may contract diseases caused by viruses, by bacteria and by fungi (Dickson, 1962; Briggs, 1978; Rasmusson, 1985). Generally control is attempted by cultivation practices designed to minimize disease carry-over from a previous crop on the soil or stubble, by the use of healthy seed and seed dressings, by the control of vectors (e.g. spraying to control aphids) by choosing disease-resistant or disease-tolerant varieties and, in some instances, by using chemicals. Bacterial diseases (e.g. that caused by *Xanthomonas*) are only locally important. Viral diseases vary in their importance. Barley stripe mosaic virus (BSMV), barley yellow dwarf virus (BYDV) and oat blue dwarf virus (OBDV), as examples, occur widely and may cause stunting and serious reductions in grain yields and quality.

The fungal diseases attract most attention. Typically, plants and grains carry numerous fungi, bacteria, yeasts, actinomycetes and other organisms that do not cause diseases. However, these organisms may have dramatic effects on grain in store and during malting (Chapters 7 and 14). Diseases caused by fungal pathogens that are of worldwide importance are powdery mildew (*Erysiphe graminis*), spot blotch (*Helminthosporium sativum*, various synonyms), net blotch (*Drechslera teres*, various synonyms), scald or barley leaf blotch (*Rhynchosporium secalis*), various diseases caused by *Septoria* spp., take-all or foot-rot (*Ophiobolus graminis*), scab (*Gibberella zeae*) and loose-, covered- and semi-loose smuts (*Ustilago nuda*, *U. hordei* and *U. nigra*, respectively). There are wide variations in how effectively fungal diseases can be controlled. Spraying with fungicides, for example to control mildew, is now commonplace. Failure to control fungal diseases can result in catastrophic yield losses and reductions in grain quality: grains may be thin, shrivelled, discoloured and infected with fungi. Grain contaminated by ergot, the sclerotia of the fungus *Claviceps purpurea*, is dangerous. The sclerotia, which are often purple-black in colour and about the same size as grains, replace grains in the ears of infected plants. The sclerotia are collected with the grain at harvest and contain alkaloids that are extremely poisonous. If eaten, they can cause hallucinations, convulsions, gangrene, abortions and death. In the UK, maltsters reject any grain contaminated with ergot. Elsewhere upper limits of contamination are specified, and presumably reliance is placed on grain-cleaning machinery or flotation treatments to purify the grain.

There is concern about possible undesirable effects that residues of herbicides, growth regulators, insecticides and other pesticides and fungicides may have on the quality, acceptability and maltability of grain. Government regulations and user requirements, specified by maltsters and malt users, dictate which agents may or may not be used and within what limits residues will be tolerated. This, in turn, limits application rates in the field. Testing programmes, backed by sophisticated analytical techniques, are used to ensure that the rules are obeyed.

In northern Europe, barley yields are usually best when a cool, damp season, which provides steady growing conditions, is followed by warm, dry weather during the ripening and harvesting periods. The fine weather facilitates harvesting, reduces the incidence of dormancy, split and pre-germinated grains, and, because the grain is comparatively dry, reduces drying costs. However, if the weather is cold and wet after the grain is milk ripe, dormancy is enhanced. Wet weather at harvest can result in delays and yield losses plus reductions in grain quality, some caused by increases in the microbial populations. In some areas, such as South Australia, grain yields are frequently limited by water supplies and even in the UK this can occur. Hot and dry conditions during the growing period can check grain filling and induce premature ripening, giving thin grains with high nitrogen contents, unsuited for making most grades of malt. The maltster wishes to buy sound, plump, fully viable and ripe grain that has not pre-germinated in lots of a single variety (i.e. uncontaminated), with an acceptable nitrogen content and which has been lightly threshed so that the corns are not chipped, split, skinned, broken or bruised. (Physically damaged or pre-germinated barley does not keep well in store or malt well.) Maltsters regularly advise farmers of their needs and suggest how they may be met.

In the past, during harvesting, near-ripe plants were cut off near the soil and were bound into sheaves. Originally, cutting was by sickle or scythe and the sheaves were bound by hand, but tractor-trailed machines called reaper–binders displaced this laborious manual work. The sheaves were set up in fours, as stooks in the fields, with the ears uppermost, to allow drying. They were then made into stacks for storage, which were thatched to keep the interiors dry. Later, the stacks were opened and the sheaves were threshed at the farmers' convenience. This system was laborious, the stacks were infested with rodents and, if they became damp, moulding and heating occurred with losses and deterioration in grain quality. As a result, the grain might be discoloured, and 'mow-burnt', or it became slimy. Conditions such as these are now almost unknown but were frequently recorded in older reports. In the older systems, sheaves were taken from a stack and threshed in a stationary machine. Prior to that threshing was by hand, using flails. Today harvesting and threshing are usually carried out in one operation, using combine-harvesters, although in some places, as in parts of the USA, near-ripe barley plants may be cut and swathed, that is laid in windrows, supported above the soil on the stubble. After a period of drying and curing, the cut plants are collected and threshed with combine-harvesters fitted with pick-up attachments. Usually combine-harvesters cut the plants when the grain is 'dead-ripe'. The plants are taken into the machine and by processes of beating, sieving and aspiration the grain is separated and collected, while straw, dust and chaff are discarded. The moisture content of the crop and the settings of the threshing machine are important parameters. If the grain is too dry it is brittle and easily damaged, while if it is too damp it is soft and is difficult to separate from the ear. Harvesting grain of 16–22% moisture has been recommended but, in South Australia for example, grain may be harvested at 9% moisture while under unfavourable conditions in northern Europe moisture contents may be 35% at harvest. While too 'gentle' threshing results in incomplete grain separation and losses with the straw, too 'close' and vigorous threshing physically damages the grain. A consequence of using combine-harvesters is that the entire crop of grain, with a wide range of moisture contents, becomes available during a short period at harvest-time. In this state, it is liable to deteriorate, and it should be handled, processed and stored in ways that maintain its quality and viability and prevent losses. Conditions of drying and storage are of critical importance (Chapters 6 and 7).

Grain must be sufficiently dry and cool to allow it to be stored in bulk without deterioration. It may need to be ventilated, attacks by rodents must be prevented, and insect and microbial spoilage must be prevented. Farmers may store grain intended for animal-feed in a number of ways that are unsuitable for grain that is to be malted. Therefore, hermetically sealed silos, chilled stores or chemical preservatives may be used. Maltsters will not trade with farmers or merchants who use their grain driers incautiously or who do not maintain their stores correctly. Indeed, in some areas maltsters will not accept farm-dried grain, preferring to carry out the critical drying procedure themselves, sometimes within hours of the barley being harvested.

2.10 The improvement of the barley crop

Efforts to improve barley have occurred since its early cultivation (Beaven, 1947; Hunter, 1952; Bell and Lupton, 1962; Carson and Horne, 1962; Briggs, 1978; Wych and Rasmusson, 1983; Anderson and Reinbergs, 1985; Nilan and Ullrich, 1993). Farmers have always saved seed from superior plants, and have experimented with varieties from any source that came to hand. No doubt this continuous selection created the many strains of cultivated barley that evolved from wild *Hordeum spontaneum*. Systematic attempts at improvement are more recent and date from the founding of the Plant Breeding Station of Svalöf, Sweden in 1886 and the efforts of the English maltster E. S. Beaven (1947). Beaven was helped in the numerical analyses of his trials by the Guinness brewer Gosset, an 'amateur' statistician who published under the pseudonym 'Student'.

Up to about 1800, the barley crop was made up of 'land races', which were mixtures of types grown from saved seed and named according to their area of origin. Increasingly, grain was saved from superior plants and was multiplied to give named varieties. For example Chevallier, an excellent malting variety selected in 1824, dominated the English crop for many years. In Sweden, selection produced Svanhals, Hannchen and Gull, while Trebi, Chevron and Peatland were selections made in the USA. Some results of mass selections are illustrated in Table 2.4. Selection from existing crops eventually ceases to be beneficial as successive generations become genetically more uniform. It is necessary to create more variability by crossing promising parental lines or, less usually, by inducing mutations then selecting superior lines from the progeny.

Breeders' more general objectives include enhancing grain yields; obtaining plants with shorter, stiffer straw, which minimizes stem breakage and lodging and, because of the greater grain-to-straw ratio, simplifies combine-harvesting; and ensuring that ears do not shatter, break off the stem or shed grain, while still allowing grains to separate cleanly during threshing. They must obtain lines that are resistant, or at least tolerant, to locally important pests, diseases, droughts and soil salination; and obtain plants well adapted to the local climate, which ripen at a convenient time and, in winter forms,

Table 2.4 The results of selections of Irish Archer compared with a Danish selection of Archer (Tystofte Prentice); comparative trials were made at 10–12 sites each year

Year	Barley sample	Yield[a] (t/ha)	Yield[a] (cwt/acre)	Grain total nitrogen (%)	TWC (g)
1906	Irish Archer (unselected)	2.95	23.5	1.51	37.9
	Tystofte Prentice	3.15	25.1	1.47	39.5
1910	Irish Archer I (selection)	2.57	20.5	1.55	34.1
	Tystofte Prentice	2.59	20.6	1.53	34.4

[a]The average yield changed markedly in successive years.
From Hunter (1952).

are sufficiently hardy. Grain characteristics of importance in malting varieties are the size, shape and uniformity of the grains; generally plump grains are preferred. Dormancy should be sufficient to prevent pre-harvest sprouting, but not so extreme as to prevent grain being malted many months after harvest. The grain should tend to have a low nitrogen (protein) content. After malting, the extract yield should be high, and the quality or composition of the extract (fermentability; total soluble nitrogen, α-amino nitrogen, *etc.*) should be acceptable. The grain should malt rapidly and uniformly with minimal losses in dry weight and should develop an adequate complement of enzymes. Attempts are also being made to minimize the contents of less desirable grain components, such as β-glucans, proanthocyanins and 'gel-proteins'.

The effective selection for such a complex mixture of characteristics is a difficult and laborious undertaking. In the early stages of a breeding programme, little grain is available for analysis and so aspects of malting quality can only be judged indirectly. Of the many early-generation tests proposed, the use of infrared spectroscopy to analyse some aspects of the composition of whole grains seems promising, as the grains are not destroyed and so can be grown. At later stages, sufficient grain is available for micromalting trials. Eventually production-scale trials are carried out.

Usually a breeder artificially cross-pollinates plants, using parents that show characteristics which it is hoped to combine. Selections are made on the succeeding, self-fertilized generations. Twelve or more generations may be needed before a variety is ready to release. To save time it may be possible to achieve two generations in a year, for example by growing alternate generations in Britain and New Zealand, or in Canada and South America. At each successive 'selfed' generation, plants are more nearly genetically uniform (homozygous) and so progeny will vary less. Selection of promising lines is a skilled job and may involve pedigree or bulk selection methods (Bell and Lupton, 1962; Briggs, 1978). The double-haploid technique has now been used to achieve genetic uniformity rapidly (Briggs, 1978). Worldwide, breeders have tens of thousands of barley lines they may choose to use as parents in a breeding programme. Some successes have also been obtained by mutation breeding, inducing genetic changes by chemical or radiation treatments. Useful characteristics for disease resistance and short, stiff straw have been obtained, and attempts are being made to modify grain composition, for example grains with reduced levels of β-glucans or proanthocyanins. At present technical problems have prevented genetic engineering being applied to barley; one objective is to insert genes for heat-stable β-glucanases for expression in germinating barley grains.

In many areas, barley varieties must now be 'pure' and 'distinct', to meet legal requirements. Therefore, when seed is being multiplied, stringent precautions must be taken to avoid cross-pollination or admixture by or with other varieties. This emphasis on genetic uniformity benefits malting, where it is advantageous to have each batch of grain as nearly uniform as possible. Two possible future developments may give problems. Breeders are seeking to take advantage of hybrid vigour by finding economic ways of mass producing F-1 seed from defined crosses. Farmers growing a crop from such seed would expect a substantially enhanced yield, but the genetic inhomogeneity of the grain produced might well make it unsuitable for malting. The second proposal is that by growing 'multi-lines', mixtures of varieties being very similar in genetic make-up except for carrying different characters for disease resistance, the spread of diseases, fungal diseases for instance, would be slowed. For such multi-lines to be acceptable to maltsters all the sub-lines would have to be closely similar in malting qualities.

Although each barley variety must be pure in the sense of not including lines that show characteristics outside the legal definition of the variety, and it is quite uniform, it is not perfectly genetically homogeneous. In the past, varieties often contained a range of forms and were genetically variable. It even happened that progeny selected from different crosses or reselections from the original mixture of lines were released under the same name. In Britain, in the early 1800s, land races were progressively replaced by selected lines and reselections, such as Chevallier, Goldthorpe, Spratt and Archer. Between the World Wars, 1918–39, the crop was dominated by hybrid selections, such as Spratt-Archer and Plumage-Archer. An imperative of the 1939–45 war was that yields should be maximized and the crop was dominated by short-strawed, lodging-resistant, high-yielding Scandinavian varieties, such as Kenia, which, however, had inferior malting quality. During 1953–60 the variety Proctor came to dominate the crop. It had been bred by Bell from a cross between Kenia and Plumage-Archer. This remarkable barley was outstanding. At one time it comprised *ca.* 70% of the crop and was still in limited cultivation 30 years later. It had smaller grains of different appearance to the older, pre-war malting varieties, but it had outstanding malting quality and was comparatively high yielding and strong-strawed compared with the older varieties. Since that time, many new varieties have been introduced. Many varieties are grown each season, but not all can produce grain of high malting quality. Of the 33 varieties recommended for use in England in 1988, only nine were recognized as good malting varieties. The fierce competition between breeders for the right to market new varieties led to a rapid succession of new, and improved, forms being grown. For the maltster, this has the disadvantage that new varieties must continuously be evaluated and experience in malting small amounts for 1 or 2 years is desirable before large quantities are purchased. Clearly, as many barleys are accepted, increase and then decline to low levels of production in 5–6 years, much of this work is wasted. Recently, the most popular spring malting barley was possibly Triumph, although this can show extreme dormancy. Maris Otter is an exceptionally good malting-quality winter barley. Unfortunately it is now outclassed, in agronomic terms, and was succeeded by other varieties such as Kascade and Halcyon (which has a blue aleurone layer). At present British malting-grade barleys are all two-rowed.

In many countries, comparative field trials are carried out annually, often at several sites, to obtain objective comparisons of the relative agricultural merits of different, new varieties. Additional variations may include different fertilizer applications and the use – or not – of fungicides. To obtain early information regarding varietal characteristics, maltsters carry out small-scale malting trials on the grain produced in these trials. In England and Wales, the field trials are carried out under the auspices of NIAB and the micromalting results are correlated and circulated. The NIAB publishes a recommended list of barleys (and other cereals) annually, including varietal malting grades. In Scotland, with its significantly different climate, the arrangements are separate. The European Brewery Convention (EBC) correlates the results of national malting barley trials carried out all over Western Europe and publishes the results.

The efforts of the breeders have improved grain malting quality as well as grain yield. Thus in the UK pre-1939, hot water extracts (HWE) of malts were around 288 l°/kg (d.wt), say 97 lb/Qr. At present extracts of around 308 l°/kg (d.wt) (104 lb/Qr) are frequently obtained, although on average grain nitrogen contents (which are inversely related to extract yields) have risen. Further increases in extract yields are expected. Why higher extracts are now obtained is not clear. One contributing factor may be a relative reduction in the proportions of the husk and pericarp, which do not yield extract (Table 2.5). Enhanced yields result from both the new, improved varieties and improved agricultural practices. It has been estimated that in the period 1947–83 the national average barley yield in the UK increased from 2.0–2.5 t/ha to more than 5 t/ha, and that new varieties contributed 38% of this increase (Silvey, 1986). Individual crop yields may reach 10 t/ha.

Plant breeders are improving crops around the world. Therefore, more than 90% of American malt is made from Mid-Western six-rowed barleys with white aleurone layers (Burger and La Berge, 1985). Here, also,

Table 2.5 A comparison between malts from two malting quality barleys, differing in their husk contents; both barleys were fully viable

Variety	Husk (%)	Grain nitrogen (%)	TCW (g)	Hot water Extract	
				(lb/Qr)	(1°/kg approx.)
Carlsberg II	12.2	1.42	37.6	99.1	294
Proctor	10.2	1.46	34.8	101.7	301

From Hunter (1962).

Table 2.6 Mid-western, six-rowed malting barleys grown in the USA, and the periods in which they predominated

	Manchuria	Kindred	Traill	Larker	Morex
Year released	1920(ca.)	1942	1956	1961	1978
Period of dominance	–	1945–58	1959–62	1964–79	1980–
Grain yield (t/ha)	2.49	2.26	2.88	3.15	3.78
Height (cm)	106	100	96	95	96
Lodging (%)	26	41	27	28	15
TCW (g)	27.1	27.9	28.3	29.6	30.2
Protein (%)	13.9	13.8	13.2	14.2	13.6
Plump grains (%)	48	38	37	67	70
Malt extract (%)	75.0	74.5	75.9	75.8	78.6
Protein ratio (Wort/grain) (%)	36.6	37.5	38.0	37.3	42.0
Diastatic power (°L)	192	225	230	251	255
α-Amylase (20°C; DU)	34	43	44	42	52

From Wych and Rasmusson (1983).

breeding has significantly improved grain yields and malting quality (Table 2.6). Two-rowed malting barleys are grown in the USA and there are programmes to improve them. The variety Moravian III was produced by Adolph Coors Co. Brewery for their own use. It seems that the malting of Western, or Coast, six-rowed types has been discontinued. These barleys were imported into the UK and malted pre-1939.

Malting barleys in Canada are predominantly six-rowed varieties, with blue aleurone layers, such as Conquest, Bonanza and Argyle. To compete better on the world market, strenuous efforts have been and are being made to produce more desirable (plump-grained) two-rowed varieties, which now occupy about 40% of the malting barley area (Burger and La Berge, 1985; Sisler, 1987). The varieties Ellice and Harrington have been introduced to replace the popular variety Klages (Table 2.7). The new varieties have higher diastatic powers than Klages, which is itself exceptionally rich for a two-rowed variety.

In terms of world trade, the greatest exporter of malt is the European Community (48%), followed by Australia (13%), Eastern Europe (11%) and North America (7%). France is responsible for 40% of the European

Table 2.7 Comparisons between Klages and the newer, two-rowed Canadian malting barleys Harrington (which began large-scale trials in 1984) and Ellice, which began its major trials a year later

	Klages	Harrington	Ellice
Agronomic data			
Relative grain yield (%)	100	105	106
Time to maturity (days)	96	93	93
Lodging (scale 1–9)	4	3	2
Analyses of commercial malts			
Extract (%, fine-grind, d.b.)	79.7	80.0	80.3
Fine–coarse extract difference (%)	1.6	1.4	1.3
Wort colour	1.62	1.54	1.55
Diastatic power (°L)	113	120	132
α-Amylase (DU)	44	46	41
Malt protein (%)	12.2	12.1	11.8
Soluble/total protein (%)	41.4	42.9	45.4
Wort β-glucan (p.p.m.)	140	75	36
Wort viscosity	1.43	1.40	1.38

From Sisler (1987).

Community's exports (*Paul's Brewing Room Book*, 1989–1990). There is also an important world trade in malting barley.

Clearly, great efforts are made to improve barley varieties and grow them well. Efforts are being made to improve all the crops that are malted, or sprouted, around the world. While it is clear that sorghum breeding in Africa does take account of malting quality, this is not true elsewhere: maltability is probably not considered in programmes for improving millets, oats, wheat, rye or beans that are to be sprouted.

3 Grain physiology

3.1 Some general considerations

The stages of malting were summarized in Chapter 1 but there are other aspects of malting which must be introduced here. Maltsters process batches of grain of up to *ca.* 500 t. Individual grains typically vary between 30 and 50 mg in weight. There are, therefore, about 25×10^6 grains per tonne. Before use, the quality of each batch of grain must be assessed, and tests are carried out on samples taken from the batch. The same applies to malt. However, grains in a batch vary in many ways: in size, in shape, in chemical composition, in germinability, in moisture content and in microbial 'load', as well as in degrees of physical damage. The mix or distribution of these properties is unlikely to be uniform throughout the batch. Consequently for laboratory assessment great care must be taken to obtain sub-samples that are drawn from all parts of the bulk and so are fairly representative. If sampling is not carried out correctly, the results of analyses on the sub-samples are likely to be misleading or useless. Sampling techniques are rigidly defined (Chapters 6 and 13). Grains in sub-samples may be analysed in mass, when average results are obtained (e.g. for moisture content, nitrogen content or malt hot water extract). However in other tests (such as those of germinability, viability, or recording the distribution of endosperm modification) relatively small numbers of individual grains are investigated. From statistical considerations, it is inevitable that these latter tests will not be very reliable (section 3.3). Other problems arise because different investigative techniques have been used and the results have been reported in different ways (Appendix, Tables A11 and A12; Chapter 13). This confuses the interpretation of the rich and valuable literature on malting, which is frequently of immense use. Anyone wishing to benefit from the literature must become acquainted with these problems and limitations.

Barley, or other grain, must be converted into malt of the best achievable quality, economically, in the shortest feasible time and in the best yield. The choice of malting schedule is guided by these considerations. When grain is converted into malt, some losses are inevitable. Losses are calculated in different ways, and these must be distinguished. In the simplest case, weighed barley is steeped and steeping losses are incurred; substances are dissolved and washed away from the husk and pericarp, together with dust, grain fragments and perhaps small grains. Subsequently, during germination there are respiratory losses in which grain carbohydrates give rise to carbon dioxide and water. Then, after kilning, the rootlets are removed and so their weight is 'lost' to the malt. Because of the wide variations in grain moisture content that occurs during malting, it is standard practice in scientific investigations to express all the losses and malt yields in terms of percentages of the dry weight of the barley initially steeped (Chapter 11). Such information is of value in assessing particular processes. However, for commercial purposes, it is more appropriate to consider the 'out-turn' of a maltings or a production run and this, sometimes confused with malting losses, is calculated in various ways. The calculation is based on the weight of barley entering the maltings (or the process) and the weight of finished malt emerging at the end. These calculations are often based on barley and malt weights 'adjusted' to particular notional moisture contents (e.g. 16% and 4.5%, respectively). In the worst cases, this is not done and malt out-turns are based on barley weight (8–25% moisture, delivered) and malt weight (1.5–6.0% moisture) with no corrections for differing moisture contents. Losses that may or may not be included in calculations of out-turn are: (i) the loss of weight when barley is dried; (ii) losses owing to the removal of dust, trash and thin corn during cleaning and grading; and (iii) the steeping, respiratory and rootlet (culms) losses. Clearly, while out-turn calculations are indispensible to a commercial operation, unless the details of the calculations are standardized they do not allow the performances of different plants to be compared and, unlike calculations of true malting losses, they do not permit a valid evaluation of the malting process itself. True malting losses are usually in the range 6–12% of the original barley (dry basis) while out-turns are often around 75%, indicating losses of 25%. In the oldest malting practice, barley and malt were measured by volume (bushels; quarters; Appendix Table A4).

In England it was assumed that 1 quarter (volume) of barley should produce about 1 quarter (volume) of malt (sometimes a little more or a little less). Then, legally, quarters by weight were confusingly defined so that 1 quarter (Qr) of barley weighed 448 lb (203.2 kg) and 1 quarter (Qr) of malt weighed 336 lb (152.4 kg). Thus the out-turn of malt from barley was expected to be 75%. Elsewhere different weight-equivalents were given to these units (see Appendix, Table A4).

Grains are normally infested with a wide variety of microbes (section 3.9). These mixed populations are difficult to control and appear to be practically impossible to eliminate. Microbes can have important effects on the grain in the field, in store and at various stages in the malting process and on the quality of beers made from 'infected' malts.

3.2 The functions of the different tissues of barley and other grains

The different tissues in grains serve varying functions (Briggs, 1973, 1978, 1987b, 1992; Fincher, 1989, 1992). Husk and pericarp serve to protect the inner parts of grains from physical damage. Both provide barriers to gaseous exchange between the living tissues and the exterior and so can limit respiration. The husks of barley and oats provide mechanical protection that is lacking in naked grains (wheat, rye, triticale, maize and sorghum) and so these husked grains are less readily damaged by movement or machinery. This is important during malting because the acrospire, which is retained beneath the husk in barley, is protected from damage by turning and conveying machinery. In contrast, when wheat is malted, the grain is easily damaged when softened by steeping and the exposed acrospire is easily damaged and is collected with the culms when the kilned malt is dressed. It is probable that the husk resists the swelling of barley grains in the steep.

The testa limits the inward diffusion of many solutes, such as ionized salts, that will permeate the husk and pericarp. It also prevents the outward diffusion of sugars, amino acids and other soluble substances from within the grain. In healthy grains, microbes never penetrate the testa, although they may occur on, in and between the husk and pericarp. Therefore, the testa separates the exterior from the interior of the grain in several important ways.

It used to be said that the embryo existed 'as a parasite', drawing its nutrients from the endosperm during germination (Brown and Morris, 1890). This view was based on the discovery that isolated embryos would begin to grow on water, but this growth soon ceased as their internal reserves were used. However, if they were provided with a solution of a suitable mixture of nutrients (e.g. a sugar such as sucrose and a nitrogen source such as ammonium sulphate, potassium nitrate or glutamine) then the embryo grew strongly and could eventually be planted and grown into a mature plant. Therefore, the endosperm and the aleurone layer could be regarded as 'mere' food reserves for the embryo. Since the embryo was shown to generate and release hydrolytic enzymes capable of degrading the endosperm cell walls and starch, it was assumed that these enzymes mobilized the endosperm reserve nutrients. This view of the passive role of the rest of the grain was reinforced by the discovery that only very minor changes occurred in hydrated, degermed grains, although if held in air for 3 days the junction between the aleurone and the starchy endosperm was weakened and the two could be separated. Nevertheless, studies by microscopists strongly suggested that in intact grains the aleurone layer also generated hydrolytic enzymes that contributed to endosperm degradation. A role for the aleurone layer was indicated when it was shown that barley embryos transplanted onto degermed grains grew well if the aleurone layer was living, but not if it had been killed. In addition it was realised that embryos cultivated *in vitro*, however well provided with nutrients, never produced the quantities of particular starch-degrading enzymes found in whole, germinating grains. Degermed grains produced virtually none of these enzymes.

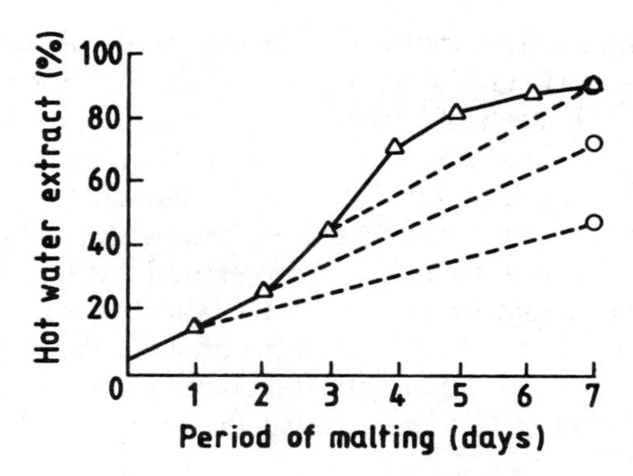

Fig. 3.1 Embryos and extract development. The continuous line (△—△) indicates the development of extract with germination time. The points (○), indicate the extract developed in degermed grains (degermed on days 1, 2 and 3) incubated under 'germination conditions' for 7 days. The extracts on the day of degerming are joined to the final extracts by dashed lines (△ --- ○). Beyond a certain incubation time, extract development continued in the absence of the embryo (Kirsop and Pollock, 1958).

The explanation for these observations came about 60 years after these early observations. Kurosawa, studying the bakanae disease of rice seedlings, found that the disease was caused by a fungus, *Fusarium moniliforme* (syn. *Gibberella fujikuroi*), and that the typical symptoms of the disease (extensive growth, chlorosis) were induced by a soluble material formed by the fungus in culture. Preparations of the active materials, 'gibberellins', were obtained and in the 1940s Japanese maltsters showed that gibberellin added to malting barley accelerated the malting process and enhanced the formation of diastatic (starch-degrading) and proteolytic enzymes. Subsequently, the possible use of gibberellins in malting became widely appreciated (Sandegren and Beling, 1959). Around 1960, Yomo showed that when separated barley endosperms and embryos were incubated under aseptic conditions in tissue culture, much greater amounts of enzymes were generated than the sum of the quantities formed when the parts were incubated separately. A soluble factor, a weakly acidic material, was produced by the embryo. This factor induced massive enzyme formation in degermed grains. Both the embryo and the endosperm (aleurone layer) had to be alive (Briggs, 1963a). The embryo factor could be replaced by gibberellic acid. Indeed, when degermed grains are incubated in water containing gibberellic acid, massive amounts of enzymes are generated; sugars, amino acids and minerals appear in solution as the starchy endosperm is extensively degraded. Within the limits of the experiments it was shown that endogenous gibberellins and not other hormones were responsible for triggering the formation of all α-amylase generated in the endosperm (Groat and Briggs, 1969). The gibberellins act on the aleurone layer, and their effects are mediated by this tissue. The addition of other

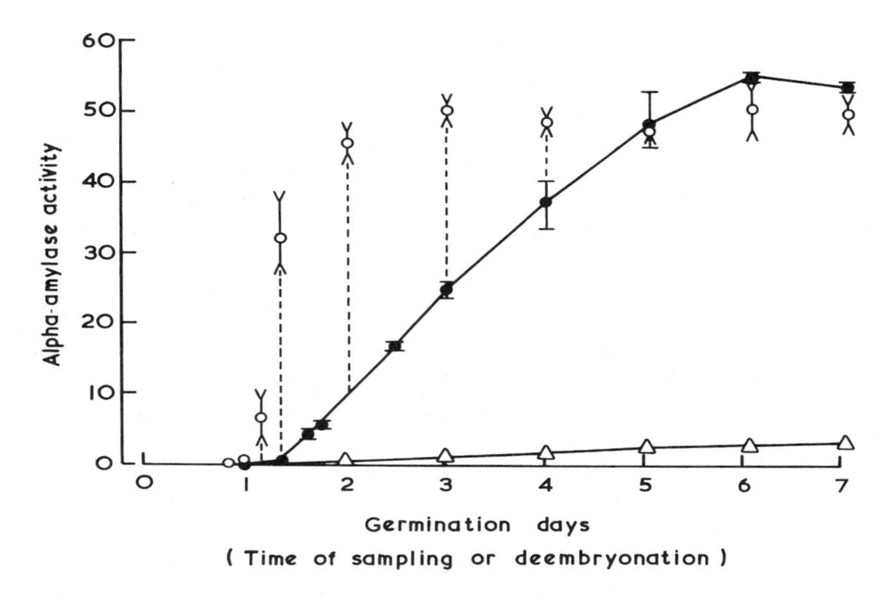

Fig. 3.2 The increase in α-amylase with germination time in decorticated grains (●—●) and in embryos (△—△), and the amount of enzyme generated after 7 days total incubation time in endosperms that were separated on the days shown (○). Alpha-amylase activity as 'starch iodine colour' (*sic*) units per corn part. The vertical dashed lines indicate the amount of α-amylase formed between the time of degerming and the end of the incubation period. Clearly the enzyme-forming potential increased between days 1 and 2. This was the result of migration of gibberellins from the embryo to the aleurone layer (Groat and Briggs, 1969).

substances to the isolated aleurone layer can alter its response to gibberellins, but it is not clear if these effects are of significance in the whole grains. These findings provided an explanation for various reports (e.g. Paula, 1932; Kirsop and Pollock, 1958) that after a certain period of germination (say 3 days) malting was not checked by embryo damage (with nitric acid) or by embryo removal (Fig. 3.1). Earlier applications of these damaging treatments checked or prevented the malting process. At the time, it was supposed that the embryos had generated sufficient enzymes for modification to continue. However, in addition to the enzyme present at the time of embryo damage, the gibberellin released by the embryo before damage was inflicted triggered continuing enzyme production and release by the aleurone layer (Groat and Briggs, 1969) (Fig. 3.2). These findings suggested the possibility of inducing grain modification with externally supplied gibberellic acid while using a variety of treatments to damage embryos and reduce embryo respiration and rootlet production and hence reduce malting losses (Chapter 14).

The response of the aleurone layer to gibberellic acid has been extensively investigated (Briggs, 1973, 1987b, 1992). The aleurone layer synthesizes and/or releases all the enzymes needed to degrade the starchy endosperm and, in some way, is able to release bound β-amylase from endosperm proteins (Chapter 4). Furthermore, this tissue releases sucrose (synthesized from the lipids stored in it) and minerals, including inorganic phosphate. Therefore, it is both a source of enzymes that degrade the starchy endosperm, so providing nutrients for the embryo, and a source of nutrients released from its own reserves. The aleurone layer's responsiveness to gibberellins is influenced by its previous history, for example how long the grain has been stored and under what conditions (Crabb, 1971; Kelly and Briggs, 1992a, 1993), and the growing conditions when the grain was forming and maturing (Nicholls, 1982). The period of time the tissue has been hydrated with access to oxygen, before the gibberellin trigger arrives is also important, as is an unidentified embryo factor (Briggs, 1987b, 1992).

The relative importance of the scutellum and aleurone layer in generating the enzymes responsible for the degradation of the starchy endosperm is uncertain (Briggs, 1992). It has variously been suggested that the embryo produces 15% or 50% of the α-amylase formed in the grain during malting, up to 40% of the endo-β-glucanase and all of the carboxypeptidase I (Briggs, 1987b, 1992). It has even been stated that the contribution by the embryo to the supply of hydrolytic enzymes is negligible and that any enzyme generated by isolated embryos originates from attached aleurone tissue. This is incorrect (Briggs, 1987b, 1992). The major considerations are as follows.

- Patterns of endosperm modification are not explicable on this basis (Chapter 2). Some of the variations in modification patterns can only be explained by assuming that the relative contributions of enzymes from the scutellum and aleurone layer differ in different grains.
- Fluorescent staining of exposed endosperms of modified, dried grains indicated that endosperm breakdown by enzymes from the scutellum differed from that caused by enzymes from the aleurone layer (Oh and Briggs, 1989).
- Isolated embryos release hydrolytic enzymes from the scutellum. This is prevented by scraping away the scutellar epithelium (which does not prevent the embryo growing on a nutrient solution; Brown and Morris, 1890). When plugs of central scutellar tissue (entirely free of aleurone tissue) are incubated in a suitable medium they generate α-amylase (Briggs and Clutterbuck, 1973; Ranki, 1990). Furthermore, enzyme production by embryos is influenced by the supply of nutrients whereas that by aleurone tissue is not.
- The use of fluorescent, specific antibody labelling has shown that α-amylase is formed in – and released from – the scutellar epithelium and this is consistent with histochemical studies of α-amylase, protease and

esterase, which indicate that the enzymes are first released into the endosperm from the epithelial face of the scutellum (Briggs, 1987b, 1992).

While contributions from the aleurone layer probably differ in different grains, there is no doubt of the importance of this tissue in malting. For example, embryos transplanted to dead, degermed grains do not induce extensive endosperm modification, and modification is inadequate in genetically faulty grains in which patches of the aleurone layer are dead (Sole *et al.*, 1987).

The current model is that when germination has started in a malting barley grain, the embryo releases hydrolytic enzymes and gibberellins. The enzymes diffuse along the grain, gradually hydrolysing the structural components of the starchy endosperm. Meanwhile the gibberellins also diffuse along the grain and *inter alia* trigger the production of enzymes in the aleurone layer and their release, together with others formed when the grain hydrated, into the endosperm. After 2–3 days under traditional malting conditions, the aleurone begins to release hydrolytic enzymes and these, together with those from the embryo, modify the starchy endosperm. This agrees with the known pattern of modification and the well-known fact that enzyme levels, malt hot water extract, cold water extract and soluble nitrogen increase first in the embryo end of the grain, then in the mid-region, and finally at the apex (Windisch and Kolback, 1929; Dickson and Burkhart, 1942; Kirsop and Pollock, 1958; Briggs, 1973, 1978, 1987b, 1992). This model provides a rational explanation for the maltsters' traditional preference for short and plump grains having embryos with wide and relatively large scutella (Mann and Harlan, 1915). Larger scutella presumably produce larger supplies of enzymes and gibberellins, while a short grain would modify sooner than a long one because enzymes and gibberellins would have less far to diffuse to reach the apex of the starchy endosperm. Plump grains tend to yield higher extracts than thin ones.

Continuing enzyme production by barley embryos, *in vitro*, is dependent on a supply of nutrients, particularly amino acids. However, moderate supplies of sugars prevent the formation of α-amylase (and possibly other enzymes) and gibberellins. Therefore, it is considered that the production of the enzymes which catalyse endosperm breakdown is regulated by sugar levels within the grain. This, in turn, is regulated by a balance between sugar uptake and utilization by the growing embryo and supplies from the modifying starchy endosperm. Thus, a feedback loop is established (Chapter 4). In some circumstances, external applications of gibberellic acid to germinating grains can increase α-amylase levels by as much as fourfold, demonstrating that endogenous hormone levels are insufficient to trigger maximum enzyme formation. Exogenous gibberellic acid gains entry at the embryo and its spread seems to be limited by the living tissues of the embryo (Smith and Briggs, 1980a). It appears to accelerate modification largely by inducing more enzyme release from the aleurone layer (Briggs and MacDonald, 1983a). As far as is known, the factors regulating modification in malting rye, wheat, triticale and oats are generally similar to those that occur in barley. When all possible transplants were made between the embryos and degermed grains of rye, wheat, barley and oats, considerable embryo growth occurred in each case. However, the chemistries of these grains differ in detail. In oats, the scutellum grows while secreting enzymes into the starchy endosperm.

In 'tropical' cereals, the situation is less clear. There are discrepant reports, but it seems that, at least in maize and sorghum, most or all of the enzymes that catalyse endosperm modification originate in the scutellum (Chapter 14; Aisien Palmer, 1983; Glennie *et al.*, 1983). Whole malting sorghum grains and separated embryos and endosperms respond only slightly, or not at all, to applications of gibberellic acid (Daiber and Novellie, 1968; Aisien and Palmer, 1983).

3.3 The statistics of grain testing

Many tests on grain and malt rely on investigations of small numbers of corns (Hudson, 1960; Pollock, 1962b; Briggs, 1978; EBC, 1987; IoB, 1991; ASBC, 1992). Examples are viability tests, germination tests, determination of TCWs and studies on the distribution of water in grains and patterns of modification. Grains from a single batch of one barley variety vary in size, shape, physical intactness, viability, dormancy, moisture content, and chemical composition. Mixing batches of grain – a necessity in large maltings – increases this variability. Evaluating a bulk of grain relies on testing small samples. It is necessary to adhere strictly to specified sampling and sample-splitting regimes to obtain samples that are adequately representative (Chapters 6 and 13). If this is not done the results of tests carried out on the samples are completely untrustworthy.

Many tests are carried out on statistically small numbers of grain, and consequently they **cannot** yield highly reliable results. For example, in determining the TCW, counting 1000 corns at random and weighing them can easily give an estimate that is up to 12% in error of the true mean value of the bulk. Weighing groups of 5000 grains reduces this error to about 4% (Hudson, 1960). The preferred method is to prepare samples of at least 50 g, and better 100 g, and to count the number of corns in them **after** they have been weighed (EBC, 1987). The results, after correcting for moisture contents, should be reported as dry weight of 1000 corns. In germination, staining or other tests in which small numbers of grain are used, the statistical uncertainty of the results is inevitably very large (Urion and Chapon, 1955). For example, if, in a sample, one grain in ten does not germinate then the range of the 95% confidence limits on the estimate of the percentage germinability of the grain bulk, on samples of different sizes, are: 50 grains, 3–22%; 100 grains, 5–18%; 250 grains, 7–15%; 1000 grains, 8–12% (Fig. 3.3). It follows that the tetrazolium staining viability test, which is often carried out on a sample of only 50 grains, **cannot** give reliable results and will often mislead if, for example, the lower acceptance limit is a low, 96% viability but the limiting test value is set at 98% or better stained (i.e. 49 or 50 grains in 50). For germination tests, often sub-samples of 100 corns are tested in triplicate. The mean value is based on 300 grains but an added advantage of using three samples is that it is possible to calculate the standard deviation of the results and so obtain an estimate of their variability. Practically, when grain has been cleaned, dried, well mixed and screened to remove small grains, the variations between sub-samples are often reduced.

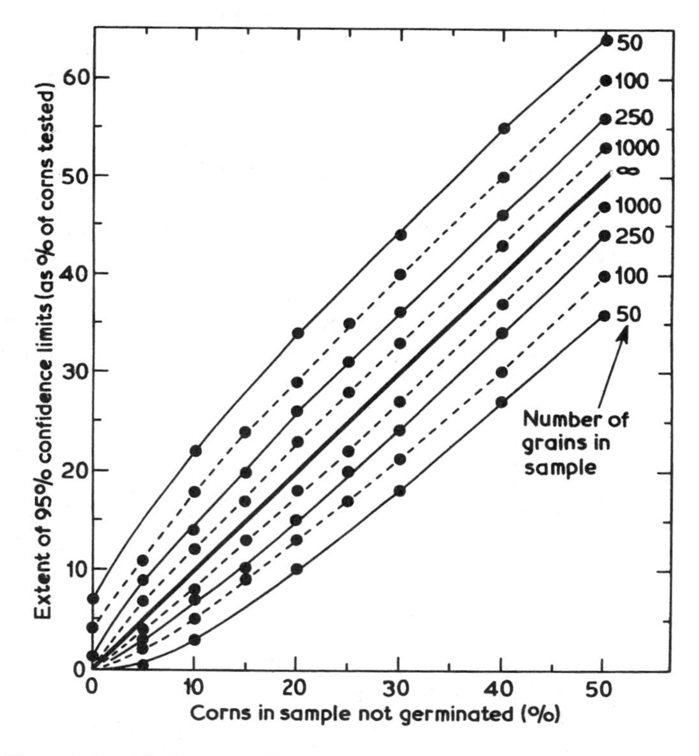

Fig. 3.3 The relationship between the percentages of corns in samples of different sizes showing a particular character (e.g. failure to germinate) and the 95% confidence limits of the estimates. The heavy control line (∞) is achieved when an infinite number of grains, or all the grains in a bulk, are evaluated so that there is no possibility of statistical error. It is seen that even with tests based on samples of 1000 grains errors are significant (data of Urion and Chapon, 1955).

3.4 Water uptake by grains

Water uptake by cereal grains (Brookes *et al.*, 1976; Briggs, 1978) has been studied extensively but, particularly in the case of scoured wheat, many of the investigations have been made with physically damaged grains. For malting purposes, grains should be intact. It seems that the processes of hydration are very similar in the intact grains of different species.

When grain is immersed in water (i.e. it is steeped) the surface layers (the husk, where present, and the pericarp) hydrate comparatively rapidly. Progressively, bubbles of air attached to the grains or retained in the crease or in spaces between the husk and pericarp disappear. If only the tip of the grain is wetted moisture spreads through the surface layers, by capillarity, and will reach the basal, embryonic end of the grain. The lodicules, situated beneath the husk around the embryo, are thought to provide a system of fine channels or spaces that, by capillarity, spread the moisture over the pericarp and the micropyle. Probably most water enters the interior of the grain via the micropyle. The embryo is the first tissue, within the testa, to hydrate. However some water may slowly gain entry over the surface of the grain, especially at the apex and near the pigment strand (Hinton, 1955; Axcell *et al.*, 1983; Briggs and MacDonald, 1983b) (Fig. 3.4). The embryo end of the grain hydrates first and remains wetter throughout steeping and germination. For example, in one instance the moisture contents of the basal (embryo) ends, the middle parts, and the apical ends of steeped barley grains were 47.1%, 38.8% and 39.1%, respectively (Witt, 1959). As the embryo hydrates it swells (Chapter 2). Presumably the water is drawn in through the micropyle by the high osmotic suction generated by the large amounts of simple sugars present (Chapter 4). Finally, the moisture content

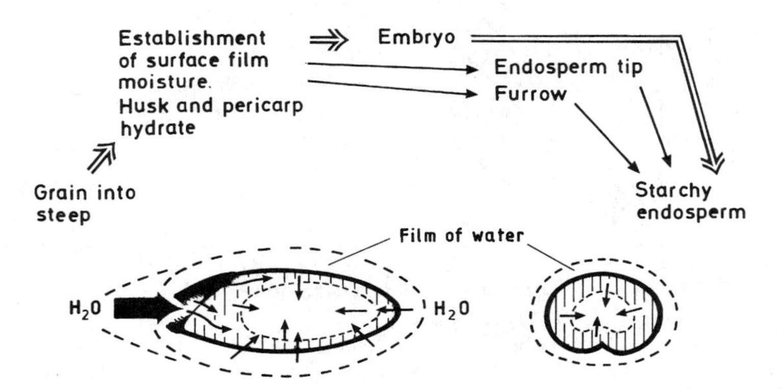

Fig. 3.4 Scheme showing the probable routes of entry of water into the barley grain (after Axcell, Jakovsky and Morrall, 1983). The embryo is fully hydrated before the central regions of the starchy endosperm (cf. Table 3.1).

of the embryo may reach 65–70% (Reynolds and MacWilliam, 1966) (Table 3.1). The endosperm hydrates more slowly.

The tissues that, perhaps locally, permit the penetration of water (the testa, pigment strand and micropyle) exclude many salts, sugars and other substances, so they are selectively permeable (section 3.5). Perforating and splitting the surface layers, including the testa, permits rapid water uptake; selective permeability is lost. Not only may dissolved salts reach the interiors of damaged grains but sugars and other small molecules can be leached from within them. Water uptake and selective permeability of the grain surface layers are purely physical processes. Both phenomena are apparent in grains killed by treatments with heat or with mercuric chloride. As air-dry grain hydrates there is an evolution of heat, the heat of hydration, as there is when, for example, dried starch or protein powders are hydrated. During steeping, there are steeping losses (section 3.1). The losses are smaller from clean, physically intact grains, when materials are leached from the husk and possibly the pericarp. Probably enzymes from microbes associated with the surface layers catalyse their limited degradation, particularly if steeping is carried out at elevated temperatures (Fig. 3.5). If appreciable numbers of cracked or broken grains are present, then nutrients are leached from within the damaged grains and the steep water becomes more intensely coloured, tends to froth more and contains more microbes. Under these circumstances, steep water should be changed more frequently. Materials dissolved in steep water include 'testinic acid' (a crude mixture of proteins and phenolic substances), uncharacterized pigments, mineral salts (including inorganic phosphate), organic phosphates, sugars, phenols, phenolic acids and amino acids. The hydrated grain metabolizes, respires (taking up oxygen) and ferments, releasing the fermentation products ethanol and carbon dioxide into the steep liquor. Microbes that are always present multiply on the grain and in the steep liquor, and they

Table 3.1 The moisture contents of various parts of barley grains steeped for short periods at 13 °C (55.4 °F). Initially the moisture content of the whole grain was probably about 15%. After 48–72 h steeping, the whole grain would have had a moisture content of *ca.* 45% and the moisture content of the embryo would have been 65–70%

	Moisture (%) after steeping at 13 °C for			
	2 h	4 h	6 h	24 h
Scutellum	18.5	36.8	50.0	57.9
Embryo axis	33.0	43.6	51.3	57.9
Pericarp + testa + aleurone	11.4	15.6	26.8	36.4
Starchy endosperm	10.0	12.5	21.4	35.0
Husk	40.5	43.1	44.9	47.1
Whole grain (by direct estimation)	21.6	25.9	28.6	38.4
Whole grain (by summation)	20.9	24.8	27.1	37.5

From Reynolds and MacWilliam (1966).

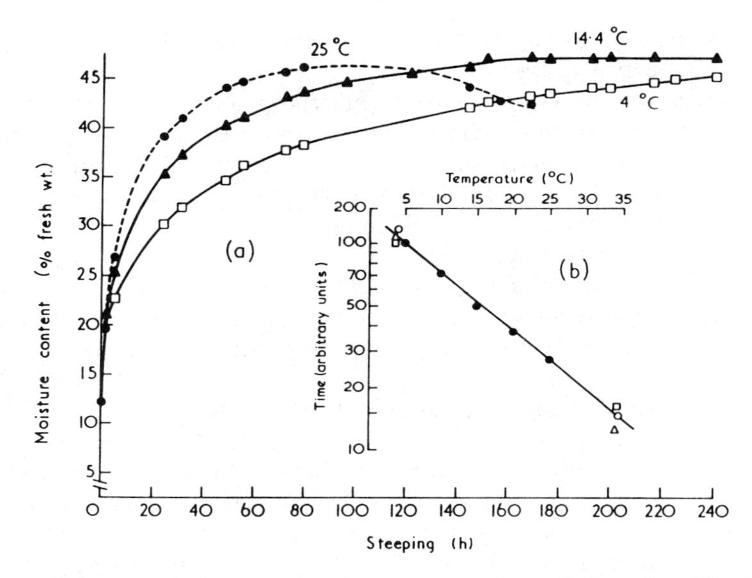

Fig. 3.5 (a) Increases in the apparent moisture contents of ungraded samples of Proctor barley steeped at three different temperatures. Samples, of known initial dry weight and moisture content, were briefly removed from the steep, centrifuged to remove surface moisture and weighed before immediately being returned to the steep. The moisture contents were calculated making the assumption that the dry weights of the samples did not decrease appreciably during steeping. However, the apparent decline in moisture content in the samples held at 25 °C (77 °F) was caused by appreciable losses of dry matter as a result of the microbial decomposition of the surface layers of the grain (Briggs, *et al.*, 1981). (b) The relationship between the relative times taken for steeped grains to attain any particular moisture content (surface moisture having been removed; note log scale on axis) and the steeping temperatures (after Briggs, 1967).

consume oxygen dissolved in the water. Among their metabolic products is acetic acid, probably derived from the oxidization of ethanol released by the grain. It is to remove these dissolved materials, the microbes growing on them and undesirable microbial metabolites, that steep liquor is changed. I believe it would be beneficial to rinse the grain at the end of each drain to remove the residual surface film of dirty water and replace it with fresh (Briggs, 1995).

Studies using autoradiography to follow the movement of tritiated water into the grain demonstrate the sequence in which the tissues hydrate (Axcell *et al.*, 1983) (Fig. 3.4). The spread of water through the starchy endosperm may also be followed by cutting grains, exposing the cut surfaces to iodine vapour and noting the colour given with the starch (von Ugrimoff, 1933) or by following the movement of tracer manganese (Davies, 1989). Endosperms of grains stained with iodine and containing less than 18% moisture are yellow; at about 18% moisture they appear to be violet and at 25% moisture or more they become blue-black. More simply, steeped grains may be briefly immersed (30 s) in boiling water, cooled and sliced apart. Where the moisture content of the starchy endosperm exceeds some fixed moisture value, probably between 32% and 38%, the starch gelatinizes and the regions where this has occurred are readily distinguished (Chapon, 1963). Using this technique, it appeared that barley grains in a single batch varied in their degrees of hydration to a surprising extent (Kelly and Briggs, 1992a), even after extended periods of steeping. The last regions to hydrate are the central parts of the starchy endosperm. Often these regions fail to modify adequately in rapidly malted grains, possibly because locally poor hydration does not permit sufficient hydrolysis of the endosperm cell walls. Irregular moisture distribution appears to be common in steeped barley grains and the practical observation that it is often preferable to leave grain immersed for several hours longer than the calculated time to reach the 'target' moisture content may be related to the need for time to allow the moisture to distribute itself more evenly within the endosperms of the most slowly hydrating grains. The poor hydration of some samples of the Australian barley Stirling seems to be associated with the formation of an unfilled channel, apparently in the region of the sheaf cells (Landau *et al.*, 1996).

When grain is steeped it swells and increases in bulk volume by up to about 25%. Individual grains swell and increase in width and depth, but not length (Table 3.2). When grain is immersed, the initial rate of water uptake is rapid, but this gradually declines until, under conditions such that the grain does not germinate, the moisture content tends to a limiting value (Fig. 3.5). Older reports referred to 'kinks', or irregularities in the water-uptake curves. These Baker–Dick breaks must have been the result of experimental errors. The rate of water uptake is influenced by corn size, temperatures, the grain sample and perhaps other variables (Figs. 3.5 and 3.6). Grain rich in nitrogen and steely grains with their tougher endosperms hydrate more slowly than grains containing less nitrogen (and, therefore, more starch) and mealy grains. Grains that have been frosted in the field are said to hydrate more rapidly than those which have not. The hydration rates vary between grains of one variety grown in different seasons. Therefore, for barleys from a particular British region the time to reach 30–35% moisture in the steep was 10–16 h in 1954, 16–24 h in 1955 and 6–9 h in 1956 (Macey, 1957). There is some evidence that grains which hydrate more rapidly, or which swell more when steeped to a given moisture content, are of superior malting quality and are those with less tough mealy endosperms (Brookes *et al.*, 1976). The increase in moisture content during a closely regulated 72 h steep (the water-absorbing power or das Quellevermögen, QV) has been regarded as useful criterion of barley malting quality. Values over 50% are very good, 47.5–50% are good, 45–47.5% are satisfactory but barleys giving values of less than 45% are inadequate (Narziss, 1976). Water uptake occurs more rapidly at higher temperatures. Many empirical rules of thumb were used to decide how long to steep grain when the steep temperatures changed. These gave different 'answers' and because of this, and in view of differences between different batches of grain, the results were unsound. The rate of grain hydration, at any particular temperature, can only be determined reliably by measuring it experimentally. However, for one particular batch of grain there is an inverse linear relationship between the temperature and the log of the time of steeping needed to reach any particular moisture content (Briggs, 1967) (Fig. 3.5). So, if the grains' hydration characteristics are known at one temperature, the water-uptake curve at any other fixed temperature can be calculated. It is likely that all barleys have about the same temperature coefficients of hydration. The energy of activation of water uptake, E, has been calculated to be 3400 cal/g per mol for the embryo and 3200 cal/g per mol for the endosperm (Sfat, 1966). The rate of water uptake is probably limited by a diffusion process. From the physiological need to obtain the most uniform germination, optimal steeping temperatures are probably 12–14 °C (53.6–57.2 °F). In practice, warmer temperatures of 16 °C (60.8 °F) or more are used to reduce the time taken to hydrate the grain adequately. It is **not** acceptable to use very warm water to accelerate steeping. Grain is easily oversteeped at higher temperatures (e.g. 25 °C (77 °F)), the regularity of germination is reduced and induced water-sensitivity appears (Lubert and Pool, 1964). The living tissues, when hydrated, are quickly damaged at steeping temperatures of 40 °C (104 °F) or more, a fact that has been exploited to reduce malting losses (Chapter 14).

In spite of irregularities at the start of steeping (Brookes *et al.*, 1976) thinner grains soon have higher moisture contents than plump grains, and this difference is maintained throughout steeping (Fig. 3.6). The irregularities at initiation of steeping are probably accentuated in commercial processes because grains are not immersed simultaneously owing to the time involved in filling steeps with water and draining them. Even moisture

Table 3.2 The effects of steeping time and moisture content on the distribution of grain widths in Kindred barley (1957 crop) steeped at 15.6 °C (60 °F). Grains of different widths were separated using slotted sieves

Slot widths of sieves retaining the grains		Barley (%) retained on the sieves			
(in)	(mm)	Unsteeped barley	Steeped 22 h (38–39% moisture)	Steeped 28 h (41–42% moisture)	Steeped 40 h (43–44% moisture)
7/64	2.78	1.8	38.2	43.4	49.5
6/64	2.38	71.6	54.5	52.2	48.3
5/64	1.98	26.0	6.7	4.1	2.0
Through 5/64	Through 1.98	0.6	0.6	0.3	0.2

From Witt (1959).

uptake and germination are important in malting. Generally thinner grains ('screenings', thin corns) are removed by sieving before barley is malted. Often the sieves used pass grain of 2.2 or 2.3 mm width or less, but the rejection size varies with the barley being used and the malt being made. In the example shown in Fig. 3.6, grains of 2.4 mm width and more vary relatively little in their moisture content, so in this case retaining grains of this size range would have been good practice. In addition to their rapid water-uptake characteristic, thin grains tend to be more dormant, to have a higher proportion of diseased corns and to be relatively richer in diastase and nitrogen (crude protein) and poorer in starch than plumper corns. Malt users prefer to have grains of even sizes if they are to be roller-milled and for the highest extracts they favour plump grains. It is expensive for maltsters to screen out and reject a large proportion of the grains they buy and to segregate, store and malt grain of different size classes.

Steeping schedules are guided by the results of preliminary, small-scale trials and experience with other samples of the season's grain. Newer malt-

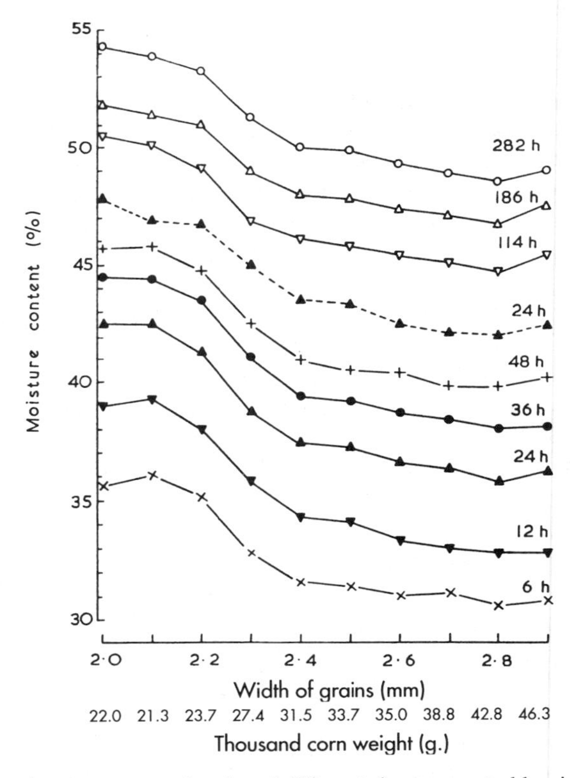

Fig. 3.6 The moisture contents of grains of different sizes, separated by sieving one batch of grain, after steeping for the times shown at about 5 °C (41 °F) (after Žila *et al.*, 1942).

ings usually have facilities to control the temperature of the steep water. In the past this was usually not the case and consequently steep temperatures and steeping times varied more widely than is current practice. To decide when grain was 'steep ripe', that is it was sufficiently steeped, traditional maltsters carried out a number of tests. First, steeps were inspected to check that the 'swell' (increase in bulk volume) was complete, that is that during steeping the level of the grain had risen to the expected extent. Second, the ends of several grains were pressed towards each other, between finger and thumb; well-steeped grain was readily compressed, tending to cause the husk to separate from the grain's surface. Third, grains were bitten to see if they could easily be crushed into two pieces, without signs of cracking or being hard. Fourth, when grains were cut across, in a farinator, they were expected to appear 'wet and macerated'. Finally, when the thumbnail was pressed into the side of a well-steeped grain, the grain was expected to bend (and not snap or break) and the husk was expected to separate from the body. These subjective tests, while useful in their time, provide no adequate alternative to the laboratory measurements of moisture content, which are used to regulate steeping in a modern maltings. In practice, it is often desirable to distinguish between the water in the grain (which remains after the grain has been centrifuged or blotted) and the film of moisture that clings to the surface when the steep is first drained. This surface moisture acts as a reserve, being taken up as the grain germinates, but is commonly rich in microorganisms and limits oxygen uptake by the tissues of the grain.

As already noted, during steeping the moisture content of grain tends towards a limiting value, provided that the grain has not started to germinate. If grains chit, as they are encouraged to do in most modern steeping schedules by using air rests, the chitted grains will take up moisture more rapidly than those which have not. If grains are drained and air-rested for short periods, their interiors will continue to hydrate at the expense of the surface film of moisture. Provided that the grain surface film is re-established, for example by reimmersing the grain, before the surface film of moisture is totally absorbed, there will not be a check in the hydration of the grain. However, with extended air-rests, this is not found since the surface film is exhausted before the end of the 'dry' period. At the end of an air-rest, when the grain is rewet, moisture uptake is resumed and, particularly if the grain has chitted, occurs at an accelerated rate.

When the grain begins to grow, the situation is less clear cut. Maltsters usually report that it is preferable to hydrate the grain sufficiently during steeping so that further additions of water, which may be applied as sprays or by sprinkling, are not needed during the germination period. Hence the old instruction to 'get all the water into the grain in the steep'. Nevertheless, some maltsters routinely sprinkle germinating grain. Possibly replacing immersions by sprayed applications of water is acceptable provided that the grains are only just chitting. During germination, the grain loses water by evaporation and by the consumption of water, which occurs when chemical links are hydrolysed. Water is also generated, together with carbon dioxide, by the respiratory processes that occur in the living tissues. As the embryo grows, the newly formed tissues (the acrospire and rootlets) need to be hydrated and the water used is, at least in part, withdrawn from the starchy endosperm. Consequently, as malting proceeds, the moisture associated with and in the grain is redistributed, and during germination the endosperm becomes drier (Fig. 3.7). Conditions that favour vigorous embryo growth may cause the endosperm to dry to such an extent that modification is slowed down and may actually cease so that undermodified malt is produced. Spraying or sprinkling grain during germination encourages embryo growth and increased malting losses. It has been disputed whether or not modification is usefully accelerated. Certainly sprinkling seems to favour embryo growth and increases malting losses more than increases in extract.

Acute problems with water distribution can occur when spray-steeping is used (Reynolds *et al.*, 1966; Kirsop *et al.*, 1967b) and, by extrapolation, might occur in extensively aerated and air-rested grain that has chitted and grown to an appreciable extent before the end of the steeping schedule. When grain was steeped by immersion for 48 h, the moisture content plateaued at

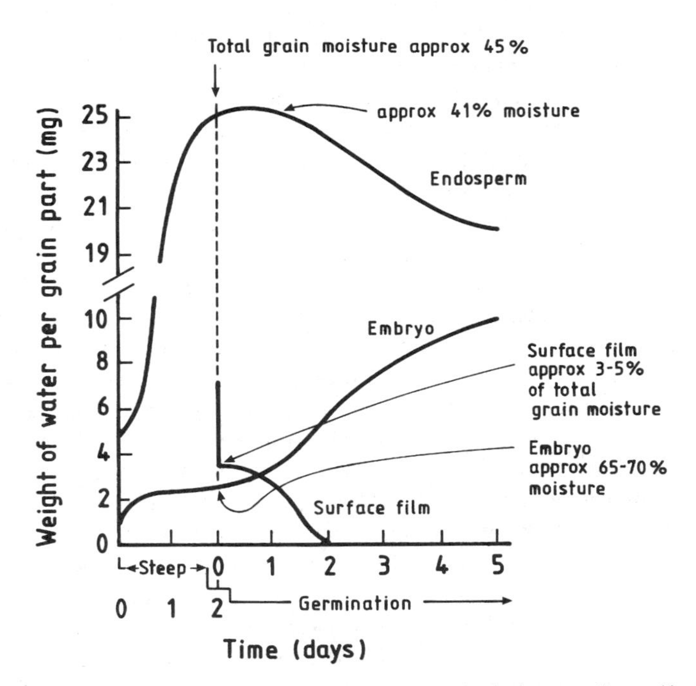

Fig. 3.7 The uptake and redistribution of water in barley grain during steeping and in the initial stages of germination (after Kirsop, 1966; Reynolds and MacWilliam, 1966; Axcell, *et al.*, 1983).

a level of 45%; the grain was regarded as fully steeped and no germination had occurred. However, during spray-steeping, the embryo grew strongly in the 24–48 h perod, so the amount of water in the embryo continued to increase. The grain (bulk) reached a water content of 45% in 36 h. However, a disproportionate amount of this water was in the embryos; the endosperms were too dry. After steeping, the embryos continued to grow strongly, abstracting more water from the endosperms. Because the endosperms were dry, an undermodified malt was produced, and because the embryo grew vigorously, the malting losses were high.

3.5 The permeability of grains to solutes

Grains have varying permeabilities to solutes (Pollock, 1962a; Brookes *et al.*, 1976; Briggs, 1978, 1987b). It seems that all the substances that dissolve in water but do not bind to grain tissues, such as the red dye Eosin, can permeate the husk and pericarp. Some substances do not reach the living tissues within the grain, while others do. The ability of grains to exclude some substances from the living tissues, but not others, is important in understanding the effects of additives on grain, and a number of other aspects of grain physiology. The permeability of the grain surface layers has been most extensively studied with barley, and the conclusions reached appear to be valid for other grass caryopses (Briggs, 1987b). Gola (1905a,b) was probably the first to recognize that the testa limited the diffusion of solutes into grains, but widespread appreciation of this important fact followed the independent work of Adrian Brown (Pollock, 1962a; Briggs, 1978, 1987b). The most striking experiments were carried out with sulphuric acid, using grains having blue aleurone layers (Brown, 1907a,b). The blue pigment turned red in contact with acid. When physically undamaged grains were placed in dilute sulphuric acid they took up water, the aleurone layers remained blue and the remaining acid became more concentrated. Thus, the grains were selectively permeable, permitting water to enter but not the acid. In strong acid (58%), water might actually be withdrawn from grain. In strong solutions of sulphuric acid (e.g. 36% or more) the husk and pericarp disintegrated but still the acid was excluded from the interiors of undamaged grains by the testa, the pigment strand and the micropyle. If all residual acid was removed by thorough washing, the huskless and pericarpless (decorticated) grains would germinate. However, in damaged or broken grains, those that had been bitten by insects, that had been pricked or cut or which had chitted (started to germinate), acid reached the interior (the blue aleurone layers became red) and the embryos were killed. Even with intact grains steeped in acid over extended periods, the acid gradually penetrated and killed an increasing proportion of corns. The selective permeability is a physical property of the grain: it is still shown by grains that have been killed – by heat or by soaking in a solution of mercuric chloride, for example. The acid was excluded by the testa (the spermoderm or seed-coat), the pigment strand and the structures located at the micropyle. The same is true for many dyes and other substances that are excluded. For example, if grains are successively soaked in solutions of silver nitrate and sodium chloride, the silver chloride that is deposited in the parts of the grain penetrated by the two salts can be located by exposing cut grains to sunlight. The black deposits of silver so produced occur throughout the husk and pericarp but not within the testa. Treated grains germinate. However, if cut grains are exposed to silver nitrate they are killed. Grains also act as small osmometers; for example, if they are soaked in solutions of increasing concentrations of substances that are excluded, such as salts or sugars, then progressively less water is taken up from stronger solutions. It is doubtful if salt concentrations in potable waters are ever sufficient to alter significantly water uptake during steeping. Ethanol, acetaldehyde, acetone, ethyl acetate, chloroform, phenols, aniline and weak organic acids, such as acetic acid, readily enter grains from dilute aqueous solution, and some of them appear to accelerate the rate of water uptake. Equally, ethanol generated in the grain during steeping readily diffuses out into the steep liquor. Ethylene glycol and urea slowly penetrate, but sugars appear to be excluded. Sugars, amino acids and many other substances are retained within intact grains during steeping. Of the strong acids, sulphuric and hydrochloric acids are excluded but nitric acid slowly enters in the region of the micropyle, and trichloracetic acid penetrates readily. Strong bases (sodium and potassium hydroxides) are briefly excluded when applied to grains in dilute solution but after a few hours they enter and damage the grains. Sodium and potassium carbonate solutions damage grains but sodium bicarbonate solutions do not. At least limited exposure to lime water does not harm grains. While limited exposure to calcium hypochloride solution does not damage grains, exposure to sodium hypochloride, which is strongly alkaline, is often harmful. These facts are of importance when solutions of these various agents are used to 'clean' batches of grain or control their associated microorganisms.

Many salts are excluded from the interiors of undamaged, quiescent grains. For example, silver nitrite, sodium chloride, copper sulphate, potassium dichromate, potassium sulphate and the strongly ionized salts mercuric nitrate and mercuric sulphate. However the weakly ionized salts mercuric chloride and mercuric cyanide readily penetrate grain and kill it. Residues of toxic salts, which may remain after grain has been rinsed, will check grain germination because as chitting occurs and the testa is ruptured they will then have access to the living embryo, even though they are excluded from ungerminated grain.

Formaldehyde slowly enters grains, limiting its use as a surface sterilant and an agent for reducing anthocyanogen levels. It is least damaging if applied to hydrated grains (Whatling, *et al.*, 1968). Iodine in solution slowly penetrates over the whole surface of the grain, as is apparent from the blue-black coloration that appears through its reaction with the starch. However, if the iodine is present in solution in low concentration in the presence of an excess of potassium iodide then it is chiefly present as the triiodide ion. This only slowly penetrates the grain (whole or decorticated, see below) in the region of the embryo. Being essentially starch free the embryo remains uncoloured, but slowly a blue collar of blue-black colour develops around the scutellum and spreads up the grain. If these grains are immersed in a solution of sodium thiosulphate, the colour remains, showing that the thiosulphate is excluded. However, if the grains are cut and the surface layers are broken, the thiosulphate gains entry and the colour is discharged.

Sodium and potassium nitrates may slowly enter grains, because they sometimes relieve dormancy. However sodium and potassium bromates (both may be used as additives in malting) (Chapter 14) and probably the corresponding iodates certainly enter grain within 4 h, mostly at the embryo end but also slowly over the rest of the grain's surface (Brookes and Martin, 1974).

Gibberellic acid is a most important additive, which is sometimes used in malting (Chapter 14). This also enters the grain in the region of the embryo since, in undamaged grain, it accelerates endosperm degradation, which still spreads from the embryo end of the grain in the normal way. This substance enters grain faster after chitting, but its ability to enter ungerminated grain is demonstrated by its ability to break dormancy. The complexity of the situation and the reasons for believing that it is the testa (not the pericarp) that excludes gibberellic acid have been reviewed elsewhere (Briggs and MacDonald, 1983b; Briggs, 1987b). When gibberellic acid is applied to whole grain, a considerable amount remains on the surface (Brookes and Martin, 1975) where it is chemically altered, possibly by association with polyphenols and their oxidation products. It seems that the embryo limits the rate at which externally applied gibberellic acid moves through the grain to the aleurone layer (Smith and Briggs, 1980a).

3.6 Assessments of husk content and some aspects of grain quality

Husk contents of barley grains have been determined in various ways. As the agents used in the different methods allow the separation of different tissues and probably leach away some of the contents of the tissues to different extents, it is not surprising that the results do not agree closely. Most of these techniques indirectly depend on the relative impermeability of testa.

The simplest method is to soak grains in alkali (for example in a solution of ammonia) and peel the loosened husks by hand. After rinsing and drying they are weighed. This is tedious, and it is not easy to collect all the husk. Various mechanical techniques have been employed for removing husk, for example by blowing grains against a wire mesh screen until the husks disintegrate. This is 'mild' pearling and there is no reason to suppose that this battering and abrasive treatment necessarily removes all the husks or removes only husk. Another approach to dehusking, applicable to barleys and oats, is to heat pre-weighed samples of grains in an alkaline solution of sodium hypochlorite. The husks separate and the residual grain is rinsed, dried and weighed. Husk content is determined by difference, having made allowances for the moisture contents of the samples. At least some pericarp remains so this method gives lower results than are obtained with the sulphuric acid decortication technique (Whitmore, 1960).

Since the observations of Brown (1907a,b), it has been known that strong solutions of sulphuric acid, acting at around room temperature, cause the disintegration and separation of the husk and pericarp of barley. Essery *et al.* (1956) devised a standardized technique for decortication. (The term decortication is preferred to dehusking, because both the husk **and** pericarp are removed.) Grains are incubated in 50% (v/v) sulphuric acid for up to 3 h at room temperature, then they are washed thoroughly in water and a suspension of calcium carbonate in water. Finally they are dried and weighed. Using appropriate corrections for moisture contents, the husk **plus** pericarp contents of grains are calculated. We routinely decorticate barley in sulphuric acid (50% (v/v)) at 25 °C (77 °F). The time taken for a given sample to be adequately decorticated varies substantially with different cultivars (Raynes and Briggs, 1985). Decortication greatly reduces the dormancy of many samples of barley. The germination of decorticated barley on a solution of gibberellic acid (50 mg/l) is a useful test of viability (germinative capacity; Chapter 6). However, if the grain has been damaged so that the sulphuric acid can penetrate, its viability is reduced and the test is then unreliable. Since the testa and aleurone layer are partly transparent, inspection of decorticated grains readily shows whether, and to what extent, endosperms are mealy or steely. The acid may cause darkening in embryos or regions of the aleurone when the grain has been damaged physically so the testa is perforated. The germinability of such grains is reduced. Alternatively, decorticated grains may be briefly rinsed in a dilute solution of iodine in potassium iodide. Blemishes in the testa, over the endosperm, are readily detected since where the iodine gains entry it binds to the starch and blue-black marks appear. If decorticated grains are soaked in a tetrazolium solution this also gains entry at any blemishes. Its presence can be detected by the red stain deposited in the adjacent living tissue (embryo or aleurone layer).

3.7 Viability, germinability and dormancy

Viability, dormancy and the ability of grain to germinate are all important properties in malting (Harrington, 1923; Bishop, 1944, 1945, 1946; Hudson, 1960; Pollock, 1962b; Belderok, 1968; Briggs, 1978; EBC, 1987; IoB, 1991).

3.7.1 Germination

In malting more than 98% of the grains must germinate promptly if a good quality malt is being made. To germinate, grains must be adequately hydrated, have a supply of oxygen, must be within a suitable temperature range and must not be exposed to harmful agents, either toxic chemicals or physically damaging machinery. A grain's failure to germinate under appropriate conditions may be because it is dead or because it is dormant (Fr. *dormir*, to sleep). Dormant grains are living but they will not germinate, at least under conditions that are suitable for non-dormant grains. During storage, grains progressively recover from dormancy. Newly non-dormant grains tend to germinate slowly. The phenomenon of dormancy must be distinguished from quiescence, the state of grains that are not germinating because the conditions are inappropriate, for example because they are dry and in store. Quiescent batches of grain may or may not contain proportions of dead or dormant corns.

Maltsters, grain merchants and farmers are all interested in germinability, but their requirements are different and so are the tests they use to characterize the physiological state of the grain. Malting is a strongly time-limited process, so maltsters must have grain that germinates promptly, in the dark, and reaches its maximum germination within 3 days under standardized test conditions. Grains that chit, that is to say when the coleorhiza has appeared, or at a later stage when rootlets are also visible, are scored as germinated. Dead grains and slow starters ('idlers', 'lie-backs') detract from malt quality. Ideally all malting grains would germinate rapidly and simultaneously. This ideal is never achieved. Grain for seed is tested differently. Grains are incubated in light and are scored for up to 3 weeks. A kernel is scored as having germinated when it has a seedling with well-formed roots and one or more green leaves. Clearly seedsmen's observations are of little value to maltsters.

3.7.2 Viability

Grain viability may be reduced in many ways: by insect or fungal attack, by excessive temperatures during grain drying, by physical damage, by deterioration caused by inappropriate storage conditions, by excessively long periods of storage under ambient conditions, and so on (Chapter 7). However, properly handled grains are surprisingly resilient. For example, well-dried grains withstand immersion in liquid nitrogen (–198 °C) or being chilled to within a few degrees of absolute zero. If carefully dried to less than 8% moisture, they will withstand short exposures to dry heat at 100 °C (212 °F). At 7% moisture and 40 °C (104 °F), grain will retain its vigour and vitality for several months, while dormancy declines rapidly.

3.7.3 Dormancy

Dormancy is both a varietal and a seasonal characteristic. Therefore, red wheats tend to be more dormant than white wheats and there are wide variations between different wheat and barley varieties. Some dormancy is desirable, at least up to and a short time beyond the normal harvest date, since without it, rain near harvest time can cause grain to germinate in the ear. Grain with any appreciable amount of pre-germination is of poor quality and is unacceptable to maltsters. Some of the older barley varieties, such as Scotch Common and Domen, show very little dormancy. However, wild *Hordeum spontaneum* and some varieties from North Africa are extremely dormant. Some modern varieties of otherwise excellent malting quality, such as Triumph, may be dormant for inconveniently extended periods. In some cases, dormancy is still intense after a year in store. Dormancy is greatly influenced by the conditions under which a crop grows. In the UK, dormancy is more common and more intense in barleys grown in Scotland than in the same varieties grown in East Anglia. Balanced applications of fertilizers may marginally reduce dormancy. However, within a variety, the main factors that influence dormancy are the weather conditions, the temperature, rainfall and perhaps the humidity. Hot, dry weather from the dough-ripe stage to harvest minimize dormancy, while cool, damp, dull weather favours it. Grain harvested promptly after a warm, dry spell may germinate much better than similar grain that has been harvested later after a delay caused by rain and high humidity. The wet weather is said to have induced secondary dormancy. At least in part this decline in readiness to germinate is caused by the build up of large populations of microbes on the grains, microbial growth being favoured by the humid conditions. Grains in a batch do not all become dormant (or non-dormant) simultaneously – so grain in bulk is heterogeneous in this, as in many other ways. There is a tendency for small grains to be more dormant than large grains. In several cereals (e.g. wheat, oats and barley) the position of the grain in the ear, or panicle, influences its size and the chance of it being dormant.

Dormancy declines during storage and the maltability of the grain improves (Table 3.3). In the past, British maltsters expected the new season's barley grain to have matured (post-harvest ripened, after-ripened, lost dormancy) sufficiently to be fit for malting 6–8 weeks after harvest. Whether this is true depends very much on the variety and how the grain is treated. As noted, some samples of Triumph from northern Britain after a bad growing season have been unfit for malting after more than a year's storage. Maltsters need to evaluate each batch of grain for its **germinative capacity** (GC), or viability, and its **germinative energy** (GE), the extent to which it will germinate in a standardized test. These tests are carried out on small numbers of grains (often three lots of 100) and so all possible

Table 3.3 The influence of post-harvest maturation on the malting quality of barley. Bright Kindred barley malted under comparable conditions at various times after harvest; the grain was grown in the northern hemisphere

Date steeped	Germination (%)	Extract (%)	Soluble/total protein ratio (%)	Wort clarity
Aug 21	97	72.5	35.5	Hazy
Sept 21	99	73.5	37.5	Slightly hazy
Oct 15	99	74.3	39.2	Clear

From Witt (1959).

precautions to ensure that the results are valid, in a statistical sense, must be observed.

3.7.4 Tests for germinative capacity

The GC value of grain is the percentage of grains in a sample that are alive. This is not easy to determine if the grains show any appreciable dormancy since, by definition, dormant (and dead) grains will not grow in standard tests. Maltsters have two approaches: (i) to use indirect tests that rely on some indication, other than germination, that the grain is alive; and (ii) to 'compel' or 'force' the grain to germinate, by overcoming dormancy using various tricks. Both approaches have their limitations. In particular, in the second group of methods, treatments that reliably break dormancy in slightly dormant samples often fail to do so when the samples are profoundly dormant.

Indirect tests. Most of the indirect tests for viability depend on the ability of reductase enzymes and their substrates, which are present in the embryos of viable grains, to reduce substances to coloured, insoluble products that are easily detected. Many tests have been used, including the reduction of *m*-dinitrobenzene or salts of selenium and tellurium. The living tissues, the embryo and aleurone layer, are coloured by insoluble reduction products. Now the reagents of choice are tetrazolium salts (MacLeod, 1967), either triphenyltetrazolium chloride or the more rapidly reduced 2-(*p*-iodophenyl)-3-(*p*-nitrophenyl)-5-phenyl tetrazolium chloride. Living tissues reduce these colourless, soluble salts – which penetrate living cells – to give bright red, insoluble deposits of formazans. Grains are split longitudinally, most conveniently with a special tool, and one half of each is immersed in a solution of a tetrazolium salt. The pressure is reduced several times at a vacuum pump to remove air bubbles then, after a period of incubation of about 45 min at a controlled temperature, the half grains are inspected. Viable embryos, aleurone layers and microorganisms are all stained. Only the embryo staining is easily seen. Wholly stained and unstained embryos are readily distinguishable as living or dead, respectively. However, sometimes staining is faint, or only parts of embryos are coloured. Recommendations are available as to how these should be scored. For example, if less than half the embryo is stained it should be scored as dead. Tetrazolium tests are rapid (1–2 h against 3 days for germination tests) and are widely used. Furthermore they detect viable grains that are dormant. However, they have two major weaknesses. First, often only 50 half grains are tested (although the EBC standard is two lots of 100 corns) and this low number makes for extreme statistical unreliability. Second, heat-damaged grains such as those that have been dried at marginally too high temperatures and/or with too low an airflow may stain yet not be able to germinate. This is a serious problem, especially in a wet harvest season when farmers have to dry the grain as fast as possible to prevent deterioration in store and so may choose conditions that overheat damp grain. An alternative test, which is believed to be more sensitive to heat damage, is based on the ability of separated viable embryos incubated on water for 8 h at 25 °C (77 °F), to synthesize starch from their endogenous reserves (French, 1959). The starch is readily detected by staining with iodine, which colours it blue-black. Such a test takes longer than those based on tetrazolium staining.

Direct tests. Direct tests of viability rely on particular treatments overcoming dormancy. Since these treatments are not always wholly successful, the results, if less than 100%, must be regarded as lower estimates. In some other tests for viability, it is incorrectly assumed that if normal germination tests are sufficiently prolonged, for example for 2–4 weeks, all living grains will germinate. It is not true that all the living grains will grow, nor is it possible for maltsters to wait for up to a month for the result of a viability test.

One frequently used procedure, first suggested about 1903, is to dry grain in a warm current of air at about 40 °C (104 °F) and then warm-store the grain at 40 °C (104 °F) for about 10 days. Then after 2–3 weeks storage at room temperature (i.e. 20 °C (68 °F)) carry out a germination test. For most moderately dormant grain this works well, but some grain samples are so dormant that not all viable grains will germinate until warm drying and storage treatments exceed a month (Briggs and Woods, 1993; Briggs *et al.*, 1994). The viability of warm-stored grain is most conveniently assessed with the hydrogen peroxide or Thunaeus test (see below). Stratification is used by seedsmen to break dormancy. This consists of placing the grain on a damp germination medium and keeping it cool, for example 4–7 °C (39.2–44.6 °F) for 4–7 days then, after the dormancy has been overcome, moving the grain to a germination chamber at 15–20 °C (59–68 °F) and subsequently evaluating the germination at intervals over a week. Other approaches, not usually employed, are to surface sterilize grains (e.g. with a solution of calcium hypochlorite and/or a mixture of antibiotics) before putting them to germinate or to set grains to germinate on substrata wetted with solutions of various agents such as potassium nitrate (0.2%); hydrogen peroxide (0.1–1.0%); gibberellic acid (0.1 g/l) or thiourea (0.2%). Usually these treatments do not wholly overcome dormancy, but some of these agents work well when applied in different ways or in combination with other treatments (see below).

Steeping grains in dilute hydrogen peroxide (0.75% (w/v), the Thunaeus test) at 18–21 °C (64.4–69.8 °F) induces chitting in all but the most dormant samples. Grains chit while immersed in the peroxide and then die. Grains that do not chit within 3 days may be rinsed and peeled (that is the husk over the embryo is stripped away with a needle, then the pericarp and testa are removed by rubbing and scratching (Bishop, 1944); this usually overcomes dormancy. These grains are then put to germinate on wet sand or filter paper. Some of these may also chit and the total number of chitted grains is used to calculate the GC value (%) of the grain. This test is widely recommended and, as two lots of 200 or 500 corns are used, the reliability is greater than seen with tests using smaller numbers of grains (EBC, 1987; IoB, 1991; ASBC, 1992). Sometimes the inclusion of gibberellic acid (2–50 mg/l) in the hydrogen peroxide increases the percentage germination found at the end of the steep. The use of this 'unofficial' modification reduces the tedious work of peeling the ungerminated grains and putting them to germinate. Peeling the husk away from the embryo, then scratching the pericarp and testa over the embryo usually overcomes dormancy (Bishop, 1944).

Grains may be decorticated with 50% sulphuric acid and the husk- and pericarp-free grain may be put to germinate. This treatment usually overcomes dormancy (Pollock, 1962b). Viable but physically damaged grains may be killed by the acid treatment. Exceptionally dormant grain may need to be both decorticated and stratified (Caldwell, 1957) and/or be wetted with a solution of gibberellic acid before they will germinate. Pricking and slitting the covering layers of grains through to the endosperm, or cutting pieces off grains reduces dormancy. The closer the wounds are to the embryo the more effective they are. In the Eckhardt test, grains are cut in half and the embryo ends are put to germinate. The test is unreliable, particularly because cut grains become heavily infected with microbes.

3.7.5 Tests for germinative energy

The germinative energy (GE) is defined as the proportion of grains (%) that will germinate under the conditions of a specified test. The difference between the germinative capacity and germinative energy of a sample of grain is a measure of dormancy, i.e.

$$\text{GC (\%)} - \text{GE (\%)} = \text{dormancy (\%)}$$

It is a notorious fact that different proportions of a batch of dormant grain germinate when submitted to different germination tests (Table 3.4) and the

Table 3.4 The germination of four different barleys in four different tests carried out in parallel

Grain sample	Germination (%)			
	4 ml	8 ml	Test tube	Bag
A	60	14	96	98
B	23	13	57	98
C	28	11	29	94
D	25	14	66	54

Macey and Stowell (1959).

results of some germination tests are much more variable than others. Furthermore, tests vary in their abilities to predict whether a batch of grain can be malted or not. These problems arise for several reasons. (i) Dormancy is *not* an all-or-none condition but is graded and its expression depends on the conditions under which germination is tested. The various different germination tests, and indeed different malting regimes, differ significantly in the conditions they provide. (ii) The conditions under which some germination tests are to be carried out are inadequately specified or controlled so that the results they give vary on different occasions. (iii) The conditions under which most germination tests are carried out differ substantially from the conditions found in malting. Therefore, although results from the tests can partially characterize the physiological state of a sample of grain they do not directly predict its behaviour in a malthouse. For this micromalting trials are required (Chapter 11).

Several germination tests are in use. Some of them have been progressively refined over the years so that the results they now give differ significantly from those obtained with the tests as originally specified. Thus, in the Aubry test, as now defined (EBC, 1987), samples of 500 grains germinate between two layers of filter paper, resting on wetted cotton wool, at 20 ± 1 °C (68 °F) and at a relative humidity of about 95%. The Schönfeld method is intended to mimic commercial steeping. The details have repeatedly been modified. Two lots each of 500 grains are separately placed in filter funnels with the necks blocked with stainless steel gauze. All operations are at 18–20 °C (64.4–68 °F). The grain is steeped for 3 h, then the water is drained away and the grain is covered with a piece of damp paper and the funnel is covered with a lid. After a rest of 18–20 h, the grain is steeped again for 2 h then drained and covered again. The first count of germinated grains is made after 72 h. Ungerminated grains are returned to the funnel, are steeped for 30 min, drained and covered again; they are finally counted after a total of 120 h. Thus scores (% germination) are made after 3 and 5 days (EBC, 1987). The Schönjan or Coldewe method is still used sometimes, although it is no longer recommended. In this test 100 grains are placed in the jar described in Fig. 3.8. This apparatus is bulky and

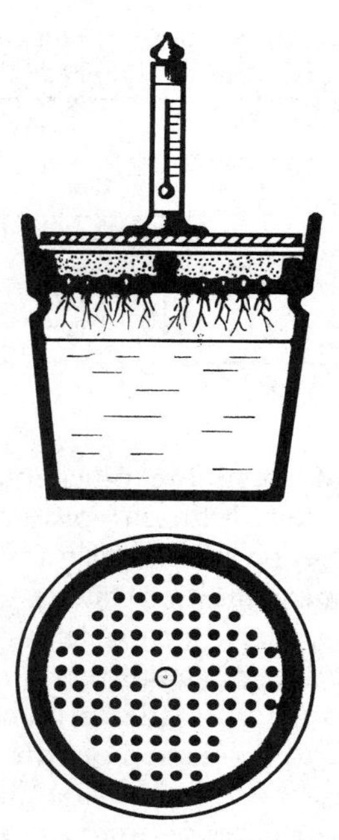

Fig. 3.8 The Coldewe apparatus. Grains are placed embryo-ends down in a perforated porcelain plate, also shown in plan view. This is supported in a jar kept in a constant temperature incubator, and the upper ends of the grains are covered with sand. The sand is watered, and surplus water trickles down and is collected in the jar, where it maintains a humid atmosphere. The sand is covered with a lid which has a layer of felt on its lower side to prevent drying. In the example shown, the lid carries a thermometer. As grains germinate the rootlets grow towards the water and protrude beyond the plate so the number of grains germinated and the vigour of their rootlet growth are easily determined. (After Heron, 1912).

awkward to keep clean but by carefully inspecting the roots, which grow from the holes in the plate, the percentage germination and the uniformity of vigour with which the grains are growing are easily judged.

Other tests are carried out by placing grains on wetted support media held in Petri dishes. Originally these tests were often carried out on graded sand (Bishop, 1944) and in some poorly characterized tests cotton wool may be used. In a widely used test 100 grains/dish (standard size) are placed on two layers of filter paper (grade specified) wetted with either 4 ml or 8 ml of water. These tests are often referred to as the 4 ml test and the 8 ml test and measure, respectively, the germinative energy and the water-sensitivity (see below). The tests should be carried out at least in triplicate. Incubation is at 18–21 °C (64.4–69.8 °F), and the numbers of grains germinated (chitted or with rootlets) are determined after a 24, 48 and 72 h incubation (Essery *et al.*, 1954, 1955; EBC, 1987; IoB, 1991). To prevent the dishes drying out, stacks of them may be placed in an incubator in loosely fitting plastic bags.

We believe that these tests should be in a water-saturated atmosphere at a more closely defined temperature. Many other supports have been used in germination tests, including cotton wool, sand, soil, layers of glass rods and porous ceramic or glass discs. We have found it advantageous to use a layer of agar as the substratum, with 1 ml or 3 ml of water on the surface and the incubation temperature held as closely as possible to 18 °C (64.4 °F). The atmosphere is maintained nearly saturated with water vapour by driving attemperated air through a bubbler in water, also at 18 °C (64.4 °F), and into the incubation chamber (Doran and Briggs, 1992).

In most seasons, the 4 ml and 8 ml Petri dish tests indicate if grain is sufficiently mature to be malted. However in some seasons, often after a very wet harvest, these tests are unduly pessimistic and indicate that grain cannot be successfully malted when, in fact, it can (Macey, 1957; Macey and Stowell, 1959). While this indicates that the tests err on the side of safety, it is a serious fault when grain is in short supply. Small-scale tests, carried out in test-tubes, have been more successful at detecting grain samples that can be malted but these, in turn, are less successful than old-fashioned 'bag' (or 'rag-doll') tests (Table 3.4; Chapter 11). In the bag test, grain samples are sewn into loose muslin bags (rag-dolls) attached to cords and labels. The bags are immersed in large bulks of grain being processed on the production unit, during steeping, germination and kilning, so the conditions of the small amounts of grain are controlled by the large bulk. The performance of a sample of barley in a bag test is a good guide to how the bulk will behave during malting. However the tests are not convenient to arrange, and it is essential to remove the bags whenever the grain is to be moved or turned, so that they are not carried away and destroyed by the mechanical equipment.

Cumulative percentage germination curves are not linear with time. Consequently there is no simple measure of 'rate of germination'. Furthermore, there is no satisfactory measure of 'final' germination (%) unless all the grains have grown in the incubation period. The best approach is to record cumulative percentage germination curves. Maltsters usually terminate these tests after 3 days, but for some purposes counts are continued for a longer period. Maltsters nearly always use 3-day germination values but are well advised to note the germination after 1 or 2 days, also. However a fully acceptable 98–100% germination value can be reached by different progressions. For example two lots of barley germinated 100% in 3 days, but while one reached 50% germination in 31 h, the other only required 15.5 h to attain this value. Clearly, the second sample was germinating more rapidly and was of better malting quality in the sense of its germinability. An exceptionally mature barley showed 80%, 94%, 98%, 98% germination (2% dead) on 4 successive days (Urion and Chapon, 1955).

3.7.6 Factors influencing germination

Many factors influence grain germination and dormancy. In the past, data were given for the minimum, optimum and maximum temperatures for grain germination. These can be grossly misleading because they are **not** constants. Germinability at different temperatures depends on the maturity (degree of dormancy) of the grain sample, the amount of water available to the grain, the duration of the test and so on. Therefore, over a period of weeks, and after a lag, samples of both barley and wheat have germinated in contact with ice (Edwards, 1932). Many samples of barley will not germinate at temperatures of 25 °C (77 °F) or more while others are reported to germinate at temperatures as high as 35 °C (95 °F). It seems that grains of barley, wheat, rye and oats that are fully after-ripened will germinate in most tests at 20–25 °C (68–77 °F) and, under exceptional circumstances, may germinate at 30–35 °C (86–95 °F). Partly dormant grain will germinate badly at 25 °C (77 °F) but, given sufficient time, most will germinate almost completely between 2 and 7 °C (35.6–44.6 °F). As grain matures, the range of temperatures in which it will germinate broadens (Fig. 3.9). The curves demonstrate that in the first day the most dormant sample germinated best at 15 °C (59 °F) but in the more mature samples most grains had germinated at 20 °C (68 °F). By 2 days the optimum germination temperature was about 15 °C (59 °F) in all the samples except the most mature, where it was 15–20 °C (59–68 °F), while in 3 days it was 10 °C (50 °F) for the immature grain and 10–15 °C (50–59 °F) for the most mature samples. After 7 days of incubation, the immature grains still showed a germination optimum of 10 °C (50 °F), but in the more mature samples complete germination was achieved by all samples over the temperature range 5 to 20 °C (41 to 68 °F) inclusive. These results are typical and emphasize that for the most rapid and uniform germination maltsters should use the most mature grain and carry out germination at 12–15 °C (53.6–59 °F). Clearly, even by increasing temperatures to 18–20 °C (64.4–68 °F) or even higher to reduce processing times, maltsters are favouring uneven germination and hence heterogeneity in their malt. It is possible that by using additives, such as gibberellic acid, or by achieving complete chitting by the end of the steeping period, this effect can be eliminated. Still, there is a clear conflict between the traditional approach of achieving high-quality malt by using long, cool germination periods, and the need to accelerate the production process by using higher temperatures. The steepness of the slopes of germination (%) against incubation temperature (Fig. 3.9) emphasizes that specifying temperature ranges in germination tests (e.g. 18–21 °C; 64.4–69.8 °F) is too lax, and that an exact temperature (e.g. 18 °C (64.4 °F)) should be specified (Doran and Briggs, 1992) since the germination percentages of the two ends of the range may be quite different.

The germination of grain may, to an extent depending on its maturity, be reduced by the presence of a water film on its surface. In traditional malting,

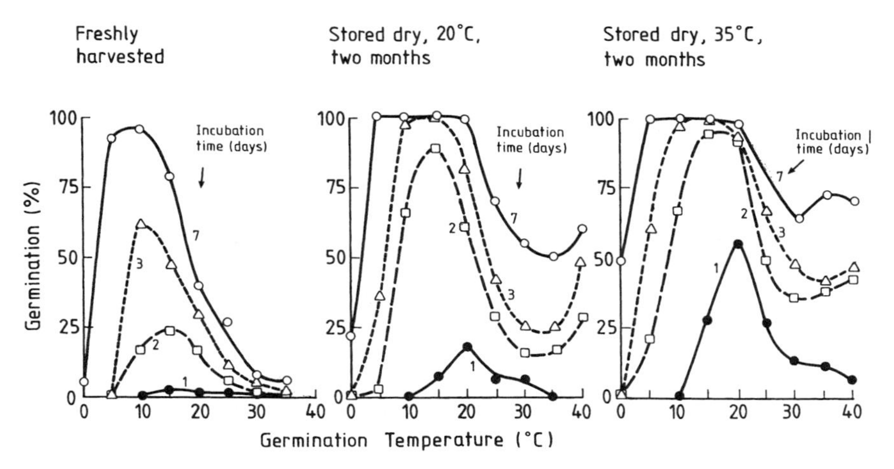

Fig. 3.9 Germination of subsamples of a batch of Sonja barley, incubated at different temperatures, on wet cotton wool, when freshly harvested and after two different storage treatments (replotted from the data of Corbineau and Côme, 1980).

grain did not begin to chit after steeping until the surface moisture film had been absorbed, and indeed an old patent describes a process for removing the film of surface water with warm air to hasten chitting. Excess water present in grain in germination tests also reduces its germination (Bishop, 1936; Pollock, 1962a). Figure 3.10 indicates the types of germination pattern that may be obtained when samples of different batches of barley are incubated at a constant temperature with different amounts of water in the germination dish. The observations are the basis of the '4 ml' and '8 ml' tests (Essery *et al.*, 1955; Pollock, 1962a; EBC, 1987; IoB, 1991). Freshly harvested grain generally germinates best with 4 ml water per Petri dish. By convention, the result of the 4 ml test is called the GE (%) of the sample. The difference between the viability GC (%) of the grain and the GE (%) is called profound dormancy (%), or merely dormancy, in this system. Grain in the presence of larger amounts of water remains covered with a film of moisture and germinates less well. In the 8 ml or water-sensitivity test, the results may be reported as WS (8 ml) %. Sometimes, especially in older reports, water sensitivity (%) is recorded as the **difference** between the scores of the 4 ml and the 8 ml tests. This practice can be misleading and should be discontinued because, as grains mature, germinabilities in the 4 ml and 8 ml tests both improve until eventually they are very nearly the same. However the 4 ml values improve faster than the 8 ml values (which are, however, also improving), so the **difference** between them **increases** and may give the erroneous impression that the grain is deteriorating. Grains placed dorsal-side (embryo-side) downwards germinate less well than if they are furrow-side downwards in the water-sensitivity test, and grains

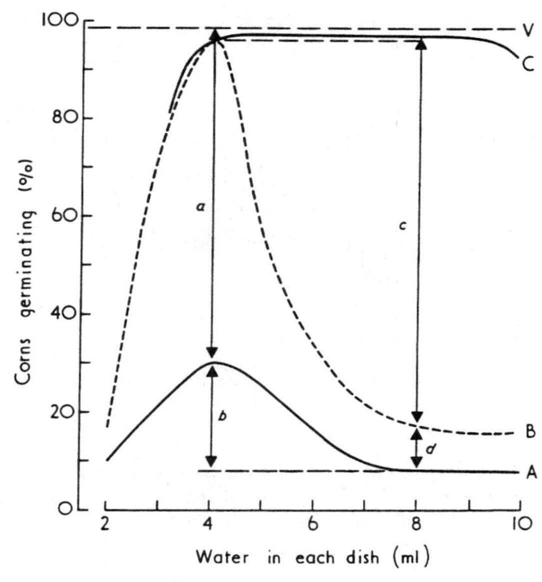

Fig. 3.10 The germination of barleys, after incubation for 3 days at a constant temperature, on filter paper in Petri dishes supplied with various amounts of water. The three curves represent results obtained with barleys of differing maturity. (A) A freshly harvested barley sample exhibiting profound dormancy (a) and water sensitivity (b). (B) Partly mature barley showing little profound dormancy, but considerable water sensitivity (c). Water sensitivity differs only by the amount between samples A and B. (C) Fully mature grain, showing essentially no dormancy. The GE (4 ml) is very close to the GC value (viability, V) of the sample. (After Pollock, 1962a.)

crowded together and touching often germinate less well than if they are spread out on the plate (Scriban, 1965). These observations can be rationalized with reference to the facts that enriching the atmosphere around the test dishes with oxygen and surface-sterilizing the grains or adding hydrogen peroxide or substances to the liquid in the dish (such as antibiotics) which are toxic to microbes, also reduce water sensitivity. The film of water on the grain presents a diffusion barrier to oxygen reaching the living tissues. Embryos of grains dorsal-side down or crowded together are more likely to be short of oxygen than those of grains spread out or where the embryos are at the top of the water film. Hydrogen peroxide decomposes to liberate oxygen and is an antiseptic, which damages and kills microbes. The microbes that are present on the grains and multiply in the film of water around them compete strongly for oxygen; they may also produce metabolites that are toxic to grains. Grains of different sizes take up different amounts of water. To allow for this it is the practice in some laboratories to include a non-standard 5 ml test when particularly large grains are being evaluated. Others have suggested that the amount of water added to the dish should be in proportion to the weight of grains. Fully mature barley will germinate when immersed in aerated water, but water-sensitive grains will not (Pollock, 1962a). This finding has important implications for malt uniformity when partly water-sensitive grain is steeped with aeration, since this favours irregular germination (Kelly and Briggs, 1992a,b).

Water sensitivity has been detected in some samples of immature wheat (Belderok, 1961). As with barley, as the wheat matures it can germinate well in the presence of a wider and wider range of amounts of water. In addition, as with barley, as the grain ripens the optimum amount of water for germination apparently declines from a 4 ml to about 3 ml/dish. In the GE (4 ml) test, which provides the optimum amount of water for the germination of water-sensitive grains, the grain attains a moisture content of around 35%, and the surface film of water is absorbed during the test. If water-sensitive grain is steeped (i.e. is hydrated by immersion to, say, 45% moisture) and then drained and set to germinate under malting conditions, germination is ragged and slow, or fails, and the moisture film only slowly dissipates because the grain is virtually saturated with moisture.

However, if this grain is steeped to about 35–37% water, the water is drained and the grain is 'rested' in a slow current of fresh air for 8–24 h (an air-rest or a dry-steep period), the surface film of water is taken up and apparently critical metabolic processes occur because if the grain is then re-immersed (to hydrate it to a suitable extent to allow successful malting to occur (42–46% moisture), the grain afterwards germinates normally. This is true even if no visible signs of incipient germination can be detected at the end of the air-rest period. When grain is steeped to a range of different moisture contents, and the samples are incubated under malting germination conditions, germination is most rapid in samples with a moisture content of about 37% (Fig. 3.11). Air-rest, or multiple air-rest, steeping schedules are now routinely used both to permit water-sensitive grains to be malted and to accelerate the rate of malting of more mature samples of grain. One of the reasons for preferring dry-casting (the dry transfer of steeped grain from steep tanks to the germination vessel), to wet-casting (where the grain is transferred, usually pumped mixed with water, and the water drains away in the germination vessel) is that wet-casting renews the film of surface moisture around the grains and so tends to delay germination in systems where chitting has not occurred and supplies 'extra' water to grains that have chitted.

Steeping barley for prolonged periods, especially if the steep water (liquor) is not frequently changed, causes steep-damage or oversteeping and subsequent germination is depressed. Oversteeping occurs more readily at high temperatures (e.g. 25–35 °C (77–95 °F)) and the steep-damaged grain appears to be water sensitive; indeed, such grain is said to show 'induced water sensitivity' (Lubert and Pool, 1964). These effects may be minimized experimentally by the use of antiseptics in the soaking liquor. Where a high proportion of damaged grains is present, the microbes grow

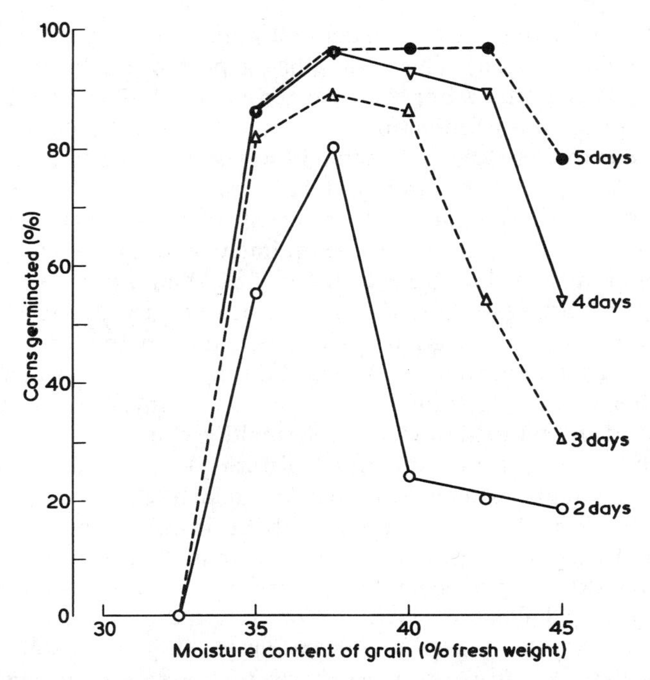

Fig. 3.11 The percentage germination, on successive days, of barley steeped to different moisture contents and then incubated under malting conditions (after Sims, 1959).

vigorously on the nutrients that are leached from the broken grains, so the steep liquor needs to be changed more frequently to prevent it becoming foul, and steep damage is particularly likely to occur. Successful germination of such grain is favoured by the removal of the film of surface water from the grains and by applications of gibberellic acid (Fig. 3.12). Frequent changes of steep water are not attractive since the water must be paid for and the discharged effluent also incurs charges for its treatment and/or disposal. Aerating steeps favours increases in the populations of microbes. While aeration favours the germination of uniformly mature grains, the germination of water-sensitive grains is irregular after steep aeration and in some cases, with immature grain, aeration actually reduces germination (Kelly and Briggs, 1992a).

Attempts to speed up steeping by elevating steep temperature above the traditional values, for example, from 10–14 °C (50–57.2 °F) to 16–20 °C (60.8–68 °F) or even higher, have been checked by difficulties caused by induced water sensitivity, or steeping damage. As with germination, compromise temperatures of about 16 °C (60.8 °F) are probably usual in steeping. At present, there seems to be no acceptable and cheap disinfectant that

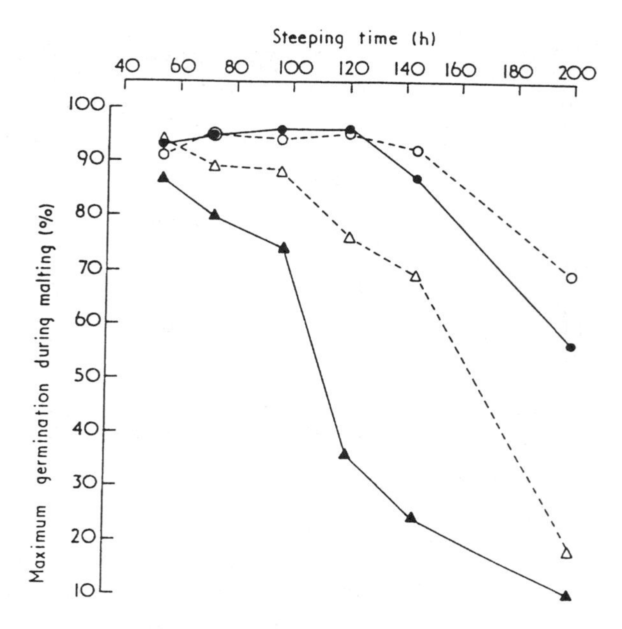

Fig. 3.12 The maximum percentage germination values achieved when malting samples of barley steeped for various periods in water and in solutions of gibberellic acid. In some samples the surface water was removed by centrifugation at the end of steeping: ▲, steeped in water; ●, steeped in water and centrifuged; △, steeped in a solution of gibberellic acid; ○, steeped in a solution of gibberellic acid and centrifuged. Both gibberellic acid and the removal of the surface water favour the germination of over-steeped grain. (After Kirsop, 1964.)

might be added to steeps to control the microbes sufficiently well to overcome this problem.

3.7.7 Overcoming dormancy

A number of treatments that reduce dormancy have been noted earlier in this chapter. During storage, under conditions that maintain grain viability, dormancy progressively declines with time. The rate of recovery from dormancy is influenced by the variety of grain, the physiological status of the sample and the storage conditions. Differences between the improvements in germinabilities of barleys of different varieties, gathered during one harvest, can be substantial (Fig. 3.13). Dormancy declines more rapidly at elevated storage temperatures, where there is also an enhanced risk of the grain being heat-damaged. Grain can certainly be stored at 40 °C (104 °F) for extended periods if it is sufficiently dry (9% moisture or less).

Experimentally, dried grains may be kept in a fully viable but dormant condition for extended periods (years) at –18 °C. Warm-drying, or sweating, grain has been used to accelerate the decline in dormancy at least since

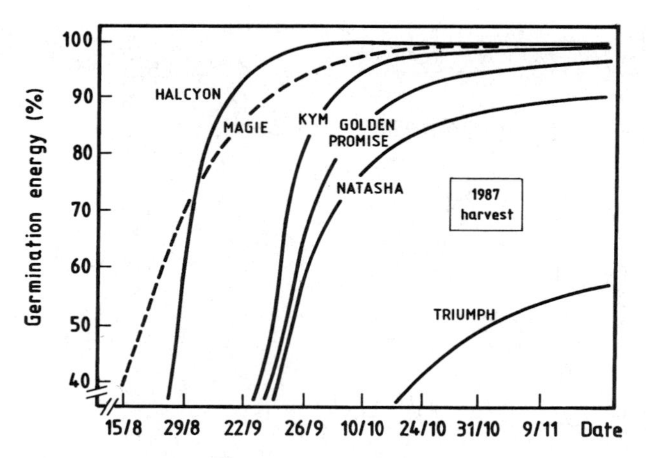

Fig. 3.13 The improvement of germinability (GE) with storage time in six examples of different barleys of the 1987 harvest. The extreme dormancy of the variety Triumph is emphasized. (After Hills, 1986.)

1759 (quoted Murphy, 1904), but there is still discussion as to how much of the beneficial affect results from warmth and how much from the removal of water. Drying at an elevated temperature and promptly cooling the grain is less effective at overcoming dormancy than allowing the grain to cool slowly (Gordon, 1968). Therefore, barley dried at 48.9 °C (120 °F) recovered from dormancy faster than grains dried at lower temperatures (48.9 > 43.5 > 37.8 > 32.2 °C) and the GE (4 ml) value increased much more rapidly than the WS (8 ml) value, so water sensitivity was relatively persistent. The energies of activation for the processes of loss of dormancy and water sensitivity were nearly the same and were calculated to be 17.5 and 18.3 kcal/mol (73.5 and 76.9 kJ/mol), respectively, suggesting that one fundamental process of maturation was being investigated under two levels of stress. Belderok (1961), working with wheat, showed that while there was no immediate effect, germinability improved faster in drier grains. Harrington (1923) had reached the same conclusion. Grains were dried over silica gel to eliminate the effects of heating. Others have shown no immediate benefits to cool drying, but found that water sensitivity declined more rapidly in drier samples of grain (Pollock, 1962a). We found that the temperature of storage was much more influential than the moisture content of the grain on the rate of decline in dormancy (Briggs and Woods, 1993; Briggs *et al.*, 1994). However, grain stored at higher moisture contents lost vigour and viability much more rapidly than drier grains stored at the same temperature.

Dormancy declines faster in grains stored warm, rather than cool, in sealed containers in which evaporation is prevented, so warmth is beneficial, independent of drying. Several reports indicate that grain dried or stored in air or oxygen recover faster than grains treated in nitrogen (Pollock, 1962a;

Briggs, 1978). We have stored grains in oxygen, air and nitrogen and have detected no significant differences in the rates of maturation (Doran and Briggs, 1993a).

Treatments that reduce or overcome dormancy have been divided into those that overcome profound dormancy (judged using the 4 ml Petri dish test), those that overcome water sensitivity (judged using the 8 ml test) and those that do both. Applications of hydrogen sulphide and other thiols, brief periods of heating (40 °C (104 °F)/3 days) without drying, and puncturing the grain coverings are said to overcome dormancy. By comparison, cool dry storage, low doses of hydrogen peroxide, elevated levels of oxygen during germination tests, and applications of a range of agents (at levels too low to damage the grain) such as mercuric, ferrous and silver salts, *N*-ethyl maleimide, iodoacetamide, mixtures of antibiotics (up to six, capable of controlling yeasts, filamentous fungi and Gram-positive and Gram-negative bacteria), washing with alkaline hypochlorite solutions and acidification of grain to pH 3 all overcome water sensitivity.

High levels of hydrogen peroxide, grain decortication, applications of gibberellic acid (especially together with hydrogen sulphide) overcome both dormancy and water sensitivity. Many other agents sometimes reduce dormancy, including brief steeps in warm/hot water, steeps in dilute sodium hydroxide, potassium hydroxide or lime water, acid ammonium fluoride, brief exposures to formaldehyde, nitrates, sodium metabisulphite, hydroxylamine, azides, cyanides, carbon monoxide, and so on. Other treatments that are reported to be beneficial at some levels of application include exposure to AC and DC electric currents, to ultrasonication, to radiofrequency electric fields and to ultraviolet radiation. Excessive exposure to many of these treatments damages grain. An interesting fact, with important implications, is that treatments that are beneficial on some occasions may be without effect on other occasions. It is concluded that multiple factors contribute to dormancy and their relative degrees of importance must be different in different batches of grain. At the fundamental level, dormancy is not well understood, but a number of factors are known to contribute to the phenomenon.

1. The physiological state of the embryo. Embryos that are mature germinate at lower oxygen levels than those from immature grains and mature embryos are harmed less by toxic agents such as tannins or abscisic acid.
2. The maturity (responsiveness to gibberellic acid) of the aleurone tissue. Poor responsiveness may be because this tissue is dormant.
3. The availability of oxygen to the living tissues. This is modulated by the amounts in the grain's surroundings, the low levels present in the steep water or water films around the grains and the competition for available oxygen by microbes on the grain surfaces or in the steep liquor.
4. The variable microbial population, which competes for oxygen, may make grain surfaces slimy and less pervious to oxygen and may generate

toxic substances or even directly attack the tissues of the grain. Microbes are dependent on nutrients leached from grains. Changing steep water removes microbes from the steep as well as removing dissolved nutrients that would support their metabolism and growth.
5. The husk and pericarp are barriers to oxygen diffusion to the living tissues of the grain. One possible explanation for the beneficial effects of abrasion is that this mechanical treatment improves the perviousness of the outer grain tissues to oxygen (Smith, M. and Briggs, 1979). It is not clear whether these tissues contain endogenous growth-inhibiting substances in addition to associated microbes.
6. Other, chemically defined substances that act as germination inhibitors when applied externally to grains have been isolated from grains and steep liquors. These include acetic acid, abscisic acid, mixtures of phenolic substances, coumarin, umbelliferone and herniarin. Their importance (if any) during malting is uncertain.

A lack of adequate amounts of oxygen reaching the living tissues is the most probable cause of the failure of water-sensitive grains to germinate. It is suggestive that when hydrated grains are held in the absence of oxygen substances derived from anaerobic metabolism accumulate, such as ethanol. When these grains are placed in air their germination is impaired and the grains appear to be water sensitive. Steeping grains in dilute solutions of ethanol, coumarin or acetic acid also appears to induce water sensitivity. Oxygen is required for the metabolic removal of all these substances.

Aleurone responsiveness to gibberellins improves with grain maturation and this is associated with endogenous changes as well as alterations in the microbes on the husk (Crabb, 1970; Kelly and Briggs, 1992b, 1993; Briggs and McGuiness, 1993; Doran and Briggs, 1993b; Bakhuizen *et al.*, 1995). Many treatments that alleviate dormancy in germination tests also accelerate the malting of grain that is 'adequately mature'. For example, periods of warm storage, air-rest steeping, applications of gibberellic acid and hydrogen peroxide, husk loosening and abrasion, grain decortication and enriching the atmosphere around the germinating grain with oxygen accelerate malting. It seems very probable that at least some of the factors that prevent dormant grain germinating also, and in a less dramatic way, slow the process of germination and modification in 'mature', malting grains (Briggs, 1987a). There are several methods used by maltsters to minimize the effects of dormancy. (i) Choosing to use grain of varieties with inherently low tendencies to become dormant and which recover from dormancy rapidly during storage. (ii) Drying grain carefully and running it into the stores warm. Apparently the use of well-insulated stores to minimize heat loss has not been adopted, nor have insulated, air-warmed conveyors been installed between the driers and stores where grain is to be held warm (Briggs, 1995). Drying grain in boxes, rather than in flow-through driers, leads to a more rapid recovery from dormancy. Probably this is because the grains are warmed in the boxes for longer periods (8–12 h against, for example, 1.5 h) and are run to store warm while the cooling section of a flow-through tower-drier chills the grain to nearly ambient temperatures. Some maltsters 'blank off' the cooling section of tower driers when they wish to keep the grain warm. During warm storage, the dormancy of grain and its microbial population both decline. (iii) Maltsters arrange to have carry-over stocks of mature barley from the previous season's harvest so that malting can continue while the new harvest's grain is held in store to mature. This is expensive both in the cost of storage and in the cost of financing grain for over a year. (iv) Moderate physical grain battering, as when grain is moved around in conveyors or is passed through abraders, sometimes (but *not* always) improves grain germination. (v) The malting process is adjusted to minimize the effects of residual dormancy (including water sensitivity). Adjustments may include the incorporation of hydrogen peroxide in the steep liquor, application of a solution of gibberellic acid to the grain, pre-washing the grain with lime water (saturated or half-saturated) a dilute solution of sodium hydroxide or a dilute mineral acid at the start of steeping.

3.8 The respiratory metabolism of grain

Grain respiration, the uptake of oxygen and the evolution of carbon dioxide (which may, in part also evolve from fermentative processes, so-called anaerobic respiration), reflects the vigour of grain metabolism and so is of great importance to maltsters (Pollock, 1962a; Briggs, 1978). Grains respire during some forms of storage (but not if the moisture content is less than 12%), during steeping, germination and the early stages of kilning. The respiration rate correlates closely with heat output and so it is of importance in deciding the ventilation rates and other cooling arrangements necessary to control grain temperatures during the air-rests in steeping programmes and during germination.

Microbes make a variable, and often highly significant, contribution to the respiration of a grain mass. At present there is no way of controlling or measuring this on the industrial scale. Another complication is that in many studies 'respiration' has been measured in terms of the rate of carbon dioxide evolution. Carbon dioxide can be generated as the result of true oxidative processes, in which sugars or lipids are completely oxidized to carbon dioxide and water (this is 'respiration' in modern terminology), as the result of decarboxylative reactions, as the result of physical desorption of gas (as when grain is first wetted) and as a result of purely non-oxidative processes such as those that convert sugars to ethanol and carbon dioxide (fermentation in modern terms, anaerobic respiration in older terms). The fermentation and respiratory processes are usually the most important. Fermentation may be measured as the rate of ethanol production, and respiration as the rate of oxygen uptake. Under anaerobic conditions, only fermentation can occur. In air, the grain may still partly metabolize by the fermentative route, the extent depending, in part, on the accessibility of oxygen to the living tissues. The respiratory quotient (RQ: the ratio of the number of moles of carbon dioxide liberated to moles of oxygen consumed in a given time) and some related measurements are sometimes used to indicate (ambiguously) the type of metabolism occurring in grain. Thus the total oxidation of hexose sugars to carbon dioxide and water gives an RQ of 1, the total oxidation of triglyceride about 0.7 and of ethanol 0.57. In the partial oxidation of ethanol to acetic acid, oxygen is taken up but no carbon dioxide is liberated. In tissues in which triglycerides are being converted into sugars the RQ is about 0.4. Under anaerobic conditions or in other circumstances where sugars are being fermented to ethanol and carbon dioxide there is no oxygen uptake and the theoretical molar ratio of ethanol to carbon dioxide is 1:1. Usually an excess of carbon dioxide is produced by fermenting grain, the extra being produced by decarboxylation reactions unrelated to fermentation. About half of the respiration of the newly wetted grain results from the embryo and half from the endosperm; plus there is a variable contribution from the microflora. As malting proceeds, the respiration rate increases, peaks and declines. Although part of this comes from reactions in the aleurone layer and part from the microflora, it is believed that most is caused by the growing embryo. The variable respiration is accompanied by variable heat output.

In store grain, respiration is low. At 12% moisture or below it is regarded as negligible but it rises very sharply with increasing moisture content (Fig. 3.14). Respiration is also temperature dependent, so moist grain must be ventilated to cool it, preventing overheating and maintaining its viability. Respiration is also influenced by the size and nature of the microbial population. Quite small changes in the moisture content can have dramatic effects on the respiration rate. For example, for various moisture contents (%), rates of carbon dioxide evolution (in mg CO_2/kg grain/24 h) have been quoted as 12%, 0.4; 14–15%, 1.4; and 17%, 100 (de Clerck, 1952). Below moisture contents of about 15%, grain in store respires steadily. However, at moisture contents of 15% (and more rapidly at higher moisture contents) the respiration of grain increases progressively with time: this is the result of the increasing population of microbes, chiefly fungi, on the grains (Oxley and Jones, 1944; Matz and Milner, 1951). At higher moisture contents, the respiration of the microbes becomes so intense that it causes grain heating and spoilage (Chapter 7). For several reasons, if grain is to be stored at low ambient temperatures, (e.g. 15 °C (59 °F)) it should be dried at least to 12.0% moisture and preferably to lower values (Chapter 7). Respiration, and heat output, increases rapidly with increasing temperatures, but this effect is less at lower (e.g. 13%) than at higher (17–19%) moisture contents

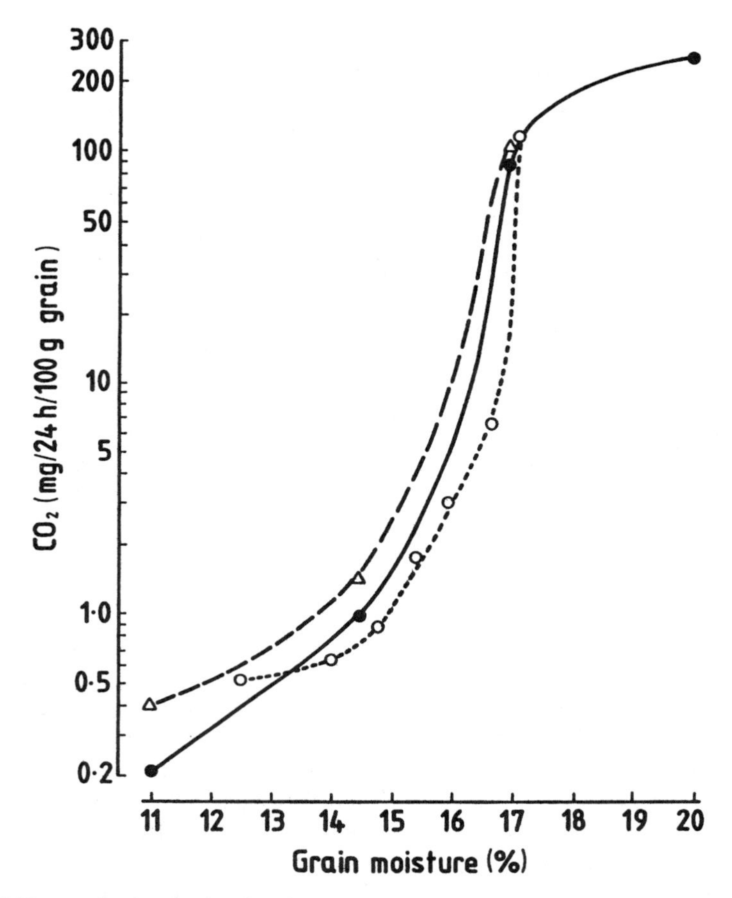

Fig. 3.14 The respiration (rates of carbon dioxide emission; log scale) of air-dry samples of grain of different moisture contents (various sources): barley A(△), barley B (●) and wheat (○). A large proportion of this respiration will have been caused by microbes associated with the grain.

(Vermeylen, 1962). To prevent deterioration, grain should be kept cool and dry (Chapter 7). Even if damp grain is 'kept healthy' and cool by vigorous ventilation, this will not prevent the multiplication of microbes and appreciable losses of dry matter. The respiration rate of grain increases as it hydrates and, in wheat and rye, a direct relationship has been demonstrated between the amount of unbound, 'freezable' water in the grain and the rate of respiration (Shirk and Appleman, 1940; Shirk, 1942). Even in air, wetted grain initially has an RQ of 2 or more, and ethanol accumulates. After the grain has chitted, the RQ falls to about 1, showing that fermentation has ceased, and the residual alcohol dissipates (Chapon, 1959) (Fig. 3.15). If the husk is stripped from the unchitted grain, ethanol accumulation ceases as

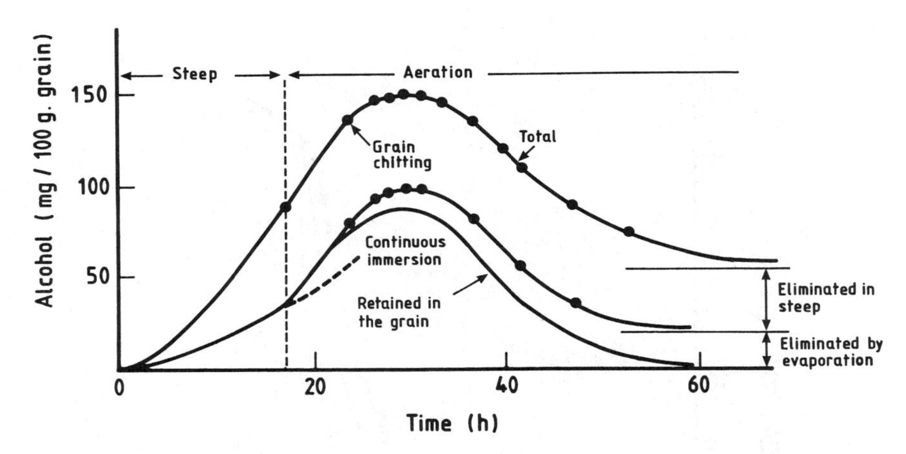

Fig. 3.15 The production of ethanol during a short steep and in the subsequent germination period. Note that initially on transfer from the steep to germination conditions more alcohol is accumulated in the grain than occurs if the grain remains immersed in the steep. (After Chapon, 1959.)

the Pasteur effect operates and the oxygen of the air switches off fermentation (Merry and Goddard, 1941). The husk is a diffusion barrier to gases and peeling away the husk increases the oxygen uptake rate by about three times, as well as depressing the RQ. If respiring grain is covered by a film of water or is immersed in aerated water, the oxygen uptake rate is immediately depressed to about 0.3–0.5 of its previous value and the RQ increases. These effects can be reversed (and water sensitivity can be reduced) by placing the grains in oxygen instead of air (Dahlstrom *et al.*, 1963; Dahlstrom, 1965; Crabb and Kirsop, 1969). The concentration of oxygen dissolved in water is much less than that in an equivalent volume of air, and the amount declines with increasing temperature (Appendix, Table A8). In contrast the demand by grain for oxygen to support respiration increases with temperature, with energies of activation estimated as 12.0–14.7 kcal/ °C per mol from water and 10.0–13.0 kcal/ °C per mol from air (Dahlstrom *et al.*, 1963). At elevated temperatures, the rates of oxygen uptake by microbes and microbial growth and multiplication are also increased. Therefore, the higher the temperature the more difficult it is to meet the oxygen requirements of grain during steeping.

Under anaerobic conditions, ethanol accumulates in hydrated grain, which cannot germinate in the absence of oxygen, and eventually the grain dies. Experimentally, very high levels of alcohol may accumulate. The addition of as little as 0.1% ethanol checks the germination of grain in air, while 0.5% severely restricts rootlet growth. Strangely, holding grain in 1–3% ethanol, in air, then transferring it to an aqueous medium may sometimes break dormancy.

Different batches of grain respire at different rates. In part this is likely to be caused by different loads of microbes. However, long ago it was reported that, even with surface-sterilized grains, smaller grains respired faster than larger grains with the same moisture and 'protein' contents and that grains rich in 'protein' respire more vigorously than grains of the same size but containing less protein (Abrahamsohn, 1911). Isolated embryos have RQ values of 0.8–1.0 as they respire their stored sugars and triglycerides. In contrast, degermed grains and isolated aleurone layers have RQ values of between 0.2 and 0.6, and in aleurone tissue stimulation by gibberellins causes substantial quantities of triglyceride to be converted to sucrose. Probably there are no significant differences between the respiration rates of dormant and mature barleys. However, in barleys that have been stored for extended periods 'extra' carbon dioxide is generated and RQ levels reach 1.8–2.2 compared with values of 0.8–1.3 for fresh grain (Anderson, 1970; Crabb, 1970). Crabb and Kirsop (1969) calculated that the effective concentration of oxygen at the embryo, beneath the husk and pericarp, is equivalent to an atmosphere of about 4%, which is about a fifth of that of the atmosphere. Embryos from dormant grains need higher levels of available oxygen to allow them to germinate than do mature embryos (Crabb and Kirsop, 1969; Doran and Briggs, 1993b). It is evident that a water film and/or a large population of microbes, which compete for oxygen, could prevent the germination of dormant grain and so cause water sensitivity. Isolated, wetted husks and husk fragments from barley and oats take up oxygen (Corbineau *et al.*, 1986; Lenoir *et al.*, 1986). The oxygen uptake rate is highly temperature dependent. The variation in uptake rate with incubation time by grain is strongly influenced by the previous storage conditions. The original investigators attributed their results to the oxidation of polyphenols and indeed the presence of several polyphenol oxidases was demonstrated. In fact the oxygen uptake is the result of microbial activity; these multiply on the husk and in the surrounding liquid and are supported by nutrients derived from the husk (Briggs and McGuinness, 1993).

Steeped grain takes up oxygen by a first-order reaction and the uptake rate increases with increasing hydration (Table 3.5). Unaerated steeps soon become anaerobic. Towards the bases of conical steeps, oxygen levels are lower and carbon dioxide levels are higher than nearer the surface of the grain bed after a period of immersion (Chambers and Lambie, 1960; Eyben and van Droogenbroeck, 1969; Cantrell *et al.*, 1981). In trials, the first steeps became anaerobic in 2–5 h, the second steeps in 0.5–1.5 h and the third steeps in about 0.5 h. In addition to supplying oxygen to the immersed grain, aeration loosens the grain during swelling and, in some designs of steep vessel, thoroughly mixes the grain. Ethanol diffuses from grains and about half is found in the steep water. This and other substances leached from the grain help to support the growth of microbes. The microbes consume oxygen and generate carbon dioxide. Quite probably the acetic acid that

Table 3.5 The gas exchanges of barley samples immersed in flowing water initially equilibriated with different gas mixtures. At least half of the dissolved oxygen was taken up as the water travelled through the bed of grain. Most of the germinated grains occurred near the water inlet. Significant gradients of oxygen concentration must have occurred across the grain bed

Circulating gas O_2 content (%)	Output CO_2 (nmol/kg dry matter per h) at				Uptake O_2 (nmol/kg dry matter per h) at				RQ ($+CO_2/-O_2$) at				Grains germinated (%)
	3 h	24 h	48 h	72 h	3 h	24 h	48 h	72 h	3 h	24 h	48 h	72 h	
5.8–6.3	0.65	1.73	2.33	2.35	0.24	0.52	0.84	1.05	2.73	3.37	2.83	2.26	12–14
10.2–11.4	0.57	1.81	2.50	2.55	0.37	0.90	1.22	1.41	1.51	2.01	2.05	1.83	13–14
21.0	0.73	1.80	2.47	2.70	0.80	1.65	2.25	2.55	0.90	1.09	1.09	1.06	89–95
37.0–40.5	0.56	2.15	2.99	3.23	0.69	1.85	2.90	3.32	0.82	1.03	1.03	0.97	94–96

From Ekström *et al.* (1959).

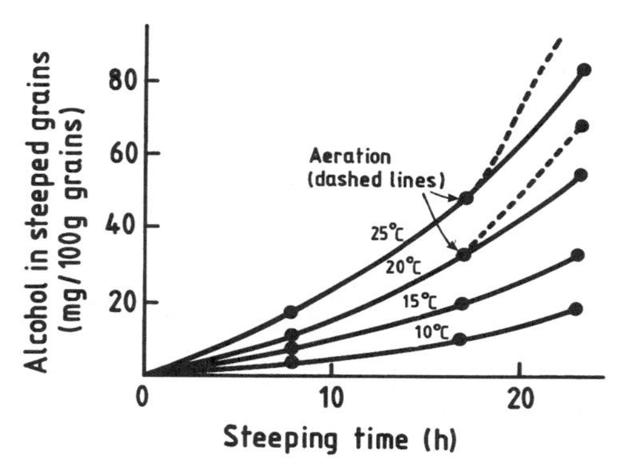

Fig. 3.16 The influence of steeping temperature on the accumulation of alcohol in barley and the influence of aeration (After Chapon, 1959.)

accumulates in steep water is formed by microbes from the ethanol which leaches from the grains. Ethanol production is influenced by temperature (Fig. 3.16). Interestingly, aeration can accelerate ethanol formation in steeping, indicating that under these conditions oxygen may **stimulate** fermentation more than it depresses it, through the operation of the Pasteur effect. Carbon dioxide continues to accumulate after the depletion of the oxygen and can reach levels sufficiently high that it effectively depresses both respiration and fermentation, as it does when it accumulates around grain in air, a fact used in Kropff malting (Chapter 14).

It was suggested, for traditional malting in which grain should not germinate until it is out of the steep, that it is preferable to steep in water saturated with nitrogen or carbon dioxide to check grain metabolism and obtain more uniform chitting after casting (Chambers and Lambie, 1960; Enari *et al.*, 1961). However, germination was more even if an air-rest was given between two immersions in water saturated with carbon dioxide. Excessive aeration or oxygenation in experimental steeping can cause mature grain to chit while still immersed (Table 3.5), when it can over-hydrate, leading to uncontrolled growth, high malting losses and the production of bad malt (Chambers and Lambie, 1960). Nevertheless limited aeration of steeps can induce more vigorous growth after steep-out, with more rapid respiration (Fig. 3.17), more rapid accumulation of enzymes and more rapid completion of modification (Eyben and van Droogenbroeck, 1971).

In traditional floor-malting, controlled even germination after steeping was the objective, particularly in the absence of air conditioning, when vigorous growth could not be checked by cooling the surrounding air. Therefore, long, cool steeps with minimal or no aeration were usual in the UK. However near-anaerobic steeping is now obsolete, since in modern commerce what is needed is **rapid**, even germination to maximize the productivity of the malthouse. Grain cooling is more simply achieved in pneumatic

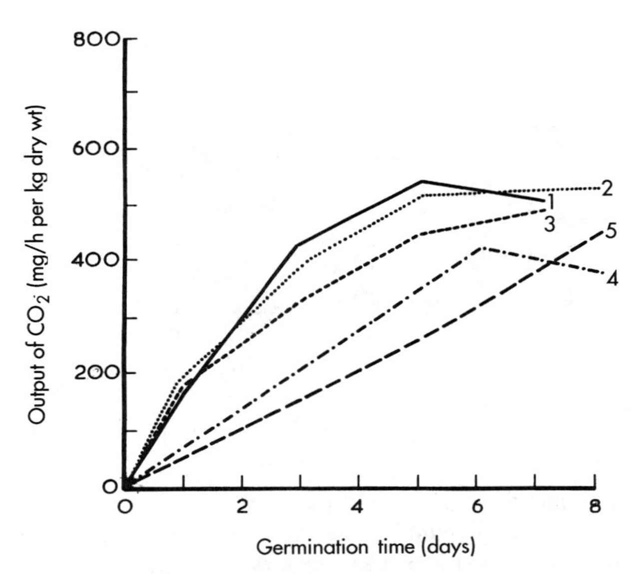

Fig. 3.17 The rates of carbon dioxide output ('respiration') by barley samples during germination following various steeping schedules. 1, 75 h steep, two aerations; 2, 75 h steep, four aerations; 3, 75 h steep, one aeration; 4, 75 h steep, no aeration; 5, 92 h steep, no aeration. (After de Clerck, 1938.)

malting plants, and so uncontrolled growth owing to over-heating is unlikely to occur. Steep oxygenation may be achieved by the use of aeration (pumping compressed air into the base of the grain–water mix), and/or by the use of air-rests (usually two). Normally during air-rests (dry-steeps) air is sucked down through the bed of grain ('CO_2-extraction') so renewing the oxygen supply and removing carbon dioxide, a heavy gas which tends to settle in the base of the steep vessel. When the grain is cast from the steep vessel, it has usually chitted and, in some production programmes, it is so advanced that roots are visible. In some floor-maltings, applications of sodium or potassium bromate have been used to check heat production (Chapter 14).

In general, aeration of steeps is insufficient to keep the water fully saturated with oxygen and, in any case, aeration is usually intermittent. A compressor must serve several steeps in turn. With continuous running, compressors become hot and warm steeps with the hot air in an undesirable way. Comparison between aerated and non-aerated steeps shows how aeration can slow the decline in the level of dissolved oxygen in the steep water and diminish the amount of carbon dioxide dissolved in the water (Fig. 3.18). The micro-environment around grain seems to be more uneven in conical than in flat-bottomed steeps. Whether this has a significant effect on malt quality is uncertain (Cantrell *et al.*, 1981) except in those cases in which excessively deep conical steeps cause pressure-damage to the grain

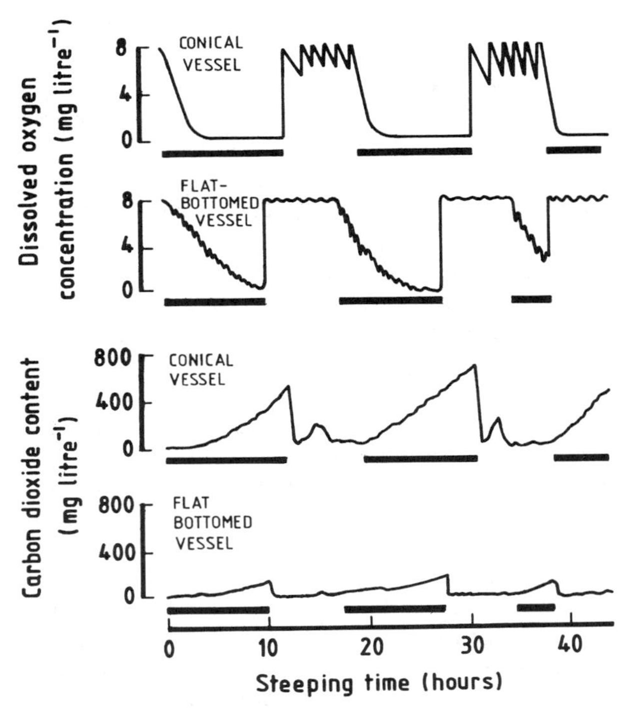

Fig. 3.18 Variations in the oxygen and carbon dioxide contents of two steeping vessels during steeping with air-rests. The flat-bottomed steep was of 154 t capacity, the conical steep was 12 t capacity. Immersion periods are shown by the black bars. During immersion periods, the flat-bottomed vessel was periodically aerated. During air-rests, both vessels received intermittent carbon dioxide extraction. The flat-bottomed vessel kept the grain in a better aerated state. This was reflected in the lower ethanol contents of the grain held in this vessel, relative to that in the conical steep. (After Cantrell *et al.*, 1981.)

(Chapter 9). During successive air-rest periods, respiration becomes progressively more vigorous as the grain moisture content increases (Eyben and van Droogenbroeck, 1971).

In floor maltings, the carbon dioxide level around the grain is often between 0.8 and 3.5% of the atmosphere. In pneumatic maltings, the value is normally about zero (Eyben and van Droogenbroeck, 1971). As a result, carbon dioxide may slow grain growth on floors. During the initial period after casting from the steep, fermentation continues and grain ethanol levels continue to rise, but as grains chit and the surface film of water is taken up, ethanol contents cease rising and then decline (Fig. 3.15). Some alcohol evaporates and some is oxidized within the grain. From then on, the gas exchange of the grain is primarily respiratory in nature. Carbohydrates are the main substrates, but small amounts of triglycerides are also used.

The intensity of respiration and the heat output change with germination time, the exact pattern depending on the grain sample, its moisture content at steep-out, the steeping schedule it has received, whether additives have been used, and the grain temperature. Thus air-rest steeping and applications of gibberellic acid favour vigorous growth and respiration and a rapid heat output, while sodium and potassium bromate suppress rootlet growth, respiration and heat generation. The respiration rate also depends on the proportion of oxygen in the air surrounding the grain so, for example, reducing the oxygen content to 5% reduces the respiration rate to 60% of that which occurs in air (Briggs, 1978). Elevated levels of carbon dioxide suppress respiration. Enzyme production in germinating grain is not directly related to the respiration rate (Fig. 3.19) and the metabolic linkages between the different processes are not simple (Chapter 4). The respiration and heat production rate increase by roughly 10% for each 1 °C (1.8 °F) increase in temperature (Nielsen, 1937). As malting proceeds, the respiration rate rises, peaks and then declines somewhat if the germination period is extended. This decline may be a result of the grain starting to dry. The moisture content of steeped grain, usually in the range 42–48% (with sprinkling), is chosen with reference to various factors including the need to achieve adequate modification but with limited embryo growth and malting losses. At higher moisture contents, grain grows and respires more vigorously (Fig. 3.20). Mellow barleys, fully mature and with nitrogen contents of about 1.4% were traditionally malted at 42–43% moisture. More steely, 'stubborn' grains and grains with higher nitrogen contents may now be steeped to 46–48% moisture, with sprinkling, while even higher moisture contents are reached in grain that has been resteeped (Chapter 14).

Within limits, grain respires faster at higher temperatures, but when grain was 'malted' at about 4–5 °C (40 °F) the respiration rate increased linearly with germination time: at 17.8–20 °C (64–68° F), the respiration rate peaked and declined while at an intermediate temperature (12–13 °C; *ca.* 54 °F), the respiration rate rose, levelled off and changed relatively little in the latter stages of the germination period (Day, 1891). While the total carbon dioxide released over a 10-day germination period increased with increasing temperatures, rootlet production increased with increasing temperature between 4.4 °C (40 °F) and 12.8 °C (55 °F) but then remained about the same at temperatures up to 21.1 °C (70 °F). Therefore, under different malting conditions, the relative sizes of the losses owing to respiration and rootlet production will vary. Peak respiration occurred sooner at higher temperatures, for example at day 2.5 at 70 °F (21 °C) and day 4 at 60 °F (15.6 °C). In traditional floor-malting, the maximal respiration rate occurred between days 3 and 5, and the rate during this period correlated well with the ultimate malting loss (Meredith *et al.*, 1962).

Traditionally, the germination stage of malting was carried out at 12–14 °C (53.6–57.2 °F), but now values of 16–18 °C (60.8–64.4 °F) are not uncommon,

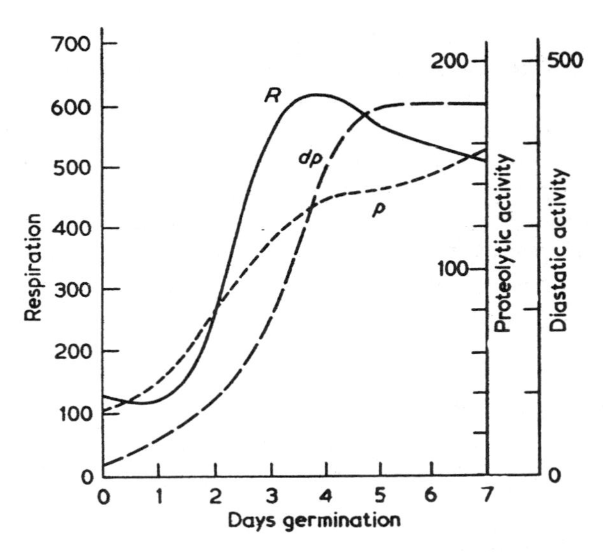

Fig. 3.19 The 'respiration' (carbon dioxide output rate mg CO_2/h per kg grain), the diastatic power (dp; Windisch–Kolbach) and the proteolytic activity (p; Idoux) of barley during a 7 day malting period. (After de Clerck and Cloetens, 1940.)

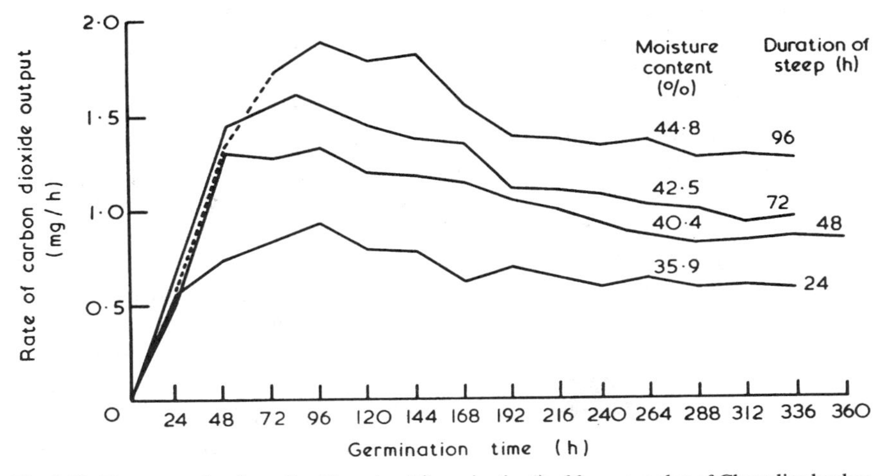

Fig. 3.20 The rates of carbon dioxide output ('respiration') of four samples of Chevalier barley, previously steeped for the times and to the moisture contents indicated, during malting at about 14 °C (57–58 °F). (After Day, 1896.)

and indeed temperatures may be allowed to rise to 25 °C (77 °F) or even more. Altering the temperature of germination causes different metabolic processes to occur at different rates relative to each other; they do not alter 'in parallel'. Consequently, malts prepared from one batch of grain at different temperatures have different compositions and enzyme complements.

When malt is kilned its temperature rises and, until it is sufficiently dry that metabolic processes cease, its respiration rate is accelerated together with many other metabolic processes. This is readily apparent on the upper floors of two- and three-floor kilns and in the upper layer of grains in deep-loading kilns. Indeed, the heat output of the green malt is significant in the early stages of kilning (Chapter 10). Fully dried malt (1.5–6.0% moisture) does not metabolize or respire although chemical changes can occur in it. Apart from the losses of aroma that occur from dark malts during prolonged storage, other changes are not understood. It is widely believed that 'white' malt gains in quality during a period of storage after kilning.

3.9 Microbes and malting

Grains always carry a population of microbes (mesophilic, thermophilic and thermotolerant) (Milner *et al.*, 1947; Hummel *et al.*, 1954; Tuite and Christiansen, 1955; Sheneman and Hollenbeck, 1960, 1961; Christiansen and Kaufmann, 1969, 1974; Briggs, 1978; Petters *et al.*, 1988). Bacteria, filamentous fungi (moulds) and yeasts are always present. Actinomycetes, slime moulds, viruses, protozoa and even nematodes have sometimes been recorded. Some will grow below 0 °C (32 °F), but optimal growth conditions occur in the range 10–45 °C (50–113 °F) and the moisture requirements of different species vary (Chapter 7). The very mixed nature of this population, the range of locations in which the microbes occur – on the grain surfaces, loose in dust, attached to, or even growing in the pericarp or husk or between these layers in barley – and the occurrence of dead microbes, viable vegetative forms and quiescent and tough spores makes this population difficult to evaluate or control. Indeed, even experimentally, when there are none of the limitations that apply to grain intended for use in foodstuffs, it requires the use of solutions of hypochlorites, salts of heavy metals (mercury or silver) and a mixture of antibiotics to control the microbes. It is doubtful if they are ever eliminated from intact grains, although this may be achieved in grains that have first been decorticated with sulphuric acid. One gram of malting barley (approximately 25 grains) may yield tens of thousands of filamentous fungi, hundreds of thousands of yeasts and millions of bacteria.

It has long been known that obviously mouldy grains, or heavily stained or weathered grains (grains discoloured because they are, or have been, heavily infected with microbes that were able to multiply as a result of wet weather around harvest time) would not malt well. In warm weather, floor-maltsters often had difficulty in preventing the grain becoming mouldy. Comparatively recently, it has been recognized that microbes often play significant roles in malting and influencing malt quality even when their presence is not immediately apparent (Kelly and Briggs, 1992b, 1993).

Obviously diseased grains that have been attacked by pathogens are avoided by maltsters (Chapter 6). However, effectively all batches of grain carry populations of saprophytic microbes. These respire, producing carbon dioxide and utilizing oxygen in competition with the grain; many produce and release enzymes, some produce substances with plant-hormone-like properties, others produce metabolites that are phytotoxins (damaging to the grain) while yet others may produce mycotoxins that are harmful to animals, including humans. In addition, some microbes influence malt so that beer made from it is prone to gushing (over-foaming; gas instability): on opening a container much of the enclosed beer is rapidly converted into foam, which gushes out. Other microbes cause the beer to become hazy (loss of colloidal stability).

The fungi found on grain are frequently sub-divided into 'field fungi', which grow on grain at moisture levels of >21% (equilibrium relative humidity (RH) over 90%) and storage fungi, which grow at lower moisture levels. Common field fungi include *Alternaria* spp., *Cladosporium* spp., *Aureobasidium* spp., *Epicoccum* spp. and *Fusarium* spp. The yeasts present include pink *Rhodotorula* or *Sporobolomyces* spp. and *Cryptococcus* and *Torulopsis* spp. The most common bacteria are Gram-negative rods, including *Erwinia herbicola* and *Xanthomonas campestris* (Flannigan *et al.*, 1982, Flannigan, 1996). Heavy populations of field microbes are detrimental to malting. They are a major, perhaps **the** major, cause of water sensitivity. During 'dry' storage, the populations of fungi and bacteria progressively decline. If grain is rapidly dried and stored dry (e.g. at 11–12% moisture) some field fungi survive well, and their presence (e.g. of *Alternaria* spp.) rather than that of storage fungi, which develop in 'damp' storage (over 13% moisture), is an indication that grain has been dried quickly and stored correctly (Christensen and Kaufman, 1974). The decline of the microbial population contributes to the decreased water sensitivity and increased maltability of the stored grain (Crabb, 1971; Kelly and Briggs, 1993).

Bacteria will grow on grains at moisture contents down to about 18%, while some fungi will grow at moistures as low as 13.5%. Since, in the UK, grain intended for malting is normally stored in the moisture range 11–16%, it is usually fungi that are responsible for grain deterioration. Various storage fungi may grow in grain stored with moisture contents of 13.5% or more, so it is best to store grain at moisture contents lower than this. Common storage fungi are the *Aspergillus glaucus* group, other *Aspergillus* spp. (including *A. fumigatus*), *Penicillium* spp., *Absidia* spp., *Mucor* spp. and *Rhizopus* spp. Various of these species can attack the grain embryo, or germ, and kill it. Fungal mycelia and spores, in grain dust, constitute a health hazard causing, for example, allergies, aspergillosis, farmer's lung and thresher's lung. Microbes on the grain compete with each other and during storage particular groups tend to become the most numerous at particular moisture levels. Their presence in grain samples can indicate how these samples have been stored. At 13.5% moisture, *A. restrictus* and *A. halophilicus* (members of the *A. glaucus* group) can slowly grow and kill grains. In the moisture range 14–16% other members of the *A. glaucus* group, *A. candidus* and *A. ochraceus* multiply. Penicillia require a moisture content of about 19% to grow. Heavy microbial attack can cause grain heating and more complex problems. As the storage fungi multiply, field fungi decline and become rare (Chapter 7).

During steeping, the microbial population multiplies and spores and microbes move around in the steep liquor so that infections spread between kernels (Haikara *et al.*, 1977; Petters *et al.*, 1988) (Table 3.6). Most types of microbe increase, but at different rates so the proportions within the population alter. Some weathered barleys are reported to take up moisture unusually rapidly. Clearly it is desirable to keep steeps clean to minimize microbial contamination. In some steeping vessels, cleaning is difficult (Chapter 9). The cleaning and anti-microbial agent most favoured seems to be sodium hypochlorite. This agent must not be allowed to come into contact with grain, however, or the final malt is tainted with a 'disinfectant' flavour.

For well over a century, maltsters have been seeking acceptable, effective agents for controlling microbes in steeps. Among the more promising agents that were used or recommended are hydrogen peroxide (0.1%, 48 h exposure; 1.0%, 24 h exposure; or 0.1% + 0.1% lactic acid), lime water (saturated) or half-strength lime water (but **not** milk of lime), sodium or calcium hypochlorites, sodium or potassium sulphites (15–20 g SO_2/hl) or bisulphites (0.02% SO_2), formaldehyde (0.05–0.2%), sodium hydroxide (0.25 M), sulphuric acid (0.125 M), phosphoric acid and salicyclic acid. For various reasons, these agents are disliked for some other characteristic, such as requiring extra plant (lime water), being costly, being corrosive (acids, bases), being harmful to grain during prolonged exposures (sodium

Table 3.6 Estimates of the numbers of microbes associated with malting barley at different stages of the production process. The data were obtained by incubating inocula on selective media at 25 °C (77 °F) for 14 days

Sample	No. of aerobic heterotrophic bacteria/kernel or ml	No. of lactobacilli/ kernel or ml	No. of filamentous fungi/kernel or ml	No. of yeasts/ kernel or ml
Barley (dry)	1.8×10^6	2.0×10^2	2.0×10^2	4.7×10^3
Barley (first steep)	6.7×10^5	4.2×10^3	8.0×10^2	4.6×10^5
Barley (second steep)	6.6×10^6	7.8×10^4	1.7×10^3	1.1×10^6
Green malt (5 days)	5.7×10^7	8.7×10^6	1.5×10^2	3.9×10^6
Kilned malt	5.6×10^6	1.6×10^5	2.0×10^2	3.2×10^4
Screened malt	5.5×10^6	5.7×10^4	8.3×10^2	1.8×10^4
Bore-hole water	1.1×10^1	0	0	0
Water (first steep)	1.7×10^7	2.9×10^6	8.0×10^2	5.7×10^5
Water (second steep)	2.0×10^7	2.6×10^6	9.0×10^2	1.2×10^6

From Petters *et al.* (1988).

hydroxide, sodium hypochlorite, formaldehyde), tainting the grain and giving malt an unpleasant odour and flavour (hypochlorites), or working only erratically. For example, older maltsters found that steeping in lime water helped to limit the subsequent development of mould on the malting floor. However in recent years, when in any case 'moulding' is rare, little benefit has been found from using this agent, and it has been reported that although its use has been followed by a decline in the fungal population the bacterial population increased (Sheneman and Hollenbeck, 1960, 1961).

Most microbes continue to increase on malting grain during germination, but the numbers decrease on kilning (Table 3.6), particularly if sulphur dioxide is applied. At the end of kilning, the population is relatively enriched in lactobacilli.

Isolates of microbes have been added to grains in micro-malting trials, sometimes with the object of swamping undesirable species such as those that can generate mycotoxins and sometimes to determine what undesirable effects particular species cause. The results are complicated, since they vary with the species and with the isolate of the species that is tested (Prentice and Sloey, 1960; Sloey and Prentice, 1962; Gjertsen *et al.*, 1963, 1965, 1973; Amaha *et al.*, 1973). Various microbes, and cell-free filtrates of microbial cultures, have been shown to reduce grain germination. In some instances, steep- and respiratory-malting losses are increased; often rootlet growth is reduced but, less usually, it may be increased. Some isolates (often *Fusaria*) led to increased levels of soluble nitrogen in the malt, so that the nitrogen indices of modification (soluble nitrogen ratio (SNR), Kolbach index) were increased, as was malt colour and the level of formol nitrogen. In particular instances, the levels of extract, of diastatic power and of α-amylase are increased. Some malts had unusual or offensive odours or flavours (molasses, burned, unclean, winey, harsh, spicey, bitter, fruity, astringent) and corns were sometimes discoloured. Some microbes altered malts so that beers made from them tended to become hazy. Often microbes that multiplied during the malting process caused beers made from them to have a tendency to gush. Isolates that produced 'gushing factors' have included *Alternaria*, *Stemphylium*, *Fusarium*, *Nigrospora*, *Penicillium*, *Aspergillus* and *Rhizopus* spp. The microbial products that destabilize the beer appear to be various and include a hydrophobic peptide and microcrystals of calcium oxalate. Experimentally, the addition of starter cultures of selected lactic acid bacteria to steep water has reduced the growth of other, less desirable microbes on malt and led to a superior final product (Haikara *et al.*, 1993; Haikara and Laitila, 1995).

Only rarely have traces of mycotoxins been found in beers, or on malts. However in particular localities, in individual seasons, barleys have been infested with mycotoxin-forming fungi and there is no room for complacency. The problem is complex. Hundreds of mycotoxins have been recognized, differing widely in their properties. These include the aflatoxins, ochratoxin A, patulin, sterigmatocystin, trichothecenes (such as nivalenol and deoxynivalenol or vomitoxin), tenuazonic acid and zearalenone. Such complexity makes analysing these substances very difficult. In various species of animals, the results of ingesting mycotoxins include death, liver and kidney damage, the induction of cancer, vomiting, diarrhoea, haemorrhaging, problems with the blood-forming system and reproductive problems. Different isolates of single species of fungi vary widely in their ability to form mycotoxins and in the ranges of toxic agents that they produce. Microbes that are capable of making mycotoxins do not necessarily do this under all conditions. In practice, mycotoxins do not seem to pose a significant problem in making European-style clear beers. It is not at all clear why this is so. An understanding of why microbes do, or do not, generate toxins might enable us to reduce the residual risks still further. The question of whether mycotoxins are of more significance in products made from 'tropical' cereals malted under less well-controlled conditions appears to be open, but it is likely since, for example, fungi readily grow on malting sorghum.

4 The biochemistry of malting

4.1 Introduction

The major biochemical metabolic pathways are known to operate in the living tissues of germinating grains (Pollock, 1962a; Harris, 1962a,b; Briggs, 1978). These pathways, which include many hundreds of compounds that are interconverted under the influence of hundreds of enzymes, interlock and provide routes by which, for example, the carbon skeletons of the carbohydrates, amino acids and lipids can be interconverted, and by which the degradation of highly polymeric substrates, such as polysaccharides and proteins, can provide respiratory fuel and building blocks for polymeric substances, such as the components of cell walls. The grain tissues undergo changes as malting proceeds. The aleurone cells are partly depleted of their contents but are metabolically active. The contents of the dead starchy endosperm are partly degraded and depleted. The embryo metabolizes and grows, chiefly at the expense of degradation products from the starchy endosperm. The degradative and biosynthetic processes proceed simultaneously; there is a net breakdown of polymeric substances, such as starch, and there is a net migration of substances from the aleurone layer and the endosperm to the embryo. Most attention is given here to the biochemistry of barley, particularly those aspects that are of most importance to maltsters and malt users and provide a background to changes that occur in malting (Chapter 5). However, some information is given on other cereal grains.

In considering the biochemical changes that occur in malting grains, it must be remembered that modification in the starchy endosperm progresses along the grain with time, so a partly malted grain retains some of its 'barley character'. Then, in each batch of grain, there is heterogeneity in the sense that individual grains vary, for example in size, shape, maturity, chemical composition and potential for generating enzymes. Furthermore, individual grains modify at different rates and may receive different treatments during steeping, germination and kilning. Consequently malts are heterogeneous; bulk analyses of grain or malt samples are averages, and analyses of small numbers of grain or of individual grains may show a wide range of values scattered around the true 'bulk mean'.

Newer methods of analysis can provide data on minute traces of substances in germinating grain. It is not always clear to what use such measurements can be put. However, a consideration of the recent advances in understanding malt flavour compounds, such as dimethyl sulphide (DMS), and the need to gain an understanding of nitrosamine formation in malting demonstrates clearly that the minutiae of grain biochemistry should not be overlooked.

4.2 The compositions of grains

4.2.1 Quiescent grain

The gross compositions of some cereal grains are summarized in Table 4.1. These are representative values, as there are wide ranges of compositions within one species (Table 4.2). Despite continuing improvements, the analyses of biological materials, like grain samples, are neither simple nor wholly reliable. In many cases, means of quantifying all of the components of a particular class of substances, e.g. polyphenols, are not available. It is clear that carbohydrates make up most of grains.

4.2.2 Changes occurring during malting

The changes that occur during malting are normally described in terms of physical alterations in the grains (Chapter 2) and alteration in conventional malt analyses (Chapters 14 and 15). It is necessary to understand the biochemical basis of these changes. The gross changes are the net result of the degradation of reserve substances, the interconversion of substances in the living embryo and aleurone layer, the net flow of substances to the embryo from the aleurone layer and starchy endosperm, the synthesis of new grain substances and their incorporation into the new, growing tissues (the acrospire and rootlets) of the embryo. Allowances must be made for the malting losses: the losses in dry matter that occur in the conversion of grain into malt. The losses result from: (i) leaching of substances from the grain

Table 4.1 Examples of representative proximate compositions (g/100g d.m.) of samples of cereal grains. Grains typically contain 10–15% water

Species	Protein	Fat	Crude fibre	Carbohydrate	Ash
Wheat	10.5–16.0	1.8–2.9	2.5–3.0	74.1–78.4	1.8–2.0
Barley	11.8	1.8	5.3	78.1	3.1
Oats (entire)	11.6	5.2	10.4	69.8	2.9
Rye	13.4	1.8	2.6	80.1	2.1
Triticale	15.0	1.7	2.6	78.7	2.0
Rice (paddy)	9.1	2.2	10.2	71.2	7.2
Maize	10.0–12.1	4.0–9.1	2.1–3.5	74.5–80.2	1.6–2.0
Sorghum	12.4	3.6	2.7	79.7	1.7
Pearl millet	13.6	5.4	1.3	77.9	1.8
Foxtail millet	13.9	4.8	9.0	68.6	3.7
Proso millet	12.8	4.0	12.7	67.7	2.8
Finger millet	6.6–11.3	1.1–7.7	0.7–7.8	61–76	2–5

From Burgess (1962); Kent (1983).

Table 4.2 The composition of barleys grown in Sweden and Montana (% (w/w) d.b.)

	Two-rowed barleys				Six-rowed barleys			
	Sweden		Montana		Sweden		Montana	
	Mean	Range	Mean	Range	Mean	Range	Mean	Range
Number of varieties analysed	81		16		11		7	
Glucose	0.3	0.1–0.8	0.2	0.06–0.3	0.6	0.2–1.4	0.2	0.06–0.3
Fructose	0.1	tr.–0.4	0.1	0.05–0.3	0.2	0.1–0.5	0.1	0.09–0.2
Sucrose	1.6	0.6–3.1	1.9	1.4–2.3	1.9	1.1–3.9	2.0	1.6–2.3
Fructans	0.4	tr.–0.8	0.5	0.4–0.8	0.3	tr.–1.0	0.5	0.4–0.7
Starch	62.2	55.9–66.6	57.2	53.3–59.8	58.9	52.9–64.1	57.2	53.0–60.6
Crude protein	10.7	8.6–13.4	14.4	12.2–16.8	11.5	8.9–14.0	13.8	12.2–15.8
Crude fat	3.0	2.7–3.3	2.8	2.3–3.7	3.3	2.8–3.7	2.7	2.1–3.3
Ash	2.4	1.8–2.9	2.9	2.3–4.0	2.4	2.2–2.7	3.1	2.3–3.8
Total fibre	19.3	14.0–24.7	19.9	17.6–22.9	20.9	17.8–23.8	20.5	17.6–23.0

Data from Åman *et al.* (1985); Åman and Newman (1986).
tr., trace amount.

during steeping; (ii) fermentative processes and the respiratory oxidation of substances to carbon dioxide and water and (iii) the removal of the rootlets. With barley, the total losses during malting are usually in the range 6–12% of the original dry weight. Losses may be much larger when some other cereals are malted (Chapter 14).

A comparison of the average composition of a European two-rowed barley and a pale malt made from it shows that during malting there is a net degradation of high-molecular-weight polysaccharides, especially starch, and protein; this is partly offset by an accumulation of simpler substances, sugars and amino acids, which contribute to the cold water extract of malt (Table 4.3). Allowing for the malting losses (*ca.* 10%), it is apparent that about 18% of the barley starch is degraded during malting. On balance there is a substantial loss of substances that could have given rise to extract during malting.

Grain tissues differ in their chemical compositions and biochemical functions. The proportions of the different tissues vary in different species or different samples of grain from one species. The distribution of the main classes of chemical substance in grain will be considered.

4.2.3 *The husk and pericarp*

The detailed compositions of the husk and pericarp are not known. Analyses have been carried out on fractions called husk, obtained by crushing malt and collecting the 'husk fraction' by sieving. This fraction certainly

Table 4.3 The chemical composition of European two-rowed barley and pale, deculmed malt

	Proportions (% d.wt)	
Fraction	Barley	Malt
Starch	63–65	58–60
Sucrose	1–2	3–5
Reducing sugars	0.1–0.2	3–4
Other sugars	1	2
Soluble gums	1–1.5	2–4
Hemicelluloses	8–10	6–8
Cellulose	4–5	5
Lipids	2–3	2–3
'Crude protein' (N × 6.25)	8–11	8–11
Salt-soluble proteins		
Albumin	0.5	2
Globulin	3	–
Hordein 'protein'	3–4	2
Glutelin 'protein'	3–4	3–4
Amino acids and peptides	0.5	1–2
Nucleic acids	0.2–0.3	0.2–0.3
Minerals (ash)	2	2.2
Other substances	5–6	6–7

Compiled from various sources by Harris (1962a).

contains husk material (palea and lemma) but it also contains lodicules, pericarp, testa, some nucellar tissues, aleurone layer tissue, fragments of acrospire and other embryonic tissue, as well as some adherent starchy endosperm. The gross composition of such a fraction is shown in Table 4.4. Much of the 'fibre' of grain is concentrated in the husk. In general plumper, 'bold' grains have lower husk contents and lower fibre contents than thin grains. Starch found in husk fractions is derived from contaminating starchy endosperm tissue; the husk itself contains no starch. The husk fraction of malt has major effects on the character, flavour and stability of beers. Malt grains from which the husk fraction has been removed by milling give bland and insipid beers, while grains enriched with husk-sieve fraction produce astringent beers that readily become hazy.

Together the husk and pericarp make up 9–14% of the dry weight of most barleys. In one instance the lemma was estimated to be 7.3% and the palea 3.1%. Six-rowed barleys tend to have higher husk contents than two-rowed grains and to contain more fibre (Table 4.2). Fibre is insoluble material but is not rigidly defined, so values vary according to the analytical method used. Malts from six-rowed barleys were used to increase the husk content of brewers' grists, to 'open out' the mash and to allow more rapid wort run-off after mashing. Estimates of husk contents differ depending on the technique of measurement used. Soaking grain in water or a strong solution of ammonia, followed by hand separation, drying and weighing is tedious and imprecise both because samples are small, and so may be unrepresentative of the bulk, and because materials are lost in solution. Pearling in a special mill, or by battering grain in a 'cage' impelled by an air-jet, does not remove all the husk but does remove small amounts of other tissues. Husk may be washed away after treating grains with a hot solution of alkaline sodium hypochlorite, after which the grains may be dried and weighed and the husk content found by difference. Decorticating grain by soaking in 50% sulphuric acid, followed by thorough washing, removes the husk and pericarp; the loss in dry weight is a measure of the

Table 4.4 The approximate composition of a 'husk' fraction of barley malt prepared by sieving. Such fractions contain true husk, together with pericarp, testa, nucellar material, aleurone layers, fragments of embryo and starchy endosperm

Fraction	Proportion (%)
Water	7.4
Protein	7.1
Pentosans	20.0
Fibre	22.6
Fat	2.1
Ash (including silicates)	10.0
Starch	8.2
Other nitrogen-free materials	22.6

Data of Geys, cited by Hopkins and Krause (1947).

combined quantity of these tissues, but the value includes the weight of substances extracted from damaged corns (Briggs, 1978; Pollock *et al.*, 1955). Husk and pericarp may be considered together since they invest grains, are fused together in most barleys and they appear to serve the same functions. They do not alter appreciably during malting, but the polyphenolic materials they contain may be of significance, and enzymes derived from the surface flora of microbes and perhaps plant enzymes surviving from when the tissues were alive may be significant during mashing. The respiration of the microbes adversely affects grain germination by competing for oxygen with the living tissues. The husk and pericarp limit the rate of inward diffusion of oxygen in respiring grain. The husk protects the grain against physical damage and protects the acrospire during malting. In naked grains, such as wheat, rye, triticale, sorghums and maize, or naked barley (which is not malted commercially), grains are more readily damaged by mechanical handling, by compression during steeping and by turning machinery during germination. The acrospires are particularly easily damaged. The husk may physically restrict barley swelling during steeping and, subsequently, may restrict acrospire growth. The pericarp, the husk and the associated microbes are of importance in accentuating or causing dormancy (Chapter 3).

The husk and pericarp have waxy cuticles. The wax on the husk has been characterized. The surface cell layers of the husk are rich in silica, giving this tissue its abrasive qualities. The husk is rich in holocellulose, hemicelluloses (containing hexose, pentose and uronic acid residues) and lignin (positive vanillin/hydrochloric acid test). It may contain about 2% crude protein. Starch is absent from the husk and pericarp in mature grains. When grain is steeped, inorganic ions, sugars, phenolic acids, amino acids and other phenols are dissolved. These seem to originate from the husk. When the grain is steeped in weakly alkaline water, the quantity of material dissolved is increased and contains a substantial amount of testinic acid, a mixture of polyphenols (perhaps including lignans) and protein. Some of the polyphenols may also originate in the testa. Briefly soaking grains in mildly alkaline solutions at the start of steeping leads to malts that give beers which are less likely to become hazy and which contain lower levels of polyphenols.

4.2.4 The testa

When the grain is decorticated with 50% sulphuric acid, the smooth, clean, acid-resistant layer that remains enveloping the grain is the testa. It is probably altered to some extent by the acid-treatment (Briggs, 1987b). Decorticated grains retain most of the proanthocyanins (anthocyanogens), but these are readily leached into steep water. In the intact grain, proanthocyanins are located almost entirely between the two cuticularized layers of the testa (Aastrup and Outtrup, 1985).

The testa together with the pigment strand make up 1–3% of the weight of the barley kernel. This tissue is selectively permeable. Its two cuticularized layers are separated by the remains of cell walls, and they stain strongly for waxy, lipidic materials (Collins, 1918; Briggs, 1978). The testa contains a cellulosic polysaccharide component, a chemically resistant estolide of fatty acids and hydroxy-fatty acids, and a wax that differs in detailed composition from the wax of the husk and contains, *inter alia*, long-chain alkane hydrocarbons and 5-*n*-alkylresorcinols. The testa, pigment strand and the micropylar structures normally limit the ingress of water and dissolved substances. Adding organic solvents, such as ethyl acetate or chloroform, to water leads to more rapid water uptake by grains. This may be owing to the effects of the solvents on the waxes of the testa or the micropyle (Briggs, 1978).

4.2.5 The embryo

Barley embryos typically weigh about 1.5 mg and represent 2–5% of the grain dry weight (Briggs, 1978). When embryos (axis **plus** scutellum) are separated from grain and are grown on agar they grow relatively little unless provided with nutrients (mineral salts plus a sugar plus a source of assimilable nitrogen) when they will grow into a mature plant. Therefore, the embryo has the full potential for development, but its own reserves of nutrients are insufficient to allow it to develop fully. The healthy embryo is waxy-yellow in colour; it cannot carry out photosynthesis. The other parts of the grain either (i) protect the inner parts (husk, pericarp and testa); or (ii) serve as an elegantly organized food store and system for mobilizing these reserves (aleurone layer and starchy endosperm) so that nutrients are made available for the embryo. When embryos are cultivated *in vitro* they release gibberellin hormones and a complex mixture of hydrolytic enzymes, which, in the intact grain, are involved in mobilizing the reserves of the grain and supporting the embryo's metabolism and growth.

Ungerminated, unhydrated embryos contain little or no starch, but traces appear shortly after the onset of germination. They are, however, rich in soluble sugars such as sucrose, 14–15% d.wt; raffinose, 5–10% d.wt; and fructans. They are relatively poor in free hexose sugars. The ash content is 5–10% d.wt; about one fifth of the grain's supply. Cell walls and polysaccharide gums comprise about 16% of the dry matter. The embryo contains a little holocellulose, some pectin (uronic acids) and hemicelluloses (approximately 7% d.wt). Initially it contains no lignin, but traces of this appear during germination. By convention, the nitrogen content $\times$ 6.25 (%) is regarded as a measure of the 'crude protein' content. For embryos this value is about 34%. It includes amino acids and other compounds. Some of the protein reserves are deposited in subcellular protein bodies in the cells of the scutellum.

The embryo is rich in triacylglycerols (triglycerides: 13–17% d.wt), which constitute about 0.33 of the grains' supply. These are located in specialized subcellular organelles, the spherosomes. Therefore, the embryo is rich in readily metabolized reserve substances (soluble sugars, lipids) that support its metabolism in the initial stages of malting. In the initial stages of germination, the raffinose is fully utilized and together with sucrose, fructans and triacylglycerols it provides the initial fuels and building blocks for the growing embryo. Later these are supplemented and replaced by soluble materials being taken up via the scutellum from the modifying endosperm. Thus during malting, substances migrate from the aleurone layer and starchy endosperm to the embryo. Here some are respired and lost as carbon dioxide and water; others are used to build the new growing tissues of the acrospire and roots. The acrospire is retained in finished barley malt, but the rootlets are removed and so, in a sense, are 'lost'.

4.2.6 The aleurone layer and nucellar tissue

Residual nucellar tissue occurs between the aleurone layer and the testa and projects into the starchy endosperm as the sheaf cells (Briggs, 1978; Morrall and Briggs, 1978; Briggs and Morrell, 1984; Fincher and Stone, 1986). The nucellar cuticle becomes fused with the testa as the grain matures. Nothing is known specifically of the composition of the nucellar tissue, but it is comparatively resistant to degradation by hydrolytic enzymes. In most cereals, the aleurone layer is one cell deep, but in barleys it is usually three cells deep. The cell walls are thick (e.g. 2.9 μm in barley, 3.3 μm in wheat). They are not homogenous but contain at least two different layers that differ in their resistance to degradation by enzymes. The aleurone, with the associated nucellar tissue and testa, makes up 8–15% of the barley grain's dry weight.

The aleurone cell walls are not obviously altered during malting, but electron microscopy shows that limited degradation does occur. Under other conditions of germination, cell wall degradation can be extensive. The cells are filled with dense cytoplasm containing prominent nuclei, mitochondria and a range of other organelles. Of these, the spheroplasts, which contain chiefly triacylglycerols, and the protein bodies of aleurone grains, which contain dense deposits of protein with inclusions of salts of phytic acid and possibly glycoproteins, are important deposits of reserve substances. During germination these reserves are utilized and the appearance of the cell contents changes remarkably. Changes begin when the tissue hydrates, but the major alterations are initiated by the arrival of gibberellin hormones from the embryo, or by gibberellic acid applied to the grain by the maltster (Chapter 14). The tissue is a source of nutrients for the growing embryo and it provides a large part of the hydrolytic enzymes that, together with those from the scutellum, are responsible for modifying the starchy endosperm.

In some barleys, the aleurone cells are coloured blue or red with anthocyanin pigments. These colours are useful varietal markers. The pigments are in no way detrimental to malt quality. The chemically related, but colourless, proanthocyanin polyphenols seem to be confined to the testa. Aleurone cells are completely devoid of starch, but they do contain about 20% triacylglycerol (which is about 90% of that of the degermed grain) and 17–20% of crude protein (N × 6.25). The aleurone is rich in inorganic materials (over 75% of those of the degermed grain) and in soluble sugars, particularly sucrose and raffinose and perhaps verbascose and fructans. About 40–45% of the tissue is cell wall, which contains approximately 85% arabinoxylan, 8% holocellulose (containing glucose, mannose and galactose) and 6% protein. The arabinoxylan chains are substituted with acetyl and feruloyl groups. The ferulic acid substituents cause the walls to fluoresce blue in ultraviolet light. The interface between the aleurone layer and starchy endosperm is rich in deposits of a laminarin-like β(1–3)-linked glucan, which stain with Aniline Blue. In some reports, the glucan is regarded as part of the aleurone layer. Pectins, if present, only occur in trace amounts.

4.2.7 The starchy endosperm

During malting, hydrolytic enzymes, many of which are secreted by the scutellum and aleurone layer, accumulate in the starchy endosperm and catalyse the partial breakdown of its structural components (Thompson and La Berge, 1977; Briggs, 1978, 1987b; Fincher and Stone, 1986). As a result, the physical coherence and strength of the tissue falls, some of the products of hydrolysis accumulate and some are used by the embryo. It is these changes that constitute modification. The materials remaining in this tissue at the end of malting provide most of the extract obtained when the malt is mashed. Consequently, and with regard to the type of malt being produced, malting is carried out in such a way as to ensure that sufficient enzymes are accumulated and that sufficient endosperm degradation is achieved, while embryo growth is minimized to reduce the 'expense' of losses of 'potential extract', that is reserves from the starchy endosperm. In contrast, in agriculture, the contents of the endosperm are all utilized by the seedling as it grows and becomes established in the field.

The starchy endosperm of the mature grain is dead. It does not respire, nor does intermediary metabolism occur in its cells. It is not uniform in composition. In steely regions there may be zones rich in protein. There is always a gradient of increasing starch content and decreasing protein and β-amylase contents moving inwards from below the aleurone layer, across the sub-aleurone region and towards the central region. The sheaf cells, above the ventral furrow, are an invagination of cell walls of nucellar origin and lack cellular contents, including starch. The depleted and crushed cell layer, immediately adjacent to the scutellum, also consists of compacted cell walls. Because both it and the sheaf cells fluoresce strongly in ultraviolet light and are relatively resistant to hydrolytic enzymes, it has been proposed that this, too, may be nucellar in origin (Briggs and McDonald, 1983a). It has not been possible to analyse these different regions separately, so analytical values for endosperm tissue are not truly just for this tissue.

The starchy endosperm makes up 75–80% of the barley grain's dry weight. The thin cell walls, *ca.* 2.3 μm thick (1.8 μm in wheat), together with water-soluble gums, make up about 10% of the tissue. The walls contain a holocellulosic, fibrillar matrix, about 6% of the polysaccharide present, which contains glucose, mannose and galactose. Uronic acids, and hence pectins, are negligible in amount (<1%). The remainder, the amorphous layer, is about 23% arabinoxylan and 71% β-glucan (mainly mixed linkage β(1→3)(1→4)-glucan). Protein is also present. Possibly the cells are cemented together by protein. The relatively insoluble cell wall components may be cross-linked by peptides and/or diferulic acid. Apparently the gums, which are soluble in hot water, are not so linked.

The cell contents are seen under the microscope to be chiefly starch grains of various sizes embedded in a matrix of protein. Protein (N × 6.25) makes up about 9% of the starchy endosperm. The strength with which the protein adheres to the starch and cell walls varies in different grains. Where the protein is tough and its adhesion is strong it may be responsible for the hardness and high density of steely grains. Much of this protein serves as a reserve of amino acids and some of it is deposited as small protein bodies, which are different in structure to aleurone grains. All the β-amylase of the grain and some of the debranching enzyme that is present before germination begins are located in the starchy endosperm.

In mature, ungerminated grains, starch is practically confined to the starchy endosperm. The content varies widely in different samples (Table 4.2) and tends to be less in thin grains. Therefore, in plump, two-rowed malting barleys starch may make up 63–67% of the grain, but in thinner, six-rowed grains the proportion may be 57% or less. In two-rowed, plump barleys starch comprises about 85% of the starchy endosperm. Starch grains are laid down in amyloplasts during grain development. The residues of the amyloplast membranes probably enclose the mature starch granules. Granular starch is not pure polysaccharide but contains traces of other substances such as phosphates, lipids, free fatty acids and proteins. In all cereals, polar lipids and/or free fatty acids are closely associated with the polysaccharide component (Morrison, 1978; 1988). The polysaccharide starch is, however, the main component of starch granules, whether from the large (20–48 μm) diameter or small (1.7–10 μm diameter granules. The exact ranges of dimensions of the starch granules are different in different barley cultivars.

Analyses of starches from some normal barleys showed amylose contents varying from 25.3% to 28.5%, lysophospholipids 630–984 mg%, B-granules of 2.9–3.1 μm mean diameter making up 3.7–11.0% (by volume) of the total starch. The mean diameters of the A-granules ranged from 12.5 to 15.0 μm (Morrison *et al.*, 1986). In 16 samples of Triumph barley, TCW values ranged from 34.3 to 46.2 g (d.wt), with starch contents from 53.7 to 63.4% and amylose contents (in the starch) of 27.5–29.4%; B-granules made up 3.7–12.6% by volume of the starch. Mean diameters of B-starch granules ranged from 2.6 to 3.3 μm, while for A-granules the range was 12.2–13.7 μm (Morrison *et al.*, 1986).

As well as the major structural components, the starchy endosperm contains minerals, some free amino acids and sugars, including glucose (**1**), fructose (**2**), maltose (**18**) and sucrose (**13**). Triglycerides are also present in small amounts. Generally cell nuclei and other sub-cellular organelles cannot be detected in mature endosperm tissue, but traces of nucleic acids, DNA and RNA are thought to be present.

4.3 The chemical changes occurring during malting

The chemical changes occurring during malting are complex (Harris, 1962a,b; Pollock, 1962a,b; Briggs, 1973, 1978, 1987b, 1995). They can only be understood by appreciating the range of, sometimes conflicting, processes that occur during steeping, germination and kilning, and the effects of deculming and dressing the malt. Polymeric reserve substances, such as starch and proteins, are partly hydrolysed in the endosperm; the low-molecular-weight degradation products diffuse through the grain. Those reaching the living tissues may be metabolized, together with the reserves of these tissues (e.g. sugars, lipids). The aleurone may release some of its metabolic products. There is a net movement of materials to the embryo where they may be respired, converted into new substances and/or be incorporated into the growing tissues of the acrospire and the rootlets. Thus the synthesis of new, complex molecules (proteins, polysaccharides) in the embryo partly offsets the degradative changes that occur elsewhere in the grain, chiefly in the starchy endosperm.

Understanding the balance of changes is complicated by the losses of dry matter that occur during malting. These losses result from: (i) leaching of substances into the steep water; (ii) oxidation to carbon dioxide and water via the respiratory process and, to a lesser extent, the loss of carbon dioxide and some evaporated ethanol formed by glycolysis (anaerobic respiration); and (iii) the removal of the rootlets and dust. Total malting losses usually amount to 6–12% of the original dry weight of the barley steeped. Malting losses cause a decline in the TCW (Table 4.5). A comparison of the average composition of two-rowed barleys and pale, derooted (deculmed) malts shows that during malting there is a net degradation of high-molecular-

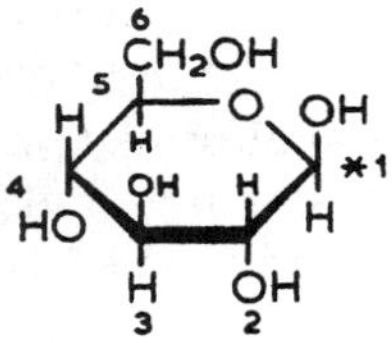

(**1**) Glucose; β-D-glucopyranose

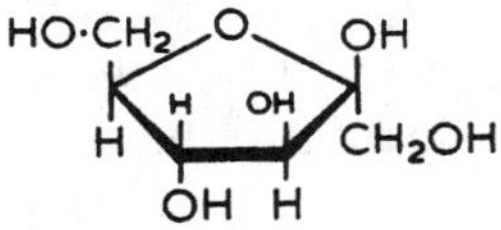

(**2**) Fructose; β-D-fructofuranose

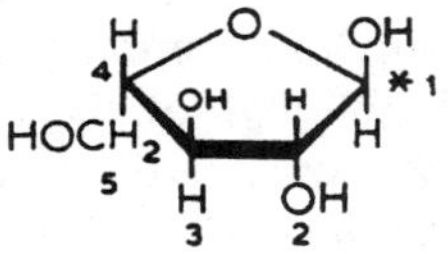

(**3**) Arabinose; α-L-arabinofuranose

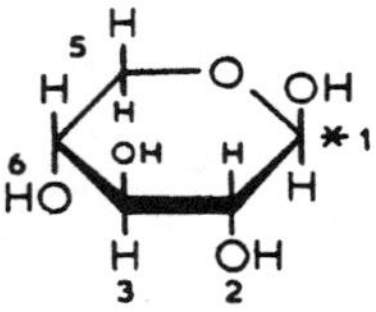

(**4**) Xylose; β-D-xylopyranose

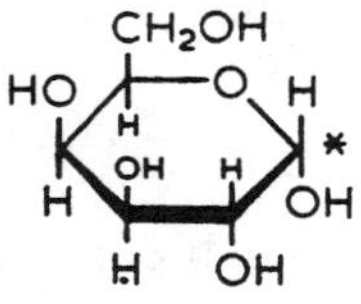

(**5**) Galactose; α-D-galactopyranose

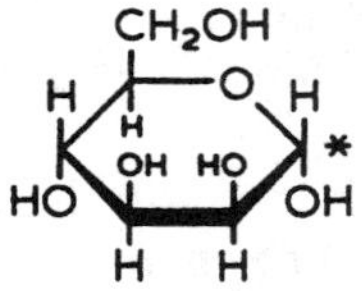

(**6**) Mannose; α-D-mannopyranose

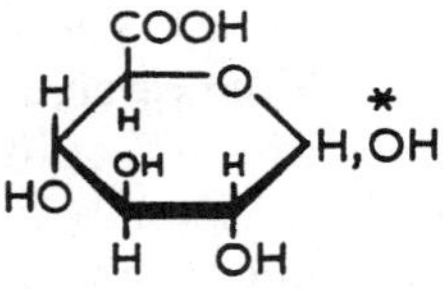

(**7**) D-Glucuronic acid

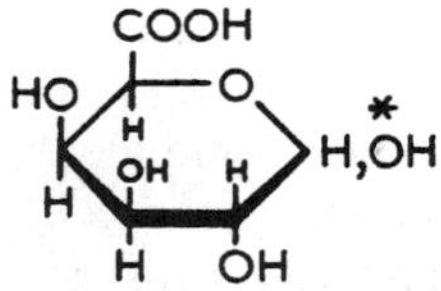

(**8**) D-Galacturonic acid

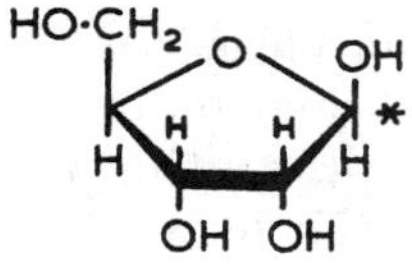

(**9**) Ribose; β-D-ribofuranose

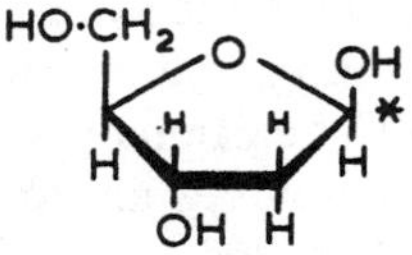

(**10**) Deoxyribose; β-D-deoxyribofuranose

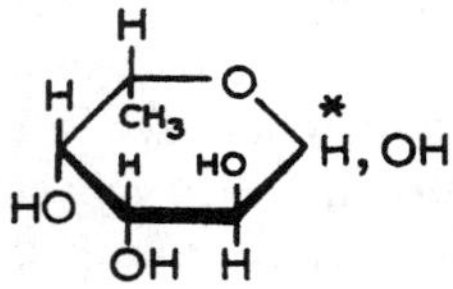

(**11**) Fucose

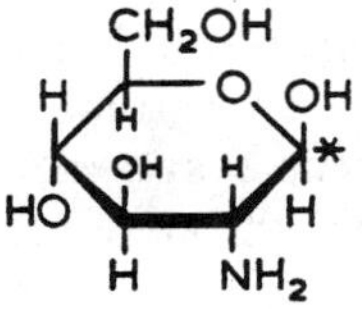

(**12**) β-D-Glucosamine

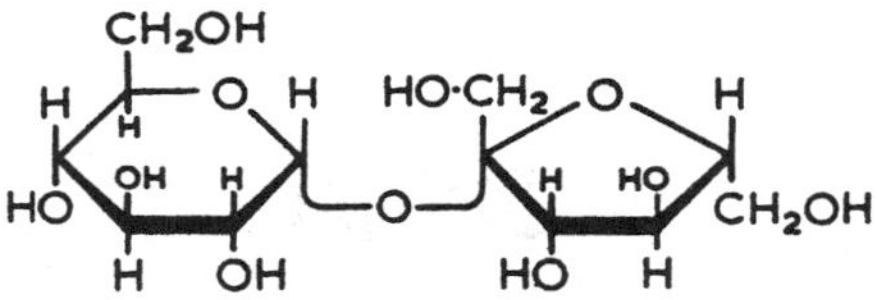

(13) Sucrose; α-D-glucopyranosyl-(1→2)-β-D-fructofuranoside

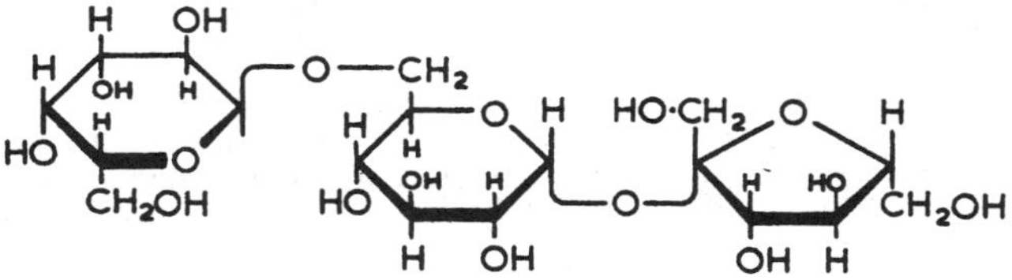

(14) Raffinose; α-D-galactopyranosyl-(1→6)-α-D-glucopyranosyl-(1→2)-β-D-fructofuranoside

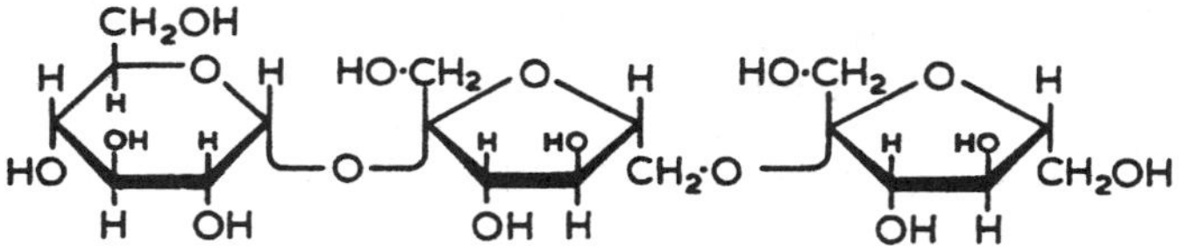

(15) Kestose (6-kestose); α-D-glucopyranosyl-(1→2)-β-D-fructofuranosyl-(6→2)-β-D-fructofuranoside

(16) *iso*-Kestose, (1-kestose); α-D-glucopyranosyl-(1→2)-β-D-fructofuranosyl-(1→2)-β-D-fructofuranoside

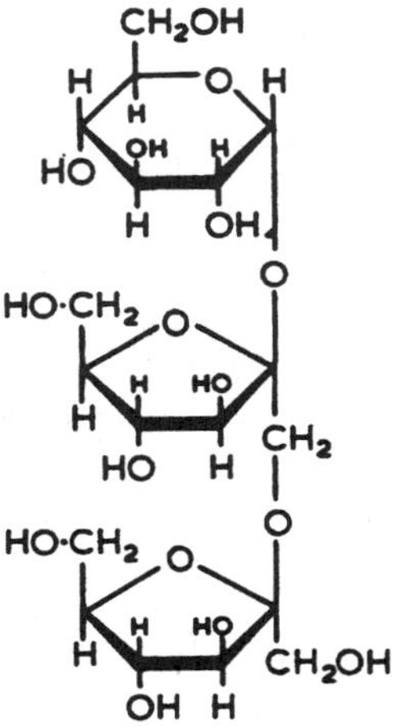

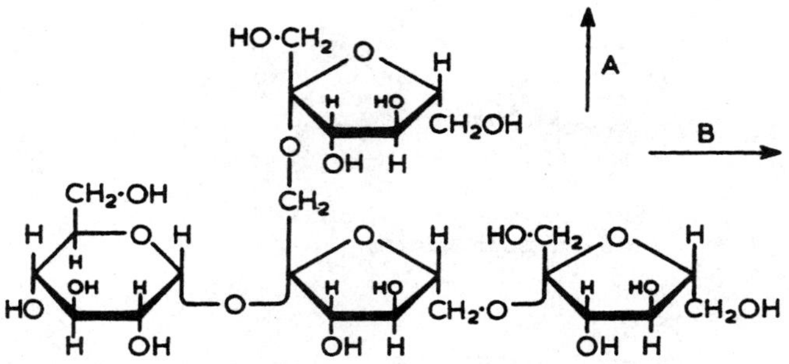

(17) Bifurcose, the basic unit of two series of fructosans. Extended in direction A, with β (2→1) linked fructofuranose units: kritesin (inulin-type) fructosans. Extended in direction B with β(2→6)-linked fructofuranose units: hordeacin (phlein type)

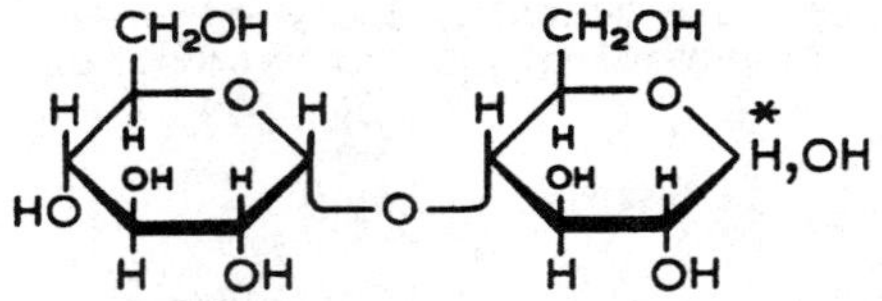

(18) Maltose; α-D-glucopyranosyl-(1→4)-D-glucopyranose

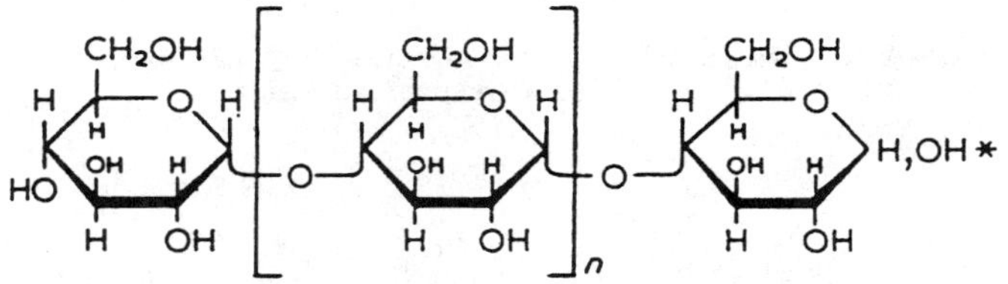

(19) Maltotriose (n = 1); maltotetraose (n = 2); maltopentaose (n = 3)

weight polymeric materials (starch, hemicelluloses, reserve proteins) and an accumulation of simpler, soluble substances (sugars, amino acids) that contribute to the malt's cold water extract (Table 4.3). Compared directly, the results, which are expressed as percentage dry weight, can be misleading. If, however, account is taken of malting losses, then the changes are more striking. Thus about 17% of the original barley starch is degraded ($(65 - 60 \times 0.9 \div 65) \times 100 = 17\%$, assuming a 10% malting loss). The loss of starch is 19.2% in the example in Table 4.5. Although this loss of potential extract is partly offset by the rise in the levels of simpler, soluble sugars, it still represents a substantial loss.

Table 4.5 Changes in the carbohydrate composition of a sample of Carlsberg barley (N, 1.43%) during traditional floor-malting. Only traces of free pentoses were ever detected. The disappearance of raffinose was real, but the apparent disappearance of 'glucodifructose' resulted from difficulties in separating it from other chromatographic fractions[a]

	Raw barley	Steeping (days)			Germination (days)				Kilning (days)				Finished Malt
		1	2	3	1	2	6	9	1	2	3	4	
Soluble carbohydrates (g/1000 corns)													
Fructose	0.027	0.012	0.027	0.042	0.100	0.097	0.132	0.145	0.189	0.267	0.297	0.262	0.194
Glucose	0.016	0.012	0.015	0.031	0.085	0.143	0.173	0.178	0.238	0.312	0.300	0.285	0.462
Sucrose	0.301	0.208	0.298	0.265	0.276	0.370	0.540	0.624	0.519	0.713	0.965	0.755	1.315
Maltose	–	–	0.008	0.012	0.026	0.147	0.179	0.123	0.116	0.109	0.103	0.109	0.257
Maltotriose	–	–	–	–	–	–	0.052	0.063	0.069	0.055	0.055	0.059	0.148
'Glucodifructose'	0.033	0.040	0.043	0.047	0.029	–	–	–	–	–	–	–	–
Raffinose	0.058	0.054	0.077	0.050	0.014	–	–	–	–	–	–	–	–
Fructosan	0.226	0.584	0.459	0.370	0.401	0.410	0.621	0.842	0.752	0.494	0.477	0.374	0.352
Insoluble carbohydrates (g/1000 corns)													
Glucosan	0.725	0.577	0.711	0.824	0.869	0.876	1.611	2.161	1.396	1.216	1.036	1.129	0.613
Araban	0.047	0.016	0.030	0.039	0.025	0.039	0.120	0.148	0 .130	0.141	0.112	0.115	0.110
Xylan	0.050	0.020	0.030	0.039	0.031	0.058	0.139	0.182	0.156	0.175	0.168	0.153	0.142
Starch	25.74	24.72	24.48	25.06	24.63	24.43	22.06	20.70	20.95	20.65	20.90	21.50	20.79
Blue value of starch[b]	0.326	0.331	0.317	0.308	–	0.310	–	0.381	0.371	0.377	0.364	0.374	–
TCW (g)	39.1	39.2	38.8	38.7	38.6	38.9	37.6	37.6	37.2	–	–	36.8	35.3

[a] All samples analysed with rootlets, except finished malt which was without rootlets
[b] Value for isolated starch: pure amylose about 1.4, amylopectin about 0.1.
From Hall *et al.* (1956).

Table 4.6 Sugars, soluble in 90% ethanol, of Ymer barley and malt

	Quantities of sugars (mg/1000 corns)	
	Barley	Malt without roots
Fructose	41	221
Glucose	34	629
Sucrose	421	2100
Maltose	22	299
'Glucodifructose'	92	180
Raffinose	203	0
Maltotriose	0	74
Other higher oligosaccharides, including fructosans	326	181

MacLeod *et al.* (1953).

4.4 The carbohydrates of quiescent and malting grains

Barley carbohydrates are usually considered in groups, defined by the techniques used to separate them. Particular attention will be paid to: (i) the starch; (ii) the non-starchy polysaccharides (NSP), hemicelluloses, gums and other polysaccharides; and (iii) readily soluble sugars, monosaccharides, disaccharides and some oligosaccharides (MacLeod, 1953; Harris, 1962a; Pollock, 1962a; Briggs, 1978; Henry, 1988). However, sugars also occur in combination in glycosides, having a range of aglycones (including phenols), in glycolipids, in glycoproteins, in nucleotides, in derivatives of nucleotides, and in nucleic acids. The hexoses glucose (**1**, 0.03–0.6%) and fructose (**2**, 0.03–0.16%) occur free in barleys and alter in amount during malting (Table 4.5). They also occur in oligosaccharides and, like the pentose sugars arabinose (**3**) and xylose (**4**), glucose occurs in substantial amounts combined in polysaccharides. The major glucose-containing polysaccharide is starch. Smaller amounts of galactose (**5**), mannose (**6**), glucuronic acid (**7**) and galacturonic acid (**8**) also occur combined in polysaccharides. The pentose sugars ribose (**9**) and deoxyribose (**10**) occur combined in nucleic acids, nucleosides and nucleotides. Small amounts of fucose (**11**) and glucosamine (**12**), which occur in glycoproteins, have also been detected.

Ungerminated barley contains 1–2% of carbohydrates that are soluble in 70% ethanol (Tables 4.2, 4.3, 4.5 and 4.6). Sucrose (**13**; 0.34–2.0%) is always the most abundant sugar, followed by raffinose (**14**; 0.14–0.83%), traces of stachyose, 'glucodifructose' (**15, 16**; 0.1–0.43%) and higher oligosaccharides and fructosans (fructans (**17**) 0.02–0.97%). As noted, traces of glucose (**1**), fructose (**2**) and maltose (**18**) (0.006–0.14%) are also present. These last occur chiefly in the starchy endosperm, where the maltose is concentrated in the sub-aleurone layer, where β-amylase (which can generate it from starch) is also abundant. 'Glucodifructose' is a mixture of two or three of the trisaccharides kestose (**15**), isokestose (**16**) and neokestose. The first two have been recorded in barley. These are the simplest members of two series of fructosans (**17**) in which sucrose is linked to extended chains of various lengths of D-fructofuranose residues joined mainly by β(2→1)-linkages in the kritesin (inulin) type, and mainly by β(2→6)-linkages in the hordeacin (phlein) type. Probably, it is the latter type of fructosan that occurs in barley. It is believed that these substances are formed by transfructosylation from sucrose to the extending chains. Oligosaccharides up to at least GF_{10} (glucodecafructose) are present. Sucrose (**13**), raffinose (**14**), glucodifructose (**15**,**16**) and the fructosans are concentrated in the embryo and the aleurone layer, the living grain tissues.

The grains of other cereal species contain a similar range of sugars and oligosaccharides to those found in barley. The seeds of the legumes contain less starch (30–50%) than is found in cereal grains. The wide range of sugars that they contain is, however, richer in the galactose-containing sugars raffinose, stachyose and verbascose. These are degraded during sprouting, when α-galactosidase activity increases. This is fortunate, as the galactose-containing oligosaccharides induce flatulence in humans.

The soluble, diffusible sugars that accumulate during malting originate partly from the degradation of starch, partly from the breakdown of hemicelluloses, gums and oligosaccharides, and partly from synthesis from triacylglycerols. Sugar interconversions occur in the living tissues. For example, sucrose is readily formed in the embryo when this is supplied with glucose or maltose. The multiple origins and fates of the soluble sugars and the practical difficulties encountered in analysing such complex mixtures make it difficult to appreciate all the factors that contribute to the changes in their levels. During malting, the quantities and proportions of the sugars change dramatically (Tables 4.5 and 4.6). The details of the changes depend on the grain sample and the malting conditions used. Maltose (**18**) and maltotriose (**19**) appear in significant amounts, being formed by the partial hydrolysis of starch. Initially raffinose and sucrose decline, chiefly in the embryo. Later sucrose is reformed in the living tissues, either from other sugars or from triacylglycerols. Raffinose disappears during the early stages of malting and does not reappear.

As the smaller molecules are diffusible, they provide the means by which carbohydrate material from the polysaccharides of the starchy endosperm, which are progressively degraded, moves to the living tissues. Here they are utilized either to form new tissues in the roots and acrospire or to be respired. Most oligosaccharides and polysaccharides (the fructosans are an exception) are formed by the transfer of sugar residues from nucleotide derivatives. The common glucose donors are UPDG (**20**) and ADPG (**21**). Considerable quantities of sugars are utilized to support fermentation (during steeping especially) and respiration during malting.

The changes in the levels of sugars that occur during kilning are very dependent on the conditions being employed. When pale, white malts are being made, using low kilning temperatures, there may be a rise in simpler

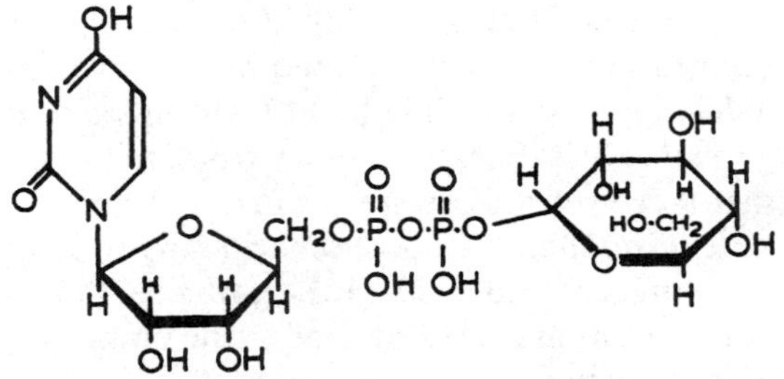

(20) Uridine diphosphate glucose; UDPG

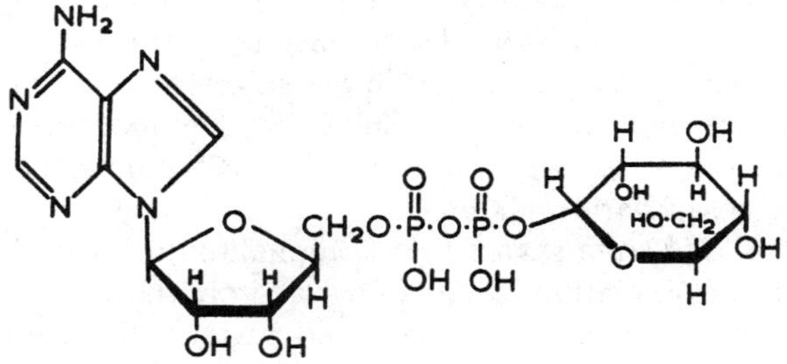

(21) Adenosine diphosphate glucose; ADPG

sugars (Table 4.5). When darker malts are being prepared, the high temperatures used may lead to a decline in some sugars, which interact with nitrogenous substances to form melanoidins. The sugars present in malt are the main contributors to the pre-formed soluble materials or cold water extract. The cold water extract is often in the range 18–21% of the dry matter of the malts.

4.5 Starch and its breakdown

Starch, the major component of the barley grain (commonly 58–65% in malting barleys but values as wide as 51.5–72.1% have been noted), and other cereal grains, provides the major part of the extract obtained when malt is mashed (Chapters 1 and 5), indirectly as the soluble sugars formed from it during malting (the cold water extract) and, more importantly, as the sugars, oligosaccharides and dextrins that arise from its hydrolysis during mashing (Harris, 1962a,b; Pollock, 1962a; Briggs, 1978; Henry, 1988). Starch granules from different cereals differ markedly in their appearances (Kent, 1983) and in their properties. These differences are significant when malts of different grains, or other grain products, are used in mashing (Chapters 1 and 5). The polysaccharides in the starch granules are deposited in an orderly, semicrystalline state. This is shown by the dark 'Maltese cross', which is apparent when granules are viewed in a polarizing microscope, and the layered structure arranged around a dark spot, the hilum. Sometimes, in dry granules or those damaged in malting, cracks appear radiating out from the hilum. The layered structure is particularly apparent after limited attack by amylases, when it is seen that some layers are rapidly degraded while others are more resistant.

The orderly structure of starch granules breaks up when they are held in hot water. At first they swell, some components leach out and then they gelatinize – that is the granules rupture and the contents disperse. On cooling, a thick dispersion may set to a gel. Granules may be battered and chipped in milling, making them more vulnerable to enzymolysis. Isolated starch granules have small amounts of various substances associated with the polysaccharide, including protein and some lipid (Morrison, 1978, 1988; Galliard, 1982, 1987). It is possible that a coat of adherent protein retards the degradation of starch granules. The lipids are divisible into the surface components, including triacylglycerols, and internal components. These internal components are polar; typically they are chiefly lysophospholipids, particularly lysophosphatidylcholine in wheat and barley, but free fatty acids occur in maize. They are associated with the amylose component of the polysaccharide and have a major influence on grain structure.

Barley starch grains occur in a range of sizes, with major groups of small (1.7–2.5 μm) and large (22.5–47.5 μm) grains. There are variations between varieties (Briggs, 1978). In one sample, the small granules made up about 90% of the granules by number but only about 10% by weight (Bathgate and Palmer, 1972). Small granules are more readily degraded than the large during malting but less readily degraded during mashing. The granules have different gelatinization temperature ranges (small, 51–92 °C (*ca.* 124–198 °F); large, 47–61 °C (*ca.* 117–142 °F)). The smaller granules are widely believed to have a higher amylose content than the large (Bathgate and Palmer 1972). However, amylose to amylopectin ratios have usually been determined by measurements of iodine-binding capacity and, as usually performed, this technique is suspect. When other techniques are used, this difference is not always found. Data on amylose/amylopectin ratios need to be checked by different techniques (Kano, 1977; Sargeant, 1982; Morrison, 1988).

Starch is a mixture of two polysaccharides, amylose and amylopectin, each of which is a mixture of chains of various lengths, all of high molecular weight, of poly-(α-D-glucopyranose). Fractions can be isolated that have intermediate properties between 'pure' amylose and 'pure' amylopectin, so starch contains a complex range of molecular types. The starches of most British barleys contain 18–26% amylose, the balance being amylopectin, but varieties are known, carrying the 'waxy' gene, in which the starches contain 0–3% amylose; in other types, the amylose content may reach 47% of the starch. So far no advantages have been found for using malts from barleys containing unusual proportions of amylose.

Amylose chains contain, on average, about 2000 glucose residues. However they do vary in length. It was thought that amylose chains consist exclusively of linear chains of poly-[(1→4)-α-D-glucopyranose]. However refined studies of cereal starches show that sub-fractions contain various small numbers of α(1→6)-links, which prevent the total hydrolytic degradation of the amylose sub-fraction by the enzymes β-amylase and phosphorylase; this defines the 'β-amylolysis limits' of these preparations. Amylose associates strongly with many substances, including polar lipids and iodine. The polysaccharide adopts a helical structure and the associated substance comes to occupy the central space of the helix. Amylose gives an intense blue-black colour with iodine. Many substances that associate with it will precipitate it from solution. The existence of complexes between amylose and the endogenous polar lipids of cereal starch granules may help to explain the ordered structure of the granules and their resistance to amylolysis. Residual materials from the amyloplast and associated protein, which surround and adhere to starch granules, may also slow amylolysis (Table 4.7). In the grain, amylases appear unable to attack starch granules until the investing endosperm cell walls have been substantially degraded. In barley, an enzyme that hydrolyses lysophospholipids accumulates during germination and may well degrade the lysophospholipid which occurs in starch associated with the amylose. In so doing, it may render the amylose more susceptible to amylolysis.

Amylopectin consists of molecules of different sizes, with very high molecular weights, estimated to be 10^6–10^8. It differs from amylose in that its α(1→4)-linked chains of glucose residues are extensively branched through α(1→6)-links so that these molecules have very highly branched structures. Thus, while a chain of amylose has one reducing glucose residue (C-1, unsubstituted) and another non-reducing glucose residue at the other end, a molecule of amylopectin contains one reducing glucose residue but very large numbers of non-reducing glucose residues that terminate its branches (Fig. 4.1). Chemical measurements of the ratio of non-reducing terminal glucose residues to the total numbers present, and separation of the shorter chains formed when amylopectin is debranched by selective, enzyme-

Table 4.7 Analyses of starches from Carlsberg barley and malt

	Barley	Malt
Moisture (%)	11.2	14.7
'Protein' content (N × 6.25%)	0.51	2.05
Ratio of glucose residues (non-terminal : terminal)	25 : 1	22 : 1
'Blue value' with iodine	0.366	0.409
Amylose content (%)	26.2	29.2
Ratio of non-terminal : terminal glucose residues in amylopectin	18–19 : 1	15–16 : 1
Amylose (g/1000 corns)	6.72	6.37

From Harris (1962a).

catalysed hydrolysis of the α(1→6)-links indicate the complex structure of amylopectin. Amylopectin gives a characteristic reddish tint when mixed with iodine solution. When the 'blue value', the colour dispersed starch gives with iodine, is compared with the values given by purified amylose and amylopectin, it allows a value of the ratio of these substances in the starch to be calculated (Tables 4.5, 4.6 and 4.7). However, unless all the lipids have been extracted from the starch before the measurements have been made, and other technical problems are overcome, the results can be uncertain. If starch is debranched (the α(1→6)-links are specifically hydrolysed) with the enzyme iso-amylase the straight chains that result can be separated. While amylose chains contain more than 135 glucose residues, the separated chains from debranched amylopectin contain two groups, one containing about 18 glucose residues, the other about 60 (Sargeant, 1982).

Barley starch is partly degraded (e.g. 15–18%) during malting, and the

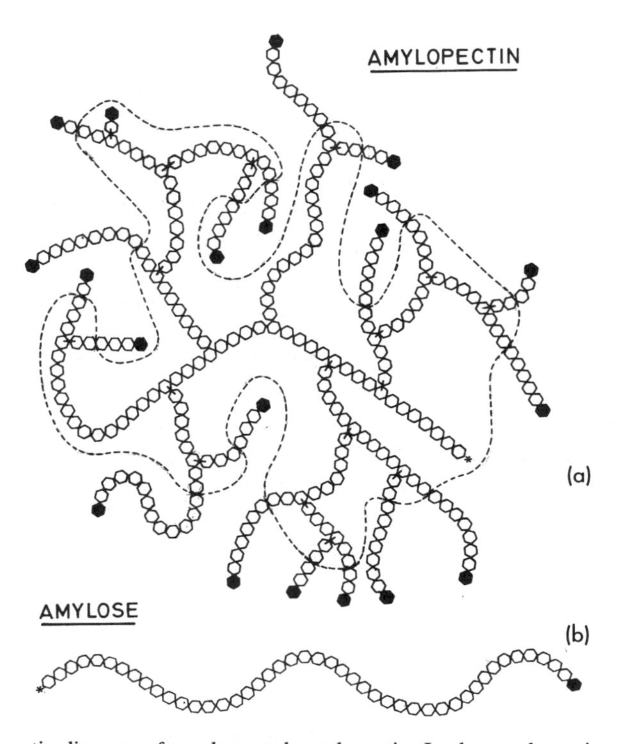

Fig. 4.1 Schematic diagram of amylose and amylopectin. In the amylopectin component of starch, the chains of α(1→4)-linked D-glucopyranose units (○) are joined via α(1→6)-branch-points (○-○). The glucose residues at the non-reducing chain ends are shown as filled in units, while the reducing chain ends marked with asterisks. The dashed line indicates the approximate margins of the β-limit dextrin remaining after the molecule has been partially degraded by pure β-amylase. Amylose, the straight-chain component of starch, has only a single reducing end and a single non-reducing end to each molecule.

composition of the residual starch is usually reported to alter (Tables 4.5 and 4.6). However, others have not detected any changes in the fine structure of the residual starch (Kano *et al.*, 1981). The extent of the changes differs a little on different occasions. The proportion of amylose in the residual starch increases (e.g. from 22% (barley) to 26% (malt)), the total amylose content of the grain declines and the average chain length of the amylopectin is reduced (Table 4.7). Therefore, most evidence points to amylopectin being more readily degraded than amylose. Malt starch is more readily degraded than barley starch by a standardized mixture of enzymes. Small amounts of 'transitory starch' are formed in the embryo while relatively massive amounts of starch degradation are occurring in the starchy endosperm.

4.5.1 *The enzymes that degrade starch*

Starch granules are not accessible to degradative enzymes until the investing endosperm cell walls are degraded. Degradation of the polysaccharide may then be facilitated by removal of investing protein and residual amyloplast materials and hydrolysis of internal lysophospholipids. Starch hydrolysis is also facilitated by granule damage, as is caused in milling, or by swelling or dispersion (gelatinization), as occurs during mashing (Chapters 1 and 5). The enzymes discussed here are those that occur in germinating barley. Those that occur in other cereals are generally similar in their properties, but the amounts which occur in the final malts may be very different.

The mixture of cereal enzymes that partially or completely degrades starch is called diastase. This mixture is not constant in its composition. Diastase from ungerminated barley grain is very different to that from barley green malt, which, in turn, is changed by kilning. Diastase has mostly been studied in connection with its activity during mashing (Chapter 5), when the temperatures used are high, e.g. 65 °C (149 °F), and most of the characteristics of the starch degradation process can be explained in terms of the catalytic activities of two enzymes: α-amylase and β-amylase. However, in the germinating grain *in vivo*, a considerable number of other enzymes are involved (Harris, 1962b; Pollock, 1962a; Briggs, 1973, 1978, 1992, 1995; Maeda *et al.*, 1979; Hill and MacGregor, 1988; Sun and Henson, 1996). Probably two systems of starch degradation occur at different locations in the grain. The microscopic etching, pitting and separation into layers (laminae) that occurs during starch degradation in the starchy endosperm differs from the steady reduction in the diameters of the transitory starch grains that occurs during starch utilization within the living tissue of the embryo (Brown and Morris, 1890).

Phosphorylase. Phosphorylase is an enzyme that catalyses the reversible phosphorolysis (with inorganic phosphate ions) of terminal, non-reducing glucose residues of starch, releasing glucose 1-phosphate (Fig. 4.2). The quantitative changes that occur in this enzyme during malting have not been adequately studied. Its activity is hard to measure because of the interfering effect of other enzymes that are present. Phosphorylase occurs in the embryo, where it may attack transitory starch. The glucose 1-phosphate liberated is then able to enter the metabolic pathways immediately. Barley contains phosphatases, which are able to hydrolyse glucose 1-phosphate to glucose and inorganic phosphate. Thus phosphorylase and phosphatase, acting in concert with inorganic phosphate, are able to sequentially degrade starch chains with the liberation of glucose. Phosphorylase is not able to hydrolyse α(1–6)-linkages in starch or dextrins, or to remove substituted glucose residues; therefore, acting alone, it can only degrade amylose and the outer chains of amylopectin to sites near to the branch points, leaving a product almost identical to β-limit dextrin (Fig. 4.1).

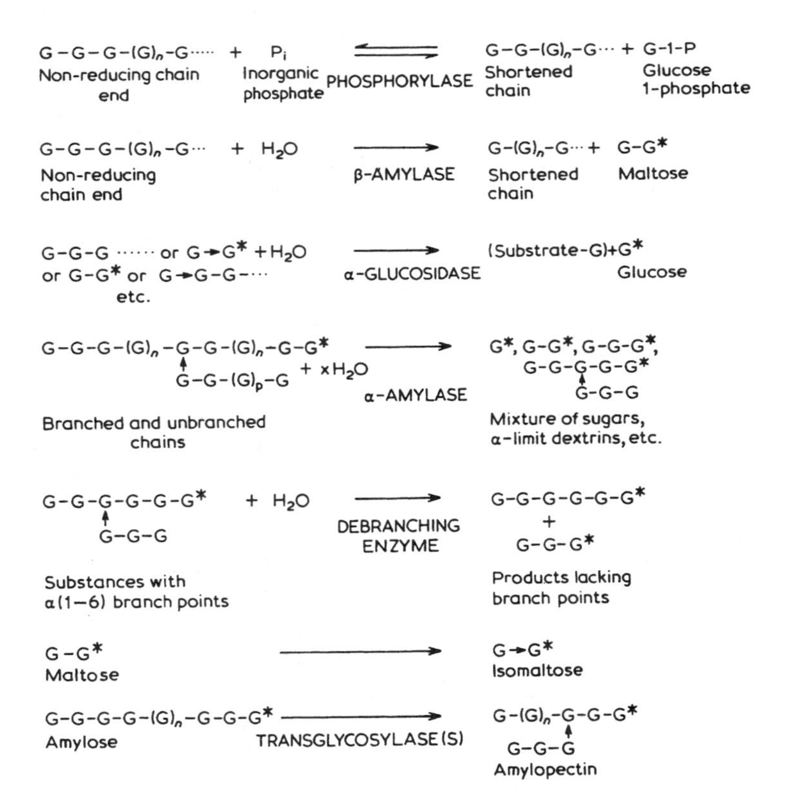

Fig. 4.2 Summary of the modes of action of some enzymes important to the degradation of starch. G, α-D-glucopyranose; –, (1→4)-link; →, (1→6)-link; *, a potential reducing group.

α-Glucosidase. α-Glucosidase (maltase) has been regarded as one enzyme that is highly concentrated in the embryo and aleurone layer. It increases in amount during germination. It has now been proposed that two or more enzymes, of different specificities, are present. A substantial proportion of the enzyme is insoluble, which makes quantification difficult. The enzyme, which has a pH optimum of about 4.6, hydrolyses either α(1→4)- or α(1→6)-glucosidic linkages at non-reducing chain ends. Thus the enzyme can hydrolyse maltose (**18**) (hence it sometimes is called maltase) and iso-maltose (**22**), in each case giving two glucose molecules. It can also remove terminal, non-reducing glucose residues from starch, dextrins or smaller oligosaccharides (Fig. 4.2). This enzyme, which has several isozymes, is probably the 'raw starch factor' that accelerates the degradation of starch granules by mixtures of α-amylase and β-amylase (Maeda *et al.*, 1979). Like α-amylase, α-glucosidase is able to attack granular starch (Sun and Henson, 1996). It is unstable in solution at temperatures of 40 °C (104 °F) and above, but it is reported largely to survive kilning. α-Glucosidase has a limited ability to catalyse transglucosidation reactions. Other transglucosidases may occur in barley. Such enzymes are able to transfer individual or groups of sugar residues from one compound to another, with the formation of a similar or a different type of linkage. Thus, an α(1→4)-link in a chain may be broken and the separated end may be joined to the same, or a different chain via an α(1→4)-link or an α(1→6)-link (Fig. 4.2). Such enzymes are responsible for the formation of the α(1→6)-branch points in amylopectin during grain development.

β-Amylase. In the absence of other enzymes β-amylase is unable to degrade starch granules. However, it is able to carry out a stepwise attack on amylose, dextrins and soluble starch chains in solution. It is an exo-enzyme and catalyses the hydrolysis of the α(1→4)-linkages penultimate to the non-reducing chain ends, releasing the disaccharide maltose (**18**) and an oligosaccharide shortened by the removal of two glucose residues. This enzyme will not attack α(1→6)-glucosidic links or α(1→4)-links immediately adjacent to them. Thus, while β-amylase, acting alone, can degrade amylose completely to maltose, it degrades the outer chains of amylopectin to maltose but the residue remains as a β-limit dextrin in which the outer

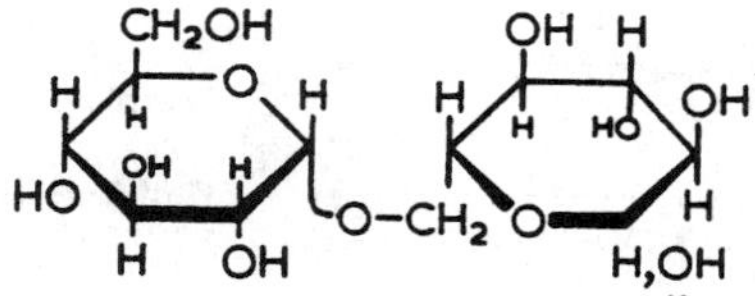

(22) *iso*-Maltose;
α-D-Glucopyranosyl-(1→6)-D-glucopyranose

chains have been degraded to stubs of probably two or three glucose residues, outside the α(1→6)-branch points (Fig. 4.1). β-Amylase has a broad pH optimum, around 5.0–5.3, and it is resistant to brief exposures to acid pH, e.g. 3.0–3.5. It is rapidly inactivated in solution at 70 °C (158 °F) and is inhibited by heavy metal ions, agents that block thiol groups and by polyphenols. In the barley grain, virtually all the enzyme occurs in the starchy endosperm. It is concentrated in the sub-aleurone layer. In barley little or no β-amylase is synthesized during germination, but in rice more is formed in the embryo. The enzyme occurs in soluble ('free') and insoluble forms. In different samples of barley, between 0.3 and 0.7 occurs in the bound, or latent, form in the quiescent grain. The latent (bound) β-amylase is attached to insoluble proteins via disulphide linkages. It can be liberated, or converted, into free soluble enzyme by limited proteolysis by chemical agents that can break the disulphide linkages and possibly by disruption of hydrophobic links (Grime and Briggs, 1996; E. Buttimer and D. E. Briggs, unpublished data). The soluble or salt-soluble enzyme is heterogeneous and, in addition to the occurrence of isoforms, some part exists as purely enzyme protein that may polymerize in solution, while another part is linked via disulphide bonds to an enzymically inactive protein, protein Z. This protein is of interest in that some of it survives malting and brewing to appear as the major antigenic protein in beer (Hejgaard, 1977, 1978). During malting, the level of free β-amylase may initially fall during steeping but subsequently, during germination, nearly all the β-amylase becomes free, and the bound form disappears (Harris, 1962b; Bettner and Meredith, 1970). The 'free' enzyme of malt occurs as several isozymes (which vary with grain variety) and is present as several slightly smaller forms than the 'parental' enzyme, as a result of limited proteolysis. Confusingly, some of the 'latent enzyme', when released by chemical treatments, is found to be proteolysed (Grime and Briggs, 1995, 1996).

α-Amylase. In contrast to β-amylase, α-amylase is virtually absent from mature barley unless it has pre-germinated. The considerable quantities that are present in malt are synthesized *de novo* in the embryo and aleurone layer, and large proportions are secreted into the starchy endosperm. α-Amylase is an endo-acting enzyme that catalyses the hydrolysis of α-glucosido-(1→4)-glucose linkages at random within starch chains, although attack is slower on short-chain dextrins, is slower near the chain ends and does not occur in the immediate vicinity of α(1→6)-branch points. Therefore, this enzyme acting on its own is able to degrade starch to a complex mixture of sugars (including glucose (**1**) and maltose (**18**)), oligosaccharides and dextrins, some of which contain one or more α(1→6)-branches. On its own and in contrast to β-amylase, α-amylase attacks starch granules and progressively degrades them. This degradation occurs more rapidly if β-amylase is present, and faster still if α-glucosidase is added to the mixture

(Maeda *et al.*, 1979). Whenever a starch chain is broken by α-amylase, a new non-reducing chain end is formed that may be attacked by β-amylase or other exo-acting enzymes (e.g. α-glucosidase, phosphorylase). Since α-amylase can cleave amylopectin chains either side of α(1→6)-branch points, it can 'by-pass' branches and β-amylase can then degrade these chains more fully. Thus, acting together, these enzymes degrade starch more rapidly and completely than they are able to do alone. In mashing, at about 65 °C (149 °F), starch granules progressively swell and gelatinize as their structure is disrupted. As this occurs, they are more readily attacked by α-amylase. In the presence of calcium ions and at pH 6.8, α-amylase is able to survive heating for at least 15 min at 70 °C (158 °F) provided it is present in a crude, unpurified grain extract. The pH optimum is 5.3, and below about pH 4.9 the enzyme becomes unstable. Agents that attack thiol groups have little influence on this enzyme. Electrophoresis of extracts of germinating barley reveal the presence of three groups of α-amylase bands having different properties (Briggs, 1992). Different varieties have different banding patterns. The first group of isozymes, α-amylase I, is only present in minor amounts in malt. These enzymes are inhibited by copper ions and seem less stable than the others. Of the two remaining sets of bands in green-malt extracts, the major (α-amylase II) results from a set of isozymes and the minor set (α-amylase III) results from these same isozymes reversibly associated with a protein, BASI, that reduces their catalytic activity (and, incidentally, inhibits the bacterial proteolytic enzyme subtilisin). The amount of this inhibiting protein declines during malting, and it is denatured when extracts are heated at 70 °C. A considerable excess of this protein is needed seriously to reduce the rate at which granular starch is degraded by α-amylase (Weselake *et al.*, 1985).

Because it rapidly generates reducing sugars (maltose) while only slowly reducing the iodine-staining capacity of soluble starch, β-amylase is often called the saccharogenic amylase. In contrast, α-amylase rapidly reduces the viscosity of starch pastes and destroys their ability to give colour with iodine, while only relatively slowly increasing the reducing power of mixtures, hence it is sometimes called the liquefying or dextrinizing amylase.

Debranching enzyme. Another important enzyme catalyses the hydrolysis of α(1→6)-linkages in amylopectin, dextrins and oligosaccharides. This is known as debranching enzyme, or limit dextrinase (Serre and Laurière, 1989). Previously the ability to hydrolyse these links in amylopectin was attributed to 'R-enzyme' or debranching enzyme and the links in dextrins were thought to be broken by limit dextrinase. These terms are still used although it is now known that both activities are catalysed by a single enzyme. Sometimes, because of its ability to degrade a particular bacterial polysaccharide in a particular way, debranching enzyme is also called pullulanase. By breaking branch points by hydrolysing the α(1→6)-interchain links, debranching enzyme facilitates the breakdown of starch by the other enzymes. This enzyme has a pH optimum of 5.0–5.5, but its characteristics are unclear. Reportedly, in solution it is unstable at temperatures over 30 °C (86 °F) and very unstable over 50 °C (122 °F). The enzyme occurs as an inactive zymogen in the endosperm of barley and various other cereals (Maeda *et al.*, 1979). Much of this zymogen may remain in the inactive state at the end of malting (Longstaff and Bryce, 1993). In malt, the enzyme appears in both freely soluble and bound forms. The bound form is solubilized with cysteine. A further complication is the existence of a heat-stable protein in barley that inhibits malt limit dextrinase (MacGregor *et al.*, 1995). The inhibitor increases in amount during malting. However, there is also evidence that some enzyme is synthesized *de novo* in the aleurone layer during germination.

Enzyme stability. During kilning diastatic enzymes are inactivated to extents that vary with the kilning conditions. It seems that α-glucosidase is destroyed more readily than β-amylase or limit dextrinase, while α-amylase is the least readily degraded. Thus the levels and proportions of the diastatic enzyme found in the final malt are influenced by the type of barley that has been malted, the way malting was carried out and the final kilning conditions. Malts intended to have maximal diastatic power (DP), for instance those used in grain distilling, are kilned at the lowest possible temperatures or, in some few instances, green, unkilned malt may be used.

The estimation of the sum total of diastatic enzymes in malt (DP) is usually carried out by measuring the increase in concentration of carbohydrate reducing groups when grain extracts containing the enzymes are incubated with soluble starch. By modern criteria, this venerable measurement method, in its various forms, lacks rigour and gives variable results (Chapter 13). The specific estimation of the individual enzymes that together make up diastase is technically difficult. The commercially used assay for α-amylase, the 'international method' based on the work of Sandstedt, Kneen and Blish (1939), relies on following the degradation of β-limit dextrin in the presence of an excess of pure β-amylase. The supposition is that only α-amylase and β-amylase occur in grain extracts. However such a method determines debranching enzyme and α-glucosidase as α-amylase so that, although it is operationally useful, it is not specific (Briggs, 1973).

4.6 Non-starch polysaccharides

Non-starchy polysaccharides include the hemicelluloses and gums (Briggs, 1978, 1992; Henry, 1988; Fincher, 1992). True cellulose, pure poly-[β(1→4)-D-glucosan], has not been demonstrated in barley, but holocellulose is present and makes up approximately 5% of the grain's weight. Holocellulose is a crude polysaccharide fraction, insoluble in dilute sodium

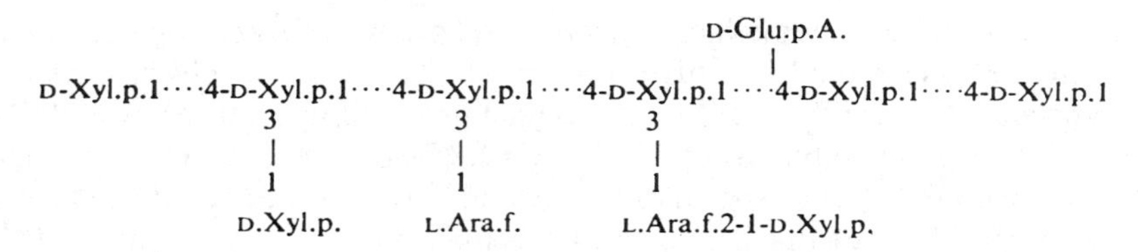

Fig. 4.3 Pentosan from barley husk hemicellulose. D-Xyl.p, β-D-xylopyranose residue; L.Ara.f., α-L-arabinofuranose residue; D-Glu.p.A., D-glucuronic acid (pyranose form) residue. The residues joined by dotted lines may be linked either directly or through a chain of (1→4)-linked D-xylopyranose residues, which probably have the β-configuration. (From Harris, 1962a.)

hydroxide solution, which contains glucose (**1**), mannose (**6**) and small amounts of galactose (**5**). It occurs chiefly in the husk where it is associated with lignin and probably occurs in the pericarp, but small quantities occur in the cell walls of the other tissues. Levels of holocellulose do not alter appreciably during malting. Traces of glucuronic (**7**) and galacturonic (**8**) acids occur, particularly in the husk and embryo, indicating that small quantities of pectins may be present in these tissues.

Hemicelluloses and gums make up 10–11% of barley grains. They occur in the cell walls of all the tissues. They have similar chemical compositions, but gums are extracted from ground grain by hot water. The distinction is arbitrary, since raising the extraction temperature, e.g. from 40 to 80 °C (104 to 176 °F) extracts progressively more gums, leaving less hemicellulose in the residue. The hemicellulose can be dissolved in dilute sodium hydroxide and some other reagents. As usually prepared, hemicelluloses make up about 8% of the grain and gums about 2%. The gums appear to grade imperceptibly into the hemicelluloses, which probably differ from the former by having higher molecular weights, closer associations with insoluble protein and possibly appreciable degrees of cross-linking and acylation of the carbohydrate material. The gums and hemicelluloses of the cell walls of the starchy endosperm must be substantially degraded during malting. They serve as a polysaccharide reserve for the grain (Morrall and Briggs, 1978).

The major components of the gums and hemicelluloses are (i) pentosans, made up chiefly of arabinose (**3**) and xylose (**4**); and (ii) β-glucans, made up of glucose (**1**). However, polysaccharides containing galactose (**5**), mannose (**6**) and uronic acids (**7**,**8**) are also present in minor amounts. Their chemical structures have yet to be defined.

Pentosans. Pentosans may have large molecular weights, *ca.* 10^6. During malting, the pentosan content may alter little or may increase (e.g. from 5.7% to 6.4%: Henry, 1985). In part this may result from the formation of new pentosans in the growing acrospire. Estimates of total pentosan levels in barleys range from 6.7% to 11.0% (Lehtonen and Aikasalo, 1987).

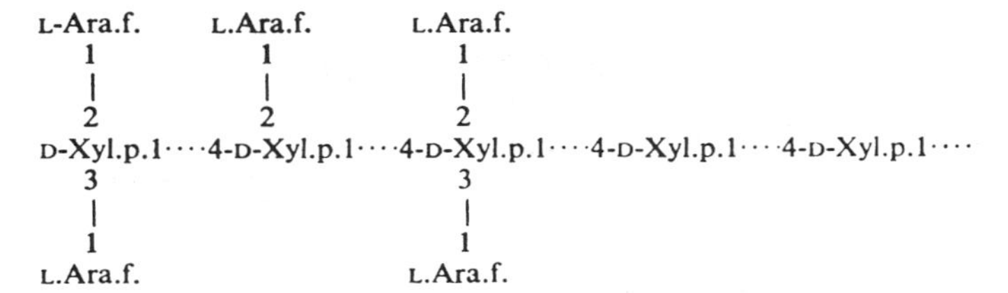

Fig. 4.4 Water-soluble pentosan gum from the barley endosperm. The symbols are as shown for Fig. 4.3. It is now known that the arabinose moieties are often substituted with ferulic acid residues joined via ester links, and the gum may also be acetylated. (From Harris, 1962a.)

Pentosans from different parts of the grain vary in their compositions. A material from the husk consisted of a chain of β(1→4)-D-xylopyranose residues (**4**) variously substituted with α-L-arabinofuranose (**3**) and D-glucuronic acid (**7**) residues (Fig. 4.3). Galacturonic acid (**8**) is also present in the husk. Pentosan gums from the starchy endosperm are simpler in structure in that the xylan chain is only substituted with α-L-arabinofuranose units (Fig. 4.4). The arabinose:xylose ratio varies widely in different pentosan preparations, e.g. from 1:1 to very low values, and there is the possibility that pure arabans and xylans are present. The realization that the cell walls of the starchy endosperm fluoresce weakly and those of the aleurone layer fluoresced strongly in ultraviolet light led to the discovery that at least some pentosans are heavily substituted with ferulic acid residues (See Fig. 4.20, below), joined to arabinose units by ester links (Ahluwalia and Fry, 1986). In addition they may be heavily acetylated (Bacic and Stone, 1981). So far there is little evidence for covalent cross-links between chains. These substituents must have major effects on the solubilities of the pentosans and their susceptibility to enzymic attack. Because of their low viscosities, pentosan gums from barley cause few problems in brewing, and so they have attracted relatively little interest. In contrast, worts from rye and triticale malts, or mashes made with unmalted wheat, rye or triticale, may be extremely viscous because of their pentosan contents (Blanchflower and Briggs, 1991c; Chapter 14). However beer hazes with high pentosan contents have been encountered from time to time. An old suggestion was that the progress of modification in malting could be followed by measuring the increasing quantities of soluble pentosan gums. This proposition was found to be unsound.

The β-glucans. The β-glucans (synonym β-glucosans) are a family of polysaccharides made up of unbranched chains of β-D-glucopyranose residues. The great majority contains mixtures of β(1→4)- and β(1→3)-linkages (Fig. 4.5). However there is evidence for a laminarin-type of β-glucan containing exclusively β(1→3)-linkages, which tends to be concentrated at the interface between the starchy endosperm and the aleurone layer, but bead-like

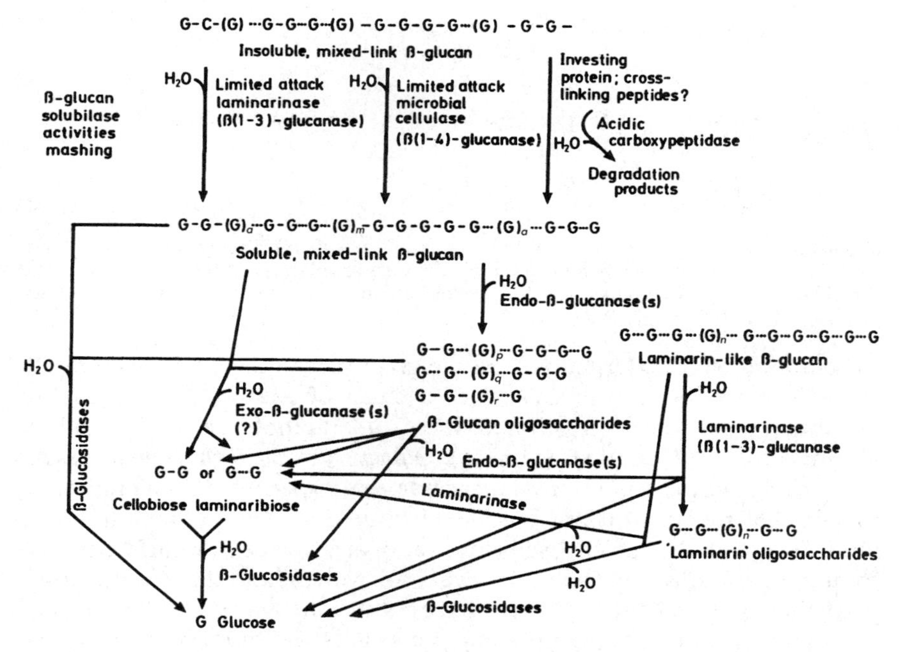

Fig. 4.5 The proposed routes by which mixed-link and laminarin-like barley β-glucans are degraded. G, β-D-glucopyranose residue; –, β(1→4)-link; ···, β(1→3)-link.

deposits occur throughout the starchy endosperm (Fulcher *et al.*, 1977; MacGregor *et al.*, 1989). In the mixed link β(1→3; 1→4)-glucans, the ratio between the bond types varies (e.g. values of 6:1 or 7:3 have been reported) and the chain lengths (molecular weights) also vary. Molecular weight distributions may peak at about 10^5 and 4×10^3 (Foldager and Jørgensen, 1984). Usually single, or a small series of (1→4)-linkages alternate with single (1–3)-linkages, but sometimes several adjacent (1→3)-linkages may occur (Thompson and Le Berge, 1977). It has been estimated that 90% of water-soluble β-glucan consists of cellotriosyl and cellotetraosyl units joined by β(1→3-links. Probably β-glucan molecules adopt a sinuous, worm-like configuration in solution (Woodward *et al.*, 1983). The β-glucans are the largest components of endosperm cell walls. They must be extensively degraded during malting for modification to be adequate (Bourne, *et al.*, 1982; Bourne and Wheeler, 1982, 1984; Anderson, *et al.*, 1989). Specific analyses of β-glucans in the presence of many other carbohydrates and especially starch have not been easy to achieve. The preferred methods of analysis are based on either measuring the fluorescence of the adduct between β-glucan and the dye Calcofluor or measuring the products of β-glucan degradation (glucose or oligosaccharides) specifically produced by carefully purified hydrolytic enzymes. Measurements of the viscosities of extracts are not acceptable for quantifying the amounts of β-glucan present,

Table 4.8 Changes in the β-glucan content and other malt analyses during the malting of Triumph barley, steeped with air-rests and germinated at 15 °C (59 °F)

	Germination time (days)				
	Barley	O (Steep-out)	2	4	6
Total β-glucan (%)	3.90	2.92	1.55	0.87	0.58
Moisture (%)	13.2	47.5	44.3	41.8	39.5
Malt modification (%)	–	19	70	93	99
Homogeneity (%)	–	50	59	63	94
Extract (f.g., %)	–	63.3	75.3	75.6	75.1
Wort strength (°Plato)	5.59	7.90	8.54	8.56	8.51
Wort viscosity (cP)	2.96	2.58	1.75	1.69	1.69
Wort β-glucan (mg/l)	–	2044	284	37	17

Total β-glucan was determined by an enzymatic method. Modification and homogeneity were determined by a sanded grain/Calcofluor staining method. Soluble β-glucan, in the wort, was determined after precipitation with ammonium sulphate. Other analyses from EBC (1987). From Foldager and Jørgensen (1984).

although they are useful for other purposes. Some estimates of β-glucans in cereals (%) are: wheat 0.5–1.5, triticale 0.3–1.2, rye 1.9–2.2, barley 3–6, oats 2.5–6.6, rice 0.6, maize 0.1 and sorghum 1.0 (Stuart *et al.*, 1987). Other estimates of β-glucan in barley are lower and yet others are higher. The content of β-glucan in barley varies with the variety and the growing season. Within a variety, there is no correlation between grain samples' total nitrogen contents (TN) and β-glucan contents (Alexander and Fish, 1984). This polysaccharide seems to be largely or entirely confined to the starchy endosperm. Hemicellulosic β-glucan may be dissolved with dilute sodium hydroxide, or in warm water following hydroxylaminolysis or treatment with selected proteolytic enzymes such as a barley acid carboxypeptidase. As no evidence for covalent cross-linking or any substitution by ferulic acid has been found, it seems likely that the less-soluble β-glucans are physically closely associated with insoluble proteins and, perhaps, insoluble pentosans and holocellulose and the treatments are responsible for disrupting these associations. Estimates of the molecular weights of β-glucans vary widely and depend, in part, on the methods of extraction employed.

During malting, the total amount of β-glucan present declines. Only a very small amount should remain in well-modified malts (Table 4.8). High-molecular-weight β-glucans are extracted when under-modified malts are mashed. As malting proceeds, the β-glucans of the endosperm cell walls are initially solubilized, and the levels of β-glucan gums increase, but ultimately they are degraded. The β-glucans in the endosperm are degraded more than the pentosans, so that while gums from raw barley contain 14–20% pentosan (really furfural-yielding materials) the gums from malt contain 70–90%. Undegraded β-glucans form highly viscous solutions that may slow the separation of wort at the end of mashing and which certainly slow beer filtration, clogging the filters and increasing the use-rate of filter aids.

Furthermore, they are sometimes associated with hazes and may even form gelatinous precipitates in strong beers. The β-glucans may conceivably contribute to beer foam and palate fullness, but they should be adequately degraded during malting or mashing. The survival of substantial amounts of β-glucan in malt indicates inadequate degradation of the endosperm cell walls, which, in turn, indicates that the cell contents, such as the proteins, have not been modified. Malts usually contain 0.2–1% β-glucan. Therefore, the determination of β-glucan in malt is an excellent indication of the extent of modification (Anderson *et al.*, 1989). The decreased viscosity of 'Congress'

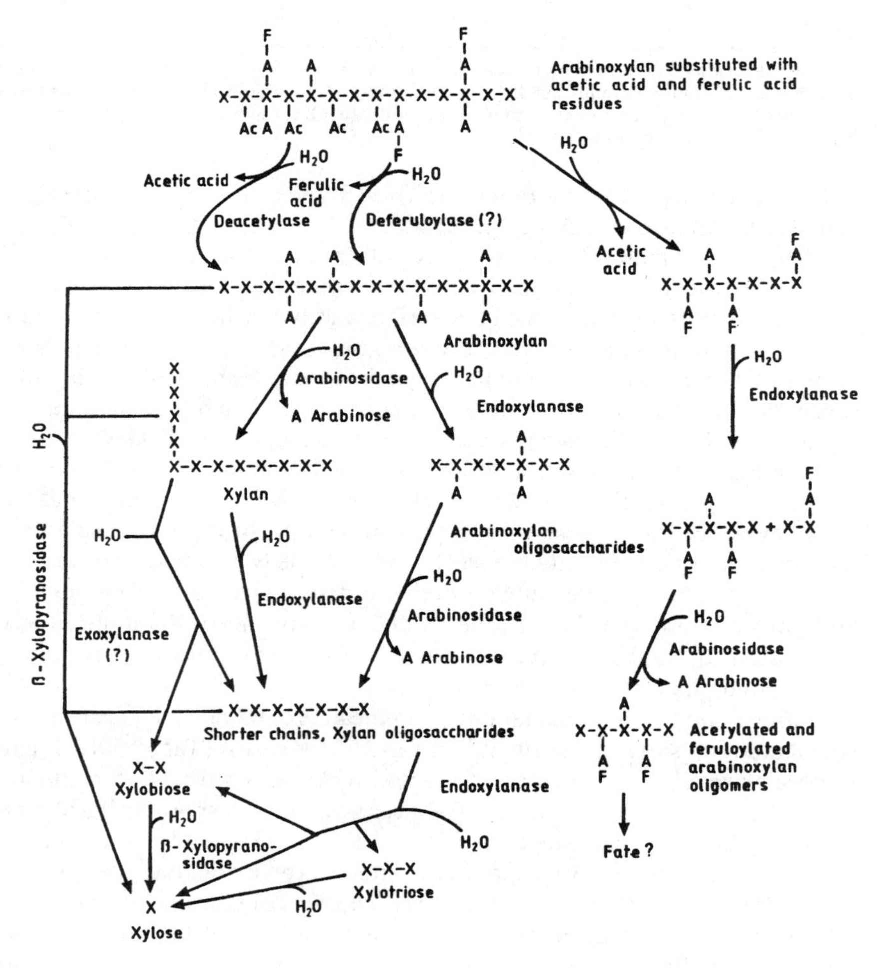

Fig. 4.6 Possible routes by which enzymes degrade arabinoxylan gums and hemicelluloses. Some of the proposed steps are hypothetical. X, β-D-xylopyranose residue; A, α-L-arabinoufranose residue; Ac, acetic acid substituent; F, ferulic acid substituent.

extracts from grain at different successive stages of malting has been a useful guide to the progress of modification in preparing traditional, under-modified European lager and North American malts. This index proved to be unreliable when long-germinated, well-modified, traditional British ale malts were being made, because viscosity stopped falling before the desired degree of modification was achieved (Chapter 13). With more quickly grown, modern malts, the viscosity of hot water extract worts is a useful guide to modification. 'Inhomogenous' malt, which contains a small proportion of unmodified corns, can yield a viscous extract. This may readily be detected using the wort from a small-scale 70 °C (158 °F) thick mash supplemented with diastase (pre-heated to destroy β-glucanase) to ensure starch degradation (Bourne *et al.*, 1977, 1982; Briggs, 1987b,c; Anderson *et al.*, 1989).

4.6.1 Breakdown of non-starch polysaccharides

The net alterations in the non-starchy polysaccharides that occur during malting are the result of their enzyme-catalysed hydrolytic degradation in the endosperm, and the synthesis of new materials in the growing embryo. Some of the latter are separated from the malt when the rootlets are removed. In some malting processes, maximum extract yields are obtained before adequate β-glucan breakdown has occurred, and hence endosperm modification is not complete. In these cases, it is desirable that germination is prolonged or the process is otherwise altered so that adequate modification is achieved.

Enzyme activity initially increases the solubility of the hemicelluloses, converting them to water-soluble gums, which, in turn, are degraded to oligosaccharides and sugars. Our knowledge of the breakdown of pentosans is incomplete. Barley enzymes capable of hydrolysing ferulic acid residues from pentosans have been detected. Enzymes catalysing the hydrolysis of acetyl groups from esters occur in barley, but it is not known if they deacetylate pentosans. It is also not known for certain if these pentosan chains are cross-linked and, if so, how the links may be broken. Various enzymes that degrade grain pentosans have been recognized (Fig. 4.6) (Harris, 1962b; Thompson and La Berge, 1977; Briggs, 1978). Arabinosidase catalyses the hydrolytic separation of α-L-arabinofuranose residues from arabinoxylan chains, or oligosaccharides. When sufficient substituents have been removed, the xylan chain is hydrolysed by endoxylanases into xylan, arabinoxylan and probably feruloyl arabinoxylan oligosaccharides. Three endoxylanase isozymes have been separated. The existence of an exoxylanase, which was said to attack hydrolytically the non-reducing ends of xylan chains with the liberation of xylobiose (**23**), was deduced from inhibitor studies. This enzyme has not been isolated and its existence may be doubted. β-Xylopyranosidase (xylobiase) catalyses the hydrolysis of xylobiose (**23**) to xylose (**4**). It is probable that the enzyme also successively hydrolyses single, unsubstituted xylose residues

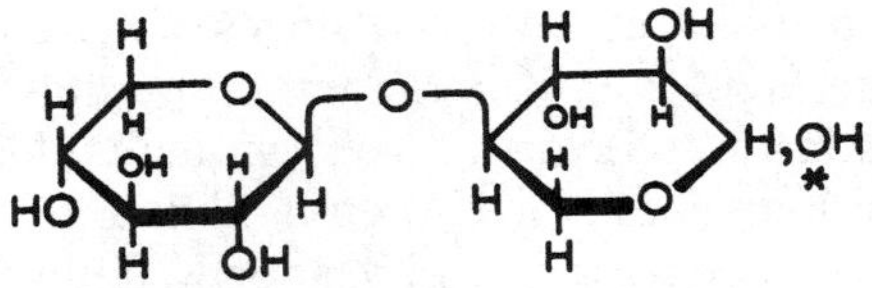

(23) Xylobiose;
β-D-Xylopyranosyl-(1→4)-D-
xylopyranose

from the ends of xylan chains. All these enzymes increase in amount during malting. As free pentose sugars do not accumulate to any significant extent in malting grain, they must be rapidly utilized by the living tissues.

The routes of hydrolytic breakdown of barley β-glucan are better characterized (Fig. 4.5). During malting there is an initial increase in the amount of water-soluble β-glucan gum at the expense of hemicellulosic β-glucan. A relatively heat-stable enzyme (or enzymes) can also increase the amount of soluble β-glucan during mashing. This has been named β-glucan solubilase (Bamforth, 1981, 1989). The enzyme(s) is more heat stable than endo-β-glucanase (Scott, 1972). The solubilase is present in barley and increases in amount during malting. Gibberellic acid accelerates its formation (Bamforth and Martin, 1981). This activity has been attributed to a grain acid carboxypeptidase, which presumably degrades protein that invests the β-glucan and reduces its solubility. However, limited hydrolysis of the β-glucan chains may also contribute to increased solubility and endo-β(1→3)-glucanase as well as a cellulase (endo-β(1→4)-glucanase) derived at least in part from microbes on the surface of the grain probably contribute to solubilase activity (Yin and MacGregor, 1988, 1989; Yin *et al.*, 1989). Various hydrolytic enzymes have been found in barley that are able to attack β-glucans or their breakdown products (Thompson and La Berge, 1977; Briggs, 1978). An active laminarinase has been highly purified. This enzyme is an endo-β-glucanase and is able to attack the seaweed polysaccharide laminarin, a β(1→3)-poly-α-D-glucan. However, this enzyme has very little effect on barley β(1→3;1→4)-glucan, only breaking chains in those regions where several consecutive β(1→3)-linkages occur. The recently discovered 'laminarin-like' β-glucan may be its natural substrate, and it may be instrumental in degrading this material and weakening and breaking the junction between the cell walls of the aleurone layer and the starchy endosperm. The enzyme may also attack the cell walls of fungi and so serve as a defensive mechanism. Barley contains one endo-β(1→3)-glucanase isozyme, but germinated grain contains two (Brunswick *et al.*, 1987). A better known enzyme is barley endo-β(1→3;1→4)-glucanase, often referred to as β-glucanase, which has a high activity on mixed-link barley β-glucans. This enzyme occurs as two isozymes that differ in their stabilities to heat in kilning and mashing (Brunswick *et al.*, 1987; Loi *et al.*, 1987). These enzymes sharply reduce the

(24) Cellobiose; β-D-Glucopyranosyl-(1–4)-D-glucopyranose

(25) Laminaribiose; β-D-Glucopyranosyl-(1→3)-D-glucopyranose

viscosity of solutions of β-glucan. Both laminarinase and endo-β(1→3;1→4)-glucanase increase very considerably during malting. Survival during kilning is favoured by slow drying at the lowest feasible temperatures. Cereals vary markedly in the amounts of endo-β-glucanase generated during germination; barley and rice produce most (Stuart *et al.*, 1987).

Germinating barley contains small amounts of a β(1→4)-glucanase, or cellulase. This enzyme attacks mixed-linked β-glucan and semisynthetic substrates containing β(1→4)-linked D-glucopyranose residues. This enzyme changes relatively little during malting and the major part, or possibly all, is derived from the microbes on the grain's surface. Its importance, relative to the other endo-β-glucanases is unknown. Early studies using inhibitors indicated that exo-β-glucanases were present. These were supposed to break the penultimate links at the non-reducing ends of the β-glucan chains, with the release of disaccharides, either cellobiose (**24**) or laminaribiose (**25**). However, in the absence of supporting findings, and since these disaccharides can be released by endo-β-glucanases, the existence of exo-β-glucanases is doubted.

Two β-glucosidases (cellobiase, laminaribiase) are able to hydrolyse cellobiose and laminaribiose to glucose, but at different relative rates. These enzymes are also able successively to hydrolyse glucose residues from the terminal, non-reducing ends of many β(1→3)- or β(1→4)-linked oligosaccharides and polysaccharides. There is some disagreement, but β-glucosidase activity appears not to increase very rapidly during malting. It seems to be concentrated in the living tissues: the embryo and the aleurone layer. Tissue extracts are able to catalyse β-transglucosylation reactions, which are probably catalysed by β-glucosidases. Cellobiose (**24**) and/or laminaribiose (**25**) do not accumulate in malting grains, so they must be rapidly degraded or utilized.

When milled barley is mashed, there are progressive changes in the viscosity of the extracts. The initial rise mainly results from gums dissolving, and the subsequent decline occurs because of their breakdown. However, the test is of no proven value as a means of deciding the malting quality of a barley. The important values are the levels of endo-β-glucanase activity and the quantity and quality of the β-glucan remaining in the finished malt. The determination of total β-glucanase has been carried out by following β-glucan breakdown using measurements of viscosity, or a radial diffusion technique. The results are not very reproducible, and semisynthetic, coloured substrates have been produced to allow alternative methods to be employed (Chapter 13).

Germinating barley contains enzymes that are able to degrade a range of other polysaccharides, including chitin, pectin and mannan. Some of these, like laminarinase, may help the plants to resist fungal invasions since they will attack the walls of fungal hyphae. Chitin has not been detected in uninfected barley grains.

4.7 Regularities in the carbohydrate composition of barley

Different samples of barley, even of one variety, may differ to substantial extents in their chemical compositions (Bishop and Marx, 1934; Briggs, 1978). Therefore, barley samples differ in their values for making malt, and this greatly complicates the task of deciding which barley varieties have superior malting qualities. The problem became amenable to rational investigation when it was realized that when the nitrogen content (hence the 'protein', measured as N × 6.25%) of the grain was greater, of necessity less carbohydrate was present, and so the grain yielded a lower extract after malting. In addition, it was shown that in certain other respects grain composition does not vary in a random manner. Therefore, within a particular variety when a range of samples is tested, it is found that there are regular relationships between the total carbohydrate content of samples and the proportions of various chemically separated carbohydrate fractions (Fig. 4.7). The recognition of this 'regularity principle' permitted rational appreciation of variations in grain composition. Thus, with increasing total carbohydrate content the extract fraction (mainly derived from the preformed soluble sugars and starch (both 'reserve' carbohydrates)) increases proportionately more than the hemicellulose or insoluble carbohydrate fractions (both of which have structural functions in the grain). There are probably wide variations between varieties (cf. Table 4.2). At least with older varieties, the six-rowed barleys contained less of the extract fraction and more of the others than did two-rowed barleys, a finding consistent with the view that six-rowed barleys are more 'husky'. Husky barleys are also richer in fibre, in insoluble carbohydrate and in crude cellulose (Table 4.2).

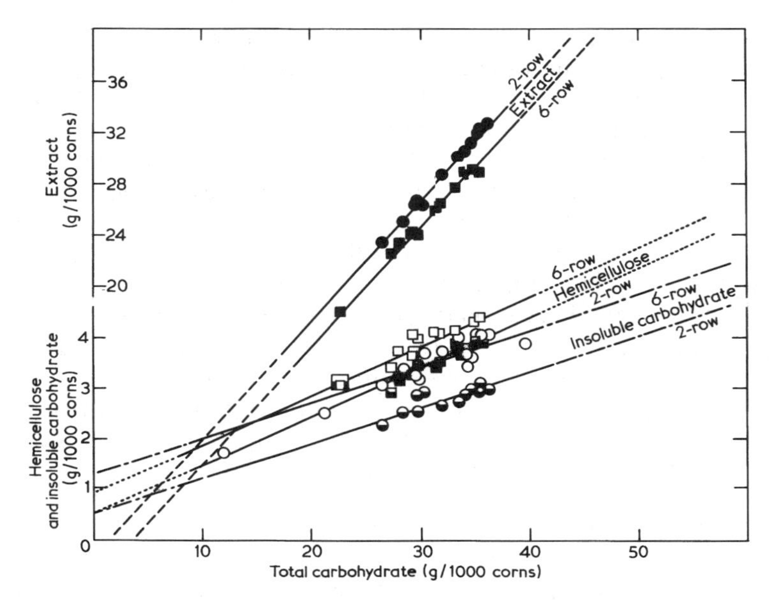

Fig. 4.7 The relationships between the amounts of three carbohydrate fractions and the total carbohydrate contents of various grain samples. The results illustrated are for samples of a two-rowed barley and a six-rowed barley and illustrate the 'regularity principle' of grain composition within one variety. (After Bishop and Marx, 1934.)

When malted in parallel, barleys having high extract-value fractions yield malts giving higher hot water extracts (Chapter 14). If a series of barley samples is malted in a standardized way, then for samples of one variety, there are correlations between barley analyses and malt analyses. Consequently, within limits, barley analyses can be used to predict the gross chemical compositions of the malts prepared from them. The nitrogenous fractions of barley also exhibit regular changes with grain total nitrogen content.

4.8 Proteins and amino acids in barley

The nitrogenous components of barley and malt are of great importance and have been extensively studied (Harris, 1962a; Briggs, 1978). As yet, much of the data appears to be of limited value to maltsters and malt users. The nitrogenous components vary with, for instance, the variety, the growing season, the soil and the amount of fertilizer applied by the farmer. Within a batch of barley, the nitrogen contents of individual grains may vary quite widely. Indeed this is true of grains taken from one ear. It is usually found that if barley is fractionated into different size (width) bands by sieving, the nitrogen content varies in the different fractions. For example,

Piratzky and Wiecha (1937) noted the following nitrogen contents (based on dry weight) for grains of different widths: <2.2 mm, 2.05%, 2.2–2.5 mm, 1.89%; 2.5–2.8 mm, 1.90%; and >2.8 mm, 2.05%. Clearly, by malting particular size fractions of a barley it is possible to select for particular characteristics. The nitrogen content × 6.25 (%) is regarded as a measure of the 'crude protein' content of grain. In the Institute of Brewing (IoB) methods, grain nitrogen contents (TN) are reported, but the methods of the European Brewing Convention (EBC) and the American Society of Brewing Chemists (ASBC) refer to grain protein (%). Different factors, for example × 5.7, have been used in calculating crude protein contents for some other cereals. Reporting grain nitrogen fractions as 'protein' is unfortunate, since grain contains many nitrogenous materials that are not proteins. This measure is extremely valuable for quality-control purposes in judging the value of barley or malt (Chapters 7 and 13). As already noted, a barley with a high TN ('protein') content will give a malt with a lower yield of extract than a malt from a barley having a lower TN content. The relationship between TN and hot water extract differs between varieties. Barley with a high nitrogen content may be slow to take up water and slow to modify during malting. Because it is likely to have a high respiration rate and will grow vigorously, adequate malt modification is only achieved at the expense of a high malting loss.

In recent years, with wide differences in the mean nitrogen content in different seasons, barley samples with nitrogen contents of 1.3–3.2% have been offered for sale in the UK. In most seasons, barleys with nitrogen contents in the range 1.4–1.9% are used in the UK for making malts for the older types of brewing practice or for malt whisky distilling, but where malts are to be used when making mashes with high proportions of unmalted adjuncts, whether for brewing or grain distilling, higher nitrogen contents are appropriate. Therefore, the malts (barley, wheat, rye, sorghum or millets) used in making foodstuffs or various beverages are selected with reference to their nitrogen contents, among other characteristics (Chapter 15).

The crude protein (N × 6.25%) or nitrogen (TN%) contents of grains are used for quality-control purposes. However modern methods of analysis (including selective extraction techniques, selective precipitations, gel exclusion chromatography, ion-exchange chromatography, electrophoresis and isoelectric focusing, often used in conjunction with specific enzyme assays and various immunological techniques) have demonstrated that cereal grains contain thousands of true proteins, many of which are linked to other chemical entities to give, for example, lipoproteins and glycoproteins. Increasingly, the importance of individual proteins, including the various enzymes that regulate modification during malting and the recovery of extract during mashing, is being recognized. However, until recently, the major advances in our understanding were made not by studying individual proteins but by studying nitrogen-containing solubility fractions, defined as the materials that dissolved in various reagents and/or were precipitated by various reagents (Hudson, 1960; Harris, 1962a,b). Each fraction contains many substances, including numerous proteins, but this was not apparent to the early investigators. Bishop (1928, 1929a,b, 1930) refined the classical Osborne method of fractionation and divided grain nitrogen-containing substances into non-protein nitrogenous substances (NPN, 'proteoses' and 'peptones' a fraction that included amino acids, amides, simple peptides and amines) and four 'protein' fractions. These 'protein' solubility fractions were named albumin, globulin, hordein and glutelin. Albumins and globulins, which include many enzymes, dissolve readily from finely ground preparations of grain in salt solutions but differ in that if salt is removed from this salt-soluble fraction, e.g. by dialysis, one group of proteins, the albumins, precipitates while the others remain in solution. Hordeins are insoluble in salt solutions but dissolve in warm aqueous alcohol, while the residual insoluble material is the glutelin. These fractionations are still used, but they have been refined (Préaux and Lontie, 1975; Shewry and Miflin, 1985). More salt-soluble protein is recovered if a reducing agent such as a thiol (e.g. mercaptoethanol) is added to the extraction medium to break interprotein disulphide bridges. It is preferable to extract the hordeins with aqueous isopropyl alcohol containing another reducing agent, sodium borohydride. To a degree, it seems that the hordein fraction represents the major group of storage proteins in the grain, while the glutelin fraction apparently contains storage and structural proteins. Therefore, while an Osborne-type fractionation scheme is empirical, the fractions obtained represent to some degree functionally different groups of proteins. Similar fractionation schemes applied to different cereals yield essentially similar fractions, but often these are given different names: for example the alcohol-soluble prolamine fraction, called hordein in barley, is termed secalin in rye. The hordeins, which are exceptionally rich in proline and glutamine, may be separated by electrophoresis (SDS–PAGE, sodium dodecylsulphate–polyacylamide gel electrophoresis) into several groups, the B-hordeins, which are rich in sulphur (80–90% of the fraction), the C-hordeins, which are sulphur-poor (10–20%) and the D-hordeins (<5%). The D-hordein content of barley has a strong variety-independent negative correlation with malt extract (Howard *et al.*, 1996). The electrophoretic patterns of the hordeins so obtained differ and serve to distinguish between different groups of barleys. The gel proteins, which consist of D-hordeins together with some B-hordeins, slow wort separation during mashing (Wallace and Lance, 1988). The suggestion that hordein electrophoretic patterns are necessarily indicative of malt quality has been disproved (Shewry *et al.*, 1980).

The classification of proteins into solubility groups is not clear-cut. For example, 'free' β-amylase occurs in the salt-soluble fraction, but bound, 'latent' β-amylase occurs as an insoluble, bound complex that only dissolves if reducing agents or proteolytic enzymes are used in the extraction medium. The amounts of β-amylase and protein Z in grains increase markedly as a series of progressively more nitrogen-rich samples are investigated. They may well serve as reserve proteins, ultimately being degraded as germination proceeds beyond the limit allowed in malting (Wallace and Lance, 1988).

The quantities and relative amounts of the solubility fractions change substantially during malting (Table 4.9). Nitrogenous substances are lost by leaching during steeping, but there are no gains or appreciable losses from the whole grain during the other stages. During germination, there is a significant decline in the amount of hordein. The final, apparent rise in the amount of glutelin and the fall in non-protein nitrogen (NPN) may be the result of the protein denaturation and chemical reactions that occur during kilning. The protein composition of the solubility fractions alters with time. For example, α-amylase and some other enzymes appear in the salt-soluble fraction while α-amylase inhibitor (BASI) and protease-inhibiting proteins decline during the germination period. The hordein fraction has been studied because its unusual solubility characteristics make it easy to separate and because it apparently represents the major nitrogenous reserve of the grain. Well-made malts often contain only about a half of the quantity of hordein originally present in the unmalted barley. Many of the nitrogen fractions that add to beer characteristics (head or foam, mouth-feel and body, tendency to form haze) appear to contain substantial quantities of hordein partial breakdown products. Polyphenolic tannins readily bind to hordeins as well as to other proteins. Inadequate protein breakdown, an aspect of modification, during malting can lead to slow wort separations in

Table 4.9 Changes in the nitrogen solubility fractions (N, g/1000 grains) of a Chilian six-rowed barley, steeped at 14.5 °C (58 °F) for 0–53 h; germinated at 20 °C (68 °F) in couch to 120 h, 14.5 °C (58 °F) to 290 h and then 17.5 °C (63 °F) to 336 h. Loaded to kiln at 340 h. Kilning finished at 460 h. All samples were analysed with their rootlets

Process time (h)	TCW (g)	Total N (g)	Salt-soluble proteins N (g)		Hordein N (g)	Glutelin N (g)	Proteose N (g)	Non-protein nitrogen (g)
			Albumin	Globulin				
Steep								
0	46.0	0.807	0.099	0.076	0.261	0.267	0.025	0.132
53	45.1	0.795	0.089	0.074	0.262	0.262	0.017	0.091
Germination								
124	44.5	0.783	0.099	0.071	0.249	0.240	0.008	0.114
170	43.7	0.788	0.120	0.093	0.207	0.207	0.003	0.159
217	44.2	0.804	0.111	0.101	0.199	0.213	0.009	0.168
288	43.4	0.791	0.102	0.111	0.192	0.190	0.023	0.167
337	42.9	0.779	0.085	0.117	0.183	0.195	0.021	0.173
Kilning								
384	42.7	0.782	0.104	0.081	0.191	0.222	0.040	0.150
460	41.4	0.758	0.060	0.114	0.176	0.228	0.033	0.147

From Bishop (1929b).

mashing. This may be caused by the mash bed being clogged by fine particles held together by 'gel proteins' (van den Berg *et al.*, 1981; Moonen *et al.*, 1987). Gel-proteins, which are assayed using a detergent extraction procedure, seem to be a hordein sub-fraction that has been polymerized by the formation of interchain disulphide links, presumably caused by the oxidation of thiols on the unlinked chains by air entrained in the mash. In well-modified malts, the level of gel-proteins are low.

Nitrogenous components of wort, derived from malt, are essential nutrients for yeast in brewing and distilling. Interactions between nitrogenous substances and carbohydrates during kilning are responsible for much of the colour, flavour and aroma of malt. The banding patterns obtained when hordein fractions are separated by electrophoresis are constant for a particular variety. Numerous banding patterns have been recognized and, while most are not unique to one variety, they can be used to distinguish between single grains of varieties belonging to different hordein groups. The method is used to check the purity of batches of grain (Chapter 6). Hordein occurs mainly in the starchy endosperm, possibly much of it in protein bodies. In some instances, the protein may adhere strongly to the cell walls and starch granules, impeding their degradation and so slowing modification.

The nitrogen solubility fractions conform to the 'regularity principle' of grain composition, as is the case with the carbohydrate fractions (Bishop,

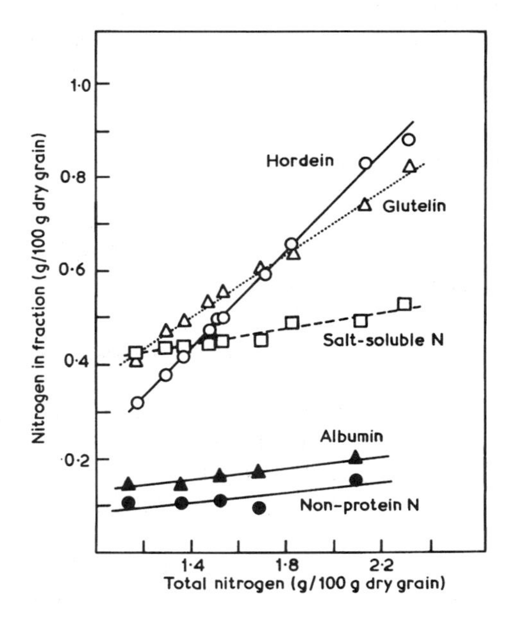

Fig. 4.8 The regularity principle of grain composition illustrated by the relationship between the amounts of the various nitrogen solubility (crude protein) fractions and the total nitrogen contents of a range of samples of the two-rowed barley variety Plumage Archer. (After Bishop, 1928.)

1928; Harris, 1962a; Briggs, 1978) (Fig. 4.8). During grain formation in the ear, the nitrogen solubility fractions increase in a regular, progressive manner, but the onset of grain maturity can 'fix' the nitrogen composition of the grain at various stages. Thus, when a series of mature, fully ripe samples of one variety of grain, having different total nitrogen contents, are analysed and the results are expressed graphically, it is seen that hordein and glutelin fractions increase most rapidly with increasing grain total nitrogen (TN%) (Fig. 4.8). It is striking that the fractions containing mostly the protein reserve substances increase most rapidly in the latter stages of grain formation, in a way that is analogous to that in which starch, the major carbohydrate reserve, increases most rapidly with increasing total carbohydrate content (Fig. 4.7). The recognition of the regularity principle of grain composition led to a rational approach to barley evaluation, and to the recognition of extract prediction equations and other predictive relationships. However, the detailed relationships between grain total nitrogen content, malt extract and the proportions of the different nitrogen solubility fractions are different in different varieties (Chapter 14).

In all cereals, the prolamine fractions decrease during germination. This is important when the germinated grains are being used in foodstuffs for monogastric animals, which besides being unable to fully digest cereal grains also require a supply of essential amino acids. Of these, lysine and methionine are present in sub-optimal proportions. Albumins, globulins and the NPN materials are richer in these amino acids than is hordein (or other prolamines), which is poor in lysine and methionine but rich in proline and glutamic acid, partly present as its amide, glutamine. As germination proceeds, prolamine levels are reduced and the proportions of essential amino acids present in the grain increase, so increasing the proteins' nutritional value for monogastric animals. The protein in malt rootlets contains about twice the proportion of lysine as does protein from whole grains.

Salt (saline)-soluble proteins may serve a range of functions. The importance of the enzymes is obvious, but it should be noted that not all enzymes are completely or even partly soluble in saline solutions (e.g. latent β-amylase; α-glucosidase; some protease and lipase activities) (see Fig. 4.13 below). Other salt-soluble proteins may serve as metabolic regulators, as structural or storage proteins, as inhibitors of particular enzymes (e.g. malt α-amylase, malt limit dextrinase, malt protease, amylases and proteases from insects, fungi and mammals). Cereals, including barley and wheat, contain lectins; these are proteins that have strong binding affinities for particular carbohydrate groupings and may cause haemagglutination (the clumping of red blood cells). The barley and wheat lectins have affinities for *N*-acetylglucosamine The thionins (hordothionin from barley) are proteins that complex with lipids and have the unusual property of dissolving in hydrocarbon solvents (petroleum ether). They are glycoproteins rich in cysteine and basic amino acids and are associated with protein bodies in the endosperm. In their native states, they are toxic to yeasts and other fungi, bacteria and mammals.

Many grain proteins are polymorphic. The isozymes of α-amylase have been extensively studied and variations in the isoproteins of the hordeins provide the basis for recognition of different hordein groups of barley varieties. As an example, a sample of Mona barley contained the following amounts of proteins (in mg/g grain): chymotrypsin inhibitors 1 and 2, 0.24 and 0.08, respectively; β-amylase, 0.98; and Z-protein, 2.20. Z-protein may comprise as much as 4% of the grain protein (Kreis *et al.*, 1987; Briggs, 1992).

Nitrogen ('protein') is not uniformly distributed throughout the cereal grain. In a barley containing 12.4% protein, the lemmae and paleae contained 1.7 and 2.0% protein, respectively, the germ 35% protein and the residual tissue 12.3% protein (Briggs, 1978). Of the protein in the grain, about 60% resides in the endosperm; about one-third is in the aleurone layer and two-thirds is in the starchy endosperm (Enari and Sopanen, 1986). In the starchy endosperm, protein is concentrated in the sub-aleurone layer (Chapter 2). The different tissues contain different proportions of the major nitrogen solubility fractions, and enzymes tend to be concentrated in particular tissues.

When grain is malted and germinated, not only do the proportions of the various nitrogenous fractions alter but also there is a massive degradation of reserve proteins in the endosperm (caused by rising levels of proteolytic enzymes) a rise in the water-soluble NPN fractions and the biosynthesis of many new nitrogenous substances in the aleurone layer and, especially, in the growing embryo. This is achieved by the embryo taking up simple amino acids and peptides from the endosperm and using them to form new substances. Thus, during malting there is a migration of nitrogenous, and other, substances to the embryo from the rest of the grain to support its growth (Table 4.10). This 'dry matter' and nitrogen is partitioned between the rootlets, which are finally removed from malt, and the acrospire and other parts of the embryo, which are retained. Changes occur during kilning that influence the nitrogenous materials, including the partial or total inactivation of enzymes and the interactions of some nitrogenous substances and

Table 4.10 Dry weights and nitrogen contents of parts of barley during malting, illustrating the migration of nitrogen and dry matter to the embryo during germination

	Dry weight (g/1000 corns)				Nitrogen (mg/1000 corns)		
	Total grain	Endosperm + integuments	Germs	Roots	Endosperm + integuments	Germs	Roots
Barley, steeped	42.32	40.81	1.51	0	543	83	–
Malt at 5 days	41.70	38.64	1.81	1.26	456	114	57
Malt at 11 days	38.46	31.77	5.00	1.69	323	228	76

After Brown (1909).

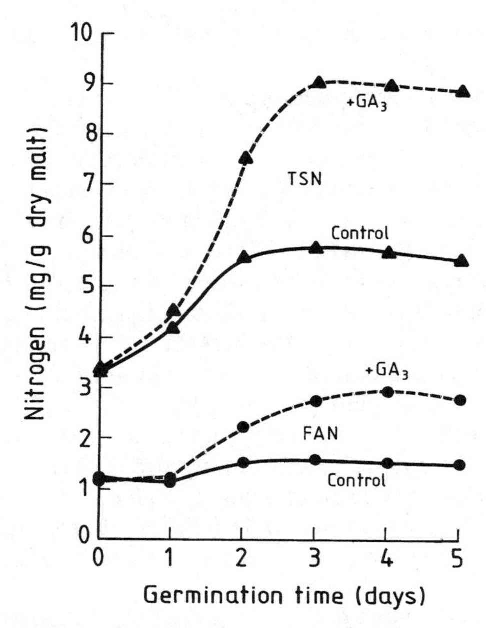

Fig. 4.9 Changes in the levels of TSN and FAN in Piroline barley malted in the conventional way (control, solid lines) or with an addition of gibberellic acid (+GA_3, dashed lines). (After Burger and Schroeder, 1976.)

carbohydrates in Maillard reactions, which generate colour, flavour substances and aroma.

Apart from substances leached during steeping, there is no net loss, or gain, of nitrogen to grain during malting. However the respiratory losses cause a reduction in the dry matter of the grain and so increase its nitrogen concentration or content (% dry basis) (Table 4.9). However, a substantial quantity of nitrogen migrates into the roots (Table 4.10), and as these are eventually removed they reduce both the weight and the nitrogen content of the finished and dressed malt. Thus, the final nitrogen content of the malt depends on these conflicting factors as well as on the nitrogen content of the original barley.

When malt is mashed, nitrogenous substances come into solution and are present in the sweet wort. Some are 'pre-formed soluble substances' in the malt and merely dissolve, while others are formed from initially insoluble proteins by the proteolysis that occurs during mashing. The extent to which this occurs depends on the details of the mashing procedure (Chapter 5). There is disagreement as to the proportion of total soluble nitrogen (TSN) that is pre-formed in malt. The amount solubilized during mashing will depend on the type of malt and the mashing regime. Perhaps 70–80% of the TSN of wort is pre-formed in many instances.

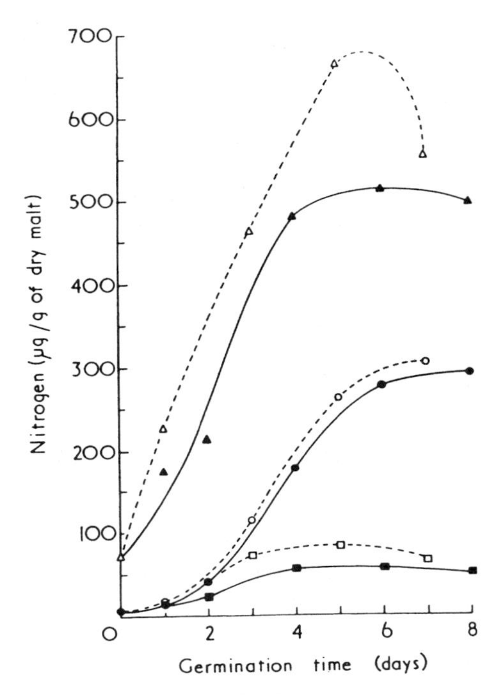

Fig. 4.10 Alterations in the levels of proline nitrogen, total amino nitrogen and ammonium nitrogen in Proctor barley malted with and without an application of gibberellic acid (GA_3). Total amino acid nitrogen control (▲) and plus GA_3 (△); proline nitrogen control (●) and plus GA_3 (○); ammonium nitrogen control (■) and plus GA_3 (□). (Data from Jones and Pierce, 1963.)

The TSN that occurs in solution in the worts of carefully standardized analytical mashes is often expressed as a percentage of the initial dry weight of the malt (Chapter 13). Formerly it was the practice to give the wort a standardized boil, collect the coagulated material and distinguish between the coagulable nitrogen fraction (coag. N) (chiefly high-molecular-weight materials) and the permanently soluble nitrogen (PSN), which remained in solution. Subsequently, for a time, there was a convention in the UK that PSN = 0.94 × TSN. This was not satisfactory because the proportion of coagulable nitrogen present in the TSN varied with the type and degree of modification of the malt (Chapter 13). Strong kilning denatures some proteins and reduces the amount of coagulable nitrogen found in the wort, but PSN values were said to be little influenced by kilning conditions. The nitrogenous components in wort vary widely in their characteristics. It is important to be able to estimate the levels of low-molecular-weight nitrogenous substances that may serve as yeast nutrients. At present, techniques are used to determine the free amino nitrogen (FAN) that occurs in worts prepared during analytical mashes (Chapter 13).

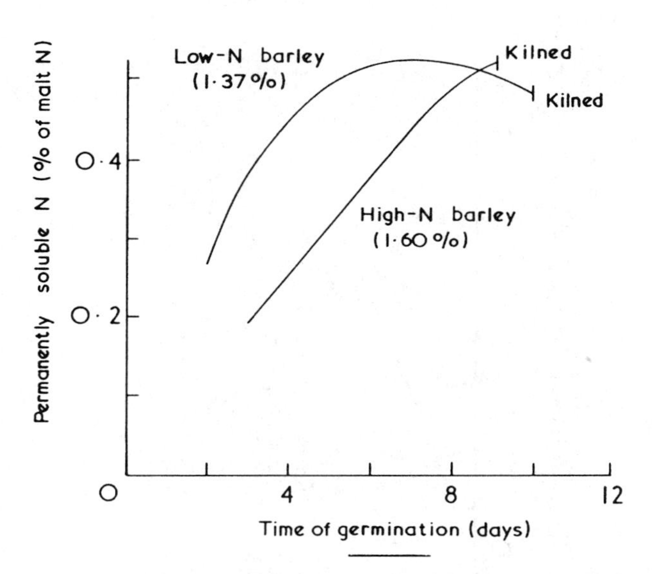

Fig. 4.11 Alterations during malting in the amounts of PSN in two samples of barley having different nitrogen contents (After Macey, 1957.)

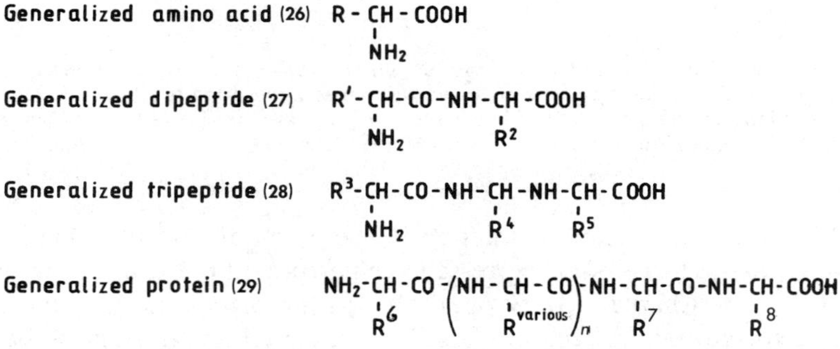

Generalized amino acid (26) $R-CH(NH_2)-COOH$

Generalized dipeptide (27) $R'-CH(NH_2)-CO-NH-CH(R^2)-COOH$

Generalized tripeptide (28) $R^3-CH(NH_2)-CO-NH-CH(R^4)-NH-CH(R^5)-COOH$

Generalized protein (29) $NH_2-CH(R^6)-CO-\left(NH-CH(R^{various})-CO\right)_n-NH-CH(R^7)-CO-NH-CH(R^8)-COOH$

Fig. 4.12 Generalized formulae of an α-amino acid, a dipeptide, a tripeptide and a polypeptide chain such as occurs in proteins. Many proteins are glycosylated or otherwise substituted.

As malting proceeds, the levels of TSN and FAN increase and plateau and may later decline. The levels attained are strongly dependent on malting conditions and are strongly stimulated by additions of gibberellic acid (Figs 4.9 and 4.10). Thus the levels of TSN obtained from a malt are not wholly regulated by the TN content of that malt but are influenced by the manner in which malt is made (Fig. 4.11). Levels of TSN can be depressed by applying potassium bromate in the early stages of malting (Chapter 14).

The ratios of soluble nitrogen obtained from malt, using a small-scale analytical mashing procedure, to the TN content of the malt has long been used as a measure of the quality of malt. When a series of similar malts are made from barley samples (of one variety) of progressively greater nitrogen contents, the TSN values of the malts also increase, but they do not increase in direct proportion to the TN values of the malts. Consequently, under these conditions, as TN values increase the ratios of TSN to TN decline. Figure 4.12 illustrates that peptides (**27**, **28**) and proteins (**29**) are formed of chains of α-amino acids (**26**) jointed by peptide links (–CO–NH–). The peptide chains in proteins are folded and packed together in complex ways and not infrequently the proteins are glycosylated, acylated or joined to other chemical groups. In addition polyphenolic tannins readily associate with proteins with the result that their effectiveness as enzymes may be reduced, their degradation by proteolysis may be slowed or prevented and their solubilities may be reduced. The formation of associations between polypeptides and tannins in beers is the major cause of haze formation.

The hydrolysis of proteins that occurs during malting and mashing has been extensively studied. The overall protease activity in grain increases during malting and declines during kilning, and enzyme activity occurs in different solubility fractions (Fig. 4.13). The pH–activity profile clearly indicates that several enzymes are involved in proteolysis (Fig. 4.14). The endogenous inhibitors of protease activity decline during malting and are absent from finished malts (Mikola and Enari, 1970). There have been major advances in our understanding of the enzymes that degrade proteins (Enari and Mikola, 1967; Mikola and Enari, 1970; Mikola, 1981, 1983; Enari, 1986; Enari and Sopanen, 1986; Wallace and Lance, 1988; Briggs, 1992; Jones *et al.*, 1993). Protease activity is the result of the activities of a very complex mixture of exo- and endopeptidases (Jones *et al.*, 1993). Germinating barley contains a mixture of at least five (and probably many more) endopeptidases, or proteases, whose overall activity increases (approximately 20-fold) during malting. About 90% of the endopeptidase activity has been attributed to two thiol-dependent proteases, which are inhibited by *p*-chloromercuribenzoate and have pH optima of 3.9 and 5.5. Of the remaining 10% of endopeptidase activity, some is caused by three enzymes that are metalloproteins, are inhibited by ethylenediamine tetraacetic acid (EDTA) and have pH optima at 5.5, 6.9 and 8.5, respectively. Germinating grains also contain five carboxypeptidases (pH optima 4.8–5.6), which are inhibited by DFP (diisopropyl fluorophosphate). These enzymes have different, complementary substrate specificities and increase in amounts during malting. They attack peptides exclusively from the carboxyl end of the chains. In addition there are four neutral aminopeptidases, with pH optima about 7.2, which attack peptides from the amino ends of the chains, and two alkaline peptidases with pH optima of 8–10. These enzymes also increase during malting, but to comparatively minor extents. The proteases and carboxypeptidases occur chiefly in the starchy endosperm, where their pH optima are close to the pH of the

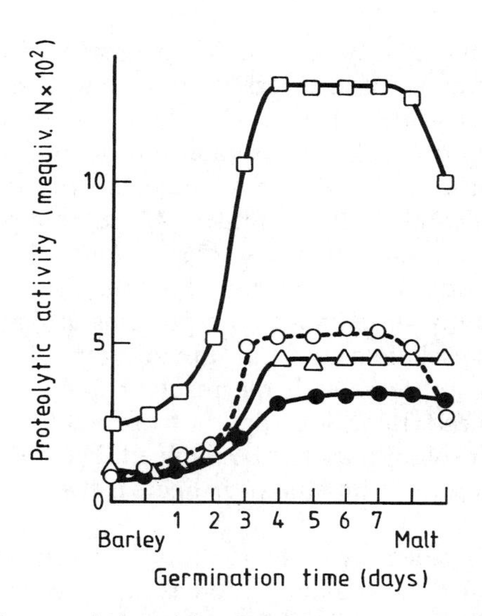

Fig. 4.13 Changes in proteolytic activity observed during malting. Total proteolytic activity (□); water-soluble activity (●) ; activity soluble in saline solutions (△); insoluble activity (○). (After Kringstad and Kilhovd, 1957.)

tissue (about 5.2). The levels of at least some of these enzymes can be greatly increased (e.g. doubled) by applications of gibberellic acid during malting. It seems that carboxypeptidases are present in excess, and that the rises in FAN and TSN that occur during malting are limited by endopeptidase activity (Burger and Schroeder, 1976). The neutral and alkaline peptidases are confined to the living tissues. The scutellum is rich in alkaline peptidases but carboxypeptidases and neutral aminopeptidases are also present.

The reserve non-embryo proteins in the quiescent grain occur approximately 70% in the starchy endosperm and 20% in the aleurone layer (Wallace and Lance, 1988). They are deposited, in part, in special bodies: the aleurone grains in the living tissues of the scutellum and the aleurone layer, and in structurally different protein bodies in the starchy endosperm. During grain germination, several stages of protein mobilization and usage may be distinguished. Following hydration, the protein reserves in the scutellum and aleurone layer are degraded and at least some of the amino acids liberated are used to synthesize hydrolytic enzymes, including α-amylase, thiol-dependent proteases and carboxypeptidases. These are released into the starchy endosperm where, once the cell walls have been degraded, the proteases and carboxypeptidases act to degrade the reserve proteins to amino acids and peptides. The aleurone layer takes up some amino acids, which it converts to glutamine (**40**) which is released back into the starchy endosperm (Fig. 4.15). The simple peptides, amides and amino acids diffuse to the embryo where they are taken up and peptidases complete their degradation

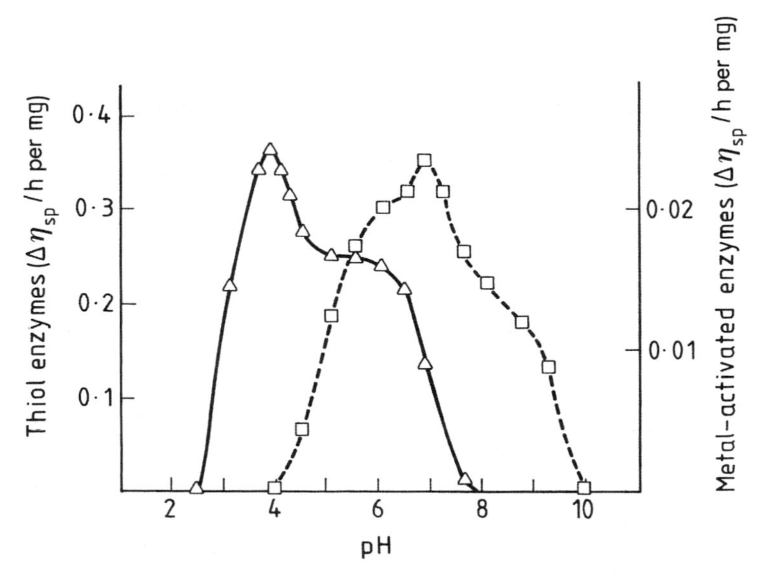

Fig. 4.14 The pH-activity relationships of two groups of proteolytic enzymes extracted from freeze-dried Pirkka malt. △–△, the activities of thiol-containing endoproteinases, resistant to EDTA, estimated by their ability to degrade gelatin; (□-□, the activities of metal-activated proteinases, resistant to *p*-chloromercuribenzoate, estimated by their abilities to degrade gelatin. Gelatin degradation was detected by falls in viscosity. (After Enari and Mikola, 1967).

to amino acids. These are interconverted and utilized by the growing embryo to form new proteins in the growing rootlets and acrospire.

The individual proteolytic enzymes vary in their heat stabilities and hence their survival during kilning. Peptidase activity is reduced more than protease activity during kilning. The insoluble protease fraction appears to be more heat resistant than the soluble fraction.

The levels of free amino acids (Fig. 4.15) that are found during malting and which are the major component of FAN have been extensively studied (Fig. 4.10). Smaller, soluble peptides also occur, but analytical methods for these are less satisfactory. Yeasts utilize nitrogenous components from wort at different rates (Hough *et al.*, 1982), so the composition of the mixture is important. At one extreme ammonium ions (from amides), the amino acids glutamic acid (**39**) and aspartic acid (**35**), the two amides glutamine (**40**) and asparagine (**36**), and serine (**51**), threonine (**52**), lysine (**46**) and arginine (**34**) are rapidly utilized. At the other extreme the imino acid proline (**50**) is essentially unused. However, proline may support the growth of microorganisms, which can contaminate yeast and spoil beer. Free amino acids differ in the ease with which they react with reducing sugars during kilning and wort boiling to yield melanoidin. In addition to the common amino acids, amides and imino acids that occur in proteins (Fig. 4.15), barley

$CH_3 \cdot CH(NH_2) \cdot COOH$
(30) α-Alanine*

$NH_2 \cdot CH_2 \cdot CH_2 \cdot COOH$
(31) β-Alanine

$HOOC \cdot (CH_2)_3 \cdot CH(NH_2) \cdot COOH$
(32) α-Aminoadipic acid

$NH_2 \cdot (CH_2)_3 \cdot COOH$
(33) γ-Aminobutyric acid

$NH_2(HN{=})C \cdot NH \cdot (CH_2)_3 \cdot CH(NH_2) \cdot COOH$
(34) Arginine*

$HOOC \cdot CH_2 \cdot CH(NH_2) \cdot COOH$
(35) Aspartic acid*

$NH_2 \cdot CO \cdot CH_2 \cdot CH(NH_2) \cdot COOH$
(36) Asparagine*

$HS \cdot CH_2 \cdot CH(NH_2) \cdot COOH$
(37) Cysteine*

$S \cdot CH_2 \cdot CH(NH_2) \cdot COOH$
|
$S \cdot CH_2 \cdot CH(NH_2) \cdot COOH$
(38) Cystine*

$HOOC \cdot (CH_2)_2 \cdot CH(NH_2) \cdot COOH$
(39) Glutamic acid*

$NH_2 \cdot CO(CH_2)_2 \cdot CH(NH_2) \cdot COOH$
(40) Glutamine*

$NH_2 \cdot CH_2 \cdot COOH$
(41) Glycine*

(imidazolyl) $CH_2 \cdot CH(NH_2) \cdot COOH$ (ring: N, NH)
(42) Histidine*

HO (pyrrolidine ring, N–H) COOH
(43) Hydroxyproline*

$CH_3 \cdot CH_2(CH_3)CH \cdot CH(NH_2) \cdot COOH$
(44) Isoleucine*

$(CH_3)_2CH \cdot CH_2 \cdot CH(NH_2) \cdot COOH$
(45) Leucine*

$NH_2 \cdot (CH_2)_4 \cdot CH(NH_2) \cdot COOH$
(46) Lysine*

$CH_3 \cdot S \cdot (CH_2)_2 \cdot CH(NH_2) \cdot COOH$
(47) Methionine*

(phenyl) $CH_2 \cdot CH(NH_2) \cdot COOH$
(48) Phenylalanine*

(piperidine ring, N–H) COOH
(49) Pipecolinic acid
(= piperidine-
2-carboxylic acid)

(pyrrolidine ring, N–H) COOH
(50) Proline*

$HO \cdot CH_2 \cdot CH(NH_2) \cdot COOH$
(51) Serine*

$CH_3 \cdot CH(OH) \cdot CH(NH_2) \cdot COOH$
(52) Threonine*

(indolyl, N–H) $CH_2 \cdot CH(NH_2) \cdot COOH$
(53) Tryptophan*

HO (phenylene) $CH_2 \cdot CH(NH_2) \cdot COOH$
(54) Tyrosine*

$(CH_3)_2CH \cdot CH(NH_2) \cdot COOH$
(55) Valine*

* Occur in proteins.

Fig. 4.15 Amino acids and some related substances found in barley and malt.

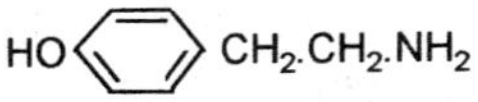

$HO-C_6H_4-CH_2.CH_2.NH_2$

(56) Tyramine

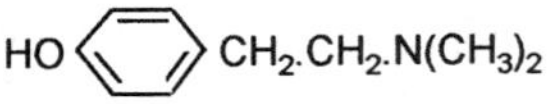

$HO-C_6H_4-CH_2.CH_2.N(CH_3)_2$

(57) Hordenine

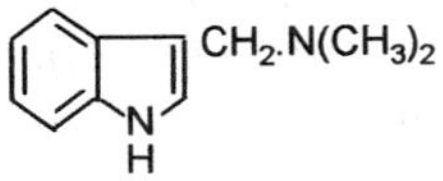

$CH_2.N(CH_3)_2$ (indole, N–H)

(58) Gramine

$HO.CH_2.CH_2.NH_2$

(59) Ethanolamine (2-aminoethanol)

NH_3

(60) Ammonia

$(CH_3)_2.NH$

(61) Dimethylamine

$CH_3.(CH_2)_3.NH_2$

(62) Butylamine

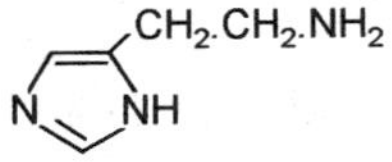

$CH_2.CH_2.NH_2$ (imidazole, N, NH)

(63) Histamine

HO, CO.NH.$(CH_2)_4$.NH.C(=NH)NH_2

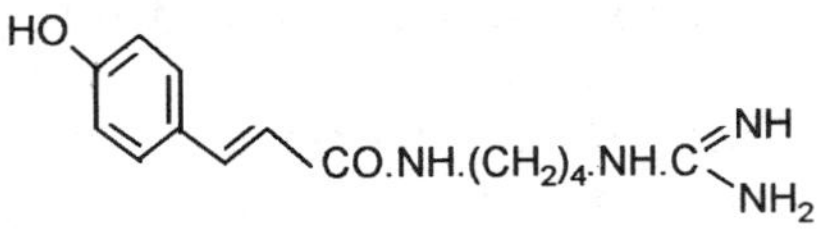

(64) Coumaroyl agmatine

$NH_2.(CH_2)_4.NH_2$

(65) Putrescine (1,4-diaminobutane)

$NH_2.(CH_2)_5.NH_2$

(66) Cadaverine (1,5-diaminopentane)

$NH_2.(CH_2)_4.NH.C(=NH)NH_2$

(67) Agmatine

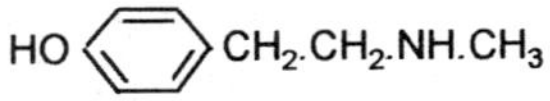

$HO-C_6H_4-CH_2.CH_2.NH.CH_3$

(68) *N*-Methyltyramine

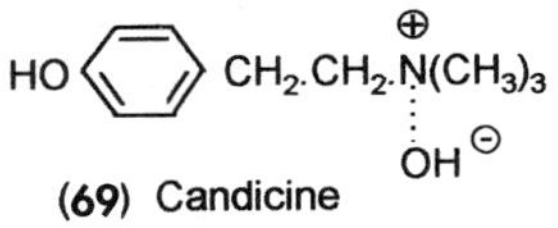

$HO-C_6H_4-CH_2.CH_2.\overset{\oplus}{N}(CH_3)_3 \; OH^{\ominus}$

(69) Candicine

$HO.CH_2.CH_2.\overset{\oplus}{N}(CH_3)_3 \; OH^{\ominus}$

(70) Choline (a B-vitamin)

$(CH_3)_3.\overset{\oplus}{N}.CH_2.COO^{\ominus}$

(71) Betaine

$CH_3.NH_2$

(72) Methylamine

$CH_3.CH_2.NH_2$

(73) Ethylamine

$CH_3.(CH_2)_4.NH_2$

(74) Amylamine

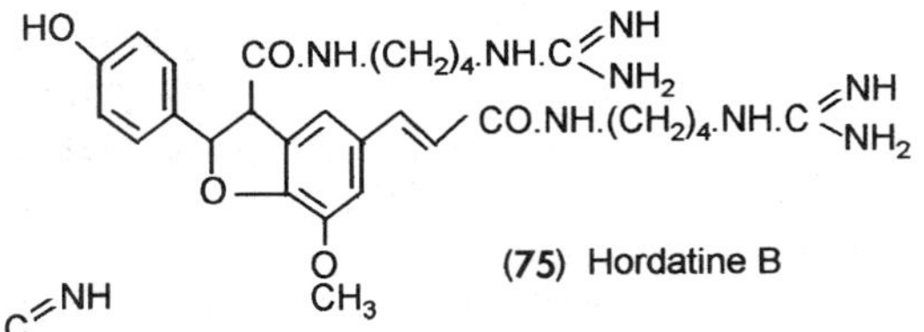

HO, CO.NH.$(CH_2)_4$.NH.C(=NH)NH_2, CO.NH.$(CH_2)_4$.NH.C(=NH)NH_2, O, O, CH_3

(75) Hordatine B

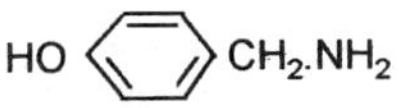

$HO-C_6H_4-CH_2.NH_2$

(76) *p*-Hydroxybenzylamine

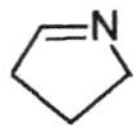

N

(77) Pyrrolidine

Fig. 4.16 Various amines and related substances found in barley and malt.

contains significant quantities of other free amino acids such as β-alanine (**31**), α-aminoadipic acid (**32**) and γ-aminobutyric acid (**33**).

Several amines, which are derived from amino acids, are also present (Fig. 4.16). The free amino acids and imino acids present in germinating malt are derived directly from reserve proteins by hydrolysis and indirectly by metabolic interconversions in the living tissues. Proline (**50**), which forms a substantial proportion of this fraction (Fig. 4.10), is a major component of hordein. Proline accumulates in barley seedlings under conditions of water stress (Briggs, 1978). As malting is germination with water stress, this may be a reason for the accumulation of this substance.

4.9 Nucleic acids and related substances

Barley contains the usual range of low-molecular-weight derivatives of the nitrogen-containing purine and pyrimidine bases (**80–85**), including the cofactors NAD^+ and $NADP^+$, ATP, and sugar derivatives such as UDPG (**20**) and ADPG (**21**) (Briggs, 1978) (Fig. 4.17). Grain contains deoxyribonucleic acid (DNA, **78**), which occurs in nuclei, mitochondria and plastids and carries the genetic information coding for the various proteins, and ribonucleic acid (RNA, **79**), which in its various forms carries the information of the code into the protein biosynthetic machinery and provides some of the molecules involved in protein biosynthesis. Nucleic acids represent 0.2–0.3% of the dry weight of the barley grain, of which about 30% is DNA and 70% is RNA. They are present and metabolically active in the living tissues, but it is not clear if they occur in the dead, starchy endosperm. During malting, enzymes accumulate that are able to degrade hydrolytically DNA and RNA (DNAase and RNAase, respectively), and other enzymes, the nucleosidases and nucleotidases, are present which are able to complete the breakdown eventually to free purine and pyrimidine bases, sugars (ribose (**9**) and deoxyribose (**10**)) and inorganic phosphate. During mashing, malt nucleic acids are completely degraded to nucleotides and free bases. As much as 6% of the soluble nitrogen of decoction wort may be contributed by the bases adenine (**80**) and guanine (**81**). The stages of hydrolytic breakdown of nucleic acids are: (i) hydrolysis by nucleases, with the release of nucleotides (various isomers; base–sugar(ribose or deoxyribose)–phosphate); (ii) the hydrolysis of nucleotides by nucleotidases to give inorganic phosphate and nucleosides (base–sugar) (ribose or deoxyribose); and (iii) the hydrolysis of the nucleosides by the nucleosidases to yield the free bases (purines and pyrimidines) (Fig. 4.17) and free sugars (ribose (**9**) and deoxyribose (**10**)). In addition, bases may be deaminated while free or as nucleosides, so adenosine may give rise to inosine, adenine (**80**) to hypoxanthine, cytidine to uridine, and cytosine (**82**) to uracil (**83**).

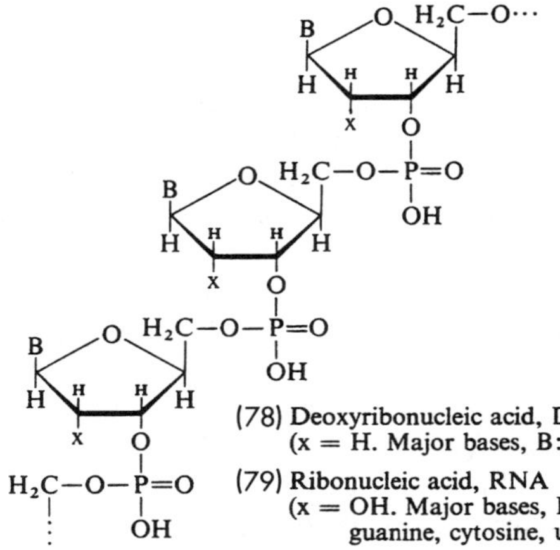

(78) Deoxyribonucleic acid, DNA
(x = H. Major bases, B: Adenine, guanine, cytosine, thymine)

(79) Ribonucleic acid, RNA
(x = OH. Major bases, B: Adenine, guanine, cytosine, uracil)

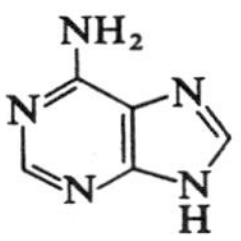

(80) Adenine (a purine)

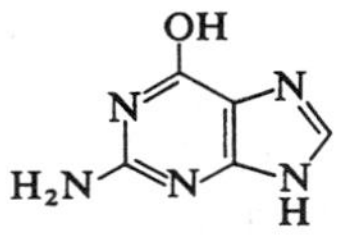

(81) Guanine (a purine)

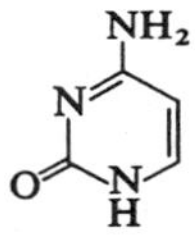

(82) Cytosine (a pyrimidine)

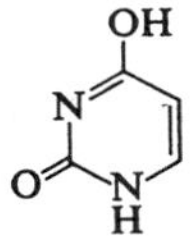

(83) Uracil (a pyrimidine)

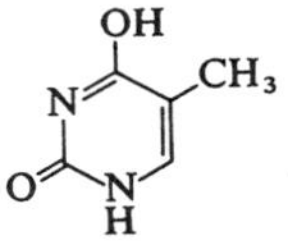

(84) Thymine (a pyrimidine)

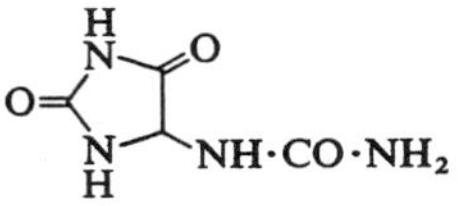

(85) Allantoin

Fig. 4.17 The structures and components of the nucleic acids.

Allantoin (**85**), an oxidative degradation product of purine bases, is found in malt.

Following mashing, various mixtures of free bases and nucleosides are found in the wort. Some nucleosides have flavour-enhancing characteristics, but it seems unlikely that their concentrations are sufficient to influence beer flavour. The products of nucleic acid breakdown may be important in ensuring that yeast growth begins promptly with a minimum lag. Worts from highly kilned malts contain a higher proportion of nucleosides than worts from moderately kilned malts, which are richer in free bases. Nucleotides are rarely or never present. It is concluded that nucleases (DNAase and RNAase) and nucleotidases (phosphatases) are comparatively heat stable, but nucleosidases are liable to be partly inactivated during kilning.

4.10 Other nitrogenous grain components

Grains contain many other nitrogenous materials in small amounts (Palamand *et al.*, 1969; Briggs, 1978; Wackerbauer and Toussaint, 1984; Izquierdo-Pulido *et al.*, 1994). Various of the vitamins containing nitrogen and some amines are present (Fig. 4.16). Numerous heterocyclic nitrogen compounds, which contribute to the flavour and aroma of malt, are formed during kilning. Traces of ammonia (**60**) are sometimes present. This may be formed by the breakdown of the amides glutamine (**40**) and asparagine (**36**). Barley contains a number of amines (Fig. 4.16), many of which increase during malting. The simple volatile amines include methylamine (**72**), dimethylamine (**61**), ethylamine (**73**), butylamine (**62**) and amylamine (**74**). Several polyamines occur, including putrescine (**65**), spermine, spermidine, cadaverine (**66**) and agmatine (**67**). Some of these may occur in combination with other substances. For example, coumaroyl agmatine (**64**) occurs and this, through oxidative coupling, gives rise to anti-fungal hordatines (**75**). The decarboxylation of amino acids is one source of amines. Thus the decarboxylation of histidine (**42**) gives rise to histamine (**63**), which is found in malt, and the decarboxylation of tyrosine (**54**) gives rise to tyramine (**56**). Tyramine (**56**), and its methylated derivatives *N*-methyltyramine (**68**), hordenine (*N*-dimethyltyramine, **57**) and the quaternary ammonium compound candicine (*N*-trimethyltyramine, **69**) occur almost exclusively in the roots. Tryptamine and β-phenylethylamine also occur in malt. During malting, root juices become spread over the surface of the malt, and these amines can be converted to *N*-nitroso-dimethylamine (NDMA) by reaction with oxides of nitrogen during kilning. Gramine (**58**), in contrast, is confined to the acrospire in malt. In seedlings, it may be an effective antagonist to some microbial pathogens. It is derived from the amino acid tryptophan (**53**), but it is not a direct decarboxylation product. Neither hordenine nor gramine are found in other cereals. However these, or their seedlings, contain nitrogenous substances that are absent from barley. For example, wheat and maize contain complex hydroxamic acids, and sorghum contains a cyanogenic glucoside, called dhurrin, which on breakdown liberates cyanide. Germinating barley contains traces of other cyanogenic glycosides.

Malting barley contains *p*-hydroxybenzylamine (**76**) and pyrrolidine (**77**). Other nitrogen-containing substances that occur, free or combined, in appreciable amounts, are choline (**70**), ethanolamine (**59**) and betaine (**71**).

Some of the amines, such as tyramine (**56**), histamine (**63**) and hordenine (**57**) may reach beer in sufficient amounts to have physiological effects on drinkers. Not all the amines present in beer originate in the malt. It is unclear if they are present in sufficient amounts to influence flavour.

4.11 Lipids

The lipids present in barley, and other cereals, are a very complex group of substances with widely varying properties, ranging from hydrocarbons from the cuticular waxes, through quinones with cofactor functions in intermediary metabolism, neutral glycerides, which are reserve substances, to polar glycerides with structural functions in the cells. The lipids of grain are reported to contain 65–78% neutral lipids, 7–13% glycolipids and 15–26% phospholipids (Briggs, 1978; Morrison, 1978, 1988).

Among the neutral lipids there are hydrocarbons and unchanged substances from the cuticular waxes, carotenoids, the tocopherols (vitamin E; **108** in Fig. 4.18), ubiquinones, sterol and wax esters, mono-, di- and triacylglycerols (mono-, di- and triglycerides (**86–89**)), free sterols such as β-sitosterol (**90**) and fatty alcohols. The diglycerides include the 1,2 and 1,3-isomers.

The polar lipids include free fatty acids (**91–97**), glycolipids (**98,99**) and phospholipids (**100–106**). Traces of 5-*n*-alkyl resorcinols are present from the waxes. In addition variously hydroxylated fatty acids occur. Recorded frequencies of total fatty acids (**91–97**) in barley are 14:0, 0.3–0.5%; 14:1, traces; 16:0, 18–27%; 16:1, 0.1–0.3%; 18:0, 0.4–1.0%; 18:1, 9–21%; 18:2, 50–59%; and 18:3, 4–7%. The high proportion of unsaturated fatty acids

(**86**)	(**87**)	(**88**)	(**89**)
$H_2C{\cdot}O{\cdot}CO{\cdot}R^1$	$H_2C{\cdot}O{\cdot}CO{\cdot}R$	$H_2C{\cdot}O{\cdot}CO{\cdot}R^1$	$H_2C{\cdot}O{\cdot}CO{\cdot}R^1$
$R^2{\cdot}CO{\cdot}OCH$	$HO{\cdot}CH$	$R^2{\cdot}CO{\cdot}O{\cdot}CH$	$HO{\cdot}CH$
$H_2C{\cdot}O{\cdot}CO{\cdot}R^3$	$H_2C{\cdot}OH$	$H_2C{\cdot}OH$	$H_2C{\cdot}O{\cdot}CO{\cdot}R^2$
Triglyceride (triacylglycerol) ($R^1{\cdot}CO{\cdot}OH$ – fatty acid)	1-Monoglyceride (monoacylglycerol)	1,2-Diglyceride (diacylglycerol)	1,3-Diglyceride (diacylglycerol)

CH_3 CH_3 CH_3 CH_3 CH_3 CH_3 HO

(**90**) β-Sitosterol

Common fatty acids.

	Formula	Name
(**91**)	$CH_3{\cdot}(CH_2)_{12}{\cdot}COOH$	Myristic acid (14:0)
(**92**)	$CH_3{\cdot}(CH_2)_{14}{\cdot}COOH$	Palmitic acid (16:0)
(**93**)	$CH_3{\cdot}(CH_2)_5{\cdot}CH{=}CH{\cdot}(CH_2)_7{\cdot}COOH$	Palmitoleic acid (16:1)
(**94**)	$CH_3{\cdot}(CH_2)_{16}{\cdot}COOH$	Stearic acid (18:0)
(**95**)	$CH_3{\cdot}(CH_2)_7{\cdot}CH{=}CH{\cdot}(CH_2)_7{\cdot}COOH$	Oleic acid (18:1)
(**96**)	$CH_3{\cdot}(CH_2)_4{\cdot}CH{=}CH{\cdot}CH_2{\cdot}CH{=}CH{\cdot}(CH_2)_7{\cdot}COOH$	Linoleic acid (18:2)
(**97**)	$CH_3{\cdot}CH_2{\cdot}CH{=}CH{\cdot}CH_2{\cdot}CH{=}CH{\cdot}CH_2{\cdot}CH{=}CH{\cdot}(CH_2)_7{\cdot}COOH$	Linolenic acid (18:3)

explains the low melting point of the mixture of barley triacylglycerols, which is an oil. The frequency of the different fatty acids differs markedly

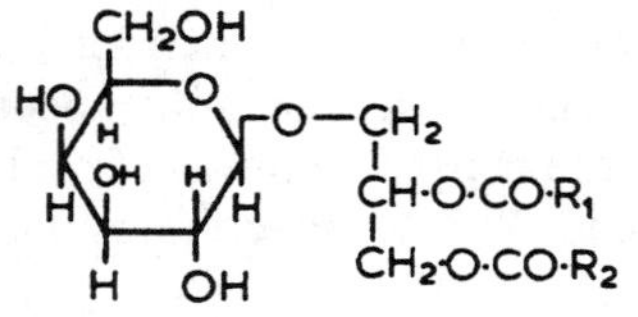

(**98**) Mono-β-D galactosyl diglyceride

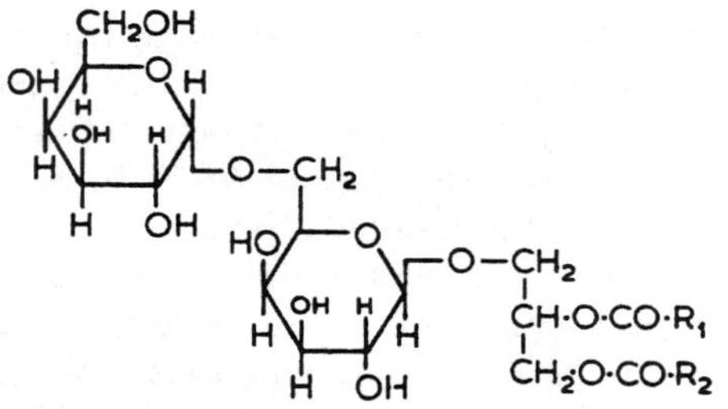

(**99**) Digalactosyl diglyceride

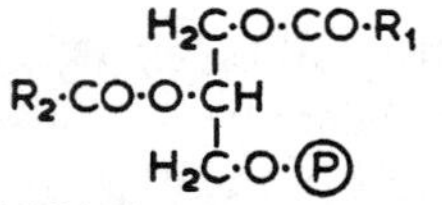

(**100**) Phosphatidic acid

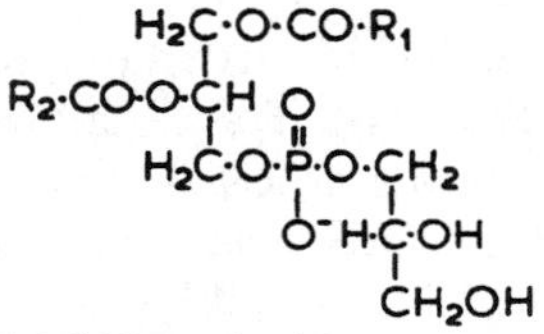

(**101**) Phosphatidylglycerol

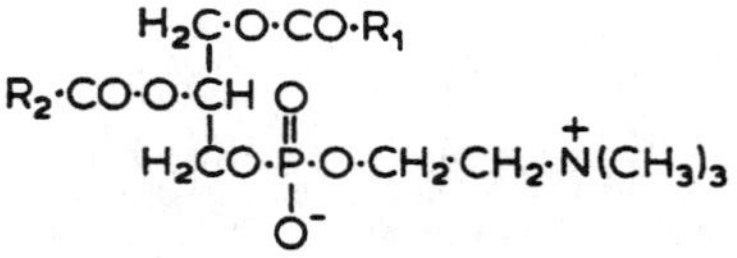

(**102**) Phosphatidylcholine (lecithin)

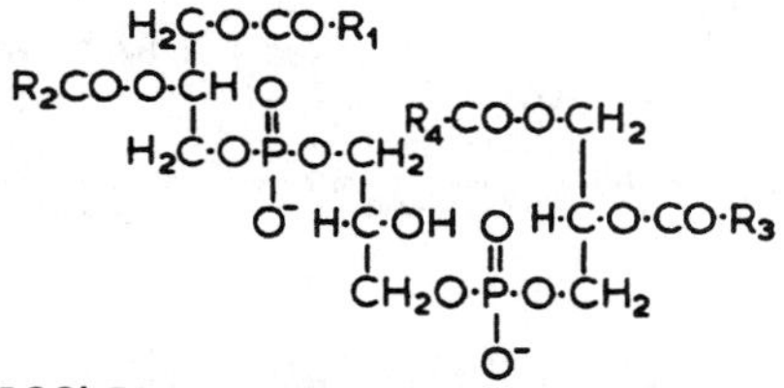

(**103**) Diphosphatidylglycerol (cardiolipin)

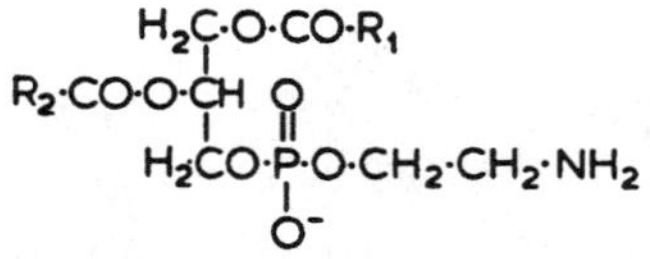

(**104**) Phosphatidylethanolamine (cephalin)

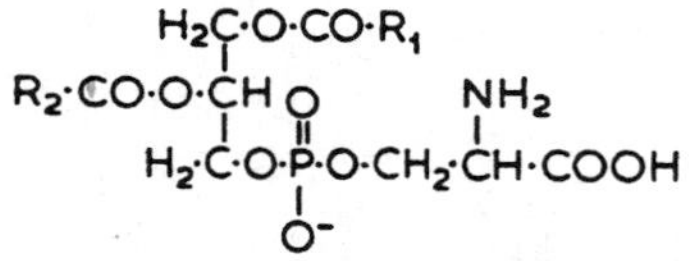

(**105**) Phosphatidylserine (cephalin)

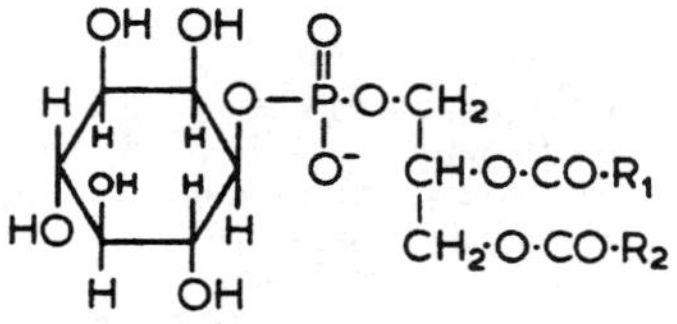

(**106**) Phosphatidyl inositol

between the different acid-containing fractions. The phospholipids include phosphatidylethanolamine (**104**), phosphatidylcholine (**102**), lysophosphatidylcholine and phosphatidylmyoinositol (**106**). The glycolipid fraction includes digalactosyl diglyceride (**99**) and monogalactosyl diglyceride (**98**) as well as glycosides of sterols, such as β-sitosterol (**90**) (Table 4.11).

As barley deteriorates during poor storage there may be a rise in the levels of free fatty acids, at the expense of other lipids. The total lipid contents of barleys are usually about 3.5% of the grain's dry weight. Older estimates of 2–2.5% were too low because the extraction techniques used failed to dissolve the more polar materials. Older estimates are of the non-polar 'petrol-soluble' materials. About one-third of the petrol-soluble lipids occurs in the embryo. Of the remaining two-thirds, more than 90% is located in the aleurone layer. The major components of this fraction are the neutral glycerides; these are largely confined in special subcellular organelles, the spherosomes, in which each oil droplet is bounded by a membrane. A variable quantity of the petrol-soluble, or neutral, lipid is utilized during malting. Estimates are often of 10–12%, but reported results differ widely. Studies of the free and combined fatty acids have been reported (Table 4.11). The triacylglycerol (triglyceride, **86**) fraction is the major neutral lipid material, which is mainly used during malting. Probably the greater part is hydrolysed by lipase via the di- and monoacylglycerols (**87–89**) to give finally glycerol and free fatty acids. These are utilized chiefly to support respiration, but a proportion may be converted into sugars via the glyoxylate cycle, which operates in the aleurone layer after gibberellin stimulation. Starch granules have lipids on their surfaces and polar lipids internally, which may be complexed with amylose (section 4.5). Most malt

Table 4.11 The fatty acids present in lipid classes of Weeah barley and malt made from it

	Fatty acid (mg/g d.wt)	
Class of lipid	Barley	Malt
Phospholipids	4.3	2.4
Glycolipids		
Digalactosyl monoglyceride	1.1	0.9
Digalactosyl diglyceride	1.7	1.0
Monogalactosyl monoglyceride	0.3	0.2
Monogalactosyl diglyceride	0.6	0.4
Neutral lipids		
Acylsteryl glycoside	0.2	0
Triacylglycerol	30.2	19.8
Diacylglycerol	1.2	1.0
Monoacylglycerol	0.3	0.3
Free fatty acids	1.7	2.0
Steryl esters	0.5	0.3
Total lipids	42.1	28.3

Anness (1984).

lipids are retained in the spent grains during mashing. However, undesirably large amounts have been dispersed into the wort when thin-bed, rapid wort separation systems have been employed with finely ground malt. It has been proposed that lipids should be extracted from spent grains and added to fermenting beer as an anti-foaming agent. The small quantities of lipid that are normally dispersed in wort are important. Non-polar lipids tend to collapse beer foam, so the presence of excess quantities is detrimental, while some polar lipids may help to stabilize it. Yeast, grown anaerobically, requires sterols and unsaturated fatty acids to maintain its viability and permit growth. Unsaturated fatty acids, or some of their metabolic products, may contribute to the development of stale flavours or other flavour defects in aging beers. Barley contains two lipoxygenase (LOX) enzymes that catalyse the oxidation of linoleic acid to the 9- and 13-hydroperoxides. LOX enzymes may survive to some extent in kilned malts. The hydroperoxides are isomerized and further changed, via ketols, to produce, among other substances, the trihydroxy fatty acids that occur in beer. It is believed that as beer ages these various products are converted to the aldehydes hexanal, *trans*-2-hexenal and other substances having off-flavours. In addition, lipids probably affect other beer characteristics, for example by favouring gushing (over-foaming) and encouraging the formation of esters by yeast, which, in turn, influence beer flavour and aroma. Barley also contains the enzyme superoxide dismutase (SOD), which occurs as several isozymes and increases in amount during malting. This enzyme catalyses the destruction of superoxide radicals, with the production of hydrogen peroxide and oxygen. In so doing it presumably helps to minimize deterioration owing to lipid oxidation (Bamforth, 1983).

4.12 Phosphates and inorganic grain constituents

After combustion 2–3% of the grain's weight remains as ash. Such a determination, or analyses of the ash constituents, gives no idea of the components' locations or roles in the living grain (Hopulele and Piendl, 1973; Liu *et al.*, 1975; Liu and Pomeranz, 1975; Pomeranz and Dikeman, 1976; Briggs, 1978). The quantity and composition of the ash depends on the variety and the soil and climate. Erdman and Moul (1982) reported that the average ash analyses for seven barleys grown on one site were: total ash, 2.2%; K, 0.46%; P, 0.24%; S (total), 0.16%; Mg, 0.14%; Ca, 0.037%; B, 1.3 p.p.m., Ba, 3 p.p.m.; Cd, 0.1 p.p.m.; Cu, 4.4 p.p.m.; F, 1 p.p.m.; Mo, 0.38 p.p.m.; Na, 100 p.p.m.; Ni, 0.21 p.p.m.; Se, 0.1 p.p.m.; Si, 2300 p.p.m.; Sr, 1.4 p.p.m.; and Zn, 27 p.p.m. Of the trace elements the high value for silicon (Si) results from the substantial deposits of silica (SiO_2) that occur in the outer layers of the husk and make it abrasive. The sulphur values

Table 4.12 Mineral analyses (d.b.) of a barley and a malt prepared from it

	Barley	Malt
Phosphate (as phosphorus) (%)	0.39	0.40
Potassium (%)	0.52	0.36
Magnesium (%)	0.16	0.13
Calcium (%)	0.14	0.13
Zinc (p.p.m.)	16	20
Iron (p.p.m.)	38	36
Manganese (p.p.m.)	16	14
Copper (p.p.m.)	11	7

From Pomeranz and Dikeman (1976).

include both inorganic sulphur compounds and organic sulphur compounds (e.g. cysteine, methionine) that were originally present in the grain. The minerals extracted from barley malt during mashing are usually sufficient to support healthy yeast growth. However, in some instances the levels of zinc may be so low that fermentation of the wort by the yeast is slow. Recoveries of zinc in worts are apparently very erratic.

Analyses of dissected grains and electron-probe microanalyses agree that the minerals in the interior of the grain are concentrated in the aleurone layer and embryo. During malting, minerals from the surface layers, especially potassium, are leached into the steep liquors. For instance, the ash content of one barley declined from 2.44% to 2.20% during steeping. Much mineral material is leached from the embryo end of the grain, especially potassium. Substantial quantities of phosphate are supposed to be lost from the grain during steeping, but not all analyses support this (Table 4.12). As malting proceeds, the minerals of the aleurone layer are rapidly mobilized and the phytin deposits in the aleurone grains (which are rich in phosphorus, potassium and magnesium) are broken down. They are released into the starchy endosperm and are taken up by the growing embryo. Thus during malting the minerals in the grain are redistributed. The rootlets and acrospires are rich in minerals. Those present in the rootlets are lost to the malt when it is deculmed.

Phosphate is a major component of the ash and it makes up about 1–1.5% of the grain's dry weight (Lolas *et al.*, 1976; Lee, 1990). Phosphate esters are

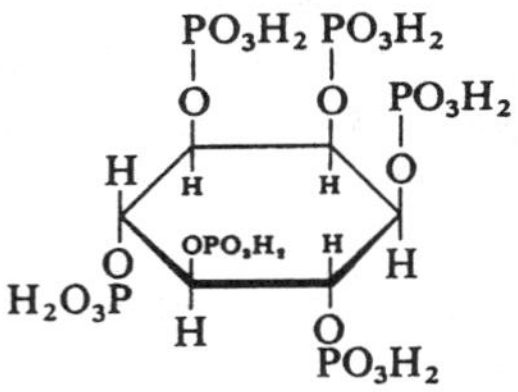

(**107**) Phytic acid, *myo*-inositol hexaphosphate

Table 4.13 Some reported vitamin contents of barleys and malts

	Barley	Malt
B_1; thiamin (**114**) (μg/g)	1.2–16.0	2.4–8.0
B_2; riboflavin (**111**) (μg/g)	0.8–3.7	1.2–5.0
Nicotinic acid (**115**) (μg/g)	47–147	48–150
Pantothenic acid (**112**) (μg/g)	2.9–11.0	4.3–12.9
B_6; pyridoxin etc. (**113**) (μg/g)	2.7–11.5	3.8–7.5
Vitamin E (**108**) (μg/g)	2.1–5.2	2.0–2.5
Biotin (**116**) (μg/g)	0.11–0.17	0.09–0.22
Folates, (**118**)		
Free (μg/g)	0.1	–
Combined (μg/g)	0.2	0.4
Choline (**70**) (mg/g)	1.0–2.2	–
myo-Inositol (**117**)		
Free (mg/g)	0.18	–
Bound (mg/g)	1.4–3.2	–

From Knorr (1952); Briggs (1978); Voss and Piendl (1978); other values are given by Scriban (1970).

involved in nearly all aspects of grain intermediary metabolism. As well as being present as inorganic phosphate (PO_4^{3-}), phosphate occurs combined in sugar phosphates, ATP, nucleic acids, phospholipids, and so on. Phytic acid (*myo*-inositol hexaphosphate, **107**) is the major reserve of phosphate of grain; about 68% of the grain's combined phosphate occurs as 'phytin', a mixed (potassium, magnesium) salt of phytic acid (Lolas *et al.*, 1976). This occurs as solid deposits within the aleurone grains in the cells of the embryo and the aleurone layer. During malting the activity of phytase, a phosphatase that is able to hydrolyse phytic acid, increases in amount. Phytic acid in the aleurone layer is hydrolysed, ultimately to inorganic phosphate and *myo*-inositol (**117**). The amount in the entire grain declines by about a quarter (Lee, 1990). The fate of the *myo*-inositol is unknown, but the inorganic phosphate is released from the aleurone layer into the starchy endosperm and is taken up by the embryo (Clutterbuck and Briggs, 1974). There is a net decrease of 10–30% of the grain's phytate during malting. The survival of the enzyme phytase is very temperature dependent. Consequently, when a lightly kilned lager malt is mashed using a rising series of temperatures and a decoction regimen, 60–90% of the phytate remaining in the malt is hydrolysed. However, when strongly kilned ale malts are mashed at a single, comparatively high temperature during infusion mashing, little further breakdown of phytate occurs. Phytate, even to a greater extent than inorganic phosphate, has a high affinity for calcium ions. Therefore, when calcium from the mashing liquor interacts with the phosphates during mashing or the copper boil, hydrogen ions are released and the pH is usefully reduced (Chapter 5).

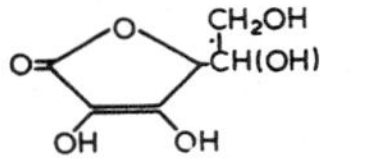

(109) Ascorbic acid

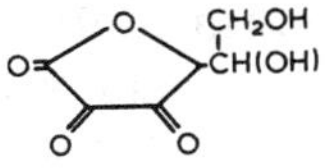

(110) Dehydroascorbic acid

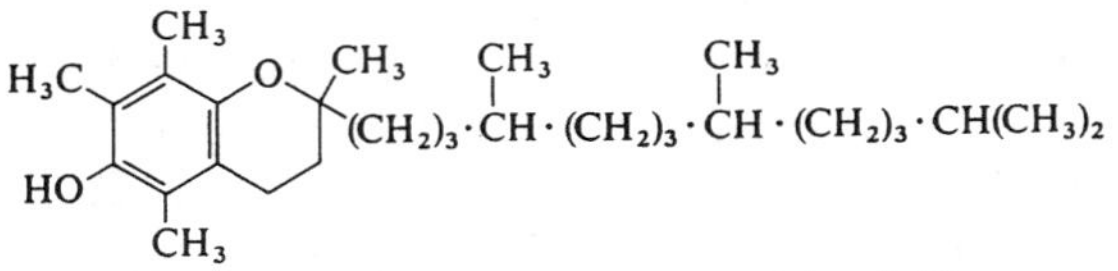

(108) α-Tocopherol (one component of vitamin E)

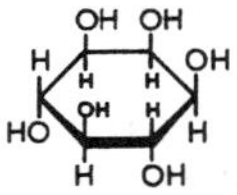

(117) *myo*-Inositol

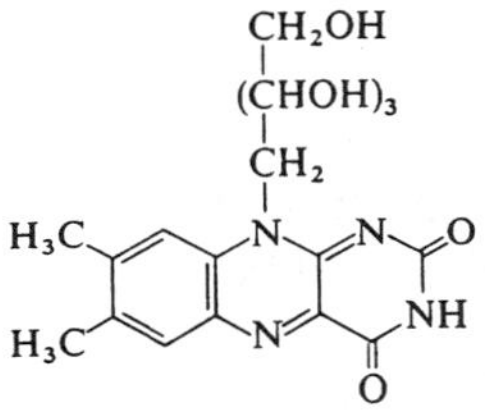

(111) Riboflavin, vitamin B_2

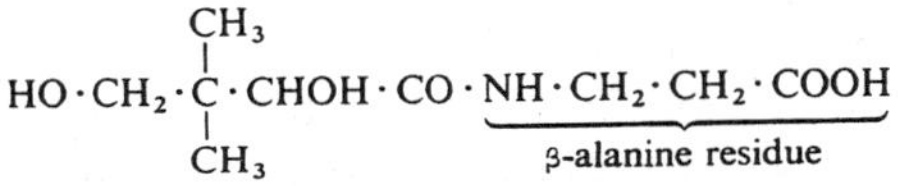

$HO \cdot CH_2 \cdot C(CH_3)_2 \cdot CHOH \cdot CO \cdot NH \cdot CH_2 \cdot CH_2 \cdot COOH$

β-alanine residue

(112) Pantothenic acid

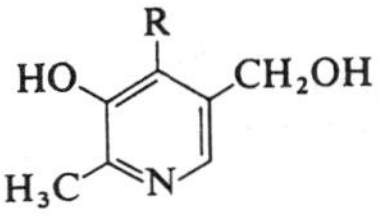

Pyridoxin ($R = CH_2OH$)
Pyridoxal ($R = CHO$)
Pyridoxamine ($R = CH_2 \cdot NH_2$)

(113) Vitamin B_6

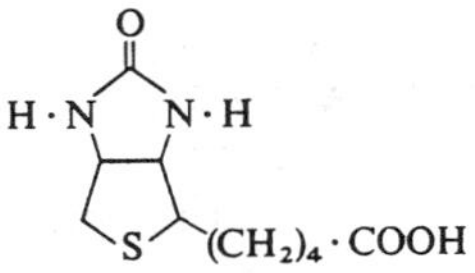

(116) Biotin

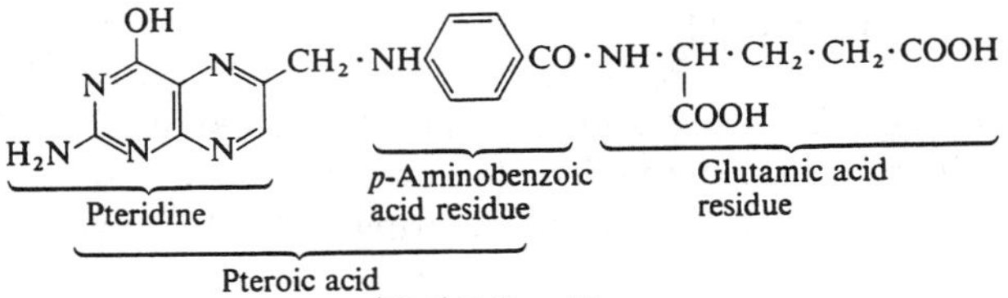

Pteridine

p-Aminobenzoic acid residue

Glutamic acid residue

Pteroic acid

(118) Folic acid

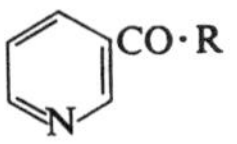

(115) Nicotinic acid ($R = OH$)
Nicotinamide ($R = NH_2$)

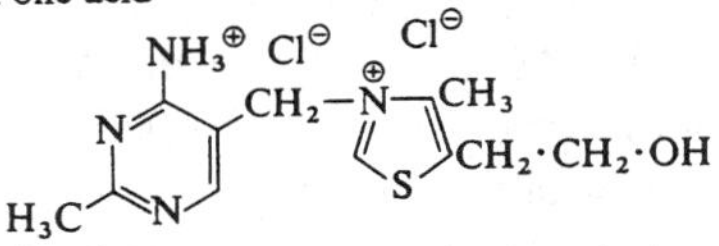

(114) Vitamin B_1 (Thiamin, Aneurine)

Fig. 4.18 Some vitamins that occur in barley and malt.

4.13 Vitamins and yeast growth factors

Barley and malt are rich in substances that are growth factors for yeast and/or are vitamins for human beings (Table 4.13, Fig. 4.18) (Harris, 1962a; Pollock, 1962a; Scriban, 1970, 1979; Briggs, 1978; Finney, 1982). Usually these substances are concentrated in the living tissues, the embryo and aleurone layer. They often occur combined in more complex molecules that serve as cofactors in intermediary metabolism. Some move into the rootlets during malting and so are lost to the malt. Rootlets are a rich source of growth factors for microbes. There are substantial disagreements regarding the levels of some vitamins in barleys and malts. In part this may be because of variations between samples and of failures to allow for losses in dry matter during malting, but often it is the result of technical difficulties with the bioassay methods used for estimating these substances.

For mammals, the fat-soluble vitamins are A, D, E and K. Vitamins A and D do not occur as such in barley or malt, but grains contain carotenoids, which may give rise to vitamin A, and sterols, which may act as precursors for vitamin D when ingested by mammals. Vitamin E (e.g. α-tocopherol (**108**), occurs in barley oil, which contains a range of tocopherols and tocotrienols (Morrison, 1978). Vitamin K is reported to increase in cereals germinated in light. Some of the sterols and fatty acids are essential growth nutrients for yeast under anaerobic conditions and are, therefore, important in fermentations.

The water-soluble vitamins C (ascorbic acid (**109**) and dehydroascorbic acid (**110**)) increase during germination. They are destroyed, apparently completely, when malts are kilned, but when malts are air-dried for use in foodstuffs significant amounts may survive (Chapter 1; Finney, 1982). It has long been known that sprouted peas and beans have anti-scorbutic properties, attributable to vitamin C, and these are widely used in foods. For human beings, it is clear that sprouting grains increase their digestibility and food value. However the increase in levels of most vitamins is, at most, moderate.

The water-soluble group of B vitamins, and other growth factors, is a complex mixture. Riboflavin (vitamin B_2, **111**), pantothenic acid (**112**) and the B_6 group of vitamins (pyridoxin and chemically related substances, **113**) are usually reported to increase during malting. Other vitamins such as thiamin (vitamin B_1; **114**), nicotinic acid (**115**) and pantothenic acid (**112**) alter very little. Other important factors include biotin (**116**) *myo*-inositol (**117**), folic acid (**118**) and related folates. Most reports deny the presence of vitamin B_{12} in grain, but the presence of tiny amounts (0.06 μg/g) has been reported. Possibly this originated from the microflora. Choline (**70**), a growth factor for some microbes, is said to increase in amount during malting.

4.14 Miscellaneous substances

Germinating barley grain contains appreciable levels of common metabolic intermediates, including glycerol and acetic, malic, succinic, fumaric, malonic, α-ketoglutaric, lactic, citric, aconitic, isocitric, oxalic and glycolic acids. Oxalic acid can give rise to hazes of calcium oxalate in beers. Traces of free cyanide (0.7 μg/100 g dry matter) may occur in barley and increase

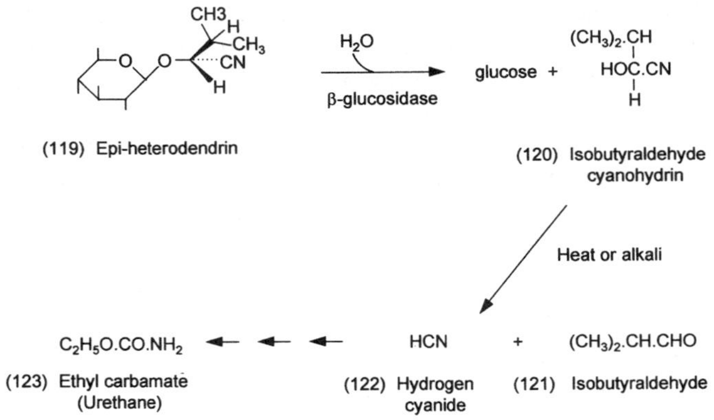

CH3 H CH$_3$ CN H O O

H_2O → glucose + $(CH_3)_2.CH$ HOC.CN H

β-glucosidase

(119) Epi-heterodendrin

(120) Isobutyraldehyde cyanohydrin

Heat or alkali

$C_2H_5O.CO.NH_2$ ← ← ← HCN + $(CH_3)_2.CH.CHO$

(123) Ethyl carbamate (Urethane)

(122) Hydrogen cyanide

(121) Isobutyraldehyde

Fig. 4.19 The generation of ethyl carbamate and hydrogen cyanide from epiheterodendrin.

during germination. Germinating barley forms traces of the cyanogenic glycoside epiheterodendrin (**119**) (Fig. 4.19). Five cyanoglucosides have been recognized in the epidermis of barley (Pourmohseni *et al.*, 1993). Grain can metabolize added cyanide to asparagine (**36**). Germinating sorghum seedlings contain substantial quantities of the cyanogenic glycoside dhurrin.

Ethyl carbamate (urethane, **123**), suspected to be a carcinogen, has been detected in barley grain-spirit and whiskeys where it appeared during storage (Battaglia *et al.*, 1990). In the spirit, a reaction catalysed by copper ions occurs between ethanol and cyanide or substances derived from it, such as lactonitrile and cyanate ion. The volatile precursor, that distills with the spirit is cyanide (**122**). This, in turn, is derived from isobutyraldehyde cyanohydrin (**120**), which is decomposed by heat. The cyanohydrin is released from its β-D-glucopyranoside, epiheterodendrin (**119**) by hydrolysis catalysed by yeast β-glucosidase (Fig. 4.19). Epiheterodendrin is found in the acrospires of malt. There are wide variations between malts made from different barleys (Cook *et al.*, 1990). Some barley malts yield as much as 800–1200 p.p.b. measurable cyanide (MC) whereas those from other barley varieties yield as little as 0–100 p.p.m. MC. Whisky malts are increasingly being made from varieties of barley that produce only small amounts of epiheterodendrin (urethane precursor, glycosidic nitrile, combined nitrile, ethyl carbamate precursor). These findings have obvious implications for the use of sorghum malts, since it is well known that some sorghum tissues are extremely rich in the cyanogenic glycoside dhurrin. Ethyl carbamate has also been detected in pot-still whiskey, and again its formation is dependent on the presence of copper. The route(s) of formation may include the degradation of copper–peptide or copper–protein complexes (Riffkin *et al.*, 1989).

Many other volatile substances have been detected in malting grain and malt. Although some may be artefacts, their occurrence is of interest because of the wide range of 'novel' compounds that are formed during kilning, with consequent implications for flavour and aroma. Ethanol, the chief product of the fermentative breakdown of sugars, is present.

Very many enzymes have been detected in quiescent or germinated grain. In barley, and a number of other plants, chitinase and lysozyme have been detected, but the grains themselves do not seem to contain substrates for

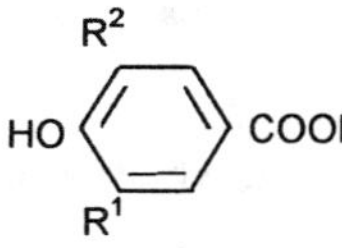

R^2 HO COOH R^1

R^1 HO CH=CH.COOH R^2

(125) *p*-Hydroxybenzoic acid ($R^1 = R^2 = H$)

(126) Gallic acid ($R^1 = R^2 = -OH$)

(127) Protocatechuic acid ($R^1 = OH$; $R^2 = H$)

(128) Vanillic acid ($R^1 = OCH^3$; $R^2 = H$)

(129) Syringic acid ($R^1 = R^2 = OCH_3$)

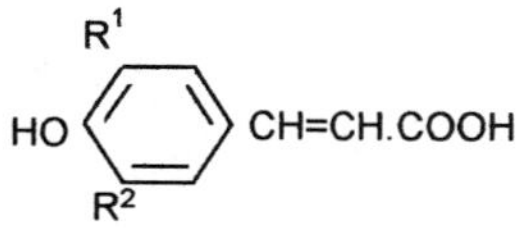

(132) Caffeic acid ($R^1 = OH$; $R^2 = H$)

(133) Ferulic acid ($R^1 = OCH^3$; $R^2 = H$)

(134) *p*-Coumaric acid ($R^1 = R^2 = H$)

(135) Sinapic acid ($R^1 = R^2 = OCH_3$)

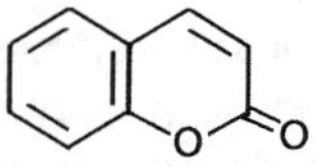

O O

(130) Coumarin

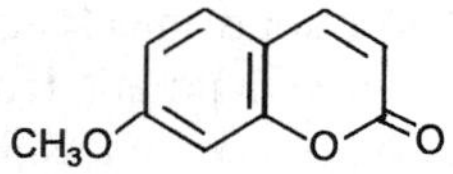

CH_3O O O

(131) Herniarin

HO COOH HO OH O.CO.CH=CH OH OH

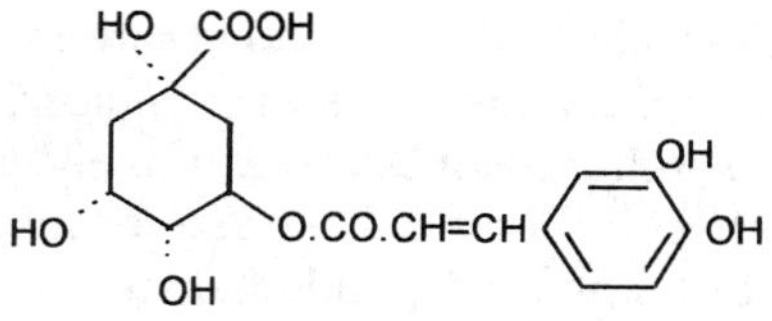

(124) Chlorogenic acid

Fig. 4.20 Some phenolic acids and related substances that occur free or combined in barley and barley malt.

these hydrolytic enzymes. It is supposed, therefore, that chitinase helps resist attack by fungi, by degrading the chitin in their cell walls, and lysozyme acts against bacteria with susceptible cell walls. The enzyme is detected by its ability to cause the lysis of *Micrococcus* cells.

Some differences between different malting grains are worth highlighting. For example, hordenine (**57**) and gramine (**58**) do not occur in malting wheat or rye. By comparison, various complex hydroxamic acid glucosides, which occur in rye, wheat, triticale, sorghum and maize and which are defence chemicals, do not occur in barley (Niemeyer, 1988).

4.15 Phenols and related substances

Numerous phenolic substances have been detected in barley (Briggs, 1978). Many of these remain uncharacterized. Phenolic substances already noted include tyrosine (**54**) and the related amines, including hordenine (**57**), tocopherols (**108**), coumaroyl agmatine (**64**) and and the 5-*n*-alkylresorcinols.

Other simple compounds include phenolic acids (Fig. 4.20), some of which occur free but the majority of which are present bound as esters or glycosides from which they can be liberated by hydrolytic reactions. Chlorogenic acid (**124**) is an ester of caffeic acid (**132**) and quinic acid. The simple acids occur in two groups: the substituted benzoic acids, such as *p*-hydroxybenzoic acid (**125**) and gallic acid (**126**), and the substituted cinnamic acids, such as caffeic acid (**132**) and ferulic acid (**133**). Other related substances, which contain internal ester links between a carboxylic acid group and a phenolic hydroxyl, are coumarins such as coumarin itself (**130**) and variously hydroxylated and methoxylated derivatives such as herniarin (**131**).

Barley malts contain the flavanol chrysoeriol (**136**) and other substances that are probably related (Fig. 4.21). Barley leaves contain complex mixtures of flavonoids, the nature of the mixture being variety dependent. These leaf flavonoids are flavone-*O*-glycosides based on isovitexin, isoorientin and isoorientin-3′-methyl ether and *C*-glycosyl flavones based on apigenin, luteolin and chrysoeriol (Fröst *et al.*, 1977).

The group of phenols that has been most studied contains the flavan-3-ols (Fig. 4.21) such as (+)-catechin (**137**), gallocatechin (**139**), and the pigments and condensed tannins that are derived from the catechins. Many barley grains have reddish colorations along the veins of the husk, and some few are quite strongly coloured, for example having red-pigmented husks or blue aleurone layers. Indeed, Canadian malting barleys were bred to have blue aleurone layers, as a useful identification aid, and the malting variety Halcyon, used in the UK, is also pigmented this way. The pigments are without any adverse effects on malting behaviour or when using malts. The colours result from the presence of anthocyanidins (**141**, **142**, **143**), or their glycosides, such as cyanidin 3-arabinoside.

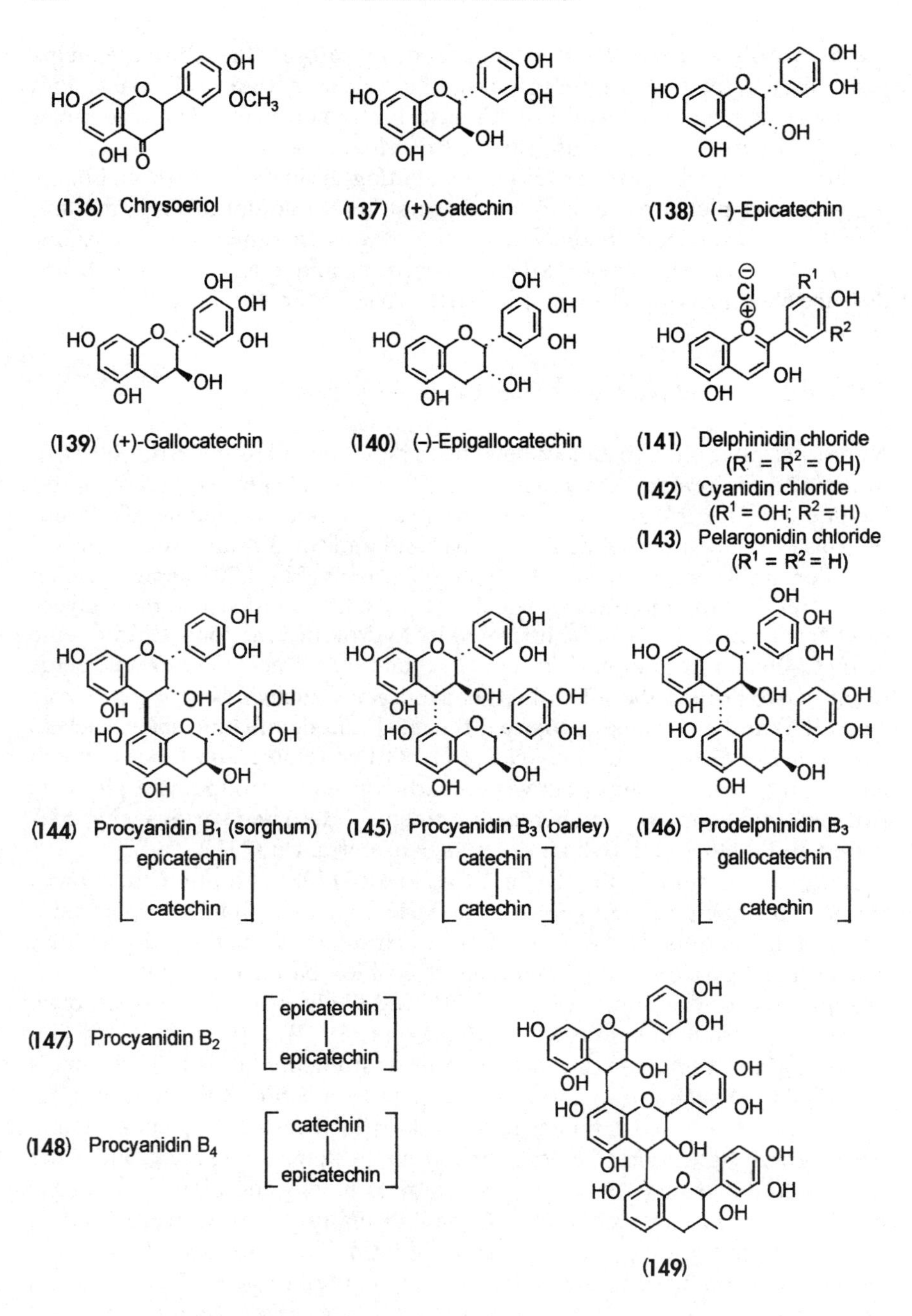

Fig. 4.21 Flavonoids and related substances found in barley. Propelargonidins have been detected in some barleys.

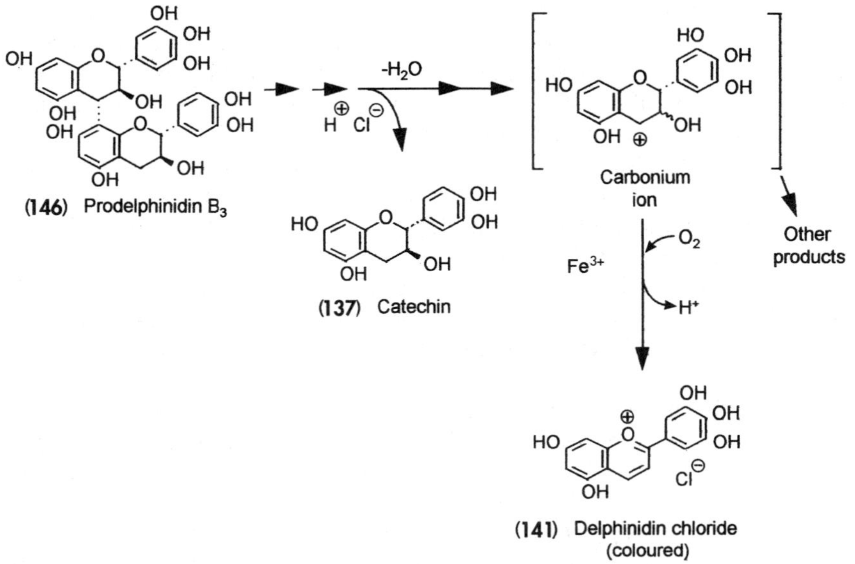

Fig. 4.22 The route by which colourless proanthocyanidin (an anthocyanogen) gives rise to the coloured anthocyanin pigment delphinidin when an acidified solution is heated in air.

The colourless, condensed tannins were originally called leucoanthocyanins, because their chemistry was unknown and they gave rise to coloured anthocyanidin pigments when heated with mineral acids in air (Fig. 4.22). It was thought that these were monomeric flavanol 3,4-diols. However, this is not the case, and the term leucoanthocyanin is incorrect. Maltsters and brewers usually refer to the condensed tannins as anthocyanogens and chemists now call them proanthocyanidins. Acid treatment of barley and malt anthocyanogens gives rise to the coloured substances delphinidin (**141**), cyanidin (**142**) and, less frequently, pelargonidin (**143**). These three substances differ in the degree of hydroxylation of one ring. Of the common cereals, apparently only sorghum contains comparable condensed tannins. The condensed tannins are polymers of flavan-3-ols, of various degrees of complexity. Bi-, tri- and tetraflavonoids occur, perhaps with substances having even higher degrees of polymerization. They seem, like the catechins, to occur free in barley and sorghum, but the relative quantities differ in the different species. For example, the dimer procyanidin B_1 (**144**) occurs in sorghum and procyanidin B_3 (**145**) is a major dimer in barley. All the procyanidin dimers are important in brewing. Procyanidins B_1 (**144**), B_2 (**147**), B_3 (**145**) and B_4 (**148**)) occur in hops. Other pro–anthocyanidin dimers occur that, on acid treatments, give rise to delphinidin (e.g. prodelphinidin B_3, **146**) and to pelargonidin (**143**) in a few barley varieties. The catechins (**137–140**) are not themselves proanthocyanidins. When the polymeric condensed tannins are heated with strong mineral acids, in the presence of air (to provide oxygen) and with catalytic amounts of iron salts (McFarlane and Sword, 1962; Porter *et al.*, 1986), the terminal catechin unit is released, and a proportion of the other monomer units is converted via carbonium ion intermediates into coloured anthocyanidins (Figs 4.21 and 4.22).

Because of the complex nature of the mixture of phenols in barley and because of their instability under a range of conditions and in the presence of air, they are difficult to characterize and methods for their quantification are only partly successful. Therefore, phenols with 'tanning' ability have been characterized by their ability to form hazes or precipitates with solutions of standard proteins, cinchonine sulphate or soluble polyvinylpyrrolidone (PVP). 'Oxidizable' phenols have been assayed, as tannins, with cinchonine sulphate after oxidation with hydrogen peroxide and peroxidase. A wide range of colour reactions has been used to determine total polyphenols, including reactions with diazonium compounds, with ferrous ions and with the Folin–Dennis and Folin–Ciocalteau reagents; however, as each phenol or phenol derivative gives a different colour yield, the results of these determinations are of limited value because of the complexity of the mixture of grain phenols. More specific colorimetric reagents are available, for example the vanillin/strong mineral acid reagents that give colours with flavonols. However, since the different flavonols have properties that differ to an important extent, this approach is also of limited industrial value. The most promising approach is the quantification of the separated substances by HPLC.

Polyphenols with tanning power bind strongly to peptide links in polypeptides and proteins and to similar chemical links in synthetic polymers. This property has been used in several assays for anthocyanogens. An extract of malt (made with 60% ethanol containing ascorbic acid as an anti-oxidant), wort or beer is shaken with a powdered adsorbent (Fig. 4.23): originally nylon-66 but now the more selective PVPP (AT, Agent AT-496), which has a higher binding capacity than nylon. The powder, with the adsorbed condensed tannins (the anthocyanogens) is separated and rinsed so interfering substances are removed. Then the PVPP powder is heated in air with a reagent containing butanol, *n*-methyl pyrrolidone (to displace the phenols into solution) hydrochloric acid and an iron salt. The anthocyanogens are degraded (Fig. 4.22) and the colour from the anthocyanidins generated in the reactions is measured. Unfortunately the colour yield varies according to the nature and degree of polymerization of the substances in the mixture of anthocyanogens. However, many useful results have been obtained using this technique. Phenols appear to occur in all parts of the grain, but particular groups are locally distributed. For example, lignin is virtually confined to husk tissues and the 5-*n*-alkylresorcinols occur in the testa wax. In particular, barley coloured anthyocyanin pigments may occur in the husk, in the pericarp or in the aleurone layer. In contrast, the anthocyanogens

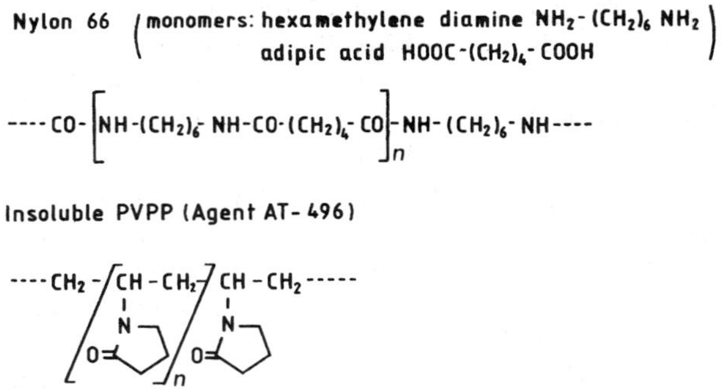

Nylon 66 (monomers: hexamethylene diamine NH_2-$(CH_2)_6$ NH_2
adipic acid $HOOC$-$(CH_2)_4$-$COOH$)

$$----CO-\left[NH-(CH_2)_6-NH-CO-(CH_2)_4-CO\right]_n-NH-(CH_2)_6-NH----$$

Insoluble PVPP (Agent AT-496)

$$----CH_2-\left[CH-CH_2\right]_n-CH-CH_2-----$$

N
O

N
O

Fig. 4.23 The formulae of the insoluble polymer nylon 66 and cross-linked PVPP. These substances adsorb polyphenols from solution.

(proanthocyanidins), located by staining the exposed surfaces of cut grains with a vanillin–hydrochloric acid reagent, are confined to the testa (Aastrup and Outtrup, 1985). This explains the observation that anthocyanogens are initially retained when grains are decorticated but are rapidly leached out in subsequent soaking. Birdproof sorghums, which are extremely rich in condensed tannins, have them concentrated in the testa–pericarp region (Morrall *et al.*, 1982).

Lignin is an insoluble, polymeric and chemically complex substance that contains hydroxylated and methoxylated phenyl propane residues. It is probably linked to polysaccharides in cell walls, but its chemistry is little understood. In barley, it is virtually confined to the husk, to which it confers rigidity and strength. It seems to remain unaltered during malting, but traces are formed in the vascular strands of the growing embryo.

The phenols are important in a number of ways. They certainly have antimicrobial properties, and so help to protect the grain and may play a role in controlling dormancy. Their importance in malting and brewing is better understood. For example, when gibberellic acid is applied to grain a proportion is retained by the surface layers, where it is associated with phenols (Nutbeam and Briggs, 1982). Polyphenols readily bind to peptides and proteins. Initially this association is reversible but with the passage of time, and particularly under oxidizing conditions and in the presence of catalytic amounts of heavy metal ions, the molecules become joined irreversibly through covalent linkages to side chains of the proteins. The more complex polyphenols that interact with proteins are termed tannins, after the various vegetable polyphenols that are used to treat hide so 'tanning' the proteins in it and converting it into leather. Not all phenols have tanning ability; the more highly polymerized materials with larger molecules are more effective. Presumably this is because they can bridge between adjacent polypeptide chains, forming large and progressively less soluble polymeric aggregates. The polyphenol–protein interactions are important in various ways. During mashing, enzymes may be inactivated; in mashes with bird-proof sorghums, with their exceptionally high levels of condensed tannins, enzyme activity may be so severely curtailed that the mashing process fails. High levels of tannins reduce the digestibilities of grains and malt preparations. This can be important with foods based on sorghum malts. In beer brewing, phenols and polyphenols are probably important for their anti-oxidant properties, their astringency, their contributions to the formation of hot and cold breaks and to the formation of the most common types of non-biological haze in beers, both chill-hazes and permanent hazes. As phenols are probably readily oxidized they will remove oxygen which has 'contaminated' beer, perhaps with advantages, for example towards flavour stability and antagonizing contamination by aerobic microorganisms. However, the oxidized products can probably polymerize and interact with other polyphenols and proteins, leading to haze formation (Fig. 4.24).

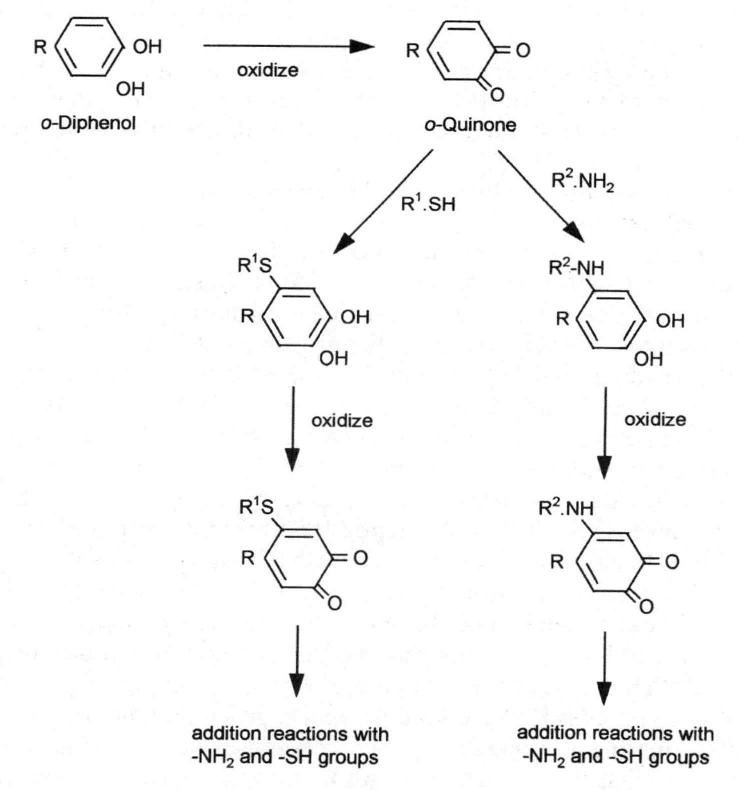

Fig. 4.24 A scheme illustrating how phenols may oxidize to form quinones, which react with other molecules to generate covalent linkages, e.g. with thiol or amino groups in the side chains of peptides and proteins. Alternative reaction schemes involve free-radical intermediates.

Some phenols are removed from grain during steeping. The quantities removed are increased by giving a short alkaline steep (long alkaline steeps can damage grains). Beers made from alkaline-steeped malts have superior, 'cleaner' flavour characteristics, tend to be less astringent and to be less prone to form hazes. The alkaline wash removes other substances besides phenols from grain, including some protein. The crude extract has been termed testinic acid and has been extensively studied (Stevens, 1958). However there is disagreement as to which polyphenols are responsible for astringency, and beers made from malts lacking anthocyanogens are said to have normal flavours.

Brewers of beer and malt-vinegar, and hence maltsters, have been most interested in the anthocyanogens (proanthocyanidins) because it is these that provide most of the tannins that are of importance in beer and vinegar haze stability. In brewing, sweet wort is heated with hops during the copper boil and is held, for example, in a whirlpool separator; a coagulum forms (the hot break or trub) that consists in part of denatured protein from the mash associated with polyphenols from the malt and from the hops. This is removed. During wort cooling a second coagulum (the cold break) may form, which may be separated before the hopped wort is inoculated (pitched) with yeast to initiate fermentation. It is desirable for these breaks to be formed promptly so that they can be separated to give bright (entirely clear) worts. The proteinaceous substances and phenols removed with the breaks are not carried forward into the packaged beer, and so they cannot contribute to the formation of non-biological hazes. Hazes, finely suspended light-scattering particles which form in bright beers with the passage of time, may consist of fine crystals of calcium oxalate or, rarely, undegraded starch, β-glucan or pentosan. By far the most frequently encountered hazes are formed primarily between polypeptide materials and phenols, with which carbohydrates and metallic ions may be associated.

It was thought that as beers age the simpler polyphenols, and especially the dimeric, trimeric and more complex anthocyanogens, polymerize by oxidative and acid-catalysed mechanisms and cross-link the peptides until they are large enough to scatter light and form a haze. However this view has been challenged (Derdelinckx and Jerumanis, 1987). Initially the polyphenol–protein associations are reversible since the haze (chill-haze) only appears when the beer is cooled, and it disappears again when it is warmed to room temperature. However, with the passage of time the number of covalent links increases until the haze becomes permanent; that is it will not disperse when the beer is warmed. Brewers take various precautions to minimize haze formation, including the exclusion of oxygen (air) from beer, the exclusion from the process stream of heavy metal ions (copper, iron) that could catalyse haze formation, the removal of proteinaceous material with the trub by vigorous copper boils, precipitation in the copper by adding calculated amounts of a tannin, adsorption from the wort onto particles of silica hydrogel, degradation by the proteolytic enzyme papain added to the beer and adsorption of polyphenols from the wort onto particulate PVPP (Fig. 4.23). Haze precursors have been removed by inducing polymerization in the mash using oxygen, hydrogen peroxide or formaldehyde. These last processes are no longer in use.

Many attempts have been made to produce barley malts that give beers with reduced haze potentials. Three approaches have been partly successful. (i) The use of alkaline steeps; these are used in some maltings, e.g. in mainland Europe, but only rarely in the UK. (ii) Steeping or resteeping grain in dilute formaldehyde solutions, or adding formalin to the finished malt (Macey *et al.*, 1966; Withey and Briggs, 1966; Whatling *et al.*, 1968). The anthocyanogen levels in worts from formaldehyde-treated malts are dramatically reduced and beers have greatly enhanced shelf life, i.e. the period before haze appears is extended. The formaldehyde probably cross-links the polyphenols and proteins yielding insoluble polymers that are not extracted into the wort during mashing. The use of these processes has been discontinued because of 'sentiment' against the use of formaldehyde. However all tests appeared to show that the treatments were safe. Similar approaches have been made to reduce the tannin levels of sorghum malts. (iii) The most promising approach is the use of barleys bred from mutants that form no anthocyanogens. Extended studies, particularly in Denmark, have led to the discovery of numerous mutants with blocks in the biosynthetic sequence leading to proanthocyanidins. For example, grain with the mutation *ant-13*, originally in the variety Foma, contains only traces of catechin and proanthocyanidins. Beers brewed with malts made from these barleys have remarkably high colloidal stabilities, that is they are unlikely to become hazy (Ahrenst-Larsen and Erdal, 1979; von Wettstein *et al.*, 1980, 1985). At present (1997) agronomically acceptable 'low-proant' barleys are in limited production. The variety Galant produced grain that, at least from some areas, had partially dead aleurone layers and did not malt well. This undesirable character is not genetically linked to the low-proanthocyanin characteristic (Sole *et al.*, 1987). The flavours from flavanless beers, perhaps unexpectedly, appear to be normal.

Birdproof sorghums are comparatively resistant to the depredations of pests and birds, with consequent yield advantages. However, their high polyphenol levels make them inherently unsuitable for making malts for producing foods or Bantu opaque beers. To overcome this problem, birdproof sorghum for malting has been leached with water, has been steeped in dilute hydrochloric acid or sodium hydroxide solutions and it has been treated with formaldehyde (0.03–0.08% for 4 h). This last treatment appears to be the most effective (Daiber, 1975, 1978; Reichert *et al.*, 1980; McGrath *et al.*, 1982; Glennie, 1983).

Various oxidase and peroxidase enzymes that are able to act on phenolic substances are present in barley, but very little is known of their technological importance. Green malt and, to a lesser extent, lightly kilned malts probably contain oxidases that catalyse the oxidative polymerization of polyphenols in well-aerated, crushed grains, giving rise to tannins that are retained in the spent grains at the end of mashing. Similarly aerated mashes made with green or lightly kilned malts give rise to worts with reduced levels of anthocyanogens and to beers with longer shelf lives (MacWilliam *et al.*, 1963; Chapon and Chemardin, 1964). Similarly peroxidase activity is present in green malt and some can survive light kilning. This enzyme can catalyse the destruction of polyphenols in mashes when these are supplemented with small quantities of hydrogen peroxide; beers from these mashes are more stable (Whatling *et al.*, 1968). As hydrogen peroxide decomposes to water and oxygen it leaves no undesirable residues. However for other reasons, connected with flavour stability and possibly ease of wort separation, mashing is increasingly carried out in ways designed to exclude oxygen and oxidizing conditions.

4.16 The regulation of modification in malting

The intermediary metabolism of the living tissues of barley is complex, both in the interactions that occur between the tissues and organs and at the subcellular level (Briggs, 1978, 1987b, 1992). Maltsters need some conception of the overall processes that occur. Molecules are degraded, others are synthesized or are interconverted and others are built into macromolecules that form parts of complex structures in new cells or cell walls. The energy required to drive the biosynthetic processes is derived in part from fermentative metabolism, but chiefly from respiration. Some lipid is used as an energy source but the main respiratory fuel is carbohydrate. Details of many of the metabolic pathways, while largely elucidated, are unimportant to maltsters, but some of the consequences of the metabolic fluxes need to be appreciated. Alterations of malting conditions and applications of additives can alter fermentation and respiration rates and hence heat generation and output, respiratory losses, rootlet growth (and hence losses owing to this), the accumulation of hydrolytic enzymes, the rate and extent of modification and changes in other malt characteristics, such as the hot water extract, the cold water extract and the soluble nitrogen ratio (Chapter 14).

All these changes depend on the biochemical processes occurring within the grain. Some understanding of the organization of these processes and how they are influenced by changing conditions is needed if malting is to be understood and regulated in a rational manner. Also it is necessary to understand the roles of the separate tissues and organs of the grain, in particular the surface layers (husk, pericarp and testa), the embryo, the aleurone layer and the starchy endosperm, and how interactions between them occur. This is an area of continuing research. The development of many, perhaps all, of the newer malting methods, such as those intended to hasten malting or to make malts with small attendant malting losses, have been inspired or facilitated by advances in our understanding of grain organization. Some aspects are considered in the following sections. Many of the changes in malt characteristics associated with altering germination conditions are considered in Chapter 14, following a consideration of malt analyses (Chapter 13).

4.16.1 The influence of the barley sample and moisture content

Barley samples differ in suitability for making malts and grains of different varieties differ in the types of malt that may be most readily prepared from them; indeed some barleys are regarded as being unsuitable for malting (Shands *et al.*, 1941, 1942; Harris, 1962b; Ballesteros and Piendl, 1977; Briggs, 1978, 1987b; Pollock, 1962a,b). This may be because they make malts with unacceptably low yields of extract, they may only modify adequately at the expense of very high malting losses, they tend to germinate slowly, endosperm modification may be unacceptably slow or irregular or they may produce insufficient diastatic enzymes. A particular sample of a genetically good malting barley may be unusable because of poor viability, heavy fungal contamination, a high level of dormancy (which may disappear during prolonged storage), small grain size, a large proportion of damaged grains, or an unduly high nitrogen content. Barleys with high TN or crude protein contents (TN × 6.25%) give malts with low extracts. They are rich in some enzymes and they respire and grow vigorously (processes that lead to high malting losses). Their endosperms are often dense and steely, and they tend to take up water slowly during steeping. Often they are 'stubborn', that is their endosperms are slow to modify. To offset this they may be steeped to high moisture levels, which favours enzyme production and high malting losses. Such high-nitrogen barleys, of specially selected varieties, are used in making enzyme-rich, highly diastatic malts in which high enzyme levels and substantial yields of soluble nitrogenous materials are more important than high yields of extract (Chapter 14).

In malting it is desirable to achieve target moisture levels exactly. At less than 32% moisture, barley will not germinate. At 34–37% moisture grain will germinate rapidly, even if it is water sensitive (and so this is the target moisture to begin an air-rest); however, this is too low a moisture content to permit adequate endosperm modification, and so the grain must subsequently be wetted more (Chapters 3 and 14). Mellow barley with low TN values (*ca.* 1.4%) may be steeped to 42–43% moisture and malted in floor-maltings. More steely, stubborn and high-nitrogen grains may be steeped and wet until they have moisture contents as high as 46–48%. Usually these higher moisture contents can only be achieved if wetting is continued after grains have chitted. In special instances, as in resteeping to achieve root kill and reduced malting losses, even higher moisture contents (of 50% and more) may be reached. The temperature of steep water is important. At high temperatures, water uptake is faster, but water sensitivity may be induced and hence irregular germination may follow (Chapter 3). During steeping, as the moisture content increases so does the grain's metabolic rate, the rate at which it generates ethanol (Figs 3.15 and 3.16) and releases carbon dioxide, or the rate at which it takes up oxygen. Enzyme levels change to a small extent during steeping, for example the diastatic power may decline. These changes are influenced by steep aeration and air-rests.

Steeping is carried out in such a way that, it is hoped, all grains will be evenly and adequately hydrated, so that germination is rapid and uniform and that rapid and even modification of the grains is subsequently achieved. At the same time, the moisture level is low enough to limit embryo growth and respiration, and hence malting losses. The metabolic rate of grain is strongly dependent on its moisture content. The malting loss is strongly correlated with the maximum carbon dioxide output rate (Meredith *et al.*, 1962). Therefore, as initial moisture contents are increased, the extracts of the final malts tend to increase up to a maximum value and may then decline a little, while their hardness declines, that is their friability increases. In a series of barleys malted in a similar way, α-amylase levels, malting losses, rootlet yields, soluble nitrogen ratios and levels of soluble polyphenols increased with increasing moisture contents, while the β-glucan levels declined (Ballesteros and Piendl, 1977).

During malting the grain gains moisture as a product of respiration; in addition, sometimes (unusually in the UK) the grain is sprayed or sprinkled with water during germination (spray steeping). Conversely, water is lost by being bound in the products of hydrolysis (e.g. of starch to sugars, or proteins to amino acids) and there are losses by evaporation to the ventilating airstream. At the same time, as malting proceeds, the moisture in the grain is redistributed (Fig. 3.7). Initially the moisture film on the surface of the grain is taken up and later water is withdrawn into the growing embryo from the endosperm. The progressive drying of the endosperm causes modification to slow or even, in extreme cases, to cease. Water sprinkled onto grain to rewet it during germination is likely to be retained by the embryo and so accelerate its growth and the rise of malting losses, without, however, improving endosperm modification (Chapter 14).

4.16.2 Temperature, germination and malting

The effect of temperature on germination and malting has been extensively studied (Shands *et al.*, 1941, 1942; Harris, 1962b; Pollock, 1962a; Ballesteros and Piendle, 1977; Briggs, 1978). At low temperatures, near freezing, grain does not grow at a rate that is in any way useful to maltsters. As the temperature increases so growth occurs more rapidly and grain metabolism rises until the temperature reached is damaging to the grain, growth occurs more slowly or the grain is killed. There are other restrictions on the choice of temperatures used in malting. Traditionally, steeping was carried out at about 13 °C (55.4 °F) but, where possible, higher temperatures of 16–18 °C (60.8–64.4 °F) are often used to accelerate water uptake. However, at these and especially at higher temperatures water sensitivity will be induced and the grain will germinate unevenly. To some extent the adverse effect of higher temperatures may be offset by the use of air-rest steeping. In the maltsters' time scale, dormant grains germinate best at 12–13 °C (*ca.* 54 °F); they germinate markedly less well at 18–20 °C (*ca.* 66 °F), and much worse at higher temperatures (e.g. 25 °C (77 °F)). Therefore, the maximum temperature at which grain will grow rapidly and uniformly is dependent on its degree of maturity, that is the extent to which residual dormancy is still present.

In samples of fully mature (non-dormant) grain, steeped uniformly and then malted at different temperatures (12–25 °C (54–77 °F)), grain held at the higher temperatures grows faster, produces roots faster (Chapter 2) and initially produces many enzymes faster. However, the rate of formation of some enzymes, such as α-amylase and protease, rapidly falls off so that after a while grain growing at a lower temperature and producing enzyme at a lower but steady state comes to contain more enzyme than the 'high-temperature' grain. By beginning germination at a high temperature, e.g. 17 °C (62.6 °F), to induce initially rapid enzyme production, then after several days reducing the germination temperature, e.g. to 13 °C (55.4 °F), to maintain a steady rate of enzyme production, it is possible to obtain malts with unusually high enzyme contents in a relatively short time (Narziss, 1975, 1976).

In traditional British malting, germination temperatures of 12–13 °C (53.6–55.4 °F) over 7–10 days were preferred, but in modern practice temperatures of up to 18–20 °C (*ca.* 64–68 °F) are common, at least at some stages of germination, over periods of 3.5 to 6 days, depending on whether additives, abrasion and germination-accelerating schedules have been used. In special instances, even higher temperatures may be used to process poor-quality, nitrogen-rich barleys. Warm-water steeping, often using novel schedules, and germinating at elevated temperatures (25–30 °C; 77–86 °F), combined with the use of gibberellic acid, can limit embryo growth and so reduce malting losses (Pool, 1964a,b; Macey, 1977). However warm water steeping is not widely used (Chapter 14). Germinating at elevated temperatures can lead to irregular growth, and it does not accelerate all the changes and metabolic processes in grains to the same degree. Consequently, malts grown at higher temperatures tend to be inhomogeneous (individual corns may vary widely in their extents of modification), and they differ in their characteristics from malts made from the same barleys at lower temperatures; they are not merely the same type of malt produced in a shorter time.

4.16.3 Grain metabolism, enzyme development and modification

When grain is immersed during steeping, the imbibing tissues begin to metabolize, and they metabolize at accelerating rates as their water contents rise (Briggs, 1973, 1978, 1987b, 1992; Enari and Sopanen, 1986). Consequently, the oxygen dissolved in the steep water is rapidly utilized and the grain begins to ferment (respire anaerobically), the major products being ethanol, a smaller amount of L-lactic acid and carbon dioxide (Chapter 2). These products are derived, via the glycolytic pathway, from stored sugars. The process is temperature dependent and occurs faster at higher temperatures (Chapon, 1959). A considerable proportion of the ethanol and probably most of the carbon dioxide diffuses out of the grains and into the steep liquor (Smith and Briggs, 1979; Cantrell *et al.*, 1981). Steep aeration reduces alcohol production and often leads to subsequent more vigorous germination and respiration during the germination phase (Chapter 3). By initiating chitting in the steep, steep aeration or air-rests may also accelerate water uptake. Even with aeration, steep alcohol contents continue to rise and may rise even faster at steep-out until after chitting occurs, when the surface layers of the grain are broken and oxygen reaches the living tissues more readily. The alcohol content of the grain may be reduced by its oxidation in the living tissues or by evaporation. At steep-out, oxygen uptake is limited by the diffusion barriers presented by the film of surface moisture, the husk and the pericarp. The microbes present compete with the grain tissues for oxygen. The oxygen uptake rate is accelerated by the removal of the film of surface moisture, artificially or when it is taken up by the grain at the start of the germination period. Decorticated grain, from which the husk and pericarp have been removed experimentally, respires much more rapidly than intact grain because of the absence of the barrier to gaseous exchange normally presented by the husk and pericarp. The microbes also respire and multiply, both on the grain and in the steep liquors. Therefore, they compete with the grain tissues for oxygen and metabolize materials leached from the grain. When the grain has chitted, ethanol formation has ceased, L-lactic acid is rapidly lost and residual ethanol has disappeared, then oxygen uptake and carbon dioxide evolution occur at approximately equal rates, so the respiratory quotient is about one, and the major respiratory process is the total oxidation of carbohydrate to carbon dioxide and water. However a little triglyceride is also utilized. Probably the major routes of carbohydrate breakdown are the glycolytic sequence and the tricarboxylic acid cycle, with the pentose phosphate pathway playing a minor role. In the presence of air, ethanol production is switched off, but if grain is placed under anaerobic conditions ethanol production is instantly resumed. The hydrogen, or equivalent electrons, that are removed from metabolic intermediates when they are oxidized are carried via a normal electron transport chain to oxygen and are oxidized to water by cytochrome oxidase. Substrate

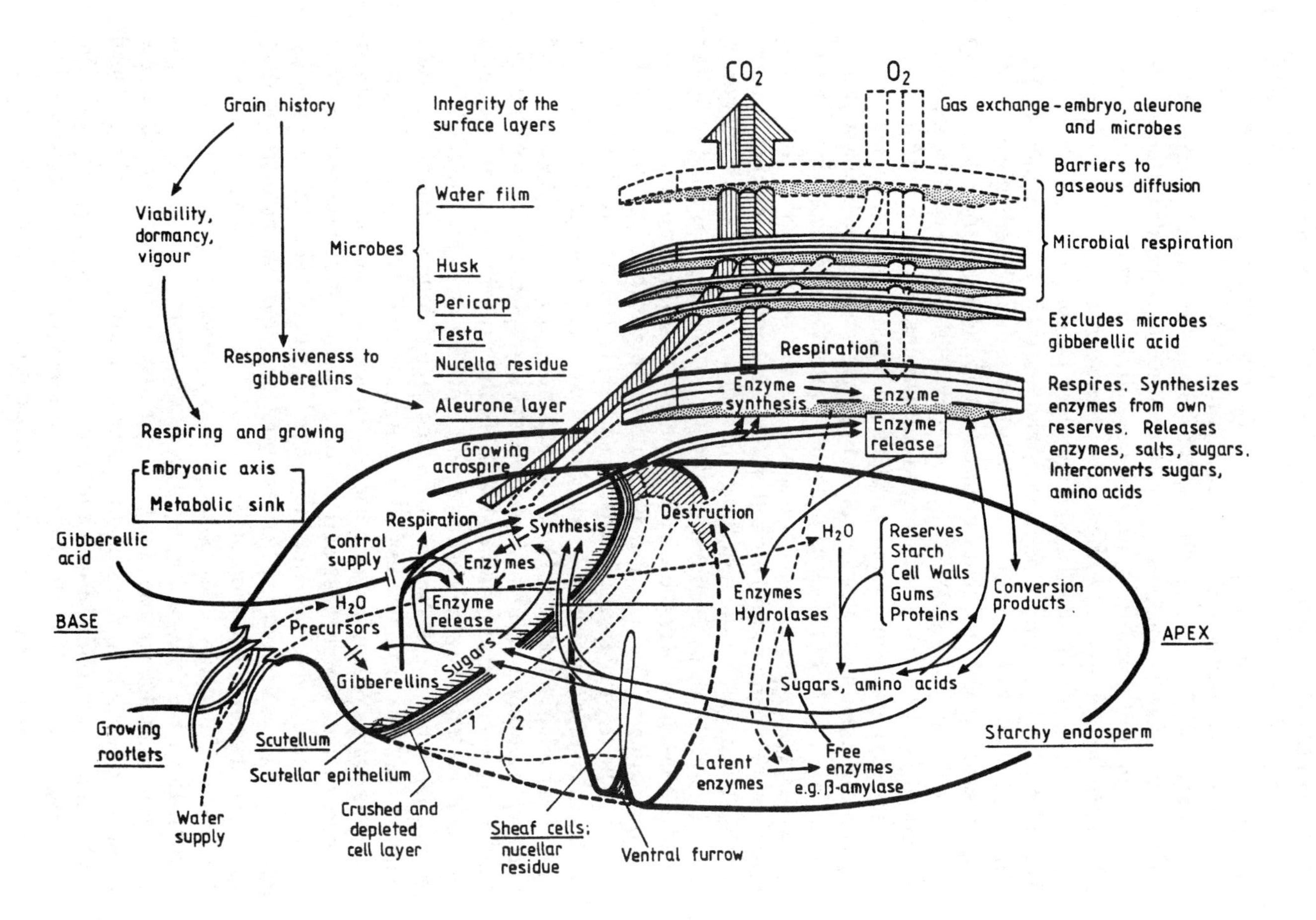

CO_2
O_2
Grain history
Integrity of the surface layers
Gas exchange - embryo, aleurone and microbes
Barriers to gaseous diffusion
Viability, dormancy, vigour
Water film
Microbes
Husk
Pericarp
Microbial respiration
Testa
Excludes microbes gibberellic acid
Responsiveness to gibberellins
Nucella residue
Respiration
Aleurone layer
Enzyme synthesis
Enzyme
Enzyme release
Respires. Synthesizes enzymes from own reserves. Releases enzymes, salts, sugars. Interconverts sugars, amino acids
Respiring and growing
Growing acrospire
Embryonic axis
Metabolic sink
Respiration
Synthesis
Destruction
Gibberellic acid
Control supply
Enzymes
H_2O
Reserves
Starch
Cell Walls
Gums
Proteins
Enzyme release
Enzymes
Hydrolases
Conversion products
BASE
H_2O
Precursors
APEX
Sugars
Gibberellins
Sugars, amino acids
1
2
Growing rootlets
Scutellum
Starchy endosperm
Scutellar epithelium
Latent enzymes
Free enzymes e.g. β-amylase
Water supply
Crushed and depleted cell layer
Sheaf cells; nucellar residue
Ventral furrow

level and oxidative phosphorylation processes provide the energy-rich substances that drive the constructive processes of the tissues. In detail, the situation is more complex than this, since a proportion of the respiration of the grain is cyanide resistant, indicating the presence of another respiratory pathway not involving cytochrome oxidase. The nature of the cyanide-resistant oxygen-consuming mechanism(s) is not clear, but part of the mitochondrial oxygen uptake is by a cyanide-resistant 'terminal oxidase'; polyphenol oxidases, lipoxygenase and ascorbate oxidases are among other enzymes that catalyse oxidative reactions. Respiration provides most of the energy that drives the grain's metabolism and the growth of the rootlets and acrospire. It also generates heat and during malting this must be utilized or removed to keep the grain bed at the correct temperature. Just as the rate of respiration changes during malting, so does the heat output of the grain. Grain respiration is influenced by the availability of oxygen. Thus, as the oxygen partial pressure is reduced the respiration of the grain declines, but not in direct proportion; for example, in 5% oxygen, respiration occurs at about 60% of the rate found in air (Briggs, 1978). The changes in the levels of the major organic acids involved in respiration have been investigated (South, 1996). Both D- and L-lactic acid accumulate in the stages of germination and kilning. Presumably the D-lactate is produced by microbes on the grain. A lack of oxygen at any stage causes the resumption of alcohol production, a cessation of growth and a cessation of enzyme production. Prolonged anaerobiosis is progressively damaging to the grain and finally kills it, so that when it is returned to air it will not grow. In contrast, within limits, increasing the oxygen partial pressure increases the rate of grain respiration and enzyme production. Elevated levels of carbon dioxide suppress respiration and, to a lesser extent, fermentative processes. Various specialized malting processes, such as Kropff malting (Chapter 14), have used anaerobiosis and elevated levels of carbon dioxide to regulate grain growth and hence to control malting losses.

At steep-out, the respiration of the grain mass has a variable, but probably substantial, component owing to the microflora. The remainder is about equally from the embryo and the aleurone layer. As germination proceeds the respiration rate of the aleurone does increase to some extent, but the major part of the increase occurs in the growing embryo. Thus, since the embryo is the most important respiring unit of the germinating grain and also produces the rootlets, which are removed at the end of malting, the embryo is the chief cause of the losses of dry matter that occur during

Fig. 4.25 Some of the metabolic inter-relationships that are important in malting barley. The approximate extents of endosperm modification after 1 and 2 days slow malting (1,2) are indicated. Gibberellin formation is shown to be oxygen dependent and influenced by sugar levels, which are controlled by a feedback loop. Microbes on the surface layers compete with the grain tissues for oxygen and may release phytotoxins. The roles of the surface layers and the ways in which hydrolytic enzymes are activated or generated are indicated.

malting. Numerous attempts have been made to devise means of checking embryo metabolism and growth during malting, in order to reduce malting losses. The intensity of grain respiration, and the accompanying heat output, is greatly influenced by malting conditions. Therefore, nitrogen-rich, high-protein grain respires faster than comparable barley having a lower nitrogen content. Although extended steeping often checks the initial rate of chitting and reduces the initial rates of respiration, subsequently grain grows more vigorously and respires faster at higher moisture contents.

Steep aeration (Fig. 3.17) and additions of gibberellic acid promote vigorous growth and respiration. Within a limited range of temperature, grain respires faster at higher temperatures. In particular trials while malting at 4–5 °C (40 °F), respiration increased linearly with germination time, but at higher temperatures of 18–20 °C (64.4–68 °F), the respiration rate peaked and declined, while at intermediate temperatures of 12–13 °C (*ca.* 54 °F), the respiration rate rose, levelled off and changed little in the later stages of germination (Day, 1891). Under traditional malting conditions, the maximum respiration rate occurs between germination (flooring) days 3 and 5 and the rate during this period correlates well with the ultimate malting loss (Meredith *et al.*, 1962).

Embryo growth, grain respiration and the development of the major enzyme systems of importance to grain modification and in the mashing process when the malt is utilized are not tightly linked in the sense that they do not all change in parallel in a sample of malting grain (Fig. 3.19). Furthermore, to a degree, they may be varied independently by changing malting conditions, with consequent alterations in the characteristics of the malt produced. However, when using one set of malting conditions to process one type of barley, the experienced maltster may use acrospire growth, rootlet production or (in principle) carbon dioxide output as indirect guides to the progress of modification in a batch of grain. The alternatives are to dry samples and analyse them using various physical or chemical techniques (Chapter 13).

The embryo orchestrates the malting process in each grain. Removal of embryos at steep-out prevents grain modifying, but transplanting an embryo onto a degermed grain causes the process to resume. Indeed embryos can be transferred between grains of barley, wheat, rye and oats with some success. For many years, it was thought that the embryo produced all the enzymes necessary for modification, but this is now known not to be the case (Briggs, 1973, 1987b, 1992). Isolated embryos produce limited amounts of a wide range of hydrolytic enzymes and ribonucleases. In addition, as shown by Harugoro Yomo in the period 1958–60, they also produce a substance, now known to be gibberellin, that triggers enzyme production in degermed grain through its action on the aleurone layer. Thus, the gibberellin is a plant hormone that carries a message from the embryo to the aleurone layer. Most recent data show that the major gibberellin in

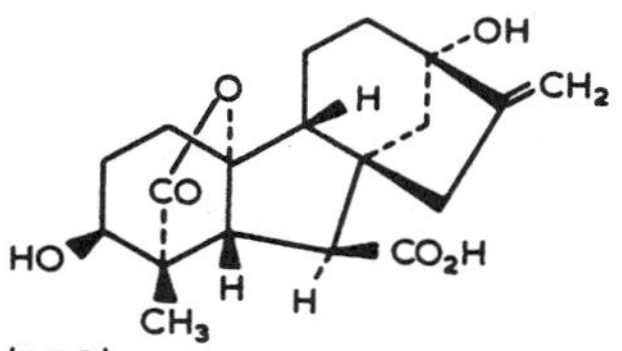

(**150**) GA_1

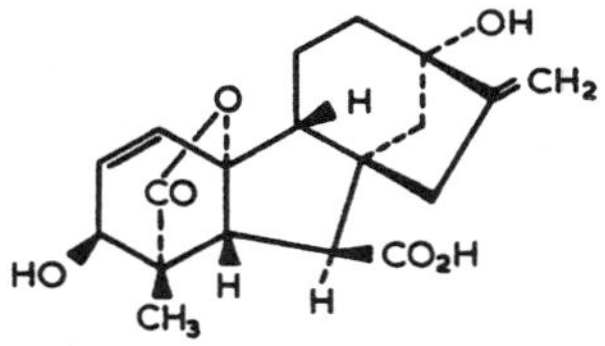

(**151**) Gibberellic acid, GA_3

malting barley is GA_1 (**150**) with smaller amounts of the closely related compound gibberellic acid, GA_3 (**151**), also being present. Gibberellic acid is available commercially, prepared from cultures of the fungus *Gibberella fujikuroi*. When legally or contractually permitted, it may be used in malting to overcome residual dormancy (Chapter 3) and to accelerate modification (Chapter 14). The initial supply of enzymes for modification originate from the embryo, specifically the epithelium of the scutellum, and later they are supplemented by greater quantities of enzymes originating from the aleurone in response to gibberellins from the embryo, and perhaps extra gibberellic acid supplied by the maltster. There is indirect evidence that embryos of different varieties of barley differ in the amount of gibberellins they release into the endosperm. The enzymes from the scutellum slowly permeate the crushed cell layer and begin to degrade it. They then attack the cells of the starchy endosperm more rapidly, initially degrading the cell walls (which lose their β-glucan component) and then attack the cell contents, most obviously the protein and the starch granules. Until cell walls are degraded, enzymes cannot spread into particular cells. This modification spreads into the endosperm with the front being roughly parallel to the face of scutellum, although it may advance faster around the sheaf cells. Later, as gibberellins trigger enzyme production in the aleurone layer, modification advances faster adjacent to this tissue (Chapter 2). Differences in the staining of the modified regions in sections of grain indicate that the mixture of enzymes from the scutellum differs from that of the aleurone layer (Oh and Briggs, 1989; Briggs, 1992). The best estimates of the particular enzymes found in the malt are that 5–10% of the α-amylase, 40% of endo-β-glucanase and about 100% of carboxypeptidase I originate in the scutellum. In addition to enzymes from the living tissues, other enzymes pre-formed in the starchy endosperm (β-amylase and some debranching enzyme) are involved in modification.

When a solution of gibberellic acid is applied to grain it gains entry at the embryo end and spreads towards the grain apex, reinforcing the normal pattern of modification and accentuating cellular breakdown under the aleurone layer (Briggs and MacDonald, 1983a).

At later stages of growth (which are not reached in malting grains), barley seedlings can make gibberellins from simple precursors of terpenes. However in the initial stages of germination in malting, the endogenous gibberellin involved in regulating endosperm modification is formed from a stored precursor, probably the hydrocarbon *ent*-kaurene (Briggs, 1973, 1987b). Gibberellin production begins when respiration has reduced the soluble sugar reserves of the embryo to a low value, and indeed its formation may be triggered by this decline. A similar decline occurs in the sugars of the embryo of germinating sorghum (Aisien, 1982b), but in this grain gibberellin production seems to be unimportant. Supplying barley embryos with sugars reduces their production of hormone and the rise in sugars in embryos of malting grains, consequent on the arrival of sugars from the modifying endosperm, also seems to cut off hormone production. Therefore, there is probably a feedback loop that regulates gibberellin levels and hence the production of enzymes and endosperm modification (Fig. 4.25).

Gibberellic acid applied to the grain by the maltster by-passes this regulatory process. The response of the aleurone to hormone may vary, and it can be regulated by the viability and maturity of the grain and the microbes on the grain surface. *In vitro* experiments show that various other plant hormones (such as indoleacetic acid, cytokinins, ethylene and abscisic acid) can influence the response of aleurone layers to gibberellins, but whether or not these hormones have significant effects in the intact grain is unknown. During grain hydration, an unidentified 'embryo factor' may diffuse down the grain and sensitize the aleurone to the subsequent arrival of gibberellin. In Galant, a barley variety in whch patches of aleurone tissue are likely to be dead in the mature grain, modification occurs only poorly or not at all in starchy endosperm adjacent to the dead aleurone, graphically illustrating the importance of this tissue (Sole *et al.*, 1987).

As gibberellins diffuse along the grain, away from the embryo, they progressively trigger responses in the aleurone layer. Its respiration rate increases and processes that began when the tissue was hydrated and became aerobic may be accelerated or be controlled. The major changes in the aleurone are dependent on respiration and are enhanced by improving the supply of oxygen, e.g. by decorticating grain or increasing the amount of oxygen in the surrounding atmosphere. In the hydrated grain, the phytate reserves in the aleurone grains are broken down and, in response to gibberellin, inorganic phosphate and other mineral ions are released into the starchy endosperm to be taken up by the embryo. Lipids, which are a major respiratory substrate of the aleurone layer, are broken down more rapidly in response to gibberellins and a proportion of the fatty acids is metabolized via the glyoxylate cycle to form the sugar sucrose. The key enzymes of the glyoxylate cycle are formed in response to the arrival of gibberellin. In addition, the aleurone can take up simple molecules, generated in the adjacent modifying starchy endosperm tissue, and convert them into other substances, which may then be released back into the starchy endosperm. Thus glucose is converted into sucrose, and several amino acids are turned into glutamine.

The most striking action of the aleurone layer in response to gibberellin is the release of a large battery of hydrolytic enzymes, including α-amylase, β-glucanase, laminarinase, pentosanases, proteases, phosphatases and enzymes that degrade nucleic acids. These are formed at the expense of the reserves of the tissue itself. Unlike the embryo, aleurone tissue is not dependent on a supply of external nutrients for making the hydrolytic enzymes it subsequently releases. Some hydrolytic enzymes are formed when the aleurone layer is hydrated (e.g. phosphatase, carboxypeptidase, β-glucanase, laminarinase), but their release into the endosperm is strongly accelerated by, or is wholly dependent upon, the arrival of gibberellins. Other enzymes, such as α-glucosidase, occur in the resting aleurone layer but their levels increase in response to gibberellins. α-Amylase is absent from the hydrated aleurone layer and its synthesis *de novo* and secretion into the starchy endosperm are independently triggered by the arrival of the gibberellin hormone. The formation, and integrity, of the α-Amylase molecules and possibly their secretion are also dependent on the availability of adequate levels of calcium ions. Various enzymes are released in sequence from the aleurone layer; they are not all formed and released simultaneously.

The observed patterns of endosperm modification (Fig. 2.21) are fully explained by this sequence of events. The enzyme that has been most thoroughly studied is α-amylase. In the finished grain it occurs, like many other enzymes, in a gradient along the grain, with most being at the embryo end and approximately equal amounts being on the dorsal and ventral sides. In contrast to barley, wheat, rye and triticale, it seems that most or all of the enzymes involved in modification when maize or sorghum are malted originate from the scutellum. There are conflicting reports, but the enzyme levels in these grains usually change little, or not at all, in response to applications of gibberellic acid (Chapter 14). A complicating factor in studying malting barley is the wide variations that seem to occur between individual grains. Inspection of the patterns of modification found in grains of one batch shows not only how widely the rates of modification can vary in different grains but also indicates that the relative contributions made in the embryo and the aleurone layer vary. When grain is germinated on wet filter-paper, α-amylase formation is followed by a period of destruction, and it is apparent that the level present at any time is the net result of the enzyme formation and destruction rates. Under some malting conditions, for example when the grain has been steeped to a high moisture content, some significant destruction of certain enzymes (e.g. α-amylase, β-glucanase) may also occur towards the end of the germination period. As a consequence of the progressive nature of modification, and the unequal distribution of enzymes along the grains, the basal or embryo ends of malt corns not only contain greater levels of enzymes than the distal ends but also contain lower levels of β-glucan, yield less viscous acid extracts (Fig. 4.26) and yield higher levels of hot water extract, cold water extract and soluble nitrogen when

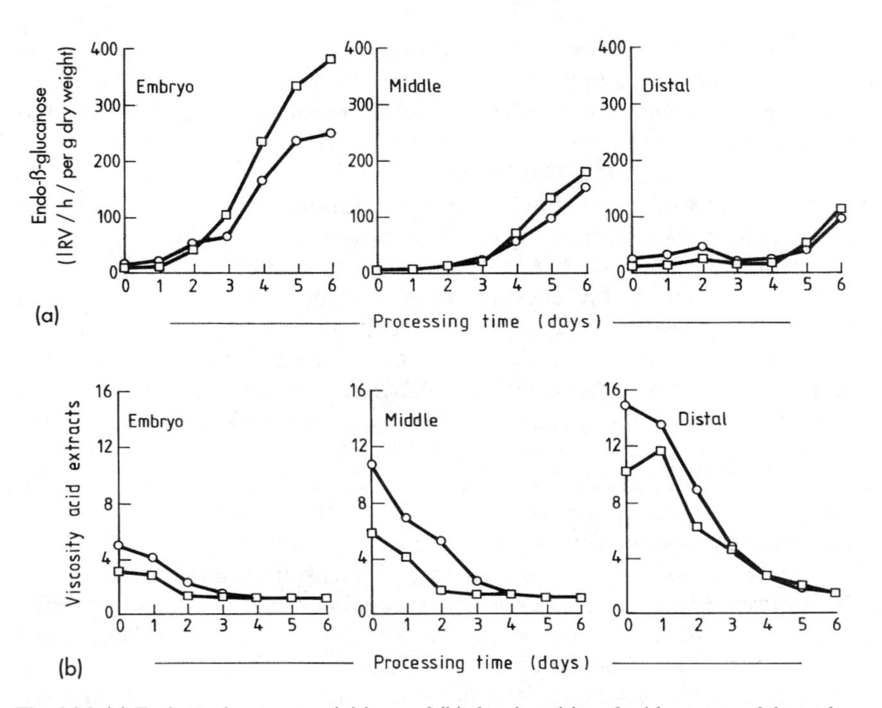

Fig. 4.26 (a) Endo-β-glucanase activities; and (b) the viscosities of acid extracts of the embryo, middle and distal portions of two barleys during malting; Ark Royal (□); and Georgie (○). (After Morgan *et al.*, 1983.)

mashed. Thus the modification of grain apices is always less than that of the basal, embryo ends (Dickson and Burkhard, 1942; Kirsop and Pollock, 1958).

Probably as a consequence of the reductive cleavage of disulphide links and limited proteolysis by enzymes from the adjacent tissues, together with other factors, the latent β-amylase of the starchy endosperm is converted to free, soluble forms during malting. The situation with debranching enzyme (limit dextrinase) is different and unusual since in addition to being formed *de novo* in the aleurone layer it also occurs pre-formed in a bound form in the starchy endosperm and a soluble enzyme inhibitor is also present.

The accumulation of hydrolytic enzymes in the endosperm causes modification and an accumulation of hydrolytic products, including sugars, oligosaccharides, amino acids and peptides as well as inorganic ions. These substances constitute the cold water extract of finished malt. During germination, some of these diffuse to the embryo, which takes them up and uses them to support its metabolism and growth. The carbohydrates taken up by the embryo restore its internal level of soluble sugars, and this seems to switch off gibberellin supplies to the aleurone layer in the regulatory feedback loop. However, when gibberellic acid is added to the grain, this control mechanism is by-passed, enzyme levels are increased and the modification rate is accelerated. One consequence is an increase in the levels of hydrolysis products (the cold water extract). Gibberellic acid also accelerates embryo growth but relatively less than it does the rate of modification. It also enhances enzyme production, the respiration rate and heat output and tends to overcome dormancy (Briggs, 1973, 1978; Chapter 14).

If barley is germinated under traditional malting conditions and at intervals some grains are degermed then held for the full malting period before being analysed, it is found that in samples where embryo removal occurs before about germination day 3 the malt extract and α-amylase levels are well below normal; however, in samples in which embryo removal is on day 3 or later, analyses are more or less normal. The explanation is that a pulse of gibberellins, together with some hydrolytic enzymes, leaves the embryo on about day 3. This triggers enzyme release from the aleurone layer and hence endosperm modification (Briggs, 1973, 1978, 1987b, 1992). Similarly it has been possible to malt degermed grains, on the experimental scale, after dosing them with gibberellic acid. These findings have important implications since by inducing grain modification by adding gibberellic acid while deliberately damaging the embryo to limit its growth, malt may be made with minimum malting losses (Chapter 14). However, the embryo is not merely a source of gibberellins and enzymes, it is also a 'sink', consuming the products of endosperm modification. Thus, when embryo growth is checked not only is the rate of increase in malting loss reduced but also the levels of soluble substances in the grain (the cold water extract) are elevated and the final malt, prepared using an embryo-damaging technique, has a slightly different composition to that of a conventionally prepared malt. The increased level of cold water extract indicates that the malt is rich in soluble sugars, amino acids and peptides. These react readily on kilning to give melanoidins and a dark colour and richer flavour and aroma (Chapter 14).

The malting of other cereals and the sprouting of seeds of pulses has been relatively little studied (Chapter 14), particularly at the biochemical level. While in **general** terms, modification in wheat, rye and triticale is probably regulated in ways similar to those occurring in barley, there are indications that there may be significant differences between them. In the case of oats the scutellum grows during the modification period, while it is releasing enzymes (Briggs, 1987b). In rice and the 'tropical' cereals, the situation is even less clearly defined. In maize and sorghum it seems the grains are largely or wholly uninfluenced by additions of gibberellic acid and most hydrolytic enzymes originate in the scutellum. These grains are regularly watered during malting and an initial period of enzyme formation may be followed by a period of destruction. The ratios of hydrolytic enzymes and their effectiveness in catalysing conversion during mashing vary widely between cereals, and between varieties of one cereal.

4.17 Some chemical and biochemical aspects of roasting and kilning

Nearly all malts are dried or kilned before use (Chapter 15). The objectives of kilning are to alter an appropriately modified green malt to produce a product, malt, from which the roots can be removed, which is stable on storage, is friable and has the correct characteristics for the particular malt type (extract yield, enzyme complement, flavour, aroma, colour, 'character') (Hind, 1940; Hopkins and Krause, 1947; Harris, 1962a,b; Pollock, 1962a; Bathgate and Brennan, 1971; Bathgate, 1973; Narziss, 1976; Palmer and Bathgate, 1976; Briggs, 1978; Moir, 1989). Kilning now consists of passing a flow of air through a bed of malt at various rates and at temperatures that are increased as the product dries. In early days, and still in tropical areas where small amounts of malts are made, products were air-dried in the sun or in a loft (Chapters 1 and 14). This is not feasible when malt is made on a large, industrial scale. During conventional kilning, a pale malt is dried by the passage of air at low initial temperatures. As the malt dries so the temperature is raised to various levels depending on the product being produced. In this latter, 'curing' phase, the malt is cooked and particular colours, flavours and aromas develop. Enzymes survive best at low temperatures and at low moisture contents. The drying strategy used for pale malts favours enzyme survival. In making some special malts, and roast barleys, very high temperatures may be applied; in manufacturing crystal malts, high temperatures may be applied before the green malt is dried (Chapter 14). Under these conditions, enzyme destruction is almost complete. Thus, during kilning, moisture is removed and the chemical and biochemical attributes of the product are altered. Rootlets (culms) are easily removed from the cooled, dry malt and the screened product, the finished malt, may be stored for extended periods provided it is kept cool and dry. However, while pale malts are believed to improve if stored for 1–3 months before being used in brewing, many special malts lose their characteristic aromas and should be used soon after being made. Unlike green malt, the kilned product is brittle and friable and is easily broken up by simple roller mills to a state suitable for use in mashing. Different types of malt are prepared by varying the grade and type of barley, or other cereal grain, being used, by germinating it in various ways to alter its enzyme complement and degree of modification and, most obviously, by kilning or roasting the various types of green malt in different ways. The number of types of malt it is possible to produce is virtually limitless. One may divide them arbitrarily into 'normal' white malts, which are pale coloured and which retain sufficient enzyme activity to permit the production of good worts when mashed alone, and 'special' malts, which are usually, but not always, dark, but which lack appreciable enzyme contents. These latter products are used in small proportions in brewers' or malt-extract manufacturers' grists to add special flavours, colour or aroma characteristics to the final products. Within these major sub-divisions there are many types of malt (Chapter 15). Among the 'white' malts one may distinguish many malt types such as 'high-DMS' lager malts, Pilsener malts, pale-ale malts, highly enzymic malts used in grain distilleries and North American breweries, peated malts (flavoured with peat-smoke), mild-ale malts and dark (Munich)-type lager malts. Among the special malts one finds relatively pale carapils, the darker crystal malts (both caramelized) and a wide range of roasted malts, amber, brown, chocolate and black in order of increasing colour intensity; roasted barley, logically an adjunct, is by tradition usually classed as a coloured malt. In the past, malt characteristics were, to various degrees, modified by the type of fuel used to heat the furnace air. Some malts were flavoured with wood-smoke or were altered by the amounts of oxides of sulphur generated when using sulphur-rich fuels. Sulphur dioxide is sometimes still added to kilning air for several purposes (see below). However, now that indirect kiln heating is becoming usual (Chapter 9), peating malts for making Scotch whisky is the only surviving process in the UK in which smoke is used to flavour a malt. However, small quantities of 'smoked malt' made from barley or wheat are still prepared elsewhere, for use in beer brewing.

Alterations in the way in which kilning is carried out can dramatically alter the enzyme complement of the finished malt. If, in the initial stages of kilning, a flow of warm, comparatively moist air is passed through grain (having a moisture content of over 40% and at temperatures below 50 °C (Bathgate, 1973)), as occurs initially on the top-deck of a three-floor kiln for example, the malt will continue to grow, endosperm modification will proceed, β-glucan levels decline and enzyme levels will be enhanced, as will the levels of soluble sugars (reflected in the cold water extract) and soluble nitrogen-containing substances. Carefully stewing malt, keeping it moist at 40 °C for 20 h, can increase amylase levels, increase the levels of reducing sugars and free amino acids and allow continued degradation of β-glucan; α-amylase is increased by up to 30% (Witt, 1945; Linko and Enari, 1966). However, as kilning proceeds and the grain temperature rises, enzymes are destroyed to a greater or lesser extent, and proteins are progressively denatured. The degree of destruction depends on the temperature regime used, the moisture levels of the malt at the different temperatures and the duration of the whole process. A proportion of the grains of very pale malts (more than 75%) may be capable of germinating, after very careful kilning. At any particular temperature, enzymes are inactivated more readily in malts of higher moisture contents. To reduce malt moisture levels to very low values (e.g. 1.5%), kilning at high final temperatures has to be prolonged – with greater risks of enzyme destruction. Enzymes vary greatly in their sensitivities to thermal inactivation, and so the enzyme spectrum present in a kilned malt actually differs to that found in the green malt. Under typical kilning regimes, some enzymes, e.g. α-glucosidase, are appreciably inactivated at 45 °C (113 °F), which is well below normal kilning (air-on) temperatures. At curing temperatures of 70–80 °C (158–176 °F), activities of enzymes such as β-amylase and, particularly, β-glucanase are reduced; this inactivation is even greater at 90 °C (194 °F) when more stable enzymes such as α-amylase and endopeptidases are also appreciably inactivated. The degree of enzyme destruction is extremely sensitive to the moisture content at any temperature. Therefore, traditional pale ale malts were dried to very low moisture levels of 2–3% moisture at comparatively low temperatures before curing temperatures of 105 °C (221 °F) were reached and final moistures of 1.5% were achieved. Nevertheless, appreciable levels of diastatic enzymes survived in such malts, although many enzymes, such as peptidases, β-glucanase and phytase, were almost totally inactivated. To be of acceptable quality, such malts had to be thoroughly modified before they were kilned because the low complement of enzymes meant that even extended mashing could not break down some unwanted substances found in undermodified malt. By comparison, traditional, older style pale European lager malts were undermodified by British standards, but the slow, low-temperature kilning regimes used in their manufacture favoured enzyme survival. Consequently, during the traditional extended decoction mashing systems then used, the conversion of the undermodified malt was complete (Chapter 5).

To prepare highly enzymic malts, kilning is carried out at low temperatures in a rapid airflow to ensure that the grain is cooled by evaporation and that no stewing occurs. In addition, driven by the need to save fuel, accepted moisture levels for 'normal' malts have risen to 4.5–6% and these levels can be achieved using lower kilning temperatures, which, in turn, favour enzyme survival. Experimentally, malts of 12–14% moisture have been prepared. In some grain distilleries and spirit factories, green malt is used and the exceptionally high enzyme complement of this material is of benefit to the users, giving highly fermentable worts.

During kilning, colour, aroma and flavour increase, the pH of the extract falls, hot water extract, cold water extract and soluble nitrogen levels tend to decline, while the level of extractable polyphenols increases. The changes that occur influence the character of the malt. In general terms, much of the chemistry involved is understood, but in detail the various stages by which colours, aromas and flavours develop are highly complex. Colour and aroma formation and the generation of a range of flavours are favoured by kilning well-modified malt, rich in soluble sugars and amino acids, at elevated temperatures while the grain is wet. These last are the conditions that favour enzyme destruction and explain why there is an approximate inverse relationship between malt colour and enzyme content.

During kilning, sucrose levels may increase, by an unidentified mechanism; reducing sugars and amino acids, while they may increase initially, finally decline in amount. Products of the interactions between them, melanoidin precursors (adducts between sugars and amino acids such as 1-valino-1-deoxy-D-fructose) and melanoidins, increase in amount. The formation of melanoidins via a series of Maillard reactions is a complex, non-enzymic process and the chemistry of their coloured, high-molecular-weight, polymeric substances is incompletely understood (Bathgate, 1973; Moir, 1989). In broad terms, reducing sugars interact with amino compounds, (amino acids, simple peptides) or proline to yield initially Schiff's bases, but in later stages aldosamines and ketosamines are formed via Amadori rearrangements. The latter products may condense with another sugar to yield diketosamines (Fig. 4.27). Diketosamines are unstable and break down to give a range of products including hydroxymethylfurfural and reductones. Some of these products interact and polymerize to yield melanoidins and a range of small-molecular-weight substances. Other reactions break down amino acids via the Strecker degradation route. Many of the lower-molecular-weight substances have important flavour and aroma characteristics, while the high-molecular-weight melanoidins are coloured. Melanoidins from darker malts have higher molecular weights than those from pale malts (Obretenov *et al.*, 1991). Reductones will consume oxygen and so stabilize beer. Thus high-temperature reactions produce high-molecular-weight substances that are coloured, such as melanoidins, and low-molecular-weight substances that contribute to aroma and flavour. This last group includes acids, alcohols, aldehydes, ketones, esters and O-, S- and N-containing heterocyclic substances

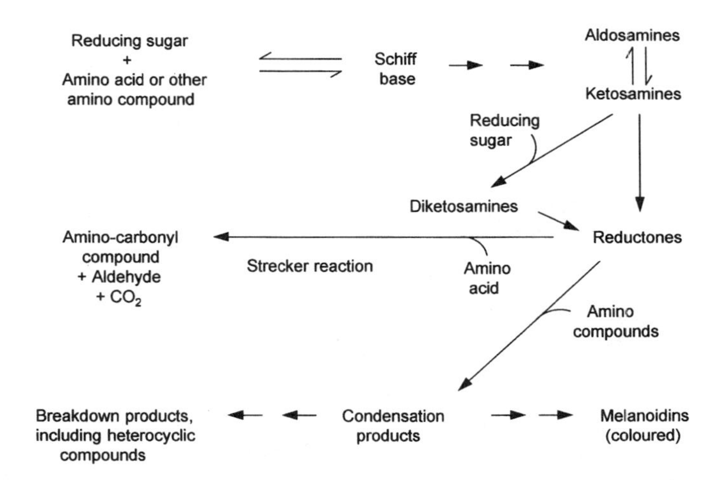

Fig. 4.27 Some of the chemical steps involved in the formation of melanoidins and some flavour and aroma substances.

(Fig. 4.28, Table 4.14). Bitter β-diketopiperazines (cyclic dipeptides) are also formed. In addition, under kilning conditions, the fatty acids of lipids are partly oxidized and degraded to yield a range of substances including unsaturated aldehydes, alcohols, lactones and other acids, with characteristic, usually undesirable, odours. Polyphenols may be converted to coloured phlobaphenes; sugars may be caramelized and, at higher temperatures at least, starch is altered so that it is less readily degraded by enzymes. Worts made with mashes containing highly kilned malts are less fermentable and contain more dextrins. In roasted barley, the granular starch is strongly pigmented, and 1,6-anhydroglucose units are generated by the roasting, together with new linkages between the sugar residues (Bathgate and Brennan, 1971; MacWilliam, 1972). The changes occurring in starch result from pyrolysis. The β-glucans are also degraded to more soluble, less viscous substances at 115 °C (239 °F); this is a lower temperature than that at which starch is appreciably depolymerized (140 °C; 284 °F) (Morgan, 1971).

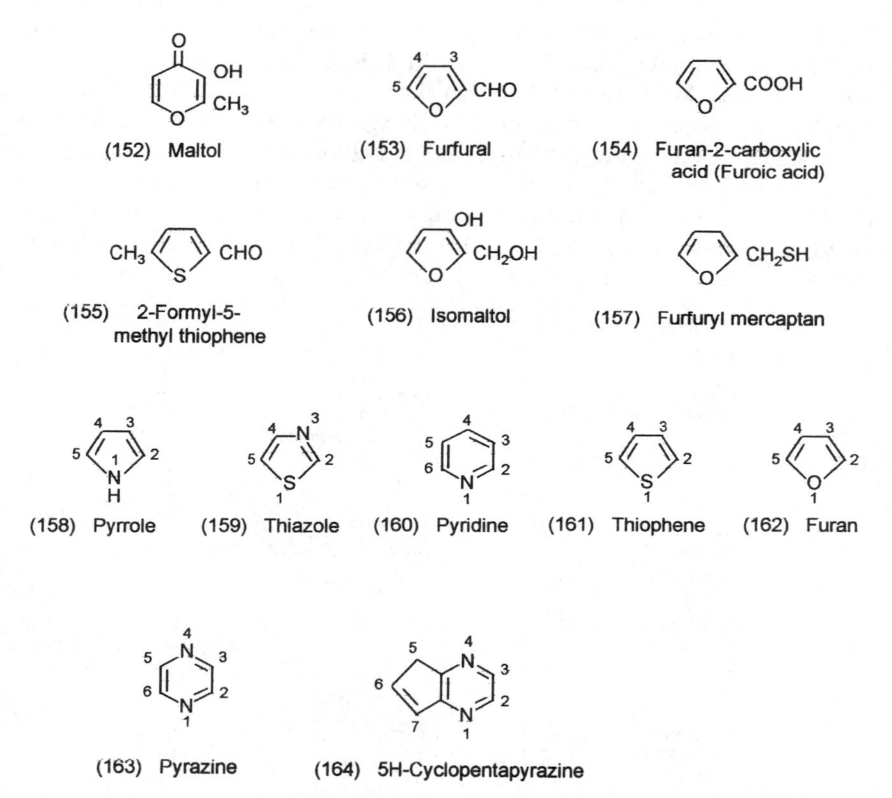

Fig. 4.28 Representative flavour and aroma substances with heterocyclic rings in their structures.

Therefore, in addition to the coloured melanoidins and possibly the phlobaphenes, the reactions which occur are able to generate a complex mixture of small-molecular-weight substances that are volatile and individually have strong, characteristic flavours and aromas. Together they contribute to the 'character' of a malt. Many substances occur below their 'flavour threshold' concentrations, but as they occur in mixtures their individual importance is virtually impossible to assess. During kilning, many volatiles are evaporated from the malt and are lost. From the brewers' viewpoint, the situation is further complicated since more colour-forming and flavour-forming reactions occur during the hop-boil, and some volatile flavour substances are lost at this stage. In addition, many substances may be metabolized by yeast and changed into different substances having their own flavour and aroma characteristics. In addition to the formation of

Table 4.14 Examples of volatile substances with flavour and aroma characteristics that have been found in dark malts and roasted barley

	Examples
From lipids	Hexanal, *trans*-2-hexenal, *trans,trans*-2, 4-hexadienal, heptanal, *trans*-2-octenal, *trans*-2-nonenal, *trans*-2-*cis*-6-nonadienal, *trans,trans*-2, 4-nonadienal, 1-hepten-3-ol, 1-hexanol, *trans*-2-hexen-1-ol, 1-octanol, 1-nonanol, *trans,trans*-2, 4-decadienol, hexanoic acid, octanoic acid
Aldehydes from Strecker degradations	Acetaldehyde, propionaldehyde, isobutyraldehyde, isovaleraldehyde, methional, benzaldehyde, 2-phenylacetaldehyde
Compounds with oxygen heterocyclic ring structures	Furfural, furfuryl alcohol, 2-pentylfuran, 2-acetylfurfural, 5-methylfurfural, 3-phenylfuran, maltol, isomaltol, furan-2-carboxylic acid
Sulphur-containing molecules	Hydrogen sulphide, methyl mercaptan, ethyl mercaptan, carbon disulphide, dimethyl sulphide, diethyl sulphide, dimethyl disulphide, furfuryl mercaptan
Heterocyclic sulphur-containing molecules	Thiazole, 4-methylthiazole, 2-acetylthiazole, various thiazolines, thiophenes
Other substances	Ethanol, vanillin, *p*-hydroxybenzaldehyde, acetic acid and other volatile acids
Amines	Ammonia, methylamine, dimethylamine, ethylamine, *s*-butylamine, isobutylamine, *n*-butylamine, *p*-hydrobenzylamine, isoamylamine
Heterocyclic nitrogen-containing molecules	Pyrrolidine, 2-formylpyrrole, 2-acetylpyrrole, 2-methylpyrazine, dimethylpyrazine, ethylpyrazine, 3,6-dimethyl-2-ethylpyrazine, indole, pyridine, methylpyridine, 2-acetylpyridine, 5-methyl-6,7-dihydro-5*H*-cyclopentapyrazine
Phenolic substances	4-Hydroxybenzaldehyde, 4-vinylguaiacol, 4-vinylphenol, vanillin
Other groups of substances	Other alcohols, aldehydes, esters, alkanes, alkenes, aromatic hydrocarbons, lower fatty acids, lactones
Diketopiperazines	Cyclized dipeptides, e.g. *cyclo*-L-phenylalanine-L-proline, *cyclo*-L-proline-L-proline

(After Bärwald *et al.* (1969); Slaughter and Uvgard (1971); Harding *et al.* (1978); Tressl *et al.* (1975, 1979, 1981); Farley and Nursten (1980); Hough *et al.* (1982); Moir (1989).

desirable flavours and aroma substances, kilning removes unwanted green-grain flavours from malt. (Experimental beers prepared from unkilned, green malt can have an appalling, long-lasting after-taste.) Experimentally, these green-malt off-flavour substances can be removed from wort by the unusual process of steam stripping during the hop-boil.

Several independent variables are involved in the formation of aroma, flavour and colour in malts. Aroma and flavour are difficult to assess, but fortunately their intensity is loosely related to the coloration achieved during kilning. From the maltster's standpoint, the factors involved in colour development are: (i) the degree of modification (the analysis) of the green malt loaded onto the kiln, (ii) the durations and levels of the various temperatures and the ventilation rates of the different stages of a kilning or cooking programme; and (iii) the moisture contents of the grain at the different stages. Skill and experience are needed to make good coloured malts (Chapters 14 and 15). Although it is technically feasible to use low-dried malts (perhaps of as much as 12% moisture), yet nearly always, and certainly for brewing purposes, malts are dried to 6% moisture or less. Even pale brewing malts need to be cured, although the chemical reasons for this are obscure. Highly kilned malts may give rise to turbid worts on mashing. Even so, cured malts seem to produce better-quality products. Even pale-lager malts are usually cured at temperatures of up to 85 °C (185 °F). For pale-ale malts, the traditional curing temperature was about 105 °C (221 °F). Curing ensures that the malt confers a particular character to beer and is said to reduce the likelihood of the formation of chill-haze or permanent non-biological hazes.

Peat-smoke is used to supply special flavours to Scotch whisky malts. Substances that are condensed on the malt include furfural (**153**) and related substances, and various phenols and hydrocarbons and related substances (MacFarlane *et al.*, 1973; Bathgate and Taylor, 1977). Of these substances, the phenols seem to be the most important and include phenol, guaiacol, *o*-cresol, *m*-cresol, *p*-cresol, several xylenol isomers and *p*-ethylphenol. Other classes of compound present in the peat reek (smoke) include alkanes, alkenes, hydrocarbons, aldehydes, alcohols, organic acids and pyrazines.

Some European, 'Continental' types of lager require malts that give rise to appreciable quantities of dimethyl sulphide (DMS; **166**) in the beer (Fig. 4.29). DMS has characteristic flavour and aroma properties. Considerable effort has been expended on finding how best to make high-DMS malts (Parsons *et al.*, 1977; White, 1977; Dickenson and Anderson, 1981; Brookes and Dickenson, 1983; Pitz, 1987). The primary precursor of DMS in malt is the sulphonium compound *S*-methylmethionine (SMM, **165**), (Fig. 4.29) which is formed by methylation of the amino acid methionine (**47**). Levels of DMS and/or DMS precursor (SMM) are often specified for lager malts. SMM is not volatile, but it is heat labile so that during kilning at elevated

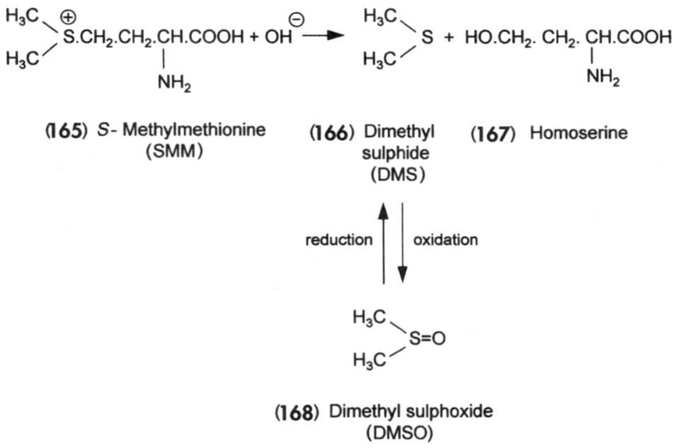

$$(H_3C)_2\overset{\oplus}{S}.CH_2.CH_2.CH(NH_2).COOH + \overset{\ominus}{OH} \longrightarrow (H_3C)_2S + HO.CH_2.CH_2.CH(NH_2).COOH$$

(165) *S*-Methylmethionine (SMM) (166) Dimethyl sulphide (DMS) (167) Homoserine

reduction ↑ ↓ oxidation

$$(H_3C)_2S{=}O$$

(168) Dimethyl sulphoxide (DMSO)

Fig. 4.29 The origins of dimethyl sulphide.

temperatures it is degraded. One of the products of thermal degradation is DMS, which, being volatile, is readily lost from the malt. Another precursor of DMS is dimethylsulphoxide (DMSO, **168**), which occurs in some malts. Presumably it is formed by the oxidation of DMS or SMM. DMSO is non-volatile and it can be reduced to DMS by yeast during the fermentation stage. In two classes of UK malt, levels of these substances (as μg DMS equivalent/g malt) have been measured. In lager malts the values were, DMS, 0–5; SMM, 1–10; DMSO, 0–2; and for an ale malt the equivalent values were 0–2; <1; and 1–2 (courtesy of C. J. Dickenson). For a range of other malts, made in different ways, the values were DMS, 0.4–3.8; SMM, 5.2–11.9; and DMSO, 1.0–3.1. Values of SMM up to 15.8 and of DMS + SMM of up to 29.3 have been recorded. For the malt to confer the DMS character to beer it is necessary to preserve and enhance the SMM levels during malt manufacture. High DMS malt is made, therefore, using a favourable barley variety, which is rich in nitrogen and which is well modified during an extended germination period. Typically six-rowed barleys give higher levels of SMM than two-rowed varieties. The use of bromates or ammonium persulphate is avoided, since these reduce SMM levels. Abrasion and gibberellic acid, like high steep-out moisture contents, favour modification and high SMM levels. Kilning is carried out in the shortest possible time, at a low temperature. Curing is avoided, to minimize the decomposition of SMM. For example, extending kilning by 5 h at 75 °C (167 °F) can reduce the SMM level by 50%. The amount of DMS found in a beer made from a particular malt is strongly influenced by details of the brewing process.

Apart from the functions of heating the grain and carrying away moisture vapour and other volatiles, the stream of hot gas passing through the bed of green malt during kilning may influence the properties of the finished

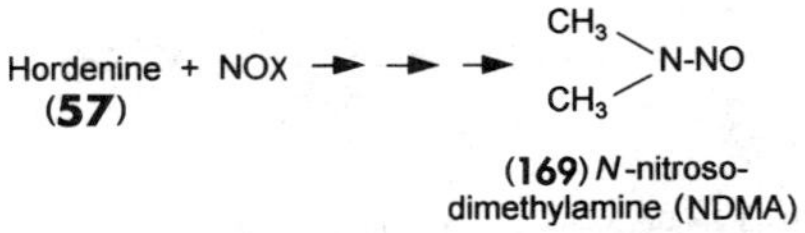

Hordenine + NOX → → → $(CH_3)_2N\text{-}NO$
(**57**)
(**169**) *N*-nitroso-dimethylamine (NDMA)

Fig. 4.30 The formation of *N*-nitrosodimethylamine during kilning.

product in various other ways. In directly fired kilns, in which the furnace gases pass through the bed of malt, the fuel chosen must meet a number of requirements and be burned cleanly without the generation of noxious or oily vapours, which would contaminate the malt. Poor quality or imperfectly combusted fuels may give off-flavours or taints to malt. In the early 1900s, some malts were contaminated with arsenic when unsuitable solid fuels were used. Now only low-arsenic solid fuels (e.g. selected grades of anthracite) may be used in direct-fired kilns in the UK. There is also a fear that incompletely burned fuels may generate carcinogenic hydrocarbons, which may condense on the malt and be carried forward into the final product. However, there appears to be no evidence for this occurring in practice. Many fuels are rich in sulphur (some coals, and many oils) and this has positive and negative aspects. High-sulphur fuels, when burned, give rise to sulphur dioxide (SO_2) and sulphur trioxide (SO_3). On the one hand, these acid gases damage and corrode the kiln structure and add to atmospheric pollution. High levels may cause local discoloured marks on malt grains, producing 'magpie' malt. The sulphur (sulphite or sulphate) on the malt is carried into the wort and may, through the metabolic activities of yeast, give rise to unwanted flavours in the beer. On the other hand, sulphur dioxide in the gases prevents the unwanted formation of *N*-nitrosodimethylamine (NDMA, **169**) (see Fig. 4.30); it reduces the colour of malt by its bleaching action and by interfering with the Maillard reaction; it sometimes increases the malt extract; it reduces the pH of mashes, enhances recovery and elevates the levels of soluble nitrogen and inorganic phosphate and it may increase fermentability (Witt and Adamic, 1957; Witt, 1972; Narziss, 1976) (Table 4.15). Sulphuring also partly 'disinfects' malt, reducing the populations of moulds, yeasts, bacteria and thermophilic actinomycetes (Flannigan, 1983). Indeed, when many North American maltsters began to use sulphur-free, natural gas as a kiln fuel they added sulphur dioxide to the kilning air (*ca.* 0.004%), or burned sulphur at the kiln air-inlet to gain the benefits.

Using super-sensitive analytical methods, such as gas or liquid chromatography employing thermal energy analyser detectors, it was discovered that some malts contained small amounts of *N*-nitrosodimethylamine (NDMA), at the parts per billion (10^9) level. This compound is suspected to be a carcinogen. Studies have largely explained the origin of this substance and have shown how its formation can be prevented (Hardwick *et al.*, 1981, 1982; Wainwright, 1981). When kiln gases containing oxides of nitrogen (NOX, a mixture of NO and NO_2) mix with malt, these interact with hordenine (**57**),

Table 4.15 Malts made from Larker barley (13% protein) with and without exposure to sulphur dioxide during kilning

Germination time (days)	Sulphured	SO_2 in malt (p.p.m.)	Extract (%)		Soluble protein (%)	α-Amylase (ASBC units)	Inorganic phosphate (p.p.m.)
			Fine ground	Coarse ground			
3	–	9	74.4	72.2	4.0	34.8	422
3	+	32	75.0	72.6	5.4	32.6	495
5	–	9	75.6	74.2	4.7	43.6	545
5	+	23	76.3	74.8	6.3	41.7	660

From Witt (1972).

present in the roots and smeared on the malt from crushed roots, and give rise to NDMA (Fig. 4.30). The oxides of nitrogen are formed by interactions between the oxygen and nitrogen of the air at temperatures of 1800 °C (3272 °F) or more. The problem is less with indirectly fired kilns, provided that these are not situated near other industrial plants or vehicles whose exhausts generate NOX, since the air heaters operate at relatively low temperatures. The problem was more acute when malt was directly kilned with the 'pure' fuel natural gas, rather than with oils or solid fuels containing sulphur. Several measures are used to prevent the formation of nitrosamines. (i) Since NDMA formation is dependent on NOX, the presence of these gases is minimized in various ways, e.g. by siting kilns away from other industries and heavy traffic, ensuring that all vehicles switch off their engines when not working, and by fitting 'low-NOX' burners to directly heated gas-fired kilns that burn the gas at less than 1800 °C (3272 °F) in the absence of excess air. Increasingly, kilns are being converted to indirect firing so that furnace gases do not come into contact with the malt and the heat exchangers operate at well below 1800 °C (3272 °F). (ii) NDMA formation is favoured by high humidity and temperatures of 60 °C (140 °F) or more, so malts are initially dried at low temperatures and these are not raised until the malt is comparatively dry. (iii) NDMA formation is favoured by high levels of hordenine and hence strong root growth. This can be limited by adjusting malting conditions and by the use of soluble bromates to check rootlet growth. (iv) NDMA formation is favoured by high pH values on the surface of the green malt. This is reduced to a pH of less than three (when amines are protonated and unreactive) until the malt is largely dried and all the free water has been removed, by supplying sulphur dioxide to the airstream, either directly or by burning stick-sulphur in the kiln air inlet. Using these various approaches, maltsters now routinely control NDMA levels to well below the legal limits, and often to below the level at which it can be reliably detected. NDMA (**169**) has been extensively studied because it is volatile. However NOX will react with other amines to generate other *N*-nitroso compounds; for example *N*-nitrosoproline has been detected in malt (Pollock, 1981). Because sulphur dioxide acidifies the surface of green malt, it does not influence the production of non-volatile *N*-nitroso compounds generated within the grains. Minimizing the levels of NOX in the kilning gases is the best single precaution to take to minimize the formation of nitrosamines.

5 The principles of mashing

5.1 Introduction

Malts are intermediates in the preparation of other products. Mashing is not part of malting but, because it is an essential part of the most important processes that use malt, it is necessary to have a knowledge of it to appreciate malt quality. In principle, ground malt (with or without other unmalted cereal preparations, starch and supplementary enzymes: the grist) is mixed with warm water to produce a slurry, called a mash. During mashing the temperature may be altered. After a period, the mash may be cooled and fermented with microbes, including yeasts, as in the preparation of some soured African porridges, Bantu beers, some grain whiskeys and some other spirits. Depending on the product, the fermented liquid or alcohol may or may not be separated from the residual solids (Chapter 1). Alternatively the hot, sugary liquid, the wort in brewing, may be separated from the residual solids and the liquid is then processed as in the manufacture of clear beers, malt whiskys, malt vinegar and malt extracts. The 'strength' of the wort is judged by the quantity of material in solution, the extract. The objective of mashing, in the latter cases, is to obtain from the grist as much extract, of the correct quality, as is economically feasible. The strength of a wort is usually given as a specific gravity, relative to water, at a particular temperature or as the concentration (% w/w) of a sucrose solution having the same specific gravity. Because the latter values are found with reference to tables, they are usually reported as °Plato or °Balling after the originators of the tables (Appendix, Table A4). By comparison, the extract obtained from a grist, or a malt, may be reported as litre-degrees of extract per kilogram of grist or malt (l°/kg) (IoB) or as a percentage of the malt or grist solids that have been extracted (EBC, ASBC) (Chapter 13). When a pale malt is ground up and extracted with cold, alkaline water, some 16–22% of its dry matter is dissolved and of this cold water extract about half is fermentable by yeast. If the ground malt is extracted with water using a specified rising temperature profile or a constant temperature of 65 °C (149 °F) then 75–83% of the dry malt dissolves and becomes extract. Better malts yield higher extracts and, using the IoB methods, have extracts of 305 l°/kg (dry matter) or more, up to 315 l°/kg. The actual yield of hot water extract (HWE) depends on the quality of the malt, the degree to which it has been broken up in grinding, the temperatures used, the duration of the mesh, and so on (Chapter 13). The wort solids are between 60 and 85% fermentable. The cold water extract (CWE) is an estimate of the soluble substances preformed in the malt. The extra substances dissolved in mashing, shown by the difference between the hot water extract and the cold water extract, are produced during the mashing process by the enzymic degradation of initially insoluble materials, chiefly starch, to give soluble breakdown products. The biochemical changes that occur during mashing are complex.

5.2 Commercial mashing processes

The raw materials used in mashing are water (liquor: usually containing controlled levels of soluble salts), ground malt, or more usually a mixture of malts, and sometimes malts mixed with unmalted adjuncts and other quantitatively minor additions such as enzyme preparations derived from microbes (Chapter 12) (Hind, 1940; Briggs *et al.*, 1981; Bathgate, 1989; Narziss, 1992). Sometimes other extract may be obtained by dissolving sugars or syrups in the wort (Chapter 12). The materials used depend upon the product being made, economics and local regulations, restrictions and laws. For example, in the manufacture of Scotch whisky or some German beers, only liquor and malt may be used in mashing. Mashing processes can vary widely in their details.

5.2.1 *Liquor*

The liquor, or water, used in mashing must be potable, that is it must be 'clean', free of toxic agents and free of materials that reduce the quality of the product, for example by adversely affecting flavour or aroma. The water must be nearly free of microbes and wholly free of pathogenic organisms (Moll, 1979b). In addition, it must have an acceptable mineral composition, which may be adjusted before use. Traditional brewing centres produced characteristic beers using their local waters. It became established that, for example, very pale lagers would be made using exceptionally soft waters containing few dissolved salts, as in Pilsen, while more bitter pale ales would be made using permanently hard water rich in calcium sulphate, as in Burton-on-Trent. Calcium carbonate and bicarbonate-rich water, having temporary hardness, are disliked as they increase mash pH in an undesirable way. Waters containing calcium sulphate or chloride reduce mash pH values to a desirable extent. Different ions, such as magnesium, sodium, sulphate and chloride, influence the flavour of the product. As well as adjustments to the salts in the liquor, mineral acids, lactic acid or small amounts of 'acid malt' may be added to mashes to adjust the pH.

5.2.2 *Grist*

The grist is usually prepared from a mixture of malts and may also contain adjuncts, cooked, pre-cooked or raw (Chapter 12). Malts and some adjuncts, such as roasted barley or torrefied wheat, must be broken up by milling before they are mashed. Cooked grits are 'doughed in' with the malt mash and so facilities for mixing are required if these adjuncts are used. Within limits, the more finely a grist is ground the more readily extract is recovered from it. However, the finer the grist the more difficult it is to separate wort from the residual solids. In traditional British beer-brewing, mash tuns are used and these only function if a relatively coarsely ground grist is used. Lauter tuns require a finer grist, while the newest mash filters can separate wort from a grist so fine that it is almost all a flour. Worts from different grists vary in their qualities. Milling must be regulated to obtain the desired proportion of husk fragments, coarse grits, fine grits and flour. Simple roller mills may be used or more complex mills may be used in which the grist is fractionated by sieving and the fractions are remilled separately before being recombined in the final grist. To produce very fine grists hammer mills may be employed. Sometimes the surface of the malt is damped with steam or fine sprays of water to soften the husk and reduce its tendency to shatter before the malt reaches the rollers. In an extreme case, termed wet-milling, the malt is steeped (wetted by immersion in water) before the wet malt is rolled, mixed with mashing liquor and is conveyed to the mashing vessel.

5.2.3 Infusion mashing

Usually dry grist is mashed-in by mixing it with warm water, at an exactly controlled temperature, just before it is discharged into a mashing vessel. A wide variety of vessels is in use. In traditional British breweries and distilleries, infusion mashing is carried out in mash tuns: vessels in which enzymic conversion of the mash occurs and which are also used to separate the worts from the residual solids (spent grains or draff). These vessels, which are circular in cross-section, vary greatly in size but they are deep, typically 2.0–2.5 m (6–8 ft approximately). The base is fitted with a false bottom of slotted plates. Usually there is also a sweep-arm to discharge the spent grains at the end of the process, after wort recovery. In older mash tuns, there were often mechanical raking devices to mix the contents. Above the mash surface in the vessel is a set of rotating sparge arms, tubes from which water at a controlled temperature can be sprinkled onto the surface of the mash. At present, rather simple mashing programmes with isothermal 'stands' are favoured. Grist is mixed with warm water to achieve a temperature usually in the range 63–67 °C (145.4–152.6 °F). After a stand of about 1 h wort is withdrawn from below the slotted plates, which hold back the residual solid materials, the spent grains. The wort is recirculated by spraying it back onto the surface of the mash via the sparge arms. When the wort 'runs bright', that is it becomes completely clear because all the fine suspended particles are held back by the filter-bed formed by the spent grains, recirculation is stopped and the wort is directed to the copper (kettle) where it will be boiled with hops. As the wort is withdrawn, the mash settles on the false bottom and contracts. Before it dries, hot water (typically at 75–80 °C (167–176 °F)) is sprayed onto the surface of the mash to wash out entrapped extract. This sparging is continued until the specific gravity of the wort being collected falls to a low value. The spent grains are discharged and are used for cattle-feed. Infusion mashing only works well with well-modified malts. Limited quantities of adjuncts may also be incorporated in the grist. Infusion mashing is usually termed an isothermal process but, in fact, the temperature increases during sparging and traditionally it was often raised during the stand by additions of hot water; as a result, temperatures might be 63 °C (145.4 °F) then 68 °C (154.4 °F) rising to, say, 77 °C (170.6 °F) during sparging. In the past, in beer-brewing and vinegar-brewing, remashing was sometimes employed and may still be used in making some whiskeys. In this process the grist was first mashed at, for example, 64–65 °C (147.2–149 °F), then the wort was collected without sparging. The draff was successively remashed (i.e. mixed with hotter water) and the wort collected, at 70 °C (158 °F), 80 °C (176 °F) and even 90 °C (194 °F). The first two worts went forward for further processing. The last two were used in the mashing liquor of the next group of mashes. Thus, a high wort concentration was maintained (s.g., 1.050–1.060) and extract recoveries were excellent. Some distilleries are now using modified lauter tuns (Bathgate, 1989; Wilkin, 1989). Distillers, like vinegar-brewers but unlike beer-brewers, wish to preserve some enzymes from the mash in the wort so that dextrin degradation continues during fermentation and alcohol yield is maximized. Thus the first cooled worts contain enzymes. Brewers, in contrast, wish to 'fix' the fermentabilities of their worts at some chosen value (e.g. 75%) and this is achieved in the hop-boil, when any enzymes remaining in the wort are denatured.

5.2.4 Temperature-programmed mashing

A more recent innovation in brewing is temperature-programmed mashing ('rising-temperature infusion mashing'). This is carried out in two vessels, a mash mixing and warming vessel and a wort-separation device, a lauter tun or a mash filter. In the mash-mixing vessel, the mash is stirred and warmed through a carefully chosen programme designed to allow optimal enzymic conversion of the grist. For example, a mash made with an undermodified malt or with adjuncts in the grist might be mashed in at 35 °C (95 °F) and after a 30 min hold it might be warmed (1 °C/min) to 50 °C (122 °F) and then, after another 30 min hold, to 65 °C (149 °F). After a further 30–45 min, the mash might be warmed to 70 °C (158 °F), held for 30 min, then be warmed to 75 °C (167 °F) for 30 min before transfer to the lauter tun or the mash filter where the grist will be sparged. At lower temperatures 45–55 °C

(113–131 °F), phytate breakdown, some β-glucanolysis and rapid proteolysis occur. At 60–65 °C (140–149 °F) starch conversion is rapid; extending the period during which the mash is held at this temperature increases the fermentability of the wort. At 65–70 °C (149–158 °F) the last of the extract is dissolved. The final high temperature accelerates enzyme inactivation and, by reducing wort viscosity, accelerates wort separation. During temperature-programmed mashing, vigorous mixing is avoided to minimize the entrainment of air with consequent oxidation of mash constituents. For well-modified malt grists, a shorter programme may be employed: for example, with mashing-in at 50 °C (122 °F) and rests at 50, 65 and 75 °C (122, 149 and 167 °F). Clearly such a system is very flexible and many different temperature programmes are used. At the end of the conversion process, the mash is moved to a wort-separation device, a lauter tun or a mash filter.

Lauter tuns vary widely in construction details but, in principle, they allow wort separation through a false, slotted bottom on which the mash rests. However, the grists used are more finely broken down than those used in mash tuns, the bed depths are much more shallow than those found in mash tuns and the slots in the false bottom are narrower. The mash may be mechanically disturbed with 'rakes' during wort separation. They permit faster wort separation from the mash than can be achieved with mash tuns. In mash filters, the mash is packed between filter-cloths supported on metal plates mounted on frames. The compartments containing the mash are shallower, *ca.* 6.4 cm (2.5 in) in some cases, and wort separation is rapid. However worts can tend to be cloudy. Various other devices, such as centrifuges, have been used to separate wort from mashes, but they are now little used.

5.2.5 *Decoction mashing*

Decoction mashing is traditional in most areas of mainland Europe. This system employs a mash-mixing vessel, a decoction vessel (or mash copper), sometimes a separate cereal cooker and equipment for separating the wort from the mash, usually a lauter tun or a mash filter. The decoction vessel may also be used as a cereal cooker (Chapter 12). Decoction mashing sequences can successfully convert poorly modified or well-modified malts, or mixtures of malt and cereal adjuncts. If adjuncts such as maize or rice grists are used, they can be mixed with 5–10% of the malt or with microbial enzymes and cooked in the decoction vessel or a specialized cereal cooker. Cooking gelatinizes the starch, which is partly liquefied by the malt or microbial enzymes. In addition, intact cell walls in the heated grist are softened and partly disrupted. Consequently, when the cooked adjunct is mixed with the main malt, mash starch conversion is readily achieved by enzymes from the malt.

Traditional three-decoction mashing is adapted to process under-modified lager malts. Improvements in malt quality and the need for shorter process times mean it is little used now, but the process illustrates some important principles. In one variant, the grist is mixed with water at ambient temperature in the mash-mixing vessel, then hot water is added with mixing to increase the temperature to 35–40 °C (95–104 °F). At this temperature, temperature-sensitive enzymes such as β-glucanase, the proteolytic mixture of enzymes and phytase act, many pre-formed soluble substances dissolve, the grist is completely hydrated and some natural acidification occurs through the activities of lactic acid bacteria. After 1.5–2 h about a third of the stirred mash is transferred to the decoction vessel, where it is heated with mixing. It is held at about 65 °C (149 °F) for approximately 20 min to permit starch liquefication, then it is heated to boiling at which point unmodified fragments of malt are disrupted, residual starch is gelatinized and total enzyme destruction occurs. Then this hot material is pumped back into the stirred mash-mixing vessel, increasing the temperature of the combined mash to approx 52 °C (125.6 °F). This series of operations constitutes the first decoction. At 52 °C, which is maintained in the main mash for about an hour, proteolysis and some β-glucanolysis continues and the starch which was gelatinized and liquefied during the decoction is converted. Again about a third of the mash undergoes a decoction, and when it is returned to the main mash after boiling the temperature of the combined mash is 65 °C (149 °F). At this temperature, large starch granules are gelatinized and readily converted but the rate of proteolysis is reduced and β-glucanolysis ceases. Finally, a third decoction increases the temperature of the main mash to about 76 °C (approximately 169 °F), when enzyme activity ceases and considerable protein denaturation occurs. The mash is then transferred to a lauter tun or mash filter and the wort is collected. The triple-decoction mashing process is slow, it lasts approximately 6 h. Mashes of this duration are only needed with under-modified malts or mashes made with uncooked adjuncts. Mashing conditions are modified with reference to the grist being used and the type of product being made. For example, boiling periods are extended when dark beers are being made.

For better-modified malts double- or single-decoction mashing processes may be used. Sometimes the initial temperature rise is achieved by heating the mash-mixing vessel directly. Sometimes only a quarter but at other times as much as half of the mash may be used in a decoction. In the Schmitz process, an interesting variant, the grist is mashed in at 50 °C (122 °F) and after a rest the whole mash is heated to 65 °C (149 °F) while being stirred. The mash is then allowed to stand and solids, including unconverted starch and unmodified fragments of malt, settle to the bottom. The supernatant liquid, containing extract and enzymes in solution, is pumped off. Then the thick residual mash is stirred and heated to boiling to disrupt residual grist particles and to gelatinize starch. It is then cooled to a suitable temperature and the enzyme-rich liquid is added back. The enzymes present rapidly liquefy and convert the gelatinized starch. The temperature is then raised to 77 °C (170.6 °F) before wort collection. When badly modified malts are used, decoction mashes may yield 1.5–2% more extract than single-temperature infusion mashes. However, when adequately modified malts are used, good extracts are obtained with infusion mashes. Maximal extracts from particular grists are recovered using temperature-programmed mashing sequences.

5.2.6 *Double-mashing*

Double-mashing is used in areas, such as North America, where large proportions of the brewer's grist (25–60%) consist of starchy adjuncts such as rice, maize or sorghum grits: materials that require cooking because of the high gelatinization temperatures of their starches (Chapter 12). Two mashes are prepared. In the first, or adjunct mash, cereal grits together with a proportion of highly diastatic malt (diastatic power (DP) 80–200 °L) or microbial α-amlyase are mixed (doughed in) at about 35 °C (95°F). After a 30–60 min stand, the cooker contents are heated. After a hold of *ca.* 20 min at about 70 °C (158 °F) to allow any available starch to be liquefied, the mash is heated further and is held at 100 °C (212 °F) for 45 min to disrupt the grits and complete starch gelatinization. In the past, the adjunct mash was boiled but now, to save heat and where a bacterial α-amylase is being used to liquefy the starch, it may be held at a lower temperature. Once the adjunct mashing sequence is underway, the second mash, with a mainly or all malt grist, is prepared. This is mashed in (doughed in) at about 35 °C (95 °F) and is allowed to stand for about an hour, when the adjunct mash is pumped into it and the two are mixed. The temperature of the combined mash is about 68 °C (154.4 °F). This temperature is held for about 15 min during which time the liquefied starch of the adjuncts is rapidly saccharified. Then the temperature of the whole mash is raised to 73 °C (163.4 °F) and any residual starch is converted. Wort separation is achieved using a lauter tun or a mash filter.

5.2.7 *Other mashing methods*

For grain whiskies, i.e. whiskies made using a proportion of unmalted adjuncts in the mash, a wide range of mashing regimes is employed. For example, in the preparation of Scottish grain whiskies the unmalted adjunct may be whole or degermed maize, wheat or (less usually) barley. The grains may be milled or used entire during cooking. Conversion may be with very pale malt rich in enzymes, or even using green malt, in which the costs of kilning and the avoidance of enzyme destruction are major considerations (Bathgate, 1989). Cereals are most efficiently cooked at high temperatures and pressures. The pressure release at the end of cooking causes substantial disintegration of the adjuncts, leading to better extract recovery and so, eventually, a better alcohol or spirit yield. Mixing the cooked cereal slurry with a mixture of malt and cold water gives a mixed-mash temperature of about 65 °C (149 °F), at which starch conversion occurs rapidly. The malt constitutes some 8–15% of the total grist. At the end of the mashing period, the whole mash is cooled. In 'all grains in' fermentations, the cooled mash is pitched with yeast. Clearly this favours the continued production of fermentable sugars during fermentation. In other systems the wort is separated and the clear liquid is fermented. After fermentation, the alcohol is recovered by distillation (Bathgate, 1989).

In the types of mashing described, the major objective is to convert all the starch and obtain it in solution either as fermentable sugars or as dextrins. In the preparation of many African porridges and beverages, such as Bantu beers, this is not an objective (Novellie, 1966, 1968, 1981; Novellie and de Schaepdrijver, 1986; Taylor, 1989; Chapter 1). The malts used may be made from sorghum or millets and adjuncts may be whole raw sorghum or maize grits. A typical commercial mashing sequence is as follows. About 10% of a low-tannin sorghum malt is ground and held in water at 48–50 °C (118.4–122 °F) for 8–24 h after inoculation with a thermophilic lactobacillus, *Lactobacillus leichmannii (delbrückii)*. During this initial mash, or souring, some limited sugar production occurs and lactic acid is produced. The 'sour' is about one-third the final volume of the beer, has a pH of 3.0–3.3 and contains about 0.8% (0.3–1.6%) lactic acid. Most of the starch remains undegraded at the end of this period. The sour is pumped to the cooker, and two volumes of water and the adjunct are added. The mixture is cooked for 2 h, then the thick, soured (acidified) porridge is cooled to 60 °C (140 °F), sometimes with the addition of a little ground sorghum malt when the temperature has fallen to 75 °C (167 °F) to liquefy the starch and facilitate mixing and cooling. Then ground sorghum malt is added and the mash is held at 60 °C (140 °F) (mashing temperature) for 1.5–2 h; the pH is usually 3.7–4.3. Starch is liquefied and the mash (14% solids) becomes less viscous. Relative to barley malt, sorghum malt is poor in amylases and most of the sorghum diastatic power comes from α-amylase activity. Sorghum amylases are more resistant than barley amylases to acid conditions, but even so they act slowly in the acid mash. It is important that sufficient free amino nitrogen be formed initially in the sour to support the *Lactobacillus* spp and subsequently in the mash to support yeast growth. At the end of mashing, the mixture has 'thinned' and is sweeter (containing about 6% fermentable sugars), but it still contains much undegraded starch. The mash is cooled to 30 °C (86 °F), coarse suspended solids are removed by straining or centrifugation (about 24% of initial solids) and the liquid is inoculated with yeast. The turbid, pink-brown product is consumed while still fermenting. It contains suspended granules of starch as well as liquefied starch and enzymes from the malt. Saccharification probably continues slowly during fermentation.

5.3 Some aspects of mashing biochemistry

Historically various mashing regimes were arrived at by trial and error. However, with increased understanding of the biochemistry involved, it has been possible to refine and improve the reproducibility of mashing with consequent improvements in yields of extract and in the consistency of products. The most widely studied are brewery mashes made with grists predominantly of barley malt, but the same principles apply to mashes made by distillers, brewers of malt vinegar, and others.

The enzymes that catalyse the important changes which occur during mashing are chiefly hydrolases, which catalyse the breakdown of starch, proteins, phytate and other substances. The conditions in a mash are extremely complex because a mixture of enzymes acts simultaneously on a mixture of substrates, some of which are substantially depleted during the mash and many of which are insoluble (at least initially) and may be embedded in insoluble particles, under conditions that are often remote from those that are best for the maximal activity of individual enzymes. Furthermore, there are changing trends in pH and usually substantial changes in temperature. Substantial degradation of starch does not occur until the temperature is high enough to cause swelling and gelatinization. At the elevated temperatures used, malt enzymes are more or less rapidly thermally inactivated (denatured), so their catalytic properties are progressively destroyed. At the end of mashing, the operator needs to have recovered as much extract as is feasible, and to have produced a wort that has the correct consistency, colour, aroma and chemical composition. As the temperature rises, enzyme-catalysed reactions accelerate. At the same time, the rates at which enzymes are denatured also accelerate. The temperature stabilities of malt enzymes vary widely. Often enzymes are partly protected from thermal inactivation by the presence of their substrates and other substances present in the mash. Therefore, as mashing proceeds, the substrate is destroyed and the enzyme becomes more prone to inactivation. Substrate protection explains why some enzymes are appreciably more stable in thicker mashes (with higher grist : water ratios) than in thinner mashes. Thus enzyme survival is influenced by many factors including the mashing programme and sometimes the presence of particular ions. Enzyme activities are influenced by pH, as are their stabilities. The pH values of mashes are adjusted to be the 'best compromise' for the various processes that are going on. Quantitatively, the most important changes that occur in mashing are the solubilization and breakdown of carbohydrates. Pre-formed sugars, particularly sucrose and some lower saccharides, are dissolved during mashing and may be partly hydrolysed during this process, but most of the carbohydrate that occurs in wort originates from starch. However in some grists, particularly those made with undermodified malts or with barley adjuncts, β-glucans should be adequately degraded to prevent problems with beer filtration (caused by the viscosity of the polysaccharide) or the formation of gelatinous precipitates during storage. Worts made from all-malt grists made with well-modified malts do not give these problems. The activity of β-glucan solubilase, whether owing to a malt carboxypeptidase, a microbial endo-(1→4)-β-glucanase, a mixture of these or some other enzymes is more heat stable than the malt mixed-linkage endo-β-glucanases. Consequently at mashing temperatures, β-glucan is dissolved from the endosperm cell walls, the hemicellulose fraction, but insufficient β-glucanase activity may be present to degrade it (Scott, 1972; Narziss, 1981a; Bamforth, 1982; MacGregor and Yin, 1990). Malt endo-β-glucanases are rapidly inactivated in mashes at temperatures over 50 °C (122 °F), therefore, if malt is the only source of enzymes, a 'temperature rest' at 45–50 °C (113–122 °F) is necessary to achieve some degradation of residual β-glucan. Unfortunately β-glucan solubilase will continue to act at higher temperatures, bringing more β-glucan into solution. However if the use of microbial enzymes is permitted, the addition of more heat-stable bacterial or fungal β-glucanases to a mash can be helpful (Chapter 12).

The conversion of starch to soluble products is quantitatively the most important change to occur during mashing (Briggs *et al.*, 1981). As much of the starch as possible must be solubilized, and the mixture of products should have the right degree of fermentability. Starch granules are only degraded slowly by enzymes unless they have been held in warm water so that they are swollen and partly disaggregated, that is they have been gelatinized. Starch gelatinization occurs at temperatures at which malt amylases are unstable. Starches from different sources differ in their gelatinization temperatures and, indeed, starch granules from a particular source gelatinize over a range of temperatures in a time-dependent manner (Chapter 12). Thus, large barley starch granules gelatinize in the range 60–65 °C (140–149 °F) and the small granules at 75–80 °C (167–176 °F). In mashes made with barley malt at 65 °C (149 °F), large granules swell and gelatinize and are readily degraded by amylases, but small granules may not be gelatinized and converted; as a result they survive mashing and do not contribute to extract. In pre-cooked adjuncts, such as flaked maize or torrefied wheat, the starch is partly or wholly gelatinized. Because the gelatinization temperatures of maize, rice and sorghum starches are high, and malt enzymes are rapidly inactivated at these gelatinization temperatures, preparations of these cereals require pre-cooking before they can be converted by malt enzymes (Chapter 12).

For hydrolytic processes to occur in mashing, the enzymes involved must survive for a significant period. The degradation of starch is initiated by α-amylase, which is usually present in excess in all-pale-malt mashes. The β-amylase attacks the non-reducing ends of the dextrin chains as these become available, so saccharifying the starch and increasing the fermentability of the mixture. These enzymes are progressively destroyed during mashing. It has usually been supposed that α-glucosidase and limit dextrinase have little influence on starch conversion in mashing, but this may not be true. The fermentabilities of distillers' worts and mashes are higher than would be expected if only α-amylase and β-amylase were operating to degrade the starch. Under mashing conditions, even when calcium ions are present, α-amylase is not completely stable; as the starch is degraded the rate of destruction of the enzyme increases as the 'protective' substrate is removed. β-Amylase is less stable than α-amylase and in mashes is largely destroyed in 40–60 min at 65 °C (149 °F). To obtain adequately fermentable wort, it is necessary to have sufficient α-amylase in the mash to 'liquefy' the starch rapidly, while there is sufficient β-amylase still present to carry out saccharification. Maximal extracts are obtained from all-malt, isothermal mashes at 65–70 °C (149–158 °F), but the fermentabilities of the worts prepared at higher temperatures are less, because these cause the more rapid destruction of β-amylase.

In temperature-programmed mashes, the extract increases and levels off at 50 °C (122 °F), as up to this temperature the changes are mainly the result of pre-formed materials dissolving; little amylolysis occurs. However, as the temperature increases towards 65 °C (149 °F) the extract increases very greatly because massive amylolysis occurs, but it hardly increases more as the temperature rises to, and remains at, 75 °C (167 °F). The optimum pH for starch conversion, at mashing temperatures, is about 5.3. In all barley-malt mashes, amylase levels are rarely limiting. However, when adjuncts are used, it is found that if the ratio of adjunct to malt is too high starch conversion is incomplete and the fermentability of the wort, which is often about 75%, declines to about 60–65%. When green malt is used in mashes with adjunct-rich grists, as in some distilleries, highly fermentable worts may be obtained, presumably because the green malts are exceptionally rich in enzymes. Nearly all the carbohydrates present in wort are derived from starch or pre-formed soluble materials (Harris, *et al.*, 1955). Traces of pentose sugars (arabinose, xylose, ribose) have been detected in some worts together with the unfermentable dextrins, pentosans and β-glucans. The high viscosities of worts from triticale malts are caused by pentosans (Blanchflower and Briggs, 1991c) but the high viscosities of worts made from under-modified barley malts and from mashes containing unmalted barley adjuncts are usually caused by β-glucans. The β-glucanases of malt are heat labile and are easily destroyed during kilning. Therefore, if under-modified malts are to be used in mashing they should be kilned at low temperatures to favour β-glucanase survival, and the mashing process should have a rest at 35–45 °C (95–113 °F) to permit some β-glucan degradation.

When sorghum malts are mashed alone it is generally found that starch conversion is incomplete and that the worts obtained are not suitable for making European style, clear, hopped beers. In part this is because of the low levels of amylases present in sorghum malts and in part because of the high gelatinization temperatures of sorghum starch (55–75 °C (131–167 °F)). This has implications for extract determination for sorghum malts (Chapter 13). Millet starches differ in their gelatinization temperature ranges: finger millet 62.5–74 °C (144.5–165.2 °F) pearl millet 56.5–72 °C (133.7–161.6 °F) and foxtail millet 56–67 °C (132.8–152.6 °F). Starches from their malts have about the same gelatinization temperature ranges (Malleshi *et al.*, 1986a).

A proportion of the complex mixture of proteases, carboxypeptidases and aminopeptidases that degrades proteins during barley malting continues to act during mashing. In worts from all-malt mashes, about 5–6% of the wort solids are nitrogen-containing materials. The total soluble nitrogen present in the wort represents 30–40% of the total nitrogen originally present in the malt. Of the nitrogen-containing materials present in wort and derived from proteins, complexity ranges from simple amino acids through di- and tripeptides to complex polypeptides. Analytically, the materials in the total soluble nitrogen are described in various ways, including the coagulable nitrogen (coag. N), which is precipitated when the wort is boiled, the permanently soluble nitrogen (PSN), which is not precipitated by boiling, the formol nitrogen (chiefly a measure of amino acids and simple polypeptides), and the free amino nitrogen (FAN) (mainly amino acids) (Chapter 13). In addition, worts contain substantial quantitities of ammonium ions, which yeasts use readily. Some 50–70% of the amino acids present in wort are preformed in the malt, the rest are generated during mashing (Enari, 1981). Since 60–70% of the malt protein remains in the spent grains at the end of mashing it appears that the supply of total soluble nitrogen is not limited by lack of substrate. Probably the quantity of proteases is limiting, since the carboxypeptidases, which degrade the products of proteolysis to amino acids, are present in excess (Sopanen *et al.*, 1980; Enari, 1981). It used to be thought that in decoction mashing proteolysis occurred at 35–50 °C (95–122 °F) and that in infusion mashing at 65 °C (149 °F) very little proteolysis occurred. In fact, proteolysis occurs over a wider temperature range so that in temperature-programmed mashing proteolysis is rapid at 45 °C (113 °F), slower at 55 °C (131 °F) and, although it is slower still, it continues even at 70 °C (158 °F). Proteolysis is maximal at a mash pH of about 4.6, a value significantly below that of normal mashes. In single-temperature, 1–3 h experimental infusion mashes, most proteolysis is achieved at 50–55 °C (122–131 °F). In brewing it is generally thought that yeasts need 100–150 mg free amino nitrogen per litre for a good fermentation (Enari, 1974). In infusion mashing, there is little room for compromise in the temperatures employed. Usually a temperature of about 65 °C (149 °F) is best for many products, but it is likely that temperatures in the range 63–69 °C (145.4–156.2 °F) are in use. A major advantage of temperature-programmed mashing is the possibility of adjusting the temperature/time programme in such a way as to favour, or curtail, the activities of particular groups of enzymes. Thus, with reference to the barley-malt enzymes, β-glucanase activity is optimal at about 45 °C (113 °F), phytase is optimal at about 50 °C (122 °F), proteolysis is most rapid at 50–55 °C (122–131 °F), the highest yield of fermentable extract is obtained at 63–65 °C (145.4–149 °F), while the highest extract recovery is attained at 65–70 °C (149–158 °F).

Nitrogen-containing substances, other than amino acids and peptides, occur in wort and, although they are present in relatively small amounts, they are important for yeast growth. Significant levels of nitrogen-containing yeast growth factors, or vitamins, are present in wort as are purine and pyrimidine bases, their equivalent nucleosides and a range of amines such as methylamine, dimethylamine, ethylamine, hordenine and choline. Nucleic acids comprise 0.2–0.3% of barley. During mashing at least 95% of these are degraded (Harris, 1962b). Bases derived from nucleic acids may account for 8–10% of the nitrogen in the wort. Some bases may be essential for yeast growth and others may enhance beer flavour (Søltoft, 1990). During mashing, nucleotides are completely hydrolysed by phosphatases to nucleosides and inorganic phosphate. Higher mashing temperatures favour the survival of nucleosides, but at lower temperatures hydrolysis to free bases and pentose sugars is favoured.

It is not clear if lipids are degraded during mashing. Small quantitites of triglycerides (triacylglycerols) are dispersed in wort together with diacylglycerols, monoacylglycerols, sterols, sterol esters, possibly lysolecithin, other polar lipids and free fatty acids. Rapid wort-separation techniques are liable to give worts containing elevated levels of lipids. Unsaturated fatty acids are precursors of the substances that impart 'stale' flavours to beers on aging. Sterols and unsaturated fatty acids are needed by yeasts that have been stored anaerobically. Lipids in beer tend to reduce foam (head) formation and retention.

Worts contain a range of organic acids, mineral ions (including phosphates) and phenolic substances. Calcium ions, including those added in the mashing liquor, cause precipitates of calcium oxalate and calcium phytate. Of the inorganic substances dissolved during mashing, only zinc ever seems to be in short supply as a yeast nutrient. The concentration of zinc in wort can vary widely and does not seem to correlate with the quantities originally present in the malt (Jacobsen, 1986). The phenolic materials include phenolic acids and anthocyanogens (proanthocyanidins). About 10% of the malt phenols are dissolved during mashing. They may contribute to astringency in beer flavour, they help to precipitate proteins during the formation of hot and cold breaks and during the formation of beer hazes during storage. The quantities of phenols extracted increase with mashing temperature and pH. The use of adsorbents to remove phenols and other haze precursors from beer is expensive. Consequently, there is great interest in making malts from anthocyanogen-free barleys. For various reasons other processes used to reduce phenol levels in wort (the addition of hydrogen peroxide or formaldehyde to the mash, mash aeration or mashing with green malt or malt prepared by resteeping in very dilute formaldehyde) are not now used. Makers of Scottish malt whisky use malts that have been flavoured with peat-smoke. The phenols, which are largely responsible for the flavour, are volatile and are carried out of the fermented wort during distillation and appear in the final product. In bird-proof sorghums, the concentrations of phenols with tanning properties can be so high that the malts made from them are unsuitable for making Bantu opaque beers because, unless they have been treated to reduce tannin levels, the tannins largely inactivate the enzymes needed in mashing.

The composition of worts is complex and no doubt the qualities of beers and other products are dependent on the amounts of many substances or groups of substances and the ratios between them. The origins of one flavour compound, dimethyl sulphide and its precursors, *S*-methyl methionine and dimethyl sulphoxide have aroused much interest (section 4.17). Wort composition is influenced by the type(s) of malt used in the grist, the adjuncts and supplementary enzymes used (if any), the brewing liquor, the fineness of grind of the grist, the temperature/time mashing programme used, the pH of the mash, the thickness of the mash (liquor : grist ratio), the system used for wort separation, how effectively air (oxygen) is excluded from the mash, and the type of sparging regime used. All-malt mashes made with distilled water have pH values of about 5.8 when cooled to room temperature. If the liquor contains temporary hardness (calcium bicarbonate), the pH of the mash is raised. However, if the liquor exhibits permanent hardness (it contains calcium sulphate) the mash pH is reduced to a useful extent as the calcium ions displace hydrogen ions from phytic acid, and perhaps other phosphates. Usually a mash pH of about 5.3 (at mashing temperatures) is desired. However salts confer flavours to the product and mashes may be acidified by additions of mineral acids, lactic acid or wort from a previous brew acidified by a *Lactobacillus* sp., or by the inclusion of a small proportion (1–5%) of an 'acid malt' (containing lactic acid) in the grist. The 'best' pH value of a mash is a compromise among the optima for the different biochemical processes that occur.

5.4 Wort separation and sparging

When the conversion period of mashing is completed, it is necessary, in most instances, to separate the solution of dissolved substances (to brewers, wort containing the extract) from the residual solids (brewer's spent grains or draff). This separation must be achieved as promptly as possible without leaving a large proportion of the extract in the spent grains. Therefore, it is necessary for the liquid to drain from between the grist particles, to be replaced by fresh, warm water (washing or sparge liquor or, less usually, re-mashing liquor) into which soluble materials from within the grist particles can diffuse. Extract is recovered most quickly from finely ground grist. The strength of a wort declines as sparging is continued.

Wort flow through a bed of spent grains is dependent on the temperature, the bed depth, the pressure difference, the viscosity of the wort and the distribution of particle sizes. Under-modified malts often give mashes with poor wort run-off properties, and this used to be attributed to the elevated levels of β-glucans increasing the viscosity of the wort. However, at the elevated temperatures used in sparging (75–80 °C; 167–176 °F), the effects of viscosity are minor, and the slow run-off rates must be caused by other factors. Experiments with grists of different degrees of fineness of grind and with flour adjuncts show that too high a proportion of fine particles in the grist clogs the mash filter-bed, retards wort run-off and may reduce extract recovery (Barrett *et al.*, 1973, 1975). Not only may fine particles pre-exist in the grist but others may form from aggregates that consist of glucans, pentosans, fragments of cell walls, lipids and proteins adhering to small starch granules. The addition of a heat-stable glucanase to the grist frequently accelerates wort separation, apparently by degrading the β-glucan and altering the structure of the particulate aggregates. Proteins from the unmodified regions of malt also contribute to the problems of wort separation. Thus, in temperature-programmed mashing, the amount of protein in solution at first increases, but as the higher temperatures are reached 65–70 °C (149–158 °F), protein precipitation occurs and the precipitated protein tends to clog the mash (Lewis and Wahnon, 1984; Lewis and Oh, 1985). A particular gel-protein fraction has been blamed for difficulties with wort run-off (Muts and Pesman, 1986; Moonen, *et al.*, 1987). This material contains a considerable proportion of disulphide-rich D-hordein, which is believed to undergo oxidative polymerization during mashing, wort recirculation and separation. The aggregated protein, with associated materials, may form a jelly-like layer of fines that severely slows, or even prevents, wort separation. Minimizing the oxidation of the mash, by avoiding air entrapment, leads to faster wort separation.

Frequently, in the first stage of wort separation, the wort is recirculated through the bed of grist, which acts as a filter. Thus, as the initially turbid wort is recirculated it becomes clear and 'bright'. Wort collection then begins and, when the volume is sufficiently reduced, the 'goods' (residual grist solids) are sparged (or remashed), and wort collection is continued until the specific gravity has declined to some pre-chosen value. As well as the strength of the collected wort declining, the composition of the later worts also changes, often in ways that reduce their quality. First worts are collected at about 65–70 °C (149–158 °F), then sparging may be carried out at 75 °C (167 °F) or even 80 °C (176 °F). The higher temperatures are efficient at removing the remaining extract from the mash. However, the elevated temperatures precipitate proteins, inactivate residual enzymes and may disperse unconverted starch and high-molecular-weight β-glucans. In addition, the separation of the wort removes the buffering substances from the mash, and there is a tendency for the mash pH to rise. As a result, the extracted solids tend to contain proportionally more polyphenolic materials, phosphates and other minerals such as silicates, as well as nitrogenous materials, dispersed starch and β-glucan. Therefore, the 'last runnings' are rich in potential haze-forming and possibly coarsely flavoured or undesirable components. In addition, the fermentability of the last runnings may change dramatically, first rising and then declining to a low level.

From the expected stabilities of the starch-converting enzymes (usually judged from experiments made with purified preparations), it would be predicted that they would be destroyed by the end of the mashing period. In brewing wort, fermentability is 'fixed', generally at 70–75%, by the total enzyme destruction that occurs during the hop-boil. However, in whisky distilleries, the wort is unboiled and fermentabilities of about 86% are reached (Briggs, 1978). The 'extra' fermentability is owing to the continued degradation of dextrins in the wort, caused by residual levels of enzymes that have survived mashing assisted by enzymes released by the 'highly attenuating' strains of yeast used in distilleries. For special purposes, where it is permitted, enzymes such as fungal amyloglucosidase, pullulanase or an enzyme-rich malt flour may be added to a wort to obtain a nearly fully fermentable product.

After wort separation from a mash, or after distilling from an 'all grains in' fermented mash, the brewer or distiller (or other user) is left with a mass of wet residual solids, which may represent about 17–22%, by dry weight, of an original all-malt grist and probably more of a grist that contained unmalted adjuncts. The draff is initially wet, often containing 70–80% moisture, and it is normally used as a cattle-food. Dried spent grains are variable in composition, but representative values are: water 9–11%, fibre 14–19% (d.m.) (dry matter), crude protein 14–27% d.m., nitrogen-free extract 38–40% d.m., lipids 5–9% d.m. and ash 3–4% d.m. The wet spent grains become mouldy within hours, and drying the product, by squeezing out the excess liquid and drying in hot air, is expensive. Furthermore the press liquid is a highly undesirable effluent. Other uses proposed for spent grains are as culture media for fungi, to feed pigs, to subject them to anaerobic digestion to produce methane and as a basis for mushroom compost.

A consideration of the changes that occur in mashing indicates that malts should yield the highest possible extracts having the desired composition under 'standard' mashing conditions. Enzyme levels must be adequate and modification should be nearly complete to minimize problems caused by β-glucans and gel-proteins. In addition, the colours, flavours and aromas of worts should be correct and the worts should be clear and not hazy. Methods for evaluating malts are discussed in Chapter 13.

6 The selection and purchase of grain

6.1 Introduction

This chapter concentrates on barley. However the principles outlined apply to other grains. More relevant information is contained in Chapter 14.

Barley varieties giving grain suitable for malting must be selected by the breeders at as early a stage as possible during a breeding programme. Because of the constraints of the breeding process some selection criteria are used that may be only weakly related to overall malting quality. Such criteria include milling energy, sedimentation characteristics of flours, grain contents of gel-proteins, β-glucans and total nitrogen, mealiness, water uptake, and so on (e.g. Bathgate, 1977; Moll, 1979a; Ellis *et al.*, 1989). These criteria are not used or are used only to a limited extent by maltsters. The best evaluations are based on the analyses of small samples of malt prepared by 'micromalting' grain samples (Chapter 11).

When new varieties are offered by breeders they are tested in national and international trials to ensure that they meet criteria of varietal purity and agronomic performance in terms of improved yields, disease resistance, and so on. At this stage maltsters carry out micromalting and larger-scale trials to decide which varieties have the most favourable malting characteristics and will, therefore, be assigned malting grades. The breeders will already have carried out micromalting tests. In England and Wales, the NIAB and in Scotland the Agricultural Colleges annually produce lists of recommended varieties, which include assessments of malting grades. Undoubtedly micromalting is the most powerful and reliable method for grading barleys (Chapter 11). Many new varieties are put into trial each year. The more successful of these progressively displace current varieties so that, in contrast to the period to the early 1960s, maltsters must continually be evaluating the new cultivars. The results of national and international trials are frequently reviewed (e.g. Smeaton, 1985; Narziss, *et al.*, 1988, 1989; Burbidge, 1989; Anon, 1990a; Sacher *et al.*, 1995; Schildback and Burbidge, 1996). In the 30-year period to 1985, UK breeders increased barley yields by approximately 30%. In 1884 extract yields of good, pale barley malts would typically have been 78–80% (d.m.); by 1984 these had increased, in the better malts, to 84–85% (d.m.) The improvements in North American six-rowed barleys since 1910 have been assessed (Schwarz and Horsley, 1995); most improvements were in kernel plumpness, malt extract, soluble protein, α-amylase and diastatic power. The criteria which may be assessed in micromalting include water-uptake rates, malting losses, hot water extracts (fine and coarse grind), wort colour and attenuation limit, the ratio of soluble nitrogen to malt total nitrogen (SNR or Kolbach index (KI)), the free amino-nitrogen content of the wort, tannins in the wort, and α-amylase and diastatic power of the malt.

Other times when grain will be evaluated are on purchase samples, on samples taken at delivery to the maltings (to decide whether loads should be accepted or rejected) and in store when samples are drawn to check whether they are ready for malting. The times available for grain evaluation at these different stages are very different. For example, grain samples from store may be subjected to germination and micromalting tests that can extend over a fortnight. By comparison, when bidding for a lot of barley or when deciding whether or not to accept delivery of a load of grain, only 30–45 min may be available. Consequently, the completeness of assessment of grain samples varies very greatly. In some instances, rapid, but 'non-standard' methods of analysis must be used. For example, standard moisture determinations take *ca.* 4 h, but rapid (but less reliable) methods are available that take only a few minutes. These rapid, secondary methods of analysis must be frequently standardized against the primary, 'standard' methods, which are the most reliable and on which commercial transactions are legally based. At the delivery point, before the grain is accepted, the operators must check that the bulk matches the 'purchase sample', which will have been thoroughly analysed. They will usually check for varietal purity, absence of damage, viability (tetrazolium test), moisture and total nitrogen contents, screenings, the absence of insects which infest stored grain, and ergot sclerotia. Every load delivered must be rigorously checked and weighed.

Techniques for assessing grain quality are arbitrarily divided into laboratory techniques, which require specialist equipment, and hand evaluation, which assesses various points and is based on judgement. Grain evaluation should be carried out in specially equipped rooms. The standard methods of laboratory analysis are based on many years of trials and, indeed, are frequently refined. They are designed for use with barley and barley malts, as well as other brewing materials. The three most frequently used are the *Recommended Methods of Analysis of the Institute of Brewing* (IoB, 1993), *Analytica-EBC* (EBC, 1987) and the *Methods of Analysis of the American Society of Brewing Chemists* (ASBC, 1992), which are targeted at ale-brewers, lager-brewers and North American brewers, respectively. In addition there are other, 'less official' collections of analytical techniques: a range of non-standard tests may sometimes be used. The inefficiency of having different sets of analyses in use is widely recognized and gradually common methods are being adopted. In general, distillers seem to use brewers' systems of analysis.

In contrast to 'laboratory analyses' there are no published systems for hand evaluation. At least some degree of hand evaluation is indispensable to obtain stocks of uniform, sound grain. However hand evaluation alone, even when carried out by trained and experienced experts, is not enough. For example, of the grain selected by hand as being of top quality at one British site in 1956–7 only 33% malted really well, and no less than 39% was finally rejected (Macey, 1957). Frequently when barley is offered for sale, the prospective purchaser will see samples of only 100 g to 1 kg. These he will analyse and then agree to take delivery, at an agreed price, of a certain number of tonnes of grain that equal or exceed the 'purchase sample' in quality. Alternatively, he will specify the quality and amount of barley he requires, and the merchant or other supplier will contract to provide this. An example of a barley contract is given by Bathgate (1989). More rarely farmers contract to grow barley for maltsters, in closely specified ways.

The largest single cost of making malt is the cost of the barley or other grain. It is impossible to turn inferior barley into good malt. Maltsters put substantial resources into encouraging farmers to grow malting-grade varieties to a high quality standard, to purchasing the best samples available and to maintaining and enhancing the quality during grain cleaning and storage, when post-harvest maturation (the disappearance of residual dormancy) occurs. Before the grain harvest begins, maltsters refer to their customers and estimate the quantities and grades of malt likely to be purchased in the following year. The sizes of carry-over stocks needed to permit malting to continue while the new season's barley is recovering from dormancy will be noted. From these data, the barley requirements will be estimated. The areas sown to different varieties in each district will be estimated, from a knowledge of seed sales and contracts with farmers and merchants, and fields will be inspected to obtain estimates of probable yields. So which varieties will be available, and the probable grades and qualities, are known before harvest time. With this information, a maltster will decide his purchase targets: which varieties to buy, how much of each and of which grade. A programme of grain deliveries to the maltings will be drawn up. Clearly the prices given for the raw material and received for the finished malt decide the economic success, or failure, of the maltings. The problems of grain purchase are made worse by the wide variations in growing conditions that occur in different seasons. For example, in a very hot, dry season when grain filling is impaired it is necessary to accept a proportion of thinner (less plump) grains having higher nitrogen (protein) contents. As a result, it is impossible to have invariable standards year after year and the target standards must vary with each harvest. However, the maltster's objective, to buy the best quality grains available for making the different grades of malt, does not vary. Under most contracts, grain delivered in the UK must be of the agreed variety, with minimal contamination by other materials, including other cereal grains or grains of different barley varieties. The grain must be in a good physical state, of the correct nitrogen (protein) content and less than a maximum (usually 16%) moisture content.

The grain size range (TCW, sieve fractions) must be acceptable, and the viability (germinative capacity) must usually be at least 98%. Germination (after recovery from dormancy) usually exceeds 98% and is frequently 99–100%.

When a load of grain is brought to a maltings, the grain is inspected and sampled. On some criteria, for example if the presence of insect storage pests or ergot contaminants is detected the load, may be rejected out of hand. Alternatively, the load may be accepted, or accepted subject to a price adjustment, e.g. for an elevated grain moisture content or a high proportion of screenings (thin corns).

In British practice, grain is purchased subject to it meeting all of a range of criteria. Elsewhere 'systematic' grain evaluation systems have been tried, in which different characteristics of the grains are given positive and negative scores and the grain is purchased on the final score (Hopkins and Krause, 1947; de Clerck, 1952; Vermeylen, 1962). This approach seems to be faulty, since grain characteristics do not compensate for each other; for example, a low-nitrogen (protein) value cannot compensate for a low germinability. The criteria used to accept or reject grain lots differ significantly among companies and often in different seasons. Other criteria are used either for experimental purposes or when grain is being selected from store to make a batch of malt. The processes of grain evaluation will now be discussed in more detail.

6.2 Sampling

Whenever grain bulks are to be evaluated, it is necessary to obtain samples on which tests can be carried out. As bulks of grain are never uniform, or completely homogeneous in their composition or characteristics, it is necessary to obtain samples that are representative of the bulk, that is their composition represents the true average composition of the bulk. In one 15 t batch of barley, different samples were found to vary by as much as 20% in their germinative capacity and by 0.25% in their nitrogen content. Therefore, **it must be emphasized** that samples must be taken so that they represent the true average composition of the bulk of the grain (de Clerck, 1952; Hudson, 1960; Wainwright and Buckee, 1977; Moll, 1979a). If this is not done, no subsequent analyses on the samples, however carefully performed, can give a reliable guide to the value of the bulk. At various times grain needs to be sampled in flat-bed stores, in silos, in sacks, in lorries and when it is being moved by conveyors. The sets of official methods prescribe how sampling should be carried out. Each part of a grain bulk should be sampled. Particularly in large stores, this can be difficult, and it may be necessary to move the grain, for example between bins or silos, so that the moving stream of grain can be sampled. In flat bulks of grain, the grain should be sampled at intervals over the whole surface and to the full depth. From static bulks of grain, samples are taken using a probing sampling device. Various manual augers, spears and 'triers' are in use. With sacks, these may be driven into the side at the top, middle or bottom; a better method, if the sack is open, is to drive them into the grain from above and take samples at all depths. The samplers may be simple pointed devices with a pocket in the side, which can be driven into the grain, rotated and withdrawn bringing out a sample in the pocket, or it may be possible to open and close one or more pockets by rotating a cylindrical outer cover. Some probes are long and have several pockets along their length. In practice they are often difficult to drive into the mass of grain. However they have the advantage that, by providing samples from the different depths, it is possible to detect areas of inferior grain layered in the bulk. Often samples are taken using vacuum probes; these are nearly always used on loads in lorries. These may be hand held and the samples collected in a container or may be mounted on a mechanically activated probe and the samples sucked into a receiver in an adjacent building. For all the convenience of the latter type, the former is superior, because the operator can inspect the grain bulk when it is uncovered and so may detect colour variations in the bulk, indicating admixture or may detect insect pests moving on the surface when the grain is first uncovered.

When sacks are being sampled and they cannot be opened and sampled to the full depth with a spear, the recommendation is that they should be sampled from the top, middle or bottom at random. Recommended sampling frequencies differ, but *Analytica-EBC* (EBC, 1987) instructs that in up to 10 bags in a lot, all should be sampled. Between 10 and 100 bags, 10 (selected at random) should be sampled and when there are over 100 bags in a lot, the number of samples should be the square root of the number, rounded to a whole number, the samples being drawn at random. Similarly, sampling patterns in open wagons or lorries are also specified, with five sampling points in lots of up to 15 t, eight for 15–30 t lots and 11 for lots of 30–50 t. Some containers cannot be inspected and sampled and so the grain must be sampled as it flows from the lorry or wagon into the grain store. This is disadvantageous, because it must be transported into a holding bin and must then be returned to the sender if it is of unacceptable quality. This procedure wastes time and creates a greater risk of importing insects into the store. Flowing grain may segregate into streams by size or shape, so it is necessary to sample the full cross-section of a grain stream, taking samples at regular intervals until the flow ceases. Manually this is achieved using various patterns of scoops, cups and long-handled samplers. If the grain is moving in a conveyor, samples may be taken using an automatic device. These should, at short intervals, momentarily divert the grain stream into a receiver. Samplers may consist of valves that open and close, or a device that moves right across a grain stream collecting grain as it does so.

The primary sample should normally weigh about 1–2 kg. Like all grain samples, it should immediately be placed in a labelled, waterproof, clean and airtight container. This should have no smell that would taint the grain or conceal its odour. The label should be of a standardized type and should give details of the time, date and location at which the sample was taken, how it was taken, from which bulk and the name of the operator. If the sample is stored cold it should be warmed to room temperature before being opened to prevent condensation occurring. At every stage, moisture losses or gains should be avoided. Often small samples of grain are transmitted in paper envelopes. These are quite useless for moisture determinations or for deciding if the bulk of the grain has an adverse odour or taint.

The primary sample must be well mixed and sub-samples taken (also without statistical bias) for individual laboratory tests. This is achieved either by tipping and mixing on a flat surface and dividing with a 'quartering iron', or by using a sample divider designed to eliminate the unconscious selection of particular grain types. Many satisfactory patterns are in use. One simple pattern has a chamber divided by partitions. The grain in alternate compartments goes to one of two outlet spouts and a receiver. By repeatedly passing grain through a divider and recombining the fractions it is well mixed, and the smaller lots, also from the divider, are acceptable for testing.

However, tests on small numbers of grains are inherently unreliable, so for example, it is desirable to grind sub-samples of *ca.* 20 g grain even though only 1–2 g samples may be needed for analysis (for nitrogen or moisture analyses, for example). Results of viability and germination tests, typically based on 100–300 grains, are notoriously unreliable (Urion and Chapon, 1995, Fig. 3.3 in section 3.3). If such samples are **correctly** drawn from a well mixed sample of grain the 95% confidence limits of test results on small samples of grain, e.g. 300 corns, may vary significantly from the true, mean value (Fig. 3.3). If the samples are incorrectly drawn the uncertainties will be even greater. Therefore, even with 'perfect' analytical techniques it is impossible to exceed certain levels of reliability.

Many tests require grains to be counted. Some operators prefer to do this by hand, keeping score with a counter. It is essential that grains be taken 'as they come', and personal selection bias, e.g. for plump or bright grains, be suppressed. There are a variety of 'corn counters' available, but they carry the risk of selecting for grains of particular sizes or shapes. Learner's (1956) corn counter, the simplest, consists of a shaped piece of plastic containing 50 carefully shaped slots. The device is pushed into a heap of grain, removed and 'sorted' with tweezers so that 1 grain occupies each slot. The 50 grains are then tipped into a receiver. With the Kickelhayn device grain is spread over a plate containing 500 slots. When every slot is filled, the surplus is removed, then a slide is pulled out allowing the 500 grains to fall into the receiver below. Finally a 'suction bed' counter, with a given number of slots (e.g. 100) may be attached to a vacuum cleaner. One grain should adhere to each perforation. After checking that this is so the operator drops the grain onto the germination plate or into a receiver by momentarily relieving the vacuum.

6.3 Hand evaluation and some laboratory tests

There are a number of accounts of hand and laboratory evaluation (de Clerck, 1952; Hudson, 1960; Bergal and Clemencet, 1962; Briggs, 1978; Moll, 1979a; Cooke and La Berge, 1988). Traditionally hand evaluation is carried out by visual inspection with and without cutting the grains, and by assessing the 'feel', the odour and the texture of the grain when bitten. At present, hand evaluation is supplemented with a wide range of other tests, and the distinction between hand and laboratory testing is now largely artificial. Evaluating grain by eye is a difficult and skilled operation, and the staff responsible need to be carefully trained. Grain of one variety grown and harvested under contrasting climatic conditions can appear rather different. The continuing replacement of current varieties with new ones means that the grain buyer must continually update his knowledge. Often buyers have 'reference collections' of grains of different varieties and types and of common contaminants. They also have reference articles, charts and handbooks describing the grains of the varieties currently being grown.

The buyer's first concern is that the grain on offer is of the stated variety, and that there is no significant admixture of grains of other varieties. Usually grain of different varieties cannot be malted well together. Grain should be inspected against a matt-black background with northern (indirect) daylight or carefully chosen artificial light. A hand lens and a low-power binocular microscope are often needed. Grain identification is based on structural characteristics such as whether husked or naked (not malted), the grain shapes (two-row or six-row), average size, plumpness (length : width ratio), aleurone colour (dark or pale blue or colourless), tendency to be pigmented (e.g. the colour of the veins in the husk), the shape of the grain base (bevelled, nicked or plain), the rachilla type (length, degree of hairiness, length of hairs), the details of the lodicules (bib or collar types, the shapes and sizes, ease of staining, etc.), the shapes of the apices of the palea and lemma, the shape of the furrow, the coarseness or absence of spicules on the veins, and so on (Chapter 2). In closely threshed grains, there will be no signs of awns and indeed the rachillae and grain bases may have been damaged or removed. Unsurprisingly, threshed grains of different varieties may be hard to identify.

In disputed cases, grain identification may be assisted by determining the electrophoretic or possibly HPLC pattern of the hordeins (Chapter 4). In extreme cases, it may be necessary to grow the grains to mature plants to check their identities. Most attempts at identifying barleys by electrophoresis involve extracting the alcohol-soluble proteins (hordeins) from individual grains and then separating them by polyacylamide gel electrophoresis, at acid or alkaline pH values, on uniform or density-gradient polyacrylamide gels. The proteins may or may not be reduced and alkylated before electrophoresis (Cooke and Morgan, 1986; EBC, 1987). The proteins are then fixed and stained. The banding patterns obtained are specific for particular groups of barleys. Others have used the elution patterns of hordeins from HPLC to characterize barleys. There are also proposals to characterize them by the different electrophoretic patterns given when isozymes of peroxidase or esterases are separated, and attempts are being made to use immunological tests for varietal recognition. Ultimately, techniques based on sequences of DNA, the genetic material of grains, will be able to provide unambiguous varietal identification of individual grains. Attempts to use RFLPs (restriction fragment length polymorphisms) for this purpose are in progress. Attempts are also being made to find electronic methods of sorting grain types using image analysis, based on shape, colour and size.

Generally grain samples for malting in the UK must be more than 95% pure in variety. Impurity may result from attempts to sell low-value grain by dishonestly mixing it with high-quality malting barley. However accidental contamination can easily occur if the farmer sows impure seed, or if the combine harvester or any of the grain handling machinery, transport or storage facilities are not completely cleaned. All these sources of possible contamination are overcome by skillful and responsible operators.

In addition to establishing varietal purity, inspection will detect contamination of the barley by foreign grains (wheat, oats, rye), by trash such as straw rachises, string, weed seeds, sand, snail shells, soil and stones. Samples containing seeds of noxious weeds such as poisonous darnel (*Lolium termulentum*), or corn cockle (*Agrostemma githago*) are likely to be rejected. Modern cleaning machinery is efficient at removing the seeds of wild oats (e.g. *Avena fatua*), which used to be a common contaminant of barley grain. In the UK, grain lots containing ergot are rejected. Ergot bodies are the sclerotia of the fungus *Claviceps purpurea.* They are purple-black bodies, with white interiors, 0.25–5.1 cm (0.1–2 in) long, which form in the place of grains in infected plants. Because they are often the size of grains, conventional cleaning machinery may fail to remove them. They are rich in alkaloids related to lysergic acid, and ingestion can cause hallucinations, St Anthony's fire, convulsions, gangrene, abortion and death. Any signs of mites or insect pests that can infect stored grains, whether entire insects or fragments, will cause grain to be rejected and sent from the maltings as quickly as possible. If insects are found on site, in stores for example, immediate action must be taken to control or eliminate the outbreak (Chapters 7 and 8). By shaking grain in a sieve over a catch tray, free-living, mature insects can easily be found. However eggs, larvae, pupae or adults inside the grains cannot be easily detected. It is desirable to inspect the surface of each lorry load of grain for insects immediately it is uncovered. Insects are relatively easily detected at this stage and, if detected, the load can be rejected at once. Samples of grain that have been properly stored will not be contaminated with rodents (rats, mice) hairs, faeces (droppings) or urine.

As well as being of the correct variety the grain must be of the appropriate quality. It should not be unduly damp. Traditionally this was judged by the feel of the grain to the hand and when bitten, but now moisture contents are measured. The grain should appear uniform, that is the grains should all be relatively plump, be evenly coloured and show no signs of being a mixture of lots of different qualities. Grains should be relatively plump for the variety (one recommendation is for an ellipsoidal ratio of 2–2.5 to 1) and should certainly not be long, thin and shrunken. The failure of grains to fill may be because of a number of factors, including lack of water in a dry season or being damaged by frost. Seed grain should be avoided absolutely by maltsters because of the likelihood of it having been treated with noxious seed dressings containing insecticides and fungicides. These dressings should contain a vividly coloured dye and, if it is the local practice to use this precaution, artificially coloured gains should be looked for. Grains of most varieties should appear bright, clean and 'straw yellow'; unripe, greenish grain should be absent. The husk should appear 'thin' and 'fine' and lightly wrinkled, and it should have a matt surface. It should not appear 'coarse', shining and unwrinkled. (These appearances are not related to the physical thickness of the husk, but the degree to which it is stretched by the grain inside.) However, grains of pigmented varieties should have their appropriate colouring. Varieties with blue aleurone layers often appear green as the blue shows through the yellow husk. The husk should not have a blotchy or a discoloured appearance, or have dark brown, stained or blackened ends. After a damp season, grain often has a bluish-grey tinge caused by a coating of 'field' microbes, probably chiefly fungi. These discolorations have various names (weathered, smudge, blighted, black-point, stained, etc.). In all cases, the adverse appearance is attributed to attack by microbes either in the field (favoured by humid and wet conditions) or in the store when grain is stored undried. In the past, attack by microbes occurred when grain was stored unthreshed in the rick. This could lead to heavy contamination and discoloration; affected grain was described as 'foxed' (reddish coloured) or 'mow-burnt' (having a scorched appearance).

Grains of some varieties of barley may have a harmless whitish bloom, of cuticular wax, especially in the crease. This should be distinguished from an undesirable coating of white fungal mycelium. As well as appearing 'bright', the grain should have the 'correct', green-grain smell. There should be no hint of sour, sickly, sweet, sharp, musty or other adverse odours owing to microbes or contaminants (agrochemicals; diesel fuel, etc.). To detect faint odours it is desirable to smell samples held in freshly opened containers. Grain that has been dried may lose its odour.

The husk should not be damaged either during threshing; this is usually caused by incorrect setting of the clearance between the concave and the drum in the combine-harvester, or subsequently during grain cleaning, handling and conveying (Arnold, 1959). Lightly threshed grains often have awn stubs attached to them, and this is a good feature, even if the weight : volume ratio is reduced. Damaged grains, which are highly undesirable, may have the husk frayed, chipped or peeled away from the grain, usually in areas around the apex or at the base, around the germ. Grain may also be split along the side or, more commonly, down the furrow (these faults arise during grain development in some varieties under particular growing conditions), or the grain may have been so battered that some corns are broken into two. Half and broken grains release sugars etc. during steeping, encouraging the growth of microbes and giving rise to foul steep water; they become mouldy during germination. In a grain batch containing many broken corns, many other corns will have been bruised.

Unbroken, but damaged, grains are not easy to detect. There are no recommended tests for physical damage. In grain decorticated with sulphuric acid or dehusked with hypochlorite, it is often possible to see damage – such as perforations into the endosperm caused by sucking insects, or areas where part of the grain has been eaten away (e.g. in North America by cutworms, grasshoppers or army worms). If the grain has been carefully washed after decortication, it is easy to enhance the visibility of damaged areas either by briefly rinsing the grain in a solution of iodine in potassium iodide, when exposed areas of starchy endosperm rapidly go blue-black, or by soaking in a solution of a tetrazolium salt, when the living tissues (embryo or aleurone layer) adjacent to perforations in the testa become red. This latter test is less satisfactory because it takes longer and the acid used in decortication may have entered the damaged region and killed the adjacent living tissue, so removing its ability to give a red colour with tetrazolium.

Pre-germinated grains, that is grains which have sprouted in the ear, do not survive storage well and indeed may be dead. Their presence in grain is undesirable for malting. Pre-germinated grains may be detected by the presence of the dried coleorhiza or rootlets at the base of the grain, or because the coleoptile (acrospire) has started to grow under the husk. The inspection may be helped by tetrazolium staining (Sole, 1994). As with malt, the coleoptile may be made more readily visible by boiling the grain in a solution of copper sulphate. In addition, the endosperm may have started to soften, or the grain may appear shrunken and α-amylase may have appeared. However it is often difficult to detect pre-germinated grains by inspection in a sample of dry grain. Several laboratory tests for pre-germination have been proposed (Pitz, 1991, Sole, 1994), including testing individual grains for the presence of α-amylase and locating fluorescein dibutyrate esterase, enzymes which only appear when germination has started, or detecting grains in which endosperm modification has started. To carry out the test for α-amylase, according to the IoB (1993), 300 pre-soaked grains are divided longitudinally and one set of halves is heated at 70 °C for 15 min in a solution of calcium chloride to destroy β-amylase but leave α-amylase intact. After rinsing, the cut faces of the grains are placed in contact with buffered agar containing starch. After a period of incubation, the grains are removed and the surface of the agar is rinsed with a solution of iodine. Undegraded starch stains dark green–brown; clear spots indicate degradation by α-amylase from pre-germinated grains. The EBC (1987) describes two other tests, adopted by the IoB (1993), using cut or sanded grains. In one, modified areas of starchy endosperm, which contain esterolytic enzyme activity, are made visible by the fluorescence of fluorescein liberated by the enzyme from non-fluorescent fluorescein dibutyrate. The other method is to use the Methylene Blue penetration test used to detect modification in malt (Chapter 13).

Barley buyers usually investigate the physical state of endosperms of a sample of grains. This is done by cutting grains (typically two to three lots of 50) transversely in half using a corn cutter or farinator, for example of the Grobecker pattern. Better, because a greater area of starchy endosperm is exposed, grain may be cut in half longitudinally using a cutter of the VLB type. It is desirable that the endosperm should appear white and mealy (opaque, floury, farinaceous, chalky). However, some grains may appear to be flinty or steely (translucent, hard, glassy) and this can be a bad sign. Grains that are persistently steely, i.e. in which steeliness survives steeping for 24 h and redrying, will malt less well than mealy grains (Hudson, 1960; Wainwright and Buckee, 1977). Often steely grains have higher nitrogen (protein) contents than mealy grains. For example, in grain samples from one variety classified as mealy, semisteely and steely, the protein contents were 9.5%, 10.7% and 12.3%, respectively. It seems likely that grains of some varieties are more likely to be mealy than those of others, and there are differences resulting from growing conditions in the field. It has been proposed that 'coefficients of mealiness' be calculated by scoring the mealiness and steeliness of 50–100 grains from a sample and then carrying out a calculation (Hudson, 1960). This process is highly subjective and the calculations are rarely used. Corncutters cut only a few grains at a time, but they are quick. By removing the surface layers of grains in a pearling machine or by decortication with sulphuric acid it is possible to inspect the endosperms of large numbers of grains for their mealiness/steeliness. Alternatively, samples of 200–500 grains can be mounted on a wooden backing, using a rapid-setting glue. Then the interiors can be exposed for inspection by grinding away about half of the thickness of the grains, e.g. with a disc or belt sander. An older approach was to mount lots of 100 grains in slots and inspect them in intense, transmitted light in a device termed a diaphanoscope (de Clerck, 1952), to score them for steeliness. This type of device does not seem to work well.

Hard grains, which are likely to be more steely, may malt less well than soft ones. Hardness has been assessed in various specialized mills (such as the Brabender Mill and the Compara Mill) to measure the energy taken to grind the grain to a given degree of fineness (e.g. Moll, 1979a; Swanston, 1990). The conditions of the test and the moisture content of the grain need to be exactly controlled. Results are altered by changes in the moisture contents of grain samples. This approach is not used by maltsters but may be used to select promising lines by plant breeders. Using a murbimeter, originally designed to study malt (Chapter 13), it has been possible to assess the hardness of individual barley grains. Various correlations between barley hardness and other barley and malt parameters have been noted (Moll, 1979a).

Related tests are based on the sedimentation characteristics of finely ground samples of grain in 70% ethanol (Palmer, 1975a). Undoubtedly differences exist between samples and these must reflect structural differences. Unfortunately the results of this test do not correlate reliably with malting quality (Moll, 1979a).

6.4 Moisture content

It is important to know the moisture content of grain accurately at many times: during purchase, during storage, for analytical purposes, during the malting process and in the final malt (de Clerck, 1952; Hudson, 1960; Wainwright and Buckee, 1977; Briggs, 1978; Moll, 1979a). Around the world, grain may be harvested at moisture contents between 8% and 25%; in Europe the range is 12–20% in most seasons. Grain is usually purchased in the UK at a notional moisture content of 16%, with penalties such as reductions in price (or allowed weight) at higher moisture contents. For example, in one Scottish company, the penalties are weight reductions of 1.2% at 16.1–17.0% moisture, 3.6% at 18.1–19% moisture, 6.1% plus a drying charge at 20.1–21% moisture and rejection at over 21% moisture (Bathgate, 1989). This approach can be criticized in that in Britain in some seasons barley may be harvested at below 16% moisture and it can pay the farmer to wet it before sale – thereby reducing its true value, while giving the farmer a better return and giving the purchaser the extra cost of drying to a 'safe' storage value. The maltster actually wants to buy barley dry matter and not an excess of associated water. The conditions under which grain can be safely stored, and how urgent it is that it should be dried, are greatly influenced by its temperature and moisture content (Chapter 7).

The moisture content of grain is variable and, for many purposes, the composition of the grain is expressed on a dry weight basis (d.b., 'on dry'). However, it is the convention in malting (unlike the usage of many plant physiologists) to express the grain moisture content on a fresh weight basis, so moisture content, $M(\%) = 100[(W_1 - W_2)/W_1]$, where W_1 is the fresh weight of the sample and W_2 is the weight after drying. The standard reference methods for determining grain moisture contents are all based on weighing ground samples of grain before and after drying in ovens under rigidly specified conditions. As grain containing more than 17% moisture cannot be ground in a satisfactory way it is necessary to carry out 'double drying'. A weighed sample of the whole grain is dried for 3–4 h, at less than 50 °C (122 °F). After reweighing, it is ground and finally dried in an oven in the standard manner (IoB, 1993). As moisture is progressively removed from grain, that remaining is more tenaciously held, as can be seen from the reduced equilibrium relatively humidity (Chapter 10). Although prolonged heating (weeks) in a vacuum, against phosphorus pentoxide, gives a good estimate of the absolute moisture content of the grain, this is too tedious and slow for routine use. The standard 'reference', oven-drying methods use arbitrary drying conditions and do not give absolute values of moisture content. Consequently the prescribed methods must be adhered to **exactly**. The EBC method specifies that barley, ground using 0.2 mm clearance in a specified disc mill, should be dried for 3 h at 105–107 °C (*ca.* 223 °F) in an electrically heated oven in metal flat-bottomed tins that rest on thick (3–5 mm) metal shelves which distribute the heat evenly. Oven performance is checked by following the weight loss from copper sulphate pentahydrate under specified conditions (EBC, 1978). The Institute of Brewing (IoB, 1993) uses two methods, one resembling the EBC method, the other specifies drying for 2 h at 130–133 °C (*ca.* 269 °F). We have found it advantageous to slowly ventilate the oven with air pre-dried by passage through silica gel and pre-warmed to the oven temperature. This apparently reduces temperature gradients in the oven and overcomes the influence of variations in ambient humidity. The standard oven-drying methods are closely reproducible and are the basis for all commercial transactions and many operational decisions. However, they require 3–5 h to complete. In many circumstances, for example at grain intake or during inspections of grain stores or samples of moving grain, more rapid, secondary methods are needed. In general these secondary analytical methods are less precise than the standard oven drying methods and they must be calibrated and regularly checked against the primary methods. Very large numbers of surprising techniques have been tried, including measuring the heat evolved when ground grain is mixed with concentrated sulphuric acid and measuring the acetylene (ethyne) released when the ground grain is mixed with calcium carbide. Others have measured the volume of water released when the grain is refluxed with an immiscible solvent (e.g. toluene), while yet others have measured the water extracted from a grist by a water-miscible solvent, using infrared spectroscopy, gas–liquid chromatography (GLC) or a Karl Fischer titration. Microwave absorption and nuclear magnetic resonance (NMR) spectroscopy seem promising in principle, but the most commonly used rapid methods of moisture determination are based on rapid drying, the electrical properties of the grain or near infrared (NIR) spectroscopy. Numerous commercial rapid-drying devices are available, for example for drying ground grain at 135 °C (275 °F) during 15 min, or for continually following weight loss under a radiant heater. Other electrical devices measure the capacitance or dielectric constants of samples of ground grain and convert the values into estimates of moisture content. Despite the high cost, devices using NIR absorbance by grists or whole grains are now widely used. When carefully calibrated, they can rapidly (1–2 min) determine the moisture and nitrogen (protein) contents of grain samples.

6.5 Water uptake by grains and distribution in the starchy endosperm

Water uptake by grains is complex and is regulated by the porosity of the grain's surface layers, the temperature, the osmotic driving force and related properties, the ease of spreading through the grain tissues and the resistance of the grain to swelling (Chapter 3). Measurements of rates of water uptake and swelling power have been made in various ways (Hartong and Kretschmer, 1961, 1968; Meredith *et al.*, 1962; Wainwright and Buckee, 1977; Moll, 1979a). It is widely agreed that barleys which hydrate quickly malt better than those which hydrate more slowly. Presumably this reflects a higher osmotic pressure in the hydrated grain and/or a less dense structure that readily permits the grain to swell to accommodate the water taken up. To a degree water uptake is influenced by variety, but it is also markedly affected by the growing season. When grain is steeped the surface layers are readily hydrated and the moisture mainly enters the interior in the region of the embryo then spreads through the starchy endosperm (Chapter 3). The completeness of endosperm hydration can be estimated by immersing the grain in boiling water for *ca.* 30 s (Chapon, 1961), then cooling and cutting or sanding to expose the interiors of the grain. Where the endosperm is 'well hydrated' the starch is gelatinized and these translucent zones are clearly distinct from other areas. More complex tests are also available, for example following the spread of radioactive, tritiated water into the grain using autoradiography of frozen samples. Various techniques are used to determine the water-uptake rates of barley lots to decide on what steeping programmes to use. These are considered in connection with experimental malting (Chapter 11).

6.6 Grain size

For many years the usual measure of grain size was the TCW expressed in grams on a dry weight basis (Hudson, 1960; Wainwright and Buckee, 1977; EBC, 1987; IoB, 1993). With plump samples of good European two-rowed malting barleys, TCW values of 35–45 g are common, but some high-quality, but small-grained, samples may give values as low as 30–32 g. Because small numbers of grains are involved (because they must be counted) in determinations of TCW, it is imperative that samples be correctly taken and, to avoid personal bias, the samples should be weighed and then counted (Hudson, 1960). Using older, less satisfactory methods, errors of 12% often occurred when samples of 1000 grains were weighed, and weighing samples of 5000 grains still often gave errors of 4%. The IoB (1993) instructs that two samples of approximately 20 g each be taken, while the EBC (1978) uses two 40 g samples. The half corns and foreign matter are removed, and the weights of these materials are substrated. The corns remaining are counted, their moisture content is determined and the TCW values are calculated.

The determination of the TCW is tedious and for many purposes it is being supplemented or replaced by a sieving test. In principle, a given weight of grain is placed on the top sieve of a set of three slotted, mechanically shaken sieves having (in Europe) slot widths of 2.8, 2.5 and 2.2 mm. In the ASBC (1992) system, the slot widths are 7/64 in (2.78 mm), 6/64 in (2.38 mm) and 5/64 in (1.98 mm). After a 5 min period of shaking, the proportions or assortment (by weight) of the grain falling into the width fractions > 2.8 mm, 2.8–2.5 mm, 2.5–2.2 mm and < 2.2 mm are determined. These data are useful in showing the distribution of grain sizes and the proportions of screenings or 'thin corn' (< 2.2 mm, or < 2.5 mm, respectively, depending on local usage) present in the sample. Maltsters remove thin corns and usually do not process them. Consequently most grain purchase contracts specify maximum percentages of thin grains (screenings) or price adjustments to allow for their presence. In the USA, thin corns normally make up 5–15% of unscreened grain. Plump corns are preferred by some malt users for ease of milling. Thin grains in a batch tend to be richer in nitrogen (protein) than the plump grains. Corns of different widths do not hydrate at the same rates during steeping, or malt at the same rates, and thin corns are often more dormant. Grains shrink on drying and swell when wetted and so allowances may have to be made for the grain's moisture content when it is evaluated by the sieving test.

In North America, screen sizes are different. For six-rowed barleys standards are: retained on 2.78 mm screen (7/64 in), no maximum; 2.38 mm screen (6/64 in) or more retained, 70% minimum; passed through 1.98 mm screen (5/64 in), < 3%. For two-rowed malting barleys the respective values are 60% minimum; 85% minimum, and < 3% (Burger and La Berge, 1985).

An ingenious device due to Eckhardt (Anon, 1907a) permits a visual estimate of the distribution of grain widths to be obtained. The grain is dropped into the top of a tapering slot formed between a glass front plate and a calibrated metal back plate. The corns fall until they wedge. The narrower they are the further between the plates they fall. Thus the grains are classified into width classes according to how far they fall.

6.7 The bulk density of grain

It is often convenient, in connection with transport or storage space, to know the bulk density of a grain sample, that is the weight of a unit volume of grain (for example as kg/hl (hectolitre weight) or lb/bu US (US bushel weight); 1bu = 35.238 l; weight (in lb)/US bushel × 1.2827 = weight (kg)/hl (Moll 1979a). Despite the complications owing to varietal differences, the effects of moisture content and the effect of gentle or vigorous threshing on the closeness of grain packing, there are still maltsters who believe barleys with high bulk densities make better malts. Officially, European maltsters have discarded this measurement as a criterion of quality. To determine the hectolitre weight, grain is usually run, under carefully standardized conditions, into a cylinder of known volume. The excess above the upper limit of the cylinder is removed, and the weight of the grain is determined (de Clerck, 1952). Values of 60–72 kg/hl were regarded as normal for European barleys (Moll, 1979a).

6.8 The specific gravity of grains

The specific gravity of barley grains has been measured by determining the volume of ethanol or toluene displaced by a known weight of grain. Alcohol has been used because it does not permit the retention of air bubbles on the surface of the grains. Toluene is preferable since it is believed not to penetrate grains. Specific gravity values vary between 1.2 and 1.4 g/ml. Mealy grains have a lower s.g. than steely grains (Brown, 1903). Mealy and steely grains can be partly separated by flotation on salt solutions of differing densities (Munck, 1993). This is not a routine determination. Some trials indicated a correlation between grain density, nitrogen content and grain steeliness. Malt has an appreciably lower specific gravity than barley. The average results for some American samples of malt and barley were 1.171 and 1.397, respectively (Moll, 1979a). Following the decline in the specific gravity of sorghum grains during malting is a useful guide to modification in this grain (Chapter 13).

6.9 The nitrogen or protein content of barley

In Britain, nitrogen content (TN, % d.wt) of barley is always reported, whereas elsewhere the nitrogen value is often multiplied by the empirical factor 6.25 and the resulting figure is reported as the 'protein' or 'crude protein' content of the grain. While hallowed by long usage, this convention is misleading since (i) grains contain nitrogen-containing substances other than proteins, and (ii) the nitrogen contents of different proteins vary, so 6.25 is only an approximate conversion factor.

The nitrogen content of grain is a most useful value. For example, when different lots of one barley variety are compared it is found in the malts made from them that DP and TSN increase and the hot water extract decreases with increasing contents of total nitrogen; higher-nitrogen samples are more likely to be steely, slower to steep and less easy to modify than low-nitrogen samples. Barley samples frequently have nitrogen (protein) contents in the range 1.5–2.2% (9.38–13.75%) and in extremes 1.3–3.0% (8.13–18.75%). Mainland European brewers prefer malts made from barleys with 1.4–1.76% TN (8.75–11.0% protein). Traditional British pale-ale malts were made with barleys having nitrogen contents of 1.3–1.4%. However, as average grain nitrogen contents have increased, more typical values for barleys for British brewers' malts are now pale-ale malts TN 1.65% (10.31% protein), mild-ale malts TN 1.75% (10.94% protein) and lager malts TN 1.6–1.9% (10.0–11.88% protein). Grain distillers' barleys, for making malts with high enzyme contents, may have a TN as high as 2.5% (15.63% protein). For North American breweries, using high proportions of adjuncts in their grists, the malts are made from barleys having TN values of 2.0–2.2% (12.5–13.75% protein).

Because of its importance, determinations of grain nitrogen content are carried out with great care, often on replicated samples, together with the appropriate blank determinations. Each system of analysis specifies one or more reference methods. Most were based on that devised by Kjeldahl in 1883. Discussion of the details of the method are given elsewhere (de Clerck, 1952; Hudson, 1960; Wainwright and Buckee, 1977; Moll, 1979a, 1985). Because of differences between individual grains, and consequent sampling problems, the grain samples analysed must not be too small, a fact which makes microanalytical techniques inherently less reliable. The basis of the Kjeldahl method is the hot digestion of grist in concentrated sulphuric acid containing a catalyst mixture such as copper sulphate, mercury salts and selenium dioxide, often containing potassium sulphate to raise the boiling point of the mixture. During the digestion organic substances are degraded and most nitrogen is converted to ammonium ions. When the process is complete, the ammonia is liberated by the addition of alkali to the mixture and is steam distilled into a boric acid solution, where it is quantified by titration. Numerous digestion procedures, catalysts and other chemical additions to the digestion mixtures (such as hydrogen peroxide) have been proposed, but these can lead to different nitrogen recoveries. Indeed it is well known that organic nitrogen in heterocyclic compounds is less readily released during digestion than, say, that from simple amino acids. Therefore, the prescribed analytical methods must be closely followed. Recently the Dumas method for determining grain nitrogen has been recognized as a standard method. Several types of commercial equipment are available. Grain samples are oxidized and the oxides of nitrogen produced are reduced to nitrogen gas, which is determined. The method recovers more nitrogen from grain than the Kjeldhal technique and is safer, since noxious reagents are avoided and there are no toxic catalyst residues to be disposed of. However, the equipment is expensive and only analyses small samples.

The Kjeldahl type of analysis is comparatively slow and laborious and numerous attempts have been made to automate it or find modifications that give results more rapidly. Automatic digestion may be followed by automatic distillation and titration of the distillate, or the ammonia in the distillate may be determined colorimetrically (manually or with an automatic flow-through analytical device) or with an ammonium ion-selective electrode. Recently there have been proposals to speed digestions using microwave heating. Another group of secondary methods is based on the principle of direct distillation. Barley grist is mixed with alkali and heated and the ammonia released is trapped and estimated. Nitrogen yields are far from quantitative, and the doubtful assumption is made that a fixed proportion of the nitrogen is released under particular conditions of heating.

Some dyes (e.g. Orange G) bind strongly to proteins under particular conditions and a variety of commercial instruments were devised that gave estimates of nitrogen contents from the dye-binding capacities of ground barley grists. The use of dye-binding methods has declined greatly in recent years, but their rapidity and ease of use caused them to be employed in barley intake units.

Recently the use of meters employing NIR absorbance have become popular. The variation of the reflectance or transmittance of NIR at different wavelengths can be correlated with the grain total nitrogen (protein) content. The performance of NIR machines has improved as more sophisticated devices have become available and calibration methods have improved. Indeed, on-line calibrations from an international data-base, updated daily, are available. Such machines are expensive but, properly housed, are robust and give results quickly so the nitrogen and moisture contents of samples of grain can be determined in a few minutes.

It must be remembered that all alternative, secondary methods of analysis must be calibrated against a primary method, and the calibrations must be regularly checked. In some instances, different calibrations may be needed for different barley varieties or different cereals. In cases of legal dispute, reference is always to the primary, standard method. The nitrogen content of grain is regarded as so important that there have been wholly unrealistic disputes over differences so small as that between 1.65 and 1.66% nitrogen, a difference so small that it cannot be reliably determined on a heterogeneous material such as grain.

6.10 Determinations of germinative capacity

Germinative capacity (GC) is the percentage of living or viable grains in a sample. When grain is dormant many living corns fail to germinate under standard test conditions, and so either germination must be 'forced' (dormancy must be overcome) in the direct tests, or it must be estimated indirectly (section 3.7).

Methods used to overcome dormancy include dividing grains into two, using a farinator, and putting the embryo ends to germinate. In this, the Eckhardt method, problems are encountered from microbes growing on the cut grains and checking germination. Grains, decorticated with sulphuric acid (Essery *et al.*, 1955) may be set to germinate on water or solutions of gibberellic acid (50 mg/l). Unfortunately the acid treatment may kill physically damaged but viable grains. Other means of overcoming dormancy, to permit the use of germination tests to assess grain viability, include drying grain and holding it at 40 °C (104 °F) for 7–10 days (or longer) or stratifying the grain, e.g. by keeping it at 5 °C (41 °F) on a wet substratum for 5–7 days before putting it to germinate at a higher temperature (Briggs, 1978). However, these tests are not routinely used by maltsters.

The direct test, which is specified by both the IoB and the EBC, is based on that of Thunaeus. Germination is 'forced' by immersing grains (2 × 200 grains, IoB; 2 × 500 grains, EBC) in solutions of hydrogen peroxide (0.75%), renewed after 2 days, and scoring the numbers of grains that have chitted after 3–4 days at 18–21 °C (64.4–69.8 °F). The residual ungerminated grains have the husk peeled from over the embryo and the brown pericarp is rubbed away over the embryo, then they are set to germinate on wet filter paper for a further day. The total score of grains chitted is used to calculate the percentage of viable grains. While this test usually works well, some samples of barleys prone to extremes of dormancy (e.g. Triumph) cannot be reliably tested in this way, even if the solution of hydrogen peroxide is supplemented with gibberellic acid (50 mg/l) and the grain has received a preliminary drying and short warm storage treatment. We have encountered barley samples that were 99–100% viable but which would not germinate completely in the hydrogen peroxide test until they had been dried to *ca.* 8% moisture and stored at 40 °C (104 °F) for 3–4 weeks to reduce the dormancy. (With most barleys treatment at this temperature eliminates dormancy within 10 days.) All these tests are likely to give low estimates of viability, a weakness that operates in the maltster's favour to the extent that, guided by these tests, he may reject malting-quality grain because of pessimistic test results, but at least he will not buy grain that has too low a viability and so can never be turned into good malt.

The indirect tests of GC rely on the ability of enzymes in living embryos to reduce various substances to coloured products. Thus living embryos become coloured while dead embryos do not. These tests have the advantage that they work perfectly well on dormant grains. However embryos of heat-damaged grains sometimes give positive results (in an extreme case 95% of embryos stained but only 48–50% of the grains germinated) and scoring embryos as living or dead can be complicated if only parts are coloured (de Clerck, 1952; Hudson, 1960). In the early tests of Enders and Schneebauer, living embryos were detected when they converted *m*-dinitrobenzene into nitrophenylhydroxylamine, which becomes purple in the presence of ammonia. In the Kipphan method, half grains were incubated in the presence of sodium selenite, when living embryos became red. These methods have now been totally displaced by others based on the reduction of tetrazolium salts to insoluble, strongly coloured formazans. In these tests, grains are accurately cut in half longitudinally with a razor blade or a commercial cutter. In practice, often only 50 grains may be checked, but for statistical reasons clearly at least two lots of 100 each should be checked, as recommended by the EBC (1978). The IoB (1993) methods suggest a practical solution – to increase the acceptance level of percentage viability as the number of corns in a test is decreased. For example, if it is desired to reject all samples of less than 94% viability, then if 100-corn samples are tested the acceptance level is set at 98% of the corns stained, while if 350 corns are tested the acceptance level is set at 96%. (Maltsters in the UK are rarely willing to accept grain with a viability as low as 94%.) To carry out a test the half-grains are immersed in a solution of a tetrazolium salt, either (3-(4-iodophenyl)-2-(4-nitrophenyl)-5-phenyl-2*H*-tetrazolium chloride (0.3%) or 2,3,5-triphenyltetrazolium chloride (1%). After removal of air, by brief exposure to a vacuum, and the appropriate period of incubation (30 min at 40 °C (104 °F) if the triphenyltetrazolium salt is used or immediately after the 3–4 min evacuation and air-readmission period if the iodonitrophenyltetrazolium salt is used) the half-corns are drained and spread on moist filter paper. The staining of the embryos is scored in a good light with the aid of a magnifying lens. Fully stained embryos are 'viable'. If staining is incomplete, an embryo is scored as 'damaged' while if less than half the embryo is stained it is scored as 'dead'. It has been claimed that heat-damaged embryos have an unstained region in the centre of the scutellum (Briggs, 1978).

Viable aleurone tissue also stains with tetrazolium salts. Particular grain samples of the variety Galant were found to malt badly. This was caused by the occurrence of large dead areas in the aleurone layer, which were detected by the absence of aleurone staining (Sole *et al.*, 1987). At present the aleurone-staining test is not used by maltsters but is used by plant breeders to select against the 'dead aleurone' trait.

A very sensitive test for viable, undamaged embryos is their ability, when separated, to synthesize starch from the endogenous reserves when incubated for 8 h on water at 25 °C (77 °F) (French, 1959). This test is not currently used by maltsters. With some samples of grain, the results of different tests do not agree well. It is for this reason that in all reports the test used to obtain a result should be specified.

6.11 Tests of germinative energy

Frequently not all the viable grains in a sample will germinate when subjected to the 'appropriate' conditions in a germination test (section 3.7.5). When the conditions of the test are altered (water availability, incubation temperature, time) the percentages of dormant grains, those that are viable but which fail to germinate, also alter (Chapter 3).

Maltsters have tried out many germination tests over the years, seeking greater reproducibility and closer correlations between test results and grain behaviour in the malthouse. Many tests have been repeatedly modified. Problems occur not with well-matured grain samples, which germinate freely under many conditions, but with immature, partly dormant samples. Therefore, when test results are reported, the method used must be recorded. When tests are carried out, the operator must not deviate from the prescribed conditions in any way. As with all tests on grains, to be of any value the samples used must be replicated, must be taken correctly to eliminate statistical bias and must each contain a sufficient number of grains.

Both the IoB and EBC currently specify that the BRFI (originally BIRF: Essery *et al.*, 1954, 1955) tests be carried out in duplicate, but carrying them out in triplicate appreciably improves the reliability of the results. Samples, each of 100 grains, are germinated on two filter papers in 9 cm Petri dishes wet with either 4 ml of water (the GE test) or with 8 ml of water (the water-sensitivity (WS) test) (section 3.7.5). In the latter case, the grains should all be arranged to be furrow-side down. Incubation is in the dark at 18–21 °C (64.4–69.8 °F). Chitted corns are removed and counted after 24, 48 and 72 h. This gives an indication of the **rate** of germination as well as the final (3 days) result. To be considered 'ready for malting' maltsters would expect grain to germinate at least 50% by day 2, and 95–100% by day 3 in the GE test. The extent to which air-rests would be used in steeping would be regulated with reference to the WS test. Results are reported as GE percentage (4 ml, 72 h), and WS percentage (8 ml, 72 h), each rounded to the nearest whole number. This method of reporting is superior to that originally proposed, in which water sensitivity was given as the **difference** in percentage germination between the 4 ml and 8 ml tests. This 'difference' value can be misleading. The reason can be seen by reference to an example in which grain, in the 4 ml and 8 ml tests, respectively, germinated 23% and 9% (difference 14%) in September, but after 2 months of storage germinated 98% and 45% (difference 53%) in November (Reicheneder and Narziss, 1989). Clearly grain germination **improved**, as judged by both tests, yet the **difference** between these tests increased from 14% to 53%, suggesting that grain deterioration had occurred.

These GE and WS tests are widely used, but problems with them include: (i) too imprecise control of temperature, the range specified (18–21 °C (64.4–69.8 °F)) is too wide (section 3.7.5); (ii) the tests are carried out in an atmosphere of unspecified relative humidity; (iii) plump corns tend to take up water and dry the paper more than thin corns (to try and overcome this last problem some laboratories include an additional 'unofficial' test using 5 ml of water; (iv) there is a tendency for the filter paper to wrinkle and displace grains. In our laboratory, we have used germination tests carried out at 18 °C (64.4 °F) in a humidified atmosphere in Petri dishes containing agar, which is exactly flat and does not dry out, and to which different amounts of surface water have been added (Doran and Briggs, 1992). The test was more robust than the filter-paper test. However, later observations showed that the test might be influenced by the concentration of the agar in the gel (Kelly and Briggs, 1992b). Other germination tests are carried out in Petri dishes on sand, soil, cotton wool or moistened porcelain plates. Reicheneder and Narziss (1989) modified the filter-paper test in a number of ways, e.g. by pre-soaking grain for 1 h, then placing it on four filter papers wetted with 4 ml of water and incubating at 20 ± 1 °C (*ca.* 68°F) and counting after 2 and 4 days.

The *Analytica-EBC* (EBC, 1987) also specifies a version of the Aubry test, in which 500 grains are held between two sheets of filter paper supported on a layer of wet cotton wool on a perforated sheet of stainless steel held in a special cabinet. Incubation is at 20 ± 1 °C (*ca.* 68°F and grains are counted after 3 and 5 days. The *Analytica-EBC* (EBC, 1987) also gives instructions for one version of the Schönfeld method in which two lots, each of 500 corns, are placed in filter funnels that have stainless steel gauze in the neck to retain the grains but permit the free passage of water. Water changes may be carried out manually or in an automatic apparatus. The grains are covered with water for 3 h at 18–20 °C (64.4–68 °F) by placing the funnels in a water-bath. Then they are drained and the surface is covered with wet filter paper and a glass plate. Steeping is repeated for 2 h, beginning 18–20 h after the start of the test. After 72 h the grain is tipped out. The number of non-germinated corns is counted and these are returned to the funnels and are steeped for another 30 min. The final number of ungerminated grain is counted after 120 h. The percentages germinated after 3 days and 5 days are reported as percentage GE (Schönfeld method), 3 days or 5 days. This method has the attraction of evaluating grain germination after a steeping regimen (as occurs in malting), and not with grain in contact with a continuously wet surface.

Other, non-standard tests are still used from time to time. In the Schönjan or Coldewe method, 100 grains are placed, embryo-ends downward, in a porcelain plate containing tapered holes (de Clerck, 1952; Chapter 3). The plate, covered with coarse sand, is watered, the excess water draining past the grains, out of the holes. Then the plate is placed in a container, containing water to keep the air saturated, and the whole is covered with a felt-lined lid. At intervals, germination is assessed by counting the number of tufts of roots which protrude through the base of the plate. The uniformity, or variability, of the length of the roots gives a good indication of the uniformity of grain growth.

To decide on the steeping schedules best suited for particular batches of grain, it may be necessary to carry out quite complex trials involving large numbers of grain samples. This may be done using 'bag malting' or complex micromalting equipment (Chapter 11). An inexpensive alternative approach, with which quite complex trials on germinability, using large numbers of samples can be performed, is to carry out tests on small samples of grain (100–200) held in boiling tubes in a constant temperature cabinet and steeped and air-rested at will. At intervals, germination values are scored (Macey and Stowell, 1959). The tubes may be closed with a nylon mesh, held in place by elastic bands. They may be filled and emptied through the top, or small holes may be blown in the base of the tubes, which can be held upright in a rack. In this case, filling and emptying can be achieved by standing the tube, in the rack, in a vessel of water or, conversely, taking it out. Other tests have been proposed to evaluate grain vigour but are not in general use. For example, it has been suggested that the rate of ethanol production during a standard steep can be used to detect heat-damaged lots of grain (Bryan-Jones, 1986).

6.12 Grain composition

Other tests are in sporadic use for investigative, rather than quality-control purposes. 'Fatty substances' may be determined by weighing the residue remaining after the evaporation of the solvents acetone and chloroform that have been successively used exhaustively to extract finely ground grain (IoB, 1993).

Numerous methods have been proposed for determining β-glucan in barley. Some rely on initially extracting the polysaccharide into solution and selectively precipitating it with ammonium sulphate or alkaline copper sulphate and quantifying the precipitate. Others rely on removing low-molecular-weight sugars by prior extraction or reduction with sodium borohydride and then selectively hydrolysing the polysaccharides to oligosacharides or glucose, using purified enzymes, and determining the liberated sugars. It is essential that the enzymes be absolutely free of contaminating enzymes that can degrade starch or holocellulose to liberate glucose. Alternatively β-glucan can be extracted and determined by the intensity of the fluorescence given when it is mixed with Calcofluor in flow-analysis equipment. These determinations are most often carried out on malts (Chapter 13). Qualitatively, grains of low β-glucan mutant barleys can be distinguished by the poor fluorescence of Calcofluor-stained sections of starchy endosperm (Aastrup *et al.*, 1985). At present, no commercial low β-glucan barleys are available. There are many reports that β-glucans can be determined by measuring the viscosities of grain extracts. While such measurements can be useful they do **not** and **cannot** give accurate, specific and quantitative measurements of the β-glucans. There are also reports that the β-glucan contents of barley can be determined by NIR spectroscopy, but trials have failed to confirm this.

There have been numerous attempts to find mashing procedures, using added enzymes, that will give a 'barley' extract value that is universally related to the extract value of malts made from it. As potential extract is lost as barley is converted into malt, and many different malting programmes are in use, it is not surprising that this aim has not been met. However, using rigidly controlled techniques and a fixed malting programme, it has been clearly demonstrated that there is a close, linear correlation between barley extract and malt extract, the barley extract being the higher (Briggs, 1978; Meredith *et al.*, 1962). There is no generally agreed method for determining barley extract (de Clerck, 1952; Hudson, 1960; Moll, 1979a). Various features have been included in the mashing programmes to obtain the extract, including fine grinding and boiling or pressure cooking to gelatinize starch, and the addition of enzymes from malt or microbes.

Methods used to estimate starch have often lacked specificity, but by using finely ground washed grists from which soluble sugars and polar lipids have been removed and by treating them with purified enzymes (α-amylase to liquefy the starch and amyloglucosidase to convert the dextrins to glucose), reliable estimates of starch can be obtained by using specific assays of the glucose produced. It has been claimed that starch contents, and barley extracts, can be determined by NIR spectroscopy, but this is not generally accepted.

Predicting malt extracts from grain nitrogen contents is only reliable when the grain is of a known variety and the details of the inverse relationship between nitrogen and extract are known (Chapter 14).

Estimates of husk content have often been made on barley (Hudson, 1960; Wainwright and Buckee, 1977; Moll, 1979a). The samples used must be picked over to remove all cracked and broken grains. The sulphuric acid (50%) decortication test of Essery *et al.* (1956) gives an estimate of the content of husk **plus** the pericarp, and has the virtues that physically undamaged grains remain viable after the removal of the surface layers, and that mealy and steely can be distinguished by inspection. Older methods involved removing husks by hand from grains that had been pre-soaked in a strong solution of ammonia (Hudson, 1960). However, the removal of the husk using a boiling alkaline hypochlorite solution is much more rapid and less laborious (Whitmore, 1960). This method has been adopted by the EBC (1987). Another version of this test enables germ damage to be detected (Weak *et al.*, 1972). Others have used limited abrasion in a pearling-type of machine to estimate husk content from the weight of dislodged material. This approach is not likely to give results specifically referring to husk, since other tissues may be abraded in turn after the husk has been removed and there is no guarantee of complete husk removal, e.g. from the ventral furrow.

A wide range of colorimetric methods has been used for estimating grain polyphenols. These tests may be applied to malts (Chapter 13) but have various weaknesses (Chapter 4) and are not normally used in barley trading. However, if attempts to produce commercially acceptable proanthocyanidin (anthocyanogen)-less barleys are successful, then it will be necessary to adopt a test to recognize flavanoid-containing grains. This can be achieved by sanding or transecting grains to expose the testa, then treating the exposed surfaces with a vanillin–hydrochloric acid reagent. When flavanoids are present, the test causes a scarlet colour to develop in the testa, which can be seen with a hand-lens. In flavanoidless grains this colour does not develop (Aastrup, 1985).

6.13 Microbes, mycotoxins and chemical residues

From time to time there are worries about the levels of microbial contamination on grains, the possibility of the occurrence of mycotoxins, and the carry-over of agrochemicals or pesticides from the field or store. These factors have not all been routinely investigated, but potentially they are important and often there are legal limits to the levels of residues that may be present on grains.

Methods for quantifying microbes on grains are available (Dickson, 1962; Moll, 1979a; IoB, 1993; ASBC, 1992). Heavy microbial populations are inferred from grain discoloration and staining; these samples are rejected by maltsters. Various *Fusarium* spp. are common sources of mycotoxins such as vomitoxin and aflatoxins, and as a routine precaution 'scabby' barley, which is likely to be heavily contaminated, is not used for making malt. However, although *Fusaria* can be discovered on most barleys in small amounts, the occurrence of mycotoxins on sound grain is rare. The IoB (1993) gives a method for detecting *Fusarium* infection on barley.

Other technical problems arising from using barley samples heavily infested with microbes include water sensitivity, the development of off-flavours during malting and the occurrence of gushing or uncontrolled fobbing in beers made from malts contaminated with particular organisms. The best protection against all these problems is to accept only clean, sound grain for use, and to keep the malting plant scrupulously clean.

Pesticide residues can only be quantified using specialized equipment. Consequently, if the presence or absence of residues is to be determined, it is normal to send samples to be analysed to laboratories that specialize in this work.

7 Grain in store

7.1 Introduction

As harvest occurs once each year but malting is carried on the year round, it is evident that large quantities of grain must be stored. Furthermore, because of the possibility of dormancy or other problems with obtaining adequate stocks of good-quality barley immediately after harvest, it is necessary to 'carry-over' barley stocks from the previous year's harvest to beyond the current harvest. Therefore, storage periods range from a few weeks to 18 months, or even longer. The cost of storing such large amounts of grain is substantial, and the value of the grain is itself very great. The objectives are to store the grain as cheaply as possible, while minimizing or preventing losses and maintaining quality, or even enhancing it by encouraging a decline in dormancy (i.e. by encouraging post-harvest maturation).

Grain respires using oxygen, generating carbon dioxide, water and heat and losing dry matter. The respiration of stored grain should be kept to a minimum. The respiration rate increases with increasing temperature and with increasing grain moisture content (Fig. 3.14, section 3.8). At 20 °C (68 °F) barley samples at various moisture contents (%) release carbon dioxide at varying rates (mg CO_2/kg d.m. grain per 24 h): 11%, 0.35; 12%, 0.4; 14–15%, 1.4; 17%, 100; 19.6% 123; 20–25%, 359; and 30%, 2000. Similarly, in grains of 14–15% moisture, the respiration rates (same units) at different temperatures were reported to be: 18 °C (64.4 °F), 1.4; 30 °C (86 °F), 7.5; 40 °C (104 °F), 20–40; and 52 °C (125.6 °F), 250. These values are unlikely to be correct for all barley samples, if only because a large proportion of 'grain respiration' results from microbes on the surface and in the surface layers, and because microbial populations change with time, depending on the storage conditions. Respiration (and hence heat- and moisture-output rates) increases rapidly at moisture contents over 12–13% and at higher temperatures, and so do the risks of insect infestation and grain spoilage caused by loss of viability and fungal growth. Hence for prolonged storage, it is desirable to keep grain cool and dry. In the UK, often 12% or less moisture and temperatures of 15 °C (59 °F) or less are regarded as good storage conditions. However, for prolonged storage (over 3 months) a moisture content of 10% or less is preferable. An indication of the rates of heat development that may occur in grain under various conditions is illustrated in Fig. 7.1.

Grain in bulk is a good insulator. At elevated temperatures and moistures, heat is generated faster than it can escape, and the temperature of the

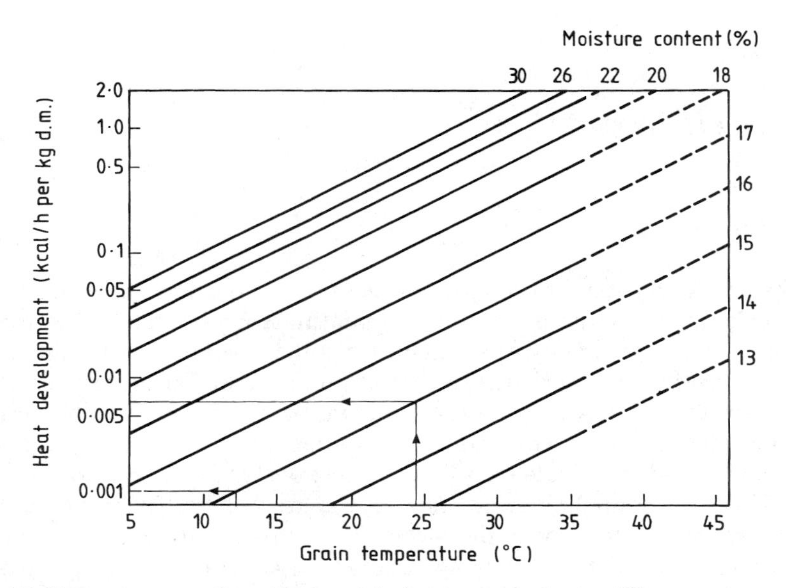

Fig. 7.1 Heat output rates (logarithmic scale) of grain samples having different moisture contents, held at different temperatures (after Brunner, 1989).

grain mass will rise – accelerating heat output and moisture generation and the process of deterioration. Such 'spontaneous heating' may result exclusively from the microbes on the grain or may be triggered by an infestation of insect pests.

Grain losses may result from faulty handling or machinery (causing grain damage and loss of dry matter), from the metabolism of the grain tissues or, more importantly, from the microbes and the depredations of insects, mites and, sometimes, rodents and birds. In addition to losses of dry matter, the germinability and quality of deteriorating and infested grain will be reduced and contamination and fouling by and with insects (entire or fragments, living or dead), insect waste, bird droppings and rodent faeces, hairs and urine will reduce the acceptability of the grain. Therefore, losses are made up of losses in weight, losses in quality (malting-quality grain may be successively reassessed as 'feed quality' and then as useless) and losses in customer trust and goodwill.

Grain is stored under a range of conditions. The so-called green grain, freshly harvested, will be at the moisture content of the grain in the field. This may be below 10% in Australia and parts of North America, is typically 16–21% in the UK, sometimes less and sometimes more, and occasionally up to 30% in some wet areas of northern Europe. Such grain must be kept cool by forced ventilation and dried at least to below 14% moisture as quickly as possible. Many farmers have their own drying facilities, but maltsters frequently distrust them because of the risks of heat damage. In some areas, maltsters take delivery of grain straight from the field to ensure that it is stored and dried with sufficiently sophisticated controls to avoid damage. After drying, cleaning and cooling, it is possible to keep grain for extended periods with minimal risk of deterioration. However, the store supervisor needs to be permanently vigilant to prevent the onset of spoilage, or to detect it and stop it if it should begin.

In the sections that follow, particular topics are discussed in more detail. However, it must be remembered that the objectives when running a grain store are to retain the largest possible quantity of grain (minimize losses) and to deliver it to the maltings in prime condition and at the lowest possible cost. Stores vary widely in the storage facilities available (Chapter 8), and in the number of grades of grain which must be kept separate. As a consequence, the problems encountered vary in type and magnitude from store to store.

Grain for malting is nearly always dried and stored in air. Experiments have shown that grain intended for animal-feed can be stored undried in an inert atmosphere, but this is technically difficult and British maltsters, at least, do not use this technique. Storing grain chilled, in a stream of refrigerated air, is unattractive to UK maltsters, since cold grain recovers only slowly from dormancy. Feed grain may be stored after treatments with propionic and/or formic acids, but as these substances kill grain this method is useless for maltsters.

7.2 The microbes associated with grains

All grains have populations of microbes, which may be larger or smaller depending on the history of the grain (Dickson, 1962; Briggs, 1978; Sauer *et al.*, 1992; Flannigan, 1996). Not all of them are pathogens. As well as occurring with the dust inevitably associated with barley, microbes occur on the surface of the husk, within the husk's surface layers, between the husk and the pericarp and probably within the tissues of the pericarp. Microbes do not occur within the living tissues of healthy grains. However pathogens may penetrate inwards beyond the testa. For example *Ustilago nuda* (loose smut) mycelia may occur within embryos and grain may be invaded, and killed, by fungi when stored carelessly under inappropriate conditions.

Under field conditions and subsequently, depending on how the grain is handled and stored, grains may carry bacteria, actinomycetes, yeasts, filamentous fungi, protozoa, slime moulds and even nematodes. More than 200 species of microbes have been identified, including some that are thermotolerant and others that are thermophilic. Bacteria are certainly important. At harvest, each kernel may carry 1.6×10^4 bacteria, 8×10^3 yeasts and 2.5×10^3 viable fungal units. To express it differently, each gram of grain may carry tens of thousands of filamentous fungi, hundreds of thousands of yeasts and millions of bacteria. Some of these will exist as vegetative forms, others as highly resistant spores.

So-called field fungi infect the grain in the field before harvest, and the microbes in the dust created during harvesting reinoculates the grain's surfaces. The extent of the infection is regulated by the weather and is greatest when grain moisture contents exceed 20%, the weather is warm and damp and the relative humidity is 90% or more. Common field fungi include *Alternaria*, *Stemphylium*, *Cladosporium*, *Fusarium* and *Helminthosporium* spp. and *Pullularia* (*Aureobasidium*) *pullulans* (Tuite and Christensen, 1952, 1955). If grains are dried quickly to a sufficient extent and are stored cool, then surviving field fungi may be found when the grain is prepared for malting. Therefore, the survival of field fungi and the absence of storage fungi (which develop under inferior storage conditions) indicate that grain has not been stored at too high a moisture content and temperature. However, if the grain is stored damp, then storage fungi may proliferate. These displace the field fungi and may cause the grain to deteriorate, to become mouldy, to heat and to die. Which fungi grow depends partly on the temperature but more on the moisture content of the grain. Because of the damage caused by storage fungi, grain for malting should be dried and cooled as quickly as possible to prevent their proliferation. The damage they cause is irreversible. Even when redried, damaged grain survives less well than sound grain during a subsequent period of storage.

In grain with a moisture content of 17% or more, spontaneous heating can occur, raising the temperature to 26 °C (78.8 °F) or more within a week; temperatures may finally reach 65 °C (149 °F), which kills the grain and greatly reduces its value, making it useless for malting. The storage fungi are probably the major cause of grain spoilage (Christensen and Kaufmann, 1974). The spores occur widely and are universally distributed in grain handling machinery and stores. They do not invade the grain to any significant extent in the field before harvest, but under moist, warm conditions they can significantly damage grain in store in 3 days. The storage fungi that grow and come to predominate on barley do so at 'preferred' moisture contents. So, for example, *Penicillium* spp. come to predominate on grain stored at 19–19.4% moisture, *Aspergillus flavus* at 17.5–18%, *A. candidus* and *A. ochraceus* at 15–15.2%, members of the *A. glaucus* group at 14–14.2%, and *A. halophilicus* and *A. restrictus* at 13–13.5%. Thus, fungi can slowly invade and kill grain with a moisture content of as little as 13% (in equilibrium with a relative humidity of about 65%). To be safe from fungal attack, all the grain in a bulk of barley must be dried to a moisture content of less than 13%, and hence maltsters frequently dry grain to 12% moisture, when grain respiration is so low that it may not be necessary to ventilate the grain. For long-term storage, even at 15 °C (59 °F), to slow grain deterioration caused by 'aging', the grain should be dried to 10% or less (Briggs and Woods, 1993). Bacteria are not thought to multiply on grains with moisture contents of less than 18%, but lactobacilli can increase on damp-stored grain. Cooling grain slows attacks by microorganisms, but even at 5 °C (41 °F) members of the *A. glaucus* group can grow slowly and others will grow below 0 °C (32 °F). The temperature optima of fungi are often about 30 °C (86 °F), but many will grow at 12–15 °C (53.6–59 °F); thermophilic organisms will grow at much higher temperatures, at 65 °C (149 °F) or more. Different species are known that will grow at temperatures ranging from –8 °C (17.6 °F) to 76 °C (*ca.* 169 °F). Therefore, for safety, grain must be stored dry (at or below 12% moisture) and cool (preferably at or below 15 °C (59 °F)). No practical, effective and acceptable fungicidal treatments for controlling storage fungi are known. Fungal 'heating' may begin in a patch of relatively damp grain in a bulk. It may also be initiated by an insect infestation, in which the heat and moisture generated by the insects allow fungi to grow and develop. As the temperature and moisture content of the grain rises, insects and microbes multiply and respire faster, and deterioration becomes 'auto-catalytic', accelerating and spreading into the surrounding grain.

In addition to killing grains and causing discoloration and heating, fungi can have a range of other undesirable effects. Malt prepared from infected grains may transmit a series of unacceptable off-flavours to beers, including flavours known as molasses, burnt, unclean, winey, spicey, harsh, bitter, fruity and astringent. The malt may be discoloured and unevenly or over-modified and the beer prepared from it may tend to become unstable, both in terms of becoming hazy and in having a tendency towards gas instability (gushing, over-foaming: an explosive release of gas, beer and froth when a bottle is opened). Most worrying is the possibility of an accumulation of one or more of the wide range of mycotoxins, which some microbes growing on grain can produce (Mirocha and Christensen, 1982; Wilson and Abramson, 1992).

Therefore, from every point of view, the maltster should obtain 'sound' barley and store it in such a way as to minimize at least or ideally prevent the activities of storage fungi.

7.3 Insect and mite pests in stored grain

Insects and mites are arthropods with, respectively, six- and eight-legged adult forms (Anderson and Alcock, 1954; Briggs, 1978; Christensen 1982; Bauer, 1984; Pedersen, 1992). Some are common pests in grain stores and others are frequently present but are not pests and cause no damage. Fruit flies and fungus beetles may occur where the grain is being germinated, and the yellow 'mealworm' (a beetle larva) may be found in pockets of damp grain dust. Worldwide, some 50 species are serious pests, but hundreds of species are encountered. Therefore, it is important to distinguish between

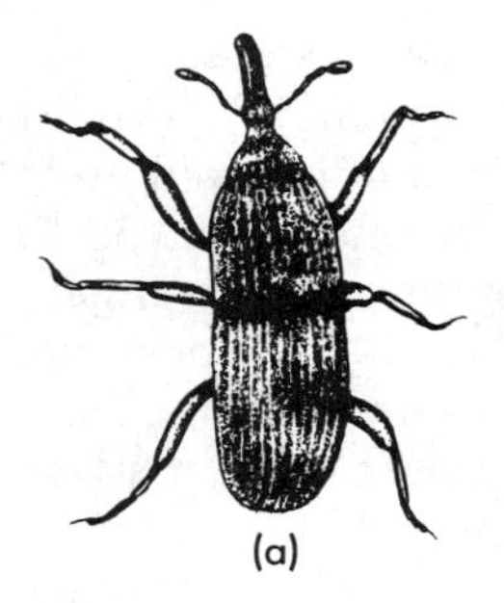
(a)

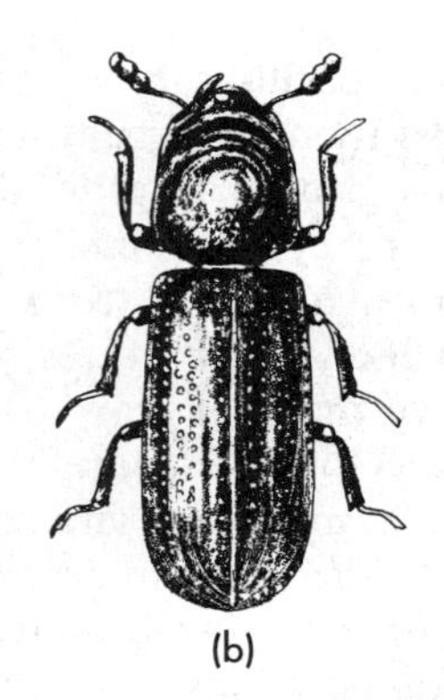
(b)

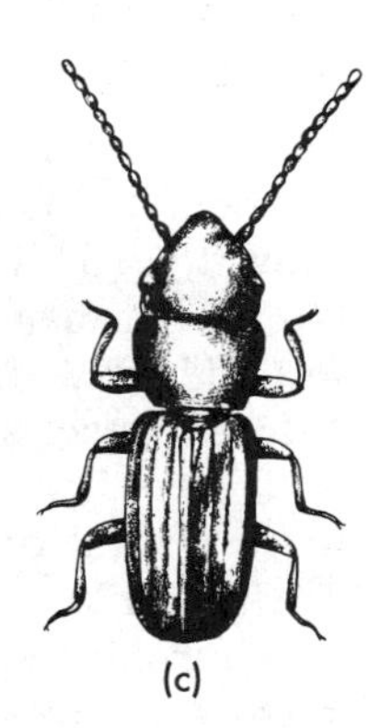
(c)

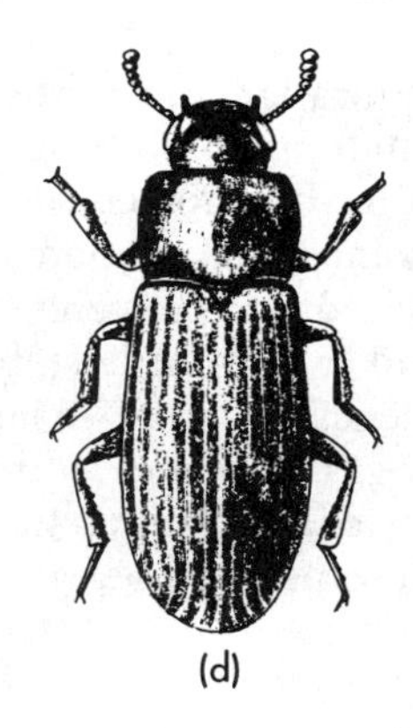
(d)

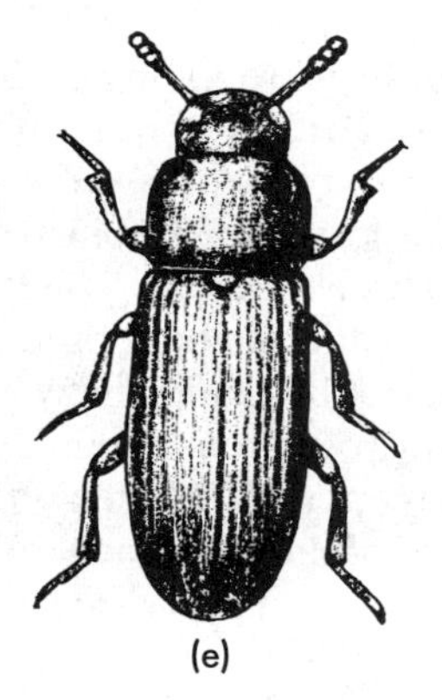
(e)

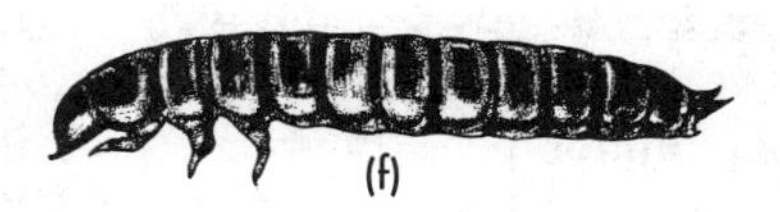
(f)

(g)

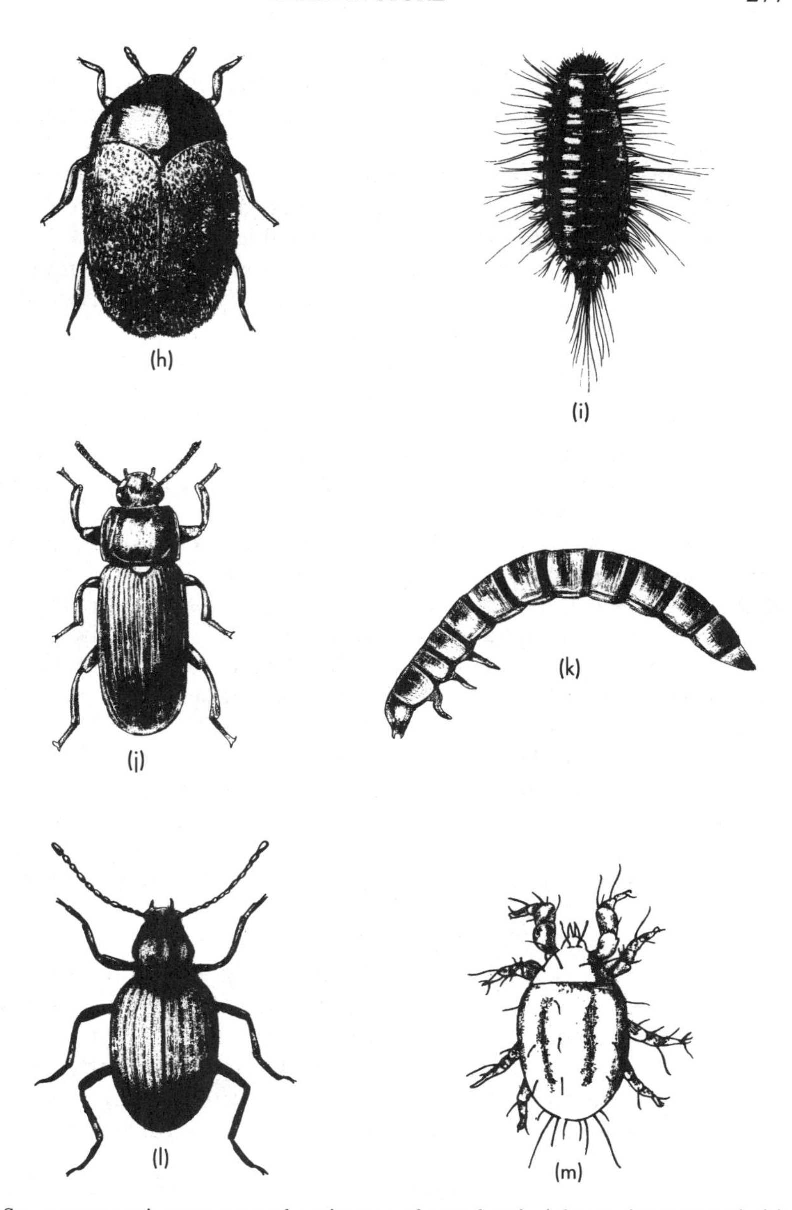

Fig. 7.2 Some common insect pests and a mite pest of stored grain (after various sources). (a) Grain weevil (*Sitophilus granarius; Calandra granaria*); (b) lesser grain borer (*Rhizopertha dominica*); (c) rust red grain beetle (*Cryptolestes ferrugineus*); (d) rust red flour beetle (*Tribolium castaneum*); (e) confused flour beetle (*Tribolium confusum*); (f) larva of the confused flour beetle; (g) saw-toothed grain beetle (*Oryzaephilus surinamensis*); (h) khapra beetle (*Trogoderma granarium*); (i) larva of the Khapra beetle; (j) adult, the yellow mealworm (*Tenebrio molitor*); (k) the yellow mealworm, larval stage; (l) the Australian spider-beetle (*Ptinus tectus*); (m) flour mite, adult male (*Acarus siro*).

the pests of stored grain (and malt), scavengers that live around the maltings eating detritus like damp dust or crushed, green malt, and adventitious 'strays' that have come from the outside. In well-cleaned maltings, detritus eaters are rare.

Some insects are harmful and some are not; they have widely different and often complex life histories, appearances, growth requirements and susceptibilities to fumigants and insecticides (Table 7.1; Fig. 7.2). Consequently it is important to identify suspected insects quickly, to determine if – and what – remedial action is required. If necessary expert advice should be sought. Some insect pests lay their eggs loose or on grains and others lay them within grains. The larvae may eat the embryos or all the inner parts of the grains, as may the adults. The larvae of some moths (caterpillars) in addition to eating grains spin silk webs that hold the grains together and – in extreme cases – create a semi-solid mass, preventing the grain flowing and so causing handling problems. In addition to direct losses, through the amount of grain consumed, insects may seriously reduce the germinability of grains. Their activity may trigger the production of local 'hot-spots' in the grain that trigger extensive fungal proliferation and spoilage; the insects themselves, their fragments and their faeces contaminate the grain. Insects and mites can induce serious allergies in those exposed to them.

Although they are not killed, most insects and mites will not multiply in grain cooled below 7 °C (*ca.* 45 °F). Grain stored with less than 12% moisture and at temperatures of *ca.* 10–15 °C (50–59 °F) is comparatively safe, that is it is not very likely to become infested. Minimum relative humidities at which insects can multiply vary for different species, but range between 1 and 65% (Table 7.1). Consequently there are insects that can multiply in grain at essentially all moisture contents and some, like the flour beetles and the khapra beetle, can even multiply in traditionally kilned malt (1.5–2.5% moisture).

The weevils are readily recognized because their heads are extended into snouts, with the mouth-parts on the ends (Fig. 7.2a). Granary and rice weevils are common pests and, in warm weather, the latter can fly, so an infestation can spread rapidly. These species lay their eggs in the grain, and the larvae and pupae remain within grains. Eventually, the adults emerge to continue the life cycle. A number of beetles are also important pests, of which the saw-toothed grain beetle is presently the most troublesome in the UK. This beetle, which gains its name from the serrations on the sides of the thorax (Fig. 7.2g), lays its eggs loose in the grain. A larva attacks a grain and burrows into the germ, killing it and often initiating heating. This species can breed in very dry grain, but it multiplies more slowly than it does when the grain is damp. The lesser grain borer is another important pest, while other slightly less important pests are the rust red grain beetle, the rust red flour beetle and the confused flour beetle (Table 7.1, Fig. 7.2b,d,e). The khapra beetle is confined to maltings in the UK and usually it is the hairy larvae, or

Table 7.1 The appearance, temperature and humidity requirements of a few common pests of stored grain: The rate of increase/4 weeks is for favourable conditions. Life cycles vary greatly in length according to the ambient conditions

Pest	Length and appearance	Optimum temperature		Minimum temperature		Minimum RH (%)	Rate of increase in 4 weeks	Comment
		°C	°F	°C	°F			
Grain weevil cycle (*Sitophilus granarius*)	1.5–3.5 mm; dark brown, black	26–30	79–86	15	59	50	15	Over 35 °C (95 °F) life not completed; adults cannot fly
Rice weevil (*Sitophilus oryzae*)	3.5 mm; dark; reddish spots on wing cases	27–31	81–88	17	63	60	25	Adults can fly
Lesser grain borer (*Rhizopertha dominica*)	2.5–3 mm; dark brown, black; cylindrical	32–35	90–95	23	73	30	20	Head 'turned down'
Rust red grain beetle (*Cryptolestes ferrugineus*)	1.6 mm; flattened, reddish-brown	32–35	90–95	23	73	10	60	Adult can fly
Rust red flour beetle (*Tribolium castaneum*)	4 mm; reddish-brown	32–35	90–95	22	72	1	70	Adults fly
Confused flour beetle (*Tribolium confusum*)	4 mm; reddish-brown	30–33	86–91	21	70	1	60	Adult does not fly
Saw-toothed grain beetle (*Oryzaephilus surinamensis*)	2.5–3 mm; dark red-brown, narrow, flattened; edges of thorax toothed	31–34	88–93	21	70	10	50	Difficult to control; adults rarely fly in UK
Khapra beetle (*Trogoderma grananium*)	Adult 2–2.5 mm; light brown Larva 3–5 mm; hairy	33–37	91–99	24	75	1	12.5	Adults hard to detect; cast skins of hairy larvae more easily seen
Angoumois grain moth (*Sitotroga cerealella*)	10–14 mm; yellow-brown	26–30	79–86	16	61	30	50	Not established in the UK
Warehouse moth (*Ephestia elutella*)	14–23 mm; variable, grey-brown, white bands on forewings	25?	77	10	50	30?	15	Larvae (caterpillars) spin webs, bind grains together
Australian spider beetle (*Ptinus tectus*)	About 2.5 mm; dark brown, globular	23–25	73–77	10	50	50	4	Nocturnal
Flour mite (*Acarus siro*)	0.4–0.5 mm; whitish; globular, eight brown legs, no division between head and body	21–27	70–81	7	45	65	2500	Taint infested grain; highly allergenic

From Howe (1965a,b); Turtle and Freeman (1967).

their caste skins, that are detected (Figs 7.2h,i). The Australian spider-beetle is another important pest, which, because of its nocturnal habits, may only be detected by night-time inspections of the grain (Fig. 7.2l).

Compared with the insects, which are commonly 1.5–3.5 mm long, or even 23 mm in the case of some adult moths, the mites are tiny, being 0.4–0.5 mm in length. Mites are arachnids, having eight legs (Fig. 7.2m). Some, like the flour mite, are pests and can severely damage grain while others are carnivores, living by eating other mites. Mites can multiply extremely quickly (Table 7.1). They breed in damp grain and are comparatively resistant to cold: they may breed even at 5°C (41 °F) and may survive freezing temperatures. In addition to the direct damage they cause to grain, mites have an oily secretion with a penetrating and unpleasant smell that leaves an offensive taint and taste. This is more easily detected than the mites themselves. Mites are highly allergenic, inducing irritations, dermatitis and gastric disorders, as well as bronchial problems. Mites will not multiply in grain having a moisture content of less than 12%.

7.3.1 The control of insects and mites

Cool, damp storage will not prevent mite infestations, and various insect pests breed in grain under a wide range of moisture and temperature conditions. As a result, many grain stores are at risk from these pests. In UK maltings, grain is normally cooled to less than 15 °C (59 °F) and is dried to 12% moisture or less (e.g. 10%) for prolonged storage. Dust and spillages of grain should be removed, since these readily become damp and are then favoured breeding sites. Not only the grain and malt stores but all the plant and surrounding buildings must be kept clean and insect-free. In particular, elevators, elevator sumps, conveyors and cleaning machinery need to be specially checked. Where possible, if it is at all at risk, the grain should be ventilated with cool air, with a relative humidity below the equilibrium humidity of the grain to dissipate heat and to even out the moisture contents of the grain without adding to it.

A major objective is to prevent the introduction of insect pests into a store. Maltsters may inspect farmers' and merchants' stores to satisfy themselves that they are being run properly before they will accept delivery. Before any load of grain is accepted, it should be carefully inspected. The lorry or wagon contents should be inspected on arrival, and samples of grain drawn from the bulk should be shaken on a sieve with a 1/10 in (0.25 cm) mesh to retain the grains and let insects fall through into a catch tray where they are easily detected. Deliveries that are detectably infested are rejected. Grain may be free of adult insects but still contain eggs or larvae. Where this is suspected, the grain must be cleaned, dried and cooled as quickly as possible and, where possible, be stored in an isolated place. To check for 'hidden' infestation, samples may be taken to the laboratory and held warm and slightly damp in enclosed containers for 2–3 weeks to encourage the development of immature insects, which can then be detected.

Combine-harvesters, trucks, wagons and lorries, sacks, stores, and grain cleaning and handling equipment can all be sources of infestation. All parts of a maltings should regularly be inspected for insect pests; this is especially true in the grain and malt stores. The temperatures of all bulks of grain should be determined regularly, at least weekly, at several sites and at different depths in the grain, and the readings should be logged or, better, recorded graphically. Any upward trend in temperature should be regarded as a warning sign. Where the grain is accessible both it and the surround structures should be carefully inspected, both in the day and at night, and grain samples should be taken and sieved to help the inspection. However, it is not easy to detect insects in a large bulk of grain where they may occur in patches in the surface layers. Increasingly, detection is being improved by the use of traps and baits to 'concentrate' the insects where they can be found (Bills, 1989). Other tests for insect infestations are available (Pedersen, 1992). Bait bags consist of plastic-mesh packets filled with insect-attractive food material. These are distributed strategically and are inspected after about a week. They are useful for detecting insects that have survived insecticidal treatments. Several sorts of traps are in use. The simplest, pitfall traps consist of simple plastic containers pressed into the grain so that the rims are level with the grain's surface. Wandering insects fall into the containers and are trapped. Alternatively probe traps consist of a tubular device with holes in the upper part and a funnel between the upper and lower, transparent part. Insects wandering through the grain, rather than on the surface, enter the holes, fall down and are trapped below the funnel. The use of insect-attractant pheromones to bait the traps does not yet seem to be practical.

Frequently grain is routinely treated with an approved insecticide, metered on as a powder or spray, as it is run into store. If an insect infestation is detected remedial action must be taken at once. The action taken will depend on the extent of the infestation, the insect(s) involved and the type of store. Suggestions that grain be stored in an atmosphere of nitrogen, be exposed to gamma-rays from cobalt-60, be exposed to percussive treatments or to infrared or microwave treatments have not found favour in maltings (Tilton and Brower, 1987). The practical options are to ventilate the grain to cool it and slow insect multiplication, to clean and dry it to remove adults and loose eggs (at the risk of spreading the infestation), or to treat the grain with an insecticide or fumigant. For various reasons, treatments that were once common have now been discontinued, and in general the agents must be applied by trained personnel wearing appropriate protective clothing (coveralls, boots, gloves, hoods, respirators) and using the correct equipment. Dose rates must be chosen that will not leave quantities of chemical residues on the grain which exceed the permitted levels. Unfortunately there are significant local differences around the world in the regulations that apply, the agents that may be used on grain intended for malting and the maximum permissible residue levels. In the UK, maltsters apply the regulations as though malting barley were a primary foodstuff, and they do not apply pesticides to malt. Consequently, as barley is steeped, germinated and kilned before being converted into malt, and the malt is processed further, there are many chances for residues to disperse; as a result, in the UK malt is likely to be an exceptionally 'safe', or residue-free material from which to prepare foodstuffs or beverages.

In general, insecticides are applied as sprays dispersed in water, as granules, dusts or powders, or as fogs. The interiors of stores are generally cleaned and 'fogged' with an insecticide before being filled. Usually grain is treated with powders or sprays while it is being moved and conveyed. Fumigants – insecticidal substances with high vapour pressures – are applied to grain when it is enclosed or covered. Where permitted, liquid fumigants, such as mixtures of chlorinated hydrocarbons, may be sprinkled over the grain before it is covered with impervious sheets, or vapours may be recirculated through the grain in a fan-driven airstream in an enclosed store. In addition spot treatments are possible by adding phosphine-generating tablets to the grain and covering the treated area. In all cases, the grain must not be approached by unprotected personnel until all the vapours have dispersed.

Numerous insecticides have been used over the years, of which many are now banned, for example, abrasive silica dusts (which damaged insect cuticles) because of the risk of silicosis; pyrethins with synergists, chlorinated hydrocarbons (too persistent, residue problems), and so on. The insecticides that may be applied to grain intended for use as foodstuffs in the UK are pyrimiphos-methyl, chloropyriphos-methyl, fenitrothion, etrimphos, methacriphos and malathion. All have maximum residue limits (MRL) of 10 p.p.m. (i.e. mg/kg) except malathion, which has an MRL of 8 p.p.m.

Fumigants have included ethylene dibromide (banned in the USA), carbon tetrachloride, carbon disulphide (explosive mixed with air), methyl bromide, sulphur dioxide, phosphine gas and hydrogen cyanide gas. The volatile insecticide dichlorovos has been used to fumigate empty stores. Phosphine may be generated *in situ* by adding pellets or sachets of aluminium or magnesium phosphide to the grain. The moisture in the atmosphere causes the phosphide to break down, forming phosphine and an unobjectionable powder. British MRL values (p.p.m.) for common fumigants on barley intended for direct use as a foodstuff are carbon disulphide 0.1, carbon tetrachloride 0.1, 1,2-dibromoethane 0.05, hydrogen cyanide 15, phosphine 0.1 and methyl bromide 0.1. While most significant levels of residues arise from treatments of grain in store, it must be recognized that there may be residues from substances applied in the field (many of which have MRLs) and agents applied to grain in store by farmers, by silo operators, by merchants, by shippers and, finally, by maltsters. Residues may consist of the substance applied and one or more breakdown products. The situation is complex and to quantify residues it is usually necessary to send grain samples to specialized laboratories where analyses for residues are carried out.

When an insect infestation is detected, it is necessary to treat the grain, the store, ventilation ducts and all handling and conveying equipment. Afterwards the grain and its surroundings must be monitored to check that the treatment has been effective. Old stocks of barley should be used up completely and each store should be cleaned and fumigated or sprayed with insecticide before a new lot of grain is allowed in. Grain stores should be made with smooth, impervious surfaces, lacking ledges, crannies or recesses where grain and dust can lodge and insects can breed undisturbed. Cracks must be filled, porous surfaces must be sealed and rough surfaces avoided. Surfaces may be sealed with paints, distemper or lacquers that contain persistent insecticides.

7.4 Other pests of stored grain

In badly designed and maintained stores, birds can enter and, apart from their direct depredations, their dirt and droppings foul the grain. However, the pests that cause most problems are rodents: various species of rats and mice. These damage the buildings, they consume considerable amounts of grain (one Norway rat may eat 9.1 kg (20 lb) of grain in a year), much of the grain is only partly eaten and the remainder is contaminated with faeces, urine, hairs and the occasional corpse. Many people become allergic to rats and mice; in addition, rodents may carry many dangerous diseases including bubonic plague, two types of typhus, Weil's disease (leptospirosis, leptospiral jaundice) and trichinosis (a form of meningitis). Therefore, it is important that these pests are controlled.

Because rodents are so common, it is necessary to attempt to eliminate them from all parts of the maltings and to prevent reinfestation from the surrounding area. This is a continuous battle. All areas should be kept clean and cleared so that rodents are not attracted and are denied places to shelter. Traditionally cats, and sometimes dogs and ferrets, were kept to control rodents, but these were largely ineffective, providing more sport than control. The most effective approaches are to gas burrows, to systematically set traps and lay poison and to design the buildings and keep them in an adequate state of repair to exclude rodents as far as possible. These pests are cautious and cunning, and a rat will often avoid a bait containing a poison if it has previously received a sub-lethal dose. Baits may be solid or liquid. Poisoned water is effective where other sources of water are absent. It is necessary to pre-bait with unpoisoned food and lay unset traps until the rodents are accustomed to them before the baits are poisoned and the traps are set. For many years rodents were controlled by warfarin and related anti-coagulant toxicants that are relatively non-toxic to humans and domestic animals. However, rodents are increasingly becoming resistant to these substances, necessitating the use of more generally toxic and more dangerous poisons. With the return to use of these more dangerous agents comes the need for extra safety precautions. Rodent control should be delegated to well-trained and well-equipped operators. All poisons must be laid so that neither personnel nor stored grain can become contaminated.

7.5 Grain heating

Grain in store may begin to 'heat spontaneously', and if the process is not checked serious deterioration may occur (Briggs, 1978). Even if the grain is loaded into store at 15 °C (59 °F) and initially at a uniform moisture content of 12%, this may still happen. It can be because the development of thermal gradients in the grain causes convective air currents through the grain and hence moisture migration from the warmer to the cooler regions. This may be minimized by ventilating the grain or by 'turning it over' (i.e. mixing it by conveying it from the bottom back to the top of a store, or to another store). Where the moisture content is above 15%, the grain is liable to heat because of the metabolic activities of the grain fungi, which generate heat and moisture and so the heating process is auto-catalytic and a hot spot forms. As the temperature rises, different classes of fungus come to predominate, and the temperature may finally reach 60–68 °C (140–154.4 °F), often killing the grain. As the temperature rises, convective air currents are established that carry heat and moisture upwards. Eventually, at the cool grain surface, the moisture accumulates and the grain may even begin to sprout over a hot spot. The heated grain may become caked and matted, it will smell musty and it may well die.

In contrast to such 'damp grain heating', 'dry grain heating', which occurs in grain stored with a moisture content below 15%, is triggered by insects, such as weevils or beetles, or even by mites. An infestation of insects locally raises the grain's temperature and moisture content, triggering the activities of fungi and creating conditions under which insect multiplication is rapid. Once again hot spots may form, but usually the maximum temperatures reached are less, e.g. 43 °C (109.4 °F).

Hot spots may initially be hard to detect, because they may be quite local and remote from fixed temperature sensors. They may be dealt with by applying fumigants (if insects are involved), by turning the grain over and, possibly, by cleaning and redrying it. Badly affected grain must be discarded for malting purposes.

To minimize the chances of heating, it is desirable that undried grain should be ventilated with cool air and that as soon as possible the grain should be uniformly dried to 12% moisture or less, cooled to 15 °C (59 °F) or less and possibly treated with insecticide as it is put into store. In addition, it is desirable that the grain be ventilated from time to time with cool dry air to cool any areas that are becoming warm, to minimize the effects of moisture migration and to try to keep the moisture content of the grain

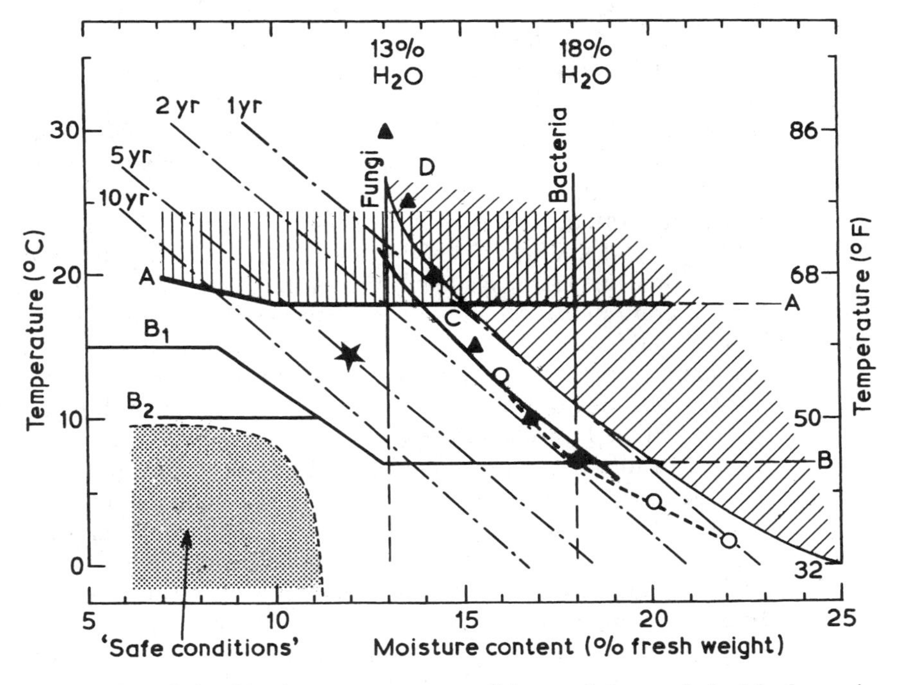

Fig. 7.3 The relationships between storage conditions and the survival of barley grains (Briggs, 1978, based on the data of Kreyger, 1958, 1959; Burges et al., 1963; Burges and Burrell, 1964; MAFF, 1966; Roberts and Abdalla, 1968). Upper region bounded by thick horizontal line; A, conditions under which insect heating is likely to occur; B, horizontal line above which conditions allow major insect and mite pests (B_1) and moths (B_2) to multiply. Region with inclined shading, bounded by line D: conditions under which heating caused by microorganisms is likely to occur. Vertical lines at 13% and 18% moisture contents: the approximate, respective moisture contents below which fungi and bacteria cannot multiply. Continuous, inclined line C, and the line marked ○ ------ ○: estimates of the conditions under which grain germinability falls by 5% in 35 weeks, an unacceptably fast rate of deterioration. ▲, Estimation of conditions under which germinability falls by 5% in 30 weeks in open storage. Inclined diagonal lines — – — –: conditions under which the germination of barley samples, held in hermetically sealed containers, fall from 100% to 95% in the times shown. Note the greater survival of grain in sealed containers. Freezing injury may occur to grains containing more than 15% water if stored below 0 °C (32 °F). The star indicates the storage conditons often, aimed for in British barley stores. Often, lower temperatures are achieved. These conditions are **not** safe unless insects and mites are controlled. The shaded zone (bottom left) indicates 'safe' conditions for 'open' storage.

uniform. However, it is apparent that even under these conditions grain spoilage can sometimes occur (Fig. 7.3). The situation is complicated in that it is sometimes necessary to store grain warm for limited periods to overcome dormancy (Briggs and Woods, 1993; Briggs *et al.*, 1994).

7.6 Moisture distribution in stored grain

It is apparent that all parts of a bulk of stored grain should be uniformly dry and at one temperature (Foster and Tuite, 1982), and often this is assumed to be true. In general this is not the case, so the variability must be reduced as far as possible because of the storage life of a bulk of grain will be limited by its dampest part. There is a belief that if two batches of grain of two different moisture contents are mixed then the moisture distribution in the bulk will quickly even out. Detailed studies have repeatedly shown that this is not true, and that both bulks should be adequately dried before they are mixed (Christensen and Kaufmann, 1974; Briggs, 1978). Furthermore, grain leaving driers is not at one uniform moisture content. In flow-through driers, variations in the grain moisture entering are reflected in variations in the moisture leaving, and in deep-bed driers the grain at the bottom, where the drying air enters, is appreciably drier than the grain from the top layers. Therefore, when grain is put into store it is desirable that it is well mixed and that it is ventilated to encourage an evening out of the moisture between damper and drier grain zones. To check the success of this approach, it is necessary to determine the moisture contents of many individual samples of grain drawn from different parts of the bulk, and not merely to mix up individual samples and obtain a single average value. When multiple samples are taken, differences of 6–8% moisture content between drier and damper parts of the grain may be found and deterioration, whether obvious or not, will be occurring in the damper regions. For this reason, as well as others, it is desirable to load grain to store with an average moisture content below that of the supposed 'safe' value.

The problems are exacerbated by the migration of moisture in the bulk of grain. When hot spots develop this is obvious. However, even in the absence of hot spots, temperature gradients in the grain drive slow, convective currents of air that transfer moisture from the warmer to the cooler regions. For example, grain is put into store in the autumn. As the temperature falls as winter advances, the outer part of a bin and the grain next to it are chilled. The air in the spaces between the grain is also chilled and tends to sink, displacing upwards the warm air that is present in the centre of the bin. At the surface of the grain this warm air is cooled and so loses moisture, transferring it from the warmer to the cooler regions of the grain. Conversely, in the summer, the surface of the bin and the outer zone of the grain will become warmer than the central parts and convection currents will flow in the opposite direction, but again they are the cause of moisture migration. To minimize these effects, it is desirable to ventilate the grain periodically. In extreme cases, it may be necessary to move it to a new bin, even passing it through a drier during the transfer.

Of course it is necessary to minimize the pick-up of moisture by grain in stores. Leaks must be prevented, but it is less obvious that draughts or ventilation with humid air must not occur. In older stores, in which grain was stored in stacks of sacks or in small piles on floors, it was particularly necessary to keep all doors and windows closed. Keeping grain dry in well-constructed silos is comparatively easy, but the conditions in flat-bed grain stores are more difficult to control (Chapter 8).

7.6.1 Adjusting grain moisture contents

In the UK grain is traded by weight, often at a notional 16% moisture content. Unless price adjustments are made for lower moisture contents, as is sometimes done, this encourages farmers to store grain at moisture contents of 16% or more, with consequent risks of deterioration. In addition, the farmer may re-wet drier grain to bring it to 16% moisture before marketing it, to gain the 'missing' weight. Indeed farming magazines sometimes publish charts indicating how much water to add to bulks of grain to adjust the moisture content to 16% (e.g. David, 1975). For example, in an exceptional year, grain might be harvested at 11% moisture. This gives a weight of about 5.6% less than if the grain was wetted to a moisture content of 16%. From the maltster's, or indeed any rational, viewpoint wetting grain to increase its moisture content and weight is a stupid system, because grain will be stored under sub-optimal conditions and so will have started to deteriorate. There will be the cost of transporting unwanted water, and it will be necessary to pay for its removal when the grain is dried. Clearly a trading system that takes account of the moisture content of each batch of grain should be agreed upon.

Normally the maltster is concerned that his grain is at or below a particular moisture content and in many countries, as in the UK, this means that the grain must be dried. Drying may be carried out using many different types of equipment (Chapter 8) and may be carried out on farm, by grain merchants or contractors, or by the maltsters. In general the last option is preferred, because maltsters understand very well how critical it is to avoid heat damage. However, this is not always feasible. The physics of drying are discussed in Chapter 10. In outline, the basis of grain drying is that warm air of a low relative humidity (RH) is passed through a layer of grain and abstracts moisture from it. The hotter the air and the lower its RH, the higher its moisture-carrying capacity and the faster it will remove moisture from the grain both because of its lower RH and because moisture migrates faster out of the grain at higher temperatures. For grain intended for malting, there is a series of important limitations. Sufficiently slow and steady drying must be maintained to avoid husk cracking, which will result from too rapid and uneven drying, and the conditions must be such that the grain viability and the vigour of its subsequent germination is not impaired. Heat damage occurs more readily in grain at higher moisture contents, and increases with exposure time. Therefore, it has been necessary to establish conditions of grain-moisture content, temperature and exposure time that divide heat damage from harmless drying conditions (Chapter 10). Wetter grain must be dried more slowly at lower temperatures to maintain its viability. In a wet harvest period, there is a temptation to raise the temperature to increase the throughput of the drying equipment. This can damage the grain. Because farmers have sometimes over-heated damp grain in their driers and because heat damage is not easily detected (the tetrazolium test for embryo viability is unreliable in this respect, Chapter 6), many maltsters will not accept grain that has been dried on the farm unless they are convinced of the drier operator's expertise and integrity. Maltsters' standards for grain quality are significantly higher than those of most other grain users. In driers, the conditions experienced by the grain are often significantly different to those used in laboratories for determining which conditions will be damaging. Individual grains in driers will receive more or less unequal treatments, and so a safety margin is required to ensure that none of the grain is damaged. The temperature of the grain in the drier is less than that of the 'air-on' (inflowing air) because it is cooled by evaporation of water from the grain; in general, the true 'in-grain' temperature is not known. Consequently the 'safe' conditions for drying differ between driers. A general recommendation is that grain having a moisture of more than 24% should be dried at less than 43 °C (109.4 °F), while drier samples should be dried at less than 49 °C (120.2 °F).

In flow-through tower driers, the moisture content of the grain may be reduced by 6% during a 2 h transit period. It is feasible to dry grain much faster, but at the risk of husk cracking, which is not acceptable (Briggs, 1978). The air-on temperatures the grain may encounter may increase from 54 °C (about 130 °F) to 60 °C (140 °F). The air is cooled by evaporation from the grain, so that the air-off (outflowing air) temperatures may be 43 °C (*ca.* 109 °F) and 52 °C (*ca.* 126 °F), respectively. In the cooling section, ambient air may reduce the temperature to about 30 °C (86 °F) or much less, if the grain is not to be stored warm initially to break dormancy. In cases where warm storage is needed, grain is preferably sent to short-term store at 38–40 °C (*ca.* 100–104 °F).

Some maltsters dry grain in large, static batches, and traditionally grain was dried in malt or barley kilns. In these methods, the grain may be dried in a deep bed for up to 12 h at an air-on temperature of 40–48 °C (104–118.4 °F) in as rapid an airflow as is feasible. The actual grain temperature may be about 30 °C (86 °F) (Bathgate, 1989). If different batches of barley are to be mixed, it may be advantageous to do it before they are dried in batch. Many other types of drier are used, including vacuum driers, drums and many sorts of equipment that are typically found on farms. In some cases, drying may be carried out slowly in store. Because the process is slow it is only suitable for use with reasonably dry grain (<17.5% moisture) and in a dry region where, for substantial periods, the air has a low relative humidity. During dry weather, air heated to 5–10 °C (9–18 °F) above ambient is passed in a slow stream through the stored grain via the ventilation ducts. The rate of drying depends on the airflow and the depth of the grain but, under British conditions, using grain bed depths of less than 300 cm (about 10 ft) it is feasible to remove 0.15% moisture in 24 h; a bin of grain may be adequately dried in 1–2 weeks in favourable weather.

As with malt kilning, there has been a fear that during drying barley might become contaminated with carcinogens from the furnace gases or formed from the oxides of nitrogen. In fact, any risks are negligible unless fuel oils are used, when indirect heating would be advisable. In addition to polycyclic hydrocarbons, fuel oils can contain considerable amounts of sulphur, and sulphur dioxide formed during combustion may be taken up by the grain (Hutt *et al.*, 1978).

7.7 Germinability, viability and grain characteristics

In northern Europe, freshly harvested grain is often dormant and this dormancy declines during storage (Chapters 3 and 6). The rate of decline is much faster at higher temperatures. However, even in grain dried to 12% moisture or less, when microbes are inactive, insect infestations may occur in warm grain. Furthermore, as storage is prolonged the grain tends to lose vigour and then die. The death rate is highly moisture and temperature dependent. Therefore, if the need to overcome dormancy dictates the necessity for a period of warm storage, the grain should be sufficiently dry that no significant grain deterioration or death occurs during this period; insects must be controlled, and the grain must be cooled as soon as dormancy is sufficiently reduced.

Often grain can be taken from driers at 25 °C (77 °F) or more, but as insulated conveyors and stores are not used, the temperature falls during transfer and subsequent storage. Gordon (1968) demonstrated that holding grain at elevated temperatures with slow cooling was more efficient at overcoming dormancy than rapid cooling. Germination in the 4 ml germination test improved much more rapidly than germination in the 8 ml test (section 3.7.5). With most older British barley varieties, dormancy virtually disappeared in most seasons after 6 weeks of storage, then the grain could be cooled to less than 15 °C (59° F) either by cleaning it and/or by ventilating it with cold air. Over winter, the grain may be cooled to as low as 5 °C (41°F).

As grain in bulk is a good insulator, it remains relatively cool when the ambient temperatures rise in the spring and summer.

However, with some newer barley varieties, such as Triumph, in bad seasons dormancy can be extremely persistent and so extra dry, prolonged warm storage is necessary to post-harvest ripen the grain and make it fit for malting. Experimentally some samples needed to be held for 15 weeks at 38 °C (100.4 °F) and dried to 10% moisture to eliminate dormancy without causing any loss of vigour or viability (Briggs and Woods, 1993).

Experimentally undried or only partially dried grain has been held at ambient temperatures in atmospheres of nitrogen, when recovery from dormancy continued, and in chilled stores where cool air was provided by refrigeration equipment. The problem with chilling damp grain is that mites can still multiply at the temperatures that are economic to use, and the chilled grain retains its dormancy. Because of this, chilling undried grain does not seem an attractive method for storing malting barley with the moisture levels found in undried barley in the UK. However in areas such as South Australia where the grain is harvested dry and non-dormant and where ambient temperatures are high, this seems an attractive system.

Figure 7.3 summarizes the conditions most likely to favour grain deterioration, fungal attack and insect infestation. If all these factors are taken into account then it is seen that 'safe' grain storage for prolonged periods is only approximately attainable in grain that is below 10 °C (50 °F) and contains less than 12% moisture. The conditions aimed for in many barley stores, about 12% moisture and less than 15 °C (59 °F), are not truly safe, and are wholly unsafe if/when moisture migration occurs resulting in some parts of the store having higher grain moisture contents. Theoretically in grain stored sealed in an atmosphere of nitrogen under these conditions (12% moisture content, 15 °C (59 °F)) germinability is only expected to fall from 100% to 95% in 5 years (Briggs, 1978). Mathematical systems for predicting the details of rates of increase in germinability of dormant grains and losses of viability in grain stored under particular conditions are being developed. These are not yet sufficiently advanced to be reliable guides to safe storage of all barley varieties on a commercial scale (Woods *et al.*, 1994).

During the storage of dry grain (less than 12% moisture), the population of microbes declines. Microbial death is faster at higher temperatures, but different species vary in the rates at which they die. However, it is likely that the decline in microbes is related to the decline in water sensitivity (section 3.7) in stored barley. Nothing certain is known about the biochemical factors that change within grains during storage and cause a loss of viability. Similarly, although biochemical studies have been carried out on deteriorating grain, nothing is known of the fundamental causes of grain aging and death in well stored, dry samples (Linko, 1960; Pomeranz, 1982).

7.8 Hazards in grain stores

Hazards in grain stores include risks from machinery (Chapter 8), risks attendant on walking on grain, risks of allergies and respiratory problems, and explosion risks from grain dust–air mixtures. Stores treated with fumigants are hazardous and must not be entered by unprotected personnel until the fumigating agents have dispersed.

Heaps of grain can be climbed over, but any grain in a silo or bin should only be inspected by a worker wearing a safety harness tethered outside the store and attended by a second person to haul the inspector up in the case of a difficulty. Should the grain begin to move (e.g. because of withdrawal or loss from the base) it will give no support, and any person standing untethered on the surface will sink and may be lost. It is important, therefore, that no grain is removed from a storage bin while the grain is being inspected. Furthermore if there is any indication that the grain has 'arched' and is not fully settled in the container, there is a risk that the 'arch' will collapse if the grain is walked on, with the likelihood that the inspector will be buried.

If the bin contains appreciable levels of carbon dioxide, fumigant or insecticide, it should not be entered unless absolutely necessary and then only if proper precautions are taken, respirators are worn, etc. Allergies and respiratory problems, including malt worker's lung, may be induced by microbes, spores, dust, insects, mites and rodents. In well-controlled stores, exposure to these should be minimal and will mostly occur through dust. Dust is generated whenever grain is moved and should be trapped by the extraction machinery (Chapter 8). Should dust arise, for example from a machinery failure, masks, respirators or filtered-air helmets should be worn. Accumulations of dust should be regularly cleared away.

Similarly all precautions should be taken to prevent the ignition of dust–air mixtures, which can cause explosions: minimim concentration 50 g dust per m^3. Dust extraction and trapping machinery should work well all the time; possible sources of ignition are eliminated and smoking and naked flames are forbidden. Plant should be designed to minimize the effects of any explosion that might occur by fitting suitable vents (Chapter 8).

7.9 Grain aeration and ventilation

Freshly harvested, undried grain should be ventilated to keep it cool and prevent deterioration (ADAS, 1985). The amount of ventilation required will depend on the grain's moisture content and temperature. Ventilation to achieve cooling may extend over months, and into the winter.

Slow drying may be carried out by passing warmed air into the grain through ventilation ducts. With dried grain (less than 12% moisture), at least in Britain, grain is usually ventilated with cold, dry air to cool it. After dormancy-breaking, grain should be cooled at least to 20 °C (68 °F) and preferably even lower as quickly as possible. In some systems, the fans that drive the ventilating air are actuated automatically when the temperature and relative humidity of the ambient air fall below pre-determined values. In hot climates, refrigeration units may be used to cool the ventilating air. It is desirable to check that airflows through the grain are adequate and uniform (no channelling) using a sensitive anemometer. Ventilation may also be used to even out the temperature and moisture content of a bulk of grain, to remove odours or to spread or disperse the vapours of fumigants. The importance of these operations to satisfactory grain storage is obvious, and consequently so is the need for careful design of the store and its ventilation ducting or other systems and fans. Clean, dust-free, loosely packed grain offers the least resistance to airflow. The logarithm of the airflow is approximately proportional to the logarithm of the pressure drop per unit depth of grain (e.g. Shedd, 1953). It follows that, with a grain air-distribution system, overloading and storing grain at too great a depth will cause too high a resistance to airflow and so prevent adequate aeration.

Grain stored in an inert atmosphere in a hermetically sealed store may keep for some months without ventilation. However, in conventional stores, grain containing 13% or more moisture will need to be ventilated; the wetter the grain the larger the volume of ventilating air required. Grain dried to 12% moisture or less and held at or below 15 °C (59 °F) has been kept for about a year without ventilation. However, the risks of insect attack (even if insecticide is applied) and moisture migration creating localized areas where deterioration can occur are significant. In some old maltings, where well-dried grain was stored on floors in layers 1.2–2.0 m (4–6.6 ft) thick, no provision was made to aerate the grain. However, the bulk had to be frequently inspected, and all was ready for emergency action if deterioration was detected. When grain was stored in sacks it was usual to ensure that these were stacked in such a way that ventilation was not impeded and the sacks could easily be inspected for signs of insect infestation.

7.10 Operating grain stores: good housekeeping

Gathering, cleaning and storing high-quality grain is an essential part of malting. The store manager must keep correct records of his stocks, ensure that adequate amounts of grain are available and that different bulks are cleaned, dried and are stored segregated by quality (variety, nitrogen content, viability and age). However, it is evident that there are many other aspects that require attention: the need to exclude vermin, to control insects, to prevent grain deterioration and to encourage dormancy breakage. To achieve the necessary results, it is necessary to adopt an orderly, active and disciplined approach to running grain stores. Many of the precautions needed have already been indicated. Often the best practices are grouped together under the heading 'good housekeeping'; this includes several general points. Standards at grain intake should be rigorously applied, and, in particular, infested or contaminated grain should be rejected. Undried, green grain and old and new stocks of dried grain should be carefully segregated, as should stocks of barley and stocks of malt. At the end of a batch, a store should be completely emptied and cleaned, together with the surroundings, machinery and ventilation ducts, and should be fumigated or sprayed before being refilled with new grain.

The stores, the associated machinery and the surrounding areas should be kept clean with no accumulations of dust or spillages of grain. All should be kept in a good state of repair and regularly inspected for signs of pests or insect infestations. Equipment should be aspirated to collect dust and built-in dust collection vacuum systems may be used as well as industrial vacuum cleaners. Rubbish must be burned or disposed of away from the site. Grain stocks should be inspected regularly, temperature readings (fixed thermometers, sensitive thermocouple probes) and moisture determinations should be made and insect traps should be set and inspected. The findings from all inspections should be recorded. From time to time samples must be evaluated for germinability and viability. Immediate action must be taken if there are any signs that deterioration is or may be about to begin. This may involve ventilating with cool air, fumigation, redrying the grain, mixing it by 'running it round' in the conveyors to the same or a different store: in fact taking whatever action is appropriate. Barley and malt stocks should be carefully segregated, even to the extent of not using the same conveyor and elevator systems. Barley contaminating malt reduces its quality. Malt contaminating barley causes major problems in steeps, which become heavily infested with microbes growing on the substances leaching from the malt corns, and during germination when the wetted malt corns will become mouldy. Malt culms (pelleted or not) are hygroscopic, rapidly become damp and can deteriorate and become infested with insects. They should be stored dry, well away from the grain store.

Personnel should be well trained, both in all aspects of running the stores and in terms of what actions to take in cases of emergency.

The cost of the barley, or other grain, is the largest single expense in making malt, and good malt cannot be made from low-quality or immature barley. With this in mind, it is clear how important it is to run grain stores efficiently.

8 Handling and storing grains and malts

8.1 Introduction

Grain, barley or whatever grain is being processed may arrive at a maltings in different sized batches. In industrialized areas, grain is usually delivered in bulk (that is loose) and is typically brought in lorries (20–25 t lots), in railway wagons, in barges or in ships. Less often it may arrive in 25 kg sacks. The grain intake arrangements must be those appropriate for the delivery system. The rate at which the grain is taken in will be rapid, to avoid holding up the transport and incurring demurrage charges. The grain will be sampled and evaluated at the point of intake and any sub-standard loads will either be redirected, or the price will be renegotiated or the load will be rejected (Chapter 6).

Normally arrangements will have been made so that only grain of one variety and one grade will be delivered to an intake in a particular period, to avoid admixtures of grains of different qualities. Maltsters prefer to hold their own grain stocks. However, storage is costly and they may contract for others to do this; if they do, then they should arrange to supervize deliveries to the store and periodically check how the store is being managed. Indeed, they may take in the grain and clean and dry it before sending it to an 'outside' store.

In the maltings, green (undried) and dried grain, steeped grain, green malt and kilned malt will be conveyed from place to place and elevated (raised to higher levels). The equipment used will be considered in this chapter. Green grain will be roughly pre-cleaned on arrival, then stored and eventually dried. The dry grain will be thoroughly cleaned and stored before entering the malting process. After malting and dressing (deculming), the finished malt will be stored until it is despatched. Batches of grain and malt are repeatedly weighed as they pass through a maltings. At various times during the last two centuries, malting has been run exclusively by a workforce or by a workforce aided by power from water, steam engines or gas engines. In the UK at present, most equipment is electrically driven and modern plants are high automated. In a few cases, the electrical power may be generated on site.

In principle, barley (or other grain), steeped grain, green malt and finished malt can be moved using the same conveyors and elevators and dried barley and malt can be held in the same stores. However it is imperative that malt, green malt and grain do not become mixed, and that different grades of grain and of malt are kept separate. Consequently, to avoid cross-contamination, it is necessary to run handling and storage facilities in a highly disciplined and orderly manner. In many maltings, the grain handling and storage facilities are largely or wholly separate from those used for malt. This, although costly, is the preferred arrangement.

Unprocessed grain is tougher than malt and is less easily damaged by handling equipment. Because of its husk, barley is more abrasive than, for example, wheat, rye or triticale, and it is less easily damaged by mechanical handling. Steeped grain, and especially green malt, is easily crushed; dry malt is friable and is liable to break up if handled roughly. For these reasons handling equipment that is suitable for grain may not be suitable for malt. Since grain and grain products are inevitably abraded, with the generation of dust, and sustain some other damage whenever they are moved it is desirable to keep the frequency and distances of grain movements to a minimum. Many different systems for handling grain are in use. Figure 8.1 illustrates a generalized system of the type used in some maltings.

8.2 Grain intake

Grain may arrive at maltings in boats, in railway wagons or in lorries. Most British maltings are not now equipped to handle sacks of grain, and so grain is delivered to them in bulk. Unloading may be through suction pipes, for example from ships' holds or from special rail wagons. In some parts of North America delivery by railway is normal, but in the UK most deliveries are by lorry.

For simplicity it is usual for all grain delivered to an intake at a particular time to be of one type, variety and grade. To speed lorry turn-round and so minimize non-productive waiting times, deliveries are 'called' to arrive in particular time slots. On arrival each lorry is weighed on a weighbridge. It is weighed again on departure and the difference is the weight of grain delivered. There must be sufficient space for the lorries to manoeuvre, park and queue ready to discharge. The weighbridge should be in sight of the grain intake so that there is no opportunity for fraud, e.g. unloading a ballast of stones from the lorry between the two weighings that would be 'scored' as grain. Arrangements vary widely but, in the best practice, when called, the lorry is driven under a waterproof roof and where possible the grain surface is uncovered and inspected for signs of insects, patches of grain of different colour, and dirt. Samples are drawn from a series of sites from the top to the bottom of each load, usually with a vacuum sampler probe and the mixed sample is taken to the laboratory to be matched against the purchase sample and checked for varietal purity, corn size distribution, freedom from insects, viability, moisture content and nitrogen

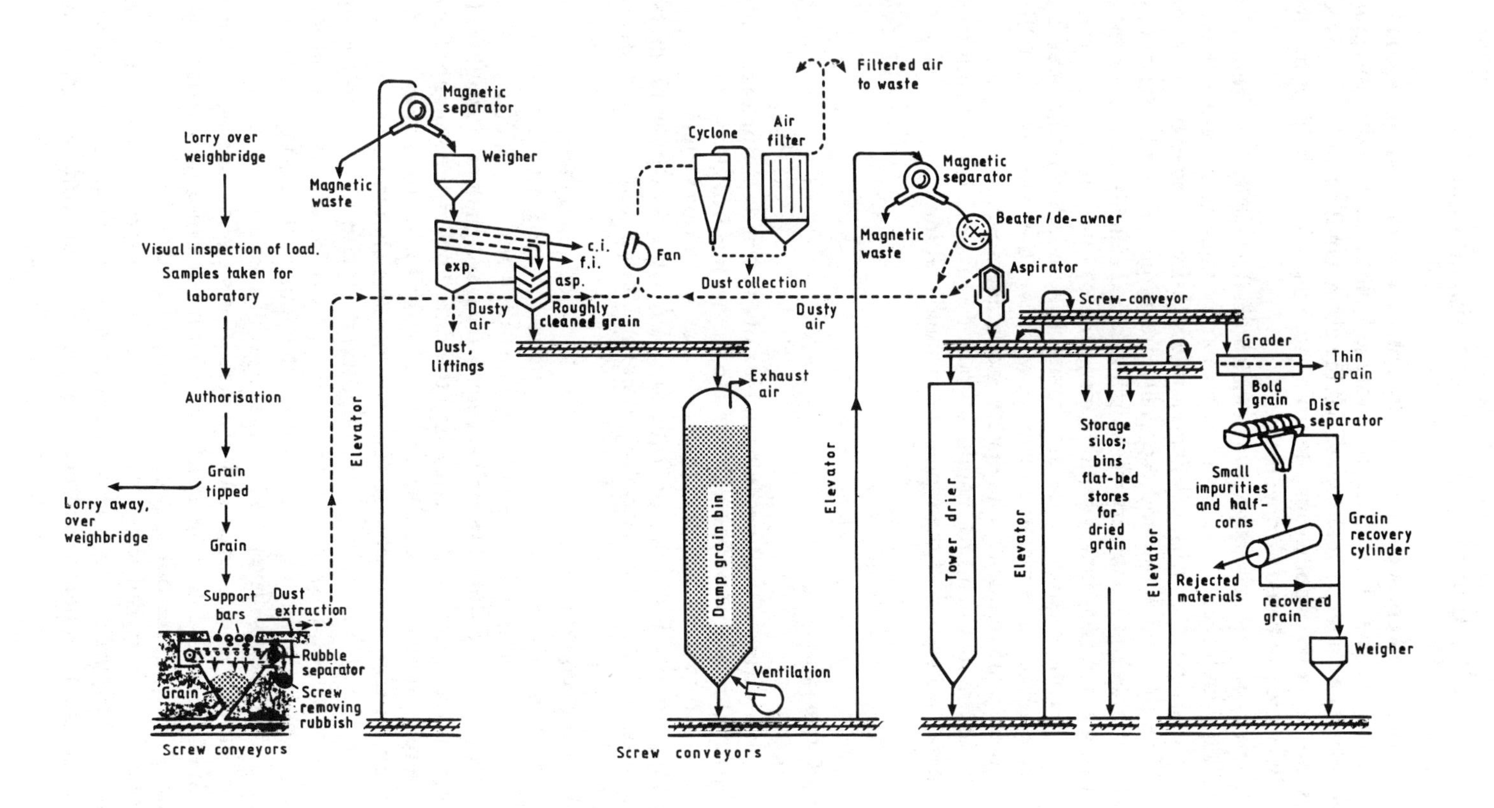

Lorry over weighbridge
Visual inspection of load.
Samples taken for laboratory
Authorisation
Grain tipped
Lorry away, over weighbridge
Grain
Support bars
Dust extraction
Rubble separator
Grain
Screw removing rubbish
Screw conveyors
Elevator
Magnetic separator
Magnetic waste
Weigher
c.i.
f.i.
exp.
asp.
Dusty air
Dust, liftings
Roughly cleaned grain
Fan
Cyclone
Air filter
Filtered air to waste
Dust collection
Dusty air
Exhaust air
Damp grain bin
Ventilation
Screw conveyors
Elevator
Magnetic separator
Magnetic waste
Beater / de-awner
Aspirator
Screw-conveyor
Tower drier
Elevator
Storage silos; bins flat-bed stores for dried grain
Elevator
Grader
Thin grain
Bold grain
Disc separator
Small impurities and half-corns
Grain recovery cylinder
Rejected materials
recovered grain
Weigher

contents (Chapter 6). Sometimes the probe is operated from the laboratory and the vacuum line delivers samples directly to the workplace. If the results are satisfactory (the tests may be completed in as little as 30 min), the lorry is unloaded. Lorry designs vary. They may be tankers with hopper bottoms, or discharge screw-augers. Some lorries are 'open', the loads usually being covered with canvas. They may be emptied either from the side if the lorry is tipped on a lifting device, or from the back if the lorry has a hydraulic ram that allows the back to tip so that the grain slides down and out at the back through an opened door. The roof covering the intake site must be high enough to permit each lorry to tip its load. Lorries can be discharged into a pneumatic handling system, but mechanical systems are much more common. Usually they discharge into a 'reception pit' in a covered hard-standing area that is usually a little higher than the surroundings so that rain or any other water drains away from the pit (Fig. 8.1). Pits vary in their sophistication. Each is covered by a metal grid capable of withstanding the weight of a lorry and is sloped so that the grain slides down to a conveyor at the bottom of a trough and is carried into the maltings. A moving woven-metal coarse mesh screen (a rubble separator) may be situated just below the covering grid. Grain falls through but larger impurities such as straw and stones will be caught and carried to one side to fall into a separate conveyor. The grain running from the lorry into the pit generates and releases dust. Lorry outlets maybe fitted with flexible canvas 'sleeves' to contain the dust and divert it, with the grain, into the pit. In addition, suction may be applied at the side of the pit, either above or below the level of the grid, to remove the dust. The dust-laden air is directed to the central dust-collection system. Rarely the whole intake area may be enclosed to form a booth, which contains the whole lorry; before lorries tip roller-blinds may be pulled down at the front and rear to help contain the dust. The dust-laden air may be sucked away to the collection system.

Individual loads of grain are likely to be of 20–25 t. Because lorries must be unloaded as quickly as possible to minimize turn-around times, the reception pits must be capable of receiving and removing grain at a rate equal to or, better, faster than lorries can deliver it.

In the UK and northern Europe, in most seasons, newly delivered grain will need to be dried. Therefore, its first destination is the green grain store, where it can be held with ventilation to minimize any deterioration before it is dried and transferred to a longer-term store.

Fig. 8.1 One layout for grain reception, conveying, cleaning, grading, drying and storage. Labelling on the pre-cleaner: c.i., coarse impurities; f.i., fine impurities; asp., aspiration section (with baffles) for dust removal; exp., expansion chamber, where the airflow drops and some dust and the 'liftings' settle.

8.3 Moving grain

8.3.1 Introduction

In the UK, the practice of moving grain by hand, in barrows or by other devices powered by the workforce has virtually disappeared (de Clerck, 1952; Vermeylen, 1962; Narziss, 1976; Briggs, 1978; Forster, 1985; Prior, 1996). Grain-handling machinery is classified as conveyors (which move grain horizontally or up slight inclines) and elevators (which raise the grain to a higher level). However some pneumatic conveying systems can act as conveyors and elevators. In addition, grain may be directed to a lower level either by allowing it to flow down a chute or spout or by allowing it to fall freely, as from a conveyor into a grain store.

The choice of grain-handling machinery must be made with reference to a number of factors (Forster, 1985; Malcolm, 1990; Prior, 1996), such as the nature of the material to be moved and how 'gentle' a machine is to the grain or malt, the quantities to be moved (often calculated as volumes) and the rates at which this must be achieved, the reliability and robustness of the equipment and how easy it is to maintain and repair, the capital and running costs (power-consumption), whether the equipment will be used intermittently or continuously, how easy it is to clean, whether it contains any dust that is generated, whether it is 'self-emptying' or if it is liable to allow cross-contamination of one sample from earlier samples, whether or not it is noisy or if it causes vibrations, and whether it is simple to operate correctly and is safe. Most conveyors are now enclosed to contain dust. They must be aspirated and the dust conveyed to the central collection facility. To minimize the risks of explosions and the build-up of pests in dust deposits, the ducting must be cleanable. In addition grain ducting must have explosion vents in suitable places to release the pressure shock should there be a dust–air explosion. These vents, typically physically weak panels in the 'walls' of ducting, are sited so that explosions cause them to burst away from other machinery or from where people may be working. These considerations do

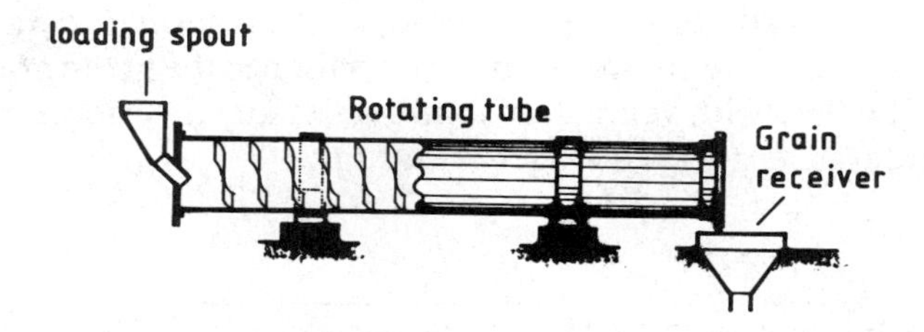

Fig. 8.2 A continuous worm-conveyor (after de Clerck, 1952). The advantages of this equipment are that it is self-emptying and it handles the grain gently. However it works more slowly than the, more common, helical screw conveyor.

not apply when steeped grain is transferred either drained or as a slurry in water (Chapter 9).

In a few places, small amounts of grain may still be moved in hessian sacks. These may be carried on the back (humped), moved on barrows or trolleys or swung in nets held by overhead cranes. Alternatively, stacks of sacks may be moved in tractor-mounted buckets or by fork-lift trucks when on pallets. Old maltings had sack elevators that lifted individual sacks vertically between floors. Sacks may be lowered between floors or often they are allowed to slide down spiral chutes or slides. For export purposes, particularly to tropical areas, malt can be exported in sacks that may be very large (1 t). Frequently these are laminated to provide strength and to exclude pests, humid air and moisture. Typically these are stacked on pallets and handled by fork-lift trucks; they may be exported in sealed metal containers, which themselves provide extra protection.

The most efficient arrangement is to handle moving grain 'in bulk' so that it can flow and, in some respects, be treated almost as a liquid. Broadly the handling equipment involved may be divided into mechanical equipment and pneumatic equipment, in which grain is transported entrained in and propelled by an airflow.

8.3.2 Screw or worm conveyors

Occasionally conveyors are used in which the grain moves down inside a rotating tube fitted with internal helical vanes (Fig. 8.2). These conveyors work over short distances and have the advantage of being self-emptying, but, because the interiors are not easily reached and the conveyors are expensive with lower capacities than conventional screw conveyors, they are not much used.

Worm, auger or archimedean screw conveyors are common in maltings (Fig. 8.3). In spite of their high power consumption, they are widely used. They carry grain horizontally or up or down slight inclines. Conveyor runs may be very long. By the use of slides, or valves, which may be automated and remotely controlled, a conveyor run may be loaded or unloaded at various points, which is very convenient. Rarely they are reversible. They are robust, simple and cheap and easy to maintain, but they are not fully self-emptying, and their power consumption is comparatively high. They are better than drag-flight conveyors for handling wet materials, but they cannot operate over such long distances. Each conveyor consists of a U-shaped metal casing, or trough, usually closed at the top by metal plates or inspection-guard grids, and in which helical metal screw blades rotate around a central shaft (Fig. 8.3). Water may be hosed into the conveyors through the grids for cleaning after conveying green malt (Forster, 1985). At intervals, the screws are interrupted and the shafts are supported by bearings, as they are at the ends. Provided they are well maintained so that

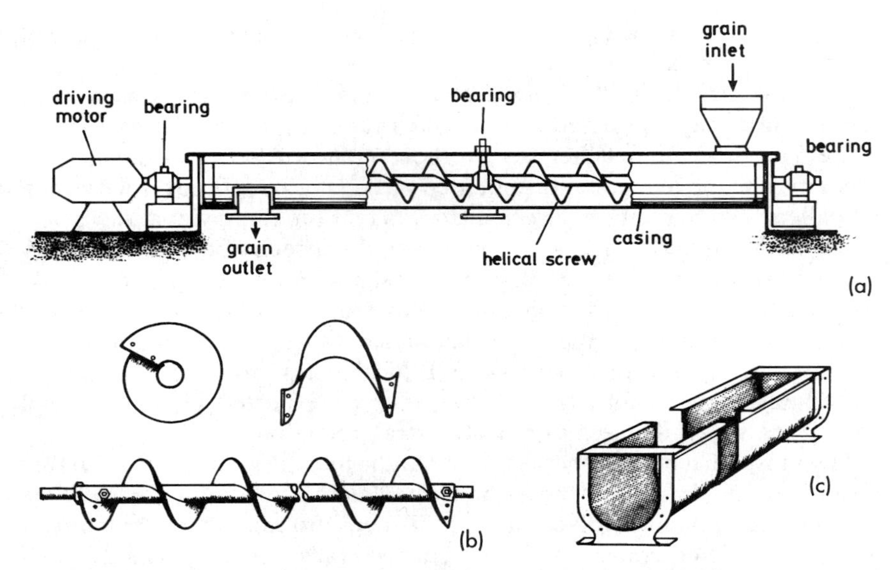

Fig. 8.3 A helical screw or worm conveyor. (a) The complete device; (b) the de-mounted 'screw'; (c) the casing. Normally the casing is closed for safety reasons and to limit the spread of dust.

the clearance between the casing and the screw is correct (often 3–5 mm; about 0.12–0.20 in) and the equipment is not run too fast or overloaded, these conveyors can move barley, steeped barley and green and kilned malt. There is a tendency for malt to be damaged by grinding between the rotating flights and the casing. The outlets must be designed to minimize grain lodgement (Forster, 1985). Such 'fixed' systems are of many capacities and some are able to convey grain at 100 t/h. Two variants of this pattern, which are unusual in the UK, are conveyors in which the continuous worm is replaced by separate segments or paddles, and others in which the normal worm is replaced by a helical metal ribbon supported by spokes radiating from the central shaft.

In addition mobile screw-augers are in use. For example, to load lorries in flat-bed grain stores, wheel-mounted augers, equipped with rubber screws in an open casing, may be pushed into grain to sweep it 'sides-to-middle'. Here it meets another screw working in a tubular casing, which lifts the grain and drops it from a discharge spout at the upper end into a lorry or other receiver.

8.3.3 *Chain-and-flight, cord-and-flight, and* en masse *conveyors*

In chain-and-flight or cord (cable)-and-flight (slat) conveyors (drag-flight, drag-link), the grain is moved along in a smooth casing of rectangular cross-section impelled by 'flights' of metal, plastic or wood positioned at right angles across the casing and moved by electrically driven chains or cords that run along within the casing (Fig. 8.4a). The flights move in one direction only. The chains and cords are 'endless' and are automatically kept at the correct tension. The flights, which rest on the bottom of the casing and sweep the grain along, take various forms and are also called drag-links, scrapers or slats. Some designs, particularly older types, can be excessively noisy. Appropriately designed, they may be quiet, may handle grain and

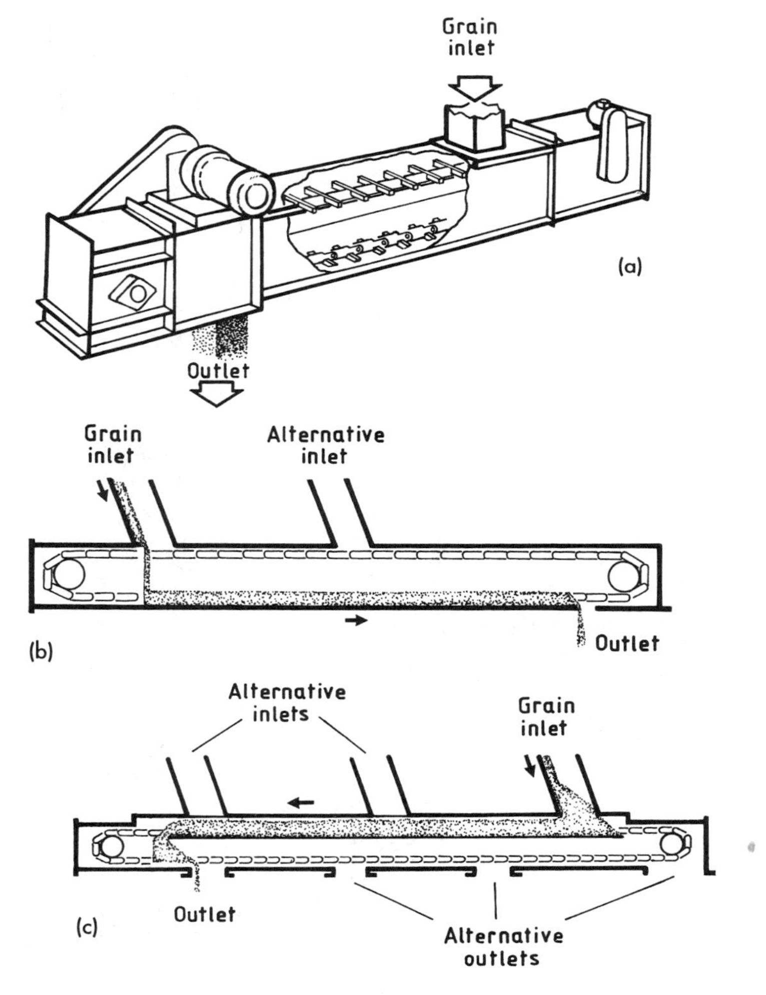

Fig. 8.4 A Redler chain and flight conveyor. (a) A schematic view of a conveyor; (b) a conveyor with multiple inlets; (c) a two-way chain-and-flight conveyor, illustrating the possibility of having alternative loading and unloading points.

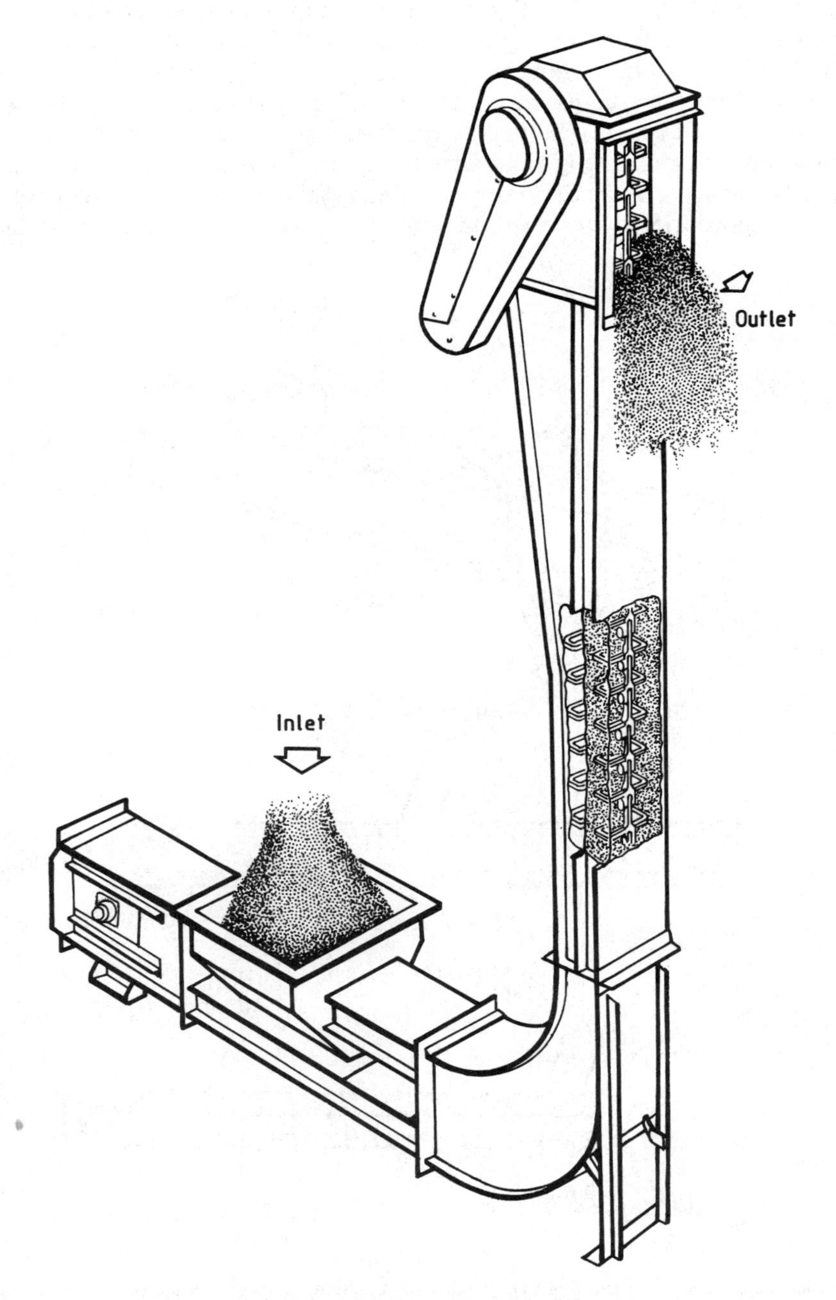

Fig. 8.5 An *en masse* conveyor and elevator (Redler). While such conveyors are used in maltings elevators generally are not.

malt gently and may so work that the grain is moved *en masse* as a steadily moving bulk. In this case, each flight may extend across the bottom and up the sides of the casing. Cable-driven conveyors with wooden or plastic flights working in a low-friction casing are relatively quiet. Such conveyors are self-emptying. They can be loaded or unloaded at various points and can work over long distances. They can be arranged to work in a number of different ways (Fig. 8.4b,c) with multiple inlets and/or outlets. They are economical in that they use relatively little power, and they may have extremely high capacities, e.g. up to 300 t/h grain conveyed. Carefully designed and when run correctly loaded, such conveyors are very gentle and may be used to handle grain or malt.

Chain-and-link *en masse* conveyors may be used to move grain horizontally and vertically (i.e. as an elevator, but as such elevators cannot be fully emptied they are not used for moving grain in maltings (Fig. 8.5).

8.3.4 Flexible-belt conveyors

Traditional band or belt conveyors move grain very gently and quietly horizontally or up slight inclines (e.g. 15°). This is the most generally acceptable conveying system. The endless bands consist of rubber-impregnated cloth or other tough, flexible material, which may have a smooth, roughened or ridged surface (to prevent grain slipping); the cloth moves over rollers, some of which are powered. Usually, but not always, the band is 'dished' by the rollers to confine the grain to the middle of the belt which should be sufficiently wide for the loads it must carry (Fig. 8.6). Grain is metered onto the middle of a belt in a thin stream and is moved gently along at 2.5–4.5 m/s (8.2–14.8 ft/s). Bands 1 m wide, capable of carrying over 250 t/h have been used. Runs of considerable length (e.g. 300 m) are possible. Devices for loading grain onto belts may be moved along or be positioned at different points, giving a degree of flexibility. Grain may be discharged by running off the end of the belt, by tipping the belt to one side and using a 'scraper' to deflect the grain stream into a receiver or by using a device to 'kink' the belt, creating a 'false end' from which the grain falls and is collected (Fig. 8.6). These conveyors handle grain and malt extremely gently and are largely self-cleaning. However, they must be washed at the discharge point when used to convey steeped barley or green malt, which may adhere to the belt and be crushed by the return rollers (Prior, 1996). They may be used open within buildings, as across the top of groups of silos, or within special casings if they are to be used outside between buildings. Dust is created at sites of loading and discharge and needs to be contained and removed by aspiration.

A recent proposal is that grain carried along on a belt could be held in place sandwiched between a ridged, supporting belt and a second, 'enclosing' belt arranged to close upon it. Sandwiched between two belts the grain

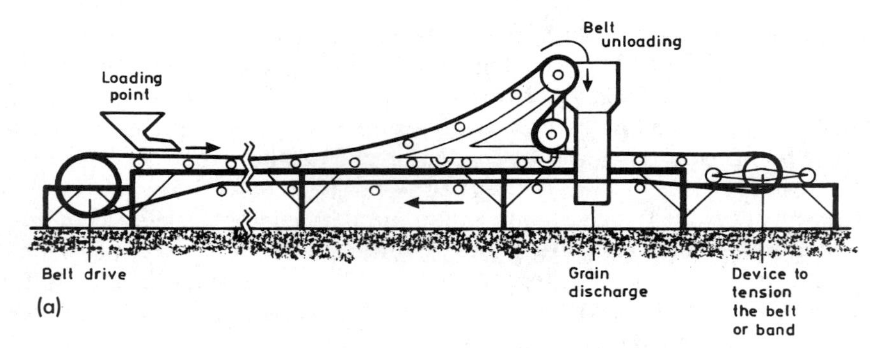

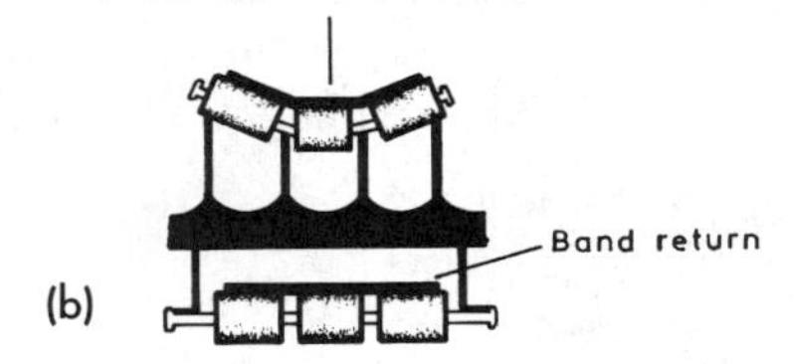

Fig. 8.6 A type of endless belt conveyor in: (a) side-view, and (b) cross-section. The belt on its return run is flattened, but while carrying grain it is 'dished' by the three rollers. In other belt conveyors, the belt may be held in a 'V' by running over two pairs of rollers set at an angle to each other. Several sorts of grain loading and unloading devices are in use.

could be conveyed without spillage and could be elevated. Another proposal is that belts should be supported by an air-cushion and run in enclosed tubes (Malcolm, 1990).

8.3.5 Jog conveyors

Jog or jump conveyors consist of metal or wooden troughs (0.2–1.0 m (0.66–3.3 ft) wide mounted on hinged legs and given an upward and forward oscillating, jerky motion through an eccentric drive (Fig. 8.7). These conveyors are inexpensive and have low power consumptions. They move green malt gently, without tearing off the rootlets. However, they are now unusual because they are cumbersome and slow (moving only 5–30 t/h) and, as they cannot exceed 15 m (*ca.* 49 ft) in length, the grain must be transferred along a series of units if the movement is over a long distance. The grain advances in jumps of 1–8 cm (0.5–3 in) and reaches a height of 5 cm (*ca.* 2 in) above the surface of the conveyor, which may give 300–350 jolts/min.

8.3.6 Grain elevators

The most usual elevators in maltings are those in which the grain is lifted in buckets, with or without bottoms, attached either to endless chains or to

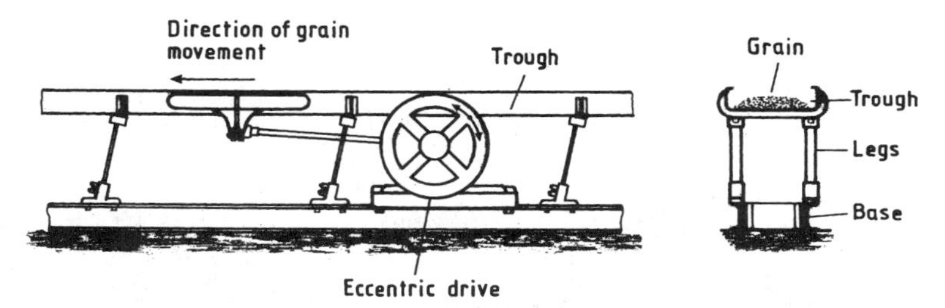

Fig. 8.7 A jog conveyor, seen in (a) side-view and (b) cross-section. The eccentric drive causes the trough to jerk forwards and upwards, moving the grain or green malt in a series of small jumps or hops.

belts that run in a casing over upper and lower pulley-wheels, the upper one of which is powered (Fig. 8.8). Elevators in which buckets are mounted on belts are now popular. The buckets, which may be of metal or plastic, are designed to avoid damage to the grain; they may be closed, or a proportion may have open bottoms. The principle of the latter is that the grain tends to fall from bucket to bucket but is prevented from doing so because friction with the sides of the bucket and with the grain in the bucket below causes it to spread out at its angle of repose; the sides of the bucket below prevent it spreading too far and spilling. In each row of buckets there is an occasional bucket with a bottom to prevent slippage. Therefore, in these conveyors continuous columns of grain are elevated, whereas in traditional elevators there is no grain in the spaces between the bottom of one bucket and the top of the grain in the bucket below. The differences between the two types are minimized in elevators in which successive buckets, with bottoms, are close together. Bottomless bucket elevators have the advantage that for a given cross-sectional area they have a higher capacity than complete-bucket elevators. The grain may be loaded into the boot on either the down leg or the up leg in these conveyors (Mills, 1979). Casings used to be of wood, but these carried an appreciable fire risk and metal casings are now usual. The grain to be elevated is metered into the lower part of the casing, the boot. The descending buckets turn round the lower wheel, scoop up the grain and, as they turn over the upper wheel, the grain is tipped into the discharge spout. The grain is metered into the elevator to prevent overloading. Normally there are devices to ensure that over-loading does not occur, grain is only metered in when the elevator is running and that the elevator keeps running until it is empty. Such elevators are not fully self-emptying and so even with carefully designed elevators with minimal dead-space, it is necessary to clean the boot after use (Forster, 1985).

Well-maintained and designed bucket elevators are very gentle. They may be made in a wide range of sizes and can easily be made to elevate grain to 100 m (328 ft) and to move 100 t/h. When handling dry goods the trunking should be aspirated, to remove dust, and explosion-relief panels should

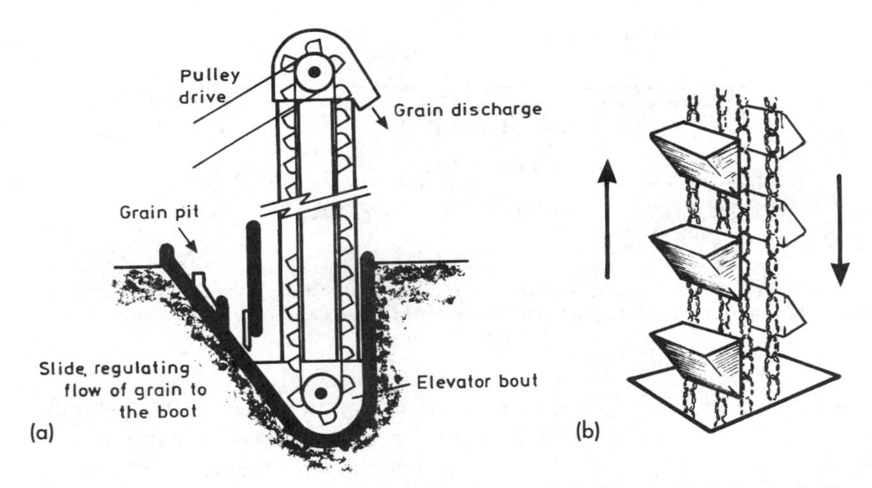

Fig. 8.8 (a) A chain-and-bucket or belt-and-bucket vertical elevator. In the example shown, grain from the pit is allowed to flow at a regulated rate into the boot of the elevator. These elevators may be surprisingly high (Fig. 8.31). (b) Ascending and descending buckets supported on chains.

be incorporated. When handling steeped grain or green malt, the specially designed elevators should be constructed of corrosion-resistant materials and be equipped with sprays for automatic cleaning (Prior, 1996).

8.3.7 Grain throwers

When some flat-bed stores or older kilns are being loaded, the grain has to be projected upwards and horizontally considerable distances to project it to its resting place. This may be achieved at the discharge spout at the end of a pneumatic transporting system, but more usually the throwers used are mechanical. Grain is metered onto a moveable, steerable device where it either meets a set of rapidly rotating, rubber-tipped paddles or a rapidly moving endless belt. These devices throw the grain, or green malt, as the operator directs. It is not easy to obtain a uniform, level surface when using such a device and the grain may 'pack-down' when loaded in this way. These are two disadvantages to using throwers to load kilns, where very even, light packing is desired.

8.3.8 Sampling devices

It is necessary to sample grain or malt when it is being moved to provide material for analysis for quality-control purposes. Various devices are used to achieve this. It is desirable that they sample the whole of the grain stream since moving grain tends to separate into streams, differing in grain size, density and shape. Numerous devices have been produced, but probably the best are those that rotate and periodically sweep across the entire grain stream, removing a sample which is discharged into a collecting container for subsequent inspection and analysis. All sampling devices should be checked to ensure that the samples obtained are fully representative of the whole grain stream.

8.3.9 Pneumatic grain conveying system

In pneumatic conveying systems, grain is swept along in a current of air. The system can collect grain from lorries, wagons, stores or ships' holds and elevate it or transfer it between sites. Such systems are advantageous when space is at a premium. Installation costs can be relatively low. Pneumatic systems are completely self-cleaning and contain the grain dust. They are flexible in being able to pick up or discharge grain at a variety of points and they can follow tortuous routes from place to place. They are able to move dry or steeped barley. However, they are mechanically inefficient and operating and maintenance costs can be high.

There is a wide range of system in use, but in 'lean-phase' (high air: grain ratio) systems airflows may be 25–35 m/s and it may need 3–6 m^3 air to move 1 kg grain. Barley in particular is abrasive, and such systems are subject to wear, particularly in bends or curves, which should be wide and gentle and kept to a minimum. Furthermore, many systems tend to damage grain and shatter malt, so they are not widely used in modern maltings, except for specialist uses, such as unloading the holds of ships or special rail wagons or lorries. However, newer 'dense-phase' or 'plug-phase' pneumatic conveying systems are said to cause much less damage to grain and malt, and move it more slowly (Malcolm, 1990). Pneumatic conveying systems may be divided into three groups, (i) the suction or low-pressure type: (ii) the compression, or high-pressure type; and (iii) suction and compression hybrid types.

In suction systems, which are powered by vacuum pumps, product leakage is eliminated since air is sucked in through any faults in the ducting. The grain may be sucked into a pick-up pipe, which may be flexible and moveable or may be fixed and have grain delivered to it from an auger or a conveyor or a hopper. The inlet to the pick-up pipe is usually guarded with a metal grid to prevent large, unwanted objects being picked up, and for safety. Valves allow the suction to be applied to one of any number of points. Grain can also be introduced from a hopper, through a valve, into the moving airstream. The grain may be moved over long distances, and at high rates (e.g. of 100 t/h). To separate the grain from the airstream, the flowing mixture is discharged into an expansion chamber (Fig. 8.9). The grain falls to the bottom of the chamber, from which it is discharged via a rotating or a tipping seal. The dirty air is passed through a cleaning system, often a cyclone and filter to remove coarse impurities and dust, then it is passed

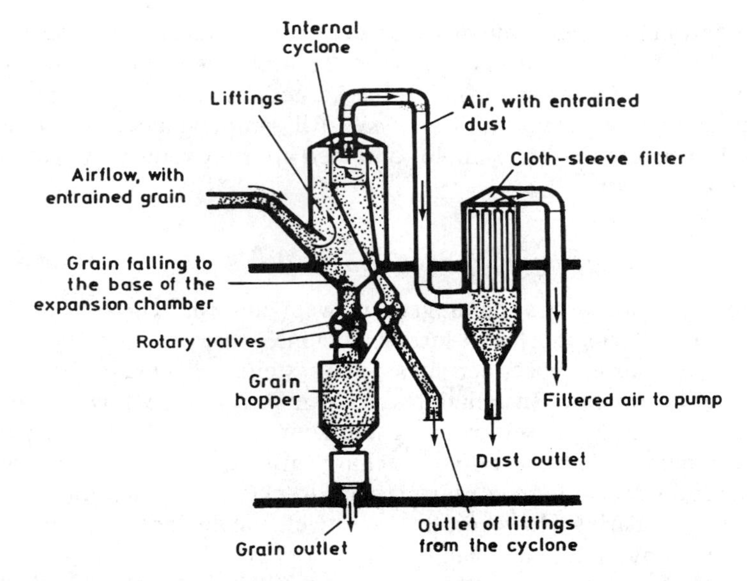

Fig. 8.9 A system for separating grain, liftings and dust from a pneumatic conveyor. The liftings are separated by an 'internal cyclone', mounted in the expansion chamber, and the dust is subsequently removed from the airstream using a textile-sleeve filter (after Vermeylen, 1962).

through the vacuum pump and discharged outside the building. Therefore, vacuum systems are well adapted to picking up grain at multiple points.

In compressed air systems, the airflow is provided by a compressor so grain must be introduced into an airflow against a positive pressure. This is best achieved using a rotating valve (Fig. 8.10). Such a system is best suited to picking up grain at one or only a few points. The grain–air stream may be directed along one of several different ducts to a discharge point where the grain is separated from the airstream in an expansion chamber, with a hopper, as in the suction system. It can be released via a tipping seal or rotating valve. The dirty air is moved into a dust-collection system then is discharged. More rarely the grain outlet is a simple nozzle that is used as a thrower. Leakage, allowing the spread of dust, can occur from faults or badly made seals in compression systems.

8.4 Mixing and blending

It is sometimes necessary to blend different samples of barley or, more usually, different batches of malt. Typically the malts are made from one variety of barley but differ marginally in their properties, so that a specification can be met. This means that grain or malt must be conveyed from different silos and be uniformly mixed in pre-determined proportions. This

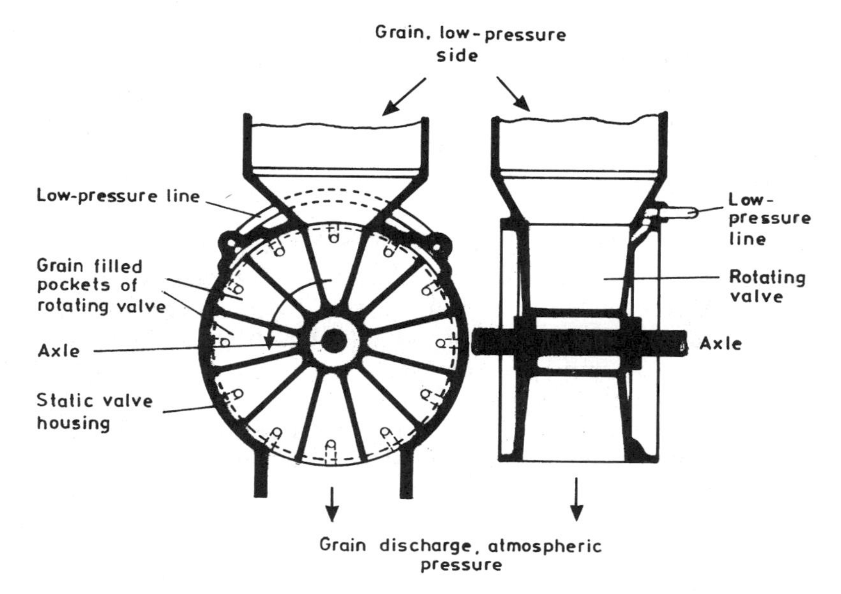

Fig. 8.10 A rotary valve used for separating grain from a conveyor working by suction. Grain falls into the pockets in the rotating valve (mounted in a static housing) and falls out when the pocket is inverted over the discharge outlet. Notice that the returning, air-filled pockets are pressure equilibrated with the suction-line before they reach a position to receive another sample of grain.

requires that the grain is delivered at different, controlled rates from the different silos to a point where they are mixed, and the mixture is collected in a garner or bin.

8.5 Weighing

Grain is repeatedly weighed during malting, at intake, into the drier, into and out of store and into steep. Similarly, malt is weighed at various stages through to dispatch.

Ideally at intake, lorries are weighed on arrival and departure and the weighbridge will commonly have to be able to weigh over 32 t in lorries 11 m (36 ft) in length. Without a weighbridge a maltster has to accept a 'certified load' document.

Within the maltings, grain may be weighed continuously on moving belts, which integrate the weight of grain moved with time and can weigh grain being moved at up to 30 t/h. These devices are said to weigh to an accuracy of within 1%. Alternatively the grain may be weighed discontinuously in 'fill and discharge' type weighers, which work well with load cells in modern equipment. The accuracy of older weighers was suspect; however, in the

UK, to meet government standards, accuracy of modern devices is claimed to be within 0.25% or even 0.1%. Hopper-type weighers, which may operate with one hopper or with two hoppers working alternately, are filled to a predetermined weight, then the grain flow is cut off and the hopper is tipped or otherwise discharged. Weight is determined by the number of discharges that occur. Automatic weighers have a wide range of capacities, for example from 1.5 to 165 t/h.

The weight of grain in sacks may be estimated approximately by weighing about 10% of the sacks, working out the average weight of one sack and hence the weight of all the sacks (average weight × number). During 'stock taking', maltsters may estimate the volumes of barley and malt they are holding in stores (an imprecise exercise), and from a knowledge of the bulk densities will estimate the weights in stock. In practice, maltsters rarely find that their grain stock figures balance precisely. This is to be expected from inevitable inaccuracies of weighing, random losses, variations in grain moisture contents, and so on.

8.6 Grain pre-cleaning, cleaning and grading

8.6.1 General considerations

Grain must be cleaned, not only to make it fit to enter the malting process but also to remove dirt and materials that would impair its keeping qualities, damage the grain-handling machinery and reduce the efficiency of driers and grading machines. However, there is a dilemma in that grain arriving at a maltings must be taken in as quickly as possible and, in many instances, it must be dried before transfer to suitable stores, but grain-cleaning machinery only works well when working at moderate rates. To overcome this problem, it is usual to avoid slowing the rate of intake by only roughly and rapidly pre-cleaning the grain on arrival, before it is dried. It is cleaned slowly and thoroughly later when the opportunity arises. At this later stage, the grain may also be graded (that is separated into size classes) and grain-sized contaminants are removed. So in principle pre-cleaning and cleaning are parts of one process; indeed the same machines may be used but working at different rates, the slower rates of working being more efficient. In practice, the two cleaning stages are separated in time and in between the grain is normally dried. At present, in the UK, the grain reaching maltings is usually relatively uncontaminated. This is because the farmers' use of selective herbicides has resulted in the near-elimination of previously troublesome weeds, and because of the care taken to prevent contamination with other grains of foreign materials. In addition, farmers and grain merchants clean grain before offering it to maltsters. The consequence is that many British maltings now use much simpler grain-cleaning processes than were once usual. Furthermore, for simplicity and to avoid unnecessary costs, grain is now graded into fewer width classes than was once the case. However, in many parts of the world, extensive cleaning is still necessary, and grain may be divided into several size classes, which are malted separately. Before and after cleaning treatments, grain is weighed to check what losses have been incurred. All the handling and grading processes generate dust, which is aspirated (i.e. carried away in a current of air) and trapped and collected centrally. The dust and light impurities may be collected from particular machines or may be collected from aspirators through which the grain flows as it is transported between locations.

Pre-cleaning may involve the removal of very large impurities on the rubble separator at intake, the removal of magnetic iron and steel impurities, the removal of ear and straw fragments and breaking awns from very lightly threshed grains, the removal of very coarse and fine impurities by sieving, the removal of dust by aspiration and (rarely) 'polishing'.

8.6.2 Magnetic separators

Magnetic separators may be used at various stages in handling barley or malt. They remove fragments of iron or steel (tramp iron) that has entered the grain stream. This may include nails, wire, nuts, bolts or fragments of metal from machinery. Besides damaging machinery, tramp iron may strike sparks, with consequent fire and explosion risks. If it is present in malt it may damage the customer's handling machinery and mills. Magnetic separators may first be applied to the grain before or after pre-cleaning. Two general types of magnetic separator are in use.

In the simplest case, the grain is spread into a thin layer by an adjustable slide, slides down an incline of 30–45° made of non-magnetic material and passes over a strong, permanent magnet (see Fig. 8.12, below). Magnetic materials cling to the magnet and must be removed by regular cleaning. Clearly there is a risk that the grain stream may dislodge metallic fragments as they accumulate.

In the other type of magnetic separator, the grain stream falls onto a rotating magnetizable cylinder or moving magnetizable belt, each moving in the direction of the stream of grain and magnetized with a strong electromagnet. The grain stream falls away, but magnetic objects are carried round out of the stream (Fig. 8.11). The cylinder or belt then becomes de-magnetized and the metal materials fall away and are collected separately. Frequently the casing enclosing a magnetic separator is aspirated to remove dust.

8.6.3 De-awners and grain polishers

If a batch of grain has been inadequately threshed, so that intact awns and sections of intact ears are present, it may be necessary to direct it through

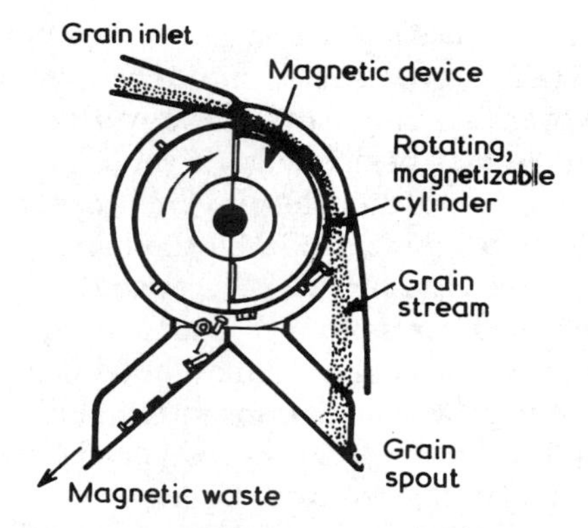

Fig. 8.11 A rotating, electromagnetic cylinder for removing iron and steel impurities (tramp iron) from the grain stream (after Vermeylen, 1962).

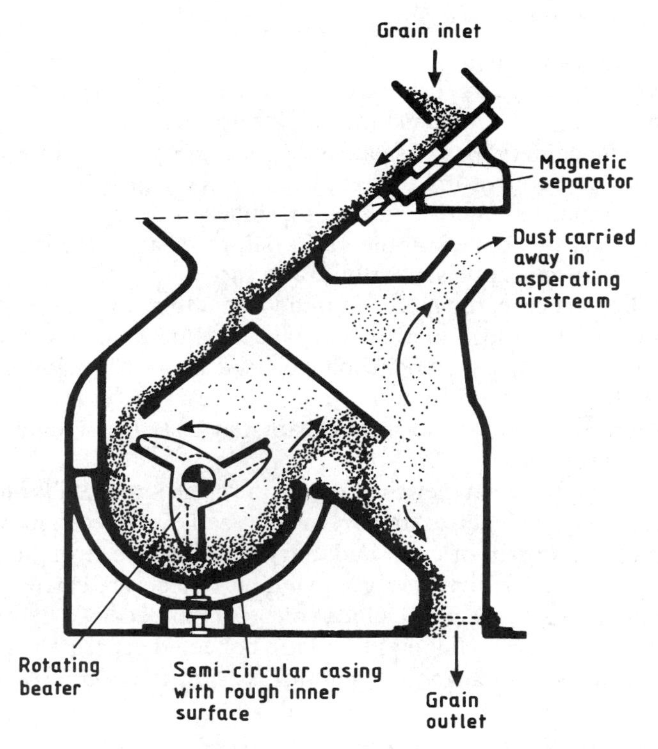

Fig. 8.12 A de-awner. This device beats grain, stirring it against a rough casing and so breaking up awns and separating grain from adherent pieces of ear. The dust created is removed by aspiration. This particular device also incorporates a magnetic separator. These machines also 'scour' grain and remove adhering dirt.

a de-awner so that the treated grain can be successfully processed. The grain passes through a semicylindrical or cylindrical chamber with a rough inner surface, in which a series of rotating paddles or beaters rub the grains together and fling them against resistant surfaces before they fall away to a conveyor (Fig. 8.12). This vigorous mechanical treatment really completes the threshing process. Clearly extra and potentially damaging physical treatments are avoided where possible and so such machines are only used when absolutely necessary.

Occasionally machines have been used to 'scour' or polish barley. This was comparatively common in the past but is probably not used at all in modern maltings in the UK. In a common design, the grain moves down a rotating, perforated metal or woven wire inclined cylinder where it is agitated and rubbed together by centrally mounted, rotating paddles or beaters. The grain is scoured and dust and dirt are dislodged. Fine dirt falls through the walls of the cylinder and is collected. Dust is removed by aspiration, which may be applied both to the grain entering the cylinder at the upper end and to that leaving at the lower end.

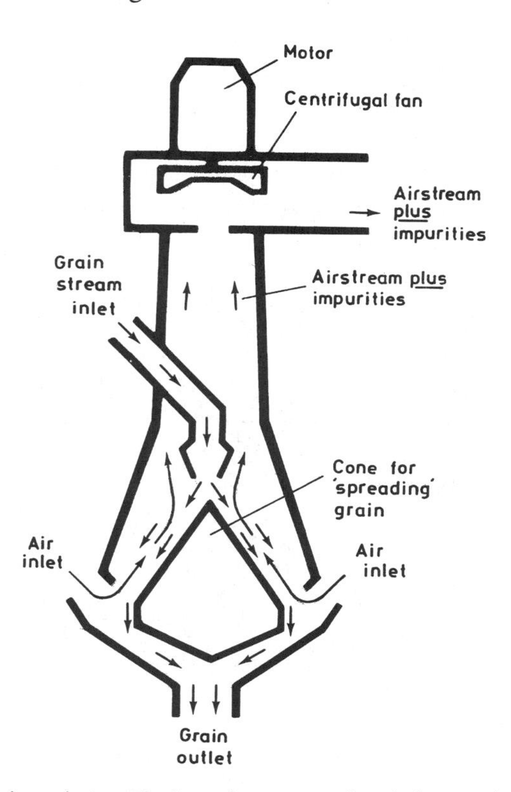

Fig. 8.13 A simple grain aspirator. The incoming stream of grain is spread over an inverted cone and meets a counter-flow of air that removes liftings and dust.

Malt polishers are much more gentle machines, which may be used after deculming. In these, the malt used to be gently brushed. Such machines are rarely or never used in the UK.

8.6.4 Aspirators

Many pieces of grain cleaning and handling machinery have built-in aspirators to remove dust. Aspiration may also be applied to moving grain or malt at many stages during handling. In principle, a thin stream of grain is allowed to fall and a current of air is fan-driven across the stream of grain, or better against it, at a rate that will carry away dust and light impurities (Fig. 8.13). In some grain-cleaning machinery, the light impurities, or 'liftings', are allowed to drop out of the airstream in an expansion chamber (cyclone) and the air is recirculated through an integral filter. In other cases the air, loaded with dust and other materials, is directed to the central dust collection point.

8.6.5 Pre-cleaning machinery

Numerous types of pre-cleaner are in use. Impurities removed at this stage include dust, sand, clods of earth, snail shells, stones, string, twigs, shrivelled or broken grains, loose husks, awns, fragments of leaves, straw, rachises and other plant debris. In a common type of simple machine, grain is metered, with or without pre-aspiration, onto the uppermost of a pair of inclined, flat

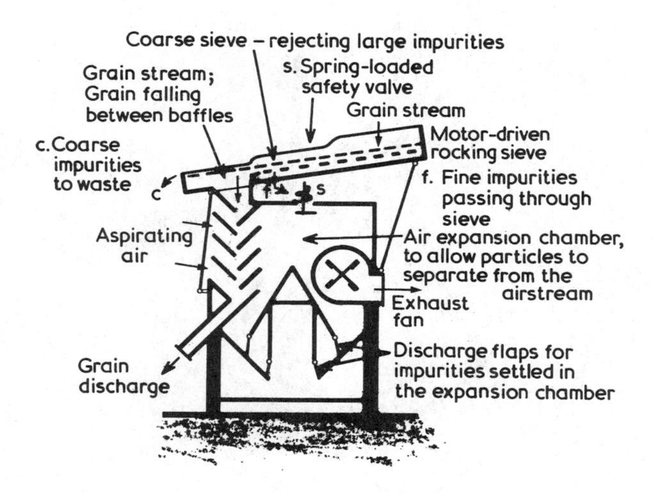

Fig. 8.14 A simple flat-bed grain cleaner. Grain, after passing through an upper sieve that retains coarse impurities, is retained on a lower sieve that passes fine impurities; it is aspirated as it then cascades down between baffle-plates (after de Clerck, 1952).

slotted sieves or screens which shake and oscillate. Frequently they are mounted in a wooden frame (Fig. 8.14). The sieves may also have a vibrating motion passed to them and be repeatedly brushed from below to keep the slots free of obstructions. The grain slides down and through the upper sieve, while coarse impurities pass over and are collected. The grain falls onto the second sieve and passes forward, while fine impurities, such as sand, fall through and are collected separately. If not aspirated on entry the grain is aspirated before it leaves the machine. To improve the machine's efficiency and prevent it being overloaded with coarse impurities, such as straw, the grain stream may first be pre-treated by passing it through a rapidly moving, woven-wire belt, a cylindrical screen or scalping reel to remove larger impurities such as snail shells, leaves and straw fragments. In some machines several sets of screens may be mounted in one casing. In some instances, the first removal of thin grains (screenings) is achieved with the sandy material (Fig. 8.15).

Other commonly used machines separate grain from impurities using rotating cylindrical screens or sieves (Fig. 8.16). The grain is fed in at an

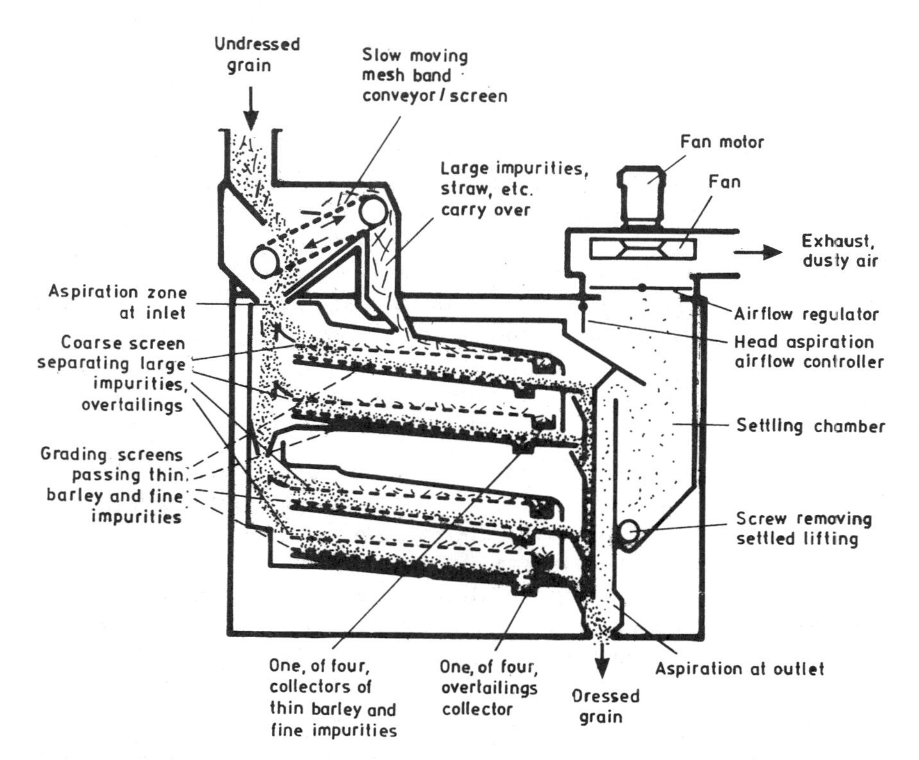

Fig. 8.15 A Kamas–Westrup type of grain cleaner, using four sets of screens simultaneously. The working rates of such machines may be 15–30 t/h on barley containing 10% screenings.

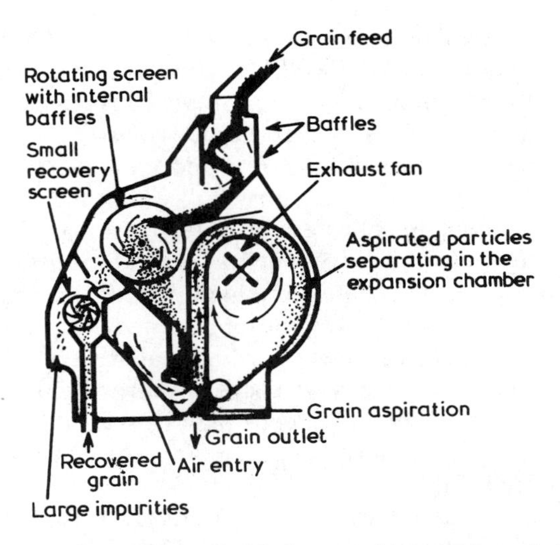

Fig. 8.16 A grain cleaner using rotating cylindrical screens (sieves). The grain falls through the first, coarse screen (squirrel cage, scalping reel) being diverted by internal baffles as it does so. Coarse impurities are carried over by the rotating scalping reel and the stream of impurities is treated with a second, smaller reel to recover entrained grain. As the cleaned grain leaves the machine it is aspirated to remove dust and liftings.

even rate and is spread to ensure that the grain-stream meets the full width of the screen. The screens may have curved internal baffles to divert the flow of materials passing through it. The grain falls into the first screen and out of the bottom while large impurities are carried over and fall onto a small recovery screen which separates any grain that is entrained and returns it to the grain stream, which is finally aspirated. Such machines are compact, have low power consumptions and do not vibrate.

8.7 Cleaning and grading

Cleaning and grading are carried out on dry grain as circumstances allow. Grain shrinks during drying and so for reproducible results grading needs to be carried out on grain dried to a particular moisture content. Cleaning and grading must be thorough, and so the machinery used must not be overloaded and, inevitably, compared to pre-cleaning the processes are comparatively slow. The more dirty, contaminated or damaged the grain is on arrival, the more complex the cleaning processes need to be.

8.7.1 Screening

In the past, maltsters often graded barley into different width classes, which were malted separately: for example 2.2–2.5 mm, 2.5–2.8 mm and above 2.8

mm. They rejected and still do reject the screenings, typically thin corns of less than 2.2–2.3 mm width or 2.5 mm in width for 'bold' or plump-grained varieties, depending on requirements and local conditions. The screenings are used for animal-feed or making enzyme-rich malt. In a bad season, as much as 20% of a barley lot will be screenings (grains less than < 2.2 mm width), but in the UK much of this will have been removed by the merchant before delivery. Usually there are maximum limits imposed in contracts that state the proportions of screenings that a lot of barley may contain.

Grains from a given lot, which differ in width, differ in their nitrogen content, water-uptake rates, dormancy and rate of germination, and the malts from the bolder grains usually have the greater hot water extracts (Briggs, 1978; Bathgate, 1989; Chapters 3, 4 and 14). While screenings are still separated and rejected for malting, for many purposes grading has been discarded and barley of a wider range of widths (all the barley above the reject size) is malted together. While laboratory screens are sized exactly (EBC, 1987) it is said that the slot sizes of some commercial screens may be so roughly controlled as to be only within ±0.1 mm of the stated width. While screening may be carried out on inclined, oscillating flat screens and these are often used for the removal of thin corn (Figs 8.14 and 8.15), other

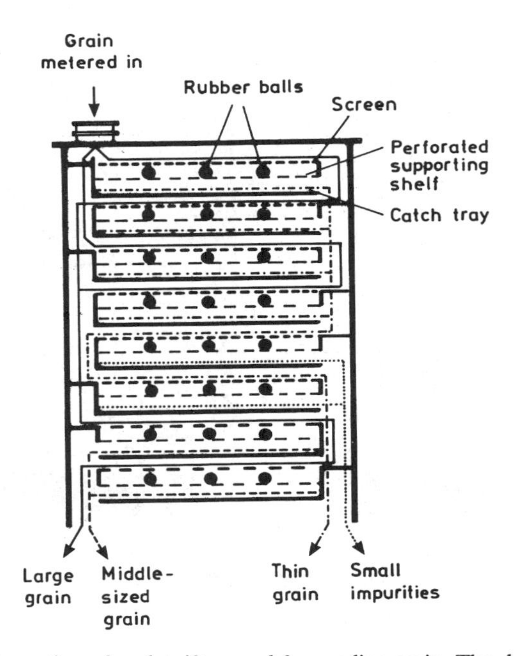

Fig. 8.17 A schematic section of a plansifter used for grading grain. The device consists of several sets of screens mounted in a frame, which all move together. The grain stream moves over the screens and is separated into size fractions that are collected separately (after de Clerck, 1952).

sorts of sieving machines are more efficient, particularly if the grain is to be graded into classes.

In plansifters, several (for example eight) sets of sieves all move together, mounted above each other in a frame that has a horizontal rotating motion (Fig. 8.17). Beneath each screen is a set of rubber balls supported on a perforated metal sheet and below this there is the metal catch tray. The balls move and bounce about and vibrate, striking the underneath of the screen and dislodging wedged grains and imparting a vibrating motion. The screens are divided into rectangular slotted areas, the slots in adjacent areas being arranged at right angles. Depending on the screens used and the routes followed by the grain, the grain may be divided into various classes, for example three grades of grain and fine or coarse impurities, or two grades of grain for malting and thin grains with small impurities. Sometimes a supporting frame is arranged to carry two sets of sieves, powered by an auto-balanced drive. Working rates of 10 t/h are quoted.

In another type of machine, grain is fed onto horizontally mounted annular screens that oscillate to and fro around a central axis (Fig. 8.18).

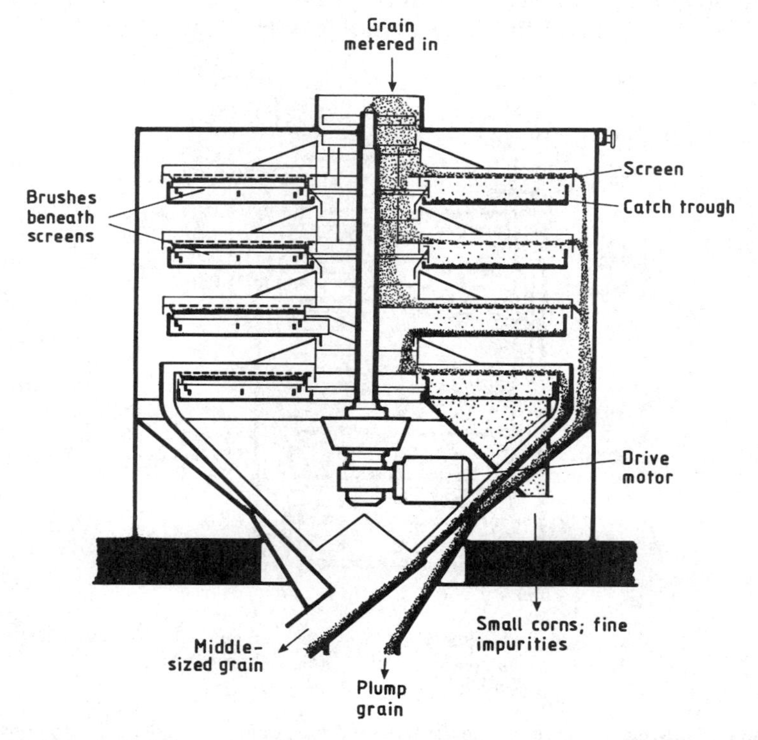

Fig. 8.18 A vertical section of a multiple-sieve grader. The screens are annular and oscillate around the centre axis. At intervals, rotating brushes move beneath the screens and dislodge any grains blocking the slots.

The larger grain makes its way, in a series of zig-zags, from the centre to the periphery, when it 'tails over'. The finer material that passes through the screen is caught on a tray below and rotating blades sweep it to a receiving slot and a finer screen. The edges of the rotating blades carry brushes that sweep the underneath of the screens and help to keep them unblocked. Grain can be processed at 12 t/h.

Recently large, rapidly rotating inclined cylindrical screens have been introduced that process grain efficiently at 50 t/h. These will reduce the screenings level of grain to less than 1%.

8.7.2 Cylinder and disc separators

The cleaning devices discussed so far separate grains and impurities by aspiration or by width. Various other machines may be used to separate broken (half) corns and other impurities by length.

In Trieur cylinders (half-corn cylinders, cylinder separators) grain is slowly fed into the upper end of a rotating, inclined metal cylinder, the inner face of which is covered with indentations or pockets, perhaps

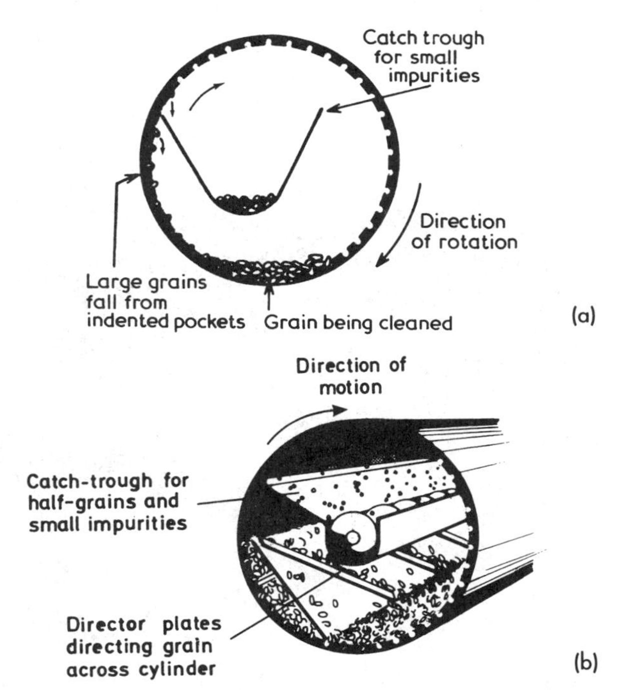

Fig. 8.19 (a) Schematic cross-section of a simple Trieur cylinder; (b) a more complex cylinder in which the 'direction plate' returns grain to the far side of the cylinder, so a greater proportion of the indented surface is in contact with the grain and is in use.

23 000–30 000/m^2. The pockets are designed to lift materials of particular shapes. Often stirrer devices are mounted within the cylinder to ensure the maximum area of contact between the grain and the inner, indented cylinder surface. In other machines internal 'plates' direct the falling grain across the cylinder (Fig. 8.19). Grains, half-grains and small objects lodge in the indentations and are lifted out of the grain stream. Large grains fall out, but smaller objects are lifted higher and when they fall they are caught in a trough suspended along the length of the cylinder. From this they are removed by a screw (Fig. 8.19). Such machines work slowly, processing only 0.2–1.2 t/h, and they must not be overloaded. Many attempts have been made to refine their designs to speed their rate of working and keep a larger proportion of the grain stream in contact with the working surface. For example, rotating vanes may be used to spread the grain over a larger part of the cylinder's inner face, or 'slides' of metal catch the grain falling from the cylinder and direct it back to the far side. Often only 20–25% of the working surface is in contact with the grain at any one time.

In high-speed cylinder separators, the cylinder rotates fast enough to generate sufficient centrifugal force to hold small impurities and half-grains in the indented pockets for a longer period, so they fall out later. Such cylinders can process 1.2–3 t/h. The materials rejected by the cylinders are re-

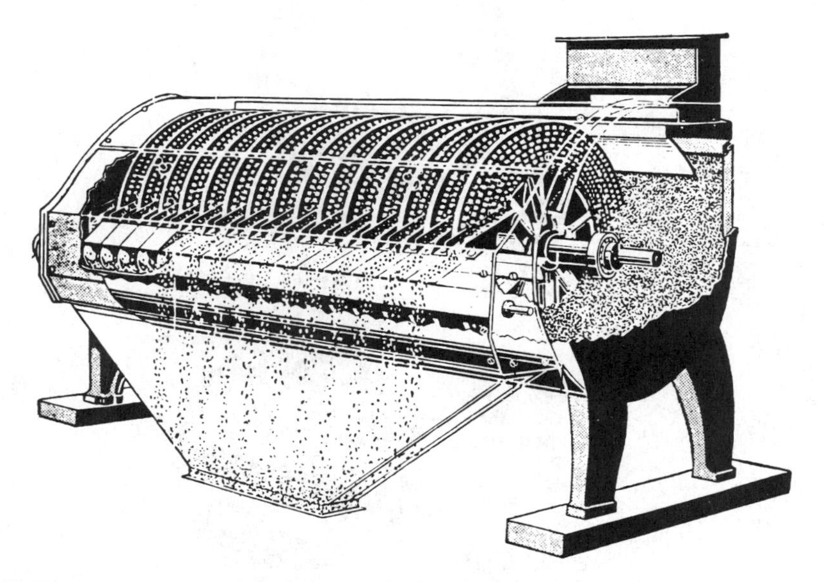

Fig. 8.20 Diagram of a Simon disc separator. The discs (Fig. 8.21) dip into the trough of moving grain, the half-corns and small impurities catch in the indentations, are lifted out and fall, being caught and directed into a separate collection stream. (Courtesy of Satake UK, Ltd.)

treated in a smaller 'corn-recovery cylinder' to recover any whole grains that have accidentally been rejected in the first treatment.

Disc separators, sometimes called Carter, Simon or Carter–Simon disc separators, are more efficient than cylinder separators because the working surface area in contact with a given weight of grain is large. They vary widely in details of construction. They must be used at moderate working speeds to prevent grain damage. In these machines a series of discs (Fig. 8.20), each indented on both surfaces and mounted on a rotating shaft, dip into a trough of grain that is slowly fed in at one end. The discs actually have an open centre, crossed by vanes, which help to impel the grain from the loading end along the trough to the unloading end. The working parts are entirely closed to contain the dust, and the machine may be aspirated. The disc types must be chosen to have indentations designed to achieve the separation being attempted. The discs are only a few inches apart in the vertical plane. As they rotate, they lift out small impurities and drop them into a catch trough (Fig. 8.20 and 8.21). The stream of impurities is later re-treated to recover whole grains. The treated grain is withdrawn from the other end of the trough. Sometimes these machines contain 're-treatment' sections, built-in screens or re-treatment cylinders. Many sizes are available and they may

Fig. 8.21 A disc of a Simon disc separator working in a batch of grain (after Lockwood, 1945; courtesy of Satake UK, Ltd.)

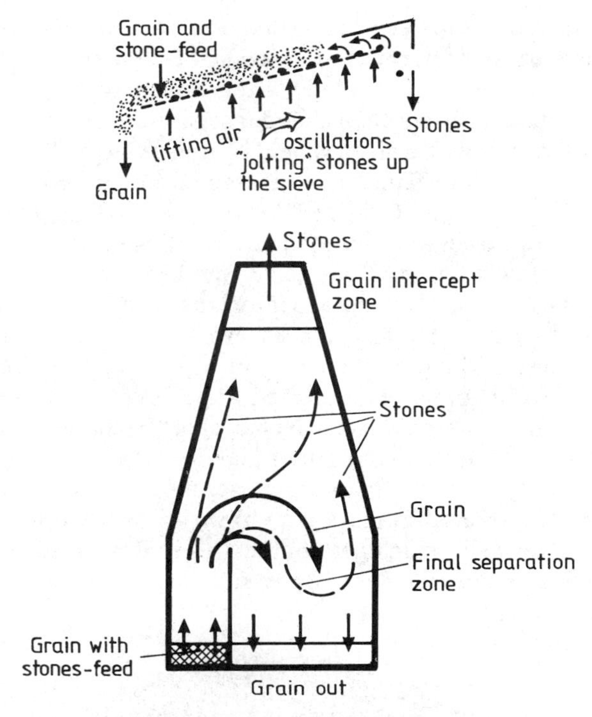

Fig. 8.22 The working principle of a table separator or destoner. (a) Section through the 'jolting' sieve or table. (b) The table in plan view. A metered flow of grain is supplied to the table by a vibrating head.

process grain at the rate of 0.7–8.5 t/h. At present in the UK, the quality of barley reaching maltings is so high that often it is unnecessary to treat the grains with cylinder or disc separators.

8.7.3 Pneumatic cleaners

At many stages, grain is aspirated to remove dust and 'liftings' (light impurities). Some devices now use carefully regulated airflows directed across a falling stream of grain to separate heavier grains from lighter materials, which are carried further sideways by the airstream and are collected separately. Such devices may be used to separate wild oats from barley grain. Alternatively, such machines may be used to separate stones from grain, the stones being heavier and less readily deflected by the air-blast than the grain.

8.7.4 Destoning

Stones of about the same dimensions as barley must be separated on the basis of their greater density. This may be achieved pneumatically, or in presteep washing devices (Chapter 9). Alternatively, the stones may be separated from barley (or finished malt) using a table separator. In one pattern, the grain is fed onto one side at the bottom of an inclined, perforated table covered with a mesh screen (Fig. 8.22). The table has a perforated surface through which an upward current of air is passed, and it is given an upward/forward oscillating, jolting motion by an eccentric drive. The lighter grain, or malt, is lifted by the airflow and slips back and down the incline table and falls over the lower edge to be collected. The stones are too dense to be lifted by the airflow, but they are thrown upwards and forwards up the table, until eventually they are propelled over the upper edge and are collected separately. A stronger airflow in the discharge zone prevents any grain being carried over with the stones. Properly harvested and handled grain should be virtually stone-free, and so de-stoning is frequently not needed. However, in some instances, the quantities of stones mixed with the barley is surprisingly large. It is essential to remove stones

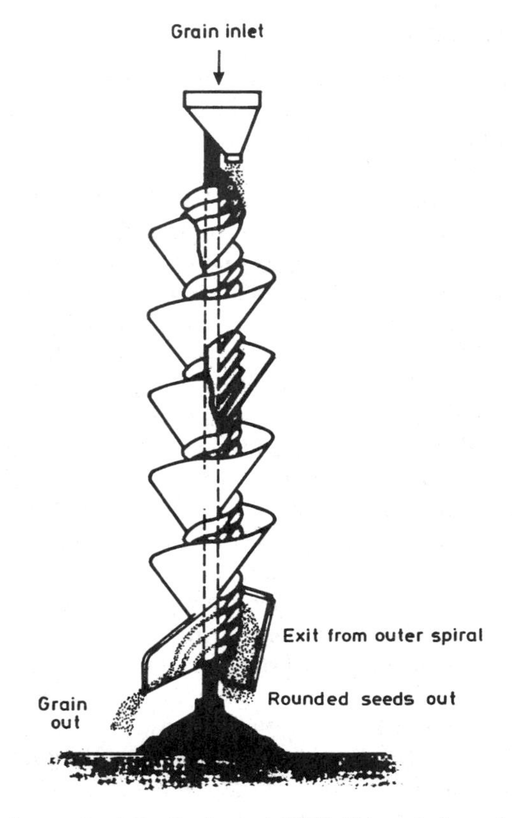

Fig. 8.23 A spiral seed separator (after Lockwood, 1945). The grain flows down the separator and small, rounded impurities move to the outside of the spiral and are collected separately.

before malt reaches the user's milling machinery. Some brewers routinely destone malt before it is milled.

8.7.5 Spiral separators

These devices are rarely, or never, used in the UK at present, but in the past the thin grain fractions with their impurities were fractionated to improve their market value, for instance for animal-feed; elsewhere spiral separators may still be used (Fig. 8.23). These devices were quite simple or could be complex (Lockwood, 1945; de Clerck, 1952). A thin stream of the material to be separated was fed to the top of a helicoidal slide. Its position in the sliding mass was controlled by its shape, density and friction. At the bottom, the different fractions (two or more) were collected separately, by dividing the stream of flowing material; small rounded seeds reached the outer edges of the spiral while wheat and barley grains held to a central position.

8.7.6 Dust and light impurities

Dust is generated whenever grain or malt is processed. It is a nuisance and a health hazard. It can cause irritation, allergic responses and 'malt worker's lung' because of its fungal content. It can explode when mixed with air and when settled it becomes damp and provides a favourable place for insects and microbes to breed. It is important, therefore, that dust is contained and trapped. In the UK, the maximum tolerated limit (threshold limit value, TLV) is 10 mg/m^3. To avoid explosions, sparks and flames must be prevented. Smoking is banned and all metal appliances are grounded to prevent the build-up of static electricity and the risk of sparks. The collected dust and liftings (small fragments of grain) can be bagged for sale or be mixed with culms and pelleted for animal-feed. As noted, dust and light impurities can be removed from moving grain or malt by aspiration with a current of air, and this may occur in the cleaning or handling machinery or when the grain or malt is flowing. In some machines, the aspirating air is ducted into an expansion chamber where its speed falls and the light liftings fall and settle in the bottom from which they can be removed by a screw conveyor, or by withdrawal through a rotating valve or double tipping valves. In some machines, the air is recirculated to contain the dust. In general the dusty air, with or without liftings, is fan driven through ducting to a central dust-collection point where the suspended solids are separated before the cleaned air is discharged. As with the grain-handling machinery, the dust–air ducting must have explosion-relief panels that, in the event of an explosion, will vent, relieving the pressure and directing the explosion away from working areas and towards the exterior of the building. Shock waves of explosions may be used to trigger the release of powder or gaseous explosion-suppressing agents into the ducting.

Various devices have been used to collect dust. In some older plants, the dusty air was allowed to pass slowly through one, or more, settling rooms and finally the air was washed by water sprays or by passage through water or wet cake filters before it was discharged. At present dust and liftings that are not trapped in cleaning machines are usually collected using cyclones and textile filters. In cyclones, dusty air with entrained solids is discharged tangentially into a cylindrical sector vessel joined to a tapering cone (Fig. 8.24). The air spirals round and the entrained particles are thrown outwards, where they strike the sides and are slowed by friction; the heavier particles slide downwards to the bottom, where they settle. The gyrating air then rises up the centre and leaves at the top carrying only fine dust. Cyclones are made in many sizes and can process, for example, 100–650 m^3 air/min. The air leaving cyclones is never completely dust free and is finally cleaned by filtration. Sometimes the air is filtered by passage through textile sleeves, under pressure, and the dust on the sleeves is periodically dislodged into a lower chamber by an automatic scraping device or by a reverse blast of air. However, usually the fan that moves the dust-laden air works by suction and so the air is sucked via a cyclone and a filter, through the fan, and to waste.

A simple suction filter is shown in Fig. 8.25. The dusty air enters an expansion/dust-collection chamber and is drawn upwards and out through cloth or textile sleeves, which act as filters. These sleeves are closed at the upper ends and are suspended in a way that permits them to be shaken

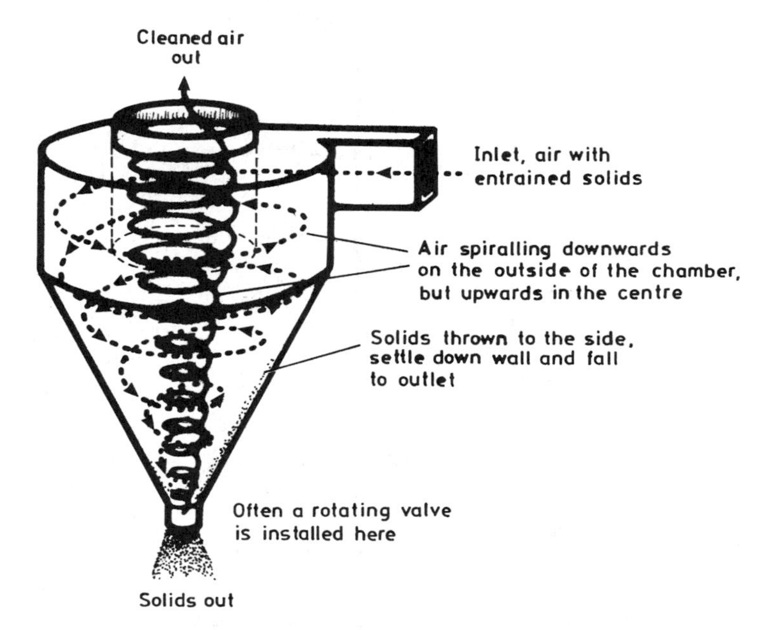

Fig. 8.24 The separation of solids from an airstream in a cyclone.

mechanically to dislodge layers of dust, which then fall back into the collection chamber. In modern machines the sleeves may be not merely circular or rectangular but may be flattened in cross-section so that they can be packed closely together and provide a large filtration area for the volume of air being filtered. In addition to mechanical shaking, it may be possible to apply a reverse flow blast of air that moves the filter cloth and dislodges the caked dust from the sleeves. The sleeves, which may be very numerous, are enclosed in an airtight chamber with an outlet that leads to the suction fan. If well maintained and the sleeves are regularly cleaned and are sufficiently fine, such machines can be very efficient. Even so the air from the extraction fan should be discharged outside the building and well away from any sensitive sites since it is not sterile. If the maltings is situated on a brewing site, the air must be discharged well away from fermenting and beer-handling areas. Ducting is earthed (grounded) to discharge static electricity and minimize the risk of ignition. Ducts are also designed with weak areas to allow explosions to vent.

Collected dust and liftings may be held ready for disposal in special holding bins above bagging points or above pelletizers.

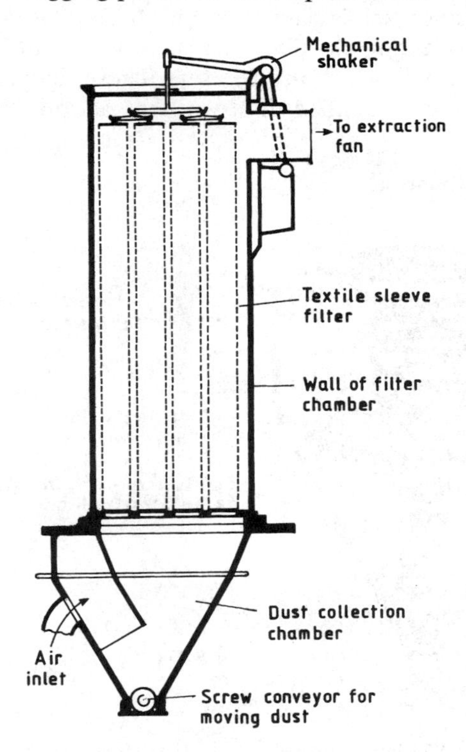

Fig. 8.25 A simple textile sleeve air filter. After a period of use the sleeves are agitated and the dislodged dust falls into the collection chamber (various sources).

8.8 Grain drying

For malting purposes, to maintain quality, grain should be adequately dry (e.g. 12% moisture or less) if it is to be stored for more than a few weeks with little or no ventilation (Chapter 7). If the grain is to be stored warm for a period to overcome dormancy, it may need to be dried to unusually low moisture levels (e.g. to 8–9% moisture if stored at 38–40 °C (100–104 °F) for, say, 3 months in extreme cases). Grain may be dried in a number of ways. However it is done, it is essential to ensure that the process is sufficiently rapid to avoid deterioration and it is carried out with care to avoid heat damage, which is very detrimental but not easily detected. In wet climates, it is essential that the capacity of the grain-drying equipment will not limit intake or, because of delays in drying, allow deterioration in stored wet grain, even in rainy seasons (Chapter 7). The physics of grain-drying are considered, with malt kilning, in Chapter 10.

Experimentally grain has been dried with air pre-dried by passage over silica gel. However, in practice most drying is carried out with heated air. The air may be heated directly and the products of combustion passed through the grain, or indirectly by passage over radiators. The process of grain drying should be carefully monitored. The instruments needed vary a little with the drying system being used but will include thermometers (sometimes recording thermometers), a grain moisture meter (carefully calibrated and regularly checked in the laboratory), a hygrometer to measure the relative humidity of air, and an anemometer to check airflows.

Some grain drying may take place on the farm, or in cooperative or merchants' stores. While it is best that the grain should be dried as quickly as possible after harvest, the risk of heat damage caused by poorly designed or improperly used drying facilities means that some maltsters prefer to dry grain themselves, and they will not purchase farm-dried grain. For many years in the UK maltsters and governmental agencies have supplied information to farmers explaining the need for care when drying grain, and how this can be carried out safely. Farm driers vary greatly in sophistication (Culpin, 1969; Briggs, 1978; MAFF, 1983). In the simplest cases, sacks of wet grain (open weave; 55–65 kg; about 120–140 lb) are placed on bars in gaps on the top of a hollow brick or concrete platform below which warm air (14–17 °C (*ca.* 25–31 °F) above ambient) is blown. The sacks remain until the weight is reduced to some pre-chosen value (e.g. to 50 kg (approximately 1 cwt)). Such devices are labour intensive and do not treat grain evenly. In more sophisticated driers, batches of grain are held in trays with perforated bases, or between perforated walls, and warm air is forced through until the grain is sufficiently dry. Then the drying compartment is emptied and refilled with the next batch of damp grain. Such devices may be semiautomatically or manually controlled. In drier areas, where the ambient relative humidity is sufficiently low for extended periods, grain may be dried in bulk. In principle, dry air is blown through the grain in store and it removes moisture until it comes into equilibrium with the grain. As time passes a drying zone moves into the grain away from the air inlets. The conditions (airflow rate, depth of grain, etc.) must be such that condensation does not occur in the layers of grain remote from the air inlets. Bulk drying may be carried out in flat-bed stores or in bins. In bins, the drying air may be forced into the base through a partly perforated base or perforated ducts and vented at the top. Alternatively the grain may be held between a perforated central metal duct and a perforated wall so the drying air moves radially from the central duct through the grain. Alternatively, the air may move through the grain between a series of intermixed inlet and exhaust ducts. Bin drying is slow and is not suitable for reducing the moisture content of the grain by more than about 3%, so the grain should not have an initial moisture content of more than *ca.* 17%. The drying air is usually warmed to 5–10 °C (9–18 °F) above ambient. The average moisture content of the grain is reduced at the rate of about 0.5% in 24 h. Because the process is slow and there is a risk of deterioration, the depth of grain to be dried should never exceed about 3 m (*ca.* 10 ft). In some areas attempts have been made to reduce heating costs by drawing air into drying bins after it has been pre-warmed by contact with black-painted surfaces heated by the sun. From the maltster's point of view, bin-dried grain is not dried sufficiently (to 12% moisture or less) and is unevenly treated; for example, the grain near the air inlet may be dried more than is necessary while that near the outlet is still damp and deteriorating.

Maltsters used to dry grain in special kilns, in metal drums (of many different types) or, experimentally, with air dried with silica gel. At present grain is dried in batches, for example in various types of germination–kilning vessels (Chapter 9), or in various types of continuous or semicontinuous drier.

In old-fashioned barley convection-drying kilns, with slow and irregular airflows and, therefore, little evaporative cooling, air-on temperatures of 35 °C (95 °F) rising to 45 °C (113 °F) as the grain dried were normal with grain bed depths of 10–20 cm (approximately 4–8 in). Drying might take one to several days, depending on air ambient conditions. In more modern batch-drying facilities (Chapter 9), with forced airflows, it is recommended that air-on temperatures do not exceed 43 °C (109.4 °F) for grain with a moisture content of more than 24%, while for grain of less than 24% moisture the air-on temperature should be less than 49 °C (120.2 °F). Because of the evaporative cooling that occurs, the grain temperature will be substantially below these values. The process may last 8–12 h. Typically a modern kiln used for drying barley is loaded to about one and a half times its normal malt loadweight (as original barley).

Very commonly, continuous, flow-through tower driers are used: often arranged to remove about 6% of moisture from grain in one pass. Thus the grain may enter at 18% moisture and leave at 12% moisture, with a transit time of 1–2 h. The moisture content of the grain may be metered automatically at the outlet and if it rises the rate of passage through the tower may be reduced to allow time for the grain to dry more. Drying capacities of 20 t/h are common, and much larger driers are also in use. It is possible to dry grain faster but this is not desirable since cracking and stress damage must be avoided. In principle grain is metered into the top of the tower and is warmed, dried by the passage of warm air, then is cooled and discharged. Heat may be supplied directly by oil- or gas-fired burners or indirectly by radiators (Fig. 8.26). Air–air heat exchangers, commonly used to pre-warm incoming air in malt kilns and economize in fuel usage (Chapters 9 and 10), are not used. The temperatures used in driers vary. Because of the short times for which the grain is exposed to the hot air and the evaporative cooling that occurs, the temperatures used are higher than those employed in batch driers. Thus as grain moves down the tower the temperature of the air-on may increase from 54 °C (129.2 °F) to 66 °C (150.8 °F); the air-off has been cooled in evaporating the moisture from the grain and may be 43–52 °C (109.4–125.6 °F). The grain is then cooled (e.g. to 30 °C (86 °F) or less) by the passage of ambient air before being transferred to store. The

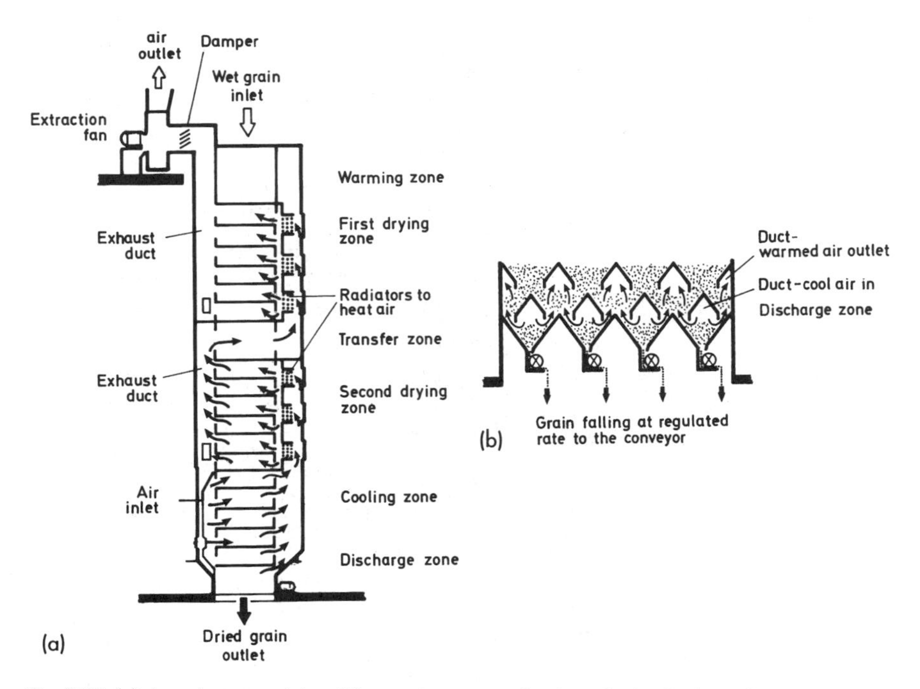

Fig. 8.26 (a) A grain tower drier; (b) a section across the ducts in the final, cooling zone showing the rotating valves that regulate the rate of grain withdrawal from the base of the tower (various sources).

warmed 'cooling air' is heated further and is used in an earlier, drying section. To speed the decline in dormancy, grain may be 'run to store' warm, leaving the drier at 30–35 °C (86–95 °F; 10–12% moisture). This grain is subsequently cooled during cleaning and grading, or by ventilating the store. The use of a special flow-through heat exchanger has also been proposed.

Tower driers come in various patterns. For example, the grain, metered in from above from a garner, may be pre-warmed by passing through radiators in a 'sweating section' before moving downwards, in columns, between louvred side walls that permit the passage of fan-driven, drying, or cooling air (Fig. 8.27). The rate at which grain is withdrawn from the base regulates the grain flow and hence the rate at which the drier works. In other types of tower driers, the incoming grain may be pre-warmed by radiators or by the air from the cooling section, which is warmed by contact with the outgoing grain (Fig. 8.26). The column of grain moves down the tower at a rate

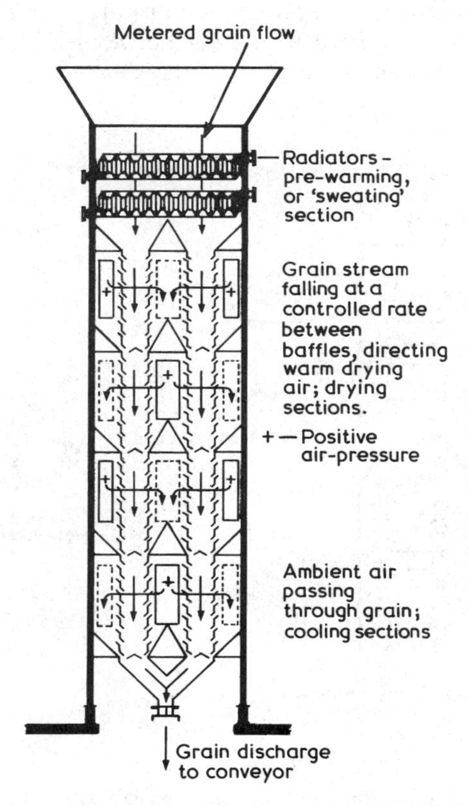

Fig. 8.27 Another pattern of grain drier in which the drying air moves from side to side across the grain column between louvres (after de Clerck, 1952).

regulated by sliding or other valves in the base. Throughout the column of grain there are open-bottomed ducts, of inverted 'V' or ∩-shaped cross-section, which alternately supply warm, dry air or carry away the cooler, damp air (Fig. 8.26). The outgoing, fan-driven airstream is often dusty and so is passed through a cyclone before being discharged from the building.

Other driers are of the inclined, flat-bed type. They come in various sizes and may have capacities of up to 60 t/h. For example, in one pattern of double drier, grain is drawn down over inclined, louvred decks, being moved by moving chains and slats (Fig. 8.28). Hot air is fan-driven up through the beds of grain. The grain is metered in at the upper end of the upper deck, moves down this deck where it is dried with cooler air, falls to the one below where it is dried with hotter air, then meets a cooling section and is eventually discharged. The rate of throughput is altered by altering the speed of the chain-and-slat conveyor. Such driers have been modified for use as malt kilns (Chapter 9).

Grain dried in a vacuum can tolerate higher temperatures than grain dried in air. Vacuum driers are complex and although they are not used in British maltings they are used in Europe (Vermeylen, 1962; Narziss, 1976). In these driers, the grain is metered into a low-pressure, 'vacuum' drying section through an airlock (Fig. 8.29). The grain, heated by radiators and heating jackets on the walls of the containing vessel, is progressively dried. It is then released through another airlock into a cooling section.

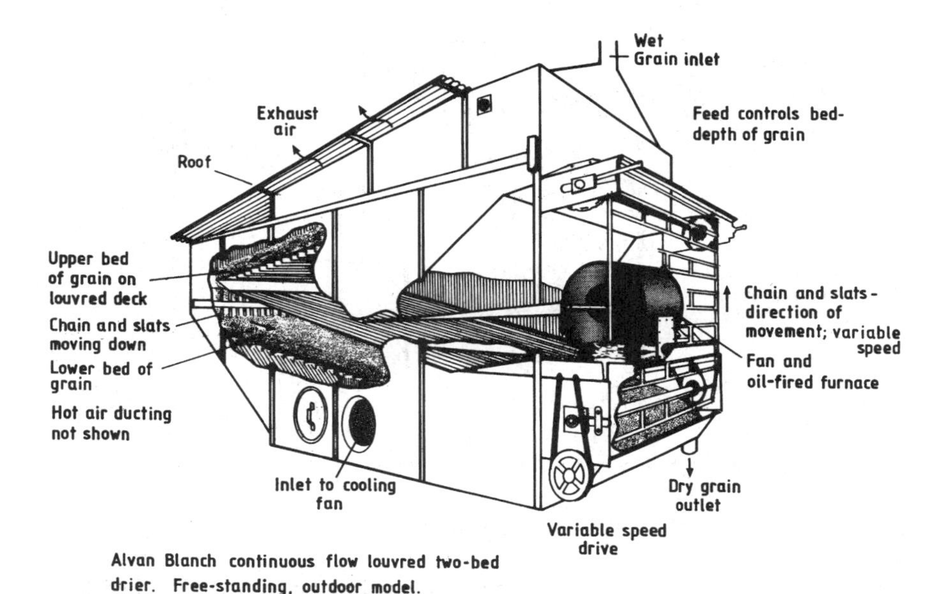

Fig. 8.28 An Alvan–Blanch grain drier. This is a free-standing outdoor model of the continuous flow louvred two-bed drier.

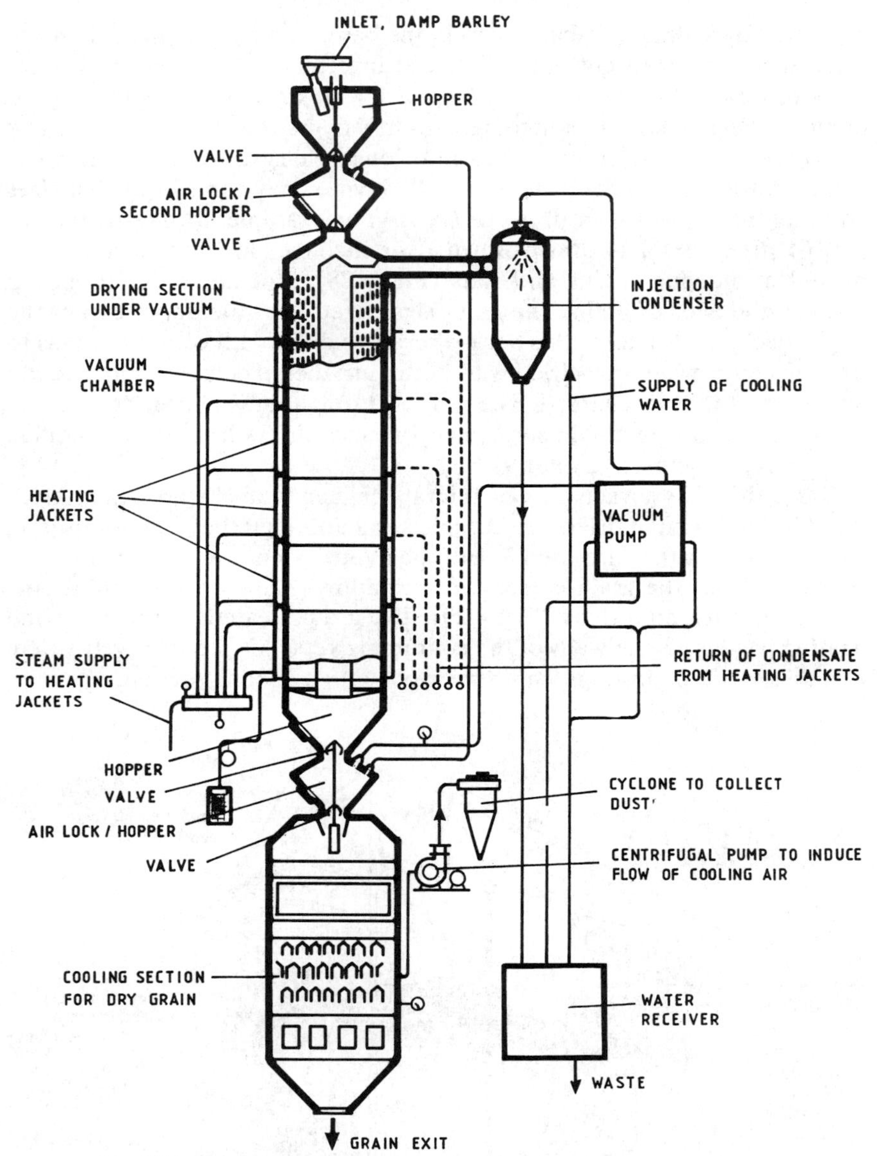

Fig. 8.29 Diagram of a Miag vacuum drier for grain (after Narziss, 1976).

8.9 Applications of insecticides and fumigants

Grain that is destined for long-term storage will need to be treated with an approved insecticide even if it is cool and comparatively dry. Insecticide may be sprayed onto the grain in small volumes of liquid or it may be applied as a powder as it is being conveyed. Application rates are carefully controlled to ensure that the insecticide is uniformly distributed and to ensure that specified residue levels are not exceeded.

Pellets, or packets, that release insecticidal phosgene gas may be mixed with grain as it is loaded into store. In some (few) enclosed stores and bins, fumigants may be recirculated in a fan-driven airstream. Fumigation should only be carried out by fully trained personnel (Chapter 7).

8.10 Grain storage facilities

The method and effectiveness of grain storage has considerable impact on the quality of the finished product (Lancaster, 1908; de Clerck, 1952; Vermeylen, 1962; Narziss, 1976; Briggs, 1978; MAFF, 1983; Gibson, 1989). Usually pre-cleaned barley is stored green, or undried, for as short a time as possible in ventilated temporary stores. Then, after drying and cleaning and possibly grading, it may be held for longer periods in other stores. In the past, when grain was stored in sacks, it was easy to segregate different lots by variety, nitrogen content and quality. Now grain is stored in large amounts in bulk, it is inevitable that each stored bulk is made up of an admixture of different batches of barley. This presents problems: the least expensive stores hold large tonnages and segregation into different lots is inconvenient or impracticable.

The best (premium) malts are made from carefully graded grains of a single variety with a uniform nitrogen content. These reasonably homogeneous samples are comparatively small lots, which must be segregated in smaller but more numerous and more expensive storage units. In the UK there is often a compromise: 'standard malts' are made from bulk-stored grain of a currently common variety. All the grain going into this store must be of the chosen variety, have less than a stipulated nitrogen (protein) content and must meet or exceed various other quality criteria. Other barleys, for top-grade or specialist malts, will be segregated by variety, nitrogen range, viability and so on. Some maltsters hold virtually 100% of their grain requirements in store, while others hold very much less, perhaps 30–40% of their requirements.

Traditionally grain was stored in sacks (fumigated before use), in heaps in the uppermost floors of floor-maltings or in other dry, draught-free places. Rarely, barley stored loose to a depth of *ca.* 1.8 m (6ft) is still to be seen in old maltings. Dried to ca. 12% moisture or less, cooled and with cool winter temperatures, this method is usually successful. Such stores are often loaded with a grain thrower and may be unloaded with a board and winch, which is used to scrape the grain into a conveyor. The different grain batches may be divided with wooden partitions. Such stores are extremely laborious to run.

In modern stores grain handling is usually fully mechanized and is often controlled from a central control panel with a wall-mounted mimic display or computer representation of the store. With the older, more traditional system, the amount and type of grain, in each storage unit is shown on a display board (often a chalked blackboard) and is recorded in stock books. In computerized units the location and amounts of stocks can be recorded in the computer. The most comprehensive of these displays indicate the stock of grain in each unit (bin, silo, etc.), if it is being ventilated, loaded or unloaded, which valves are open and which shut, which conveyor is running, if the drier or cleaning machinery is running, the grain temperature(s), when the grain was last inspected, when the next inspection is due, and so on. Such systems also indicate when grain is being received, dried, cleaned, or is being transferred to the malt production facilities.

Using British terminology, grain storage units may be divided into silos, bins or flat-bed stores. They are designed to be weatherproof and damp-proof, to exclude birds and rodents and to be devoid of nooks or crannies that can harbour grain residues, dust and insects. When empty, they should be easy to clean and spray or fumigate. As far as possible wood and porous building materials are avoided. Metal and smooth concrete structures are preferred.

Bins and silos are not distinct types of structure. In the UK, bins are usually squat structures of metal, circular or (less commonly) square or rectangular in cross-section and free-standing. Capacities of 500–800 t are common, while capacities of 5000 t or more exist. Bins are normally arranged in rows and are served by common loading and unloading conveyors. They may be up to 20 m (*ca.* 66 ft) in diameter and 15–20 m (*ca.* 49–66 ft) to the eaves (Fig. 8.30). Silos are relatively taller and tend to be grouped in blocks (silo blocks, 'elevators' in North America) for mutual support (Fig. 8.31). Each unit may be hexagonal, square, rectangular or circular in cross-section. Steel silos can have capacities of up to *ca.* 3000 t, but capacities of 300–400 t are more common. Steel bins are cheaper to construct than reinforced concrete silos of equal capacity and, in the UK, these are nearly always built at present. With changes in the ambient temperature, steel transmits abrupt temperature changes to the outside of the grain. This is undesirable because this encourages convection currents and moisture migration within the grain mass (Chapter 7). Silos and some bins are hopper-bottomed to facilitate emptying. Usually the bottom angles are about 45° from the horizontal, but this angle may need to be increased if, for example, malt in culm is being stored, since this does not flow as freely as barley or dressed malt.

Both bins and silos are filled from above by weather-proof conveyors. Silos usually have hopper-bottoms, and so when the valves are opened they are self-emptying into a conveyor. Silos, which are relatively tall and narrow, are often not equipped to allow the grain to be ventilated, but

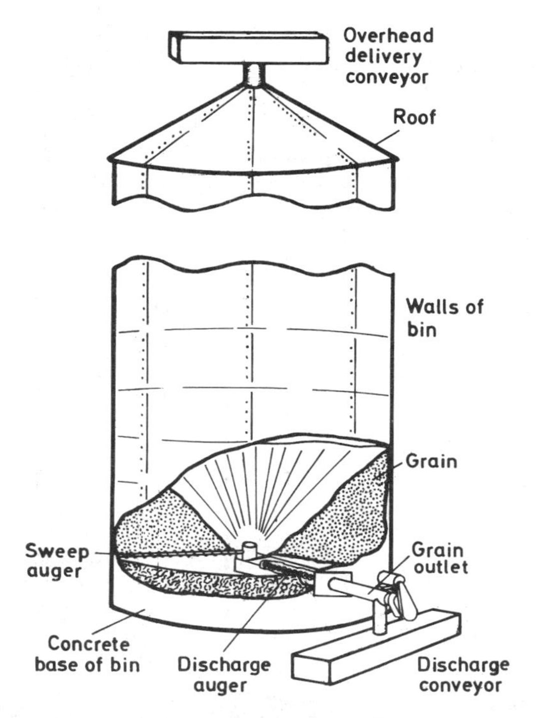

Fig. 8.30 Diagram of a metal bin for storing barley, loaded from above and emptied with the help of a rotating sweep auger. This particular bin has no aeration ducting.

special designs do permit the grain to be aerated: the air may flow between side-mounted ducts across the column of grain, or it may flow from the base to an extraction duct in the top. In some designs, the grain can only be ventilated by 'running it round', that is by withdrawing it from the bottom of the silo and either returning it to the top or discharging it into another silo. Bins with larger capacities are usually flat bottomed. The base may be solid, contain ventilation ducts or may be 'fully perforated' to permit the grain to be aerated. Bins holding green (undried) grain must often be ventilated. The noise of the fans can cause a nuisance to those nearby. Normally emptying flat-bottomed bins is by gravity into a central access-point to a conveyor (Fig. 8.30). When the grain has emptied as far as it can in this fashion a sweep auger is brought into operation. This screw auger moves around the central discharge point, moving the residual grain from the sides to the middle, where it falls into the discharge conveyor and so away. When grain is emptied under gravity, for example from a hopper-bottomed silo, it tends to funnel down the middle, so the centrally placed grain is discharged before that against the wall. A consequence is that the last grain loaded into a silo or bin is not necessarily the last out. Consequently it is not possible, even

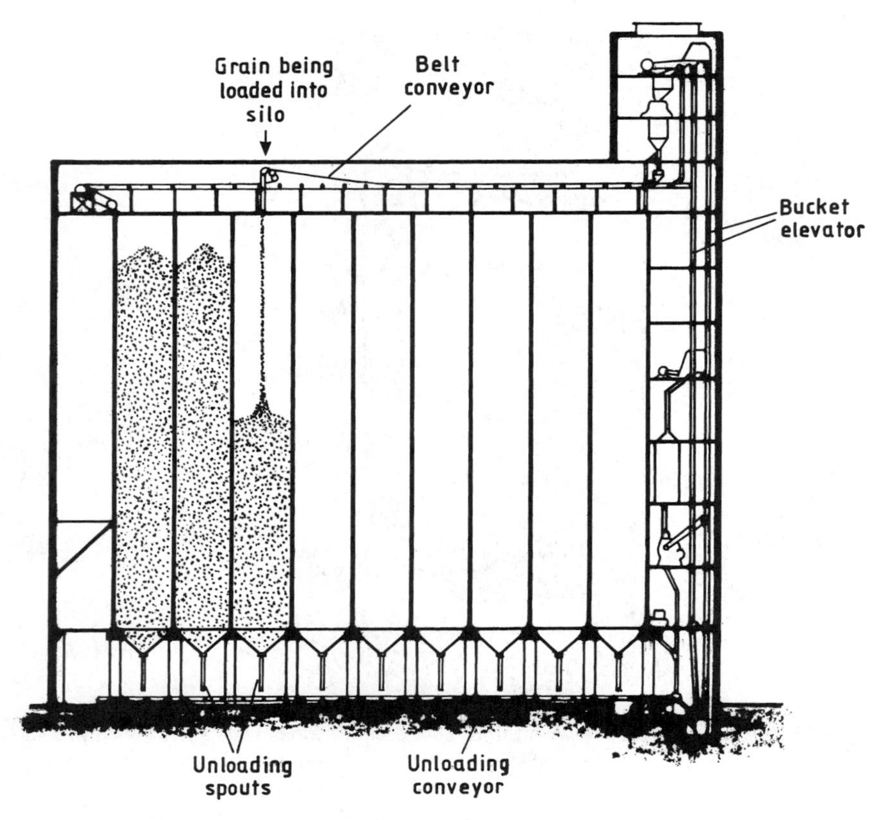

Fig. 8.31 A diagram of a block of grain silos (after Lockwood, 1945). Note the presence of pre-cleaning machinery and a weigher.

approximately, to segregate grain in a single conventional bin or silo. However in some specially designed bins, in which the unloading sweep auger runs continuously when the bin is being unloaded, the grain is removed 'in layers', funnelling is avoided and some degree of vertical segregation is possible.

Grain in bins and silos is difficult to inspect properly. Therefore, it is desirable to have temperature probes and level indicators mounted within the storage chambers. Electrical, remote-indicating thermometers are best mounted on one, or more, 'ropes' hanging into the grain mass. The temperature-sensing units may be mounted at intervals on the rope, or spaced out around it on 'candelabra'. The thermometer readings should be taken and plotted or logged at least weekly to detect any change in temperature indicating the onset of heating. Enclosed bins and silos are convenient if a fumigant has to be applied to grain. However, conventional stores of this type are never completely airtight so, for example, they cannot be used to store grain in an inert atmosphere to control fungi and insects.

Flat-bed stores are the least expensive to construct per tonne of grain stored although they occupy a relatively large ground area. Such stores are not suitable for holding many segregated batches of grain, the grain cannot easily be turned over, and unloading them cannot be fully automated. Stores designed to hold up to 25 000–30 000 t grain have been built. Flat-bed stores are constructed on smooth, waterproof, reinforced concrete platforms. They are of light construction, typically of asbestos cement or other lightweight sheeting, and preferably have a single-span roof with a single, central ridge. This avoids the inconvenience of internal supporting pillars and the risk of leakage in the roof valleys. Doors allow the passage of vehicles. Within the outer structure are side walls, or 'retaining units', designed to take the lateral thrust of the grain. These extend to a height of 2.4–3.0 m (8–10 ft). The lateral thrust can be very considerable. The grain is most conveniently loaded from above from a central, overhead conveyor with multiple outlets, sometimes fitted with directional spouts and adjacent to a walkway to facilitate inspections and maintenance. The grain fills the space below and slopes up from the retaining walls towards the conveyor. The angle of repose or barley is about 23° (see Appendix, Table A11) and the angle of the roof is often *ca.* 30°. If the loading conveyor is not centrally placed, the grain heap is asymmetrical and storage space is wasted. In another system grain loaded from above is levelled by a suspended cross conveyor that may be moved along the length of the store and is raised to accommodate the increasing depth of the grain bed. Alternatively the grain may be unloaded from lorries and projected up onto the heap using grain throwers.

Ventilation ducts may be recessed into the concrete floor and covered with perforated metal grids. These are connected to fans situated outside the main building. On the one hand, this arrangement is advantageous in that the floor is flat and unobstructed and so vehicles can move over it freely. On the other hand, these ducts are comparatively expensive and the need to clean them and treat them to control insects must not be overlooked (MAFF, 1983). The alternative is to have perforated ducts of metal or metal covered with cloth that may be demountable or telescopic and usually extend at intervals from each side of a central duct, which sits on the surface of the floor (MAFF, 1983b). As lorries move into the store to remove grain these ducts have to be disassembled. They must also be cleaned and inspected for insect pests. Centrifugal or axial flow fans may be used to provide ventilating air. If fan heaters are to be used to dry grain in bulk it may be desirable to level the grain heap to obtain equal airflows through all the different parts. This may be achieved using mobile, horizontal chain-and-slat conveyors. Flat-bed stores may be unloaded into lorries using front-loading buckets on tractors, with pneumatic devices or with mobile augers. An expensive but partly automated system is to have an underfloor conveyor system. If such a system is well designed, the floor can be as much as 70% self-emptying and a mounted sweep conveyor system can move into the face of the remaining grain heap and sweep it into the under-floor system (Gibson, 1989). Overhead walkways are too distant from the grain surface to allow the grain to be inspected adequately, and so it is necessary to walk over the grain heap to inspect it and obtain temperature and moisture readings. Adequate safety precautions must be taken when this is done.

8.11 Handling malts

8.11.1 Green malt

In many types of malting plant, green, undried malt must be conveyed from the germination compartment to the kiln. The material is wet (40–50% moisture) and has the fresh rootlets in place, so it is easily damaged and must be handled gently. Jog, *en masse* belt and screw conveyors may all be used. They must be carefully adjusted and run, without overloading, at the optimal speed. The machinery must have suitable access points for regular cleaning to dislodge sprouting and crushed grains. Green malt may be allowed to fall or run down slides or chutes to reach a lower level. It is preferred practice to elevate it with belt-and-bucket elevators equipped with plastic or stainless steel buckets. The elevator trunking may be equipped to be ventilated with warm, dry air, or it may be equipped with fixed in-place sprays for automatic cleaning. One of the advantages of germination – kilning vessels (GKVs) is that in these the green malt is not moved (Chapter 9).

8.11.2 Dry malt

Finished malt is brittle and can easily be damaged by machinery, impact, buffeting, abrasion, or compression, which cause chipping, cracking, crushing or husk separation (Brain, 1983). Therefore, malt must be handled as little as possible, the equipment used must be chosen with care and well maintained and handling or movement must be gentle. Malt is highly hygroscopic and must not be allowed to pick up moisture (become slack) so it must not be allowed to come into contact with humid air or damp machinery. In addition, it must not be contaminated with raw barley or accidentally mixed with other grades of malt. Therefore, malt-handling machinery should not be used for handling barley, and it should be well maintained and regularly cleaned to minimize physical damage and cross-contamination. Most of the grain-handling machinery described above may be used to handle malt, including bucket elevators, screw, chain-and-flight or cable-and-flight and *en masse* conveyors. Sandwich-belt conveyors/elevators, with belts having ribs to prevent slippage, are also in use.

8.11.3 *Malt storage*

Malt storage must be in moisture-proof structures, containers or bags (Gibson, 1989). Silos and metal bins are frequently used, sealed as far as possible to exclude moisture and draughts of moisture-laden air. Obviously aeration must be avoided. Many small containers may be employed so that different batches of malt may be kept segregated. Closely similar malts may be blended to meet customers' specifications, as required. Insecticides cannot be used on malt. It is desirable to minimize storage times and to 'turn stocks over'. Malt has only about three-quarters of the bulk density of barley, so a given weight of malt occupies a large volume. To avoid the damage caused by conveyors, malt bins are hopper-bottomed so they can be emptied by gravity. The angle of the hopper should be at least 5° steeper than for barley, especially if the malt is to be stored 'in culm', i.e. with the rootlets still attached, when the valley-angle may be 50° or more.

8.12 Changes in grain weight and volume during processing

For European two-rowed malting barleys, bulk densities (fresh weight) of 65–75 kg/hl (1.33–1.54 m^3/t) have been reported, but values of 68–72 kg/hl (1.39–1.47 m^3/t) are common (de Clerck, 1952; Narziss, 1976). The angle of repose of European barley is about 23–35° from the horizontal. The bulk density range of normal European two-rowed malts is 45–55 kg/hl (1.82–2.22 m^3/t), but undermodified chit malts have values of about 60 kg/hl (1.67 m^3/t) (see Appendix, Table A11). The angle of repose of malts is about 26° from the horizontal. These figures are important in calculating either the volume needed to store a given weight of material or the weight of a given volume stored. North American values for various cereals are: barley 60.5 kg/hl (1.65 m^3/t), maize 72 kg/hl (1.39 m^3/t), oats 43.8 kg/hl (2.28 m^3/t), rice 57.9 kg/hl (1.73 m^3/t), rye 72.1 kg/hl (1.39 m^3/t), sorghum 73.3 kg/hl (1.36 m^3/t) and wheat 77.2 kg/hl (1.30 m^3/t) (Hoseney and Faubion, 1992). Presumably the lower bulk-density of the North American barley, relative to the European, is owing to the presence of six-rowed grain.

8.13 Malt delivery

Most malt deliveries are now made in bulk. Delivery in small sacks is unusual but may be required by small users. The malt is weighed into a hopper that discharges into a sack; this is sewn up using an upright sewing machine. Modern malt sacks are laminated in construction with an outer woven material to provide the strength and an inner film of plastic that is impervious to water and water vapour. For export, particularly to tropical countries or where no bulk handling facilities are available, giant 'sacks' may be used. These normally have 1 t capacities and are laminated. They are filled from above and may be emptied from below when the 'spout' is untied. They are moved using fork-lift trucks, the truck prongs being inserted through the four loops attached to the top of each sack. Alternatively, malt may be exported in sealed containers.

Most malt is conveyed in bulk. Rarely brewer–maltsters or distiller–maltsters may be able to convey malt directly from the malting store to the unit where it is used. Indeed, in some grain distilleries, green malt may be conveyed from the maltings directly to the mashing plant. Normally malt has to be moved between the maltings and the customer by road, rail or water. Enclosed tipping lorries, or bulk-container lorries may be used or the malt may be placed in containers that can be sealed and transported by any available means: lorries, trains or ships. Transport costs are significant, so it is usual to deliver malt to the customer from the nearest maltings and in the largest feasible loads. If possible a return load is obtained so that the return journey of the lorry or wagon pays. For there to be a chance that this can be achieved, the lorry or other conveyance must be adaptable and able to carry materials other than grain or malt. Lorries often carry loads of 20–25 t, rail wagons 50–53 t, and boats may carry batches of malt of 300–8000 t. Whatever system is employed the malt must be kept clean, dry and out of contact with moist air.

8.14 Organization

The arrangements of grain reception, handling, cleaning, drying, storage and despatch vary a great deal. Possible types of arrangement are indicated in Figs. 8.1 and 8.31. Usually, as the years pass, plant is progressively renewed and modernized and extra facilities are provided. Thus, the grain stores should be arranged with some thought for ease of expansion, alteration and modernization. It is logical to place them close to the steeping end of malting plant, to minimize the need for long runs of conveyors.

In recent years, there has been a move towards increased automation and remote control of plant. This has been highly successful and has resulted in great reductions in workforce requirements. However, it has been necessary to ensure that, while all the processes may be remotely controlled from a central control-room, the equipment, the plant and, above all, the grain itself is inspected regularly by trained personnel.

9 Malting technology

9.1 Introduction

The origins of malting are lost in time (Chapter 1). From the limited evidence available, in the 18th and 19th centuries malting was carried on in conditions not dissimilar to those used in the 16th century (Harrison, 1587; Markham, 1615; White, 1860; Anon., 1877; Stopes, 1885; Corran, 1975). These were probably similar to those used in Norman and Anglo-Saxon times, which, in turn, probably resembled the malting carried out by Celtic and possibly Germanic tribes, both within and without the Roman Empire. There seem to have been few major differences between the malting technologies used throughout the beer-brewing regions of Europe in mediaeval times. In the UK, around 1800, the demand for malt was growing and was being met by the output from larger and larger malthouses (Brown, 1983), and the numbers of smaller malthouses declined. Home malting and brewing had been carried on in the UK, but the introduction of the malt tax ensured that only commercial maltsters willingly admitted that they did it; later the beer tax meant that few admitted to carrying out home brewing until recent years, when the laws were relaxed. In the UK and elsewhere throughout Europe and North America, ways were sought to increase the efficiency of malting. Many ideas originated – and were often patented – in the UK. However, until 1880, the crushing number of rules and regulations that had to be met by British maltsters, subject to a tax on the volumes of grain processed (White, 1860; Stopes, 1885), meant that almost all modernization was prevented. After that time, adverse economic conditions and the availability of cheap casual labour from farms to run maltings in the winter, combined with an ingrained conservatism in maltsters and malt users, meant that, although mechanized, pneumatic maltings were well understood in the 19th century, widespread adoption of more mechanized techniques did not occur widely before about 1948. Even now (1997), there are still a few floor-maltings being profitably run in the UK. Most of the early mechanical maltings were erected by brewers and distillers who had some spare capital (Causton, 1965). For example, Bass (Burton on Trent) had a large drum maltings operating in 1905 (Brown, 1983). In some respects, the malting technology used in the UK was beginning to differ from that used in North America and Continental Europe by the mid-1800s. Outside the UK, the introduction of mechanization and pneumatic malting was rapidly adopted. Despite some 'false leads', most of the practical types of malting plant that have remained in use were developed in Continental Europe.

It is convenient to consider the malting process in stages. Cleaning and grading (Chapter 8), sometimes preliminary abrasion, steeping, sometimes 'squeezing' and applying additives such as gibberellic acid, germination (which was traditionally divided into couching, flooring and withering) and kilning (which was also divisible into stages, sometimes beginning with withering but always including drying and curing). Finally there is removing the rootlets and dust from the malt (deculming and dressing) before storage. To these now must be added pelletizing the rootlets (culms), dust and broken grains for animal-feed. Obtaining supplies of water and disposing of effluent (chiefly surplus steep water) are also essential processes. The traditional key stages are steeping, germination and kilning. In the earliest technologies the stages were separate, and so used separate vessels. At present there are plants in which steeping, germination and kilning are carried out in single units or vessels (SGKVs) plants in which steeping is carried out in 'conventional' steeps but germination and kilning are carried out in the same units (GKVs), and plants have been built in which steeping and germination have been carried out in single vessels (SGVs) but kilning has been in a separate unit. A wide range of vessel shapes and types is in use, often arranged more or less horizontally but not infrequently one above another in towers. Although commonly carried out as a batch process, in some plants malting is semicontinuous in that the grain, in batches, is moved repeatedly as germination progresses; in yet other processes malting is truly continuous in that barley is fed in continuously at one end and malt emerges in a steady stream from the other. It is impossible to develop a logical, simple progression of ideas to characterize all the widely varied forms of malting plant. However, with the passage of time there have been certain strong trends in malting. Malting has become progressively more rapid, batch sizes have become larger (200 t batches of barley steeped are now common in the UK; 300 t are common in Europe and the largest batch size known to the author is 520 t, used at Buckie in Scotland). Workforce requirements are being progressively reduced, so that each operator makes more malt per annum than his predecessors. New maltings are fully mechanized and increasingly are automatically controlled.

In the UK, the sales-maltsters, who sell to a wide variety of customers, make malts to many different specifications for brewing, distilling and food use. They choose plants able to make numerous types of malt to meet customers' requirements. User-maltsters typically choose plants optimized for making the relatively few types of malt that they mainly use. They then purchase the small amounts of 'special' malts that they require. Malting plant capacities are described in various ways. As far as possible they will be described in terms of the weights of barley as-is (usually *ca.* 12% moisture content) being steeped. As the barley, or other grain, is steeped, germinated, kilned and deculmed, its weight and the volume it occupies changes. These changes must be allowed for when the plant is being designed. In addition, measurements will be given using the metric system or SI units wherever possible, but equivalents in other units will also be used (Appendix, Table A4). Older British maltings were described, for example, as 50 Qr maltings, the volume that could be steeped at one time. More recently the capacity would be given as the weight of grain. (1 quarter (Qr) barley = 448 lb = 0.2 ton = 0.2032 tonne = 203.209 kg.)

In the UK, malthouses were once to be found on major farms and estates and in practically every town and village. As malting became more industrialized, so maltings tended to become more localized, either in the best barley-growing areas or adjacent to breweries or distilleries. There are arguments in favour of both locations. 'Wet' barley weighs substantially more than the 'dry' malt made from it, so it is cheaper to transport finished malt. On the one hand, the maltster in a barley-growing area is better able to control the quality of his grain by establishing a rapport with the farmers, ensuring that they understand his requirements in terms of choice of varieties needed, gentleness of threshing and exceptionally careful drying and storage. Building sites and local taxes are often cheaper in the countryside, and labour is often less expensive. On the other hand, the maltster at a brewery or distillery site has the advantages such as easier access to labour, sophisticated central laboratory facilities, perhaps a lorry pool and the use of less expensive power or steam.

The site must have good supplies of clean water, means of dealing with effluent, good communications with the barley-growing areas and customers' plants and access to roads, railways or waterways (canals, rivers) and a port if malt is to be exported. Fuel and electrical power must be available. To minimize problems with oxides of nitrogen (NOX) and the formation of nitroamines, the maltings should be situated away from industrial areas or regions with heavy traffic. Attempts to diversify on malting sites and utilize by-products and waste heat (recovered with heat pumps) have included keeping cattle, fish-farming and producing ornamental plants in greenhouses.

9.2 Floor-malting

Floor-malting is the oldest technology used. Since it could be carried out in many types of farm building, maltings are not readily recognized in archaeological remains, but a maltings dating from 250–380 AD with a kiln containing a cross-shaped heat disperser has been recognized near Rochinne (Gocar, 1983) and the plan of the 8th century malthouse of the Monastery at St Gallen, Switzerland, with its floors arranged in a cross, is famous (e.g. Brown, 1983). The survival until recent times of small-scale malting techniques in the countryside of Norway allows us to imagine how small-scale mediaeval maltsters proceeded (Nordland, 1969). The earliest comprehensible English accounts of malting, on the farm, were given by Harrison (1587) and by Gervase Markham (1615). The processes of malting, except for increases in scale and mechanization, changed relatively little over the years (White, 1860; Anon., 1877; Scamell and Colyer, 1880; Stopes, 1885; Lancaster, 1908, 1936a,b; Leberle, 1930; Schönfeld, 1932; Northam, 1971) and remain essentially the same in floor-maltings today. None of the early English writers doubted that the Saxons and their predecessors used essentially the techniques that they describe.

A brief outline of floor-malting is given here (Fig. 9.1). The units (the steeps, kilns, etc.) that are used in other types of malting are described later in the chapter. Well-cleaned barley was steeped under water in a cistern, often for 3 days and nights. After draining, the batch was shovelled onto the floor and arranged in a heap or couch 0.61–0.91 m (*ca.* 2–3 ft) deep in English practice, 20–50 cm (*ca.* 7.9–19.7 in) in northern Europe (Vermeylen, 1962). If the weather was cold, the couch might be covered with sacking or a tarpaulin to encourage a rise in temperature and accelerate the uptake of the surface film of water and chitting. Barley remained in the couch until it started to 'come' (chit), then the heap was 'broken down' and the piece (batch) was spread on the floor, thicker or thinner according to the state of growth and the temperature. Originally the piece might be 30.5–38.1 cm (*ca.* 12–15 in) deep, but if very warm it might on occasion be reduced to about 7.6 cm (3 in). The piece was spread thinner to allow heat to escape and so prevent a rise in temperature. It was turned four or five times each day in England and remained on the floor for up to 21 days. The stage of the process was judged from the appearance of the grain. Before the advent of thermometers and for some time after, the temperature of the piece and hence the need to turn, thicken and thin it, was judged by feel using the hand, or the bare feet of the workmen (White, 1860). When growth was sufficiently far advanced, and before the grain could shoot or 'spire' (overgrow: the acrospire appearing at the end of the grain) the piece would be transferred to the haircloth on the single-deck kiln, where it would be slowly dried in a thin layer 'on the haircloth'.

The malthouse was to be 'conveniently placed and arranged' – Markham preferring to have the buildings around a courtyard. The advice to have the germination floors in a suitable cave was generally impracticable. Water for the steeping vessels (steeps, cisterns, vats, fats, ditches, troughs, uiting-fats) might be pumped or taken from a well, but 'standing water' was disliked. Originally partly filled sacks of grain might be steeped in streams, but cisterns or vats were, at various times, made of wood (tubs), stone, slate or brick, variously faced and sealed. Arrangements were made to allow the water to drain away through a bung-hole, the barley being retained by a grid, usually of wood. Markham advised that barley garners should be

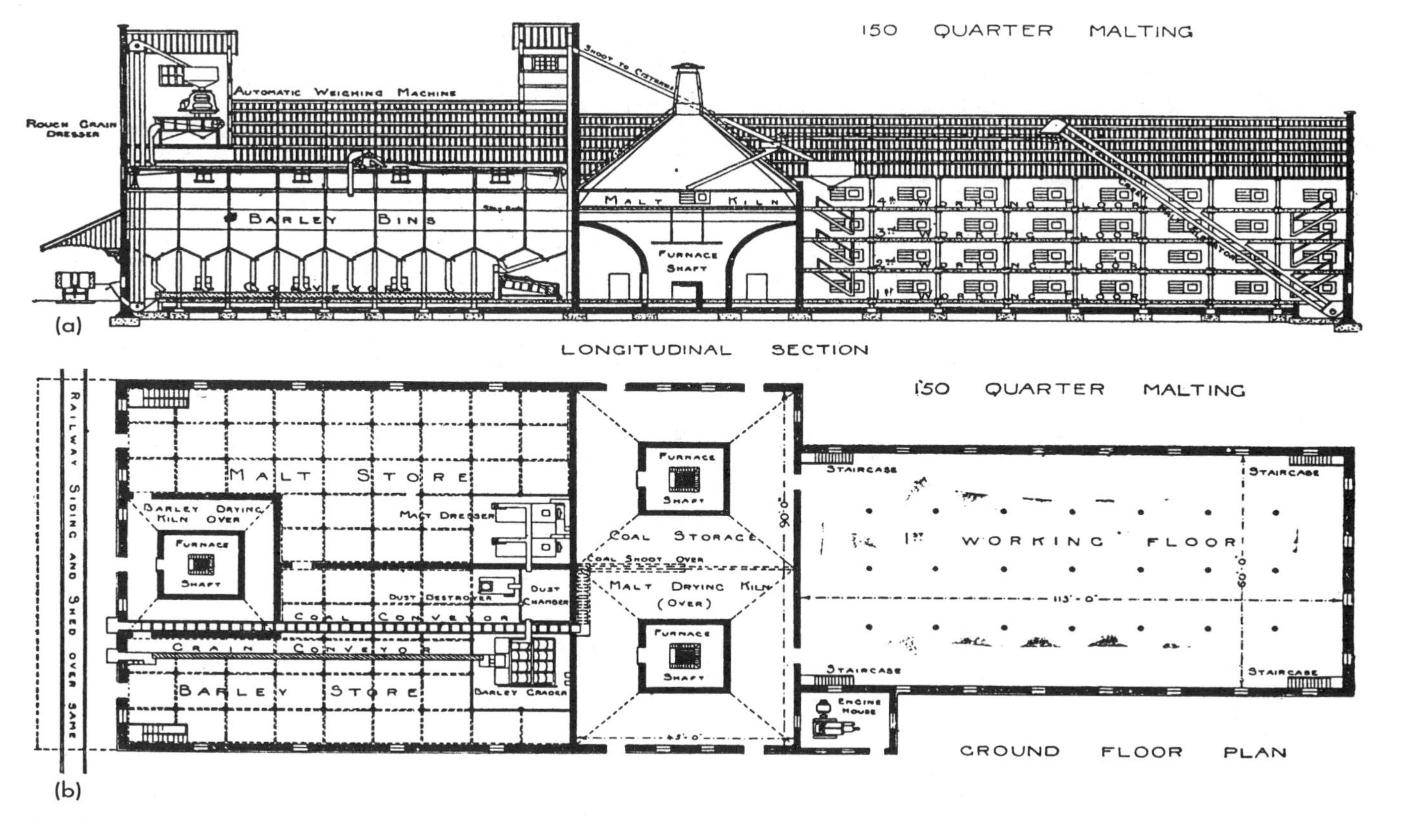

Fig. 9.1 A longitudinal section (a) and ground-floor plan (b) of a 150 Qr (22.9 t) English floor maltings (Saunders, 1905). The barley and malt stores, the barley-cleaning machines and drying kilns are to the left. The germination floors are to the right. Note that the positions of the supporting pillars are indicated, as are the window shutters. The two malt kilns are centrally placed.

above the cisterns so that grain could conveniently be run directly from the garner into the steep. In the USA, early records indicate that steeping was carried out in wooden tubs, kept sweet by cleansing with hot, slaked lime, a powerful disinfectant. At intervals during steeping the water would be changed. Originally, steeped grain had to be shovelled from cisterns. It was undoubtedly a major labour-saving advance when rectangular, flat-bed cisterns were replaced, in some instances by self-emptying hopper-bottomed metal steeps (e.g. Stopes, 1885). However, even now some cisterns are in use, being emptied by electrically driven elevator augers fed by sweep augers at the base.

In England the law came to enforce the use of a couch frame, a rigid rectangular and flat-bottomed structure in which the steeped grain had to remain until it was 'gauged' for tax purposes (White, 1860). This tax, and the oppressive regulations relating to it, remained in place until 1880 (Stopes, 1885). Germination would begin in the 'couch'. The main germination stage of malting was carried out on floors. The numerous disagreements on what sorts of floor were best emphasize the problems early maltsters faced. Beaten earth was dirty and encouraged growth of moulds on the grain; wood was too 'warm' and porous, tending to dry the grain. Also it was difficult to keep clean. Stone, brick, tiled or slate floors were used but irregular joints and irregular surfaces were unsatisfactory. In later years, the preferred material has been smooth, waterproof concrete, cement or some similar material, sometimes laid with a slight slope to allow water to drain to a gutter placed at one side. Well-laid tiles were liked, but the surface had to be perfectly flat, with no cracks or crannies to hold grain or dirt (Lancaster, 1908). The floors were often arranged one above another, often with a headroom of about 2 m (6 ft 6 in), but in some maltings the ceilings were so low that men could not work standing upright. Typically the structure was strengthened by rows of cast-iron pillars, which, by convention, came to be used as a crude unit of measurement dividing each floor area into bays. The dimensions of the floors were chosen by experience. Because temperature control was by ventilation from the side windows, wide rooms were judged to be unsuitable. The walls were whitewashed and the bricks at floor level were glazed or coated with waterproof paint. The pillars were also painted. While one- and two-floor maltings were probably originally the most common type, by the early 1990s, two-, three-, four-, five- and even six-floor maltings were built, this being facilitated by the introduction of mechanically driven conveyors, elevators and power shovels (Saunders, 1905, 1906). Floors had to be cleaned regularly. Often they were swabbed, with a solution of hypochlorite when this became available, scrubbed if needed and the surplus water was pushed aside into a drain.

Originally malting was a cool-weather occupation and the temperatures of the pieces were regulated, as far as possible, by opening and closing the side windows and by thickening or thinning each piece to retain or lose heat.

The designs of the windows varied a good deal. Often they had coloured glass and sometimes sacking covers, with various types of wooden shutter and slides with which to control ventilation. From time to time suggestions were made to have ventilation ducts in the walls and ceilings. This idea was not widely adopted until air conditioning (forcing the circulation of cooled damp air over the floors, distributed by ducts suspended from the ceilings) allowed floor-maltings to be used in warm weather. Stead (1842) patented the ideas of heating and cooling the air over floors and humidifying it by the controlled release of steam from steam pipes. Even now, not all the surviving floor-maltings are air conditioned. In air-conditioned floor-maltings, the side windows are usually sealed and insulated. Preferred air temperatures ranged from 10 to 14 °C (50 to 57 °F). In addition, it was necessary to maintain the humidity of the air over the floors. This might be done by sprinkling the grain and floors with water (which also wets the grains) using watering cans or hoses with rose sprays. Sometimes sacking 'hangings' over the windows were wetted to humidify the air. Originally, temperatures in the grain were judged by feel. However for about two centuries it has been usual to have thermometers hanging from the roof and embedded in the grain to determine the air and 'in-grain' temperatures.

Grain was originally moved with wooden shovels, or scuppets, of various patterns. The workers might be expected to work bare-foot to minimize damage to the grain. Rubber-soled boots are now probably normal (Fig. 9.2). In recent years, some maltsters have been 'reduced' to using shovels with aluminium blades. The pieces on the floors were 'turned' with shovels, the grain being thrown up and forward with a flick of the wrist to separate and mix the corns. The pieces were also lightened, ridged and mixed using wooden forks, rakes, 'stick ploughs' and multiple-pronged 'ploughs' that were pushed or pulled through the grain (Fig. 9.3). In addition, brooms were used to sweep scattered grains back into the edges of the piece (Fig. 9.2).

Manual turning has been replaced by guiding electrically powered turners, with rotating paddles or 'blades' that throw the grain in the air, separating the roots and letting it fall gently back onto the piece. Apparently these were introduced into the UK in 1947 from Germany, where they were in use in the late 1920s (Schönfeld, 1932; Cherry-Downes, 1948). Modified rotating-tine 'cultivators', driven by little two-stroke engines, have also been used. Pieces were also turned by electrically powered 'ride-on' machines equipped with rotating paddles.

Traditional floor-maltings were immensely labour intensive and little thought was given to minimizing workloads or workforce requirements. Originally, steeps were loaded and cast by hand, then the couches were made and broken down, and the pieces were laid out, thickened, thinned, turned and loaded to kiln and the kilns were levelled, turned and stripped all by hand. The introduction of metal, hopper-bottomed, self-emptying steeps allowed grain to be dropped directly onto a floor, down chutes to

Fig. 9.2 Working on a germination floor in the 19th century. The gas flares gave light for night working (Anon. ca.1877).

lower floors or into an elevator with little manual effort. It was an advance when two-wheeled tipping barrows, latterly with rubber tyres, and baskets suspended from trolleys running on overhead rails were used to move grain from the steeps to the floors and from the floors to the kilns. Power shovels were introduced in the 19th century in parts of mainland Europe and North America (Thausing *et al.*, 1882; Wahl and Henius, 1908). Originally, shovels pulled by a rope running around permanent pulleys mounted on the floor and walls were powered through belt-drives by steam or gas engines. Those used now, for stripping floors into a conveyor for example, are pulled and returned by wire ropes activated by electrical motors. The operator 'steers' the shovel which pushes the malt across the floor into the receiving chute or conveyor. Then, in response to a switch, the shovel is pulled back, rides over the grain and is positioned to take the next 'bite'.

Grain intake, cleaning and storage facilities were usually placed at one end of the maltings. Barley garners were often placed on the top one or two floors so that the grain could be run down to the steeps. Usually the floors were arranged so that as the pieces were 'worked' they moved from the steeping end towards the kilns. It will be appreciated that as a piece was thinned, the area of floor covered increased very greatly and allowance had to be made to ensure that sufficient floor area was always available.

White (1860) reckoned that for every 40 Qr (8.12 t) barley steeped a floor area of 8000 ft^2 (226.5 m^2) should be allowed. More recently, from Belgian experience, Vermeylen (1962) reckoned that for each tonne of barley steeped a floor area of 33 m^2 should be allowed. There should always be

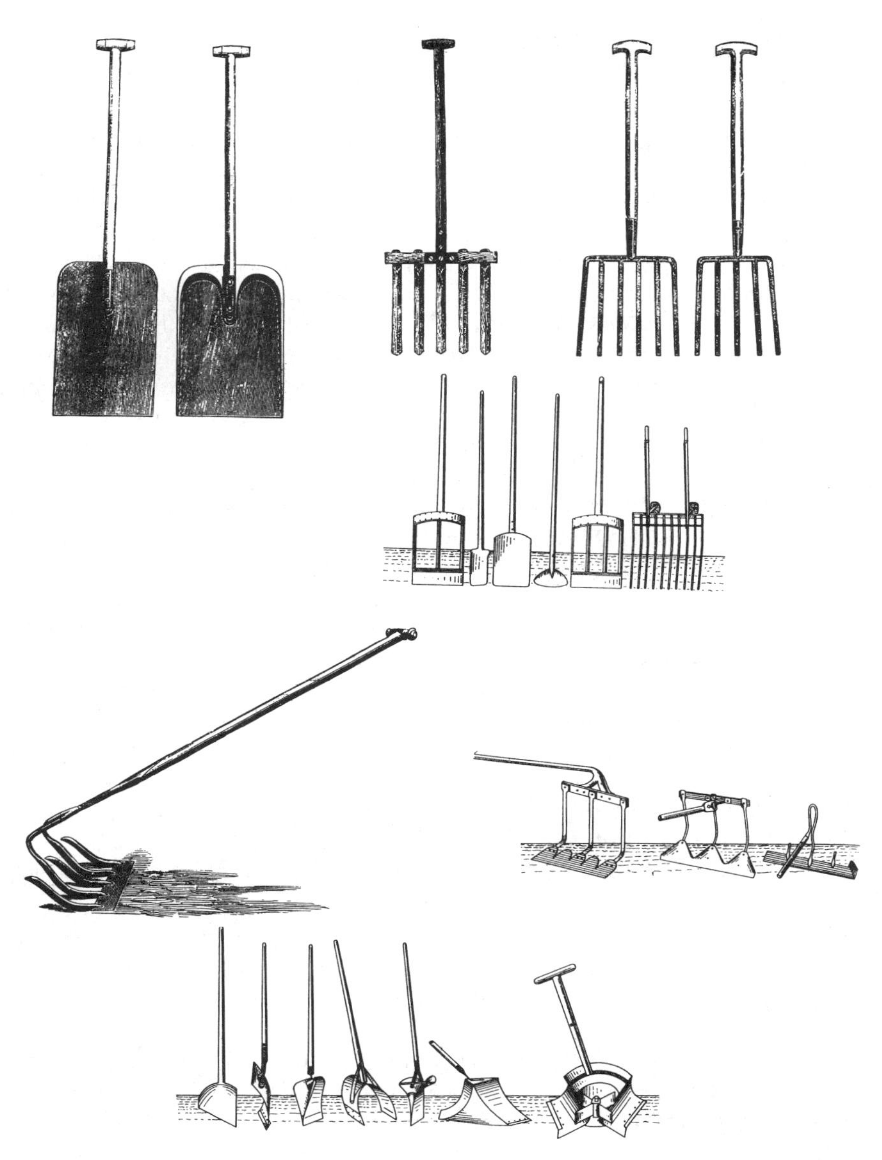

Fig. 9.3 Various sorts of shovels, forks and ploughs, once used in English and German floor maltings (Stopes, 1885; Schönfeld, 1932).

walkways between the malt pieces to minimize treading damage. Gibson (1989) allows that a piece may be thinned to 7.5–10 cm (3–4 in) when a tonne (before steeping) of barley will cover 28–37 m^2 (*ca.* 300–400 ft^2) of floor. Floor areas/unit weight barley steeped were less in cooler areas, since there was less need to thin pieces to extremes to dissipate heat. Saunders (1905) gives plans for English-style malthouses ranging from 50 to 300 Qr (10.2 to 60.9 t) steeping capacity (e.g. Fig. 9.1).

Air conditioning allowed malting to be carried on under controlled conditions the whole year round, maximizing the productivity of the malting equipment and giving regular employment, rather than casual employment during the colder months. Various modernization stages of old English floor-maltings have been documented (Hynard, 1947; Cherry-Downes, 1948; Anon., 1953, 1963a; Northam, 1962, 1965a, 1971). A few of these maltings still survive because there is a steady specialist demand for floor-malts, because the costs of the buildings were paid off many years ago and because modernization has reduced production costs. In addition the mild and equable English climate permits malting to be carried out for much of the year without the need to run costly plant to heat or cool the air (Hudson, 1986a,b). Improvements include replacing handling sacks of grain with automated bulk storage, handling, elevating and conveying equipment, arranging for steeps to be self-emptying or mechanically emptied, making arrangements for steep aeration and/or downward ventilation and the control of steep water temperatures. Malting grain is now turned with mechanical devices, the floors are stripped with powered tools and kilns have been modernized to run efficiently. Kiln improvements include the use of pressurized hot-air chambers and wedge-wire floors, semiautomated or automated firing, mechanized loading and stripping, the use of air recirculation, better insulation and better temperature control. These improvements, coupled with the use of better-quality barleys, the use of sophisticated steeping regimens to accelerate germination and, where permitted, the use of additives have increased malt outputs by 300–400%.

Mechanization of floor-maltings was well advanced in Europe long before it was tested in a limited way in England, (Anon., 1907b, 1953, 1963a; Leberle, 1930; Schönfeld, 1932; Thomson, 1950; Vermeylen, 1962; Brown, 1983). The last mechanically-turned floor in England was closed in 1992. Various patterns of 'mechanical floor-maltings' were in use (Anon., 1907b). For example, in one system referred to as the Maffei system in England (introduced in Germany before 1914), the grain was malted in semicircular 'tunnels' made of bricks. These were ventilated via channels that, for example, admitted fresh air at intervals along the apex of the tunnel and withdrew air from just above floor level where the carbon dioxide concentration was highest (Fig. 9.4). In England, in the period after the Second World War, Swedish equipment was installed in two maltings. Machines

Fig. 9.4 Diagram of a section of a mechanized floor maltings used in Munich early in the 20th century (Anon. 1907b). Note the flat-bottomed steeps, which discharged grain through tubes to the germination floors, the steerable spout used for loading green malt onto the upper deck of the two-floored kilns, and the mechanical kiln stripper.

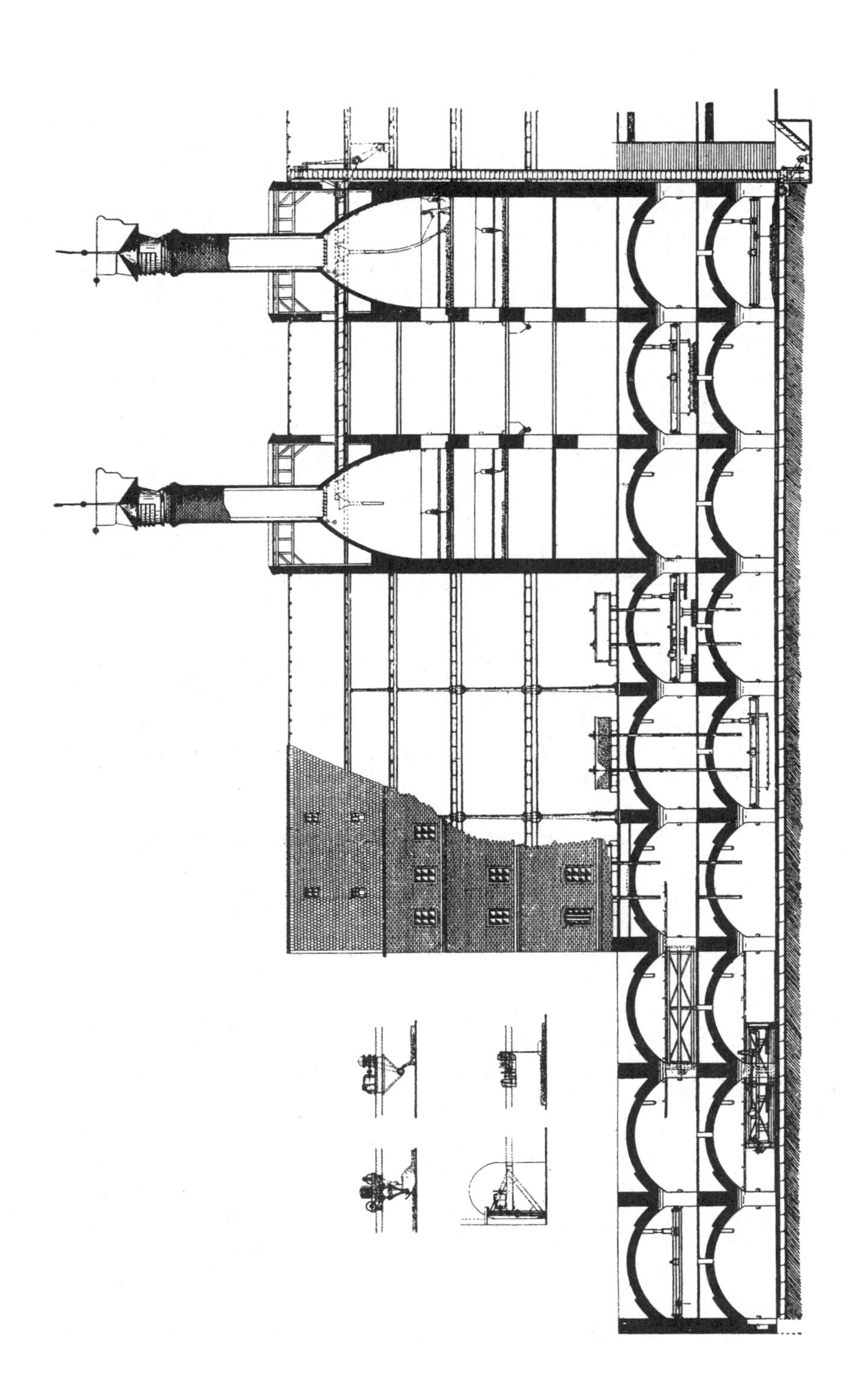

(tank cars) accepted the steeped grain from the steeps; these cars ran along rails supported on side walls and spread the grain on the floor. Other machines turned, thickened and thinned each piece as required and eventually moved it to a conveyor for transport to the kiln. The Grimsby Maltings were well insulated, equipped with 75 Qr (15.23 t) barley garners and with 6 × 75 Qr (15.23 t) circular, closed-in hopper-bottomed steeps (Anon., 1953), with facilities for pumping between the steeps. The floors, air conditioned at 11.1–12.2 °C (52–54 °F), were divided into 24 bays, with three bays each side of a central corridor on each of four floor levels. Each bay was 18 ft wide × 120 ft long (5.49 m × 36.56 m). Rails ran down the side of each bay to support the tank cars and other machines. The central corridor, cross corridor along one end of the bays and lifts between the floors allowed the machines (three turning machines and two tank cars) to be moved on mobile carriages between the bays and floors and so service all the bays. Steeped grain was directed via spouts into a tank car on a particular floor. The loaded tank car was moved to a bay and, using a rotating cross-feed and levelling device, spread the grain to the full width of the bay. One tank car took three journeys to load a bay. The 75 Qr (15.23 t) original weight of barley steeped was spread in strips in each of two bays, by one operator working for 1.5 h. Other machines turned the piece, lifting and allowing samples to fall from giant metal 'combs' actuated and tipped by metal arms supported by an overhead carriage running on the side rails. Turning, thickening and thinning was carried out with minimum grain damage. The machines stripped the floors by pushing successive strips of grain into a conveyor running across the end of each bay. Two workers stripped two bays (75 Qr (15.23 t) barley steeped) in 1.5 h. As built, this malting produced 40 000 Qr malt (6096 t) each year, with about one-third of the workforce needed by a conventional floor maltings.

9.3 Abraders

Barley that has received a limited degree of 'battering' or abrasion before steeping frequently malts more rapidly than untreated grain particularly when it is also treated with gibberellic acid and malting conditions are adjusted (Palmer, 1969). A range of treatments has been used in laboratory studies (Briggs, 1987b). Two types of equipment have been used in malting practice (Northam and Button, 1973; Brookes, 1980, 1981; Briggs, 1987b). The Henry Simon impact abrader was developed from an entoleter-type grain scourer/aspirator (Fig. 9.5). In this machine, a controlled flow of grain is delivered into the central space of a rotor, consisting of two steel discs held apart by 56 hardened steel pins. The rotor spins, usually at 1000–1500 r.p.m., or rarely 1650 r.p.m. (Northam and Button, 1973; Halford *et al.*, 1976; Brookes, 1980, 1981). The grains are thrown outwards, striking

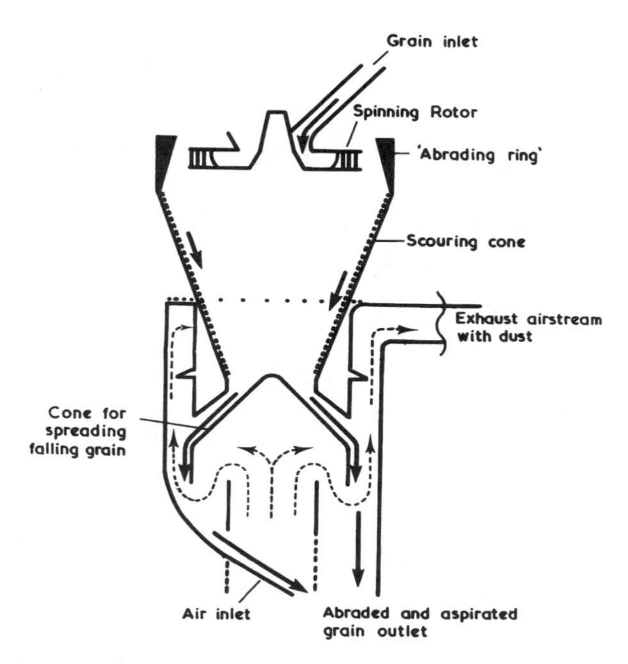

Fig. 9.5 A Simon impact abrader (after Northam and Button, 1973).

each other and the steel pins; as they leave the side of the rotor, they hit the downward-angled face of a silicon carbide abrading ring. The grains fall downwards, sliding over a scouring cone as they fall, then they are spread by a cone. Falling around the cone the grains meet an upward flow of air which carries away the dust created by the abrasion process. Weight losses depend on the working conditions and are commonly 0.5–1.0% of the grain's original weight. Each machine can process 9–12 t/h. These machines are often called 'single-hit' abraders but clearly this is not correct. Sometimes more even results are obtained if two machines are used working in series. Treatments must be moderated to avoid dehusking, skinning and grain breakage. The Richard Sizer abrader works on a different principle (Fig. 9.6; Brookes, 1980). Grain is fed into the ends of two tapered chambers, with circular cross-sections. Here it is stirred by paddles, with finger-like projections, extending from a central shaft rotating at 800–1000 r.p.m. The grains are rubbed against each other and against blocks of silicon carbide mounted in the chamber walls. The grain leaves the chambers and falls down a common shaft where it meets an upward airflow that aspirates it, removing dust and husk fragments. Each machine can process 10–12 t/h and removes 0.3–0.5% of the grain's weight. Treatments can loosen the grain's husk (Brookes, 1980, 1981; Briggs, 1987b). The device should only be used gently on malting barley, to avoid damaging grains. Another type

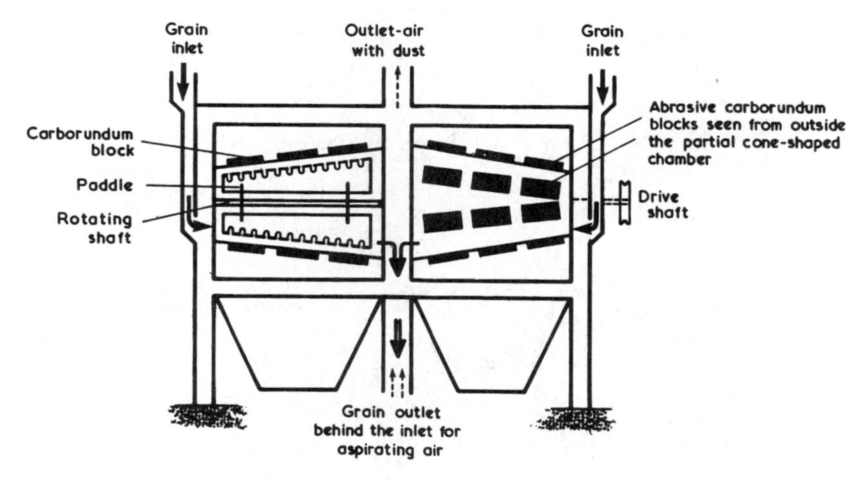

Fig. 9.6 A Sizer 'stirrer' abrader (after Porteous; personal communication; Brookes, 1980).

of machine, which was used experimentally to abrade or – in extreme cases – dehusk grain, used a rotating propeller to stir the grain in a bowl lined with an abrasive material.

Because of uneven grain treatments, giving malt of irregular appearance, and because some customers will not allow gibberellic acid to be used in making their malt, abrasion has not been widely accepted. It appears that the process is now only in limited use.

9.4 Grain washing and destoning

Once it was common to use devices to wash and destone grains before they went to steep. The practice dropped out of use, but washers are now being installed in many modern maltings (Thausing *et al.*, 1882; Stopes, 1885; de Clerck, 1952; Vermeylen, 1962; Narziss, 1976; Maule, 1996). They pre-clean grain before it is steeped and so eliminate dust problems in the steephouse. The throughputs of most devices are slow. Various steeps (e.g. with propeller agitation (Wild), vigorous aeration and pumping over and into separation devices) have been used to wash grain. Washing is also achieved in the steeping spiral used in the Domalt plant (see below). Dry destoning, although possible with barley (Chapter 8) is usually applied to the less dense malt. In 1879 a device was described in which barley, immersed in water, was passed between rotating brushes to remove surface dirt.

Vermeylen (1962) distinguished three types of washer. In the first, a thin stream of grain is metered into a reservoir of slowly moving water. The grain, being comparatively light, is moved across, some floating and some

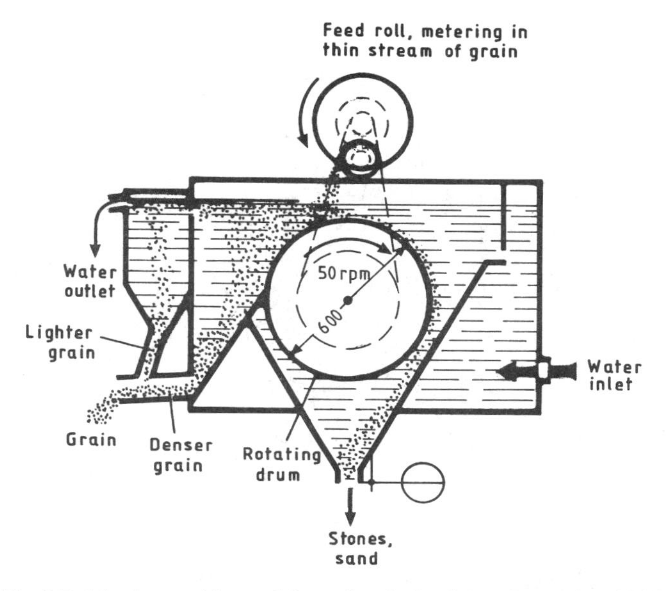

Fig. 9.7 A barley washing and destoning device (after Vermeylen, 1962).

sinking downwards. The two groups are collected, separately if desired (Fig. 9.7). In contrast, the stones and sand, being denser, sink immediately onto a rotating drum which carries them round against the water stream and deposits them in a collecting hopper. This device can treat 2–4 t grain/h and uses 110–225 l/min of water. Presumably the water could be filtered and recirculated. In a simpler device, a metered stream of grain falls onto a dispersing cone and meets a slowly rising stream of water (Fig. 9.8). The grain is swept upwards and outwards into a double settling chamber. It is possible to collect the lighter and denser grains separately. The stones and sand particles fall through the water stream and into a collector, which can be emptied at intervals. This device can clean 1–3 t grain/h with the usage of 50–150 l/min water. The third type of washer and stone separator exists in many variants (Fig. 9.9). The grain is metered into a trough of flowing water, in which it tends to float. A horizontal helical screw moves it across the trough. The stones and sand fall to the base and are moved in the opposite direction into a catch trough by a small screw. At the end of the trough the grain is lifted by a large, inclined screw conveyor working in a perforated casing. Here it is drained and sprayed with fresh water before being directed to the steep.

Another old device was a screw conveyor fitted with sprays, which conveyed grain during transfer between steeps. Such a device dislodged the dirt that had been loosened by the previous steeping and replaced the film of dirty water with fresh. Recent maltings have often been equipped with grain

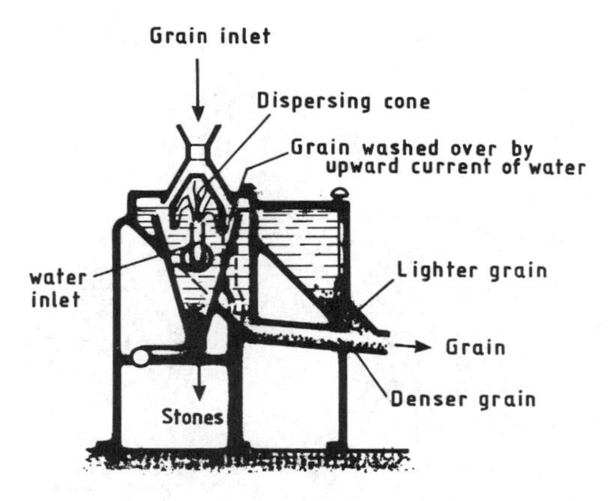

Fig. 9.8 A barley washing and destoning device (after Vermeylen, 1962).

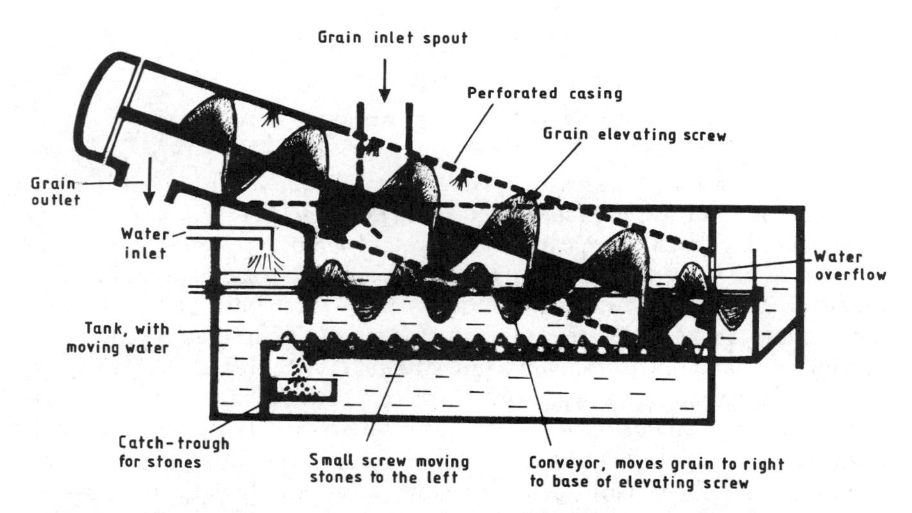

Fig. 9.9 An old type of barley washer. A stream of grain is metered into a slowly moving stream of water. The grain is moved to the right by the horizontal screw and is lifted from the water trough by the inclined screw. Dust and floating material is removed with the overflowing water. Stones sink to the bottom and are conveyed to the left into a catch-trough (various sources).

washing 'spirals' (helical screw conveyors) that remove some soluble materials, dust, husk, chaff and light grains. The grain dwell-time is often only about 15 s. Washing drums have been installed in some German maltings with throughputs of 60 t/h and residence times of 1 h (Anon., 1991; Kunze, 1994; Maule, 1996). By transferring the washed barley slurried in water to the steep vessels and by returning the transfer water the process saves water. The grain receives a second wash in the transfer water. Briggs

(1995) proposed that grain be 'plug-rinsed' at the end of a steep by allowing fresh water (with or without additives) to follow the dirty water through the grain as the steep is drained (section 14.8.2).

9.5 Weights, volumes and capacities

The number of units used to describe grain during malting, and the plant in which malting is carried out, is confusing (Appendix, Table A4). In the past, a maltings in the UK might be described as 'a 100 Qr House' in which 100 Qr barley was steeped to yield about 100 Qr malt (malt quarter = 336 lb = 152.41 kg), all measurements being on a fresh weight as-is basis (Saunders, 1905, 1906). However not only were two different quarters (of weight) in use, but in earlier times 'quarter' had been a measure of volume and outside the UK quarters were assigned different weight values (Appendix, Table A4). For most of this book, where appropriate, weights will be given in kilograms with equivalents in pounds (lb) (1 kg = 2.204 lb) and in metric tonnes (t; 1 t = 1000 kg = 0.9842 long (British) ton).

A widely used modern convention is to quote malting capacities etc. in terms of the numbers of tonnes of barley steeped (fresh weight (as-is), often about 12% moisture) in each batch. However as barley is steeped it gains weight and swells. During germination it grows roots, which take up space, and during kilning the rootlets shrivel and the bulk shrinks. At each stage there are preferred depths of grain, limited in steeps by pressure and airflow (during carbon dioxide extraction). During germination and kilning, the relationship between economically achievable airflows, malt quality and the depth of the beds of grain and drying malt are critical. Different batches of barley grain, processed in different ways, will weigh relatively different amounts compared with the original weight of barley and will occupy different amounts of space. Consequently there is relatively poor agreement between various estimates (de Clerck, 1952; Vermeylen, 1962; Narziss, 1976; Forster, 1985; Gibson, 1989). Gibson (1989) gives graphs indicating the relationships between bed depth and specific loading for barley, steeped barley and green malt. A tonne of barley (moisture content 12%) occupies about 1.42 m^3 (14.2 hl; 50.15 ft^3). On steeping, the grain swells and increases in volume by 30–50%. If agitation is provided, to prevent the grain becoming compacted, at the end of steeping it will occupy 2.1–2.2 m^3 (*ca.* 75.9 ft^3). If the grain is violently agitated during steeping, extra steep volume must be allowed, perhaps 10% depending on the design of the steep, leading to steep volumes of 2.3–2.4 m^3 (*ca.* 83 ft^3) per tonne barley steeped. During germination, rootlets will develop and after turning green malt will occupy about 2.46 m^3/t (86.9 ft^3/t) of **barley originally steeped**. Average malting losses seem to be assumed in these calculations.

For example, when planning to process a 200 t batch of barley, it is proposed to have a specific loading on the steep deck of 850 kg barley/m^2, which will give a depth of *ca.* 1.2 m. The barley occupies 1.42 m^3/t, so the 200 t batch will occupy 284 m^3. With an initial grain depth of 1.2 m, it follows that the area of the steep will be 237 m^2. This dictates that a circular flat-bed steep tank will have a diameter of 17.5 m. During steeping, the grain swells, occupies approximately 2.08 m^3/t original barley, and so the bed increases in depth to *ca.* 1.8 m and in volume to 416 m^3.

In germination compartments, bed depths and specific floor loadings are smaller than this, the optimum values being an initial depth of about 1.4 m, with a specific loading of about 670 kg/m^2 (weight original barley/unit deck area). With the volume occupied still being 2.08 m^3/t of original barley, giving 416 m^3 steeped grain, the germination vessel needs an area of 279 m^2. If the vessel is circular it will have a diameter of about 19.5 m. If the vessel is rectangular, with a width: length ratio of 1:7, which is regarded as satisfactory, its measurements will be 6.5 m × 46 m. As the grain germinates, the green malt will come to occupy *ca.* 2.46 m^3/t, and so the bed depth will increase to 1.65 m. It is interesting to compare these data with those of Thausing *et al.*, who, in 1882, reckoned that, on steeping, 100 volumes of barley increased to 133 volumes, which, during germination, increased to 200 volumes. However, on kilning the volume of the malts, with rootlets, declined to 100 volumes.

The acceptable depth of green malt in a kiln varies greatly depending on the choice from the options of accepting the extra electrical costs of powering a fan for driving large volumes of air through a deep bed of grain, using a larger, more shallow-loaded kiln or extending the kilning time (Chapter 10). Therefore, depending on the requirements of the production cycle and the choice between larger capital and larger running costs, a wide range of specific loadings is used in modern kilns. Assuming that the choice is for an initial bed depth of 1.0 m, needing a specific loading on the kiln deck of, say, 400 kg/m^2 relative to the original barley, the 200 t batch that occupies 2.46 m^3/t will occupy 492 m^3; so a circular kiln will need to be 25 m in diameter to accommodate the batch under the chosen conditions.

As the barley is steeped, its actual weight increases. For example, 200 t barley (12% fresh weight) contains 176 t dry matter. If, during steeping, its moisture content increases to 45% then (ignoring the small losses of dry matter which occur in the steep: 0.5–1.0% usually) its weight will increase to 320 t. Malting plant must be built with this in mind.

Considerations such as these are implicit in all stages of designing malt-houses. In addition, large amounts of grain take considerable times to process or convey from one part of the plant to another. Since there are always time constraints to take into account, it is important that conveying machines have sufficient working rates (t/h).

9.6 Water for steeping

The aims of steeping are to permit barley to take up water to a chosen level (imbibe), to wash and clean it, and to have the grain in the correct physiological state so that at or towards the end of the steeping period it will grow steadily and uniformly. Some steeping is carried out by spraying, when run off is minimized and little water is wasted, but most is carried out by immersion. The amount of water needed per tonne of grain varies with the design of steep, 'usual' values being 0.8 m^3/t barley per wetting for conical steeps and 1.3 m^3/t barley per wetting for flat bottomed steeps (Gibson, 1989). (For reference 1 m^3 equals approximately 220 gal (British) or 264.2 gal (US)). The extra water usage of the latter is caused by the need to fill the large under-bed plenum chamber (see below). It is now common to cover each batch of grain with water three times, with air-rests in between, so the water volumes become 2.4 and 3.9 m^3/t barley processed. In earlier times, barley steep liquor might be changed as often as four, five or even six times. In addition, water may be used to transfer the grain and will certainly be used to wash down and clean the steeping vessels, so total water volumes of 3.5–4.5 m^3/t barley processed are often allowed. It can be seen that each batch of grain processed in a modern maltings (typically 200–300 t) will use immense volumes of water (e.g. 700–1350 m^3) and will generate very large volumes of effluent, which must be disposed of. Both purchasing water and disposing of effluent are increasingly costly. A maltings needs a reliable supply of large volumes of uncontaminated water, steeping liquor, free of pathogenic organisms, chemical contaminants or substances able to taint the malt, and iron salts, which give the malt a dull, grey appearance caused by the colour given by the interaction of the ferric ions and the polyphenols in the husk (Moll, 1979b).

The nature of the steep water can influence malt quality. Highly saline waters are unsuitable, but generally the small amounts of salts dissolved in potable waters have minimal or no detectable effects on steeping. Chalky, alkaline waters are said to limit microbial growth in steep, and washing with short exposure to half- or completely saturated limewater (water saturated with calcium hydroxide, slaked lime) was once common. The use of these, and other modified steeping liquors, is discussed in Chapter 14.

It should be possible to regulate the temperature of the steep water. For hundreds of years, British maltsters relied on the temperature of the well water being relatively constant, temperatures of 10–13 °C (*ca.* 50–55 °F) being preferred. Maltsters may draw their water from wells or from domestic supplies. The latter is almost always the more expensive choice and is not desirable if, for example, the water contains traces of chlorine, which can taint the malt. It has long been recognized that by warming water and steeping at higher temperatures the duration of steeping can be shortened (Stopes, 1885). However, there are physiological limits on how warm steep liquor can be before germination is reduced and, at even higher temperatures, grain is killed. At present steep temperatures of up to 18 °C (64.4 °F) are common, and the temperatures are closely controlled. To achieve this, the water is normally stored in large, insulated holding tanks warmed by heat exchangers. Sometimes, final temperatures are achieved by blending water from hot and cold sources. Another advantage of having a water-storage tank is that it can be filled at a moderate rate using small mains and pumps, but using gravity feed and large mains it can be used to fill steeping vessels very rapidly indeed, saving 'turn round' time and ensuring that all the grains in a batch are wetted at nearly the same time.

Violent aeration, or other agitation of the grain in water, is an excellent way of cleaning the surface of grain, dislodging dirt and microbes (e.g. Heslop, 1911). Frequent changes of steep water, washing grain while it is being transferred between steeps and spraying water into the tops of a grain bed to rinse it while water is being withdrawn from below are other techniques that have been used to clean grain. These treatments also remove water containing carbon dioxide and replace it with water saturated with air and so containing oxygen. However these procedures are expensive because of extra water requirements and extra effluent production. Various suggestions have been made to permit the re-use of steep water. The ideal would be to process effluent to achieve such a degree of purity that it could be re-used in steeping. This approach has not been adopted, at least in the UK. It seems worth investigating since large effluent treatment plants receive sewage and discharge water which is so pure that it is potable. Many maltings already process their effluent to reach a standard that permits it to be discharged to water courses. Other suggested approaches have been to re-use steep liquor from a late steep at an earlier stage, with or without the addition of an antiseptic agent (Chapter 14), or to recirculate steep liquor from the base of the steep back to the top via a unit that will partially cleanse it. This involved drawing grain and water from the base of the steep, pumping it round and discharging it into the top of the steep over a conical 'grain–water separator' (Kauert, 1955; Vermeylen, 1962). The grain falls over the sides of the cone and back into the steep with a little of the recirculating water. Most of the water falls through a bed of granular magna, a mixture of magnesium and calcium hydroxides. This agent removes organic acids and carbon dioxide from the steep liquor, which is aerated by mixing during its passage from the grain separator to the bed of magna, and as it falls back into the steep. The treated liquor is not completely purified and as it is alkaline it can only be used for a limited part of the steeping programme.

9.7 Maltings' effluent

Very large volumes of water are used in malting, mainly in steeping but also in sprinkling the grain, in air conditioning and in cleaning. Maltings have variously used 22–90 hl/t malt produced, and a recent recommendation is to allow 35–45 hl/t barley used (Gibson, 1989). Other figures given are 1.25–8.75 m^3/m^3 grain steeped (Simpson, 1970). The cleaning liquor may contain detergents and hypochlorites and be strongly alkaline. Other toxic wastes may come from the laboratory (e.g. heavy-metal catalysts), oil-spills, etc. These effluents should be segregated and disposed of separately. Efforts are made to reduce water use and prevent waste. Costs are high, so water consumption and effluent production should be monitored and representative samples of effluent should be analysed to check waste, to ensure that the treatment plant is working well and that the maltings is not being overcharged. In the UK, it is often economic to divide waste into separate drainage systems, to deal with (i) storm water, (ii) domestic effluent and sewerage, and (iii) trade effluent. Spray steeping can be arranged to give minimal water usage and effluent production but barley does not malt well unless it has been well washed or has received at least one immersion steep (Briggs *et al.*, 1981; Maule, 1996).

Waste steep water is yellow-brown, readily putrescible, contains microbes (but few or no pathogens), suspended organic matter, colloidal materials and a range of low-molecular-weight substances: dissolved salts, hexose and pentose sugars, amino acids, organic acids (including phenolic acids) and phosphates dissolved from the surface layers of the grains and from the interiors of broken grains. The liquid is prone to froth. Initially it may contain floating grains, awns and rachises. Steeping losses vary in the range 0.5–1.5% (d.wt) so a 200 t batch of barley may add 1–3 t solids to the effluent stream. The longer the grain is in contact with steep liquor the more material is extracted and the higher the 'strength' of the effluent. For

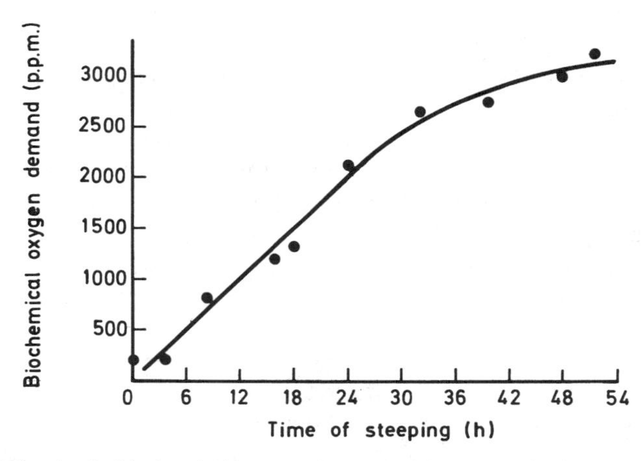

Fig. 9.10 The rise in biochemical oxygen demand with increase in time of contact between grain and steep water. In practice no one immersion is likely to last longer than 24 h (courtesy of O. T. Griffin).

Table 9.1 Analyses of waste water from a small, older type of American drum maltings. In this plant four steeps were given during 3 days. After 3 days of germination the grain was sprayed for 1.5–3 h and the excess water was discharged. At the end of the germination period the drums were washed. The volumes of steep water used were large by modern standards

	Time in contact with malt (h)	BOD (5-day) (mg/l)	Total organic nitrogen (mg/l)	Free ammonia nitrogen (mg/l)	Total solids (mg/l)	Total volatile solids (mg/l)	Total suspended solids (mg/l)
First steep water	24	960	69	16.5	4856	1968	176
Second steep water	12	920	72	16.2	2372	1382	104
Third steep water	12	185	4	3.5	418	194	32
Fourth steep water	16	254	7	3.1	452	244	24
Germinating drum water	–	50	12	7.2	534	274	80
Germinating drum wash water	–	–	9	2.4	456	222	136
Water supply	–	–	–	–	328	125	–

From Ruf *et al.* (1935).

example, initially the biological oxygen demand (BOD) of the steep liquor increases roughly linearly with time (Fig. 9.10). Using larger volumes of liquor results in effluent that is less strong (Tables 9.1 and 9.2 compared with Tables 9.3 and 9.4). Changing steep liquor leads to later effluents that are weaker and grain that is 'cleaner' (Tables 9.1 and 9.3). The maltster must choose between using larger volumes of water and so having cleaner steeped grain but having to pay heavier water and effluent charges, and economizing more with water and reducing costs. 'Dirty' grain will often only malt well if it is washed by using several changes of steep liquor. There have been proposals to use effluent to supplement mashing liquor or to make, by growing microbes, single cell protein (scp) for animal-feeds. These ideas are not used. Heat has been recovered from effluent using heat pumps and used in secondary businesses, for example to heat greenhouses or warm the water used in fish-farming.

Effluent has a high BOD. In most cases it is not acceptable to discharge it directly to local water courses, which rapidly become anaerobic or microaerophilic and become foul and lose the higher life forms. Sometimes untreated maltings effluent can be pumped out to sea. Maltings effluent has been partially purified by allowing it to flow through ozier beds or reed beds or it has been sprinkled onto grassland at the maximum rate of 37 cm (14.6 in)/ha (2.48 acres) per week. Reed beds are sometimes used (usually in sequence) to remove residual nitrogen and phosphorus and some suspended solids from partly treated effluent (Walton, 1995; Maule *et al.*, 1996). Increasingly stringent regulations are being introduced and in most areas maltsters must ensure that the effluent is fully treated either by discharging it into a public foul sewer (with or without preliminary purification) so that it is treated together with domestic sewerage, or by treating it themselves

Tabke 9.2 Quantity and characteristics of effluents from a comparatively small, older type of malthouse. Data relate to the maltings referred to in Table 9.1

Run No.	Total effluent per 24 h (gal)[a]	Barley malted (bushels)[b]	Waste per bushel of barley (gal)[a]	Waste BOD (5-day) (mg/l)	Total organic nitrogen (mg/l)	Nitrogen as free ammonia (mg/l)	Total solids (mg/l)	Total suspended solids (mg/l)	Oxygen consumed (mg/l)
1	443 000	6165	72	385	16.0	6.9	1069	80	278
2	515 000	7245	71	330	12.0	3.5	872	52	220
3	482 000	7190	67	460	14.0	4.1	1000	110	330
4	602 000	6780	89	370	10.0	5.5	864	62	312
5	527 800	7245	73	330	13.0	9.4	876	56	314
6	565 800	7350	77	455	–	–	–	–	–
Average	–	–	75	390	13.0	5.9	935	72	291

From Ruf *et al.* (1935).
[a] American gallons
[b] American bushels

Table 9.3 Typical analyses of steep water

Sample	BOD (mg/l)	Organic nitrogen (mg/l)	Nitrogen as free ammonia (mg/l)	BOD per available nitrogen[a]
Steep 1	2800	132	32	29
Steep 2	2250	75	17	42
Steep 3	1900	64	16	40
Steep 4	490	12	2	61
Composite sample	1860	71	17	36

[a] Half organic nitrogen plus all of the free ammonia nitrogen.
Water use: 13 gal/bushel (1.6 m^3/m^3); BOD: 0.24 lb/bushel (2.9 kg/m^3).
From Simpson (1970).

Table 9.4 Some analyses of malting effluents

		pH	PV[a] (4 h) (mg/l)	PV[a] (30 min) (mg/l)	BOD (5-day) (mg/l)	Total solids (mg/l)	Suspended solids (mg/l)	Total acidity as $CaCO_3$ (mg/l)
A	Steep liquor	6.35	250	157	650	2910	218	87
B	Steep liquor 1st drain	5.90	451	355	2350	3990	145	284
	Steep liquor Final drain	5.65	404	280	4050	3070	112	508
C	Steep liquor Composite	6.40	30.7	16.6	170	408	27.5	26
	Steep liquor Sample 1st drain	6.65	13.6	9.5	80	341	16.0	14

[a] Permanganate value COD.
From Isaac (1976).

to a specified standard. The choice made depends on costs and local circumstances. Some sewerage works welcome effluent from maltings because domestic sewerage has a BOD/N ratio of *ca.* 17 : 1, which is less than optimal for treatment, while malting effluents have a ratio of *ca.* 36 : 1. Increasingly it seems probable that it will become economic to purify the effluent to a sufficient extent to permit at least some re-use in the maltings. Data are available on malthouse wastes and their treatment (e.g. Ruf *et al.*, 1935; Isaac, 1969, 1976; Simpson, 1970; Isaac and Anderson, 1973; Wilderer and Fedder, 1985; Smith, 1986; Lloyd, 1990; Vriens *et al.*, 1990). Problems of treating the effluent include the high volumes of liquid involved, the fact that the volumes and temperatures may be variable, the flows are intermittent and the composition of the liquor is likely to be extremely variable. Variations in composition are compounded by differences in cleanliness of different lots of barley, and the use of different steeping temperatures, contact times or the number of water changes. BOD (5-day) values can range from very low values indeed up to around 4000 mg/l (p.p.m.). Values for all the other parameters also vary widely (Tables 9.1–9.5). Although the

Table 9.5 The effect of treating a malting effluent using a Pasveer ditch

	Screened effluent to Pasveer ditch (mg/l)	Final effluent to stream (mg/l)	Removal (%)
Suspended solids[a]	313	18	94
BOD[a]	1140	12	99
Dichromate value[a]	1501	109	93
Permanganate value[a]			
3 min at 27 °C (80.6 °F)	53	10	81
4 h at 27 °C (80.6 °F)	304	30	90
Total nitrogen[a]	36	7	81
Ammoniacal nitrogen	0.4	0.2	–
Oxidized nitrogen	0.1	0.1	–

[a] Average values over about 3 days.
Briscoe, quoted in Isaac (1969).

values for individual steeps are variable the total BOD/unit weight barley steeped is more uniform and approximates to 2.68 kg/m^3 grain (0.22 lb/bu) (Simpson, 1970). Recently reported volumes of effluent of 1.9, 2.5 and 2.7 m^3/t barley steeped illustrate the large differences occurring between maltings (Wilderer and Fedder, 1985). The levels of suspended solids and the oxidizable organic materials (variously measured as the BOD, and various types of COD (chemical oxygen demand: determined using the ability of the liquor to reduce potassium permanganate or potassium dichromate) also vary widely. The BOD is determined by bioassay and is the amount of oxygen needed to totally oxidize the susceptible organic materials in 5 days under standard conditions. The COD tests include organic materials that are not oxidizable by microbes. Usually COD refers to values found using dichromate but other results refer to permanganate values (PV).

The costs of effluent treatment are high and in the UK charges are based on the volume and the costs of handling, pumping and receiving each cubic metre ('overhead' expenses), giving a 'reception charge' (A), and a volume charge (V). Additional costs per 1 m^3 are the cost of the biological treatment (B), in which the BOD value is of crucial importance, the cost of sludge separation, treatment and disposal (C) and the cost of primary screening and settling. There are variations, but in the UK charges for sewerage treatment are usually based on the Mogden formula or some variant of it:

$$\text{Cost} = \text{total treatment charges/m}^3 \text{ (230 gal)}$$
$$= A + V + B\,(ME/MS) + C\,(SE/SS)$$

where ME and MS are the average strengths (oxygen demand of the settled liquid) of the effluent and the combined sewerage received at the treatment (sewerage) works, respectively. Similarly, SE and SS are the values for the suspended or 'settleable' solids contents of the trade effluent and the mixed sewerage and effluent, respectively.

In addition there are likely to be restrictions on the rate of flows, particularly peak values, the total daily volume and peak values of BOD, suspended solids, temperature and pH. All these points must be agreed upon and are specified in appropriate 'consents' or 'directions'. Maltsters attempt to reduce effluent charges by paying attention to all the factors in the Mogden formula. Often there are penalties for operating outside the agreed limits. There are many ways in which malting effluents may be treated and the technology chosen will depend on the size of the maltings, the local conditions, whether the effluent is mixed with other effluents (e.g. from a brewery or distillery) and how thoroughly the effluent must be purified (Vriens *et al.*, 1990).

If maltsters can avoid carrying out effluent treatment, for example by using the facilities of the local sewerage works at a moderate cost, then they will usually do so. Sometimes partial purification is economically advisable. Complete sewerage treatment is undertaken with reluctance since continually operating plant to produce a high-quality treated effluent is a specialized operation. It is neither desirable nor economic to dilute effluent with fresh water to reduce its strength. Most maltsters that deal with their own effluent use aerobic primary treatments and, in a few cases, may treat the sludge that is produced anaerobically. However, anaerobic systems are frequently proposed, and are said to be cheaper (Anderson and Saw, 1986; Reed, 1990). Treatments begin by screening the effluent to remove floating corns, husk, rootlets and other debris. The sieves may be flat or cylindrical, or bands or discs. The solids are continually moved from the screens by wipers or brushes. Often the liquid is received in a holding or mixing buffer tank, from which it can be withdrawn steadily, so evening out the flow, and in which mixing occurs leading to a 'feed' of more uniform composition for the biological treatment system. Some primary settling may occur in this tank and so it must be desludged from time to time. The next objectives are to remove biologically oxidizable materials and the suspended solids remaining after screening. This is achieved by a combination of biological oxidative treatments and settling or precipitation. (In principle, anaerobic 'digestive' processes might be used instead.) Under the aerobic conditions provided, microbes convert the dissolved susceptible organic materials into carbon dioxide, water and extra microbial mass, which accumulates and is recovered as sludge. The microbes tend to clump together and colloidal and particulate materials adhere to them so that they too are removed. The flow of liquid through settling tanks must be slow and steady so that solids can separate efficiently. Sudden surges of liquid cause carry-over of solids and so the necessary purification is not achieved. Initially the sludge may only contain 0.5–2% solids. Settling tanks all tend to be large and comparatively shallow. The disposal of sludge is potentially difficult and expensive. Sometimes it is accumulated in holding tanks and may be thickened to 4–5% solids by settling, and then is transported to be dumped or spread on fields as a soil conditioner and fertilizer. In contrast to some others, sludges from maltings **do not** contain significant residues of toxic heavy metals. In large treatment plants, the sludge may be 'dewatered', often with the addition of lime, under vacuum, by pressure filtration or by heat to reduce the volume. Under most conditions, it is not economical for maltsters to do this. Perhaps the most satisfactory arrangement is to biologically digest the sludge anaerobically at 30–35 °C (86–95 °F): the volume is substantially reduced (e.g. by 50%), the sludge is stabilized and swells less and carbon dioxide and methane are generated. The last can be burned as a source of heat and may be used to keep the digester warm.

The common oxidative treatments used in maltings are either based on trickling filter beds (bacterial beds, percolating filters, fixed film reactors) or 'activated sludge' (suspended growth) systems. The advantages and disadvantages of some different systems have been tabulated by Walker (1981). Trickling beds have the effluent sprayed over a porous bed, often 2–3 m (*ca.* 6.6–10 ft) deep, of granular material (e.g. *ca.* 3.8 cm (1–5 in) diameter) such as gravel, broken rock, coke or blast-furnace slag. The beds are usually circular but may be rectangular in cross-section. The liquid percolates downward over the solid surfaces and is in contact with an upward stream of air. The biology of such beds is extremely complicated but, in principle, a 'zoogleal' bacterial slime forms on the solid support and it is here that oxidation occurs; suspended material is retained and the bacteria accumulate. Sometimes it is necessary to add ammonium salts and phosphates to the effluent to maintain the population of microbes. Dislodged solids are carried away in the liquid stream and are collected in a separating tank. The bed must not be allowed to dry and the effluent must not be too concentrated, so some recirculation of treated effluent may be used (Fig. 9.11). The daily BOD load probably should not exceed 0.178 kg/m^3 per day (*ca.* 0.3 lb/yd^3 per day). Sometimes it is preferable to use at least two beds working in sequence, with facilities to alter the order of use (Fig. 9.11b). Such beds may be more than 98% efficient in reducing the BOD; for example, in one case a filter with recirculation reduced the BOD value of an effluent from 1500 to 20 mg/l. Such trickling beds are low filtration-rate devices and are large. Overloading must be avoided since this can lead to 'ponding'. There can also be difficulties with offending odours and flies breeding in the filter beds. More compact, high-filtration rate devices have been used, for example as a preliminary treatment preceding activated sludge treatment. In these devices, a plastic material with a high surface area, used to support a film of microbes, is packed in a tower covering a relatively small ground area, and the effluent is sprayed in at the top. Such towers are able to tolerate higher BOD loads than trickling filters. In other devices, rotating biological contactors, supporting a mass of biological slime, are partly submerged in the effluent to be treated. The 'captive biomass' and its support is rotated, exposing the wetted biomass to the air

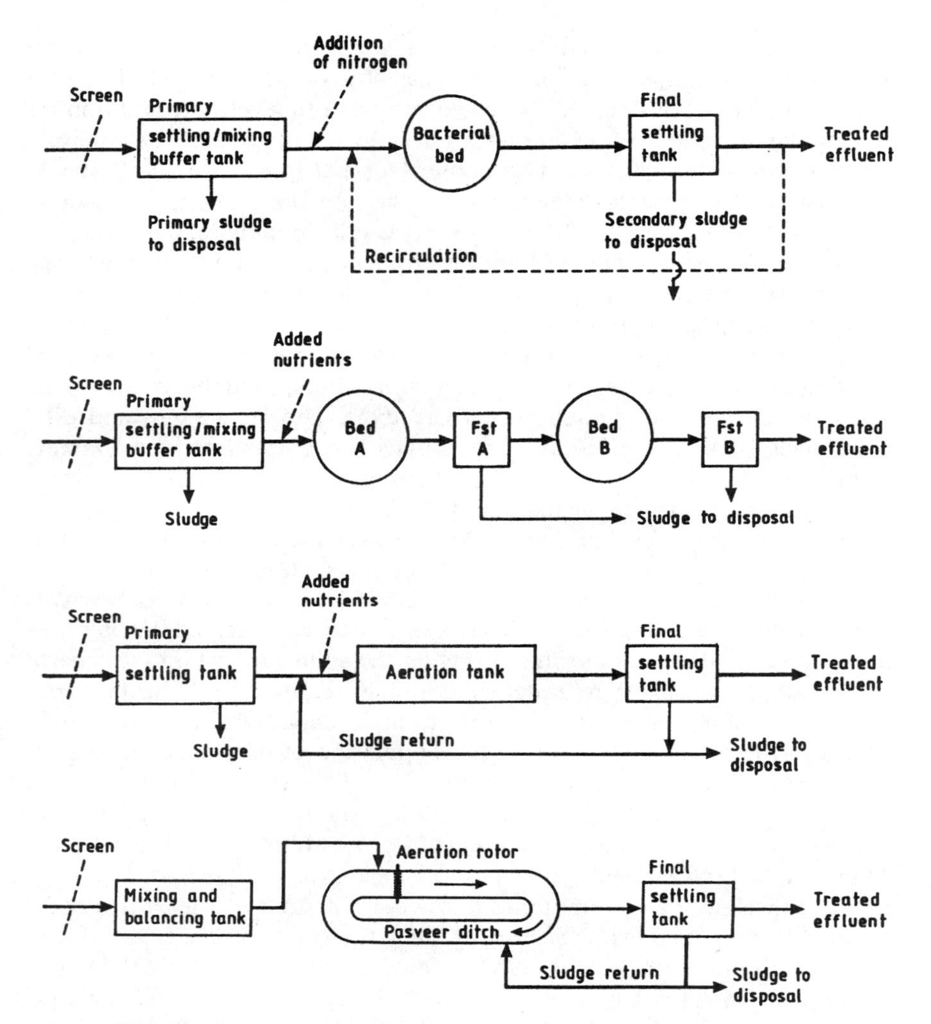

Fig. 9.11 Various plants used by maltings for treating effluent. (a) Single filtration or percolating bed system with facilities for recirculation of water; (b) double filtration, it is possible to alternate between the two beds A and B; (c) a standard activated sludge process; (d) a Pasveer ditch type of activated sludge plant. (After Isaac, 1969.)

and carrying air under the liquid. The supports may, as examples, consist of a series of high surface-area discs or plastic 'forms' packed in rotating, perforated drums. Such devices are thought to be economically attractive, they consume little power and may be in use for treating malting effluents (Smith, 1986; Vriens *et al.*, 1990; Kühtreiber, 1995).

In activated sludge systems, the effluent is aerated and agitated, the bacterial mass being kept in suspension. Oxidation occurs in the liquid. Solids are separated by subsequent settling, and a proportion of the sludge, containing microbes, is returned to the oxidation vessel to maintain a high concentration (Fig. 9.11c). A number of different types of plant is used. In 'fill and drain' plants, aeration tanks are filled in turn, supplementary sludge may be added or surplus removed and the contents are aerated and mixed. When adequate oxidation has occurred, agitation ceases, settlement is allowed to occur, the cleaner water is removed, the sludge amount is adjusted and the tank receives the next charge of effluent (Wilderer and Fedder, 1985; Vriens *et al.*, 1990). Such plant is not affected by variations in effluent flow. Another arrangement is to have two aeration/sedimentation tanks working in sequence (Eyben *et al.*, 1985). In other systems the effluent may be treated by aeration in deep shafts, or the effluent may be gassed with pure oxygen. A widely used plant is the Pasveer or 'oxidation ditch' (Fig. 9.11d). Each ditch, which is a channel, elliptical in plan, has a trapezoidal cross-section and is about 1 m deep. A 'balanced' mixture of effluent is slowly and continuously added and treated effluent is continuously withdrawn; it is usually transferred to a settling tank, although in some plants settling is allowed to occur in the ditch. Sometimes the effluent is pretreated using a 'rapid flow' filtration/oxidation system. Rotating brush aerators or rotors continuously aerate the effluent and cause it to circulate around the ditch. At intervals, sludge from the settling tank may be returned to the ditch to maintain the concentration of suspended, oxidizing microbes. Average retention times are 2–3 days. Such plant is comparatively inexpensive to build but may have problems with 'sludge-bulking' (Isaac, 1969). One plant, having three ditches each with 178 000 UK gal capacity (total 534 000 gal (approximately 24 300 hl)) treats a BOD load of 783 kg (1725 lb), equivalent to a population of *ca.* 14 000, with variation in BOD concentration of less than 1000 to more than 1600 mg/l. The results of tests made with the first ditch are shown in Table 9.5.

9.8 Steeping

9.8.1 Early steeping systems

Supposedly, steeping was originally carried out by immersing sacks partly filled with grain in running water and leaving it exposed to the air to drain between immersions (Chapter 1). However, while this process may have continued to be used on small farms, it became normal to steep in cisterns (ditches, vats, fat(t)s, uiting-fats, wetting-fats) (Markham, 1615; White, 1860). Such steeps were shallow, often being less than 1 m (39.4 in) deep. Sometimes they were of wood, being referred to as tubs and being circular in cross-section. However, more usually they were rectangular in shape (by law in the UK until 1880) and were often placed on the ground floor.

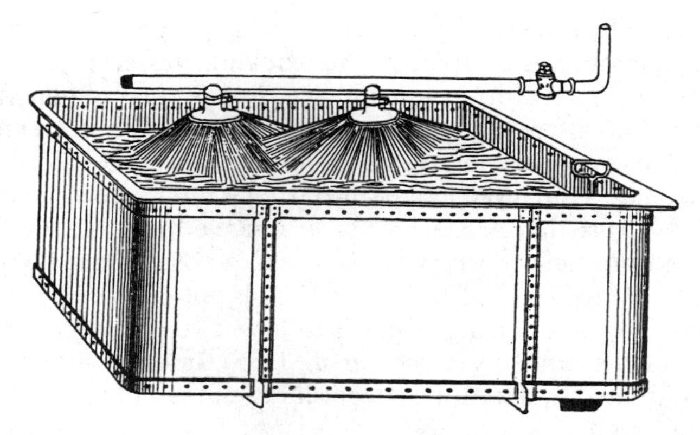

Fig. 9.12 A cast iron, flat-bottomed steep tank of 1875 (after Schönfeld, 1932).

(Subsequently the wet grain had to be thrown or pulled, using baskets and pulleys, or (later) elevated to higher floors.) Cisterns were constructed of stone, cement or concrete, or they were faced (including tiled) brick, lead or (latterly) iron (Fig. 9.12). The base had a slight fall, often to a gulley covered with a perforated grid to retain the grain while allowing the passage of water. This gulley in turn sloped down to a drain-hole, which could be closed with a bung. In later designs, water could be run in from above or (unusually) from below, and also water could be sprayed or sprinkled from above so that as water was drained from below the grain could be rinsed with well-aerated water from above. Originally, the steeped grain had to be thrown out (cast) by men using shovels. In recent years, it has been unloaded using an electrically driven elevator auger, with a 'sides to middle' sweep auger feeding the elevator at the base.

To reduce labour, simple cisterns were, in many instances, replaced by hopper-bottomed steeping vessels, of either square or circular cross-section. These vessels, often called self-emptying and self-cleaning steeps, were originally made of iron sections bolted together (Fig. 9.13). Such vessels could be filled with water from above or below (when the grain was agitated and mixed) and could be quickly filled with grain by running it down a chute from an overhead garner. The original pattern is the direct ancestor of a wide variety of more complex steeps. At the end of an immersion period, the water was withdrawn through a perforated zone at the base of the cone, so the grain could be drained. Opening a valve allowed the grain to fall either directly onto the floor or via chutes to floors further down the building. Any few residual grains were dislodged with a broom or water from a hose. In order to permit the grain to slide out, the slope of the basal cone must have an angle of at least 45–50° from the horizontal; this sets limits on the acceptable size of conical-bottomed steeps because as the steep increases in size the cone becomes deeper and the increased pressure at the

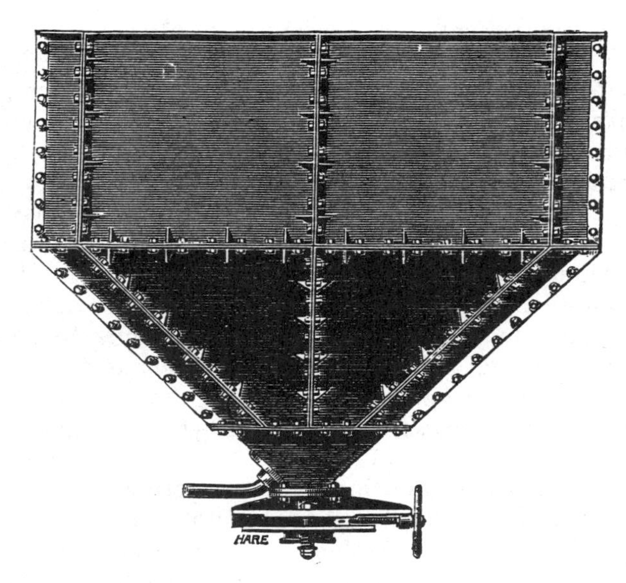

Fig. 9.13 A simple iron, hopper-bottomed, self-cleaning and simple self-emptying steeping cistern (Stopes, 1885).

base of the cone damages the grain. In the 1950s the maximum capacity was regarded as about 15 t barley (de Clerck, 1952) but the consensus now is that up to 50 t is acceptable (Arter and Olsen, 1986). A simple but costly alternative (because of the need for multiple valves, etc.) is to have a vessel with multiple hoppers in the base (Fig. 9.14). This overcomes the problem of excessive vessel depth.

The original hopper-bottomed, self-emptying steeps and the flat-bed cisterns have given rise to long pedigrees of steeps. With the exception of some spray-steep systems (e.g. the Domalt and Solek systems) flat-bottomed and conical-bottomed steep systems are almost universally used.

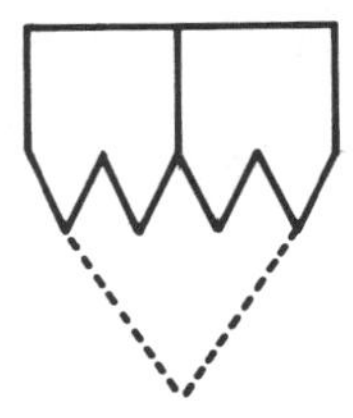

Fig. 9.14 The advantage of multiple hopper bottoms. The angle of the cone is defined by the need to have the steeped grain flow freely during emptying. By having multiple hopper bottoms rather than single cones, the depth of the steep, and hence the pressure on the grain in the lower layers is reduced.

9.8.2 Novel steep designs

During more than 100 years of malting developments, numerous steep designs have been tried. Sometimes it is obvious that the complexity of a design must have made it costly and difficult to clean and maintain, or some simpler means of achieving a result was found. In other instances, the reasons for a particular approach being discontinued is unclear.

The process of sprinkling grain – using watering cans or spray hoses in old floor-maltings, or sprinklers on turners or within drums – is, illogically, often not considered to be part of steeping. Yet spray steeping is practised and is an integral feature of Solek and Domalt plants. By spraying only slightly more water than the grain can take up, water usage is minimized, as is effluent production. However, the grain is not thoroughly washed and a preliminary wash is a necessary step, preceding spray steeping.

Hruby patented a device, in Prague in 1879, which consisted of a series of perforated cylinders mounted in an endless series on a paternoster-like device. By cranking a handle, the cylinders were rotated around their longitudinal axes, mixing the grain within, and the paternoster moved, so each cylinder moved downwards, around, up, over and down. Steeping, and washing, was carried out by filling a trough with water into which the chain of cylinders dipped, so each was immersed in turn. The grain was germinated and kilned in the same cylinders. Presumably the grain was well washed and aerated during steeping and the crank system provided a simple way of turning small quantities of grain (Anon., 1903; Heinzelmann, 1907).

The idea of steeping in drums has been tested many times. Modern drums are usually equipped to spray the grain with water during turning but they cannot be partly filled with water for steeping. To allow steeping in a drum, it must be built to withstand a large weight of water in addition to that of the grain. The alternative was to place the grain in drums with circular (1885) or hexagonal (1886) cross-sections and perforated walls. These rotated around their long axes while dipping in troughs of water. The grain would have been well cleaned and aerated during this type of steeping (Heinzelmann, 1907). A washing drum used in northern Germany processes each batch of grain for 30–60 min and treats 180 t barley every 5 h.

In 1885, Hauschild suggested using a large propeller, rotating horizontally around a central shaft, to agitate grains in circular, flat-bottomed steeps (Heinzelmann, 1907). An alternative form of propeller-mixed steep was described by Wild (1910) (see below). In 1878 Saladin suggested that steeping might usefully be accelerated by periodically placing steeps under vacuum. This idea is unsound, as wide fluctuations of pressure caused by using alternate exposures to vacuum and to air or when pumping grain–water mixtures frequently impair subsequent germination.

9.8.3 Hopper-bottomed steeps

Self-emptying (and largely self-cleaning) steeps have developed in several ways. In the UK, steeps are increasingly often made of corrosion-resistant stainless steel or another alloy, but most of those extant are probably made of mild steel and as such they need to be painted, usually with epoxy or vinyl polymer products, to minimize corrosion. In mainland Europe they are sometimes made of concrete, fitted with metal hoppers. It has been proposed that successive layers of paint should be in contrasting colours so that wear is readily detected (Adamic, 1977).

Steeps are loaded from overhead conveyors or from a garner. The running grain generates dust, and so it often falls through a ring of sprays, which 'quench' the dust, and into water. When minimizing water use is not essential, fresh water may be admitted to the base of each steep and the water 'weired over', through overflows at the sides of the steep, carries away dirt, dust, floaters or light corns. These overflows often have slides to regulate the depth of water over the grain. The overflow pipes lead to grain traps or a screen to remove solids before the effluent water is discharged. When steeps are emptied, or 'cast', the grain may be transferred after draining, using suitable conveyors. Such dry casting subsequently leads to more rapid chitting, if the grain has not already been induced to chit in the steep. Indeed, an old patent described blowing warm air over the grain as it was conveyed from the steep to dislodge the film of surface water and hasten the onset of germination. The alternative method (wet casting) involves transferring the grain slurried in water. This is simple and the piping is kept clean. Provided excessive pumping pressures are avoided (<15 psi) and the

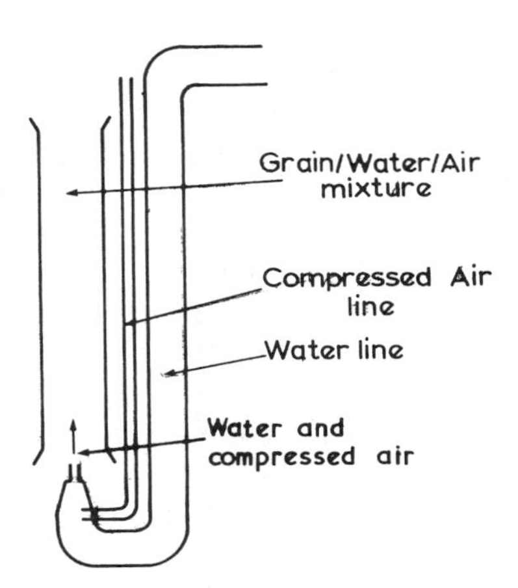

Fig. 9.15 A small, portable steep rouser introduced in *ca.* 1904. When immersed in the steep, the upward jet of water and the stream of compressed air induced an upward flow of grain suspended in water (after Heinzelmann, 1907.)

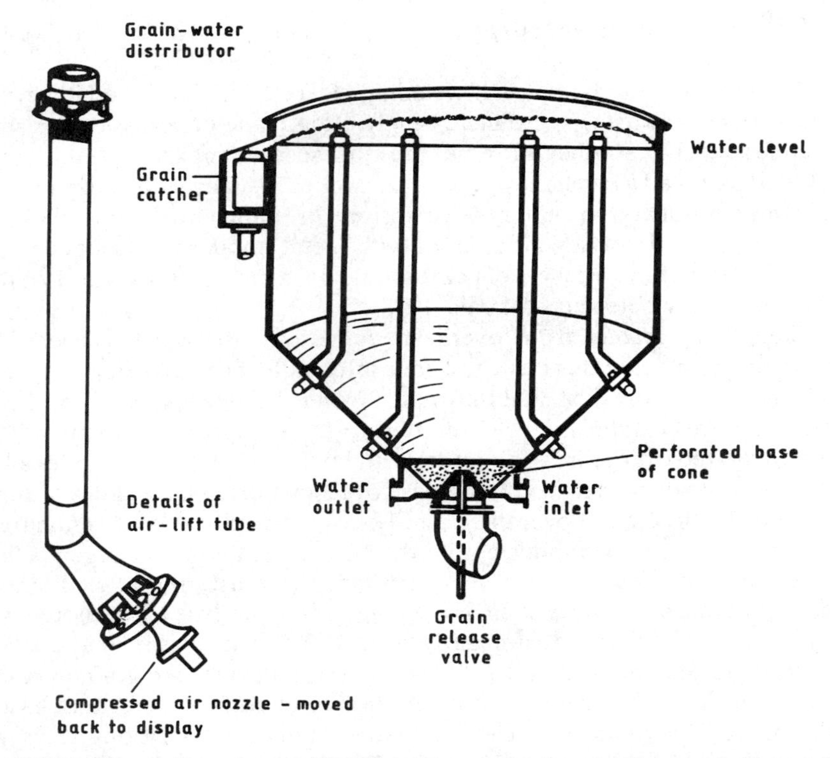

Fig. 9.16 Details of a Doornkaat air-lift tube and how such tubes could be mounted in a conical steep (*ca.* 1904). The flow of grain, water and entrained air could be regulated by altering the setting of the distributor.

impeller design is carefully chosen this works well, but it can lead to irregular chitting and is considered undesirable with grain that has already chitted. Most British maltsters prefer dry-casting.

As the desirability of mixing and aerating the contents of steeps was recognized, a number of devices were tried, several of which are still in use. In 1885 Bothmer, in Leipzig, pointed out that flat-bottomed and hopper-bottomed steeps could be aerated by introducing compressed air through a web of perforated tubes placed in the base, an approach still widely employed (Heinzelmann, 1907). Later air-lift tubes were introduced. Advancing, and more expensive, technology created an investment problem for maltsters working on a small scale. In 1914, the suggestion was made that air be drawn through grain, during air-rests, using a draught created by a chimney with a small fire at the bottom. Another device was a small, portable steep rouser, introduced in *ca.* 1904, which could be moved around within a steep or between steeps (Fig. 9.15). With this device immersed in the steeping grain, water and compressed air were directed upwards into

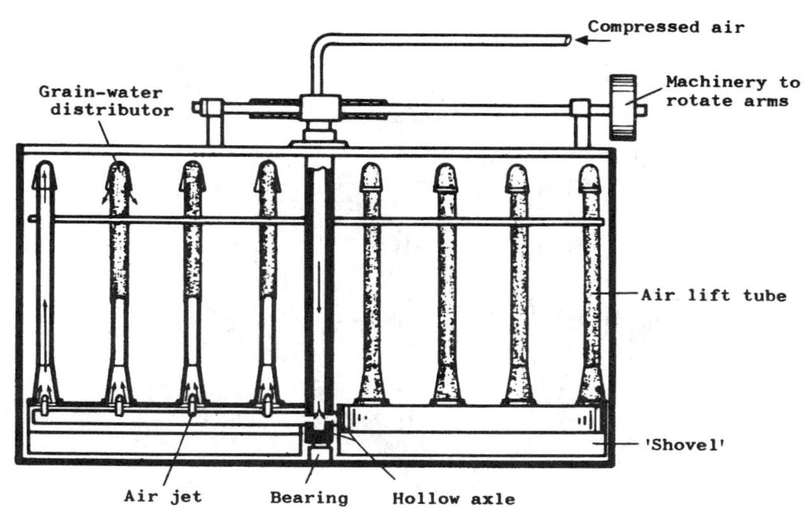

Fig. 9.17 A flat-bottomed Doornkaat steep in which the air-lift tubes were carried around on a moving support (pre-1905). The latter was equipped with shovels or scrapers that ensured that even the grain at the base of the steep was disturbed and mixed (after Schönfeld, 1932; Briggs, 1987a).

the base of a wide tube. The air–water–grain mixture rose up the central tube and out of the top. Fresh grain and water entered the air-lift tube at the base. Thus the steep contents were turned over and aerated. Recently it has been suggested that water be pumped from the base of a steep and sprayed over the surface to aerate it and (by recirculating the water) equalize the temperature in the steep.

Around 1900–5 a series of flat-bed and hopper-bottomed steeps were introduced associated with the name Doornkaat (Heinzelmann, 1907; Schönfeld, 1932; Vermeylen, 1962; Briggs, 1987a). The key device was an air-lift tube either mounted in conical-bottomed steeps or in flat-bed steeps (Figs 9.16 and 9.17). Compressed air, introduced by a nozzle at the base, carried grain and water upwards and out of the top; thus a grain–water circulation was set up. A distribution device on the top could be adjusted to regulate the spread of grain and water as they fell down onto the surface of the steep. If this device was screwed shut then circulation was prevented and compressed air spread out from the base of the tubes and agitated the steeping grain.

The concern to completely mix and aerate steep contents led to some complex designs. For example, in a flat-bed steep (Fig. 9.17) and in the conical-bottomed Germania design (Fig. 9.18), sets of aeration tubes were mounted in rows on mechanically driven, rotating frames. The tubes were over scoops or shovels that moved the grain from the base and lifted it into the rising stream of air–water–grain which was ascending the air-lift tubes and dropping back into the surface of the steep.

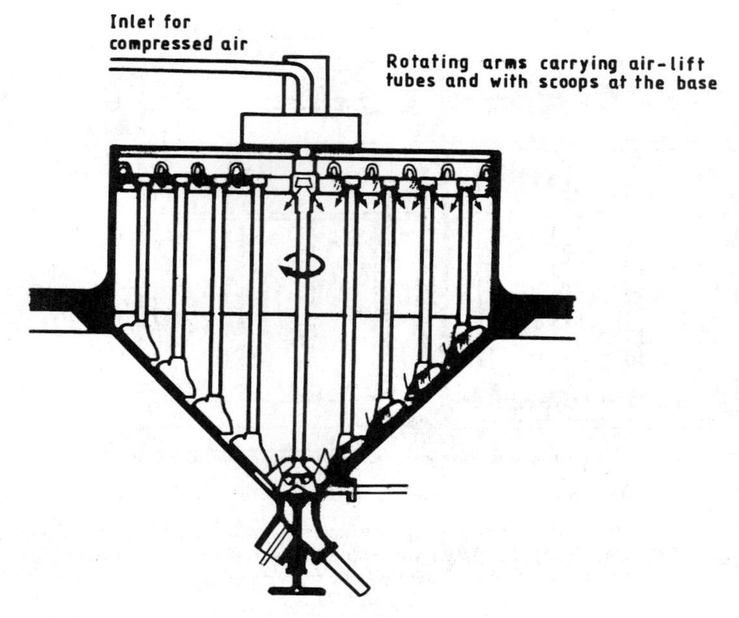

Fig. 9.18 The Germania conical steep of *ca.* 1905 in which each of the air-lift tubes, which rotated in sets, was equipped with a scoop to ensure that the grain was moved from the base (after Schönfeld, 1932).

Other complex designs were proposed, such as that of Topf und Sohne in 1907 (Schönfeld, 1932), in which the conical steep was divided into sections with radial walls. Slides allowed access of grain and water from a particular section to the base of a large, central air-lift tube, which elevated it and dropped it into another compartment. Thus the grain could be turned and transferred between compartments.

Another, simpler system had a wide air-lift tube bent over at the top, which drew in grain and water from the base of the cone. As the tube rotated, the grain was redistributed around the top of the steep (Fig. 9.19). In this steep, water could be let in or withdrawn from the base of the steep via a perforated section of cone that held back the grain; compressed air could be pumped in through the cone at the base as well as into the air-lift tube. This steep was equipped with an overflow that permitted water, added at the bottom, to 'weir over' at the top, removing floating dust, husk and light corns. A more common design, with features that are still used in many steeps, had aeration rings (perforated tubes) in the cone of the tank and a centrally positioned air-lift tube. In the example shown in Fig. 9.20, this led to a rotating 'geyser'. The grain–water–air mix elevated up the tube moved sideways down the two rotating arms and out of the two inclined nozzles. The reaction of the discharge caused the arms to rotate so that

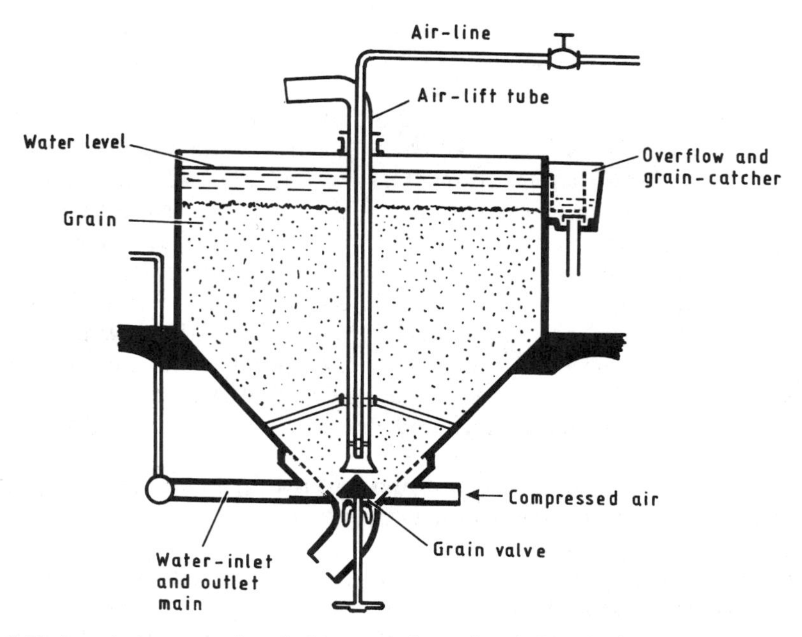

Fig. 9.19 A conical steep equipped with a central, rotating air-lift tube (after Schönfeld, 1932).

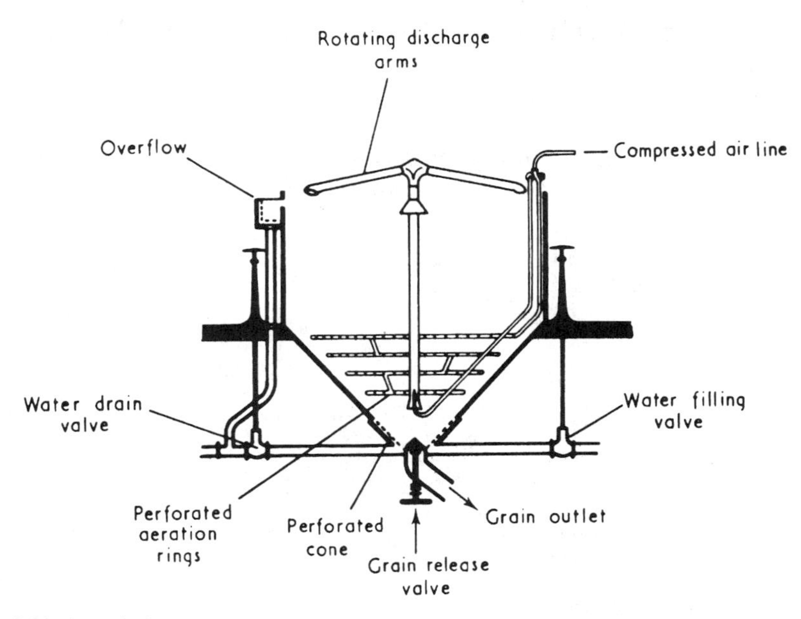

Fig. 9.20 A conical steep-tank equipped with aeration coils and a central air-lift tube fitted with rotating discharge arms (a 'geyser') (after de Clerck, 1952).

grain lifted from the base of the cone was spread around over the surface of the steep.

To ensure that grain was well mixed, it might be transferred from one steep tank to another, either pumped in water (pumped over), blown over with compressed air or 'dropped' under gravity. In other cases, grain and water were drawn from the bottom of a steep and pumped back into the top of the same steep (pumped round). Suggestions that grain should be washed by changing the water during transfers between steeps had been made by *ca.* 1884. Many variants of aerated, hopper-bottomed steeps have been used. For example, instead of having internal aeration tubes, which provided an obstruction to the free flow of grain, necessitating hosing to finish the emptying process, some steeps had aeration holes with inlet pipes that ended flush with the inner face of the cone (Fig. 9.21). The complexity of the external tubing and complicated maintenance discouraged this construction.

When pumping grain and water round or 'over' (between steeps) distribution cones or water separators may be used (Fig. 9.21). With distribution cones, the incoming stream of grain and water hits the apex of an inverted metal cone and is spread outwards to be aerated and uniformly distributed as it falls back into the steep. With water separators the cone is perforated, to permit the passage of water but not grain. The water from the cone is carried to drain while the 'de-watered' grain falls into fresh water in the steep tank below (Fig. 9.21). A water-separating cone may have a metal ring around it to prevent splashes of water and grain going beyond the limits of the steep tank.

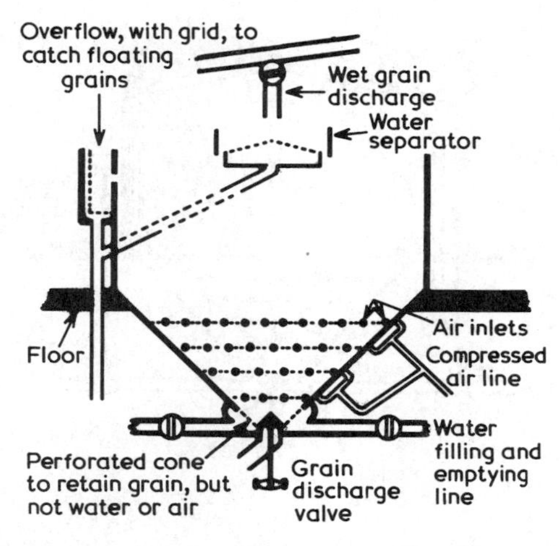

Fig. 9.21 A self-emptying steep equipped with a conical water separator for removing excess liquid from a grain–water stream; air-inlet holes are flush with the inner face of the cone (Briggs, 1978).

Aeration of grain by 'blowing' during 'dry' periods, between immersions, was found to be inefficient and, as the compressed air was heated by the compressor, the grain was warmed and dried. By the mid-19th century, maltsters had realized that as water was drained from the base of a steep fresh air was drawn in from above. By the end of the century, it was understood that supplying a continuous stream of fresh air during the dry air-rest periods between immersions might be beneficial, an idea that had occurred to Galland by 1873 (Hudson, 1986a,b). A complex aeration system was used in the Söding-Winde 'Reform' steep (*ca.* 1909); one version of which is shown in Fig. 9.22. Two sets of tubes were present. Compressed air was pumped through air-lift tubes to mix and aerate the grain while it was under water. During the dry air-rest periods, air could be sucked down through the bed of grain and out through the large inverted cones at the base of the large air extraction pipes. These were often mounted concentrically around the air-lift tubes. This complex design was probably expensive to build and difficult to clean. Apparently no British maltsters used these steeps. At the end of the 1970s most steeps in the UK were comparatively simple conical-bottomed vessels often with arrangements for aeration while the grain was

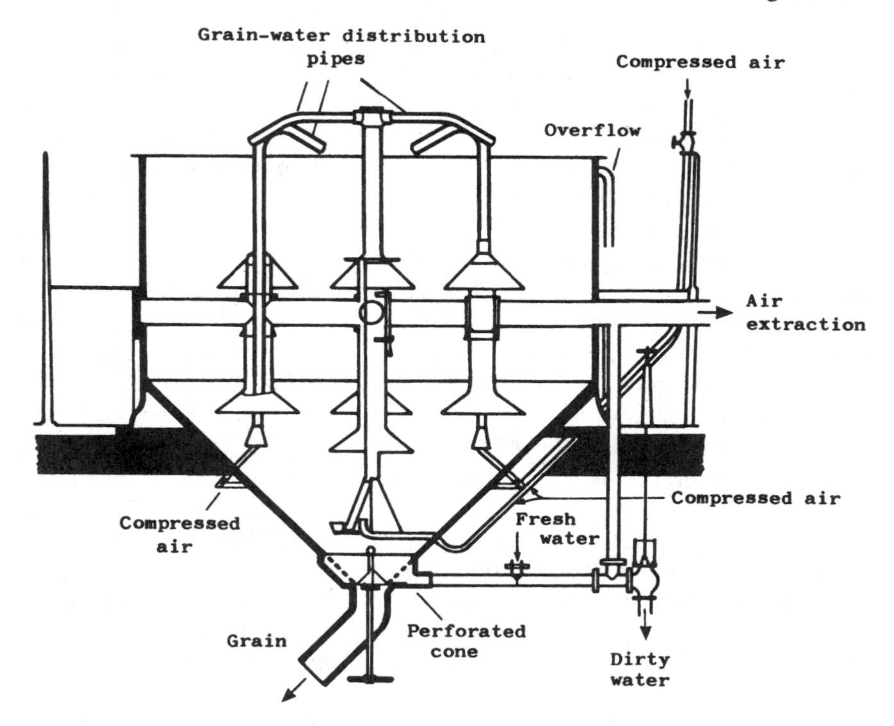

Fig. 9.22 One version of the Söding-Winde 'reform' steep, of *ca.* 1909, in which grain under water could be aerated and in which, during dry periods, air could be sucked down through the grain (various sources; Briggs, 1987a).

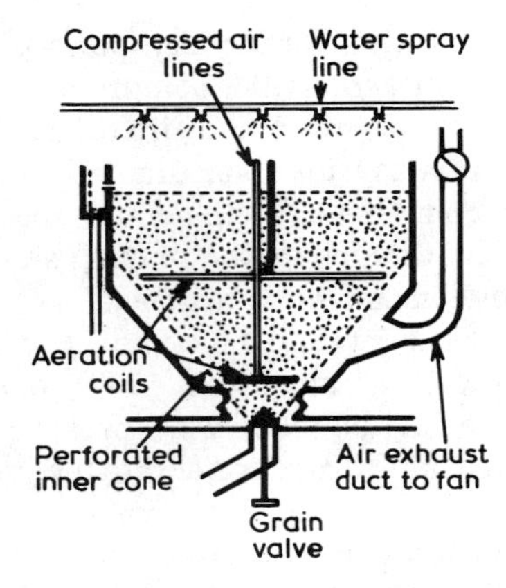

Fig. 9.23 A conical steep with a fully perforated cone for carbon dioxide extraction, and aeration coils for use when the grain is immersed. The overhead water sprays are to damp the upper layers of grain to prevent drying during periods of downward ventilation (Briggs, 1978).

submerged. Frequently the grain was transferred between sets of steeps at particular stages of the process to ensure that thorough mixing had occurred. Recognition of the desirability of carrying out grain ventilation by suction, or 'carbon dioxide extraction' as it is frequently called, led to a number of makeshift devices and then the appearance of steeps where the equipment for downward ventilation was built in. Downward ventilation was adopted because (i) it avoided pumping warm air into steeps from a compressor, and (ii) because as carbon dioxide is a heavy gas it tends to drain downwards and can be withdrawn from the base of the steep. It is not thoroughly displaced by an upward stream of air introduced at the bottom of the grain bed. In some instances, suction was applied to the base of the steep, the carbon dioxide-laden air being withdrawn from the perforated base of the cone. In other cases a wide tube with a perforated 'cage' at the base (to exclude grain) hung down into the cone from above and suction was applied to this during air-rests. Similar devices have been used, with some success, in some old, shallow, floor-malting cisterns. Two problems became apparent. Firstly, the airflow through the grain was uneven so that, although subsequent germination was more vigorous than in non-ventilated steeps, it was uneven. Grain from poorly ventilated parts of the steep grew less well than that from well-ventilated parts. Secondly, the downward flow of air into the steeps tended to dry the top layer of grain. A third problem, not initially apparent, was that the grain could be unduly warmed in summer and chilled in winter by the ambient air. Attempts to overcome this are now being made.

The first widely adopted steep tanks, designed for carbon dioxide extraction (downward aeration), had fully perforated cones in the base to ensure that suction was applied across the complete base of the grain mass

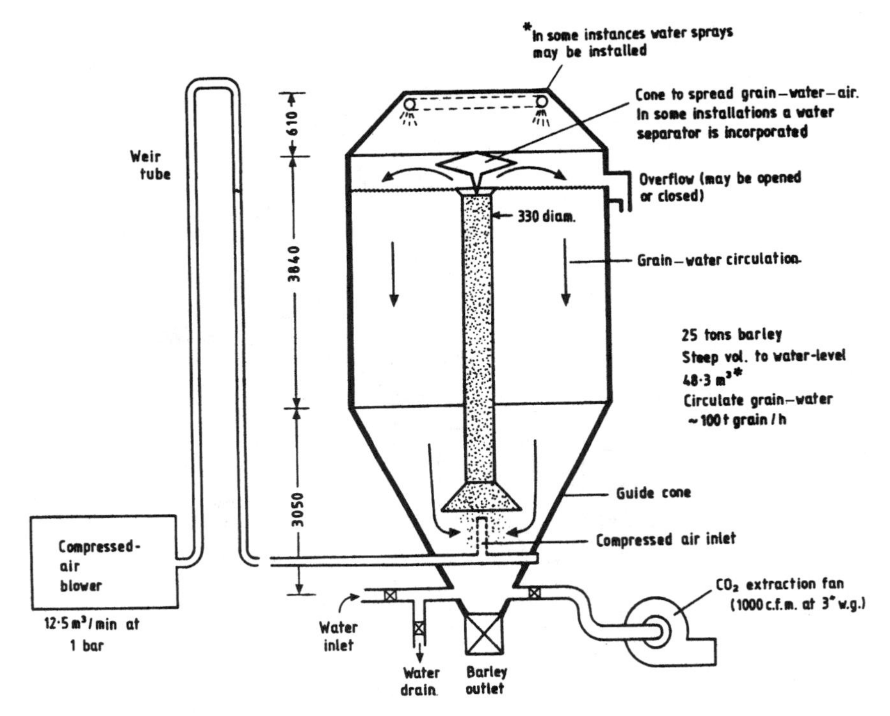

Fig. 9.24 A conical steep equipped with a central air-lift tube, facilities for carbon dioxide extraction at the base and overhead water sprays. Agitation and grain–water circulation is vigorous in these steeps. Heights given in mm (after Aalbers et al., 1983.)

to achieve a more even airflow. In addition, fine sprays were mounted above the steep to humidify the air and damp the top layer of grain as the air was drawn downwards (Fig. 9.23). The wide air outlet tube was usually carried upwards to act as a weir to hold back the water, before bending over in a 'swan's neck' and going down to an air-extraction pump. A common refinement was to divide the plenum, between the large perforated cone and the outer water-retaining solid cone, into horizontal zones, commonly three, by partitions. Different degrees of suction were applied to each zone to obtain a more uniform airflow through the bed of grain. A major problem was cleaning the air plenum to an adequate standard. Often these steeps were used to make a malting plant more flexible: either to extend steeping or germination times or to increase overall capacity.

A refined form of conical-bottomed steep has been described (Aalbers *et al.*, 1983) and steeps of this type are used in the UK (Fig. 9.24). Vessel capacities may vary between 25 t and 50 t barley, the latter being regarded as an upper value, in vessels 8.3 m deep and 5 m in diameter. The angle of the basal cone is 60°. The large, central air-lift tube is telescopically adjustable and has a spreader cone at the base to ensure that the entire grain bed is circulated. Often steeps are aerated for 10 min in every hour. In contrast to

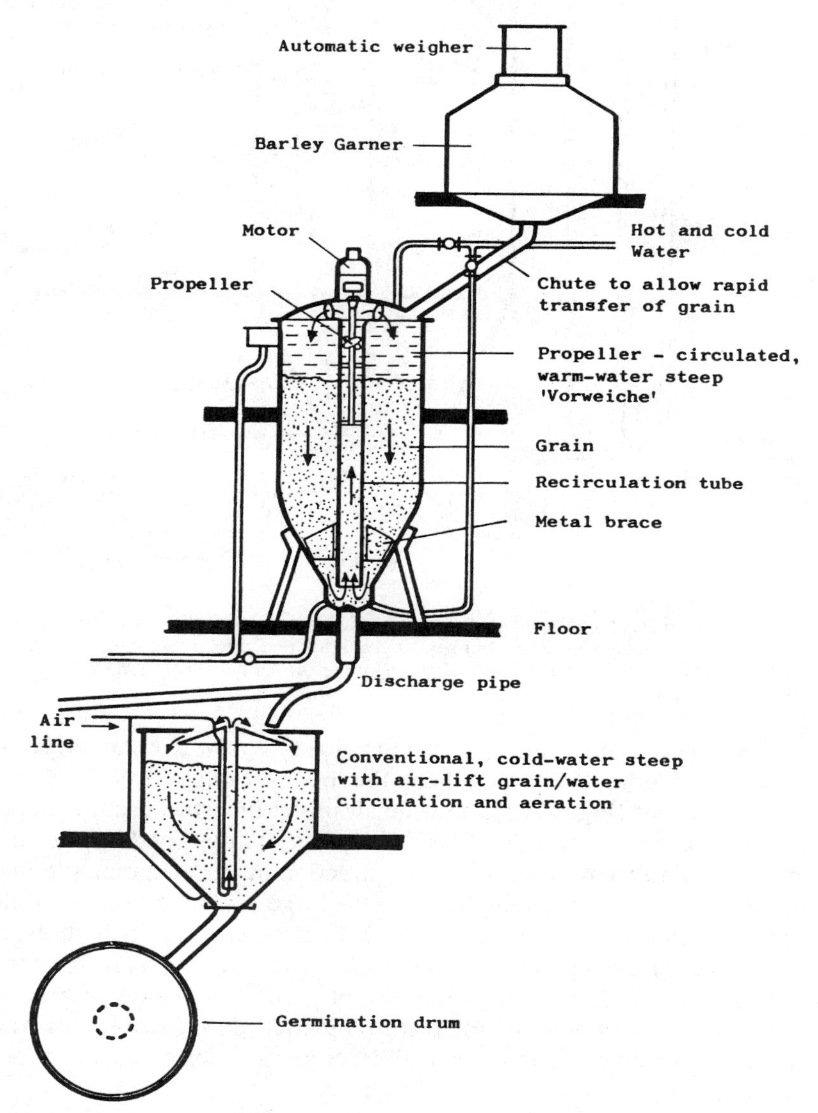

Fig. 9.25 A warm-water steeping system designed by Wild (1910) (Briggs, 1987a). A weighed load of barley from the garner was quickly discharged into the warm-water steep, which was continually mixed and circulated by the propeller working in the central lift tube. At the end of the warm steep, the drained grain was quickly transferred to a cool steep where it was rapidly mixed in cold water with an air-lift tube. The fully steeped grain was transferred to a germination drum.

older steeps, the circulation is vigorous, the grain being moved at the rate of 100 t/h. Air sprays may be provided to damp the surface of the grain during periods of carbon dioxide extraction. The quite exceptional vigour and turbulence of the 'aeration' results in thorough grain cleaning. It may be preferable, and about equally costly, to build six × 50 t conical steeps, rather than one 300 t flat-bed steep, with the advantage of 20–30% less water usage, fewer mechanical devices to maintain (because conical steeps are self-emptying) and much easier cleaning (Aalbers *et al.*, 1983). Others have indicated that the difference in water usage between conical and conventional flat-bed steeps is much greater (e.g. 0.8 against 1.3 m^3 water/t barley steeped/wetting; Gibson, 1989). There is a trend towards building steeps in corrosion-resistant or stainless steels. With these materials, a wide range of cleaning agents can be employed without risk of causing structural deterioration. In some maltings, the steeps are equipped with covers to allow automatic in-place cleaning (CIP), using internally mounted spray systems.

Two unusual types of steeping are warm-water steeping to achieve rapid water uptake (e.g. at 40–50 °C (104–122 °F)), which was widely investigated around 1900 and again in the 1960s, and resteeping, the reimmersion of

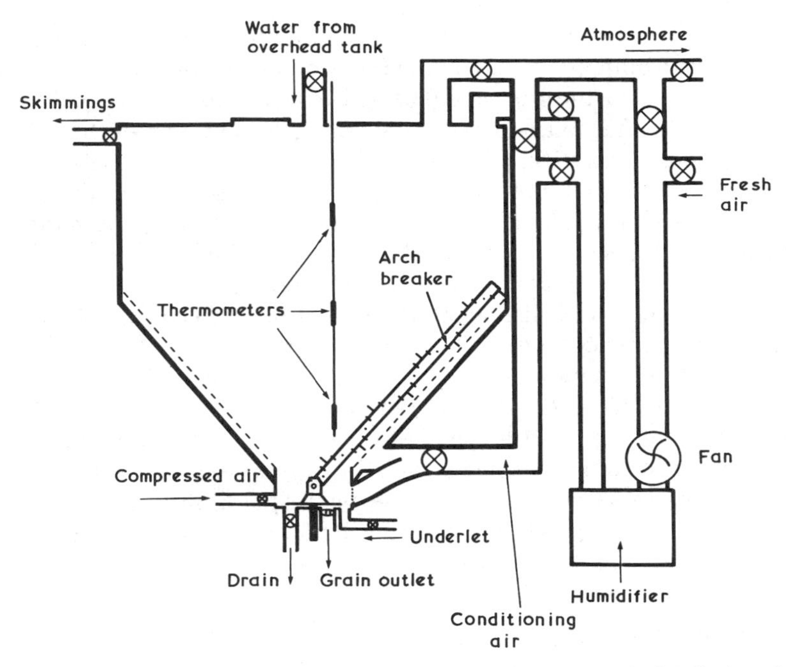

Fig. 9.26 The warm-water steeping, germination and resteeping vessel of Pollock and Pool (1967). To ensure that all the grain left the cone of the vessel at the end of steeping or germination when it was discharged, a rod carrying a number of smaller metal rods and mounted on a universal joint was moved around the cone. This arch-breaker dislodged any residual grain.

germinating grain to kill the growing rootlets, to minimize malting losses and to reduce or eliminate the need to turn grain. The intention was to achieve 'static' malting. For warm-water steeping, it is necessary that the temperature of the steep water should be controlled exactly, that the grain and warm water should be rapidly and thoroughly mixed, and that at the end of the warm period the water should be rapidly withdrawn and the grain be cooled (Briggs, 1987b). This was achieved in the system designed by Wild (1910) (Fig. 9.25). The first steep was carried out by quickly running grain from a garner into a special tank (Vorweiche) containing water at the correct temperature and a wide recirculation tube equipped with a propeller. This permitted rapid mixing and temperature equilibration of the grain–water mixture. At the end of the warm steep, the water was drained and the grain was rapidly transferred to a 'conventional' steep where vigorous mixing with fresh water using an air-lift tube ensured that the grain was cooled and an even temperature was achieved. At the end of the steep, the grain was transferred to a conventional germination drum.

A more modern design of a warm-water steeping, germination and resteeping tank (30 t capacity) was built by Pool and Pollock (1967) (Fig. 9.26). The grain could be rapidly loaded into the hopper-bottomed vessel and mixed with water at the chosen temperature. During 'dry' periods, conditioned air could be pumped through the grain bed through the fully perforated base. Compressed air could be blown into the base of the vessel. If turning was needed, the grain could be withdrawn from the base of the steep and conveyed either to a separate vessel or back to the top of the tank. To prevent the wet grain sticking or 'arching', and so not flowing, the vessel was equipped with an 'arch-breaker', a device that is also used in the base of flour hoppers. This is a long rod reaching from a universal joint in the base to the top of the cone and from which a series of small rods project. Driven from the universal joint at the base, the rod rotates around its long axis and moves round the cone, dislodging grain as it goes. Such vessels, used as steeps, have been incorporated into 'Imamalt' plants.

In principle resteeping can be carried out in correctly built SGVs and SGKVs (see below). Mendlesham-type boxes, the first commercial-scale resteeping plants and Clova type vessels may be used. Other resteeping plants have been designed, some of which are quite complex (Narziss, 1976). One includes a vessel with a conical base, but a tipping perforated deck. The weakness of this design is the very large amount of 'waste' water used to fill the 'plenum' between the tipping floor and the cone.

9.8.4 *Modern flat-bed steeps*

In contrast to hopper-bottomed steeps, in which a capacity of 50 t is considered to be the upper acceptable limit, huge flat-bottomed steeps with capacities of 200–320 t are common (Narziss, 1976; Arter and Olsen, 1986;

Gibson, 1989). The large capacities of flat-bottomed steeps are acceptable because the depth of the grain bed can be limited. Capacity is increased by increasing the diameter of the vessel. A common pattern is shown in Fig. 9.27. In these vessels, a false perforated deck supports the grain. The depth of the plenum chamber between the deck and the true base of the steep is usually *ca.* 0.5 m (1.64 ft). Aeration with compressed air is from a network of perforated pipes running beneath the deck. Carbon dioxide extraction is by suction from below the deck. Grain is loaded in from one or more spouts discharging near the centre of the steep. The grain runs in through one or more rings of water sprays to quench the dust. Grain is usually loaded to a depth of 1.5–1.8 m (*ca.* 4.9–5.9 ft), giving an initial specific loading of 1.05–1.3 t/m^2. A steep for 200 t barley might have a diameter of 15m. The most usual grain levelling and stripping device is called a giracleur. This consists of three arms equipped with downward-extending fixed and inclined blades. The arms rotate around a central support and can be controlled to rise or descend. During loading, the giracleur rotates and the blades move the grain outwards. The device rises as the steep fills, leaving the grain with a perfectly level surface. To strip the grain after draining, a discharge port is opened and the giracleur is set to sweep the grain outwards if the opening is in the side of the steep, or to draw it into the centre if the steep is equipped with an unloading port in the middle of the deck. Water is admitted and waste water is withdrawn from below the deck.

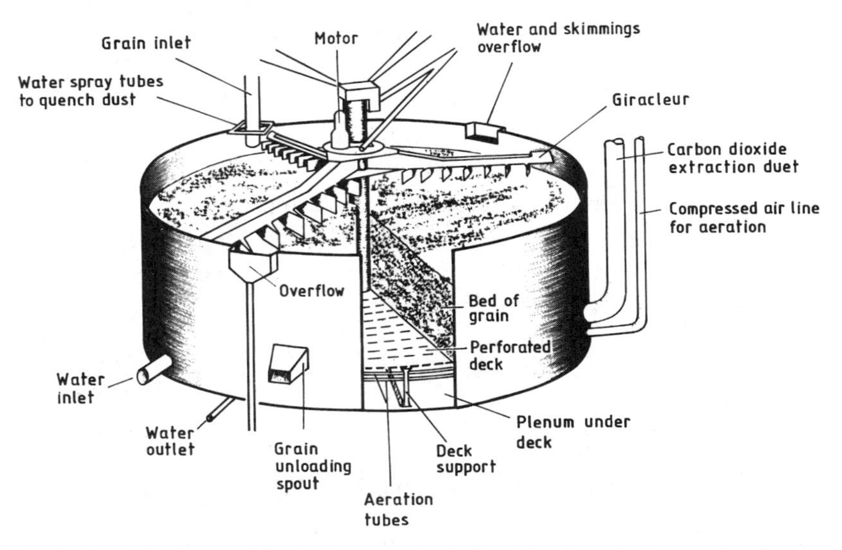

Fig. 9.27 A 'Nordon' type of flat-bed steeping tank. Provision is made for aeration from perforated pipes mounted below the perforated deck and for carbon dioxide extraction. The giracleur, which may be raised or lowered, is used to level the grain and empty the steep. In some steeps the grain is discharged at the sides, in other instances discharge is from a central duct (e.g. Fig. 9.84).

In addition some giracleurs carry water sprays and/or cleaning sprays. Such flat-bed steeps work well, but they have two main problems. Firstly, the large under-deck plenum must be filled with water, and this means extra water usage and extra effluent production. Secondly, the steeps, and in particular the air-extraction ducts and the under-deck spaces with their networks of supports and aeration pipes, are very difficult to clean. Using stainless steel construction, which is costly, permits the use of more extreme cleaning agents, which may be applied hot, e.g. at 60–80 °C (140–176 °F). Although downwardly directed high-pressure jets may be fitted on the giracleur to wash the perforations in the deck and CIP sprays and sprayballs may be installed below the decks, it still seems that deck plates must be lifted and the plenum must be cleaned by hand; operators use high-pressure hoses and wear protective clothing. This is difficult and dangerous work (Arter and Olsen, 1986). Some newer versions of these steeps are equipped with devices that lift the entire deck, giving more ready access for cleaning. Instead of giracleurs, some flat-bed steeps are equipped with sweep augers, essentially open helical-screw conveyors that can be raised and lowered and which rotate about the centre of the steep and level the grain or move it outwards or inwards as required.

Another ingenious device, to save water but retain a large air plenum with its advantage of even airflows, was installed at Crisp Maltings. The water-filled under-deck space was shallow, so saving water, but during dry periods a set of large mushroom valves were opened connecting the plenum to a large, secondary plenum situated beneath the first, which led to the air-extraction fan. This type of steep has the advantages of a shallow under-deck space for saving water and a large air plenum, which permits uniform airflow.

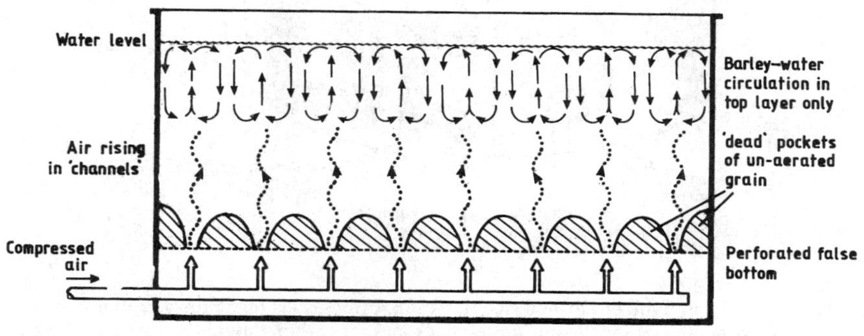

Fig. 9.28 Scheme illustrating the supposed state of grain, during aeration, in a flat-bed steep. Some sections of grain are not aerated or agitated and only the surface grain is mixed (after Aalbers *et al.*, 1983).

Commonly steeps are aerated and sparged with air at a rate of about 1.5 m^3/t per h at high pressures. During the dry periods, carbon dioxide extraction is at the rate of *ca.* 300 m^3/t per h, at a relatively low pressure (Gibson, 1989). Aeration may be sporadic or continuous. While blowing compressed air into the steep undoubtedly loosens the swelling grain and helps to prevent compaction, a consideration of the details suggests that aeration is uneven and that the grain is not well mixed (Fig. 9.28). Whether this is important is a matter of disagreement.

9.8.5 Air conditioning for steeps

Traditionally the control of steep temperature was by mixing grain with well water at a temperature that was hoped to be constant. The desirability of properly controlling the steep temperature was recognized about 150 years ago, and regulating the temperature of steep water is now common; as a result, insulated, thermostatted water tanks are now commonplace. However, only recently have some steep rooms been built with adequate insulation and often there is no attempt to control their temperatures, on the grounds that the large thermal inertia of the grain and water will be sufficient to prevent significant temperature fluctuations. In practice, however, it has been found that drawing large volumes of unconditioned air through grain, during periods of carbon dioxide extraction, can cause the upper layers of grain to dry and unwanted temperature gradients to occur. A partial solution is to mount fine water sprays over the surface of the steep to humidify the air being drawn into the grain. Other approaches are to duct humidified air from the germination units to the tops of the steep tanks or to recirculate the air from the steep, at least until carbon dioxide concentrations rise to unacceptable levels. The only reliable solution is to run the steeps with an air-conditioning unit, as is used in germination compartments (Fig. 9.29).

Aerated steeps are comparatively expensive and may not be needed until the later stages of the steeping process. In several plants 'simple' steeps have been supplemented by aerated steeps in which germination is initiated. This has the added advantage of increasing the capacity of the plant or lengthening the steeping or germination times.

9.8.6 Spray steeping

It is estimated that spraying the required amount of water onto grain so that it is taken up with little or no run-off can reduce water usage by 50–70% and nearly eliminate effluent production. Steeping by recirculating water through deep beds of grain, withdrawing the water from a sump at the base and spraying it back on the top has been tried (while downward aeration was applied continuously) but this has usually been discontinued. In practice, it seems that grain requires a first, immersion steep, presumably to wash it, and then later applications of water may be made as sprays provided that grains

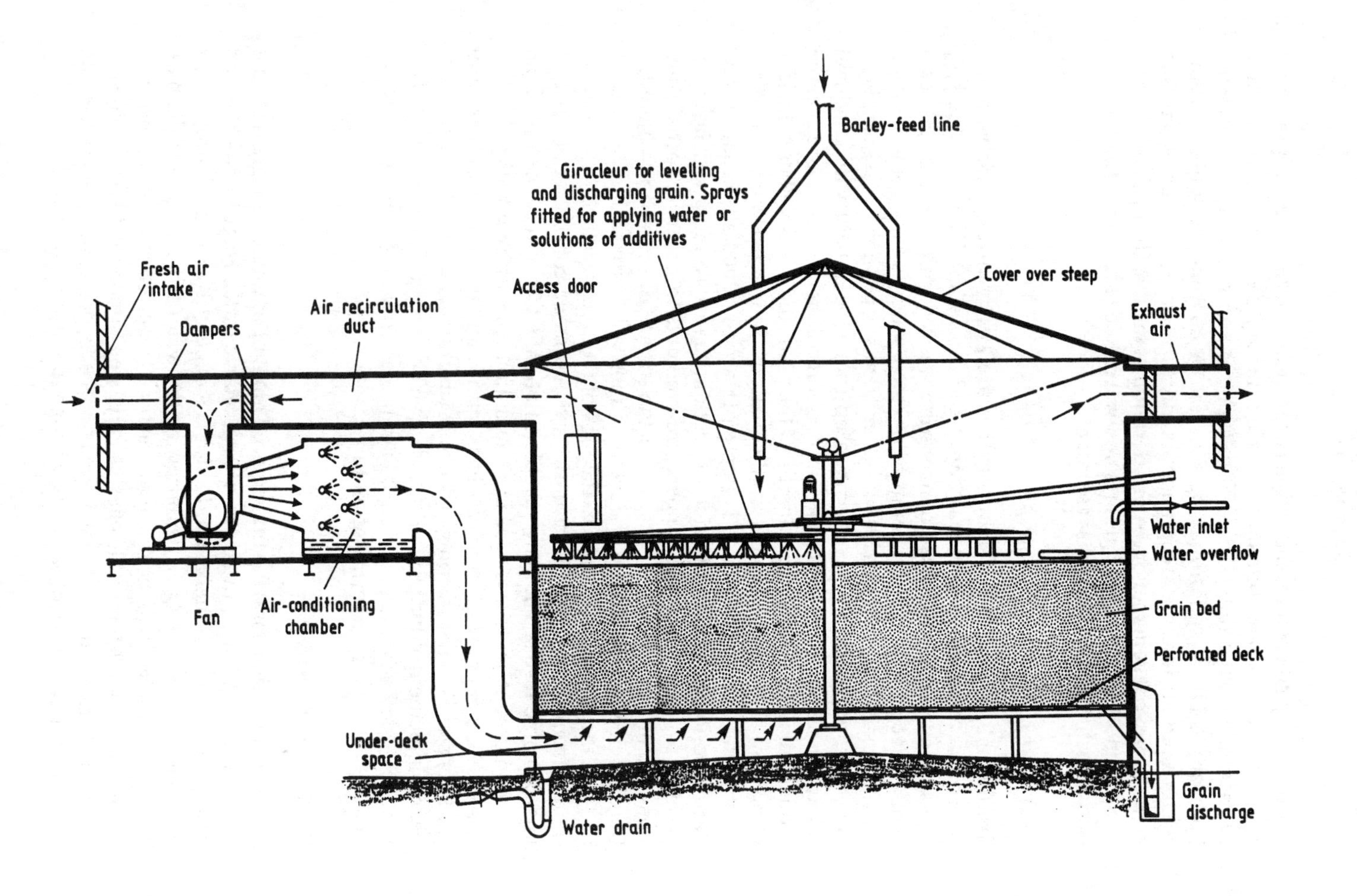
Barley-feed line
Giracleur for levelling and discharging grain. Sprays fitted for applying water or solutions of additives
Fresh air intake
Access door
Cover over steep
Dampers
Air recirculation duct
Exhaust air
Water inlet
Water overflow
Fan
Air-conditioning chamber
Grain bed
Perforated deck
Under-deck space
Grain discharge
Water drain

receive equal treatments and 'channelling' does not occur. This is the process carried out in the continuous Solek and Domalt malting systems. Most malting systems have provisions for 'sprinkling' water onto the grain after it has left the steep. In old floor-maltings, the water might be applied from a hose equipped with a rose or from a large watering can. In Saladin boxes, spray bars are often fitted to the front and rear of the turner carriage, as they may be on Wanderhaufen turners, or groups of sprays may be mounted on special carriages. Sprinkling bars may be installed in drums. All these devices permit the maltster to apply regulated amounts of water to the grain and so adjust its moisture content. Applications are most regular if they are made when the grain is moving, for example when it is being turned.

9.8.7 Applications of additives

Various additives have been or are employed in malting (Chapter 14). Some, such as lime water or other antiseptics, or hydrogen peroxide, have been applied during steeping. However, for some substances, this is wasteful and expensive.

Two additives that may be used either separately or together are gibberellic acid and sodium (or potassium) bromate. Both were applied in the steep. This was wasteful and now, when used, they are commonly applied by spraying solutions onto the grain as it is being conveyed from the steep to the germination unit. As the object is to replace the water film around each grain with a film of solution and as all grains should be treated equally, considerable ingenuity has been applied to achieving rapid and uniform treatments. The grain may be vigorously sprayed as it is moved along in a conveyor or as it 'toboggans' down a metal slide. The well-wetted grain moves over a perforated section that allows excess liquid to drain to a sump from which it is returned to the sprays.

9.8.8 Squeeze-malting

A recently introduced system, which allows grain to be malted with a low moisture content, so reducing kilning costs, involves steeping grain to 36–37% moisture and then squeezing it by passing it between smooth, Teflon-covered rollers set 1.7–2.0 mm apart. The squeezing created faults in the surfaces of the grains which permits the rapid uptake of gibberellic acid solution, when this is applied (section 14.12).

Fig. 9.29 A flat-bottomed steep equipped with a giracleur and an air conditioning unit (a Redler design; courtesy of Redlers Ltd).

9.8.9 The regulation of steeping

In older maltings, all the steeping operations were controlled by hand. In modern maltings, all operations are fully automated, with manual over-ride controls for use in emergencies. This approach leads to better timing and control of the system, fewer errors owing to human failure, lower workforce requirements and the avoidance of working unsociable hours at night-time or – often – at weekends. Automation pays for itself in many ways and is economically attractive. The entire steeping procedure – loading and casting, water filling and emptying, aeration and carbon dioxide extraction – must be programmed, instruments must be installed (and must be monitored) to check that all components of the system are working correctly, and controllers and actuators must be installed to work the conveyors, water-lines, pumps, valves, *etc*. Furthermore automatic alarm systems must be installed – on site and in the operatives' homes – so that in the event of an emergency personnel are quickly to hand.

The types of controller used have varied from simple mechanical timers through a range of electromechanical devices and programmable logic controllers, to computerized systems. Because of their flexibility and comparative cheapness, PCs are presently most in favour. Generally each zone, such as the steephouse, germination unit, *etc.* will have its own microprocessor or computer that regulates the various processes and 'reports back' to a central control room where the data are recorded and the process can be monitored by an operator, usually with the aid of visual displays called up onto screens. The use of colour graphics, combined with numerical data, has become popular. The operator will choose which barley, and how much, is to be steeped and what steeping programme is to be used. Normally one of several pre-set programmes is chosen, but these may be modified as required. Sometimes security 'blocks' are used to prevent unauthorized personnel altering programmes.

Details of these systems will not be considered. However, not only must the correct weight of barley be delivered into the correct volume of water in the correct vessel at the right temperature and at the correct time, with automatic water changes, aeration, carbon dioxide extraction and casting occurring at the correct times and for the correct periods, but safety and monitoring devices must be installed and monitored to check that electrical motors on the fans and conveyors have not overloaded, run out of control or stopped, valves are not sticking, water is not rising above pre-chosen levels, temperatures are correct, and so on. All the valves, fans, *etc.* must be automatically controlled and monitored with appropriate switching devices. The systems used must be robust and be able to withstand grain dust and high humidity. It appears that electropneumatically operated valves and slides are coming to be preferred to directly motorized units on grounds of cost and resistance to water, but their use does require the installation of compressors and compressed-air mains.

Obviously the steephouse does not operate in isolation. The main control room may also control barley and malt stores and will certainly control the barley garners, conveyors, germination and kilning units and probably the malt-cleaning plant. The result is a malting plant that functions with all its parts fully integrated.

9.9 Malting systems

9.9.1 Pneumatic malting systems: an introduction

For about two centuries attempts have been made to improve malting systems, to mechanize malting in order to reduce labour requirements and to eliminate unpleasant jobs; to find ways to malt the year round, without reference to ambient weather conditions; to reduce the size of malting plant; to increase the flexibility of malting plant so that grains of different qualities can be processed and malts of different types can be produced; and to minimize production costs (Thausing *et al.*, 1882; Stopes, 1885; Anon., 1903; Corran, 1975). In general, successful plants are moderately sophisticated, but all have substantial power requirements. Increasingly, reliability and ease of installation of automatic controls are becoming important.

One early approach was to air condition and mechanize floor-maltings (section 9.2). A whole range of other approaches have been tried. In some of the failed approaches, the plant appears to have been excessively complex, while in others some essential component was missing, for example no fans were used to force air through a grain bed. Other problems seem to have been the occurrence of excessive rust and corrosion, problems with maintenance, difficulties in keeping the plant clean and inadequate control of air temperature and humidity. Tizard, in 1854, patented a malting device in which steeping, germination (in a stream of conditioned air) and kilning were to take place in one vessel. Wooden germination drums were used by Lecambre and Persac in Brussels in about 1850–2, and at about this time Vallery, in France, used metal drums to hold the germinating grain. In 1855, Fleetwood proposed making malt in a perforated cylinder containing perforated shelves, and in 1862 R. Walker proposed using a cylinder divided into compartments by perforated walls. The Hruby (Prague) multiple-cylinder system of 1876 has already been mentioned (section 9.8.2). Stead, in 1842, had proposed air warming and humidification for the floors of malt-houses. A. B. Walker, in 1866, suggested that in summer air be cooled with ice to permit malting to continue. These, and many other ideas, were current when the trials and proposals which are often regarded as 'ancestral' to

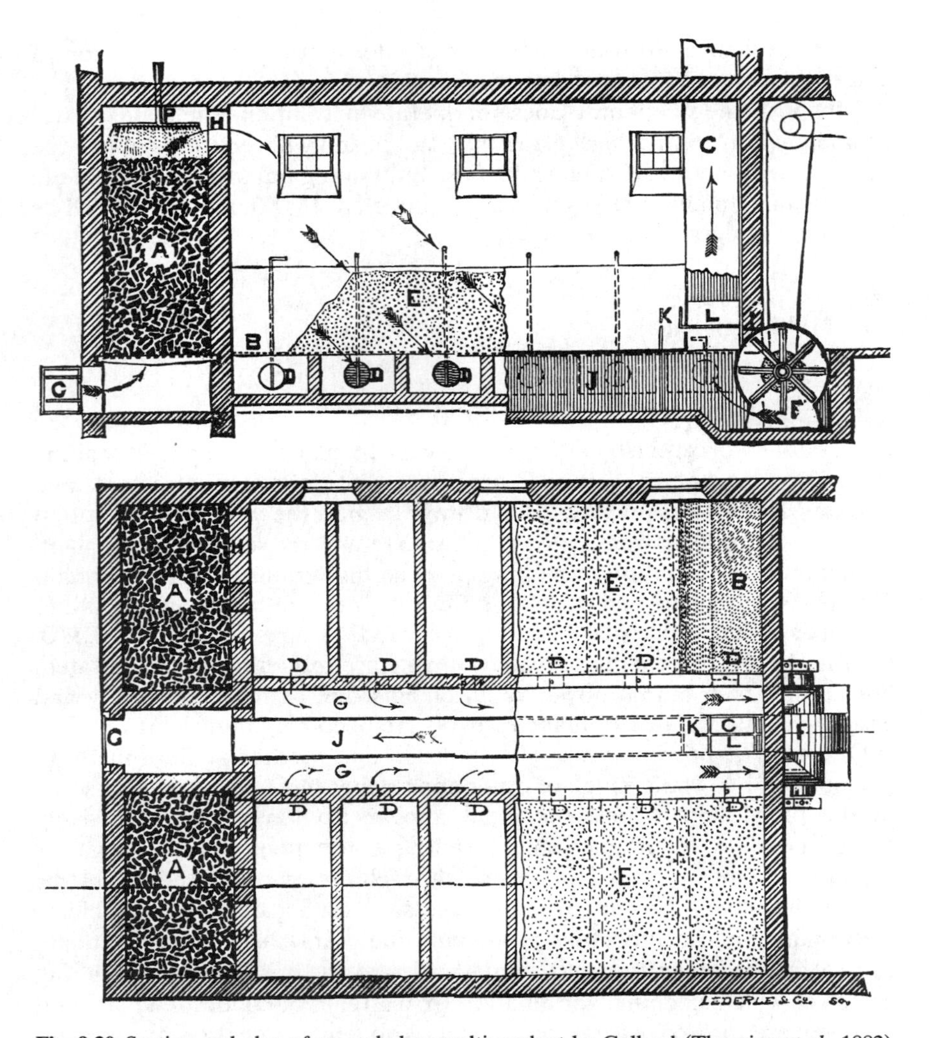

Fig. 9.30 Section and plan of an early box-malting plant by Galland (Thausing *et al.*, 1882). The air was sucked through air-conditioning towers (A) down through the grain resting on perforated decks (E) and was discharged to waste.

modern malting plants were described. One mechanical malting system that seemed to have achieved some success was that of Geçmen. In this equipment, around 25 'metal' floors each 5.2 m (17 ft) × 2.4 m (8 ft), were each formed by 20 plates mounted on a spindle and 25.4–30.5 cm (10–12 in) apart; these were loaded with grain. Grain was turned by tipping all the plates of a floor by rotating their spindles so the grain fell to the floor below. The grain was cooled in a fan-driven airflow (Scamell and Colyer, 1880). Another, reportedly successful, apparatus was the Rice malting system, used in Chicago (Anon., 1906a; Wahl and Henius, 1908), in which barley was slurried in water in the store and pumped to the steeps. The barley, washed in transfer, was steeped conventionally. The steeped grain was germinated in a flow of conditioned air and rested on perforated steel shelves supported on wheels. The layers of green malt were *ca.* 30.5 cm (12 in) deep and had a 15.2 cm (6 in) airspace above, to the next shelf. Turning was by dumping the grain into a garner, then re-forming the bed on the same shelf. Both the Geçmen and Rice systems had drying and kilning systems that resembled the germination units but in which the cool damp air was replaced by kiln air.

In 1873 Galland built his first 'pneumatic' malting plant at Maxeville in France. In this system, steeped grain was formed, by hand, into a bed in rectangular compartments with perforated floors. Air was drawn through towers packed with stone wet with sprays of water to cool and humidify it and was then drawn down through the bed of grain by a fan, which discharged it to waste. Ice could be used to cool the air and the direction of airflow could be reversed if desired. Wagons ran on rails between the boxes to aid transport of the grain. The original plant was a failure because no arrangements were made for turning the grain and the deep beds were difficult to turn by hand; as a result the rootlets grew together and felted. The first successful pneumatic maltings, built on the Galland pattern, were built at Perry's at Roscrea in Ireland, in 1876. The maltings was built by a river, which supplied cooling water and power, from a water-wheel. The crucial difference to the Maxeville plant was the provision of space in the boxes to permit workmen to turn each piece twice each day by hand (Fig. 9.30). Interestingly, the perforated false-bottom plates were 'of pure zinc'. The depth of grain was around 0.61 m (2 ft). In boxes used by Kogsbölle at Carlsberg in Copenhagen, cooled and humidified air was sucked down through the grain held in beds 30.5–50.8 cm (12–20 in) deep in iron boxes. Eventually each box was loaded with about 6.25 t (125 cwt) of grain. Originally 0.64 m (2.1 ft) deep, the grain increased to 0.78 m (2.56 ft) deep and was turned twice a day and transferred between compartments by hand (Anon., 1881; Minch, 1960). In this sense, the plant was ancestral to the semi-continuous malting systems now in use. At about this time, similar plant was used by Völkner, in Graz, Austria. Thausing *et al.* (1882) describe Marbeau's system, used in Chicago, in which malt, held in boxes with lids, was ventilated by suction, the air first being humidified by passing through several layers of cloth sprinkled with water. The grain was turned by shovelling it from one box to another.

A major weakness of the Galland boxes was the need for manual grain turning. His sometime collaborator, Saladin, overcame the problem by inventing a rope, belt-and-pulley system for actuating a system of vertical, helical screw turners (generally six, turning alternately in opposite directions) mounted on a moving carriage (Fig. 9.31; Stopes, 1885). As the

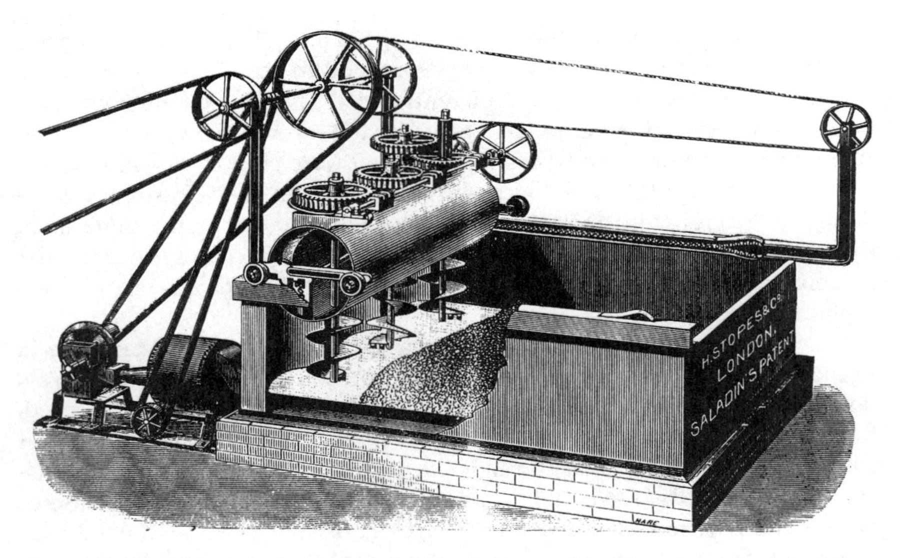

Fig. 9.31 'Saladin's turning screws in a germinating case, with echangeur and gearing' (from Stopes, 1885). The echangeur was an air-humidifying device. This is a diagram of a 'model' plant displayed by Stopes at a Brewers' Show in 1883.

carriage moved along the box, the rotating turners lifted and stirred the green grain. The same principle is still used in the turners of most compartment maltings.

Galland moved on to develop the system of drum malting, for which he is best known. The early versions, and there were several (e.g. Stopes, 1885), evidently had faults that were largely overcome by the engineer Henning, consequently the later more successful types are often called Galland–Henning drums. These drums relied on the passage of humidified air between a central duct and other ducts around the periphery of the drum to condition the germinating grain. The grain was turned by rotating the drum around its long axis.

Numerous other drum designs were once well known but their use has been discontinued. For example, Tilden's drum (Anon., 1900a, 1903), developed in the USA, consisted of inner and outer cylinders, concentrically arranged and each of perforated steel. The steeped grain filled the space between the cylinders. During the germination period, moistened air flowed from the inner cylinder outwards through the grain and to waste. At the end of the germination period, dry air was supplied to wither the grain, and then hot, dry air was passed to kiln the grain *in situ*: the drum was a GKV. Because the drum was rotated during kilning, the malt was deculmed and was free from dust. The culms fell through the outer cylinder and were collected from the floor, beneath the drum.

An early problem with drums was that when attempts were made to increase their size, e.g. to 100–200 zentners (2–4 t), uneven airflows occurred in the bed of grain. Where low airflow occurred, carbon dioxide accumulated and a rise in temperature occurred (Bleisch, 1900). The Schwager drum, described in 1898, was intended to overcome this problem (Bleisch, 1900; Schönfeld, 1932). In this drum, the central airshaft was divided in the centre and air was exhausted from both ends. The drum was longer and narrower than usual to minimize the length of the passage for the air from the outer casing. The air was admitted to the grain from ducts formed by having a solid outer cylinder and an inner cylinder, areas of which were perforated. The space between these cylinders was partitioned to form ducts. The arrangement was complex, but in principle the conditioned air entered the outer drum mantle from both ends and was exhausted from the opposite end of the drum to that in which it entered.

The attractive ease of turning grain in a drum, but doubts about uniformity of the ventilation; led to a number of proposals. For example, Sleeman (1892, 1908) (Anon., 1904), who believed that floor-malting had not changed since Norman times and was carried out by a 'wooden-headed workman with a wooden shovel', advocated a range of devices including steeping and germination drums and withering cases. The novel arrangement with one type of drum was that, while at rest, the grain, which rested between perforated walls 1.07–1.22 m (3.5–4 ft) apart, was ventilated from side to side. When the drum rotated the grain fell into a wider chamber, normally above the perforated walls when the drum was at rest, where it 'spread', the extra space permitting good mixing and turning, before it flowed back between the perforated walls. Unlike floor-maltings, which he derided, Sleeman's systems did not survive.

For many years floor-malts were regarded as being superior to pneumatic-malts, and many still believe this to be true. A major difference between floors and other plants is that much of the heat of germination is lost by conduction to the floor and by convection and radiation to the space above (Lundin, 1936). Ventilation is minimal. Jonsson devised drums that may be compared to Plevoet's box system, in which a substantial proportion of the heat was removed by conduction so that less ventilation was needed and the green malt was less prone to drying out (Lundin, 1936; Jonsson, 1937). In the Jonsson drums, the cylindrical shell (the heads) and various internal cylindrical and radial elements were cooled by the circulation of cold water (Fig. 9.32). The spacings and the gaps between and the gaps in the elements ensured that the grain was mixed well when the drums were turned. Instead of rotating in one direction, the drums oscillated, turning 1.25 revolutions in one direction, then 1.25 revolutions back (1 rev./45 min). Sufficient space was allowed to permit unloading and cleaning. The drums, 10.5 m (34.5 ft) long and 3.6 m (11.5 ft) in diameter, were housed in humidified and attemperated rooms. Limited ventilation of the

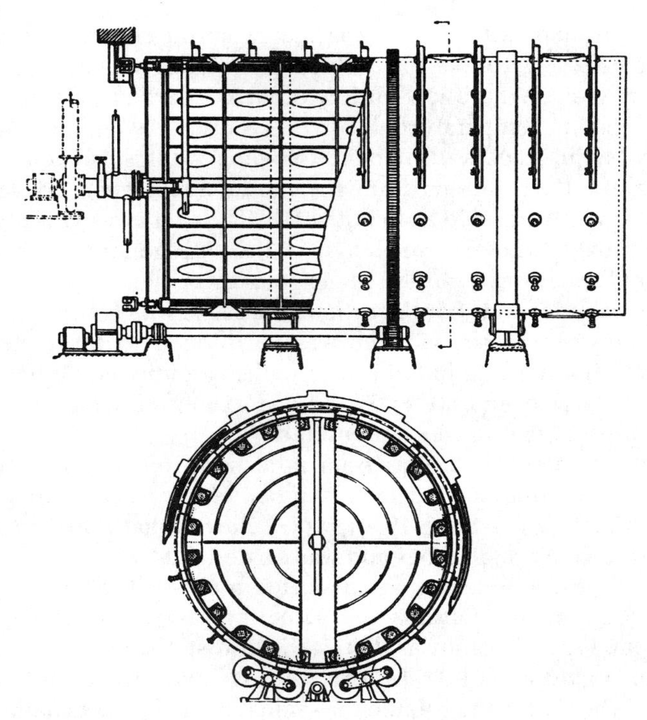

Fig. 9.32 The Jonsson drum in which the grain was partly cooled by contact with water-cooled metal surfaces within the drum (Lundin, 1936).

grain was arranged, but about 70% of the heat generated by the grain was withdrawn via the cooling surfaces. Presumably, these complex devices were expensive and hard to maintain.

The boxes and drums mentioned above were the direct antecedents of the equipment in most modern malting plants. Other devices, which have left no recognizable descendants, include the 'suspended perforated floor' patented in 1901, in which it was optimistically hoped that rocking the bed would adequately mix the green malt; the malt was to be aerated or kilned by air from 'a hopper-shaped hood capable of being raised or lowered on the top of the bed' connected by a flexible pipe to a fan. Other devices, in which the grain was moved on endless perforated webs on belts, or moving perforated floors, were described several times and were the forerunners of the Solek continuous malting system (Krause, 1892; Plischke and Beschorner, 1903; Adlam, 1907; Koch, 1911; Jakob, 1913; King and Hopper, 1936). The devices varied in whether the belts were moved by hand-cranking or electrical power, and whether the air was adequately conditioned. The Plischke system was called a Hordenwender or Wanderhaufen (moving piece) device. Originally hand-cranked, the later machines were

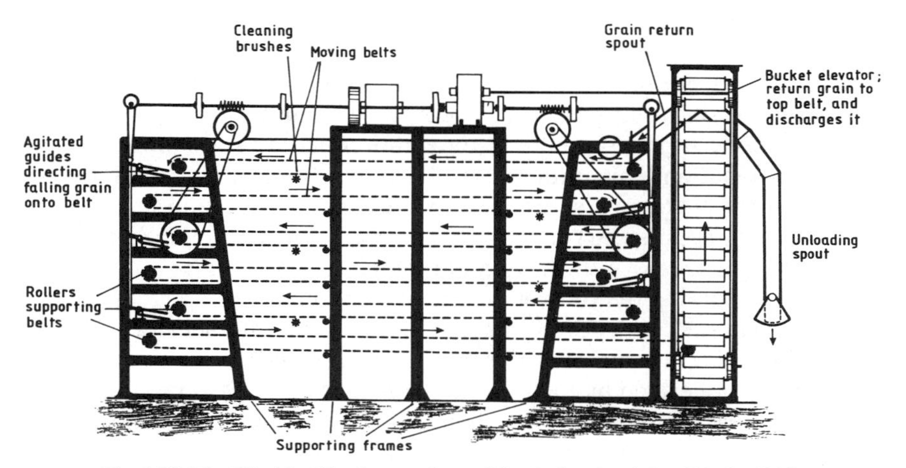

Fig. 9.33 The Plischke Hordenwender or Wanderhaufen (after Jakob, 1913).

electrically powered (Fig. 9.33). The device was designed to save space and labour and consisted of six woven-wire endless belts mounted above each other; these were offset and moved in opposite directions so that the barley falling from the end of an upper belt fell onto the belt below, moving in the opposite direction. In falling it was mixed and turned. Steeped barley was loaded onto the upper belt from a hopper. In the original system the grain might be advanced 1 belt/day, so that there were 6 days of germination with one 'turn' each day. In the later system, the grain from the lowest belt could either be discharged to kiln or to a small bucket elevator that returned it, via a slide, to the top belt. The second arrangement achieved better turning since grain movement could be continued until the operator was satisfied. The device illustrated (Fig. 9.33) had perforated mesh belts 10 m (32.8 ft) long, 2 m (6.6 ft) wide and supported a grain bed *ca.* 10 cm (3.94 in) deep. The bands were advanced 40–50 cm/min (15.7–19.7 in/min) during turning. Turning took 15–20 min. Unloading took *ca.* 5 min. Although rotating brushes were installed to clean the belts while 'inverted', the machinery had to be thoroughly cleaned between each use.

9.9.2 Pneumatic malting systems: general considerations

The different types of pneumatic plant have various features in common (de Clerck, 1952; Vermeylen, 1962; Narziss, 1976; Briggs, 1978; Gibson, 1989, 1992). Pneumatic systems must have satisfactory arrangements for loading, turning and stripping (discharging) the grain. The devices must work in as nearly a quantitative manner as possible to minimize the need for manual 'tidying up' operations and must be gentle, minimizing damage to the grain and the rootlets. While the grain is growing it must be supplied with an adequate supply of cool, humidified air to remove heat and carbon dioxide and to supply oxygen. In traditional malting, where grain was cast wet, unchitted and germination temperatures were *ca.* 15 °C (59 °F), the maximum rate of respiration and heat output was on germination days 3–5. In many modern plants, where steep aeration and air-rests are usual and where germination temperatures are higher and the grain is cast chitted or even with the first roots showing, maximum respiration and heat output occur by day 2 (post-steep) or earlier. The ventilation and air-cooling system must be able to remove this heat and to hold the grain temperature at the chosen value during the time of maximum heat output.

Grain beds vary in depth from about 0.61 m (2 ft) (old Saladin boxes) to 2.44 m (8 ft) (Popp plants). The deeper the bed of grain the higher must be the pressure generated by the fan and the higher the airflow per unit cross-sectional area to maintain an adequately low temperature differential between the top and bottom of the bed. Often the target used to be a differential not exceeding 1 °C (*ca.* 2 °F), but in practice differentials of up to 4 °C (7.2 °F) have been allowed to occur in modern plants for limited periods. Such temperature differentials lead to unevenly grown malt. In the past and when using longer germination times, mean bed temperatures were in the range 12–16 °C (53.6–60.8 °F); now temperatures of 15–19 °C (59.0–66.2 °F) are common and, not infrequently, temperatures of 21 °C (69.8 °F) may be reached or exceeded. The use of higher temperatures, deeper grain beds, larger temperature differentials and shorter germination times are dictated by commercial considerations but can lead to the production of less uniform, poorer quality malts.

Even when the air-on entering a bed of grain is fully saturated, the grain still loses water to the airflow. Therefore, daily declines of 0.5% moisture are encountered. During germination, water is **bound** in hydrolytic reactions in the grain; it is **generated** by the grain's respiratory processes, and it is **evaporated** from the grain. Air entering a grain bed saturated with moisture (100% RH) and leaving the bed saturated with water (100% RH) is removing water from the grain because it is warmed by the grain bed (from which it is removing heat) and so its moisture-carrying capacity is increased (see Fig. 10.11). Inadequate air humidification in a pneumatic malting plant cannot be tolerated. For example, when the air-on RH fell to 95% in one particular plant, the grain moisture content fell by about 2.5%/day, and modification virtually ceased in 2 days.

Not surprisingly, different maximum airflow rates are used. An old guide was that for a grain bed 1 m (39.4 in) deep, the linear airflow rate should be about 3.7 m/min (*ca.* 12 ft/min), giving about three air changes every minute. Modern germination units may use maximum airflow rates of *ca.* 0.15–0.2 m^3/t barley steeped per s (Gibson, 1989), but some certainly use less, particularly during germination periods then the grain is not respiring strongly and so the heat output is less than maximal. Airflows of up to

0.4 m^3/t per s may be used for rapid cooling. The airflow is driven by fans, sometimes duct-mounted axial-flow fans but more usually centrifugal fans fitted with multi-bladed impellers and powered by electric motors. Variable airflows may be achieved either by running various numbers of fans or by using variable-speed fans. The old, energy wasting method of allowing fans to run at speed but 'throttling' the airflow to reduce its rate is not now used. The power required varies in a complex way with the volume of air to be moved and the resistance to airflow; this, in turn, largely depends on the depth of the grain bed and the degree of bed compaction. After turning, the rootlets are disentangled and the bed is less closely packed (the piece is lightened); although the depth is increased, the resistance to airflow is less. It is important that grain beds should be evenly loaded and turned so that airflows are as uniform as possible. This may be checked using a sensitive anemometer mounted over an inverted cone that rests on the surface of the grain bed.

The conditioning air should be as nearly saturated with water vapour as possible, but it should not carry a mist of suspended water droplets and it should be at a chosen temperature below that of the grain. Depending on the ambient conditions, the air may be attemperated (warmed or cooled) by passing it through humidifying water sprays, without the use of heat exchangers. This is economically attractive and may be used in mild weather, as for much of the year in England. However in many regions in winter, the air must be heated and in summer it commonly needs to be cooled. Cooling is largely carried out by the evaporation units of refrigeration plants. Where feasible, the heat collected is used to save fuel by discharging it into the kiln inlet airstream by siting the condenser there. In some kilns fitted with sophisticated heat-recovery systems, including heat pumps, the air may emerge from the kiln at *ca.* 16 °C (60.8 °F), saturated with water vapour. This may be used, as 'conditioned air', by diverting it to the germination units via the air inlet. In most malting plants the airflow is directed upward through the bed of grain. However, in other maltings, the direction of airflow may be reversed at will and in still others the airflow is always downwards. Because the temperature of different batches of malting grain and the required air-on temperatures and airflows will differ, it is desirable that each piece has its own individual airflow circuit and air-conditioning unit so that recirculated air needs to be cooled at most by about 2 °C (3–4 °F) per cycle. At one time, germination units were housed in large 'halls' so the air over the grain beds was mixed.

The air space over the grain and in the 'air-circuit' should be minimized to reduce the volume of air that needs to be conditioned. In compartment maltings, a downward flow of air may be advantageous since it displaces downward the heavy gas carbon dioxide, which is generated by the grain, and it carries downward organic volatiles from the grain. The high humidity of germination compartments causes condensation and dripping, with rapid deterioration of paintwork, corrosion of susceptible equipment and the growth of moulds and algae, particularly near lights. Downward airflow directs some of the grain volatiles away from the germination compartment (where they would support microbial growth) and into the air plenum, which, in any case, gets dirty from drips and broken rootlets. The walls and ceilings of such germination compartments remain much cleaner in consequence. Walls may be tiled, painted with water-resistant paints (which may contain anti-microbial agents) or they may be constructed of corrosion-resistant metal. Air ducts have been illuminated with ultraviolet light (which kills microbes but which can also damage the eyes), and small amounts of chlorine gas have been metered into the airstream to check microbial growth. Chlorine is, of course, highly toxic; in addition, it catalyses the breakdown of gibberellic acid. Usually maltings are cleaned with preparations based on sodium hypochlorite and with high-pressure water hoses, so the structural materials chosen should be resistant to these treatments. Until recently all cleaning was by hand, but increasingly devices for automatic CIP are being installed. It seems, however, that no CIP systems have yet wholly replaced manual cleaning.

Nearly all germination plants are equipped to permit the grain to be sprinkled, that is to be sprayed with controlled amounts of water. To obtain even applications this is applied when the grain is being turned. Many maltsters prefer to avoid sprinkling where this is possible. The old maltster's dictum, that it is best to get all the water needed into the grain during steeping, is still widely accepted.

A wide range of air-humidifying equipment has been used in pneumatic malting (Stopes, 1885; Schönfeld, 1932; de Clerck, 1952; Narziss, 1976; Vermeylen, 1982). The humidifying systems also wash the air, and the dirt must be separated from surplus water if this is to be reused. Because they are being continuously wetted, the ductwork and spray chambers require regular cleaning to prevent the build-up of microbes and slime. Humidification systems can be divided into five classes, of which possibly the first and certainly the last three are still in use:

1. humidification towers;
2. exchangeurs;
3. hydraulic spray heads;
4. pneumatic 'ultrasonic' sprays;
5. spinning disc humidifiers.

Humidification towers were used by Galland in some of his early maltings. These were towers packed with pebbles, or coke (or Raschid rings or other 'high surface area packings' in later examples). Air was sucked up through the tower; water sprayed on at the top trickled down the tower, coating the packing and running to waste. Thus the air emerging from the top had been washed, was saturated with water vapour and was the same temperature as the water. These structures, which were often very large, even 8–10 m (26.2–32.8 ft) high, wasted water and were prone to the growth of slimy organisms. Pressure drops could also be large. It has been suggested that CIP could reduce the problems (Chumbley *et al.*, 1986). A second humidification device, apparently introduced by Saladin (Fig. 9.31), was the exchangeur (also called Linde humidifiers, echangeurs, changeurs, alternators and rotating disc humidifiers). This consisted of a horizontal, cylindrical chamber partly filled with water and containing a series of internal cylinders and rotating discs that dipped into the water and so were continually wetted. The air, passing through the cylinder and over the numerous wetted surfaces, became humid.

The last three classes of device generate fine sprays of water that either evaporate into the conditioning airstream, thereby cooling and saturating it, or form droplets, which fall to the floor. The water is collected in a sump. After settling and filtration, and possibly cooling and purification, this water is reused. Various chamber systems are in use. Usually sprays are mounted in specialized parts of the air ducts, known as the spray chambers. However, in some maltings, the sprays are mounted where the air enters the underfloor plenum of the germination compartment. Often in the spray chamber the airflow is directed alternately up and down by baffle plates. It meets a series of sprays; the objective is to have the air leave the chamber saturated with water but not carrying entrained water droplets, since these increase the moisture content of that part of the grain bed which receives the airstream and causes excessive growth. The quantities of water used can be very considerable, especially in summer; to economize on this and reduce the amount of attemperation required, a proportion of the water-saturated air from the germination compartment is recirculated. Narziss (1976) noted that a germination compartment of 35 t capacity might need 72 sprays, each discharging water at 1.0–1.5 l/min, that is up to 6480 l/h. Using a 144 h period of humidification, the amount of water to be sprayed would be up to 933 m^3. De Clerck (1952) reported water usages of 30–60 g/m^3 air in winter and up to 200 g/m^3 air in summer. With airflows up to 100 000 m^3/t barley (in summer), this gives a water requirement of 20 000 l.

The third class of humidification equipment is hydraulic sprays. Water, pumped at 2–3 atm pressure, is forced to swirl and emerges from a specially shaped nozzle, which causes it to break up into droplets, much as in a garden spray (Fig. 9.34). The objective is to have a fine spray that will present the airflow with a large surface of water from which evaporation readily occurs. These sprays are inefficient in that a range of droplet sizes is produced. Many of the larger droplets settle too quickly. In general, it is preferable to have a larger number of small nozzles (rather than fewer larger nozzles) to achieve most droplets of a small size. Droplets of 50–1500 μm diameter are obtained. Many different patterns of nozzles are in use. Some discharge ‘hollow cones’ of spray, others solid cones. The sprays should discharge into the airflow and **not** impinge on a solid surface. Within limits, the higher the

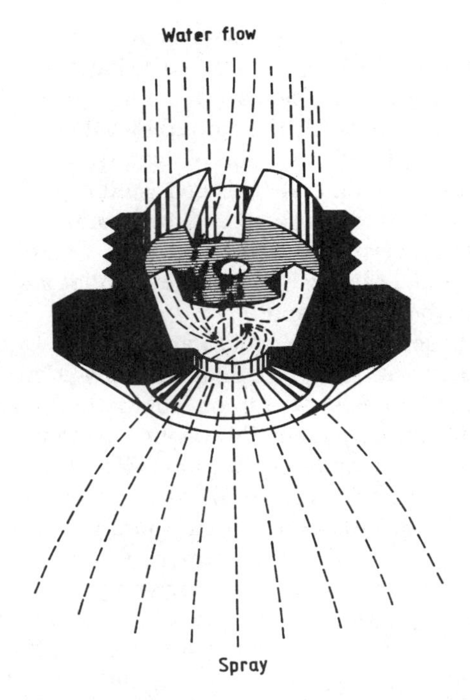

Fig. 9.34 A simple spray nozzle (after Narziss, 1976).

pressure the finer the water spray. Hydraulic sprays are simple and reliable, provided that precautions are taken to prevent fine particles or material that may have accumulated in the pipes causing blockages. Most can easily be unscrewed and cleaned.

Pneumatic, ultrasonic sprays are more complex, as compressed air is pumped into the nozzle system and, therefore, a compressed air supply is needed. The extra energy added to the system breaks up the water spray into very fine droplets indeed, so fewer sprays are needed to achieve air saturation. In the device illustrated in Fig. 9.35, the compressed airstream meets a reflector plate and this creates an acoustic oscillator. The water pumped into the airstream is broken into very fine droplets by the shock waves in the resonating chamber. Droplets with diameters of 10–100 μm are produced.

In the fifth class, spinning disc centrifugal air washers and humidifiers, mechanical energy is used to break the water into a spray or fog of very fine droplets (e.g. 2 μm diameter). Water is pumped at a metered rate to both sides of a disc that is rapidly rotated by an electrical motor (Fig. 9.36). The surfaces and edges of the disc are so textured that the water is thrown outwards by centrifugal force and leaves the edge of the disc as very fine droplets. These devices are easy to install and, because the evaporation

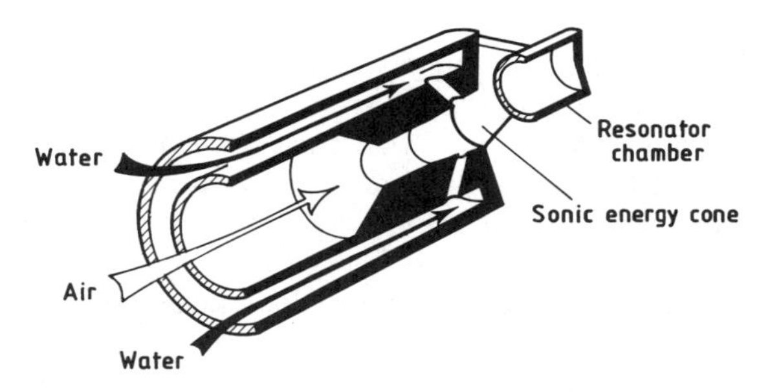

Fig. 9.35 A Lucas–Dawe ultrasonic water atomizer (courtesy of the maker).

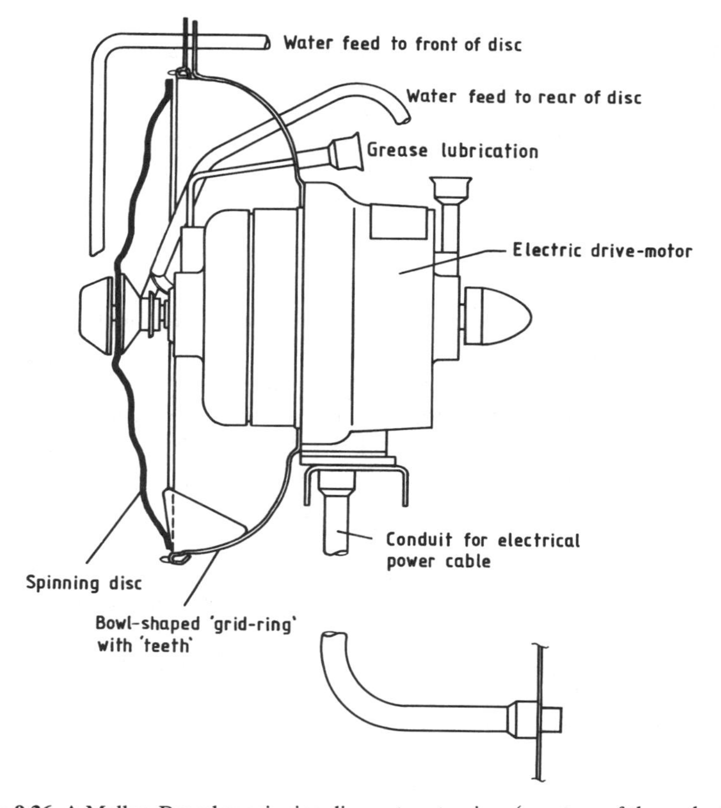

Fig. 9.36 A Mellor–Bromley spinning disc water atomizer (courtesy of the maker).

from the tiny droplets is so efficient, a relatively few units can replace numerous sprays. However, in older installations, the electric motors were not adequately protected from the high humidity, resulting in frequent electrical failures. This is easily overcome by housing the motors apart from the discs, for example by mounting the electrical motor outside the spray chamber and driving the spinning discs, from the motors, by toothed belts that pass through the chamber wall.

Chumbley *et al.* (1986) estimate that each 300 t batch of barley requires an airflow of 205 000 m^3/h (or 240 t air/h). Saturating this amount of air is technically challenging. Furthermore, the sprays will wash dust from about 40 000 t of air each week, which is likely to appear in the water-sump as about 10 kg of fine, grey sludge. To ensure that entrained water droplets and wet dust particles are separated from the airstream, spray eliminators are used. The wet airflow follows a tortuous path, and entrained substances impact on the surfaces of baffles to drain or be carried down to the water-sump. For a particular maltings, with a 240 t batch size, the maximum airflow through the spray chamber was 300 m/min. The water spray flow was 4.95 m^3/min through 480 spray nozzles operating at 3.5 bar. With an air residence time of 1 s, the air was fully saturated. The target water droplet diameter was 500 μm, droplets that are easy to generate and separate. Calculations show that each kilogram of air was exposed to 1.5–3.5 kg water, with a surface area of 18–42 m^2. Each cubic metre of the spray chamber volume contained a total water surface area of 24–36 m^2.

9.9.3 Drum malting plant

Because of their relatively small batch sizes and higher costs, drum plants are not now being built in the UK, but several are still in operation. Drums have been designed for barley drying; for steeping and germination; for steeping, germination and resteeping; for germination and kilning; and for kilning (Leberle, 1930; Schönfeld, 1932; Hynard, 1947; de Clerck, 1952; Vermeylen, 1962; Macey, 1963; Narziss, 1976). Other specialized drums are used for roasting. All those in use at present are used for germination only or for roasting.

The first practical drums were built by Galland at Pankow, near Berlin, in 1883. The first drums to be built in the USA were in Milwaukee in 1889. The design was improved by Henning. Small Galland–Henning drums were installed in the Speyburn distillery, Rothes, in 1897 and continued in use for about 90 years. H. A. and D. Taylor installed drums at Sawbridgeworth in 1894–6 (Prior, 1996) and in 1905 Bass installed 15 drums of 30 Qr (6.09 t) capacity each at Burton-on-Trent. All drums consist of long, horizontal, metal cylinders (usually steel but sometimes aluminium), supported on rollers. They are rotated by electrically driven helical gears, engaged with teeth around the drum. Drums normally have two speeds of rotation, slow (2–4 rev/h) for turning the grain and faster to adjust the positions of the doors to be opened for loading and stripping. Galland–Henning drums were loaded from an overhead spout and unloaded onto a conveyor from a ring of doors around the centre of the drum. Many of the early drums were not fully self-emptying and were laborious to empty. Apparently drums have never been built insulated. They are often equipped with sprinkler bars, which allow the grain to be wetted while it is being turned. In the Galland–Henning drums, air was withdrawn from the central, perforated duct and conditioned air was allowed in through peripheral ducts (Fig. 9.37). Because the space above the grain (needed to allow the grain to move during turning) would allow undue air leakage past the grain, a counter-balanced sliding plate sealed off the upper air passages. Usually the drums had air ducts with circular cross-sections; semicircular ducts mounted on the side walls were also used (Fig. 9.38). 'External' ducts were also made by having a fully perforated cylinder enclosed in a longitudinally 'rippled' metal cylinder. Where the ripples were away from the perforated cylinder, or between radial partitions, the spaces formed ducts (Fig. 9.38). Rectangular or triangular central ducts and other shapes were used. In 'fully perforated' drums, the air entered the grain from the central duct. To reduce the tendency of the air to move directly to the above-grain space, a baffle

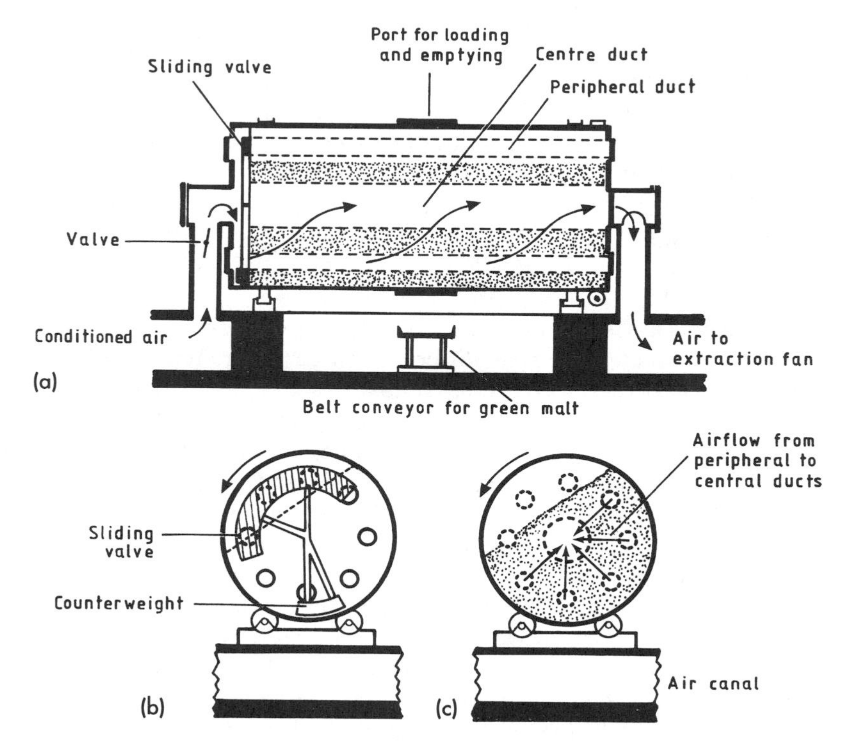

Fig. 9.37 Schematic diagrams of a Galland–Henning drum. (a) A longitudinal section of a drum; (b) a sliding valve cuts off the air supply to the space above the ducts in the space above the grain; (c) the direction of airflow from the peripheral ducts to the central duct (after de Clerck, 1952.)

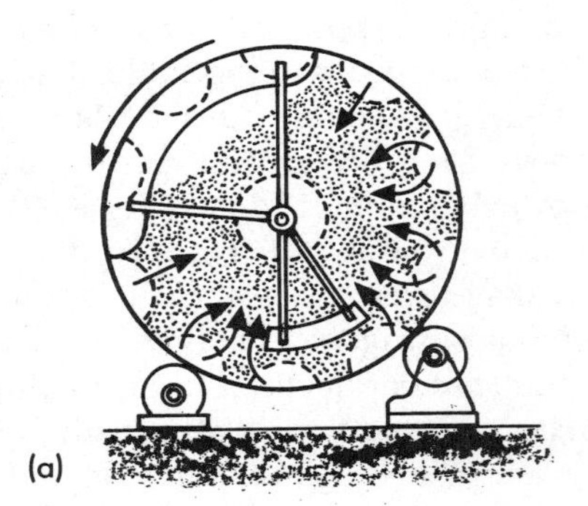

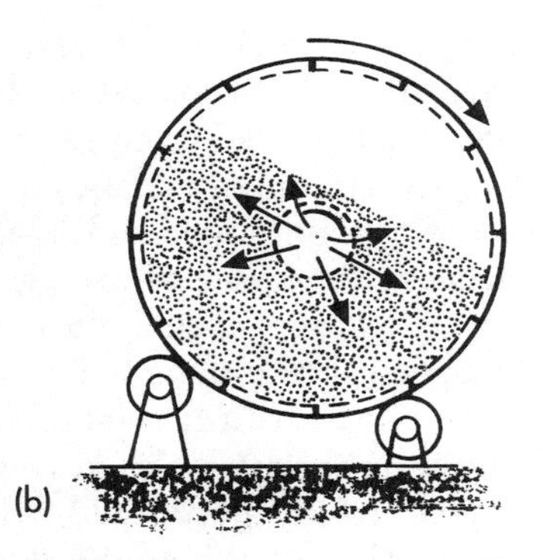

Fig. 9.38 Alternative arrangements for air-ducts in drums: (a) the airflow is from the peripheral, semicircular ducts to the central duct; (b) the airflow is from the central duct to the 'fully perforated' periphery. The baffle plate in the central duct is to minimize the movement of air through the shallow layer of grain covering the central duct.

plate was installed within the duct (Fig. 9.38). Some drums were built with longitudinal projections extending inwards from the outer wall to minimize slippage of the grain mass as the drum was rotated. In these older types, the drums were about two-thirds filled with grain and, in contrast to box drums (see below), they could only have their 'charge' (amount of grain loaded) varied in a narrow range because the air ducts had to be covered to the correct depth. The grain gently rolls over as a drum is rotated, achieving a very gentle turning action. Older Galland–Henning drums had capacities of 2.5–20 t, with values of 10–12 t being most usual. However, newer more modern drums have been built with capacities of 45 t or more and an upper capacity of 100 t has been suggested.

Modern drums are all of the box or 'decked' type. These have the advantages that they can easily receive different charges of grain and the airflow through the grain bed is more even than that which occurs in ducted drums. Modern box drums can be operated with few personnel. In the older box drums, such as the Topf type (Fig. 9.39), the grain, which was loaded into and emptied from doors in the sides, rested on a flat, perforated deck and attemperated air was passed through from below. During turning, grain was prevented from entering the air outlet by a perforated metal cover. In box drums, the grain is aerated when the deck is horizontal. As the grain is inclined by the turning process the drums are stopped at the end of the turn so that the deck is also inclined and parallel to the surface of the grain bed; the drum is then turned back so that the surface of the grain bed and the deck are both horizontal. As a result, the grain bed is of uniform thickness, say 1.22–1.37 m (4–4.5 ft) in a particular plant. Turning is usually carried out every 8–12 h. The Boby type of decked drum became popular in the UK in the mid-1960s (Fig. 9.40). Such drums are self-filling, self-levelling and self-emptying. In addition to spray bars for wetting the grain, provision may be

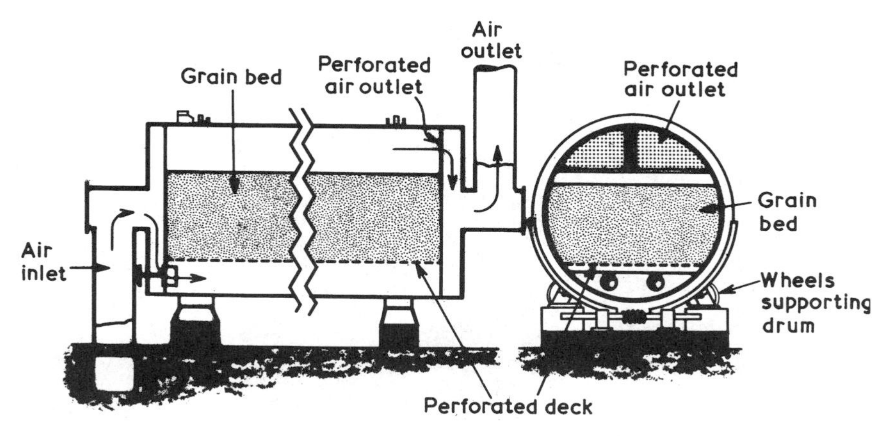

Fig. 9.39 An older pattern of a box-drum of the Topf type (after various sources; Briggs, 1978).

made for automatic CIP, with deck washing and under-deck cleaning (Palmer, 1965; Anon., 1964, 1967a,b). In several installations, pairs of steeps, each holding about 15 t (75 Qr) barley discharge into a drum (30 t). Often these are about 6.86 m (22.5 ft) long and 3.66 m (12 ft) in diameter. Other quoted dimensions of drums of various capacities, are: 15 t, 3.66 m (12 ft) diameter, 8.85 m (29 ft) length; 35 t, 3.82 m (12.5 ft) diameter, 15.2 m (49.8 ft) length; and 45 t, 4.43 m (14.5 ft) diameter, 15.8 m (51.8 ft) length. Drums are 'worked' in pairs, so each piece is about 60 t (as original barley), i.e. 300 Qr. Continuous helical blades of about 12 cm (4.7 in) project in from the walls of the Boby drums, running in 'opposite senses' from the ends to the middle (Fig. 9.40). For loading, the steeped grain is spouted in at one end while the drum rotates. The grain levels itself and forms a bed about 1.4 m (4.5 ft) deep. After each turning period, the grain bed is re-formed in an even layer, as described above. When the drum is to be stripped, a ring of doors, ports or hatches round the centre of the drum is opened. As the drum revolves the grain falls out of the hatches and into a collecting hopper, which guides it to a conveyor that carries it to the kiln. The spiral blades within the drum are so arranged that as the drum rotates they continually guide the grain from the ends to the middle and out of the doors. This continues until the drum is empty. (In some other types of drum the emptying hatches are arranged in a line along the length of the drum. During discharge the green malt falls into a hopper over a conveyor that runs the entire length of the drum.) Thus filling, turning, sprinkling (if used), emptying and cleaning are largely mechanized and, apart from opening and closing the doors, are push-button or timer controlled; as a result the amount of grain one person can malt is chiefly governed by the size of the plant. In older plants, several drums might be connected to the same air-conditioning and fan unit. In

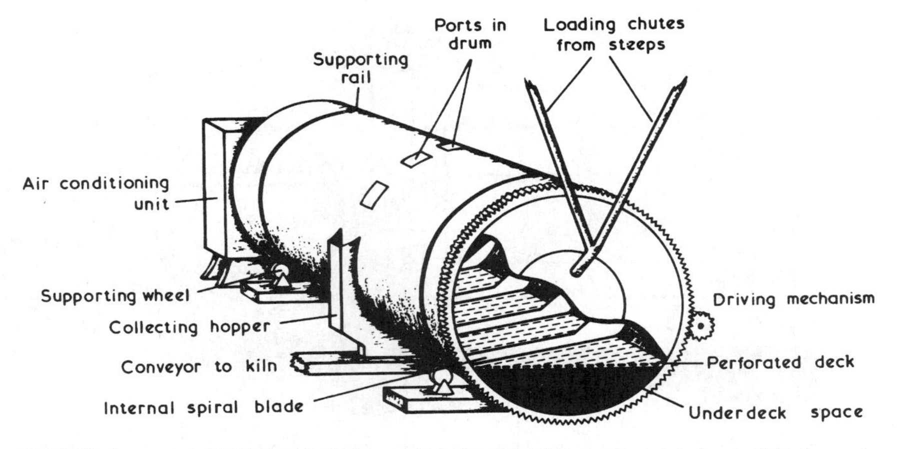

Fig. 9.40 One type of Boby decked drum. Each drum has its own fan and air conditioning unit, which is situated at the end opposite the loading spout. The inside of the drum is fitted with internal 'spiral blades', which as the drum rotates, move the grain to the discharge ports during stripping

more modern installations each drum is equipped with its own air-conditioning unit and fan; and the temperature of the air and the proportion being recirculated may be varied automatically during the germination sequence.

The Graff drums were 12.8 m (42 ft) long and 5.49 m (18 ft) in diameter with capacities of 3800 American bushels. Steeping, germination and kilning could be carried out in the same drum. Air was admitted from a central duct and exhausted through perforated regions in the casing.

9.9.4 Compartment maltings

Rectangular germination compartments, still often referred to as Saladin or Saladin–Prinz boxes, are widely used for making malt. The original boxes and ancillary equipment have been greatly refined (Leberle, 1930; Schönfeld,1932; de Clerck, 1952; Chubb, 1954; Vermeylen, 1962; Leaton, 1963; Macey, 1963; MacWilliam, 1963, 1965; Anon., 1965a,b, 1967c, 1968, 1969a,b; Green, 1965; Griffin and Pinner, 1965; Palmer, 1965; Philbrick, 1965; Pool and Pollock, 1967; Narziss, 1976; Williams, 1976; Adamic, 1977; Gibson, 1989; Kunze, 1994). The equipment is robust – an original Saladin installation of 1891 was photographed, still working, in 1967 (Crisp, 1967) and the same plant was in use for many years afterwards. Modern compartments are equipped with high-performance air-conditioning units, and the turners, fans and other ancillary machinery is electrically driven. Often boxes have been built in large halls, with walkways in between. This is advantageous in some ways; for example turners and bulldozers can easily be moved

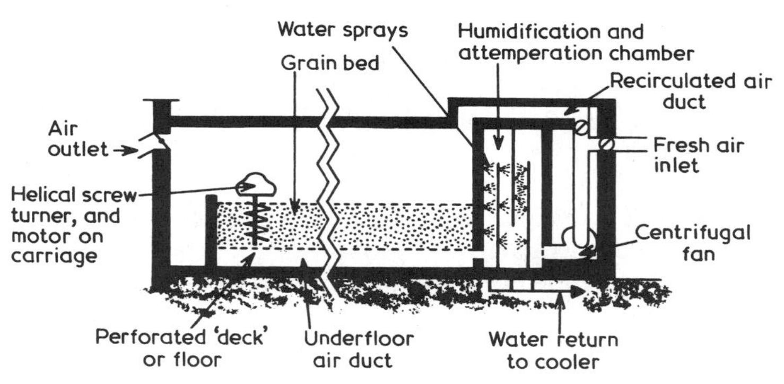

Fig. 9.41 A schematic, longitudinal section of a rectangular germination compartment, or Saladin box. A fan-driven airflow is driven through a humidification and attemperation section and up through the perforated deck and bed of grain (after de Clerck, 1952; Briggs, 1978).

between the boxes and so be shared. Unfortunately, the airstreams get mixed. This means that much of the mixed air needs to be attemperated, more than would otherwise be the case. The preferred arrangement is to have each box in a closed room, with its own air circuit (Fig. 9.41). Equipment is moved between the rooms through sealed doors. In the UK, air conditioning is often achieved using only water sprays. The water may be attemperated. Elsewhere the air may be warmed using injected steam or radiators, and cooling may be achieved using refrigerated heat-exchangers as well as the evaporation of the water from the humidification sprays. A walkway is provided at the side of each box to allow the grain to be inspected and to give access to the machinery. Usually carbon dioxide concentrations do not increase greatly because only a proportion of the air from the respiring grain is recirculated and because much of the carbon dioxide is washed from the airstream by the air-conditioning water sprays. However, it is desirable, indeed it may be mandatory, to check that toxic levels of carbon dioxide are not reached. Appropriate respirators must be readily available for use in emergencies. Other safety precautions relate to the installation of the electrical equipment in the highly humid environment and the provision of safety switches or cut-out devices so that turners, bulldozers, conveyors, *etc.* may be electrically isolated and emergency switches on the devices themselves, including a 'cut-out lever' across the front of the turner, so that machines can be stopped immediately in case of an accident.

The high humidity causes problems of condensation, plant deterioration and corrosion. In many plants, the walls and ceilings are curved so that condensed water runs down and does not drip onto the grain, walkways or machinery. In some older maltings the ceilings were warmed with embedded electrical cables to prevent condensation, but this system is too expensive to run. Walls and ceilings are usually painted, the paint being resistant to damp and to cleaning agents. Sometimes the paint also contains anti-microbial agents. In a few cases walls (both of the boxes and of the rooms) and the ceilings may be tiled. Cleaning is usually with high-pressure hoses and water, but cleaning agents containing sodium hypochlorite are often used.

The chief characteristics of traditional, rectangular boxes are illustrated in Fig. 9.41. In practice, designs vary widely in detail. A box consists of a fan, an air-conditioning unit and air ducts, which include a recirculation pathway and valves, as well as loading, turning and stripping equipment. The air duct may be beneath the perforated deck, constituting the plenum, or it may run under the walkway beside the box and be connected to the plenum chamber by doorways regulated with valves, often simple sliding valves, to allow the airflow to be controlled. Each box is a rectangular, open-topped container with a false bottom or deck of slotted metal plates or wedge-wire (giving about 20% free space) on which the bed of grain rests. The plates can be lifted for cleaning. 'Wedge-wire' of stainless steel is the preferred decking material. It is readily cleaned and, because of its large 'free space area' it offers minimal resistance to airflow, so keeping down fan power requirements and costs. The space between the deck and the true bottom is the plenum chamber. Many designs of boxes had plena that were awkward to clean. Usually the conditioning airflow is upward through the bed of grain, but in some plants the airflow is sucked downward and in others the direction of airflow is reversible. The long, vertical side walls of the boxes are exactly parallel and are built to support a load. They carry, usually on the top, rails and a rack that give support and traction to various pieces of equipment, such as a turner, plough or bulldozer. Boxes vary widely in their capacities. In the 1970s, boxes with capacities of 5–150 t (unsteeped barley) were in use, while in the UK the 'usual' size in the 1960s was about 40 t. In different plants, using typical values, the walls would have been 2–5.5 m (6.56–18.1 ft) apart, 1.4–1.6 m (4.59–5.25 ft) high and 18–32 m (59.1–105 ft) long. Grain depths were 0.9–1.4 m (2.95–4.6 ft) at steep-out. The depth increases during germination. Plenum depths were 0.4–0.6 m (1.3–2 ft). Boxes have become larger, with deeper grain beds. Sometimes they are GKVs. The plenum chambers have tended to become deeper (0.4–2 m (1.3–6.56 ft)) to achieve more uniform air distribution and airflows. Side air ducts seem less usual. Gibson (1989) stated that for good air distribution the length-to-breadth ratio should not exceed 8:1. With practical restrictions on the size of turner assemblies, the largest box should be about 6.5 m $\times$ 52 m (21.3 ft $\times$ 170.6 ft), giving an area of *ca.* 338 m^2 (3638 ft^2). If the loading depth is 1.4 m the total volume of steeped barley is 473 m^3 which, at a bulk density of about 2.08 m^3/t barley steeped, gives a batch size of about 227 t original barley (unsteeped). He concludes that the 'best' batch size in a rectangular box is 200–250 t barley. Many newer boxes in the UK fall in this range, but larger boxes of *ca.* 300 t are used in Continental Europe. Ease of construction and improved airflows have led to newer malting compartments, often with larger capacities, being not rectangular but circular in plan.

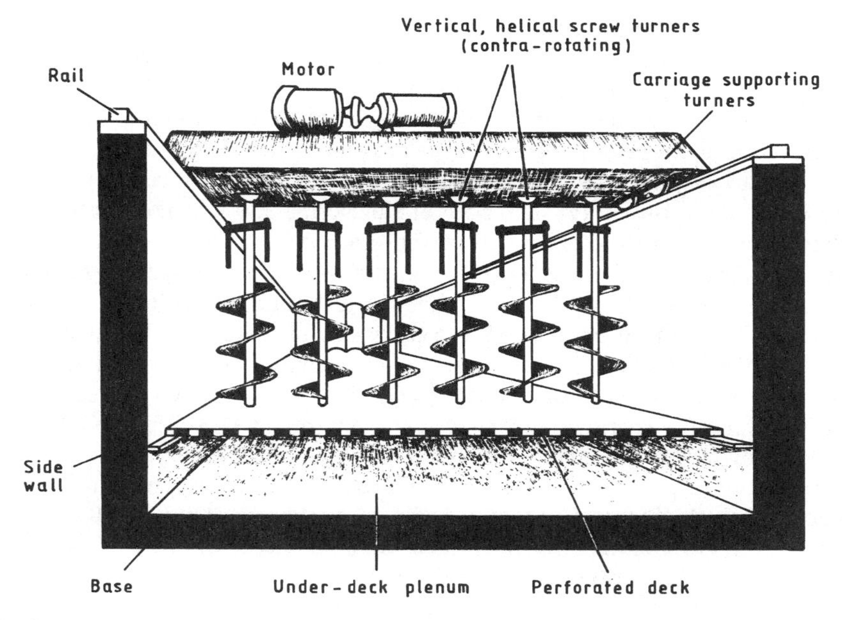

Fig. 9.42 A set of contra-rotating helical screw turners, of the Saladin type, mounted on a carriage in a rectangular germination compartment. Often the row of turners can be lifted to a nearly horizontal position so that the carriage can be moved above the bed of grain without the turners touching it.

Boxes may be loaded 'wet' or 'dry'. In wet loading, a slurry of grain in water is pumped to the box and is poured into it, loading it as uniformly as possible from an overhead conveyor using a steerable spout. The surplus water percolates through the perforated deck and escapes from a drain in the base of the plenum chamber. This is simple, the delivery system is practically self-cleaning and the grain arrives washed. However, if the grain has been allowed to chit in the steeps, or even if it has not, excess pumping pressures and physical damage in the pumps must be avoided, making gravity-feed from steeps mounted above the boxes preferable. If the grain has **not** been allowed to chit in the steeps, then arriving in the compartment with a film of surface moisture may delay the onset of chitting and slow down the onset of modification. British maltsters tend to prefer dry casting and usually the grain has been induced to chit by the end of the steeping period. The steeped grain is conveyed 'dry', using chutes and overhead conveyors, and is loaded into the germination compartment. At the end of loading the grain is usually in the box in a series of conical heaps. It is levelled by passing the turner through the box, sometimes after preliminary levelling by hand.

Turners are used to 'lighten the piece', lifting the grain to separate matted rootlets and so reduce the airflow resistance of the bed of grain. Grain increases in volume as it germinates and the rootlets 'hold the corns apart'.

After turning, the volume of a piece may be increased by as much as a third, so the depth of the grain bed is increased; this must be allowed for in the height of the walls of the box and the design of the turners. Good turners gently lighten the piece without damaging rootlets or breaking or crushing the grains. 'Root pruning' encourages the production of secondary rootlets, which can cause husk separation, spoiling the appearance of the malt. Moulds may grow on crushed rootlets and crushed or broken grains, and they may spread to infect much of the grain batch. The counter-rotating, vertically mounted, helical screw turners, of the type introduced by Saladin, are still those mostly used (Figs 9.31, 9.42), although they differ in details of design. The screws are mounted on a carriage which is supported on the rails at the sides of the box. The carriage, which is electrically powered, is driven forward by a cogwheel that meshes with the rack, next to the supporting rail. Usually the carriage advances at 0.4–0.6 m/min (1.3–2 ft/min). It must not be advanced too rapidly. In older boxes, the metal helices were solid. In many modern plants, the helix 'flights' are partly of a ribbon of metal supported from the central axis by a series of projecting struts. As the turner works, the grain is raised towards the carriage and to break up this 'wave' many turners have downwardly projecting rods mounted on the axis (Fig. 9.42). These turners really 'lighten the piece'; they do not invert it or turn it thoroughly. Where the turner has passed, the surface of the grain bed is slightly uneven, having a series of slight mounds separated by valleys running the length of the box. These are caused by the contra-rotating turners. Where they rotate towards each other, the grain is piled up and where they rotate away a valley is formed. The resistance to airflow is less through the valleys than through the hills, and the grain must be more compressed in the hills, since raking the surface flat does not overcome the problem. Therefore, airflow through a grain bed turned in this way is not truly uniform.

From time to time the capacities of malting plants are increased. With boxes, this is usually done by increasing the height of the side walls. This needs the installation of new turners and, often, more powerful fans. Attempts to use boxes loaded deeply may be frustrated by the formation of the 'bow wave' by the turners which lift the grain and deposit it onto the top of the walls, the racks and rails and even cause it to spill over the sides. To minimize bow-wave formation, the rate of turner carriage advance and turner rotation speeds are carefully adjusted. An ingenious device called a 'chinese hat' has been mounted on turners; this limits the upward movement of the grain. At the appropriate height on each turner shaft, there is a metal disc physically supported and strengthened by a cone of metal attached to the rim of the disc and the shaft. The grain in the box is slightly more compact than it would be were the 'hats' not in place, so the resistance to airflow is higher. With 'hats' the effective capacities of boxes may be increased by 10%.

Two other types of turner that have been used in rectangular boxes both turned grain extremely thoroughly and gently, but the boxes were not used to full capacity because they were 'open ended', having no end walls. The older types of machine were known by various names, such as paddle, shovel or Van Caspel turners. These are not now used. They worked like old kiln turners. A carriage moved along the box supporting a horizontal, rotating, metal axis, which carried projecting arms that carried blades or scoops. As the device moved into the germinating grain, the blades cut down into the bed of green malt, which was scooped up, lifted and dropped behind so that it was thoroughly mixed. As the turner passed, the grain bed was moved along the box by about 1.83 m (6 ft). The diameter of the circle drawn by the rotating blades was *ca.* 1.83 m (6 ft). However, if the grain was beginning to mat or the box was loaded too deeply, these turners tended to 'walk up the malt' and became derailed. Consequently, in some instances, they were replaced by 'single throw' Wanderhaufen-type turners (see below, Fig. 9.55), which also moved the grain along the box. With both of these types of turner, alternate turns had to be from opposite ends of the box, and the grain mass was moved to and fro. A small area of perforated deck, at one or other end of the box, was not covered by grain and the grain bed 'sloped' at the ends. To conserve attemperated air, valves were closed to cut off the air to the uncovered areas of decking. The open-ended boxes have usually been converted to closed-ended boxes equipped with helical screw turners, thereby increasing their capacities since they can be completely filled to the vertical end walls. Other types of turning device are considered below in connection with semicontinuous malting (section 9.10.3).

In the proposed 'Vilain system', short-steeped grain was to be loaded into a box or bin with a perforated base (MacWilliam, 1965). Turning was achieved by a device, triangular in cross-section, suspended from above and reaching to the bottom of the box. Endless series of buckets and chains moved around the device, lifted the germinating grain from in front and deposited it behind as it moved along the box. Steeping was completed by spraying the grain being moved in the buckets.

The number of systems used to unload ('strip') rectangular boxes is quite large. In many boxes, the end wall may be moved, sometimes being lowered into a slot; as in open-ended boxes, the grain is discharged from the end into a hopper and a conveyor. The conveyor is designed to minimize the damage to the green malt. In some early end-discharging boxes, the green malt was pushed out by a manually operated power-shovel. Now the green malt may be pushed by a 'bulldozer' plate mounted on a carriage in the same manner as a turner (Fig. 9.43). The device moves backwards (when the blade is raised) or forwards (when the blade is lowered) taking 'bites' and pushing successive parts of the grain load to the conveyor. In other systems, the grain is pushed using the turners, which are not rotated and which are backed by a detachable metal plate that is profiled and remains attached to the end

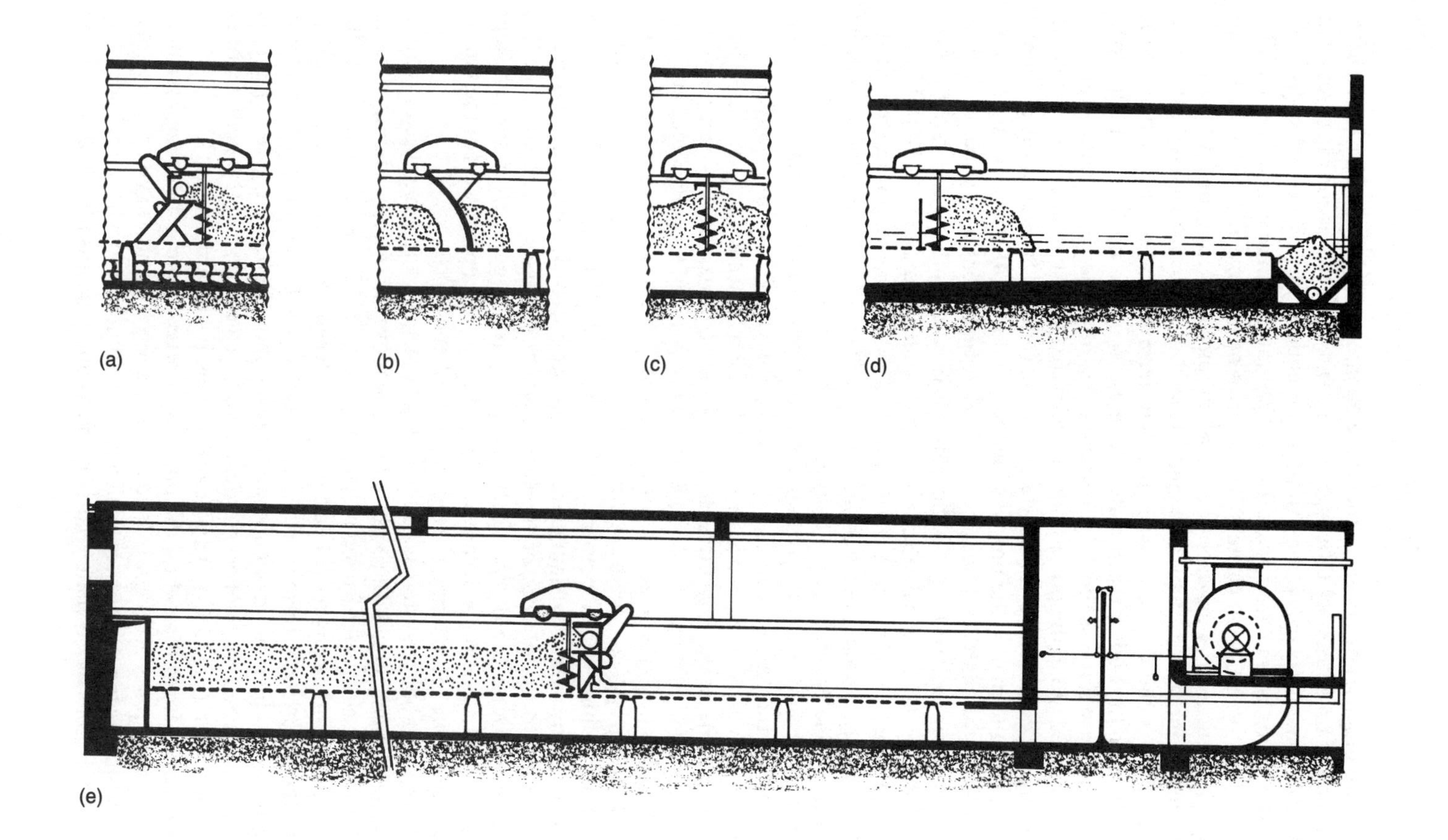
(a)
(b)
(c)
(d)
(e)

wall when not in use (Fig. 9.43d). There are also references to a 'traction discharger', described as resembling a hay rake. Some boxes were built with louvred floors. To discharge the grain, the floor sections were tipped and the grain fell into a hopper and conveyor in the base of the air plenum chamber. Other boxes are unloaded pneumatically, by suction. In a simpler but more laborious system, an operator moved the flexible suction tube about in the box by hand. In a more sophisticated system the turner, with ancillary machinery attached, moves into the grain and lifts it to a suction head, where it is taken into an extending, telescopic tube (Fig. 9.43e). In yet another system, the turner, attached to extra machinery, moves into the grain and lifts it to a helical screw cross-conveyor, which moves it sideways and drops it over the side of the box, into a gulley in the walkway, containing a conveyor (Fig. 9.43).

Plevoets (1974) proposed a rectangular box that is novel in a number of ways, including the manner of discharge. As in the Jonsson drum, some of the heat from the germinating grain was withdrawn by contact with a series of long metal walls running the length of the box and cooled with internally circulating, attemperated water. Each section was turned by a single vertical helical screw. In addition cool, humidified air was drawn downwards through the grain bed. The deck of each box consisted of a perforated moving-band conveyor. For filling, steeped grain was loaded in at one end as the conveyor advanced. For stripping, the end wall was moved, the conveyor moved forward and the grain was discharged into a conveyor.

9.9.5 *Kropff malting boxes*

In the Kropff malting system, partly germinated grain was transferred to special, heavily insulated boxes (Fig. 9.44) in which a series of valves allowed the boxes to be sealed, the grain bed ventilated, or the air under the deck to be changed (de Clerck, 1952; Schuster, 1962; Vermeylen, 1962; Narziss, 1976). After loading the box was sealed; the temperature at first rose and the oxygen was progressively used up. Sporadically, gentle aeration was applied, but for practical purposes the 'carbon dioxide rest' checked rootlet formation and respiration losses, while permitting the

Fig. 9.43 Various turning and stripping systems used in compartment maltings. (a–c) In the Koch systems, grain is either (a) lifted by the mechanical turner, cross-conveyed and delivered into an open conveyor moving to the kiln along the walkway or (b) bulldozed along the box to a discharge conveyor. (c) Note the bow wave around the helical screw turner. (d) In the Steinecker–Lenz stripping system, the turner, with a 'backing plate', is used to bulldoze the green malt into a cross-conveyor at the end of the box. (e) In the Koch pneumatic unloading system, the advancing turner lifts the green malt into a suction unit attached to an extending pneumatic tube, which, in the diagram, extends to the right through the spray chamber and past the centrifugal fan (after Vermeylen, 1962.)

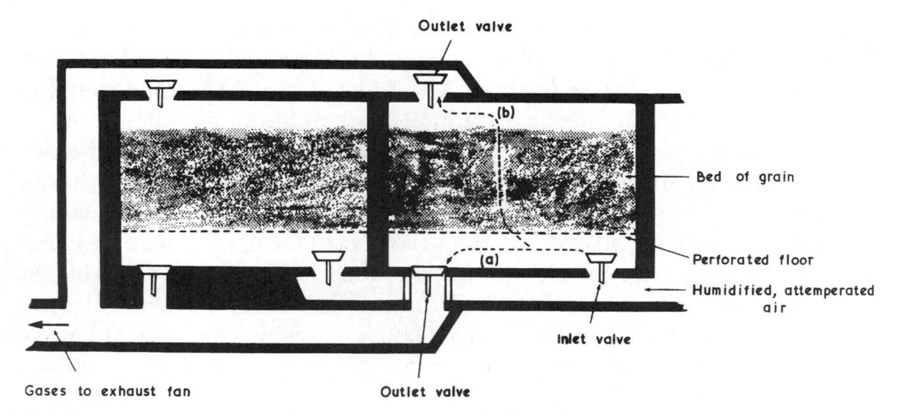

Fig. 9.44 Two adjacent Kropff boxes (after various sources). These boxes were well insulated and were gas-tight when the valves were closed. (a) The route of the air during limited ventilation; (b) the route of the airstream during more vigorous ventilation and attemperation. When the boxes were sealed the temperature rose, oxygen was used up and carbon dioxide accumulated until growth ceased, so reducing malting losses.

enzymes to continue to modify the endosperm. Thus this system was used for reducing malting losses (Chapter 14).

9.9.6 *Popp malting plants*

Various versions of this plant, designed by Dr Popp, were built (Vermeylen, 1962; Leaton, 1963; Macey, 1963; MacWilliam, 1963; Reitzel, 1964; Green, 1965). A novel means of turning the grain bed was used (Fig. 9.45). In a particular unit two vessels, each of 25 t capacity, 11.6 m (38 ft) high overall and 5.64 m (18.5 ft) diameter, were served by a single compressed air unit with a reservoir 12.2 m (40 ft) long × 2.13 m (7 ft) in diameter (Green, 1965). The heavily lagged, cylindro-conical vessel had inspection windows in the sides and a tipping floor made of two semicircles of wedge-wire. Tipping the floor discharged the grain into the hopper bottom and thence into a conveyor. Steeped grain was loaded into the vessel to the unusually large depth of 2.1–2.4 m (about 7–8 ft) and attemperated air, partly recirculated, was passed in the normal way. The temperature differential across the grain was about 2 °C (3–4 °F). When the grain was to be turned (two or three times each day), the air-conditioning ducts were closed and the upper cover or lid was raised about 0.6–0.9 m (*ca.* 2–3 ft). An air blast was then suddenly released beneath the bed of grain, from the compressed air reservoir (at 4 atm) using a special valve. The bed of grain was lifted and the piece was thoroughly lightened without root damage. The air blast escaped from under the lid and any fast-moving grains were retained by the perforated grain catcher. Afterwards the lid was closed, other ducts were opened and ventilation with conditioned air was resumed. On the one hand, if the grain

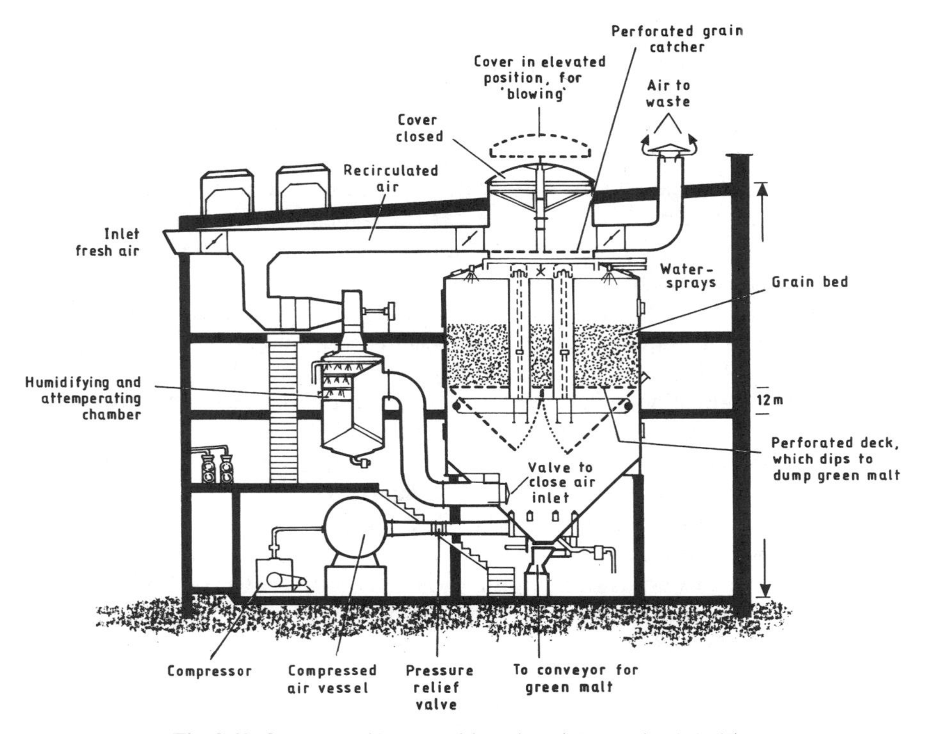

Fig. 9.45 One type of Popp malting plant (after Reitzel, 1964).

was 'blown' too infrequently, the rootlets grew together, the pieces matted and turning was ineffective. On the other hand, if the piece was 'blown' too often the blast made channels through the loose bed of grain; the bed then presented too little resistance to the airflow and could not subsequently be turned.

In principle one compressor system might serve several, say six, germination units and the potential advantages were seen as ease of automation, gentle grain turning and the possibility of carrying out resteeping in the vessel. However the feasible batch size was small.

9.10 Multi-function systems

9.10.1 Germination and kilning boxes

Various types of rectangular compartment in which both germination and kilning could be carried out have been built (Vermeylen, 1962; MacWilliam,

1963; Haman and Hartzell, 1972; Kassenaar, 1974; Adamic, 1977). The advantages are that such boxes may be used to dry batches of barley at harvest time and the transfer of green malt between the germination compartment and the kiln is done away with, removing the need for green malt conveyors and simplifying the integration of the germination and kilning cycles, giving the maltster more operational flexibility. Furthermore, the high temperatures of kilning, and sometimes the inclusion of sulphur dioxide in the hot airflow, disinfects the plant, 'sanitizing' it and reducing the need for cleaning. However, the structures must be capable of withstanding repeated temperature cycles, and the thermal efficiency of some early plants was low. The plant must be cooled before steeped grain is loaded into the box and in kilning the entire structure needs to be warmed up and any pools of water in the plenum must be evaporated. In the UK, the most successful plants are those built with low thermal capacities. Many of those built with concrete are now used as germination vessels only and the green malt is transferred to and kilned in separate thermally efficient kilns.

In the UK such boxes, also called GKVs or germination–kilning vessels, typically have capacities of about 200 t and the grain, which has been steeped in various types of steep tank, is distributed into the box using an overhead conveyor. The below-deck plenum chambers are usually deep (e.g. 2–3 m (6.56–9.84 ft)). The conditioned air is admitted from one end of the box and the kilning air from the other. The air ducting of the system not in use is sealed off with heavy doors. In the USA, this type of malting is referred to as the Fleximalt system, emphasizing its operational advantages (Fig. 9.46). Recently GKVs of circular cross-section have been installed. Typically these are well-insulated structures internally clad with corrosion-resistant metal. They often receive partially grown green malt from a circular germination vessel. By installing a GKV in this way the germination period can be extended, if this is needed, but at the expense of a shorter and, therefore, more costly (in terms of fan power) kilning cycle. The Albert Schwill (Chicago) system seemed rather similar except that, unusually, the box was unloaded by having a moving, perforated deck that discharged the kilned malt via a removable end door into a conveyor (Vermeylen, 1962; MacWilliam, 1963).

9.10.2 *Multifunction vessels: SGVs and SGKVs*

SGKVs have frequently been proposed (i) to overcome programming problems associated with malting in separate vessels, so that the durations of the steeping, germination and kilning periods can be varied at will while keeping the vessels fully occupied; (ii) to overcome the need for transferring grain between steeping and germination units and the germination units and the kilns; (iii) in some cases, to permit resteeping, that is the

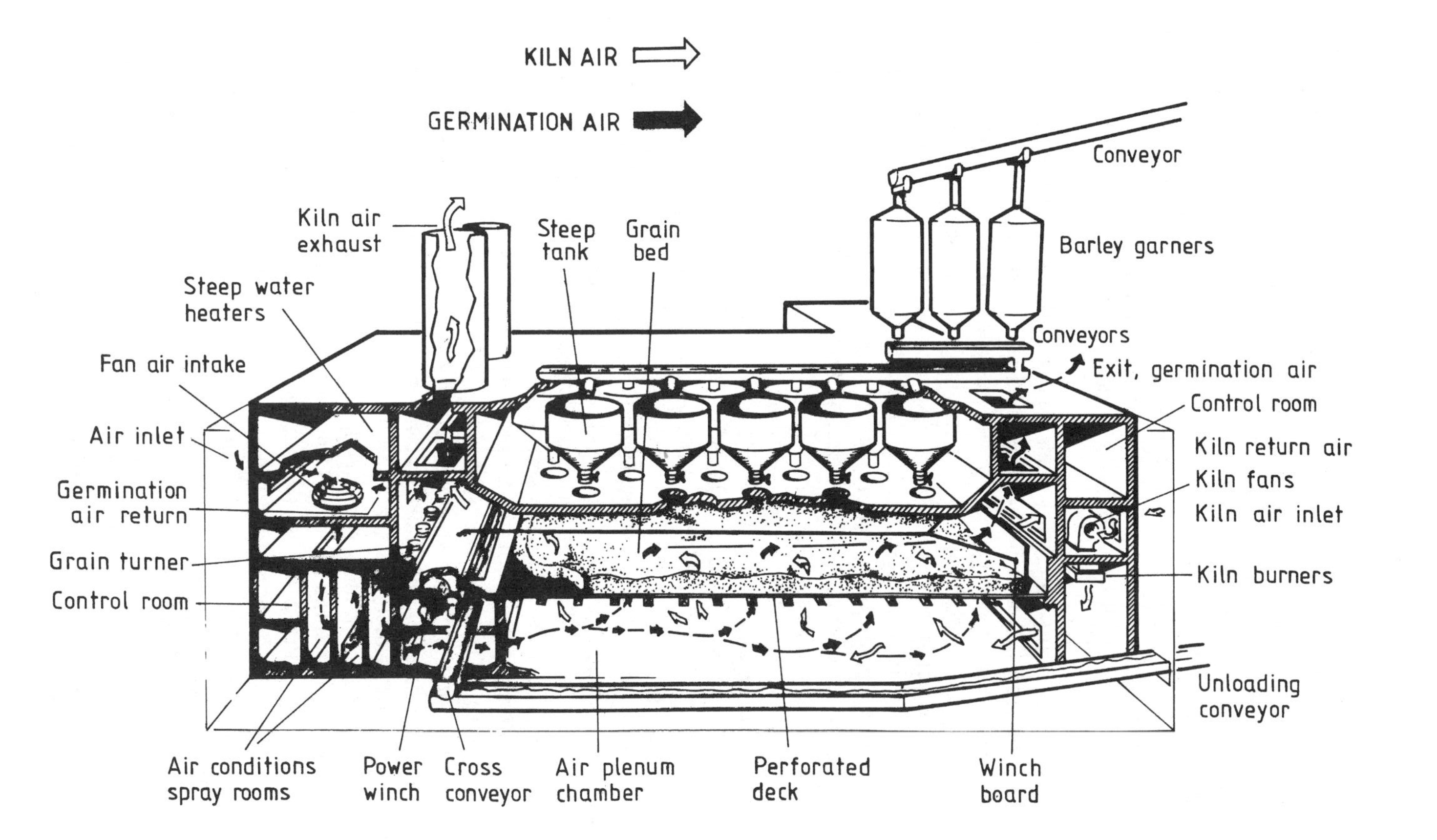

Fig. 9.46 A rectangular germination and kilning box (after Kassenaar, 1974).

reimmersion of germinating grain to check rootlet growth and reduce malting losses (Chapter 14). Thorough root killing means that the green malt does not mat and so need not be turned; such plants were sometimes called 'static malting systems'. Commonly such plant is also used for barley drying during the harvest period. Steeping and germination vessels may be built to permit resteeping in a single vessel but kilning takes place in a separate, thermally efficient kiln. Alternatively they may be used as a well-controlled, ventilated steep that is sufficiently air conditioned to control the grain's temperature well yet avoids the need to transfer the grain to a germination vessel.

Problems with multi-function vessels include: (i) design and engineering compromises must be made in the construction, as a result efficiency in the different operations is compromised; (ii) the units will almost certainly be thermally inefficient; (iii) because of the large batch sizes that are now common the air chamber beneath the perforated deck must be large to obtain even airflows during germination and kilning. The air chambers must be filled with water during steeping, or resteeping; as a result water usage and effluent production are very large, and the cost can only be sustained under special conditions where water is unusually cheap and the effluent can be disposed of cheaply, e.g. by pumping it out to sea.

With the *tout en cases* system (de Clerck, 1972), 300 t barley was loaded into rectangular boxes, 63 m × 10 m (207 ft × 32.8 ft) using a chain conveyor. The grain was levelled using a horizontal, moving, helical screw conveyor working 'middle-to-sides'. Turning was with Saladin-type turners, and stripping was by opening a row of doors in the middle of the floor over a conveyor and running the moving cross conveyor so that the screws moved the grain 'sides-to-middle'. A novel aspect of this system was that the grain was spray steeped using overhead sprayers.

The first attempt to carry out resteeping on a commercial scale was made in specially strengthened drums (Riviere, 1961). While this demonstrated the feasibility of the process, it was clear that purpose-built plant was needed. In the first commercially operational 'static' malting system, the Mendlesham box (which was also used for barley drying), steeping, germination, resteeping and kilning were carried out in a special vessel (Fig. 9.47). As originally designed the plant was inexpensive to build. The air duct ran beside the box and the drying air flowed over a retaining wall or weir (used to hold back the steep water) under the false bottom of the box, with a comparatively shallow plenum with a sloping base, out through the grain bed (*ca.* 1.5 m (*ca.* 5 ft) deep) and to waste. This elegant design ensured a good airflow but, by achieving a small plenum volume, was economical with water use and minimized effluent production. The axial-flow fan was reversible and attemperating air could be drawn downwards through the grain bed during the germination period. The original boxes were used for processing 30 t batches of barley using multiple steeping and resteeping malting

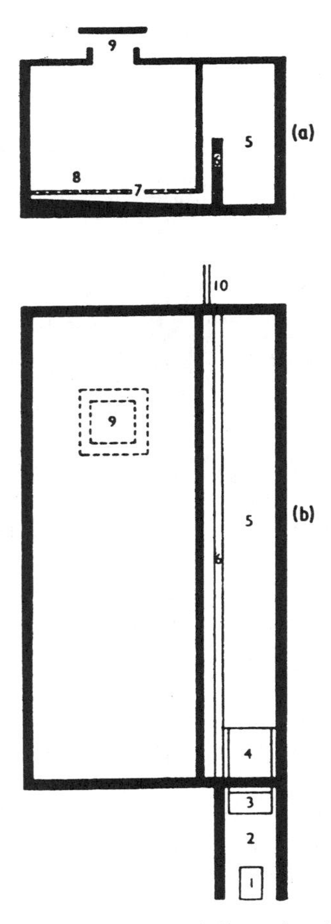

Fig. 9.47 The layout of the first type of commercially used multi-purpose SGKV (Griffin and Pinner, 1965). These Mendlesham boxes were used for barley drying and for steeping, germination, resteeping and kilning. (a) Vertical Section; (b) plan. 1. Oil burner; 2. furnace; 3, silencer; 4, fan; 5, air duct; 6, wall or weir to prevent escape of water through the air duct; 7, perforated wedge-wire floor, or deck; 8, the position of the grain bed; 9, air vent; 10, drain cock.

regimes at controlled temperatures to achieve root kill, so no turners were provided. The boxes were 15.5 m × 5.45 m (51 ft × 18 ft) and were loaded using revolving band thrower–spreaders. Unloading was by a winch and plough board through a removable end wall. Overhead sprays were used to humidify the air above the grain bed. Ventilation was with a mixture of fresh and recirculated air. However it was found that being confined to the non-turning method of operation was too restricting, and so in later versions of these boxes the grain bed depths were deeper and vertical helical-screw turners were installed. Rising fuel costs, the inherent thermal inefficiency of

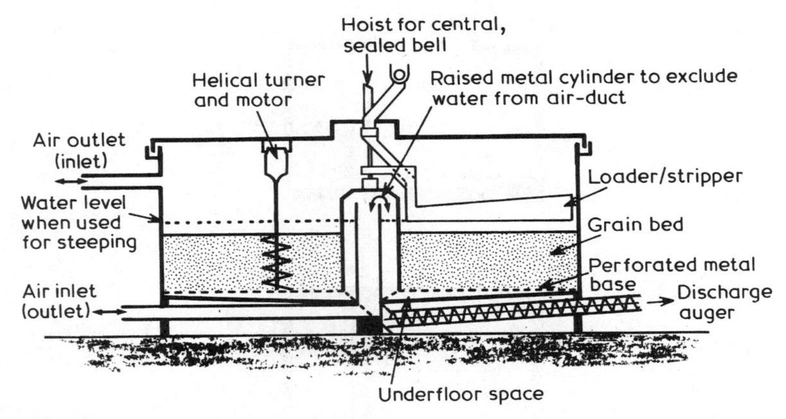

Fig. 9.48 A Boby flat-bed, circular cross-section steeping, germination and resteeping unit (after various sources; Briggs, 1978).

this type of plant and the impracticability of installing heat-recovery devices rendered the plant too expensive to use for malting, but for a period it was still used for barley drying. In the original plan, the plant was used for drying barley for about 8 weeks each year. By drying slowly, for periods of up to 24 h then running the grain away to store while still warm, dormancy in the grain was usefully reduced.

In the Huppmann system, grain was steeped in a conventional way then transferred to a special vessel which was, in effect, a hopper-bottomed steep equipped internally with a tipping floor and which could be ventilated from below with conditioned air (Holbein, 1969; Haman and Hartzell, 1972). The first germination stage and resteeping were carried out in the specialized steep. At the end of the resteep root-killing period, the floor was tipped and the piece was wet-cast into a conventional germination vessel.

The use of modified hopper-bottomed steep tanks (4.27 m, *ca.* 14 ft deep) equipped with arch-breakers and perforated inner cones (by Pool and Pollock, 1967) for steeping, resteeping and germination of 30 t batches has already been mentioned (Fig. 9.26). In commercial practice, these steeps have been used in conjunction with a flat-bed, circular cross-section vessel that was originally designed to permit germination and resteeping in the same vessel (Fig. 9.48). The equipment was used with a circular kiln and was known as MTI (Malting Technology International) plant (Anon., 1967c, 1968; Pool, 1967a,b, 1970) (Fig. 9.49). When resteeping was to be carried out, the grain was returned from the germination vessel to the steep and was subsequently transferred back to the germination vessel for the 'autolysis period' (Chapter 14). These vessels, with circular cross-sections, are typical of one type of plant that is now frequently built. The grain bed is static but the levelling, turning and stripping machinery rotates around a central mounting, the outer end of the carriage having wheels supported on a rail

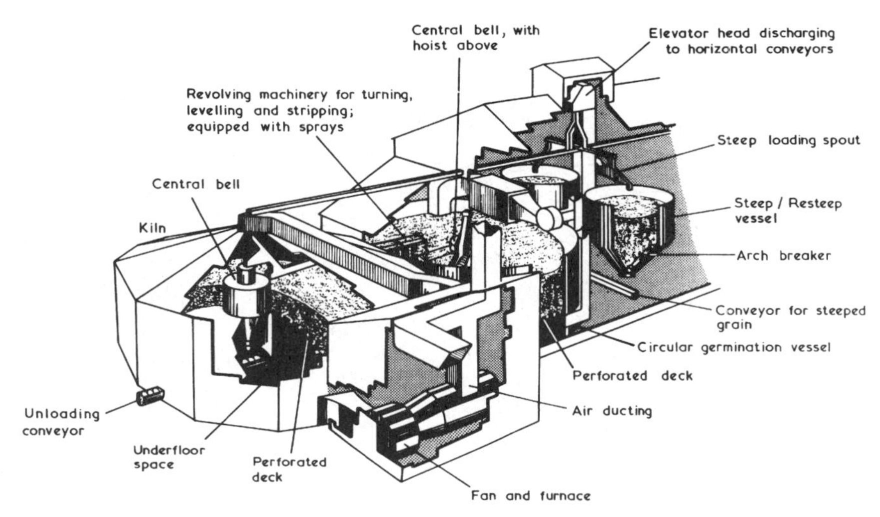

Fig. 9.49 An MTI malting plant with conical bottom steeps fitted with arch-breakers (for steeping or resteeping), a circular flat-bed germination unit with rotating machinery for loading, levelling, turning or stripping and a kiln with circular cross-section. (Courtesy of Rockley (1968); Vickers Ltd, Malting Division.)

and driven by cog wheels which mesh with a rack. For loading, steeped grain is delivered into a spreader-arm. As this rotates and rises a helical-screw cross conveyor moves the grain from the centre towards the periphery, creating a level bed. For stripping, the central bell is raised, exposing the central hopper to the conveyor, and the cross conveyor rotates, moving downwards and sweeping the grain inwards to the hopper. Turning is by a series of vertical helical screws, which are moved around the vessel on a rotating carriage. One 90 t (as barley) vessel has an outer diameter of 14 m (46 ft) and a floor area of 140 m^2 (1500 ft^2), giving a specific loading of 646 kg barley/m^2 (132 lb/ft^2). Where space is at a premium, germination vessels can conveniently be built in pairs, one above the other (Fig. 9.50).

Increasingly, germination vessels are being built of corrosion-resistant or stainless steels. By designing the vessels to have relatively unobstructed plena, it is possible to use automatic cleaning using high-pressure sprays fitted to the turning machinery to clean above the deck, and high-pressure sprays mounted beneath the deck to clean the plenum and underside of the deck. There are no obvious limits to the size of these circular vessels. The general design has been applied to kilns (Fig. 9.49). Recently, circular germination and germination-kilning units, built in the UK, have capacities of *ca.* 200 t (as barley) and are built of stainless steel or other corrosion-resistant metal (e.g. Mulford, 1992). Such vessels have superior airflow distributions to those of rectangular vessels. Also they are cheaper to construct than rectangular vessels in that they take less constructional material to enclose a particular amount of grain.

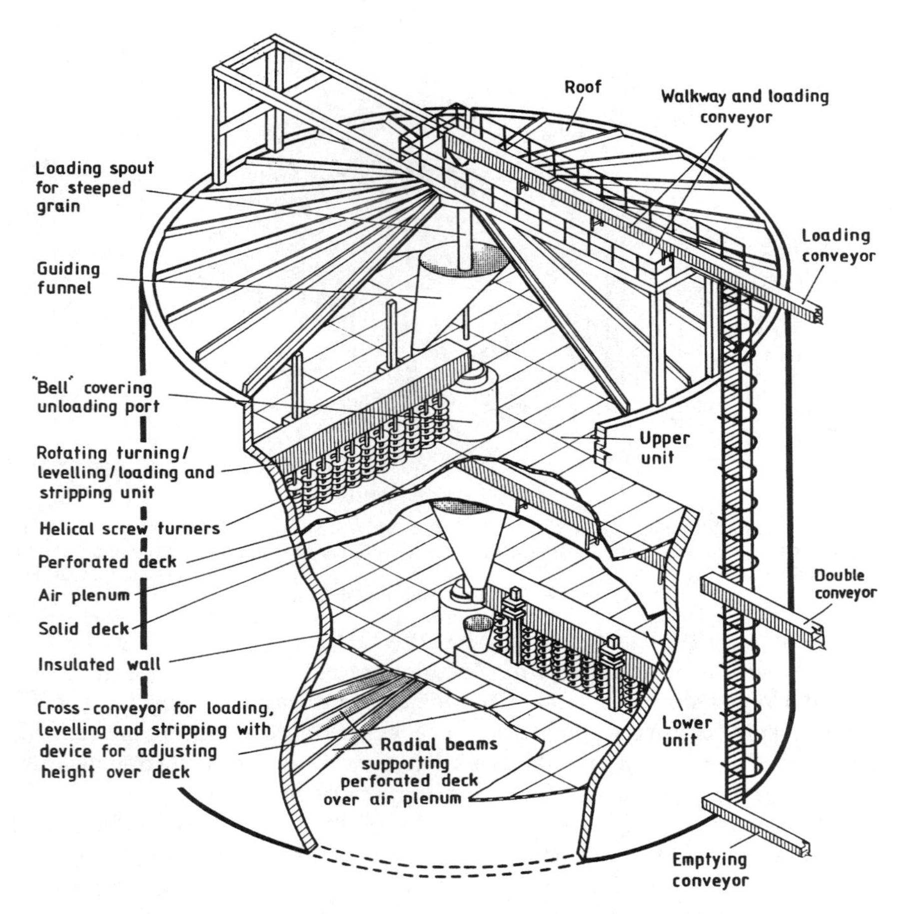

Fig. 9.50 Two germination units, circular in plan, with fixed floors and rotating machinery, built as a pair, one above the other (courtesy of Redlers Ltd).

In several modern maltings, annular germination vessels have been followed by GKVs. These vessels are essentially similar to the germination units except that while they are equipped with loading, levelling and stripping screws, they do not have turners and the specific loading of the decks is less so that the layer of green malt is less than that used in GVs. Such vessels must be well insulated and of low thermal capacity if they are to be thermally efficient. By extending germination time into the GKV, potentially at the expense of kilning time, it is possible to make malts from barleys differing in their germination-time requirements without altering the overall processing time of a batch.

The other type of circular or, more properly, annular malting vessel is that in which the 'bridges' supporting the grain levelling, loading and unloading

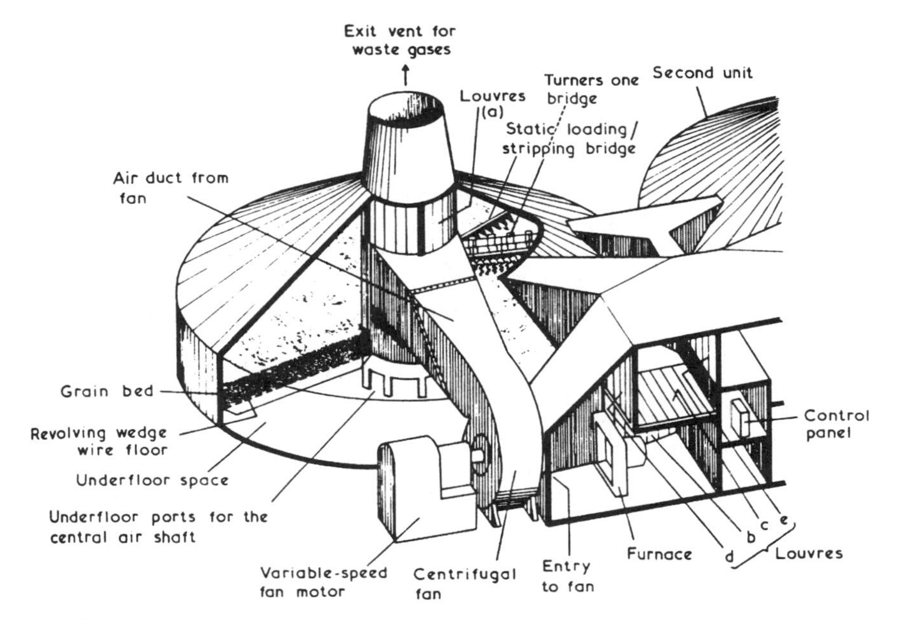

Fig. 9.51 One Clova malting SGKV unit in which barley drying and steeping, germination, resteeping and kilning may be carried out. Since their inception, all these units have been equipped with air-to-air heat exchangers, which are not shown here. The perforated floor, or deck, is rotated during loading, turning and stripping. The bridges supporting the loading and stripping machinery and the helical screw turners are stationary. By adjusting the settings of the louvres (a–e), it is possible to arrange airflows for barley drying, for germination, for kilning with fresh air or partly recirculated air, or to take 'curing' air from one unit and direct it to another, so linking kilning operations (after various sources; courtesy of Griffin (1982), Moray Firth Maltings Ltd).

machinery, and the set of helical-screw turners, are fixed, but the floor supporting the grain can be slowly rotated, carrying the bed of grain with it. The turners towards the outside of the vessel rotate more rapidly as they must turn more grain. There are mechanical advantages to this arrangement (Maule, 1995). Vessels of this type have been used as germination units and kilns, sometimes arranged above each other in 'towers'. They have also been used as SGKVs. As originally proposed, these SGKVs, which are a conceptual extension of the Mendlesham box (Fig. 9.47), were sometimes intended to work in groups of four, giving so-called Clova plants (Griffin, 1982). Usually they work in pairs (Fig. 9.51). Each group has a common control-room, services and ducting, is able to exchange air between the vessels during kilning and has common heat-recovery equipment for the kilning stage of operation. The plant and all the ancillary machinery is microprocessor controlled. These vessels are used for barley drying, steeping, germination and kilning and, if required, resteeping. The grain is supported on a rotating, annular wedge-wire floor, which is driven from below by several small electric motors. As the grain is loaded onto the rotating floor it is levelled, in successive layers, by a helical-screw cross conveyor. Conversely the box is stripped by the screw moving the grain sideways from the bed of grain, moving slowly below it, into the exit chute. The grain may be sprayed as the bed is carried round through a row of static turners. In some plants, water may be sprayed onto the grain from many overhead points. An ingenious arrangement of ducts and louvres allows germination air to be partly recirculated or furnace air to be directed through the grain bed and to waste; alternatively it may be partly recirculated or be directed to another vessel. It is usual for the hot, wet 'used' kilning air leaving the plant to be used to supply heat to the incoming air, commonly with an air-to-air heat exchanger. Fan speeds, and hence airflows, are variable in these plants.

These SGKVs are suitable for producing large amounts of similar, pale malts. However, because of the large under-deck plena, they use large volumes of steep water and produce large volumes of effluent. Consequently they can only be operated profitably where water supplies and effluent disposal are cheap. In the case of the largest unit, at Buckie in Scotland, the batch size for barley drying is 880 t max., allowing 1760 t to be dried each day. For malting the design batch size is 520 t of barley (Turcan, 1982). The diameter to the inner wall is 10.5 m (34.45 ft) and that to the outer wall is 30.5 m (100.1 ft), the annular floor between these walls having an area of 644 m^2 (6932 ft^2), giving floor loadings up to 1400 kg/m^2 for barley drying and 807 kg/m^2, as barley, for malting. Turning is by a bank of 17 vertical screw turners, which operate to a maximum depth of 2 m (6.56 ft). Those nearest the centre rotate at 3 r.p.m., those at the periphery at 12 r.p.m. The floor, of galvanized wedge-wire, is over a plenum with a depth of 0.75 m (2.46 ft) on the outside and 1.2 m (3.94 ft) towards the middle. The floor, which is driven by five 10 hp motors, rotates at the rate of one revolution in 2.75 h during loading and discharge, and more slowly (1 rev/3.5 h) during turning and loading the larger batches of barley for drying. Water, which can be attemperated, can be added to the grain from below and from fixed overhead sprays at the rate of 136.4 hl/min (3000 gal/min) and will cover the grain bed in 45 min. A wide range of other designs of static or resteeping malting systems has been proposed (Narziss, 1976). The Huppman resteeping system has been noted earlier.

9.10.3 Semicontinuous malting plants

Many proposals have been made for turning grain by transferring it to another germination compartment or by taking it out of the base of a compartment and returning it to the top. This idea could be applied to the static malting system of Pollock and Pool (Fig. 9.26) and is the basis of the Wanderhaufen system of Plischke and Beschorner (Fig. 9.33).

As early as 1889, the proposal was made to hold germinating grain on superimposed, perforated and ventilated floors in a tower. Turning was to be by dumping the grain through opening floor sections onto the floor below so that, as germination proceeded, the grain moved progressively down the tower (Gent, 1889). In the Engerth system, grain was germinated in ordinary boxes but was turned by dropping it into a drag conveyor, elevating it, then conveying it back into the top of the box (Vermeylen, 1962). In a sense the Engerth system resembles some of the early box-malting systems, except that the grain is moved mechanically and not by hand. In another type of transfer malting system, germinating grain, which is steeped in conical units, is held in one of a series of closed cylindrical germination vessels (Simpson, 1979). The vessels are unloaded using a planetary discharge unit. While the grain may be turned by conveying it horizontally to an elevator, elevating it and conveying it horizontally back to the top of the same vessel, normally the grain would be transferred to another vessel, so that at intervals of 23 h the grain 'moves on' from one vessel to the next and eventually is transferred to a kiln. Thus the malting uses batch transfer to operate in a semi-continuous fashion.

The Lausmann transposal system also relies on batch transfer (Fig. 9.52). The system turns grain very gently and ensures that, in each move, the grain is transferred from the top of one compartment to the bottom of the next (and also 'bottom-to-top' so each piece is inverted each time it is moved) (Anon., 1969a; Narziss, 1976). The compartments (rectangular boxes) are arranged in a row (in a 'street') and the deck or perforated floor of each may be raised or lowered at an exactly controlled rate. This is achieved through a series of pistons moving in cylinders, activated by oil pressure. The grain, steeped conventionally, is loaded into the first compartment with the floor in the fully lowered position and remains here, being ventilated with cool and humid air, until the first transfer and turn, usually after 24 h. At this stage the floor of the first box is fully lowered and the floor of the next box, which has just been emptied, is fully raised. The scraper transfer device, a comparatively light structure that spans both boxes and consists of endless moving belts carrying rubber or plastic scraper blades, is set in motion. Then the floor of the full box slowly rises, while the floor of the next box slowly descends. As this happens the grain is gently skimmed from the surface of the bed in the first box and is deposited in an even layer in the second box. Rubber wipers ensure that the last few grains are transferred. Each transfer may take 20–25 min. Arrangements may be made for spraying the grain during transfers. Boxes, the contents of which are generally transferred daily, may have capacities of 15–40 t or more; there are the same number of boxes as the number of transfers required, often six, and they may be loaded to a depth of *ca.* 1.5 m (4.92 ft). For stripping, the same machine may transfer grain into a hopper and conveyor or move it through a sealable and insulated door into a kiln, which, after kilning is complete, is

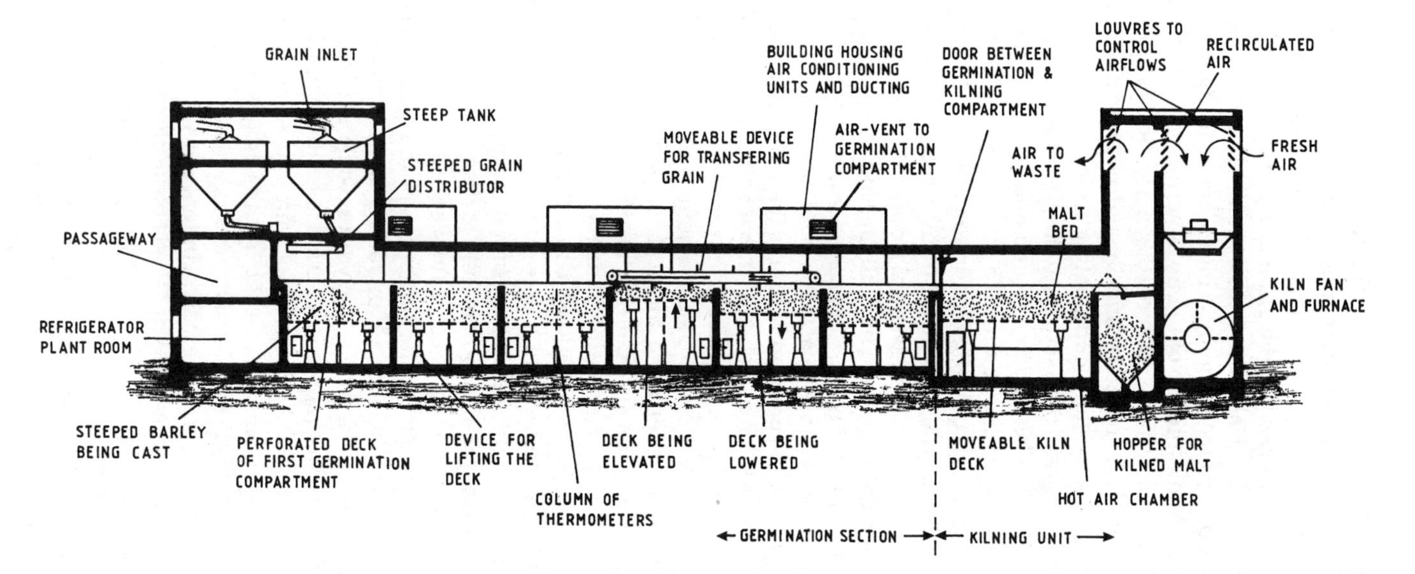

Fig. 9.52 A section of a Lausmann transposal malting plant. The first compartment, at the left, is being loaded with steeped grain. Green malt is being transferred from the fourth to the fifth compartment. The kiln is situated at the end of the street behind airtight and thermally insulated doors (after various sources).

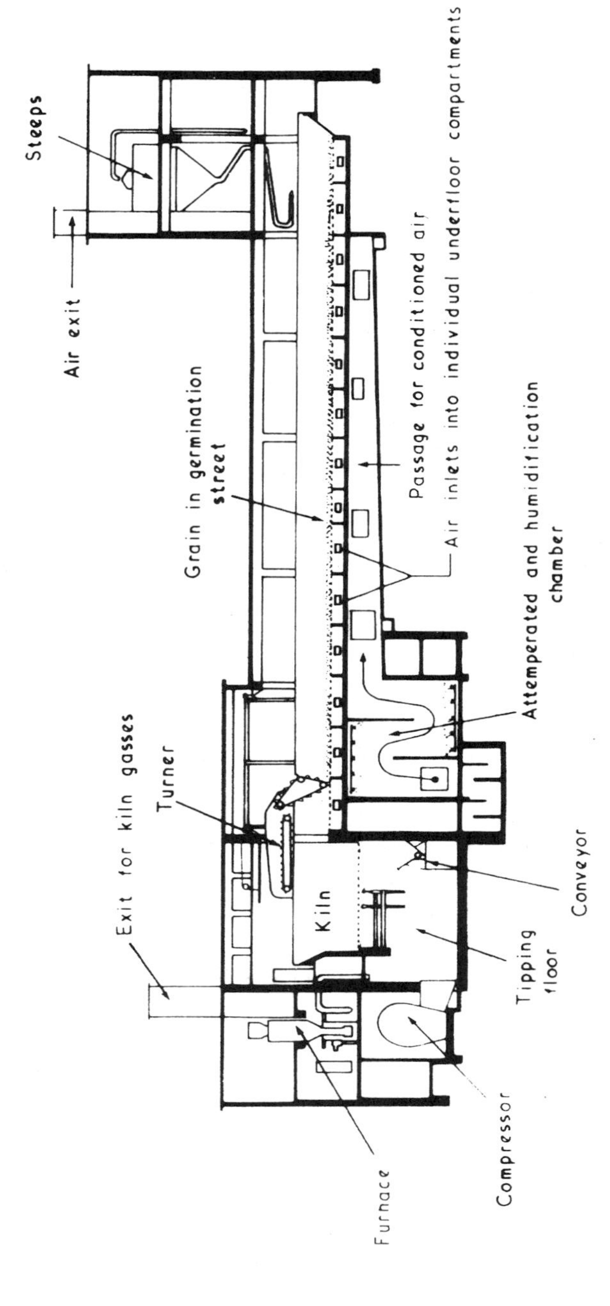

Fig. 9.53 A section of a complete Seeger Wanderhaufen malting plant, with a kiln (Ostertag patent). A bucket-and-chain, Ostertag double-throw turner is in position to move green malt from the street onto the kiln.

stripped into a hopper using the same machine (Fig. 9.52). It is feasible to build one of the boxes to hold water and so permit it to be used for resteeping. Normally a maltings has a number of streets operating in parallel.

The Ostertag Wanderhaufen or 'moving piece/heap' system may be regarded as an extension of the open-ended box system or be a version of a transposal system in which the different batches of grain are not physically separated (Anon., 1959, 1963b; Minch, 1960; Vermeylen, 1962; Leaton, 1963; Philbrick, 1965; Narziss, 1976). Various versions of this system have been widely used and a number are still in operation. In all of them the grain rests in a long box or street with side walls bearing rails and a perforated base through which attemperated air is admitted in the normal way (Fig. 9.53). The steeped grain is loaded in at one end of the street and is levelled. In some plants the grain may be spray-steeped from a trolley that moves along the box, supported on the side rails. At intervals of 8–12 h, a special turner moves along the box, turning the grain very thoroughly and moving it along. In contrast to the operation of open-ended boxes, in the Wanderhaufen the grain is always moved away from the steep towards the kiln. Thus each batch of grain, or piece, is progressively moved along the street. Fresh grain is always at the steep end of the street, the oldest is at the unloading or kiln end. Because of this, the surface of the grain is visibly higher away from the steep, since the volume of each piece increases with growth. It follows that the grain's heat output differs in different sections along the street, and this is allowed for by arranging that different sections receive different volumes of conditioned air, which, therefore, operate as different temperatures and airflows. At the end of the street, the turner may unload the grain into a conveyor (Fig. 9.54) or it may load it directly into a kiln, to which access is gained via a fireproof, well-insulated door (Fig. 9.53). Several arrangements are in use. For example, the kiln may be simple (Fig. 9.53) or it may consist of two linked kilns, side by side, separated by heat-proof doors except when grain is being transferred between them. The grain is dried in the first and cured in the second. This arrangement, which is known as linked kilns, is suitable for achieving high thermal efficiencies. Alternatively, the grain may be conveyed from the street to one of a variety of separate kilns.

A street is divided into sections called bays and most types of turner are capable of moving the grain along by half a bay or one bay each time they move along the street, although single-throw turners also exist (Figs 9.53–9.56). Thus the rate of movement of the grain along the street is controlled by the number of passes of the turner and whether each pass moves the grain by half a bay or by one bay. The plants can be operated in a number of ways. For example, in a particular Irish plant each street delivered 20 t green malt (as barley) if 4-day germination was being used or 10 t/day if 8-day germination was employed. Of course, the delivery of grain from the steeps and the ability of the kilns to process the green malt need

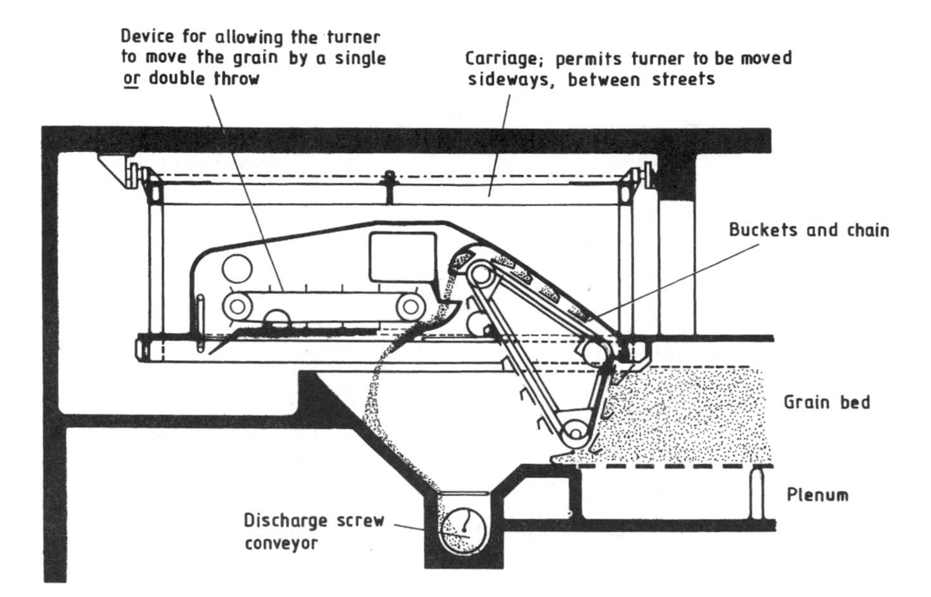

Fig. 9.54 An Ostertag bucket-and-chain single- or double-throw turner, moving green malt from the end of the street into a discharge conveyor moving it to a separate kiln.

to match the working rates of the streets. In the 1960s, when several Wanderhaufen plants were built in the UK, 8-day germination periods were common. With the faster processing periods now used – 4–5 days are common – the effective working capacities of the streets have increased. The streets are frequently built in pairs or in groups of four so that machinery can be shared. Street sizes are variable; for example one was 38.1 m long × 4.9 m wide (125 ft × 16 ft) divided into nine bays (Anon., 1963b), another was 45.7 m × 2.44 m (150 ft × 8 ft) (Minch, 1960) and another was 50 m × 3.6 m (164 ft × 11.8 ft) (Vermeylen, 1962). The successful operation of this plant is dependent on the turners causing minimal mixing between successive pieces as they advance along the street. Mixing is said to be as little as 5%. The turners used achieve very complete and gentle grain movement, completely inverting the grain bed, and are so gentle that the system is capable of making wheat malt. Two classes of turner are used, each with single- or double-throw options. In the older patterns, the grain is lifted by a series of buckets moving on endless chains and either is dropped directly behind the turner (single throw, Fig. 9.55) or the operator has the choice of using a single throw (Fig. 9.54) or moving the grain further back and dropping in a double throw. Plastic brushes alternate with the buckets and sweep the decks. The turners may be equipped with water sprays and have a safety cut-out device reaching across the width of the street, which, when touched, stops the machine. Such turners work well but frequent maintenance is

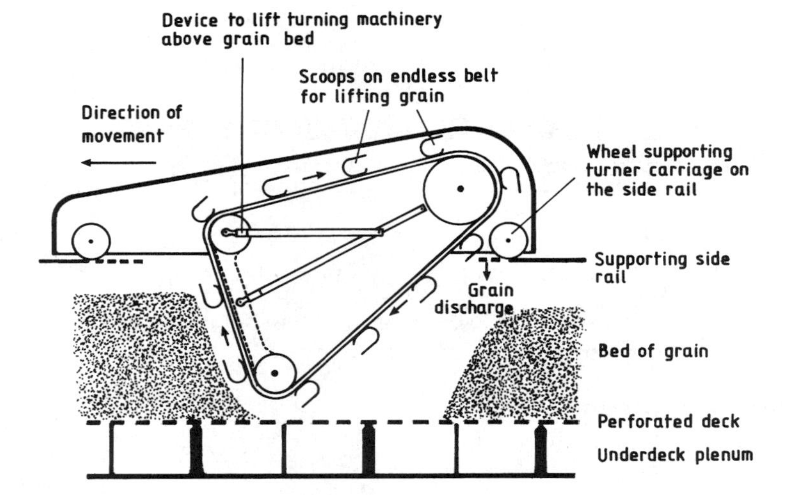

Fig. 9.55 A simple single-throw Ostertag bucket-and-chain turner.

needed and many users have replaced them with others in which grain is elevated by banks of inclined helical screws, working side-by-side, which discharge the grain either in a single-throw or a double-throw device (Fig. 9.56). As turners must always work along the box in the same direction, provision is made to retract the working parts, lifting them above the grain beds, so that the machinery can be run back to the kilning end ready for the next pass. Turners are complex and expensive and usually serve two or more streets. To permit this, they are run from the rails on the side walls of the street onto the rails of a carriage or transporter that can move sideways, moving the turner to another street. When the street is to be cleaned, a piece is missed, and so a 'space' moves down the street, allowing each part to be cleaned in turn. The air supply to the empty space is cut off to prevent waste of conditioned air.

Various other semicontinuous malting systems have been proposed in which the grain is steeped at the top of a tower and is turned each time the grain is transferred to a lower compartment. Finally it is transferred to a kiln in the base of the tower, or to one side of it. The Gent (1889) tower worked on this system, and so do the towers described by Schlimme (1972), the Kling tower (MacWilliam, 1963), the Frauenheim system (Frauenheim, 1959; Vermeylen, 1962; Witt, 1970a) and the Neubert system (Vermeylen, 1962). In the Optimälzer tower (Narziss and Kieninger, 1967; Haman and Hartzell, 1972; Narziss, 1976), the germinating grain is supported on complex tipping floors that tip in sections. Each section is so arranged that it forms a duct for the conditioning air. The grain is steeped normally before it is transferred to the first compartment. The tipping floors are said to seal so well that they can hold water and so can be used to resteep the grain. The

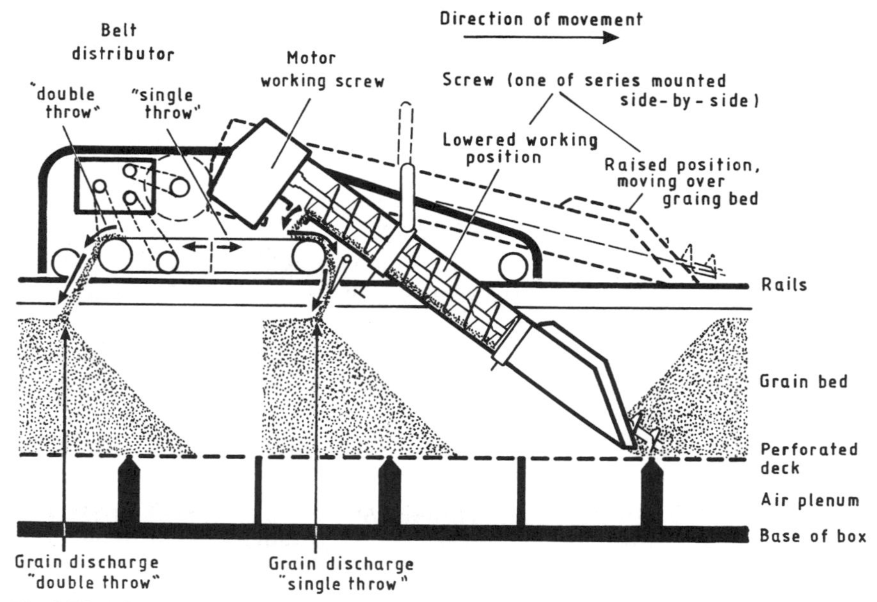

Fig. 9.56 A Seeger pattern of Wanderhaufen turner in which a row of helical screws is used to elevate the grain to a device that enables it to be discharged a short way (single throw) or a longer way (double throw) behind it in the street. Simpler, single-throw-only devices, in which the elevating screws drop the grain immediately, are also used.

Graff system, which was used in Chicago, consisted of a complex tower system in which a series of vertical, cylindrical malting chambers, ventilated horizontally, surrounded a large, central duct that distributed conditioned air downwards (Vermeylen, 1962; MacWilliam, 1963; Witt, 1970a). Turning was achieved by moving grain from the base of one vessel to the top of the next. In two other proposed semicontinuous systems, steeped grain was held in trucks with slotted bases and open tops. The trucks were moved forward at appropriate intervals to allow the grain to be turned and to position the trucks at successive 'stations' that provided a supply of cooled and humidified air during germination and hot air during kilning. In the Morel system, the trucks (nine or ten) moved forward in a straight line (Vermeylen, 1962; MacWilliam, 1963) and in the proposed Saturne system they moved on a circular track. However, the sliding seals gave problems and the modern Saturne system is rather different and is fully continuous in operation.

Some reasons given for using the somewhat complex semicontinuous, or continuous, malting systems include: (i) they are easily automated; (ii) they can be adjusted to provide optimum malting conditions for producing a particular grade of malt by maintaining the 'correct steady-state conditions' in the different sections; (iii) their workforce requirement, relative to their capacity, is low by traditional standards; (iv) they have relatively steady demands for their water, power and other services; and (v) heat is saved because kiln cooling and reheating can be avoided and heat recovery is simplified. However, with increased computer-controlled automation and increased batch-sizes, new malting plant in the UK is all designed for batch production processes.

9.10.4 Continuous malting plants

The best known truly continuous malting plant is the Domalt system, named after the Dominion Malting Co. of Canada, and used in various parts of the world (Stoddart *et al.*, 1961; Vermeylen, 1962; Leaton, 1963; Anon., 1965b; MacWilliam, 1965). The plant was developed to allow quick-growing Canadian barleys to be malted rapidly, by telescoping steeping and germination using spray-steeping. The plant used at Haddington, in Scotland, was modified to permit a longer processing time (Laing, 1986) (Fig. 9.57). The clean grain, held in a hopper, was metered into water via a dust-extraction unit, using a vibrating feed device. The grain, entrained in water, was pumped into the base of a washing and wetting unit: a rotating, upwardly inclined cylinder, *ca.* 18.3 m (60 ft) long, containing an internal, perforated spiral. The contents of the spiral were sprayed with jets of water from a centrally mounted water pipe. Thus the grain, which remained in the spiral for 1.5 h, ascended against a counterflow of warm, attemperated water and was thoroughly washed. The water drained from the lower end of the spiral. The grain emerged with a moisture content of *ca.* 30%. Originally the grain was transferred directly to the 'apron' and steeping was continued by spraying. In the modified system, the grain was transferred sequentially to a pair of novel spray-steeping vessels (Fig. 9.58) in which the moisture content of the grain was raised to *ca.* 46%. The grain was fed steadily into the top of the vessel, where it was sprayed. The downward movement of the grain was regulated by an ingenious spreader and valve system so that it moved in order, 'first in, first out'. The residence time was 24 h in each vessel. The grain was continuously ventilated from below and all the water sprayed onto it was taken up, eliminating effluent production. The steeped grain was transferred and formed into a level bed about 0.9 m (36 in) deep on an apron, flat-deck conveyor or floor of perforated metal sheets that moved forward in an enclosed tunnel at a speed of 69–91 cm/h (27–36 in/h). The speed could be varied. Conditioned air was passed through the grain bed from below. As the grain moved forward it might be sprayed. At intervals it met simple permanently mounted turners that rotated about horizontal axes, which gently scooped the grain up and dropped it behind, thoroughly turning it. Different sections of the germination tunnel could be adjusted to run at different temperatures and airflows. The grain passed through two germinating–conveyor units and into a kilning–conveyor section, where, in the Scottish plant, it remained for 24–28 h. Originally this plant was 'direct

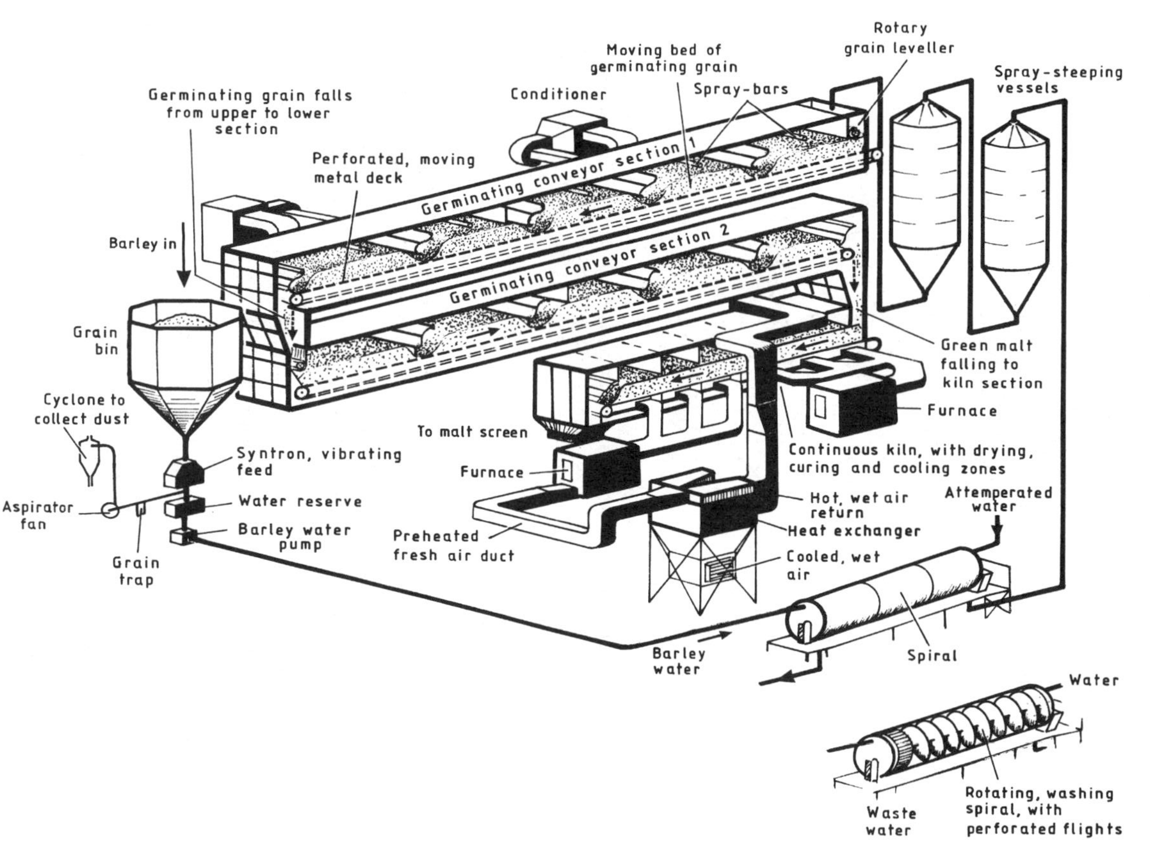

Fig. 9.57 The Domalt plant that operated at Haddington, in Scotland (after Laing (1986); and other sources).

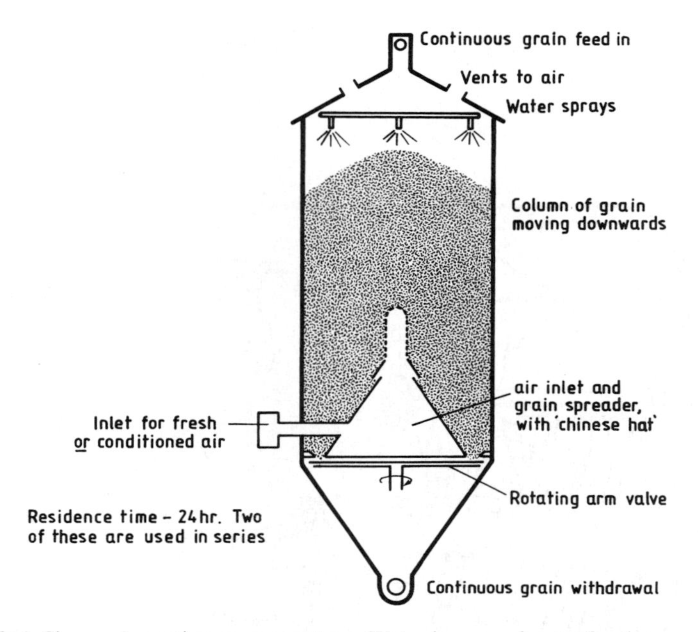

Fig. 9.58 A Simpson's continuous spray steep. Water is sprayed onto the downward-moving column of grain at such a rate that its uptake is essentially complete in the 24 h dwell period. The grain is continuously ventilated upwards. The shape of the spreader-cone at the base ensured that the grain that leaves the steep first is the first in. In the plant at Haddington residence time was 24 h and two vessels were used in series. (Courtesy of A. P. Maule, Simpson's Malt Ltd).

fired', but afterwards it was indirectly fired and a heat-recovery system was installed. From the kiln, the malt was conveyed to a deculming plant. The equipment was automatically cleaned and had built-in monitoring and safety devices that sounded alarms and stopped the plant if necessary. The original plant, built in Canada, was 70.1 m long × 12.2 m wide × 15.2 m high (230 ft long × 40 ft wide × 50 ft high), converting fast-growing Canadian barley into malt in 80 h and producing 5693 t (37 500 Qr) per year.

In the Czech 'Solek tunnel' malting system, as in the Domalt plant, the first processing step is to wash the grain by immersion; it is likely that this is necessary to clean the grain before subsequent spray-steeping as well as to provide a considerable part of the water needed by the grain (Fig. 9.59). The grain is then carried through different sections, spray-steeping, germination and kilning, on woven metal belts or bands (Vermeylen, 1962). Turning is achieved when the grain falls from one band to the next. There are seven bands in the spray-steeping section (Fig. 9.59), four sets of seven in the germination compartment and a long belt that transfers the green malt from the germination tunnel to the top of the kiln, which has six bands, in a separate building. The belts move at a rate of *ca.* 4 cm/min (1.57 in/min).

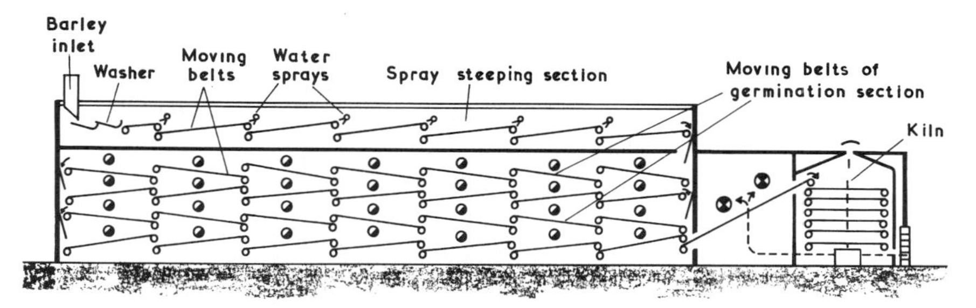

Fig. 9.59 A Solek tunnel continuous-malting plant. Grain is moved on a series of endless belts through the successive processing sections.

The tunnel is 55 m × 10 m (180.5 ft × 32.8 ft). A plant can produce about 7500 t malt per year.

The Saturne system, as finally built, is a continuous malting plant and uses continually moving, annular floors (Anon., 1965a; Cauwe, 1965; Narziss, 1976; Minch, 1976) (Fig. 9.60). Presumably the name was suggested by the annular rings of the planet. Several Saturne plants have been built in Europe. Originally steeping was by spraying only, but initial immersion steeps are now employed. An extremely large version of this plant in Metz processes 200 t barley/day; 8.33 t/h (Mauclaire, 1977). The barley is steeped by successive immersions in twin rectangular tanks, 20 m long, 2.5 m wide and 3 m deep (65.6, 8.2 and 9.8 ft, respectively), through which the bed of grain (1.5 m (4.9 ft) deep, 2.5 m (8.2 ft) wide) is moved using a very wide chain conveyor. The maximum dwell time in each steep is 7 h. After the second steep, the grain is pumped to the outer, germination ring, a vibrating device dislodging excess water. The annular germination ring has a diameter of 68 m (223.1 ft), is *ca.* 12 m (39.4 ft) wide and forms a perforated deck 1650 m^2 (17761 ft^2) in area. It is moved by five hydraulic jacks spaced around the circumference and completes a revolution in 6 or 7 days (2–15 days is the possible range). The barley is spread to an even depth the full width of the deck and wetting is finished by spray-steeping. As the grain is moved forwards it passes through rows of helical turners. It is possible to regulate the ventilation of four sections of the germination ring separately. When the grain has germinated, it is transferred sideways to an inner, annular-floored, continuous kiln, in which the moving deck is 7 m (23 ft) wide and has an area of 460 m^2 (4952 ft^2). The malt passes through a series of four process-zones during kilning. Normally kilning is completed in a 24 h cycle but may take 18–36 h. The malt from the kiln is passed through a cooler. Such plant has no peak loads and the separate units mean that loadings are optimized. Because the kiln and its heat recovery system are operating under steady-state conditions, its thermal efficiency is as good as, or better than, that of a double-deck kiln.

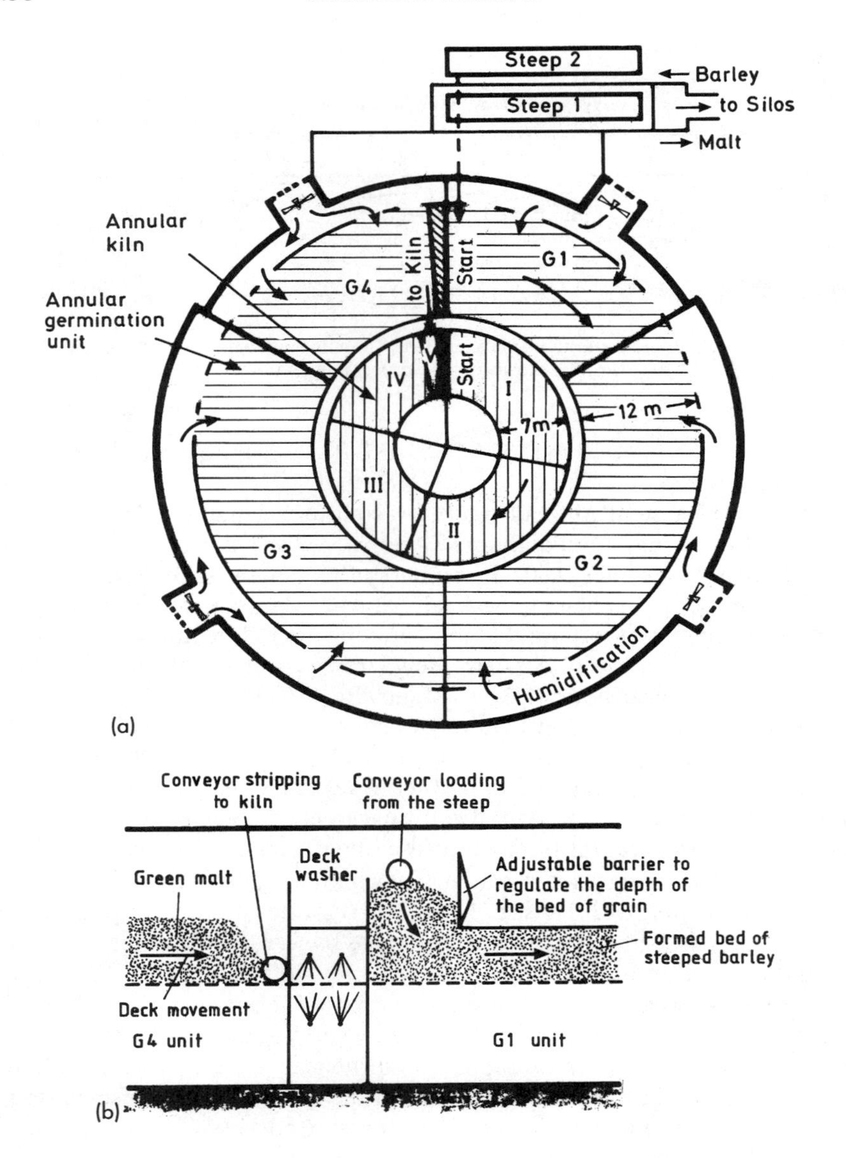

Other, less well-known, continuous malting processes have been proposed. The Kardos system used an annular grain conveyor (Vermeylen, 1962). In the Chevalier Martin system (Vermeylen, 1962; MacWilliam, 1963) malt was germinated on beds formed on inclined, perforated decks and ventilated from below. The decks were swept by 'steps', which were parts of endless-belt conveyors that received the steeped or youngest grain at the upper end of the box and discharged the older grain at the lower end. The

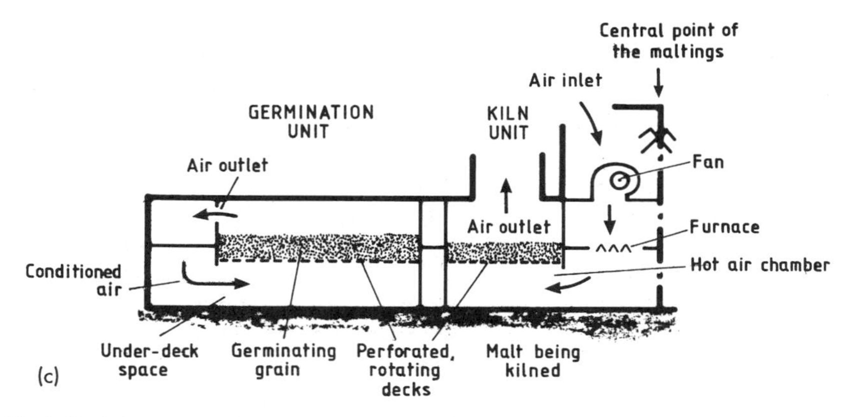

Fig. 9.60 A Saturne continuous-malting plant. (a) The unit in plan, with the four successive kilning zones (I–IV) in the inner annulus, and the germination zones (G1–G4) in the outer annulus. (b) A section through part of the outer annulus, showing the steeped-grain loading and green-malt stripping arrangement. (c) A radial section showing the relationships of the kilning and germination regions of the plant (after Mauclaire, 1977.)

grain took a day to move through each box and it was turned and mixed during its transfer to the next. Similar boxes might be used as kilns (in effect one- or two-deck), or for barley drying. In the Sogemalt continuous/semi-continuous system, barley is processed at 85 t/day. Barley is steeped in six 35 t cylindrico-conical vessels and is transferred into the first of a series of germination cases (three small, 35 t capacity, and four large, *ca.* 150 t capacity) (Houtart and Delatte, 1979). After 48 h of steeping, the first vessel is filled, a process taking 8 h. Later (16 h) the vessel is emptied and the grain moves to the next vessel. The emptied vessel is reloaded with freshly steeped grain. Each transfer takes 8 h. The grain is finally transferred to a continuous kiln. In the proposed Rotzler system, the grain is intented to move continuously downwards between zigzag perforated walls, designed to allow side-to-side aeration (MacWilliam, 1965). The grain-filled spaces between the walls are progressively further and further apart to allow for grain growth. It is unclear how the grain was intended to be mixed to prevent rootlet matting.

Many ingenious types of equipment for dealing with malting grain have been proposed and some have been tested. Experimentation seems to be continuous, but it is likely that most malt is now made using 'conventional' rectangular or circular germination compartments.

9.11 Kilning

9.11.1 Introduction

In a few instances, as in some distilleries, green, undried malt is used directly in mashing. This has the advantages that the considerable cost of kilning is avoided and the full enzyme complement of the green malt is preserved. However, the green malt cannot be stored and it is not readily transported; as a result, it must be produced very close to where it is to be used and it must be used immediately. Most malt is kilned before it is used.

Kilning is the process in which malt is dried in a stream of warm to hot air. The soft endosperm becomes brittle and friable and the rootlets become brittle so that they are easily broken and separated. Malt, the dried material **less** the roots, is readily transported and may be stored, sometimes for extended periods. Depending on the kind of malt being made, and hence the kilning schedule, the malt is partially cooked, some or considerable enzyme destruction occurs and the product develops flavour and aroma. In fact, considerable 'character' is added to the malt during kilning and the population of microbes on the grain is substantially reduced. In the past, virtually all malts were finished on kilns, but now many special or coloured malts are made using roasting drums.

In primitive times, and even now in African and Indian 'village' malting, malt was dried in the sun or in thin layers in dry places under cover. In Belgium wind malt, used in the manufacture of Old Louvain beers, there is green malt that has been air-dried in special lofts. Very large spaces are needed for this (de Clerck, 1952). However, simple air drying is impractical when large amounts of malt are being produced and is an uncertain process in many climates. It is not known when kiln drying and curing were invented. Presumably the origin in Europe is very old (Hinz, 1954; Talve, 1960). The monastery at St Gallen had a kiln at its maltings in *ca.* 800 AD and the remains of a kiln from Roman times has been found in Belgium. However, in many areas it is necessary to dry grain at harvest time and the archaeological remains of the structures used for this cannot be distinguished from malt kilns – or indeed they may have been used for drying grain and for kilning malt. In addition, in Scandinavia and many other areas, especially of northeastern Europe, 'dry bath houses' and 'hot rooms' have been in use for centuries, and baking ovens were universally employed. All these could be used for drying and curing malt.

The earliest purpose-built kilns described were of a type widely used in northern Europe, and, like many kilns (kells) built later, they were prone to catch fire. Not for nothing did Tusser (1580) write:

Take heede to the kell
Sing out as a bell
Be suer no chances to fier can drawe,
the wood, or the furzen, the brake or the strawe.
Let Gillet be singing, it doth verie well,
to keepe hir from sleeping and burning the kell.

In short, by keeping the workmaid singing you could check that she was awake and, hopefully, tending the kiln correctly and not letting it catch fire.

In the older English kilns (sometimes called smoke-kilns), a set of rafters was set at about 8 ft (2.44 m) above a fire. A 'bedding' of some material, usually straw, was arranged on hurdles in a layer across the rafters, then on this was spread 'hair-cloth', a cloth of woven horse-hair that supported the malt (Markham, G. 1615; Harrison, 1587; Anon., 1903). A fire, of furze, straw or wood (carefully chosen and dried) was lit, and the rising hot air and smoke moved up through the malt spread on the haircloth and escaped from the roof via a hole. In such old kilns, the depth of malt might be 7–15 cm (3–6 in) deep. Temperatures might be in the range 27–100 °C or even 105 °C (81–212 °F, even 221 °F) (White, 1860; Stopes, 1885). Markham (1615) mentions the French kiln, which was superior and less prone to burning down, but he does not describe it as its form 'is so well known'! Wahl (1944) noted that simple, direct-fired Brabant kilns were still in use in Germany in 1750.

9.11.2 More recent, shallow-loaded kilns

A wide range of kilns has been employed and many older types, often 'modernized', were in use within living memory; a few still are. The earlier types had a fire basket of iron in a stoke-hole, dunge or dungeon (Fig. 9.61). The space above the fire widened out into a hot-air chamber, above which was a perforated floor or deck to support the malt. To prevent sparks rising into the malt and to spread the rising hot air more evenly beneath the malt, there

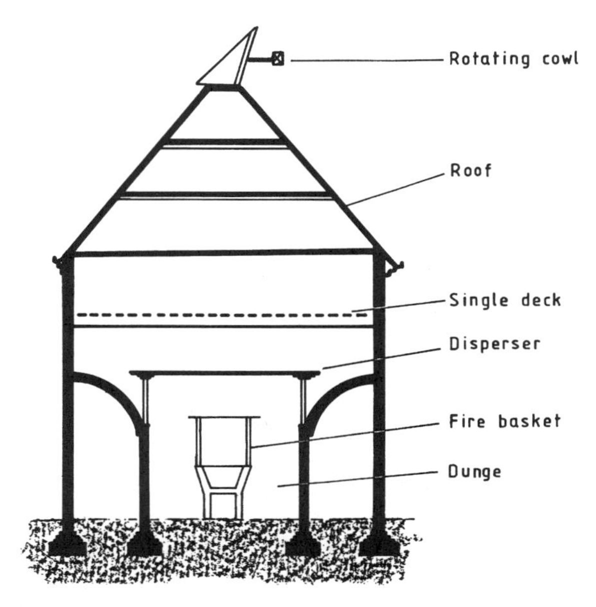

Fig. 9.61 A simple direct-fired English shallow-loaded kiln of 1800–80, with a fire basket.

was a stone, a structure of tiles or more often an iron sheet (a canopy, a spark-stone, spark plate or disperser, also called the fly-plate, the baffle-plate, the shield, the reflector, the guard and the apron) mounted above the fire (Stopes, 1885). In Diderot's *Encyclopaedia* of 1751, the desirability of arranging the disperser to shed malt dust and culms and so minimize fire risk, was pointed out. 'As far as possible' the use of wood in the construction of kilns was avoided. The kiln floor may have been of perforated kiln tiles of pottery (glazed or unglazed), perforated sheet iron, cast iron plates, woven wire or (later) wedge-wire (White, 1860; Stopes, 1885; Lancaster, 1908). The kiln floor was roofed over. In the worst case, the wet air escaped through a hole, but refinements were the use of cupolas, 'mushroom-caps' and louvred 'pagodas' or 'chinese hats' to keep out the rain; at best, rotating cowls moved with the wind and prevented down-draughts (Fig. 9.61). In some very large kilns, the roofs at the apex had 'ridge-vents' which extended some way along the top of the roof to allow the escape of the hot air (Saunders, 1905). Such kilns were loaded either by barrowing or bringing the green malt on over-head-suspended hand-propelled buckets, then tipping the green malt onto the kiln deck and levelling it by hand using forks, shovels or ploughs. The malt was turned by hand and the kiln floor was stripped by hand. Later some kilns were turned mechanically and a steam- or electrically powered barge-board was used to sweep the kilned product into a chute.

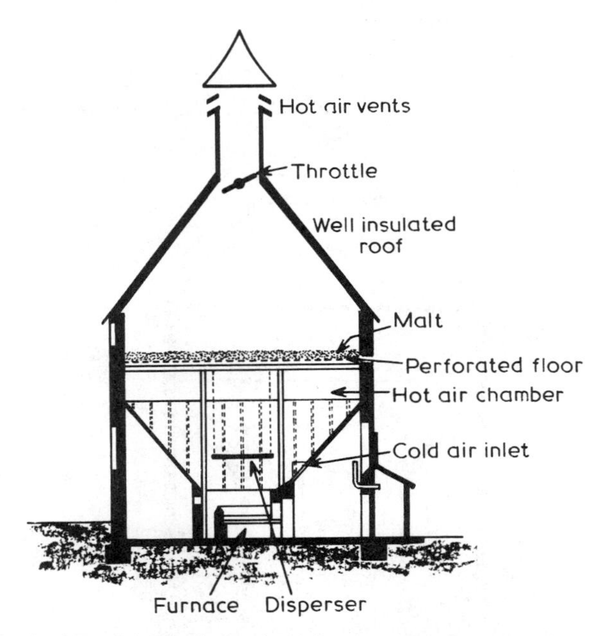

Fig. 9.62 A direct-fired English kiln (around 1880) using a furnace and a draught-regulating throttle (Briggs, 1978).

At various times, the fuels used included furze, bracken, straws, woods, peats, coal, coke and anthracite. The preferred modern fuels are various oils and gases. From the earliest records to the present, maltsters have attempted to reduce fuel usage and so keep down their costs (Chapter 10). Two early modifications to minimize fuel costs were (i) to burn the fuel in a furnace in which the rate of combustion could be better regulated and more complete combustion of the fuel could be achieved; and (ii) to have a 'throttle' in the 'throat' of the kiln, which regulated the airflow or draught (Fig. 9.62). All early kilns worked on the principle of the chimney and were greatly affected by the wind and the outside temperatures. Consequently their performances were erratic.

Another problem was that heat was wasted when hot air left the kiln less than fully saturated with water, so some of its drying power was wasted. This was particularly likely to happen in the latter stages of kilning when the

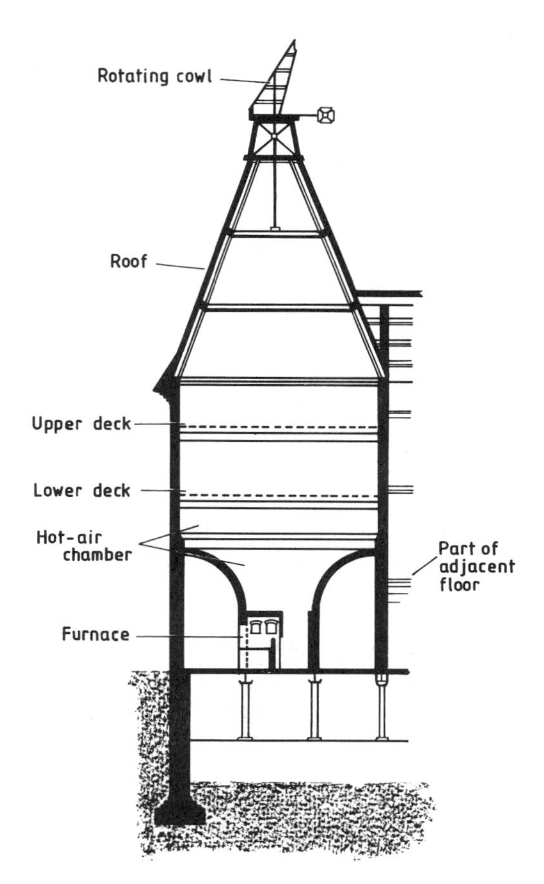

Fig. 9.63 A Stopes design for a simple double-deck kiln (Stopes, 1885).

comparatively dry malt was being cured at elevated temperatures. A solution was to arrange for the hot dry air from a batch of malt being cured to dry a younger batch of malt. Stead (1842) patented this idea and tried it in a 15.2 m (50 ft) high, fan-assisted five-floor kiln (White, 1844) but apparently this was unsatisfactory (White, 1860). Nevertheless, two- and three-floor kilns were widely used in Continental Europe, as were more complex kilns, and simple two-floor kilns were strongly championed by Stopes (1885). A Stopes double-decked kiln is illustrated (Fig. 9.63). It is virtually identical to a German kiln of about the same date (Scamell and Colyer, 1880). The first British Stopes-type double-decked kiln was erected in Brighton in 1881 (Stopes, 1885). Such kilns remained in use in Scotland for a period, where they were used to make pale, lightly cured distillers' malts. However, double-decked kilns were not successful in making highly cured brewers' ale malts, almost certainly because the small volume of very hot air needed for curing on the lower deck was not compatible with the need for a high volume of much cooler air needed for drying the green malt on the upper deck. The solution, to use ducts to supply more cool air to mix with the hot air between the decks, was not adopted in the UK (Ross-MacKenzie, 1927) but was in mainland Europe.

The English single-floor kiln was improved in a number of ways (Lancaster, 1908; Chubb, 1911). The kilns were constructed to empirically discovered standards. By increasing the height of the structure and incorporating a chimney (the chimney effect), the difference in density between the hot air within the kiln and the cold air without gave a greater driving force to the draught. This enabled greater depths of grain (about 10–40 cm (4–16 in)) to be kilned, increasing the capacity of a kiln of a particular deck area (Fig. 9.64). The furnace hot air rose up a central shaft into which cooler air could be admitted from without to adjust the below-grain temperature. The dispersers were often elaborate metal or brick tunnels or radiating tubes with occasional holes, designed to spread the hot air evenly into the hot-air chamber below the deck. The draught to the furnace was often thermostatically controlled. For preference, the kilns were situated on rising ground, away from trees or buildings that could cause turbulence and down-draughts, with the lower door positioned to face the prevailing wind, in fact roughly to pressurize the kiln air inlet. Rotating cowls on the tops of the chimneys were the best means of preventing down-draughts and minimizing the effects of changing wind direction. Sometimes the air was delivered from the outside to the kiln via an underground duct. In the best kilns, the roof was of double thickness, the roof and walls were insulated and as few doors and windows as possible were used, which limited leaks.

The old horse-hair cloth supporting the green malt was replaced with floors or decks of perforated pottery tiles (glazed or unglazed), cast-iron tiles (patented 1783), perforated iron sheet 'tiles', woven wire or wedge-wire. Perforated plates had 15% or less of free space. Strong woven wire

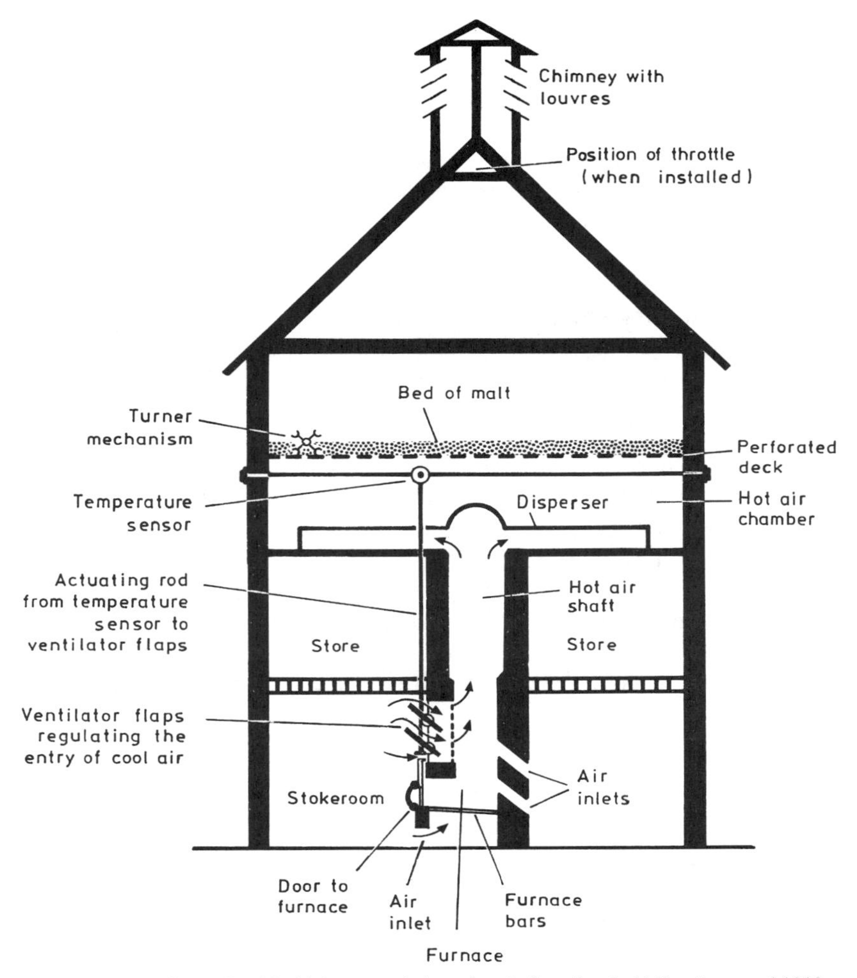

Fig. 9.64 A direct-fired British, natural-draught, shallow-loaded kiln of around 1910.

had a free space of *ca.* 50%. Metal tiles were needed when the temperature had to be very high, as when making dark amber malts. The tiles were readily blocked by dust and culms and needed regular, laborious cleaning. Wedge-wire has come to be preferred, because it has a large 'free area' (30–40%) so that it presents minimum resistance to airflow; the wire is profiled so it is not easily blocked; it is easily cleaned, and, being strong, it can be used in modern tipping or louvred floors. It was recommended that for pale malts for making India Ale the deck should be at least 14 ft (4.3 m) above the fire so 'the air chamber will be capacious and lofty' (Tizard, 1843).

Turning malts by fork or shovel as often as every 2 h in temperatures up to 105 °C (221 °F) in a smoky, fume-filled and dusty atmosphere was not a popular job, even when linked with free beer. Even in the 1920s in some places in England, people were still turning kilns by hand, often working naked except that their feet were covered by canvas 'bag-boots' to stop them burning. A mechanical kiln turner was patented by Johnston of Blackburn in 1849, but the first successful design seems to have been that of von Schlemmer, described in 1874 (Wahl, 1944). Many variants of this turner were used (e.g. Levy, 1899; Schönfeld, 1932; de Clerck, 1952; Vermeylen, 1962) (Fig. 9.65). A turner consisted of a long rod stretching across the full width of the kiln deck and supported at the ends on trolleys that ran on rails along the kiln wall. As the trolleys moved forward, cogs on the machinery engaged on a rack mounted on the wall by the rail, and the rod was caused to rotate. Mounted on the rotating axial rod were short arms that usually carried counter-balanced swivelling scoops which successively picked up, lifted and deposited the grain, so mixing it. Sometimes simpler, cross-bar stirrers fixed rigidly to the supporting arms were used (Schönfeld, 1932). Screw turners were used in some kilns and found favour in some deeper-loaded North American kilns (see Fig. 9.70, below). Often safety devices were incorporated to switch off the turner before access could be gained to the kiln deck. In the past, it was the usual practice to have one, or more, thermometers embedded in the layer of malt, to obtain the in-grain temperature. Turners were developed with devices to automatically lift the

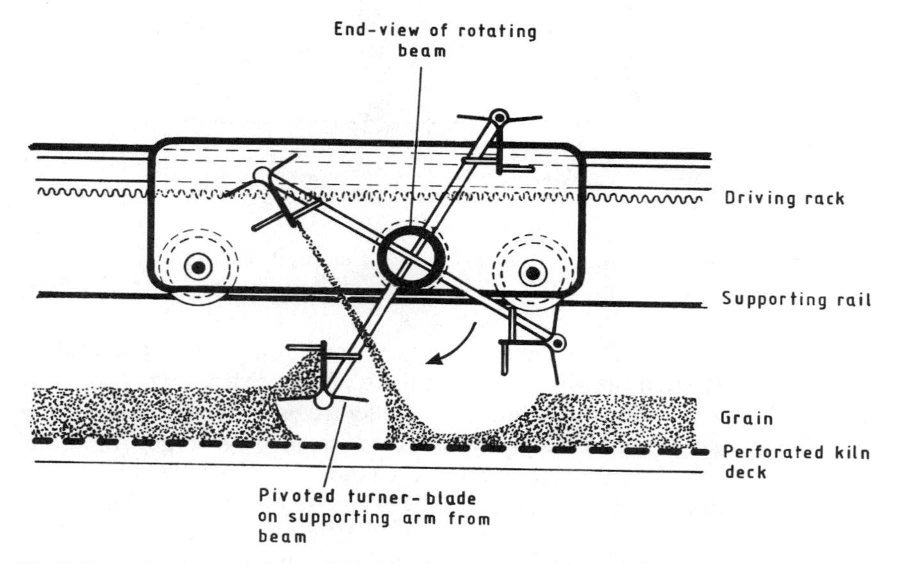

Fig. 9.65 A turner for mixing malt during kilning on a shallow-loaded kiln (after Vermeylen, 1962).

thermometers out of the way as the turner worked. In modern kilns in Europe, malt turning is no longer employed.

One of the problems with natural-draught kilns is that the airflow is so irregular. In unfavourable weather, kilning might take 5 days to complete. Stead pointed out the advantages of a steady fan-driven airflow and patented the use of fans to suck air through the malt (suction kilns) and to force air through the malt by pressurizing the air below the beds of malt (pressure kilns) (Stead, 1842). In 1885 Stopes was championing the use of suction fans to assist the kiln draught. He described the Farcot kiln in which both pressure and suction were applied using two separate fans. Suction fans were also in use in Europe (Fig. 9.66). At first most of the British kilns that were modernized had suction fans installed near the kiln outlet, but pressure fans, which drove air into the hot-air chamber, were also installed (Jaques and Mungham, 1964). The steady draughts obtained with the fans permitted kilning to be carried out according to a regular schedule and, since the fan overcame the resistance of the grain bed to airflow, the loading on the kiln floor was often increased to *ca.* 60 cm (24 in).

Other 'modernizations' occasionally carried out on these older kilns included the installation of an internal, or external, heavily insulated duct from above the kiln floor via a fan to the hot-air chamber to permit a degree of air recirculation (and hence heat saving) during curing. Loading green malt onto the deck came to be carried out by overhead conveyor, the malt being delivered by a steerable, gravity-fed chute, or by a steerable, endless-belt grain thrower. With both of these devices the grain bed must be finally levelled by hand, the bed becomes compressed and so the airflow is restricted. Different devices are used in the best modern kilns. In addition to barge-boards or power shovels used to strip kilns by moving malt into a spout or a conveyor in the floor that is usually kept covered, the floor might be stripped using the turner. Different kilns were used to make pale and dark malts. The British maltsters were well aware of the possibilities of producing malt using indirectly fired kilns, in which the gases from the burning fuel are used to heat heat-exchangers (direct hot air, hot water or steam)

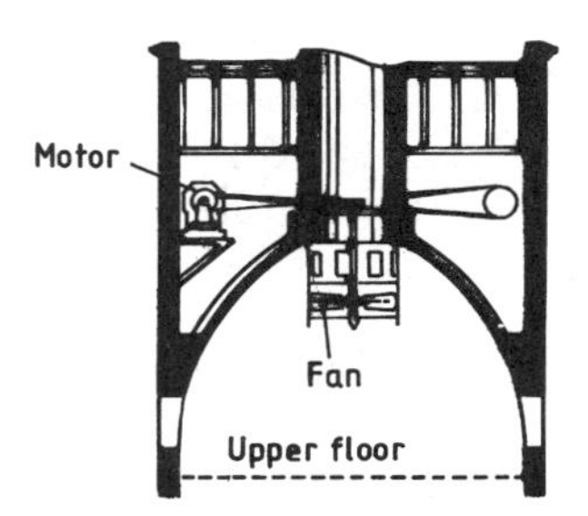

Fig. 9.66 An extraction fan mounted in the base of the chimney of a German type of kiln (Delbrück, 1910).

that heat the air which passes through the malt (Stopes, 1885; Saunders, 1905–6). They were also well aware of the penalties, in terms of thermal inefficiency, of indirect heating and, provided good fuel properly burnt was employed, they regarded the disadvantages of direct firing as of marginal importance or non-existent. Directly fired kilns were variously fuelled with straw, furze, wood, bracken, peat, charcoal, coke, coal or anthracite. Following an outbreak of arsenic poisoning among beer-drinkers (chiefly caused by copper sugars containing arsenic) in the early 1900s, analyses for arsenic in fuels for directly fired kilns became a routine practice. To trap fumes that contained arsenic (volatile arsenious acid) it was proposed either to mix the fuel with chalk, lime, slaked lime or soda ash, or to pass the hot furnace gases through a maze of channels in a stack of chalk blocks (Beaven, 1904). The heat converted the chalk to lime, which retained *ca.* 90% of the arsenic fumes and much of the sulphur oxides. It was claimed that this device worked well. Now fuels containing significant amounts of arsenic are not used. Extant direct-fired kilns are fuelled with anthracite, some special oils or natural gas. Oils and gases are easily automatically metered. To make smoked malt (in Germany) or peated malt (in Scotland) either beechwood or peat is burnt in a supplementary furnace and the fumes are fed into the stream of drying air.

One problem with directly fired, solid-fuel kilns was temperature control – and this was dependent on the skill of the kiln operator and simple thermostatic airflow regulators. Another problem was the need to stoke the furnace more-or-less continuously, which meant that the kilns had to be manned day and night. Several sorts of automatic stoker were designed to overcome this problem (Northam, 1965a). In the Prior type of stoker, coal from a hopper was slowly carried forward by a screw, and was forced upwards and fell outwards onto fire-bars where the fuel was burning in a fan-driven airstream (Fig. 9.67). The rate of firing was varied by altering the screw speed. Every 6–8 h the clinker had to be removed by hand. In principle the Hodgkinson 'low-ram' automatic stoker was similar, except that the fuel was moved from the hopper to the combustion zone by a slowly oscillating ram. In addition this equipment had moving fire-bars, which made it nearly self-cleaning. Perhaps the best known of the automatically stoked, solid-fuel furnaces is the Suxé burner, a few of which are still in use. This device consists of a horizontal, cylindrical, water-jacketed furnace. Fuel is fed from a hopper mounted over the furnace, using a chain-feed and gravity. Combustion air is provided through water-cooled tubes in the base of the furnace cylinder and secondary air enters via the hollow furnace door, so cooling the door and pre-heating the air. The rate of the fan-driven airflow is controlled by a thermostat. The furnace is easily clinkered. The hot furnace gases enter the kiln and are mixed with cool air and air that has been pre-warmed by passage through a radiator that is heated by the water circulating through the cooling jacket of the furnace.

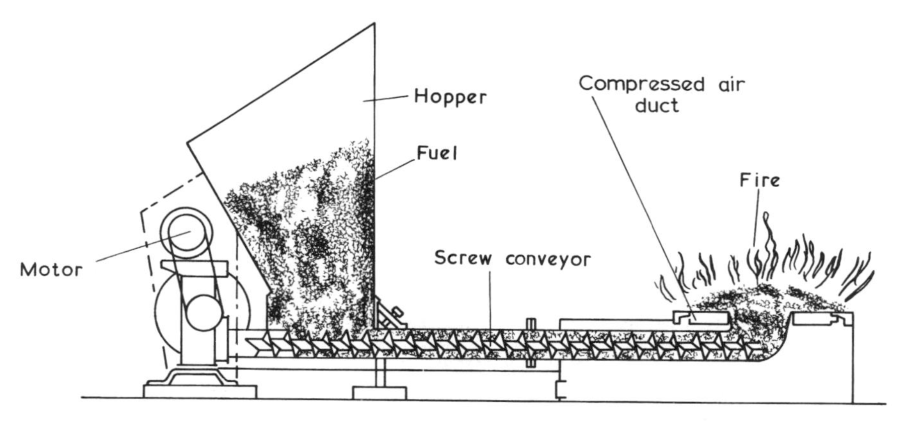

Fig. 9.67 A pattern of automatic stoker used in some older malt kilns.

Oil- and gas-fired kilns are often equipped to burn either fuel. Light oils may be burnt using simple spray and combustion chambers, but heavy oil sprays require longer dwell times in the combustion zones for complete combustion and this requires the use of single- or double-toroidal combustion chambers. Solid fuels and oils usually contain sufficient sulphur that nitrosamine formation on the malt is not a problem. However natural gas is extremely pure, and so NOX formation in the kiln and nitrosamine formation on the malt must be overcome. Usually low-NOX burners, which burn at a low temperature, are employed and sulphur dioxide (from cylinders of the gas, or from burning sulphur) is added to prevent nitrosamine formation (Chapter 4).

Until around 1790 the European kilns were similar in type to those used in the UK, but gradually their development diverged (Delbrück, 1910; Fernbach, 1928; Leberle, 1930; Schönfeld, 1932; Wahl, 1944; de Clerck, 1952; Schuster, 1962; Vermeylen, 1962; Narziss, 1976; Briggs, 1978). The idea of indirect heating was introduced in 1802 (Wahl, 1944) and gradually heat exchangers, heated with the furnace gases, hot water or steam, were improved until they were sufficiently thermally efficient to be acceptable. To encourage natural draught, the furnace gases were directed via a separate duct around the malt floors and were released into the base of the chimney or were passed up an iron tube within the chimney. The heat from this tube warmed the outgoing air and helped to maintain a draught. Natural-draught kilns were difficult to manage and to improve the draught large chimneys of 8–10 m (26–32.8 ft) were often used (Vermeylen, 1962). One-, two- and three-floor kilns (some indirectly fired) became common. The details of the designs varied considerably, depending on locality and whether light or dark malts were to be made. Multiple-floored kilns, such as that of Geçmen (Scamell and Colyer, 1880; Thausing *et al.*, 1882) and

Galland (Galland 1874; Wahl, 1944), were tried but fell out of use. It was concluded that single-floor kilns were simple to use but were thermally inefficient. Double-floor kilns were more efficient but were difficult to manage. Three-floor kilns were most efficient but were essentially too difficult to manage until the use of fans was introduced to obtain steady draughts, and 'intermediate floors' were installed (see below). Kilns for making very pale Pilsen malts were very lightly loaded with green malt (Schuster, 1962) but (depending on the kiln and the type of malt being made) loading might be as deep as 0.15–0.2 m (0.49–0.66 ft) (Vermeylen, 1962). Very many variants of kilns were in use in Europe (Saunders, 1906; Delbrück, 1910; Leberle, 1930; Schönfeld, 1932; de Clerck, 1952; Schuster, 1962; Vermeylen, 1962; Narziss, 1976).

In a simple, indirectly heated, two-deck Continental European type kiln of a well-developed pattern, a number of refinements relative to the Stopes

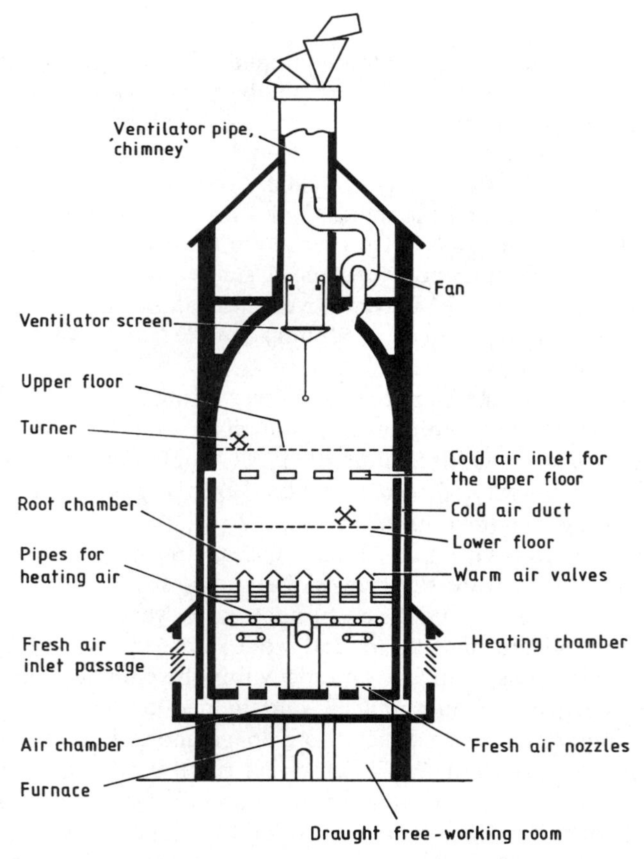

Fig. 9.68 A traditional two-floor, shallow-loaded, indirectly heated Continental European kiln (after Schuster, 1962).

double-deck kiln can be seen (Fig. 9.68). Sometimes the upper deck was loaded using a spout fed with green malt from an elevator. In other cases the malt was barrowed onto a bridge above the floor, then the barrows were tipped, dropping the green malt onto the floor, to be levelled by hand. Stripping was by hand or by using a barge-board. In two- and three-floor Prinz kilns, used in the USA about 1900, tipping floors were used to dump the malt onto the floor below. In an interesting example of heat-saving, the air entering the kiln was pre-warmed by passing it over the exhausts from the steam-engines and pumps then in use. The chimney, or ventilator pipe, is opened or closed by a moveable 'screen'. The chimney receives the hot gases from the furnace and the fan outlet (a not-entirely logical installation). The centrifugal fan normally received air from a circle of slots in the domed ceiling (Schuster, 1962) (Fig. 9.66). Admission of air into the heating

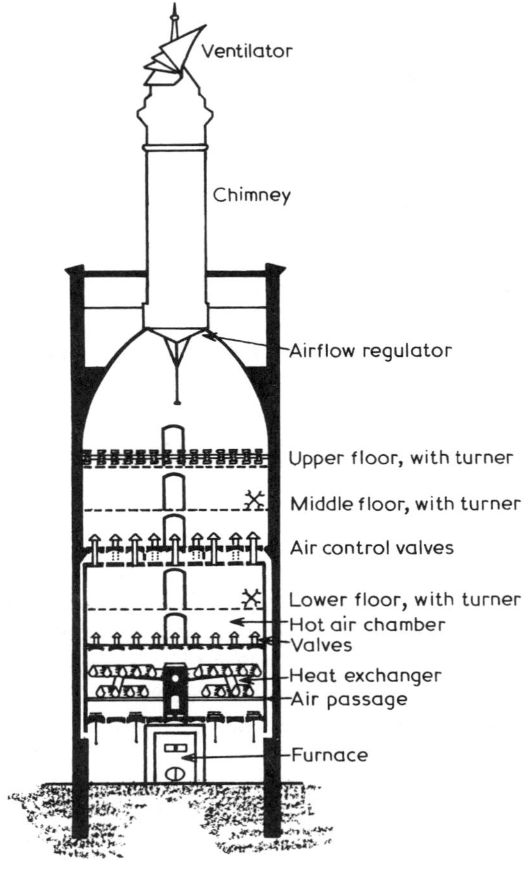

Fig. 9.69 A Continental European three-floor, shallow-loaded kiln with an intermediate floor between the lower and middle floors (after various sources).

chamber, where it meets the hot pipes of the heat exchanger, is regulated by valves. The hot air is admitted below the first deck via capped tubes or valves which distribute it and deflect falling dust and culms, trapping them in the below-deck space and minimizing the risks of fire. In addition, and most importantly, more cold air can be admitted beneath the upper deck to mix with the air rising through the curing malt on the lower deck so now a larger volume of cool air can pass through the upper deck, where green malt is being dried. At appropriate stages the lower deck is stripped, the dried malt is transferred from the upper to the lower deck and the fresh green malt is then loaded onto the upper deck. In some, more controllable, two-floor kilns, an intermediate floor was placed between the upper and lower decks (Vermeylen, 1962) (Fig. 9.69). In this three-floor kiln, the upper 'withering' and middle drying floors or decks receive the same airflow; however, the lower curing floor or deck is separated from the middle deck by an intermediate floor containing cool-air-ducts and valves. Here the hot air from the curing floor is mixed with cool air in a well-controlled way before it proceeds through the middle and top decks. The heat exchanger for heating the incoming fresh air consists of crossed layers of pipes. In this case, the heat-exchanger pipes are pear shaped in cross-section, giving them a greater surface-to-volume ratio; this gives them a greater heat-transfer efficiency and allows them to shed dust and culms more readily. Among many variants, some Bavarian kilns for making dark malts had the heat-exchanger pipes very near the lower deck so that the malt would be more strongly heated and so coloured by the radiant heat (Schuster, 1962). A diagram of a North American kiln of the 1970s (Lubert and Thompson, 1978) shows another possible arrangement in which a double-deck (two-floor) kiln, deeply loaded and using helical-screw turners, has the facility to direct extra hot air, as well as cold air, to the space below the upper deck, where it passes together with the air from the lower curing deck up through the drying malt (Fig. 9.70).

9.11.3 Deep-loading kilns

Most modern kilns are 'deep loaded' and the grain is not turned during kilning. In 1899, Saladin built a pneumatic kiln in France in which the green malt was 46 cm (18 in) deep and was turned as in his germination compartments. The iron floor was constructed in sections, so the kiln was stripped by dumping. Kilning was completed in 48 h. This pattern was not widely adopted.

Kilns working in a substantially different way are now widely used. These were developed by the frères Winckler in France (*ca.* 1918) and Müger in Germany (Fernbach, 1928; de Clerck, 1952; Schuster, 1962; Vermeylen, 1962; Narziss, 1976). In these deep-loaded, single-floor pressure kilns the malt is not turned during the drying process and so a drying zone

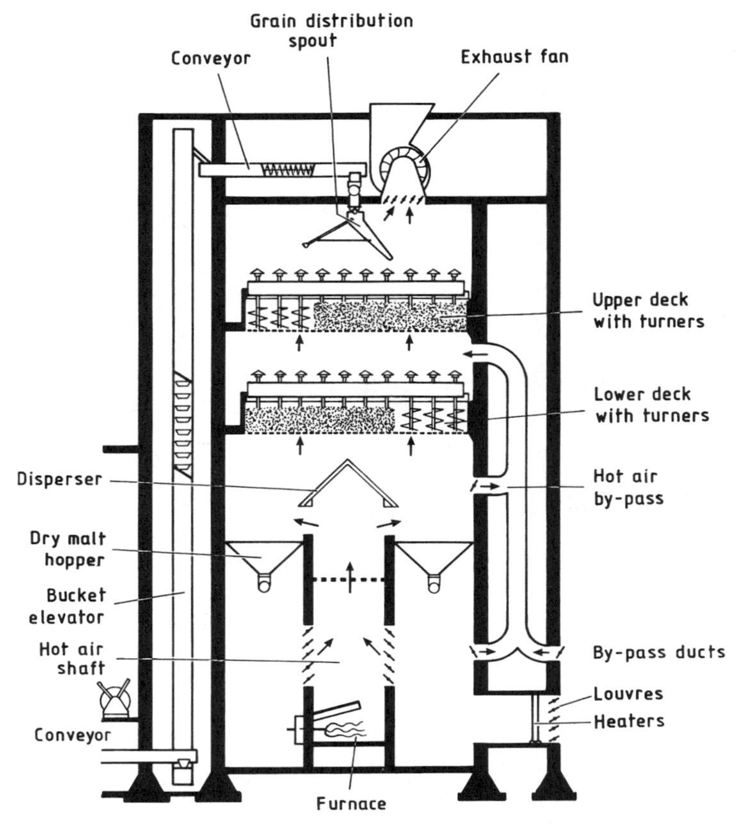

Fig. 9.70 A North American two-floor, deep-loaded kiln (after Lubert and Thompson, 1978).

spreads up through the grain (Chapter 10). Such kilns are completely satisfactory for making pale malts. In these kilns, which were all originally rectangular in cross-section, the hot air is driven by an axial-flow or a centrifugal fan and (in all the original patterns heated by direct firing) passes into a hot air chamber and up through the deck and the bed of malt (Fig. 9.71). The bed of malt must be **exactly** level. The malt beds may initially be 0.6–1.9 m (1.97–6.23 ft) deep, depending on the kilning cycles to be used. Shorter kilning cycles require shallower beds, to avoid the need for fan power capable of driving large volumes of air quickly through a deep bed of grain (Chapter 10). Such kilns function virtually independently of the weather. Fan power consumptions vary widely depending on the kiln loading and drying cycle used (Chapter 10). Quoted airflows are 12.2–15.3 m/min (40–50 ft/min) and 40–70 m^3 air/min per t; in practice, the values used must vary widely. The hot air, forced from below, moves the drying zone upwards from the bottom of the grain bed to the top. Draught

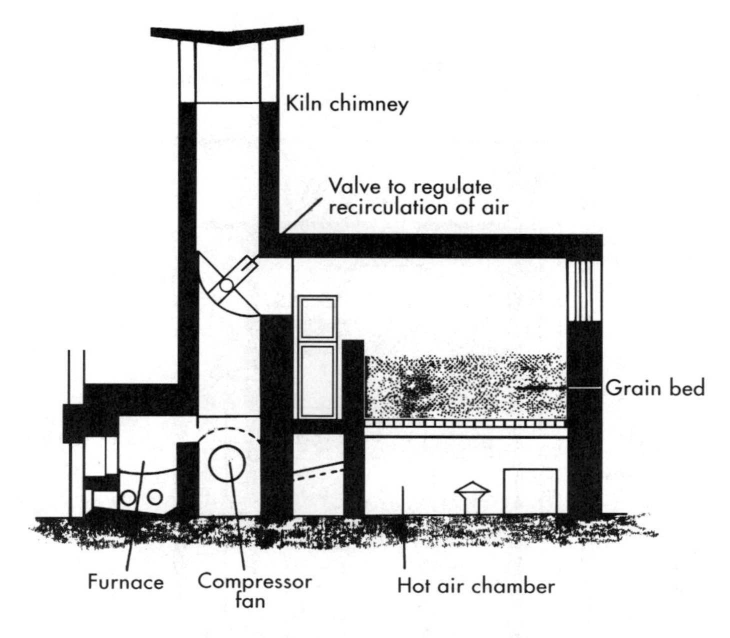

Fig. 9.71 A simple deep-loaded, direct-fired kiln of the Winckler type.

and temperature are regulated so that the air emerges from the bed of malt 90–95% saturated during the 'free-water' drying stage. Modern kilns generally have initial floor loadings of 350–500 kg (original barley)/m^2 floor area, or green malt loading depths of 0.85–1.20 m (2.79–3.94 ft) and use cycles (turn-around times) of 18, 24, 36 or 48 h, including 'down times' (Gibson, 1989). In these kilns the decks are of wedge-wire, which is strong, has 30–40% free space, conducts heat well and is readily cleaned. In the early kilns, loading might be from overhead spouts or grain throwers and levelling was achieved by hand, but these methods can lead to compaction, irregular airflows and high fan-power requirements. The deck should be loaded evenly and lightly and this is achieved by mechanical loading. For example, in rectangular kilns grain may be dropped from an overhead conveyor into the hopper of a moving loading and levelling machine, which 'meters' the flow of green malt and spreads it evenly with horizontal screw conveyors. Such kilns may be stripped by pushing the finished malt to one end and out of a door, or by uncovering a conveyor on the floor and having a moving, horizontal screw conveyor move the malt to it, usually 'sides-to-middle'. However, in many rectangular, deep-loading kilns, stripping is by tipping the floor, either the whole floor or two sections or in a series of sections. The falling malt is caught in a hopper, or hoppers, and guided to a conveyor.

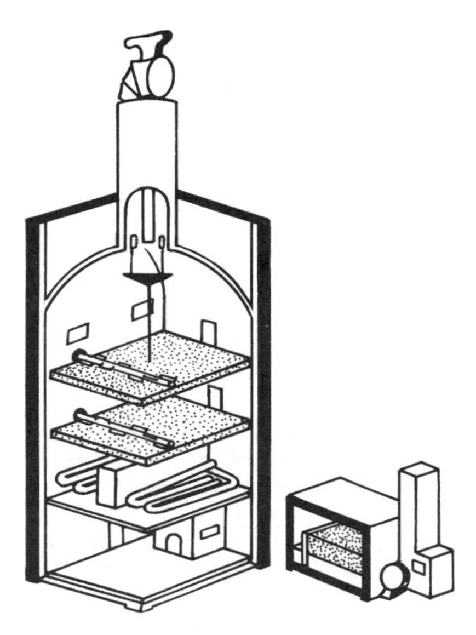

Fig. 9.72 Relative sizes of a two-floor, shallow-loaded Continental kiln and a deep-loaded kiln having about eight times the throughput capacity (after Fernbach, 1928).

Such kilns became popular because they are compact and have high capacities, saving a great deal of space compared with older kilns (Fig. 9.72), they are simple and the labour requirements are much less; indeed, they can be highly automated. Furthermore, the efficiency of fuel utilization is high. For example, Fernbach (1928) pointed out that four, three-floor continental kilns could dry 20 t malt in 48 h, while a single Winckler kiln could dry 20 t in 14–15 h. Furthermore, while each old kiln was *ca.* 29 m (95 ft) high, the Winckler kiln was *ca.* 10 m (33 ft) high. Many deep-loading kilns were built into older types of kiln, saving construction costs (Fig. 9.73). Deep-loading kilns vary widely in design (Figs 9.74 and 9.75). Many modern deep-loading kilns are circular in cross-section. This gives the advantages of better air distribution and lower construction costs. Grain may be loaded and unloaded onto a stationary floor with a spout and giracleur or with a moving, horizontal screw conveyor. Stripping may be achieved using the same devices moving the grain to a central outlet (Fig. 9.49). Alternatively, the floor may move and may be loaded and stripped using 'fixed' horizontal conveyors, as in comparable germination compartments (Fig. 9.51). The principle applies in vessels that are SGKVs and in those that are GKVs.

During the free-drying stage, the air emerging from the malt is 90–95% saturated and leaves the kiln. Now it is usually vented via an air-to-air or air-liquid-air 'run-around' heat exchanger and, more rarely, a heat-pump cooler to reduce the waste of heat by causing the warm outlet air to directly

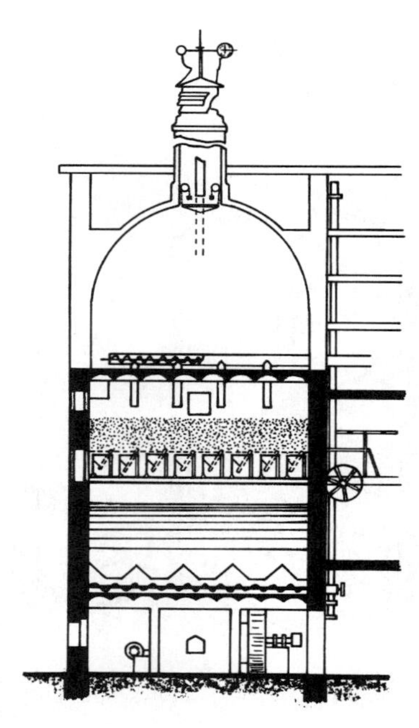

Fig. 9.73 Example of a deep-loaded kiln equipped with a tipping, louvred deck, built into an old kiln (after Fernbach, 1928).

or indirectly pre-heat the incoming kiln air, with a substantial saving in fuel consumption (Chapter 10). The forced draught of hot air warms the bed of malt and the evaporation of water from the malt cools the air. When the drying (and warming) zone has moved up through the bed of grain and reaches the top, the 'break' occurs, as is seen when the relative humidity of the air leaving the bed (air-off) falls below *ca.* 90% and the temperature begins to rise. At this time, the drying capacity of the air is not being fully utilized and so the thermal efficiency of the kiln falls. Normally at this stage the airflow through the kiln is reduced and the temperature of the air-on is increased to initiate curing. Later the valve in the kiln duct is moved to permit a proportion of the hot air to be recirculated. As curing proceeds, so less and less drying occurs and a larger and larger proportion of the air may be recirculated, until as much as 75% of the air from the bed of malt may be recycled. Very substantial savings of heat, and so of fuel, may be achieved in this way, but because the 'damp' recirculated air does not have the drying or cooling capacity of the freshly heated air there is a tendency for malts made using recirculation to be darker. Such kilns are not as easy to manage as those in which recirculation is avoided. The avoidance of recirculation and using linked kilns or modern double-decked kilns allows better control of the kilning process and better heat recovery.

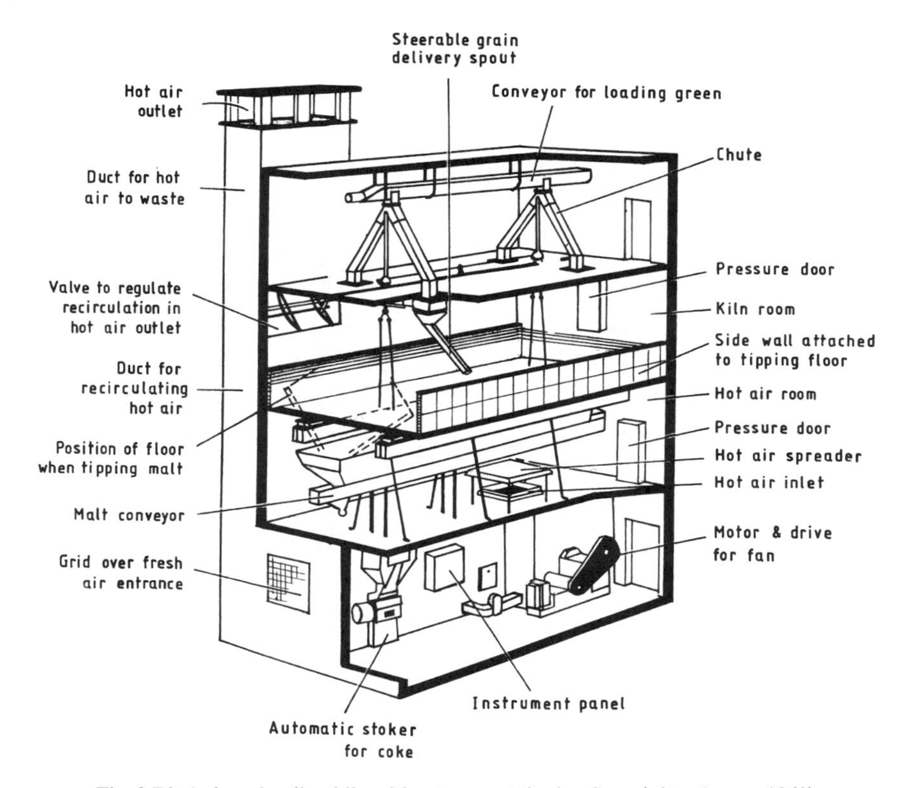

Fig. 9.74 A deep-loading kiln with a two-part tipping floor (after Anon., 1963).

9.11.4 Less common types of kiln

Kilning has been carried out in drums built essentially like the various types of germination drum but using hot air. This, indeed, is an old idea. In some older plants, the drying stage was carried out in a large drum and curing in a smaller one (Delbrück, 1910; Leberle, 1930; Wahl, 1944). The contraction of the malt, which occurred during drying, meant that often the ducting became exposed, hot air travelled above the malt rather than through it, and these drums were inefficient (de Clerck, 1952). Kilning in drums was never popular in the UK although 'roasting cylinders' were – and are – in use. An extension of simple drum kilning was kilning in drums under vacuum. This process found some acceptance in Germany and was used in the USA (Wahl, 1944). Presumably the malts manufactured with vacuum drying were all pale and enzyme rich.

The idea of 'vertical' kilns is an old one. White (1860) described a 'very pecular' upright, or tower kiln in use in a London distillery. It consisted of two concentric cylinders of wirewall, 40 ft (12.2 m) high separated by 6 in

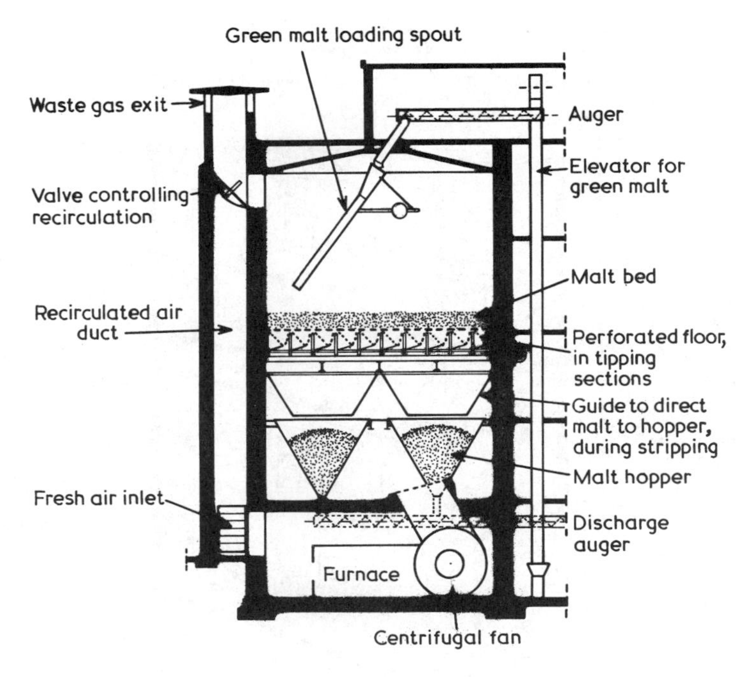

Fig. 9.75 Another pattern of deep-loading kiln with a louvred, tipping floor.

(15.2 cm), and of *ca.* 8 ft (2.44 m) diameter. Fresh wet malt was continually supplied into the 6 in (15.2 cm) gap between the wire walls at the top of the structure, and finished malt was continually withdrawn from the bottom. A fire of anthracite provided hot air, which ran up the centre of the column and outwards through the annular column of malt.

In a more modern pattern of vertical kiln, used in mainland Europe, the malt is held between vertical, perforated walls *ca.* 20 cm (7.9 in) apart and the hot air is passed from side to side (Fig. 9.76) (Leberle, 1930; Schönfeld, 1932; de Clerck, 1952; Vermeylen, 1962; Narziss, 1976; Kunze, 1994). The air spaces are *ca.* 80 cm (31.5 in) wide (Vermeylen, 1962). Each rectangular column of grain is divided into two or, more usually, three sections by transverse doors. The top of each column is wider than the bottom to allow for the contraction of the green malt that occurs during the initial drying stages. The upper and middle stages may each be 2.70 m (8.86 ft) high, while the bottom stage may be 2.40 m (7.87 ft). By using air valves and additional air inlets, the sections are arranged so that drying occurs chiefly at the top, intermediate treatment occurs in the centre, and final curing occurs in the bottom section. Because of the way the air is used in these kilns (like more usual three-floor kilns), they should be thermally efficient. In practice, malt shrinkage, which can be around a third, may result in the malt columns

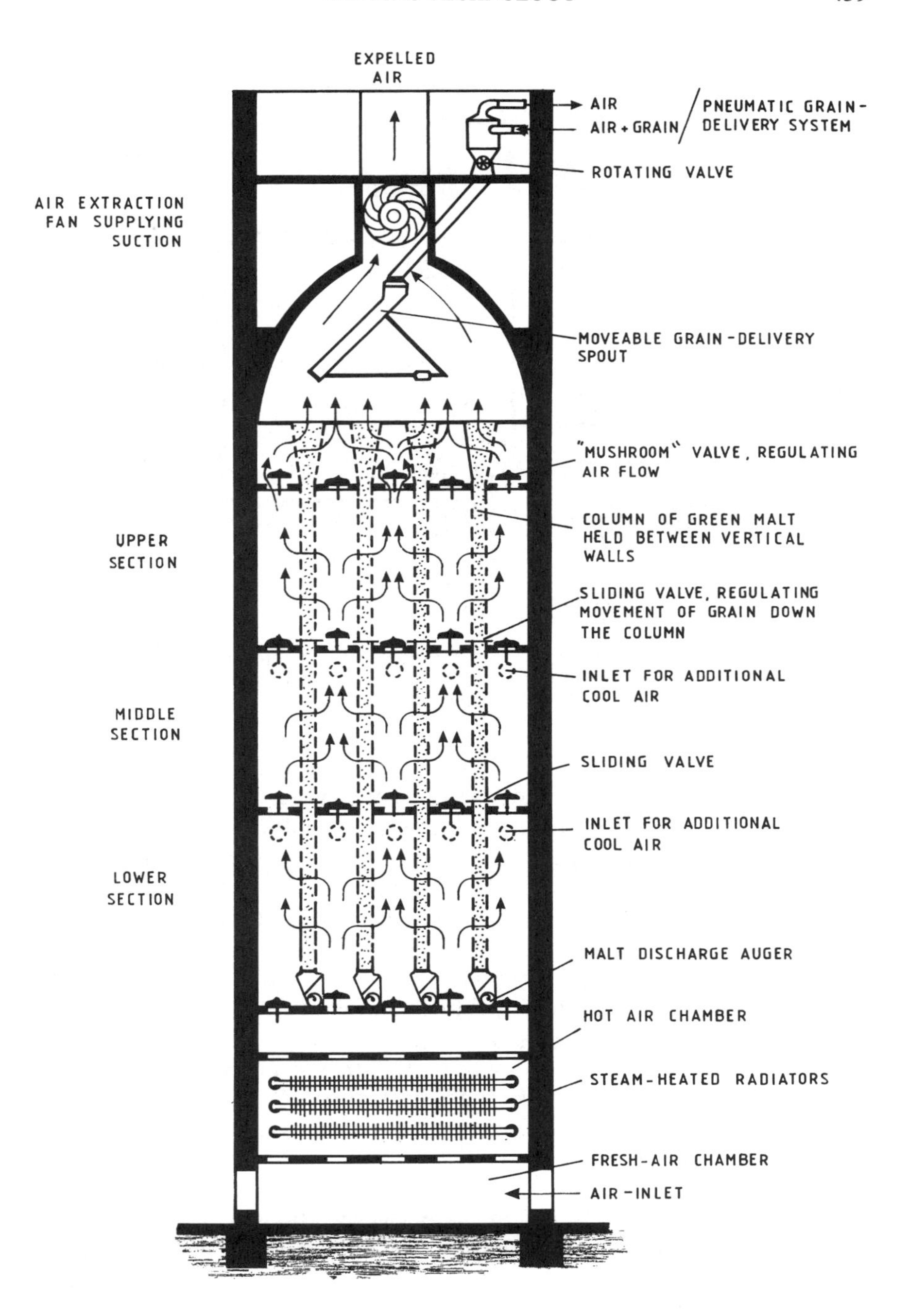

Fig. 9.76 A three-section, vertical kiln (Topf pattern; after Narziss, 1976).

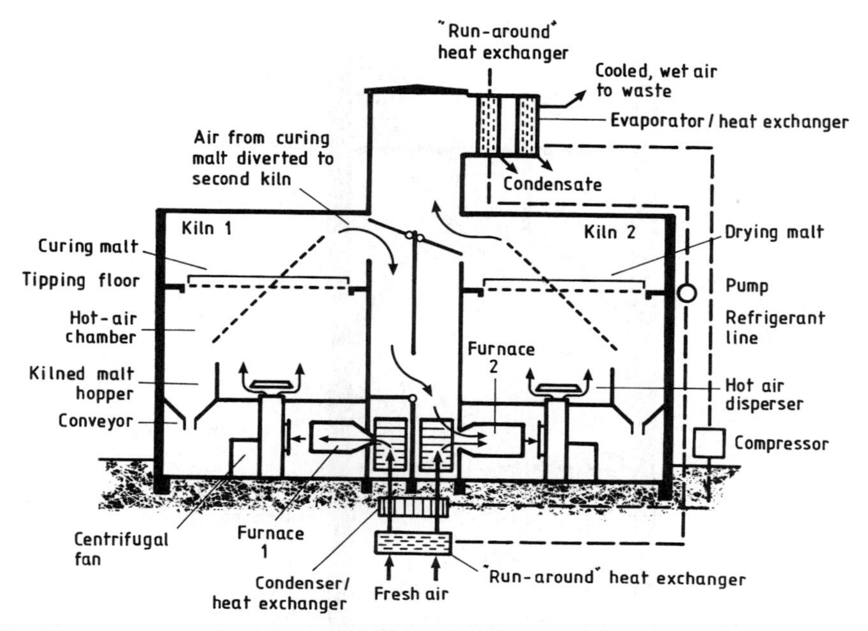

Fig. 9.77 Two deep-loading kilns, with tipping floors, linked so that the hot dry air from curing (kiln 1) is used to supplement the air in the drying kiln (kiln 2). Two heat-recovery systems are shown, an air-liquid-air 'run-around' heat exchanger and a heat pump.

becoming shorter than is desirable so that gaps develop beneath the horizontal doors and, because of the low resistance in these regions, the hot air passes here in preference to passing through the malt (de Clerck, 1952). In an alternative mode of operation, the intermediate doors remain open, green malt is added to the top of the columns as required and finished malt is withdrawn from the bottom either in batches or continuously.

As noted, continuous kilns are in use in the Domalt (Fig. 9.57), Saturne (Fig. 9.60) and Solek (Fig. 9.59) plants. Other continuous kilns have been proposed. Potentially the thermal efficiency of these kilns, working in a steady state, can be very high and even the air heated by cooling the malt as it leaves the kiln can be re-used. In 1875 Habich (quoted by Narziss, 1981b) described continuous two- and three-floor kilns in which the green malt was carried across each floor on three moving belts. A continuous, compact and highly thermally efficient 'triflex' kiln has been described by Schuster (1985). This kiln can be used with normal batch malting. A batch of green malt, 320–350 t, is held in a two-section storage vessel, which may be ventilated and cooled. Some drying occurs in this unit. The green malt is fed into the kiln and as it moves it is progressively dried and cured. The drying air is indirectly heated by furnace gas–air heat exchangers, exit gas–air and flue gas–air heat exchangers and heat pumps.

9.11.5 *Linked kilns and modern double-decked kilns*

Heat recovery in kilning is discussed in more detail in Chapter 10. Where kilns are grouped together, it has often been possible to link kilns using appropriate ducting so that by scheduling malt production the small volumes of hot, dry curing air from one kiln contributes to the larger volumes of cooler air used in drying in the other kiln. This is readily achieved in double kilns used in some Wanderhaufen plants, for example. Figure 9.77 shows two deep-loading, single-deck kilns that have been linked. The curing air from kiln 1 is directed to kiln 2, where drying is occurring. In many kilns, heat from an air–air heat exchanger (often made of glass tubes or, less usually, stainless steel) is used to pre-heat the cool air entering the kilns. In the example, the wet air leaving the kiln first gives up heat to a run-around heat exchanger, in which the warm, wet air is cooled by liquid that is pumped to the kiln inlet, where it gives up its heat via a radiator to the incoming air. The partly cooled air from the kiln outlet is cooled further by passage through a second heat exchanger, that of a heat pump, in which refrigerant is evaporated. The refrigerant is compressed and condenses in a heat exchanger situated in the kiln air inlet. Here it gives up its heat to the cool air. Maltsters are divided about the economics of using heat pumps. The extra complexity of the units and the extra resistance to air flow provided by the heat exchangers (requiring the expenditure of more fan power) are undesirable. However, some maltsters in the UK have used heat pumps with great success and short pay-back times (Watson, 1987).

One maltings has modified Alvan–Blanch barley driers as double-deck kilns (see Fig. 10.15, below). Coupling the kiln to a heat-recovery system allows the kiln to be used with high thermal efficiency; in addition, as the cooled air emerges at 15 °C (59 °F) and saturated with water vapour, it can be used in the germination compartments, reducing the amount of refrigeration and conditioning needed for the air going to the germination units (Watson, 1987).

Deep-loading, two-floor kilns are becoming popular. They can be thermally highly efficient and take up less space and ground area than linked kilns. Various designs of deep-loading, two-floor kilns have been described (Vermeylen, 1962; Schuster, 1985; Anon. 1986, 1990b; Tschirner, 1986). In all these kilns the lower and upper floors are separated by intermediate floors, drying is on the upper floor and curing on the lower. The floors are mechanically loaded and stripped. The differences between the various designs include the number (one or two) and relative sizes of the fans, and the disposition of the ducting (Tschirner, 1976). One such kiln is illustrated in Figure 9.78 (Anon., 1986). This kiln works on the suck-and-blow principle, air being drawn through the lower, curing deck and blown through the upper, drying deck. The kiln proper is circular in cross-section, has a concrete shell and is heavily insulated, the fans, ducts, etc. being housed in

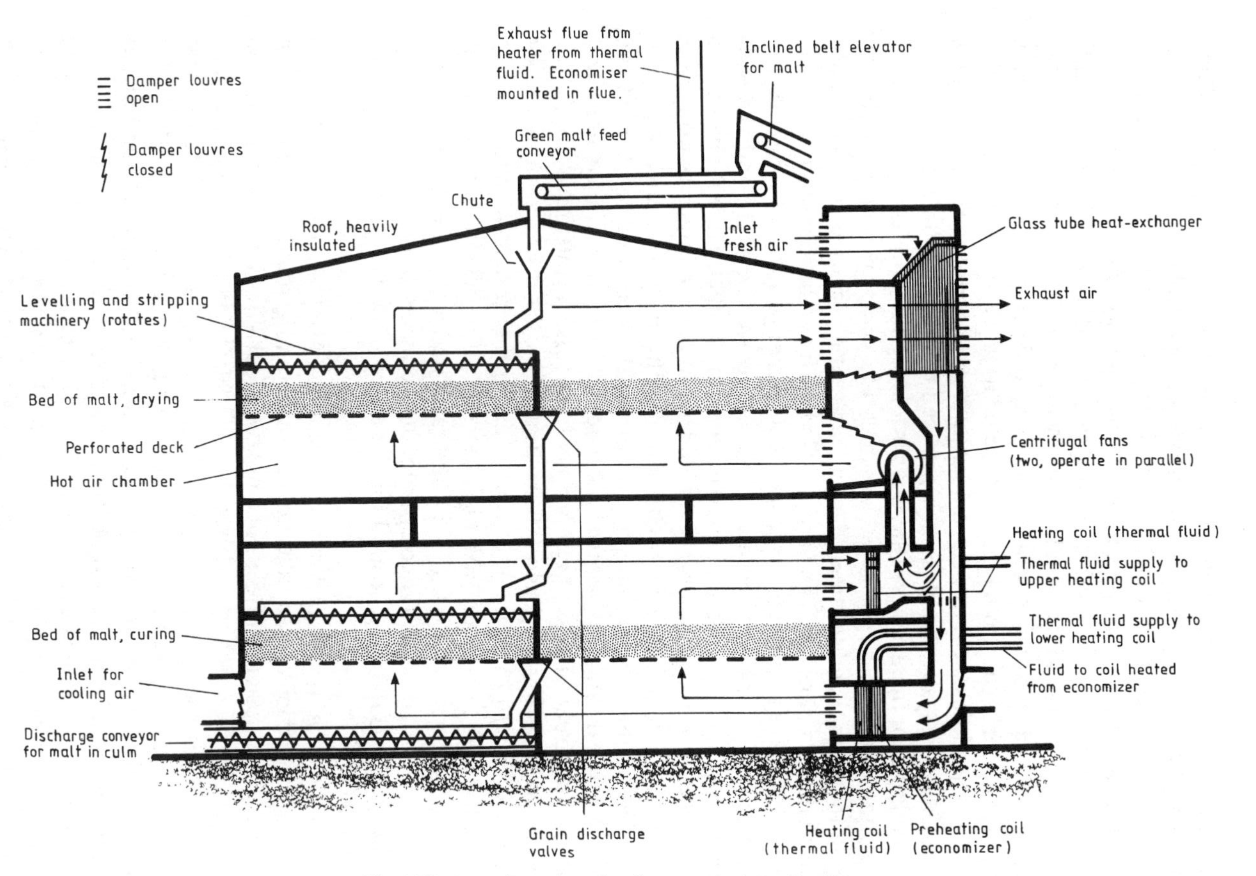

Fig. 9.78 A modern, deep-loading two-deck Redler kiln.

an associated structure. Two fans, 132 kW each, run in parallel. Each floor holds the equivalent of 30 t green malt. The total kilning cycle is 36 h, allowing 18 h on each floor, including loading and stripping. The kiln air is indirectly heated. The thermal fluid, a special oil that carries the heat to the finned-tube heat exchangers, is heated either by burning natural gas or, as a standby, by burning gas oil. A run-around heat-recovery 'economizer' system (water/glycol) recovers heat from the flue gases when natural gas is being burnt, but this is by-passed when gas oil is the fuel, because then the flue gases are corrosive. The wet air leaving the kiln passes through a glass tube heat exchanger, which pre-warms the incoming air. A heat pump has not been installed. The kiln is thermally highly efficient. Turn-around is rapid. Operations are completely automated, all mechanical, drying and curing functions being computer controlled. Proposals have been made to produce even more thermally efficient, 'continuous' kilns, some of them being of the 'vertical' kiln type (Anon., 1985a).

9.11.6 Kiln automation

By automating kilns and other malting plant, it is possible to maximize their performance and minimize workforce requirements. By loading and stripping kilns rapidly, quick turn-round times are achieved. By controlling fan operations and heat outputs and monitoring air-on and air-off temperatures (and air-off relative humidities) it is possible to time the kilning stages rather exactly and so reduce product variability. Furthermore, safety checks and interlocks are installed, for example to prevent machines operating out of sequence or in opposition, and to prevent operators from entering kiln compartments when this is undesirable, e.g. because sulphur dioxide is being used. Automatic warning systems are also installed, with the alarms in offices and in operatives' houses, so that any faults are quickly detected and rectified.

Many types of semimechanical controlling systems have been used, but now microprocessors and computers, which may be used individually or controlled from a centre regulating the processes in the entire maltings, are generally employed. These frequently contain a selection of programmes for different steeping, germinating and kilning cycles and record the conditions used in various previous cycles. In addition to essential numerical information, it is often possible, using computer graphics, to 'display' a vessel and its state, and to inspect the records or details of the ongoing and some previous production cycles.

The principles applied to kiln automation and control are now regularly used in all parts of maltings from intake to despatch. Computerized records of production conditions and malt analyses give maltsters the chance to establish correlations between them and to use the data so obtained to 'fine-tune' their production processes. Modern control systems are compact, reliable and user-friendly. The different levels of control may be made secure; for example, only managers may have the code that permits them to alter programmes regulating processing parameters, while all operators can monitor ongoing and recorded processes, initiate particular stages, etc. Usually control is exerted from a central station, but it is possible (in emergencies) to devolve control to 'outstations' that control particular stages or groups of stages of the production process. It may also be possible to over-ride the electrical controls with manual controls. On the one hand, such sophisticated automated and computer-controlled systems lead to improved repeatability in operations, the reduction of the number of unpleasant jobs that have to be done by hand and a reduction in workforce requirements. On the other hand, the staff employed must be knowledgeable and technically well-trained.

9.12 Cooling and deculming

In older practice, and rarely now, at the end of kilning in simple kilns the furnace is extinguished and the warm malt is 'rucked-up' (piled up) and allowed to develop more colour and aroma without the expenditure of more fuel (de Clerck, 1952; Vermeylen, 1962; Narziss, 1976). When the malt is cool enough the kiln is stripped. In general, modern kiln cycles do not permit this time-consuming approach, and in single-deck kilns the fans may be kept running to cool the malt. It is more efficient to strip the kiln quickly before it has a chance to cool, thus avoiding the expense of rewarming the kiln structure. The malt may be cooled separately by passing cool air through it in a special device, or during deculming. This reduces turn-around times and saves fuel, because it avoids the need to cool and then reheat the structure of the kiln. In continuous kilns, and in any sets of kilns that are working in-phase, the air that has been warmed while cooling the malt can be fed into a kiln air inlet, so saving fuel.

Deculming is the process in which the culms are broken and separated from the malt. With barley malt, the culm is virtually all rootlets, but in malts made from 'naked' grains, such as wheat, rye or sorghum, shoots are also present. Barley culms are separated because they give bitter flavours to beer, they are high in soluble nitrogenous substances, under unfavourable conditions they can contain high levels of nitrosamines and they are highly hygroscopic. While the malt is warm and very dry, the rootlets are brittle and can easily be separated. The deculming process also cools the malt. Historically, deculming was achieved by trampling, that is walking all over the malt while it was still on the kiln, then it was thrown, a shovel-full at a time, against an inclined wire screen or sieve and was brushed against

Fig. 9.79 Deculming malt (Anon. *ca*. 1877).

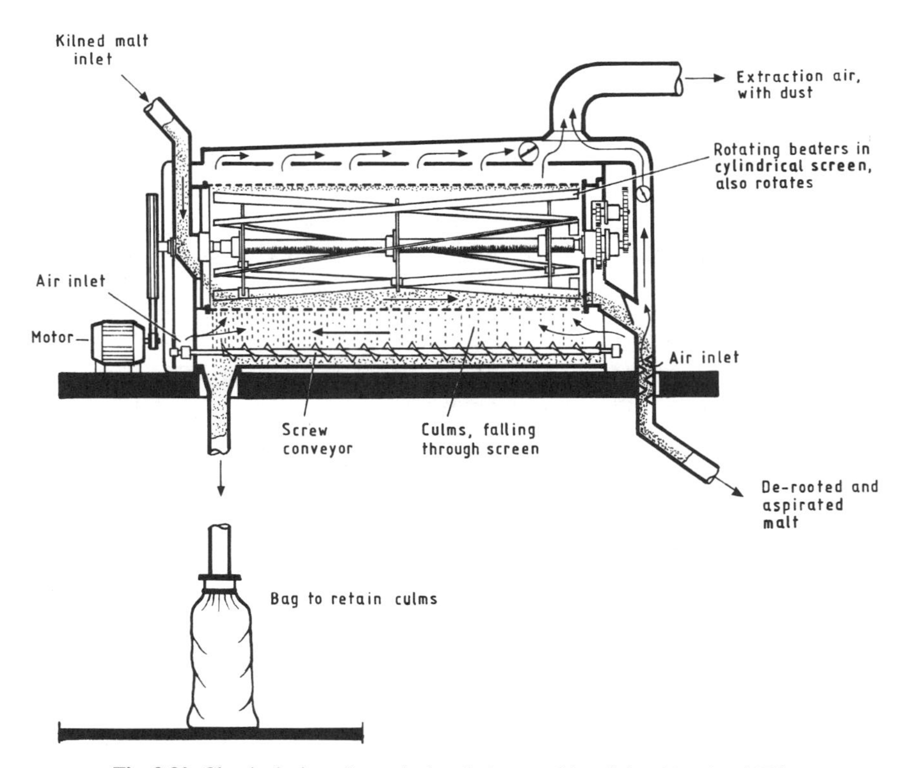

Fig. 9.80 Classis design of a malt deculming machine (after Narziss, 1976).

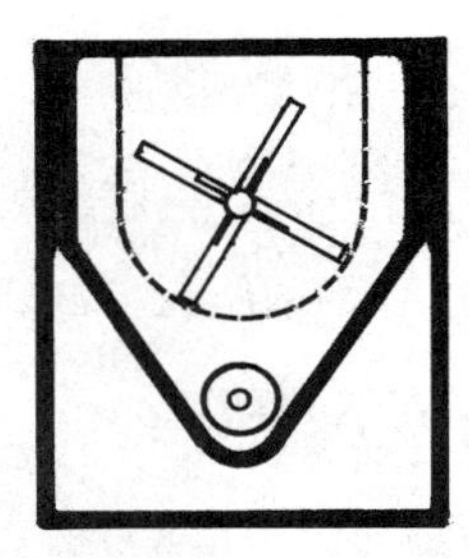

Fig. 9.81 Cross-section of a malt-deculming device. Roots dislodged by the beaters fall through the perforated metal conveying trough and are collected by the screw conveyor below (after Narziss, 1976).

it (Fig. 9.79). The broken culms fell through and the malt slid down the screen and accumulated in a heap. This filthy, dusty job was mechanized at the start of the 20th century. The 'classic' deculming machine consisted of a slowly rotating, cylindrical screen within which a set of wooden beaters rotated in the same direction, but faster. Malt 'in culm' was fed into the cylinder at one end. The angled beaters broke up the culms and gradually impelled the malt along the cylinder to the end, where it was discharged (Fig. 9.80). The screen rotated in an enclosed housing over a trough. The space around the cylinder and the malt leaving it was aspirated with an air current to remove dust. The culms fell through the screen into a trough from which they were moved to a collecting sack. Such machines worked very slowly, processing only 0.5–3 t/h. Two other types of machine, with much higher working capacities, are now usually used for deculming. In one the malt, in culm, is moved along by angled beaters in a U-shaped trough with perforated walls. The broken rootlets fall through and are caught in a trough in the outer housing from where they are moved by a screw conveyor (Fig. 9.81). The deculmed malt leaving this device is aspirated to remove dust. The other device is a pneumatic deculmer (Narziss, 1976). The malt, with any separated roots, is entrained in an airstream and propelled into a vertical cylindrical chamber. Impacts break off any rootlets that are still attached. The heavy malt falls against the airstream, past a separation cone, but the rootlets and dust are lifted and carried in the airstream. The malt is withdrawn from the bottom of the cylinder. The rootlets and dust may be separated in one or two cyclones, the rootlets being relatively easier to separate from the airstream. The air may be recirculated through a fan.

9.13 Malt storage

The cooled, deculmed malt is stored. It must be stored cool and dry in sealed stores: (i) to prevent moisture being picked up and the malt becoming damp

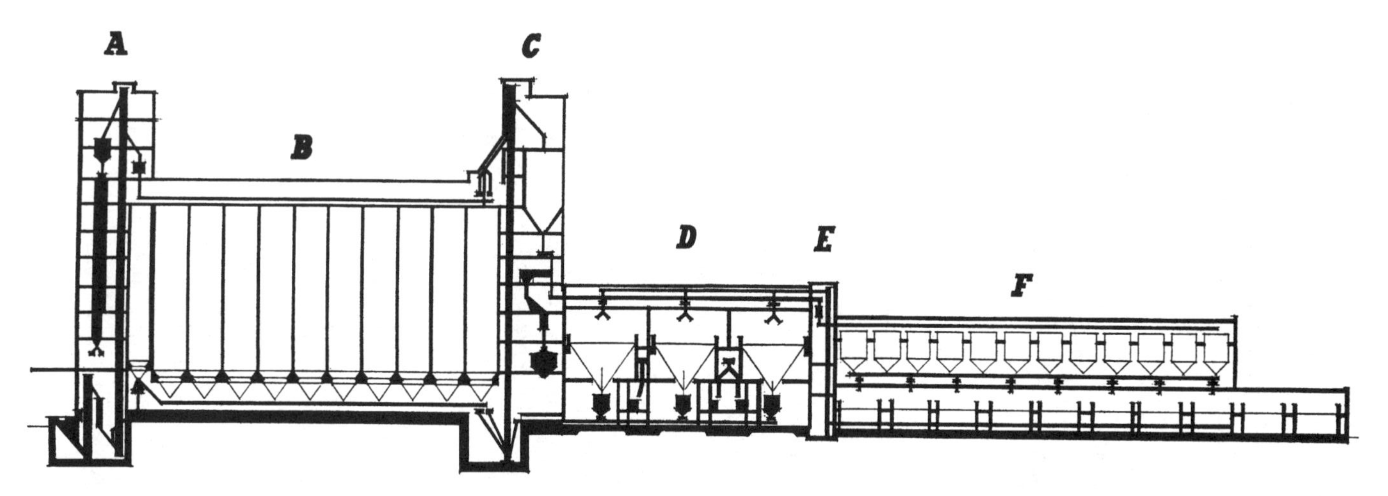

Fig. 9.82 A section of a maltings equipped with pneumatic germination Saladin boxes. Originally the boxes were open ended and turned by van Caspel turners, then by Ostertag–Wanderhaufen turners. The boxes, modified by the installation of end walls, are now turned with vertical helical screws. A. barley intake, drier and green barley bins; B, barley storage; C, malt cleaning and dispatch; D, three kilns; E, stair-hall, conveying and elevating equipment; F, conical steeps (12) above the germination boxes (8). This plant was built in 1961. (Diagram courtesy of ABM Ltd.)

(or slack) and spoiling; (ii) to reduce the chances of insect infestation or rodent depradation; (iii) to check the further development of colour and flavour; and (iv) to arrest the decline in enzyme levels. Kilned pale malt is usually stored for 4–6 weeks before it is used in brewing. There is some evidence, and a strong belief, that the quality and yield of extract obtainable from the malt improves during this storage period. Historically, malt was often stored with a layer of culms over it, to take up moisture from the air and prevent the malt beneath becoming slack. Stored in sacks, it frequently gained moisture from the air and sometimes needed to be rekilned. In modern plant, with the malt handled in bulk, these problems have vanished. The bins and silos used for bulk malt storage are described in Chapter 8 (Fig. 9.82).

9.14 By-products: culms and malt dust

Some 3–5 kg culms are produced with every 100 kg malt. Culms are collected from the deculming plant and from the kilns. The kilns should be regularly cleaned to collect culms and dust and so minimize the risk of fire. Because culms from the kiln tend to be dark, they may be kept separately. Malt and barley dust comes from many sites in the maltings, wherever cleaning is carried out or wherever grain is being moved and aspirated. Dust and culms were collected separately. They were bagged using bags with moisture-proof linings and then were usually sold to animal-feed merchants. To some extent this is still done. More rarely, they were burnt to provide heat or were composted, e.g. to make a medium for orchid cultures. Sometimes culms were extracted to make a medium for culturing yeasts. However, the dust and particularly the culms are hygroscopic and readily spoil, and their bulk densities are low, so small amounts, by weight, occupy large volumes of storage space.

Increasingly, culms and dust are transported, often pneumatically, to a ‘pelletizing’ unit. In some cases, thin barley is also transferred to this unit and is hammer-milled before being processed. Here the materials are wetted and thoroughly mixed and then converted into pellets. In one type of machine, this is achieved by compressing the mix with a screw and forcing it out of a multi-holed die. The emerging ‘worms’ of material are divided into short lengths by a rotating cutter. The pellets are cooled and are ready for sale. The advantages of pelletizing the dust and culms is that the bulk density of pellets is much less, so that storage space is saved and the pellets are more readily stored. Also pellets are more easily handled and are less dusty. The bulk densities of culms and pellets are about 224.3 kg/m^3 (14 lb/ft^3) and 561–641 kg/m^3 (35–40 lb/ft^3), respectively, depending on moisture contents and pellet sizes.

9.15 Roasting plant

In the past, dark malts were made on kilns having decks of woven wire or perforated iron plates. A layer of malt less than 3.8 cm (1.5 in) was used. Heat was provided by a wood fire that – at some stages – was extremely fierce so that kilns and their charges of malt were apt to catch fire. The advice given was to have filled water containers ready to quench this in the early stages (White, 1860). At this period, some dark malts were also finished in roasting cylinders about 4 ft (1.22 m) in length having diameters of about 2 ft (0.61 m). The cylinders, holding a charge of 3–6 bu, were rotated at about 20 r.p.m. and were heated over a coke fire. The charge was frequently inspected so that the operator knew when roasting was complete (White, 1860). When the cylinder was unloaded the steam and vile-smelling 'empyreumatic oil' that was released was extremely unpleasant and could be smelt for a long distance (White, 1860). Valentine (1920) suggested that the fumes from the cylinder should be burnt by passage through a coke fire before being vented through a chimney.

Modern roasting plants are much better controlled, but even so the production of quality products still rests on the skill of the operator, and the problems of fume disposal and fire and explosion hazards are always present. In particular, the escape of fumes from roasting plant in the UK is an offence and may lead to the plant being restricted or closed. Products now made in roasting drums include crystal malts, Munich malts, caramel malts, amber, brown and chocolate roasted malts, roasted barley and some special wheat and oat malts (Anon. 1963c; Reed, 1965; Bemment, 1985; Blenkinsop, 1991; Cantrell, 1992) (Chapters 14 and 15).

Modern roasting drums (cylinders) range in capacity from 0.5 to 5 t, with 1 t and 2 t units being usual in the UK. Special and coloured malts lose aroma on storage, so it is usual to make small batches for immediate delivery. A 1 t unit is about 3.05 m (10 ft) long and 1.52 m (5 ft) in diameter. Speeds of rotation are 20–25 r.p.m. Each drum is fitted with internal vanes to ensure that the contents are constantly mixed and turned and so are heated evenly (Fig. 9.83). Typically the furnace is fuelled by light oil or natural gas. The drum contents may be heated directly, when the hot gases pass through the drum, as when drying and roasting are occurring or, by resetting the valves in the ducting, they may be heated indirectly, when the hot gases pass round the outside of the drum. In the latter method, the contents are 'stewed', that is they are being heated while moist and evaporation is prevented. The different treatments are appropriate at different stages and for different types of product. In some instances the drums are loaded (charged) with green malt, in other cases kilned malt or even ungerminated grain is used depending on the product being made. Increasingly, roasting cylinders are better instrumented, with good control of temperatures and airflows, and they may have temperature programmes regulated

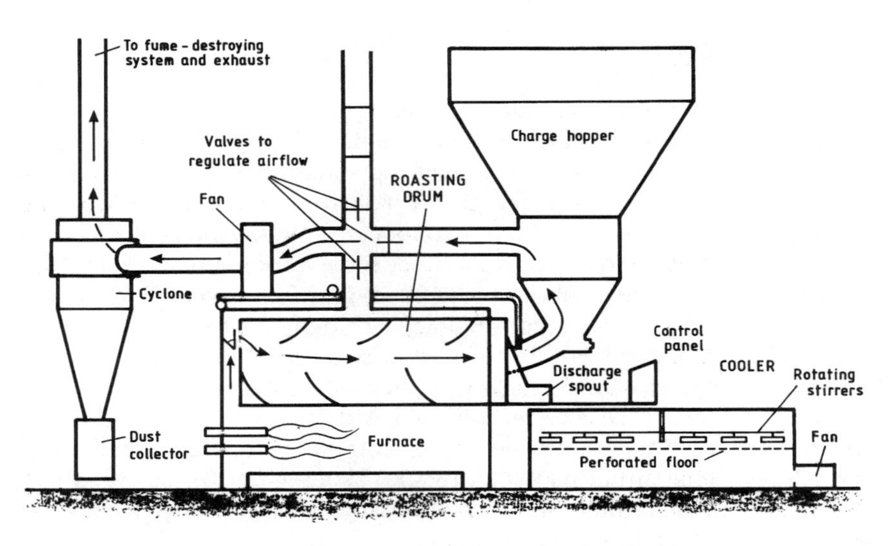

Fig. 9.83 A modern roasting drum, or cylinder, and a cooler. The drum contents may be heated indirectly or directly (after Bemment, 1985).

by a computer. Even so production is dependent on the skill of the operator, who frequently samples the charge with a device at the front of the drum in the stationary end plate; the operator cuts and inspects the grain, particularly towards the end of the of the process and judges when to cool the load. The dust from the drum, carried in the airflow, is trapped in a cyclone and the fume-filled air passes through a device (at one time scrubbers were tried, but afterburners, with or without catalysts, are now used) to destroy the fumes. In a particular pattern of thermally efficient afterburner (or 'thermal oxidizer'), the gases leave the drum at *ca.* 150 °C (302 °F) and are pre-heated in a heat exchanger to *ca.* 630 °C (1166 °F); they are then heated to 760 °C (1400 °F) by the burner, which destroys the organic volatile substances (Anon., 1985b). The gases then flow out of the device via two heat exchangers, the first of which pre-heats the gases from the drum; the second pre-heats the clean air going to the furnace that heats the drum. The gases are sucked through the device by a fan, which discharges them through a stack at about 200 °C (392 °F). Because the device is 'under suction', it is easily fitted with doors that can act for explosion relief and, when the device is not working, for access and inspection. Despite all precautions, fires still occur in these plants, so they are usually housed in isolated buildings often equipped with fire-controlling systems such as sensors and fixed overhead sprays. Some drums may be automatically flooded with water if the intended temperatures are exceeded to a dangerous extent. Drums are loaded, with a pre-weighed charge, from an overhead bin or hopper delivering through a chute that enters the stationary end plate that closes the drum.

At the end of the roasting process, it is important to cool the product quickly to reduce the risk of fire and to check further colour development. The furnace is switched off, and some products are sprayed with water; after a period when cool air is passed, the drum contents are discharged into coolers. Each cooler consists of an open-topped, circular container in which the product rests in a thin layer on a woven-wire support (Fig. 9.83). The product is continually levelled and mixed by brushes and vanes borne on centrally mounted, rotating arms. Cool air is drawn down through the mass by fans. When the material is sufficiently cool it is moved to a holding bin.

9.16 Malt blending and dispatch

Before dispatch, malt is screened and aspirated to remove small corns, broken grains, dust and impurities. It is also passed over a magnetic separator to remove iron or steel contaminants. Less usually, the malt may be passed over a destoner. In the past, malt was polished, that is it was brushed while passing between a rotating brush and a corrugated plate (peripheral speed 6–7 m/s (21 ft/s), or between contra-rotating brushes or between a rotating conical brush and a housing (de Clerck, 1952; Vermeylen, 1962; Narziss, 1976). The dust generated was removed by aspiration and was collected. This purely 'cosmetic' exercise was wasteful since the dust is rich in starch and yields **more** extract, on a weight basis, than the polished malt. This process probably survived because much malt was bought 'on inspection' and, one suspects, to remove mould. Polishing is unnecessary with well-made malts and it is not now used in the UK. In the UK, maltsters often store each batch of malt separately. Often each batch of one type of malt has a marginally different analysis. Then, when the user or customer orders a delivery, it is possible to blend malt from different batches to meet the customer's 'malt specification'. Practices differ, but while some delivered malts are blends of batches made at different times with one variety of grain, others may be mixtures of different varieties or, if the customer requires it, different grades of malt. Mixing grain lots thoroughly and in the correct proportions is not simple and the preferred means seems to be to simultaneously transport streams of malt (two or, rarely, three) at carefully controlled rates from the batch bins to the bin from which delivery will be made. Thus the delivery bin is loaded with a mixture that is uniform throughout.

For export purposes, malt may be transported in sacks of 50–55 kg or even up to 1 t capacity. These sacks are laminated, with plastic and hessian or other supporting layers, and are waterproof. Small amounts of malt may also be delivered in sacks, for example to micro-brewers, but large deliveries are in bulk. Problems with using sacked malt are that it is labour intensive to pack, transport and use and the sacks are easily damaged. Sacks are conveniently transported in sealed containers. Lorries or freight wagons may carry airtight container units or they may have built in 'tanks', which contain loose malt. This bulk malt is treated like a fluid, the containers are typically filled by a flow of malt from an overhead hopper. They may be emptied by suction into a pneumatic system, by gravity using tipping or through a hopper-bottom installed in the base. Obviously the maltsters' loading and malt users' unloading systems must be compatible with the transport being used. Every effort is made to keep raw grain separate from malt. Usually different storage containers and conveying and elevating systems are used. To prevent contamination, the transport must be rigorously inspected and cleaned before malt is loaded.

9.17 The organization of maltings

This topic is very complex; the purpose of this section is not to attempt an exhaustive treatment but to emphasize that a maltings or malt factory is not a collection of independent units but has to function in an integrated way. Ultimately the design of the factory will be indirectly regulated by the types of customer being served and their requirements. Therefore, a brewer-maltster or a maltster supplying distillers will, traditionally, have relatively few grades of malt to supply, allowing him to make relatively few, but large, batches of malt. A sales-maltster supplying breweries and distilleries of different sizes and types, and perhaps the food industry also, must be able to provide many different grades of malt in differing batch sizes. For the latter, the maltings will have to be flexible and allow more types of processing. The designer of a maltings must allow for this, not only in the type of production plant chosen and the batch sizes, but also in the capacity to store different amounts of different barley and malt grades in bins or silos.

The qualities of barleys available differ between harvests. These differences may be so great that different steeping and/or germination times may be needed to make the same grade of malt in successive years. Since economies dictate that it is desirable to have '100% vessel utilization', that is all parts of the plant should be continuously in use, considerable problems can arise in practice. Different philosophies have been chosen in balancing vessel capacities and the residence times of the grain samples passing through them. For example, in the tower malting illustrated in Fig. 9.84 in which 206 t batches of barley are processed, the stages, including loading and stripping (unloading) times, are intended to be 2 days steeping, 4 days germination and 2 days kilning, so that the steep tank delivers steeped grain to one of two germination compartments every 2 days, and the kiln receives green malt from a compartment every 2 days. All the machinery needed for transfers must be able to make transfers in the shortest possible times. A supposition is that, by choice of grain, grain maturation and the use of

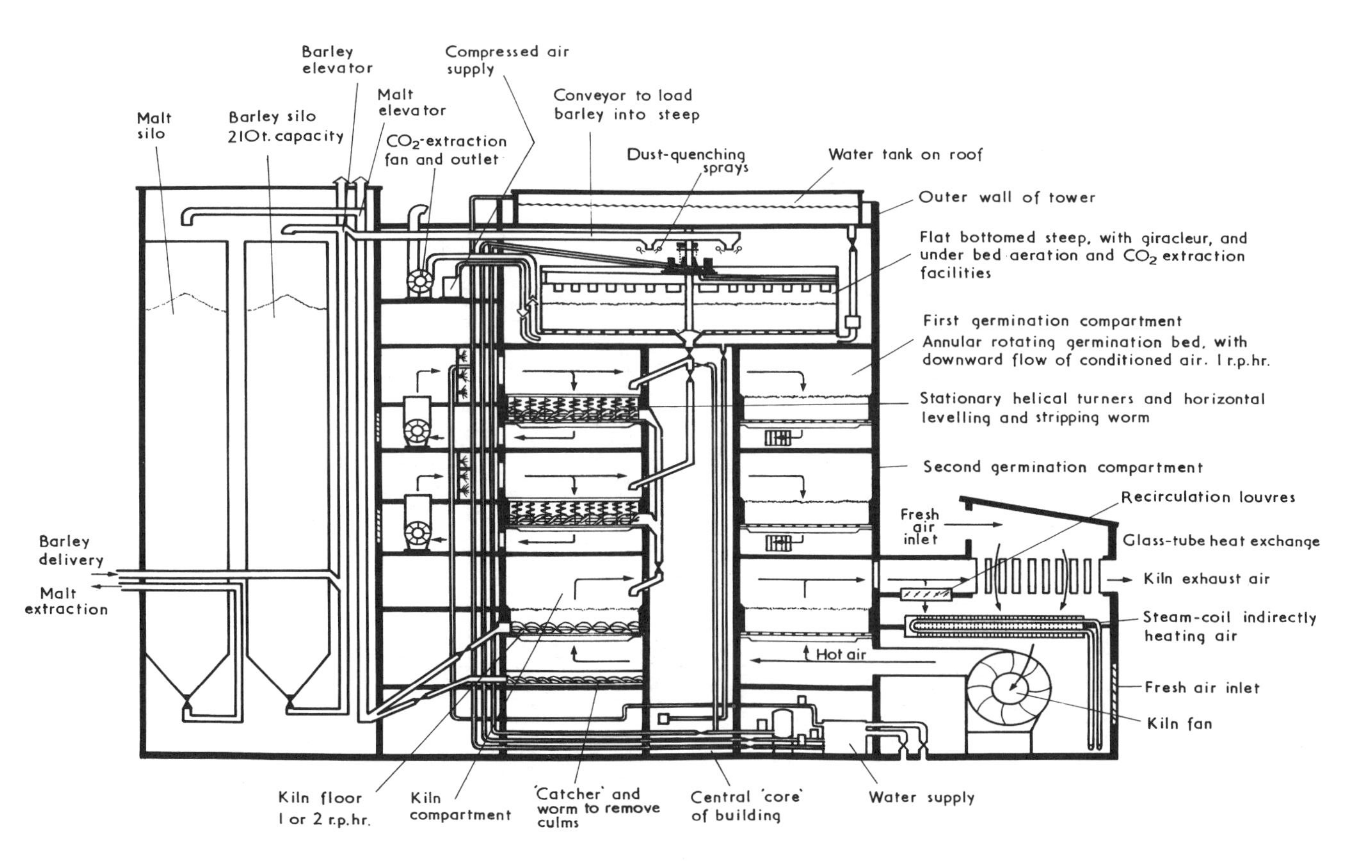

Fig. 9.84 The Carlsberg-Tetley Maltsters' tower maltings of Burton-upon-Trent constructed to process batches of 206 t barley. The steep, the germination compartments and the kiln are all circular in cross-section. (Courtesy of Carlsberg-Tetley Maltsters Ltd.)

different malting techniques, it is possible to prepare malt according to the chosen programme. Any necessity to depart from this programme would reduce the efficiency of working.

This may be compared with another tower maltings with a 200 t (barley) batch size in which three flat-bed steeps serve four germination compartments that, in turn, are served by a single kiln that is separate from the tower. In this arrangement, steeping, and the initial stages of germination because the steeps are ventilated, can take up to 3 days; the divisions between these stages are variable, giving some flexibility to the process. The germination period, in the germination vessels, is 4 days, leaving a 1 day kilning cycle. This short cycle is economic because the kiln, being outside the tower, has a large diameter and is relatively lightly loaded, the shallow bed of green malt avoiding the need for excessive fan power. In another North American tower malting, barley is first steeped in six conical-bottomed tanks, then it is dropped to another six tanks where steeping is continued. The steeped grain is discharged into a ventilated pre-germination vessel and later it is moved into one of four germination vessels. The green malt is moved half to each of two kilning floors and when it is hand-dried is moved to a third kilning floor where it is cured (Anon., 1994).

A new plant has been built at ground level to replace a drum-malting plant in which each drum had a capacity of 26 t. Barley batches of 210 t are processed using five lidded steeps (out of a total of 10) with conical bottoms (using a 2 day turnround time) four circular germination vessels (maximum germination time available 4 days) and two GKVs. In the GKVs the 48 h (max.) dwell time can be used for extended kilning or for a shorter kilning period if extra germination time (up to a maximum of 1 day) is required. Thus flexibility in processing is available. Germination time can be extended at the cost of using a more rapid, and expensive, kilning schedule. An advantage of malting plant built at ground level is that it is relatively easy subsequently to modify or extend it.

10 Energy used in malting

10.1 Introduction

For a century and more maltsters have tried to minimize expenditure on fuel and power. The swinging increases in the cost of energy, both fuels and electrical power, which began with the 'energy crisis' of 1973, have concentrated maltsters' attention on energy-saving strategies. The large amounts of money to be saved have made it economic to install energy-saving devices, having high capital costs, which were not previously worth considering. Capital expenditure is usually considered worthwhile if the 'payback period' (the time to save the expenditure, which is numerically equal to the capital cost) does not exceed about 3 years.

Fuel is used in maltings to warm steep water, to warm incoming germination air (in cold areas), in drying barley and, above all, in kilning. Electrical power is used in instrumentation, lighting and to move grain (elevators, conveyors) but especially to drive the fans used in the driers, steeps, germination units and kilns and in the refrigeration units used to cool the germinating grain at warm periods of the year. In the early 1970s, approximate power and heat usages in maltings (per tonne of malt produced) were for drying barley (18% to 12% moisture content (m.c.)), 4 therm/t malt; steeping, 1 therm/t; air heating for germination units (in UK), less than 1 therm/t; refrigeration for germination units (depending on season), 0–60 kWh/t; fans for germination units, 25–40 kWh/t; kiln fans (if installed), 0–75 kWh/t; fuel used to heat the kiln, 30–60 therm/t (1 kWh ≡ 0.03412 therm ≡ 3412 Btu ≡ 3.6 MJ) (Walker, 1978). Other estimates were given by Marsh (1975) (Table 10.1). It should be noted that the cost of power was three to ten times an equivalent amount of fuel, so figures of equal energy consumption can give rise to very different costs. Electrical power met about 10% of the total energy requirements. Of the electricity used in malting, about 60% is used in driving the kiln fans, 25% is used in driving the fans and refrigeration plant of the germination units and 15% is used in conveying, steep aeration and ventilation. In 1979, electrical costs were about equal to 40% of the fuel costs. Gibson (1989) estimated that, ignoring profits, interest and capital depreciation, the operating costs in making malt in the UK were £30–50/t malt produced. These costs were divided as fuel 25–30%, electricity 15–20%, wages 15–25%, repairs and maintenance 10–20% and miscellaneous costs 15–25%. At that time, malting barley, cleaned and entering a steep after storage etc., was probably worth £110–150/t.

Table 10.1 Estimated fuel and power consumption in UK maltings in 1975[a]. For comparative purposes all values are expressed as kWh/t

	Consumption (kWh/t)
Barley drying and storage	
Grading, material transfers, etc.	5–10
Cleaning and drying	
Air movement	5–10
Air heating (fuel)	90–120
The malting process	
Steeping	1–2
Germination	
Turning	2–4
Air movements	25–40
Refrigeration	0–60
Kilning	
Air movement	25–75
Air heating (fuel)	900–1200
Process-material transfers	2–4
Miscellaneous	
Packaging and dispatch	1–3
Dust extraction	2–6
Total	1058–1534

[a] For comparative purposes all values are expressed as kWh/t. 1 kWh ≡ 0.03412 therm ≡ 3.6 MJ. From Marsh (1975).

Attempts were made to improve the efficiency of fuel and power usage. To achieve this, maintenance was improved and it became justifiable to make substantial capital investments, for example in heat-recovery equipment. In some cases, power charges were reduced, for example by avoiding peaks in the demand and trying to use power in the off-peak periods. In addition, some kilns were fitted with burners so that they could burn different fuels, e.g. oil or natural gas, whichever was the cheaper, and so be able to accept a cheaper, interruptable gas supply.

Power and fuel use is carefully monitored and any change that suggests a fall in efficiency of usage is investigated. Staff training to encourage economies of energy consumption and openly displayed records of performance are also used. Some aspects of energy saving are discussed in the following sections. Gibson (1989) estimated that the specific energy usage per tonne of malt produced ranges from 2.48 GJ (23.5 therm, 689 kWh) to 6.81 GJ (64.5 therm, 1890 kWh) with a weighted average of 3.74 GJ (35.4 therm, 1038 kWh). There has been a considerable saving in energy usage in recent years in some malting plants, and other major savings remain to be achieved. Of the average value (3.74 GJ), the use of fossil fuel in kilning accounts for about 3.19 GJ (30.2 therm, 885 kWh) and electricity represents 0.55 GJ (5.2 therm, 152 kWh). To this must be added the cost of barley drying: about 0.5 GJ (4.8 therm, 141 kWh) for 'average' drying between 18% and 12% m.c. For the most modern and sophisticated kilns, fuel/energy usage of as little as 2.0–2.4 GJ/t (19–23 therm/t) are claimed (Tables 10.8 and 10.9, below).

10.2 Sources of heat for barley drying and malt kilning

Although several alternative sources of heat have been proposed, in practice barley drying and malt kilning are carried out in currents of heated air (Gluckman and Kidger, 1984; Kidger, 1986a; Quayle, 1987). The air may be directly heated, that is at least some of it passes through a furnace and so carries with it the products of combustion (in particular carbon dioxide, water vapour and, with some fuels, traces of sulphur dioxide, sulphur trioxide and oxides of nitrogen). Care is taken to avoid fuels containing appreciable levels of arsenic. In indirect heating, the air is warmed by radiators or heat exchangers, which may be heated by furnace gases, water, steam, thermal fluids (special oils) that are heated in furnaces, or with 'refrigerant' liquids heated by heat pumps. In addition, 'recovered' heat from the kiln outlet gases may be transferred to the incoming air using air–air or air–liquid–air heat exchangers or heat pumps. Therefore, with indirect heating, furnace gases do not come into contact with the grain or malt, but the systems are more complex; in most instances, indirect heating systems are less thermally efficient than direct systems.

To understand questions of heating efficiency and apparent major discrepancies in the literature, it is necessary to understand the calorific values of fuels and the basis of drying with warm air. As air is warmed so its moisture-carrying capacity increases. If the air is heated directly then water vapour, a product of combustion, is added to it and so some of its water-carrying and drying capacity is 'wasted'. However, if the air is heated indirectly using a heat exchanger, then no moisture is added to the airstream. Heat and moisture are carried away in the waste furnace gases. If heat is recovered from these furnace gases so that they are cooled below the dew point water condenses on the heat-exchange surface and, in addition to other heat, the latent heat of condensation is also recovered. If this recovered heat is transmitted to the drying airstream, the drying system can be more thermally efficient. The latent heat of evaporation, or vaporization, is the amount of heat needed to evaporate a unit mass of water at a particular temperature. The latent heat decreases as the temperature is increased. At a pressure of 1 atm and at 100 °C, 1 g steam has a latent heat of 539 cal. This heat is recovered as the latent heat of condensation when the steam is condensed. The heat outputs of fuels completely burned under standard conditions are given as gross and as net calorific values. These must be distinguished, or calculations based on the figures are useless. The gross, or high calorific value (HCV), of a fuel is the total amount of heat liberated, including the latent heat of evaporation of water generated in the

Table 10.2 The gross and net calorific values of various fuels; many fuels vary in composition, so values vary widely[a]

		Calorific Values (MJ/kg)	
		Gross	Net
Solid fuels			
Typical wood (dry)		15.83	14.35
Anthracite		29.66	28.94
Coking steam coal		30.70	29.75
British coals (9–10% H_2O, 8% ash)		32.84	29.55
Peat		15.91	14.47
Lignite (brown coal)		21.45	20.19
Coke		28.63	28.35
Liquid fuels (relative density)			
Light distillate	(0.68)	47.80	44.50
Gas oil	(0.83)	45.60	42.80
Light fuel oil	(0.93)	43.50	41.10
Medium fuel oil	(0.95)	43.10	40.80
Heavy fuel oil	(0.96)	42.90	40.50
Gaseous fuels			
North Sea natural gas (94% methane)		38.62	34.82
Groeningen natural gas		33.28	30.00
Carburetted water gas		18.95	17.31
Coal gas		17.97	16.14
Producer gas		4.99	4.77

[a] Theoretically the latent heat of vapourization of water is 2.25 MJ/kg. Cooling water from 373K to 288K recovers 0.354 MJ/kg. Therefore, the theoretical heat allowance is 2.61 MJ/kg. In practice, this is found to be too large and 2.454 MJ/kg is allowed.
Data of Quayle (1987).

combustion, per unit weight of fuel burned and when the products of combustion are cooled to 25 °C. Under most circumstances not all this heat is usable, as the heat in the water vapour is not recovered. For many purposes, the net, or low calorific value (LCV) of the fuel is of more use. Numerically LCV = HCV – latent heat of the water vapour at room temperature. Representative fuels and their calorific values are shown in Table 10.2. Attempts to use barley dust and straw as cheap fuels for grain drying seem to have proved impractical on the industrial scale. Inspection shows wide variations between the gross and net calorific values of different fuels, and these are related to the hydrogen contents of fuels, the hydrogen giving rise on combustion to the water that must be vaporized. Therefore, fuels with high carbon contents contain little hydrogen (e.g. anthracite, *ca.* 2.4% H), oils contain intermediate amounts of hydrogen (e.g. light fuel oil), 11.7% H) while methane, the predominant component of natural gas, contains 25.1% H. The worst case is, of course, hydrogen gas itself, but this is not used as a fuel in malting.

Increasingly maltings are changing from direct to indirect kiln firing. By placing economizers (heat-recovery heat exchangers) in the flue gases, it is feasible to recover some of the latent heat of the water vapour in some instances, as when natural gas is burned. However, when sulphur-rich fuels are burned (e.g. heavy fuel oils, *ca.* 4% S) it is important not to cool the gases below the dew point as the oxides of sulphur dissolve in the condensed water and the acidic solution obtained is highly corrosive. A temperature of about 135 °C (275 °F) is regarded as critical; in practice, flue gases are allowed to escape at *ca.* 200 °C (392 °F) to avoid corrosion problems (Kidger, 1986a). With natural gas as a fuel and the flue gases cooled to less than 53 °C (127 °F) (45 °C (113 °F)) (which is easily achieved), the combustion efficiency is in excess of 100% of the **net** calorific value but, of course, cannot exceed 100% of the **gross** calorific value of the fuel. It is ironic that when direct-fired kilns were heated by burning sulphur-rich heavy fuel oils it was necessary to take great care with the airflow and combustion conditions to minimize sulphur trioxide formation (and hence speckled magpie malt). On occasions it was necessary to inject small amounts of ammonia into the airstream to neutralize the oxides of sulphur (Macey *et al.*, 1975). Since the introduction of natural gas it has often been necessary to inject sulphur dioxide into the kiln airstream or to burn sulphur in it to minimize the formation of nitrosamines in the malt from NOX in the airstream (Chapter 4). A consequence has been considerable corrosion of kiln structures. Following some arsenic poisoning, which occurred at the end of the 19th century, fuels used in the UK for direct firing have to have low arsenic contents. Other precautions for reducing the arsenic content in the kiln airstream were used. For example, furnace gases were directed through a maze of chalk blocks, which were converted to lime by the heat and which retained acidic gases including those containing arsenic (Beaven, 1904). Others added a proportion of lime to the solid fuel and this also trapped arsenic and oxides of sulphur.

In many ways, direct-firing kilns with 'pure' natural gas appeared ideal, until the risks of nitrosamine formation from oxides of nitrogen were appreciated. Two approaches were made to overcome this problem (Kidger, 1986b). Firstly, sulphur dioxide was added to the airstream (either as a metered supply of the gas or by burning solid sulphur sticks, powder or droplets) with consequent corrosion problems with the kiln structure. Secondly, 'low-NOX' gas burners were installed. These can only utilize natural gas, so standby oil burners are usually installed. Below 750 °C (1382 °F), nitrogen and oxygen do not combine to form NOX, and below 1100 °C (2012 °F) NOX production is still low. The low-NOX burners, which require careful adjustment, operate at a low temperature with a near stoichiometric mixture of air to natural gas. This can lead to unstable flames and so, to be able to achieve reductions in heat output, it is necessary to have several burners so that one or more can be switched off. Direct heating systems have the disadvantage that the hot furnace gases, usually mixed with cool, diluting air, are passed through the fan. Fans approximate to constant volume devices; because hot air is less dense than cool air, a fan moves a smaller mass of warm air in a given time and this is less efficient than having the fan drive cool air before it is heated.

Indirect air heating has the advantage that the products of fuel combustion do not come into contact with the malt. Such systems are more expensive to install than those involving direct firing, but their thermal efficiencies have been raised to fully acceptable levels. In indirect heating systems, the kiln air is heated by finned-tube, or other, heat exchangers containing steam, water, 'thermal fluid' or, in one system, furnace gases. These heat exchangers may be installed downstream in the airflow from the fan, which, therefore, can work more efficiently. Steam can be raised by heating water with a wide range of fuels. It is only economic to use steam if the kiln is on a brewery site or some other location where steam is widely used. Thermal efficiency is likely to be about 80%, based on the gross calorific value (Gluckman and Kidger, 1984). Steam heating is easy to control and is comparatively safe. Heat may also be transferred from a furnace using special oils (thermal fluids) that may be heated up to 350 °C (662 °F) or water under pressure, which may work at 140 °C (284 °F), for example, under 4 bar pressure. Because the latter system operates at lower temperatures, it requires a larger heat exchanger to transmit a given amount of heat to the airstream. However, concerns about malt contamination with oil and possible fire risks should a heat exchanger leak favour hot-water systems working under pressure. Both systems can work at about 80% efficiency (gross calorific value).

An interesting flue gases–air heater is the ANOX unit (Allemann, 1981; Mutter, 1981). This unit is normally fired with natural gas or, for short periods, gas oil. The flame is contained in a stainless steel combustion tube and the hot gases pass successively over stainless steel heat-exchange tubes and borosilicate glass heat-exchange tubes, being progressively cooled. The cool gases go partly to waste and are partly recirculated, driven by a fan, to the combustion tube, ensuring that the surface of the tube is at less than 750 °C (1382 °F) and so nitrosamine formation is prevented. The incoming air is heated by passage over the glass, then the steel heat-exchange tubes and finally over the combustion tube before entering the kiln; thus it flows counter-current to the furnace gases. The exhaust gases are cooled to about 45 °C (113 °F), below the dew point, so water is condensed and some of the latent heat of condensation is recovered. Thus the efficiency of this heating unit is *ca.* 94% of the gross calorific value of the fuel burned and 104% of the net calorific value. Such units are large and expensive but their heat output can easily be varied and the fan driving the airstream to be heated works efficiently as it is in cool air. Used with a heat-recovery system to prewarm the incoming air, this provides a very efficient heating system. Safety devices must be installed to prevent the heat-exchange system overheating when low airflows are being used through the kiln.

10.3 The physics of water removal from grain

By convention, the moisture content of barley and malt is usually expressed as percentage fresh weight (f.wt) or on an as-is basis. Thus:

$$\text{m.c. (\%, as-is)} = \frac{\text{wet weight} - \text{dry weight}}{\text{wet weight}} \times 100$$

However, in some engineering work and some botanical studies a dry weight basis, (d.b.) is used, in which:

$$\text{m.c. (\%, d.b.)} = \frac{\text{wet weight} - \text{dry weight}}{\text{dry weight}} \times 100$$

Numerically, the values can be very different indeed; for example, a green malt could have a m.c. (f.wt) value of 45% and an m.c. (d.b.) of 81.8%. In this book, the maltster's usual convention is followed, and moisture content (% f.wt) is used throughout.

At any particular temperature water has a nearly fixed water vapour pressure, and air in equilibrium with water at that temperature will be saturated with water and have an equilibrium relative humidity (RH) of 100%. At higher temperatures water has higher vapour pressures, so air in equilibrium with the water at higher temperatures contains more water vapour, while the equilibrium RH is still 100%. The absolute numerical values of vapour pressure change slightly with atmospheric pressure, but this is ignored here. At any temperature, a given sample of barley or malt also has a characteristic equilibrium water vapour pressure which will usually be **less** than that of water at the same temperature because water in the grain is bound or more or less firmly associated with the grain's constituents. The vapour pressures of samples of grain or malt or water increase sharply with temperature and it is more convenient to work in other, RH units. RH is the quantity of water present in air at a particular temperature, expressed as a percentage of the quantity of moisture needed to saturate the air at the same temperature. Therefore, with an RH of 0% the air is perfectly dry, while with an RH of 75% it is three-quarters saturated. However, at higher temperatures, air with a given RH contains more water vapour than air having the same RH at a lower temperature. It is usual to express the water vapour pressure of the grain as the RH of the air in equilibrium with the moisture in the grain at the prevailing temperature. This convention is helpful, because as the temperature is increased so the water-holding capacity of the air increases, as does the equilibrium vapour pressure of the grain; as a result, the equilibrium RH of the grain alters much less than the equilibrium vapour pressure. However, the equilibrium RH is not completely independent of the temperature (see below). Measurements of the vapour pressures of grain samples have been made under 'static' (no

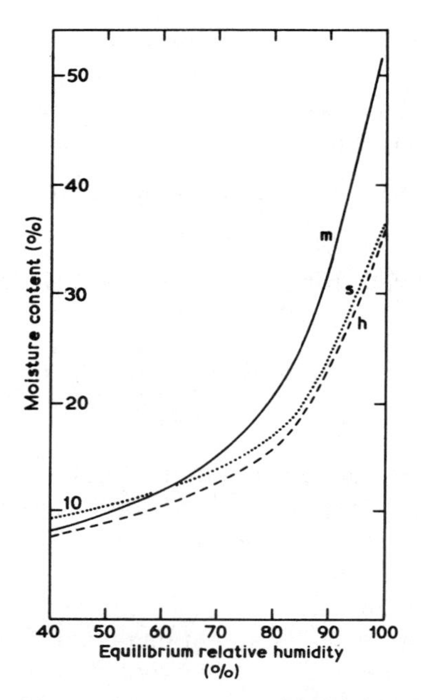

Fig. 10.1 The relationships between moisture content and equilibrium RH for malt culms (m), straw (s) and hay (h) in the range 15–27 °C (59–80.6 °F) (after Snow *et al.*, 1944).

airflow) conditions and at fixed temperatures where the moisture in the air was intended to be in equilibrium with the moisture in the grain. This has not always been achieved and some of the older data, much used in older calculations, are in error. Furthermore, different substances (starch, protein, husk) have different equilibrium RH–moisture content relationships (Snow *et al.*, 1944). Therefore, barleys and malts having different chemical compositions (e.g. higher nitrogen contents: richer in protein, poorer in starch) will have different equilibrium RH–moisture relationships also, a fact that is never allowed for. Inspection of the equilibrium RH–moisture curve for malt culms shows a relationship that is quite different for those of hay, straw, barley or green malt (Figs 10.1 and 10.2). The highly hygroscopic nature of culms is well known. Many apparent equilibria are, in fact, only approximate because of the excessive periods of time needed for a true equilibrium to be obtained (Hart, 1964).

Moisture movement between grains, and probably within grains, is slow at moderate temperature (e.g. 25 °C (77 °F)). This is illustrated by the fact that moisture–equilibrium RH curves differ (exhibit hysteresis) depending upon whether the grain has picked up moisture from the surroundings (adsorption curves) or has lost moisture to the surroundings (desorption curves) (Fig. 10.2). The failure of these curves to coincide is at least partly owing to the non-attainment of true equilibrium which might take a month

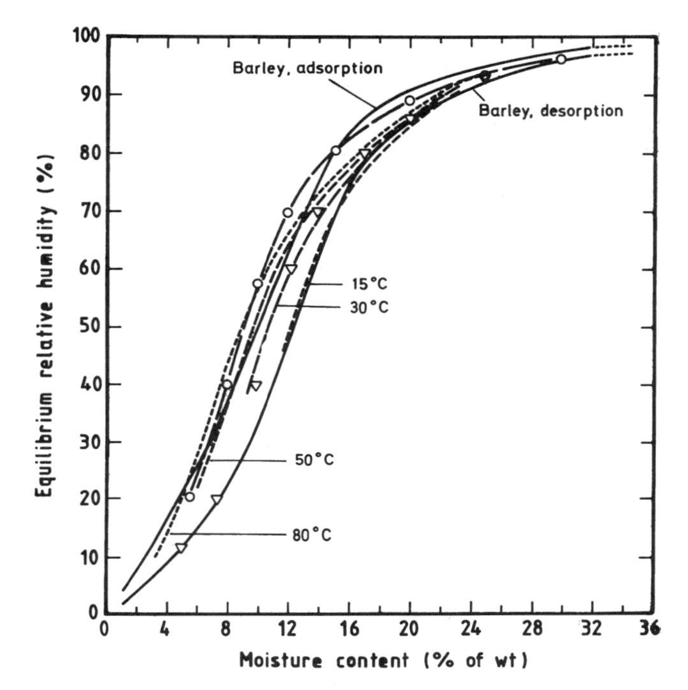

Fig. 10.2 The relationships between moisture contents (f.wt) and equilibrium RH values for barley and malt at different temperatures. The barley adsorption and desorption lines bracket a great deal of published data for grains at ambient temperature (Briggs, 1978). The failure of these lines to coincide, illustrating the phenomenon of hysteresis, is probably owing to the failure to obtain true equilibrium between the water in the grain and in the surrounding air within the experimental period. The lines labelled 15, 30, 50 and 80 °C are those calculated by Tuerlinckx *et al.* (1984) for malt. Values obtained experimentally for malt at 5 °C (41 °F) and at 45 °C (113 °F) (o–o) were reported by Pixton and Henderson (1981), who also noted that a green malt with a moisture content of 43.5% had an equilibrium RH of 96.6% at 45 °C (113 °F) and 98–99% at 5 and 25 °C (41 and 77 °F).

or more (Pixton and Warburton, 1968, 1971; Pixton and Griffiths, 1971; Pixton, 1982). The curves for barley shown in Fig. 10.2 enclose many published experimental results (Briggs, 1978). Nomograms relating equilibrium RH, temperature, grain moisture content and grain vapour pressure for Sultan barley (and some values for Capelle wheat, which are similar but relate to low moisture contents are given in Fig. 10.3

The relationships between malt moisture content and equilibrium RH have been contentious. Statements that malt of 12% m.c. or more has an equilibrium RH of 100% are certainly wrong (Tuerlinckx and Goedseels, 1979). Measurements of malt equilibrium RH–moisture values at two temperatures are shown in Fig. 10.2 which also shows values calculated using these equations (originally expressed on a moisture content (d.b.) for four temperatures. In detail the curves do not agree. In part this might be

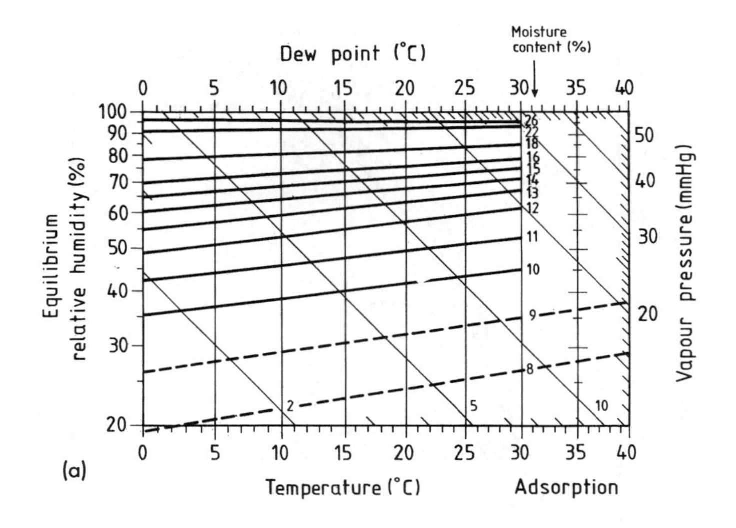

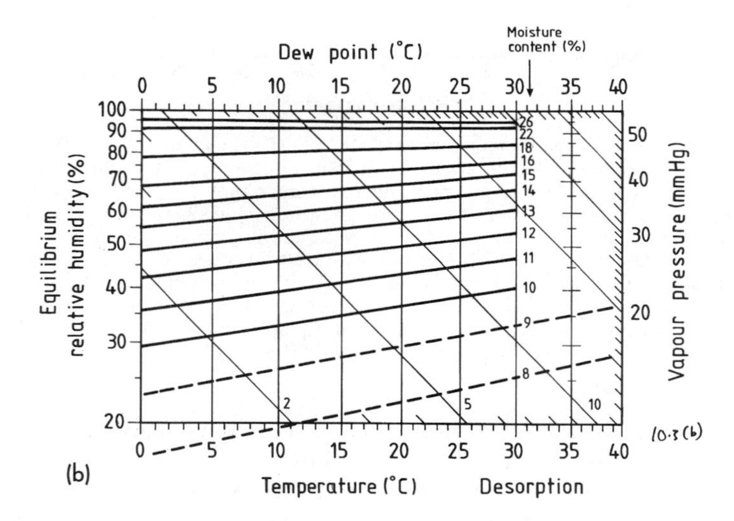

Fig. 10.3 The relationships between moisture content, equilibrium RH, vapour pressure and temperature for barley (and wheat) under adsorptive (a) and desorptive (b) conditions (after Pixton and Warburton, 1971). The solid lines are for Sultan barley and the dashed lines are for Capelle wheat. The values for the two species, where they overlap, are similar, so the data for the wheat should be a good guide for drying barley to low moisture contents, in preparation for warm storage. To relate the moisture content of the grain to the equilibrium RH at a given temperature, draw a horizontal line from the point where the temperature and moisture content lines intersect to the RH scale. The diagonal lines indicate the vapour pressure of the atmosphere in equilibrium with the grain. To obtain the dew point, i.e. the temperature required for condensation to form in an atmosphere in equilibrium with grain at a given temperature, a line parallel to the nearest vapour pressure line is drawn from the point where the moisture content line intersects the temperature line to the dew point temperature scale.

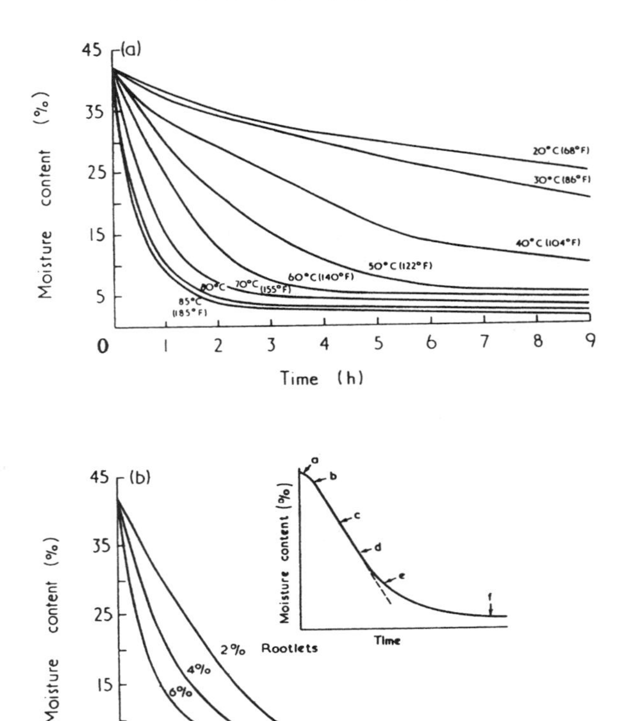

Fig. 10.4 (a) Experimental drying curves for barley in a fixed air flow, at the constant air temperatures indicated. (b) Drying malt grown to different extents in a fixed air flow at a fixed temperature (70 °C (158 °F)). Inset: the components of an 'idealized' isothermal drying curve: a, the initial phase in which the grain bed is warming and coming into equilibrium with the air stream; b, onset of the period of steady-state drying; c, the (approximately) linear steady-state stage in which 'free water' is removed; d, the transitional stage, leading to; e, the intermediate stage of drying; and (f), the slow drying period in which bound water is being removed (after Schuster and Grünewald, 1957).

because the different malt samples have different percentages of roots (Figs 10.1 and 10.4; see below). However, even at moisture contents of 34%, equilibrium RH values are significantly less than 100%; one estimate of the equilibrium RH of malt (m.c. 43.5%) at 45 °C (113 °F) was 96.6% (Pixton and Henderson, 1981). Indeed, as both malt and barley contain sugars, amino acids, salts and other soluble substances, and as the solution of any of these substances in water will reduce the equilibrium vapour pressure, it follows from the laws of physics that the equilibrium RH of grain or malt must **always** be less than 100%, although the difficulties of measuring RH at high values may lead to experimental errors that conceal this.

When grain (or malt) is ventilated and the RH of the air is above the equilibrium RH of the grain, moisture will be picked up by the grain. Conversely, if the air is 'dry' and has an RH below the equilibrium RH of the grain, water will be lost to the air and the grain will dry. This is a dynamic situation in which the conditions are deliberately maintained away from equilibrium. The rate of water loss will depend on the difference between the RH of the air and the equilibrium RH of the grain plus the time taken for water to migrate from the interior to the surface of the grain from which evaporation occurs. As the grain dries, the difference between the RH of the air and the equilibrium RH of the grain declines and so the rate of drying falls until, in a deep bed of grain for example, the RH values are equal and drying ceases. Inspection of drying curves shows that these expectations are justified (Fig. 10.4). It can be seen that at higher temperatures drying is more rapid. This is because at higher temperatures: (i) the RH of the air is less; (ii) the rate of migration of moisture from the interior of the grain to the surface is faster, so evaporation is accelerated; and (iii) the vapour pressure of the water in the grain is increased, as is the equilibrium RH, which also favours evaporation. It is noticeable that the rates of malt drying increase with the amounts of roots attached (Fig. 10.4). It has been suggested that this is because moisture is less firmly bound in better-modified malt, but inspection of Figs 10.1 and 10.2 show that this is incorrect (see below). It is probable that the large surface area of the rootlets, from which evaporation can freely occur, is responsible.

The temperatures **inside** the grains are important for the maintenance of grain viability in barley drying and in causing colour formation and enzyme destruction during the kilning of malt. As the hot drying air passes through the grain bed so the grain temperature will be reduced by an amount depending on the latent heat utilized to evaporate the water that is removed (539 cal/g (970 Btu/lb) at 100 °C (212 °F)). The latent heat of water evaporation from green malt rises as the malt dries and the residual water is more firmly bound (Tuerlinckx *et al.*, 1982, 1984). This cooling is important; for example, the in-grain temperature of damp barley should never exceed about 38 °C (*ca.* 100 °F) for any substantial period of time. Often higher air temperatures are used for drying barley; for example, it is often recommended that the temperature of the drying air should not exceed 49 °C (120 °F). This air-on temperature is only acceptable if free evaporation is occurring from the grains so that the in-grain temperatures are reduced below this value. If, for any reason, evaporation is restricted, for instance when grain is heated in a closed container, the in-grain temperature may rise to a damaging extent and grain vigour, and possibly viability, may be reduced. When green malt is kilned, enzyme survival is enhanced by using a cool airstream. (Experimentally, enzyme survival is maximal when green malt is freeze-dried and water is removed from grain at less than 0 °C (32 °F).) In experimental malting, the major proportion of diastatic enzymes survives if kilning is carried out in a rapid airflow at a temperature not exceeding *ca.* 40 °C (104 °F) (Dickson and Shands, 1942). When idealized drying curves are compared with those obtained in practice, discrepancies are observed (Fig. 10.4). When drying begins, there is an initial period of adjustment during which the bulk of grain is warming up. Green malt (*ca.* 45% m.c.) has a specific heat less than that of water (about 0.7). Barley and other grains have specific heats that decline with decreasing moisture content (Fig. 10.5). The specific heat of dry malt has been given as 0.38 (Hopkins and Carter, 1933). Often the initial adjustment period, or lag, is not seen in small-scale drying experiments where the amount of grain is small and the airflow is relatively large. After the lag, it is often stated that a linear period of drying follows because it is assumed that the evaporation of water from the grain is not restricted below that of free water. The moisture content–equilibrium RH curves (Fig. 10.2) show that this assumption is not true, so that at a moisture content of around 24%, for example, the equilibrium RH of barley is only 90% of that of water, and at lower moisture contents the equilibrium RH is lower still. Furthermore, in a constant airflow at a fixed RH, the difference between this RH and the equilibrium RH of the drying grain is declining. This in turn ensures that the rate of drying will decline, as will the likelihood that under these dynamic drying conditions moisture migration within the grain will be able to maintain the 'equilibrium vapour pressure' at its surface. Inspection of drying curves obtained under closely controlled conditions confirms that there is no true linear phase of drying (Fig. 10.4). Rather, the rate of moisture removal declines with drying, slowly at first and increasingly more rapidly until it becomes very slow indeed as equilibrium between the moisture in the airstream and that in the grain is approached. At higher temperatures, the equilibrium moisture contents of drying barley are lower, as predicted

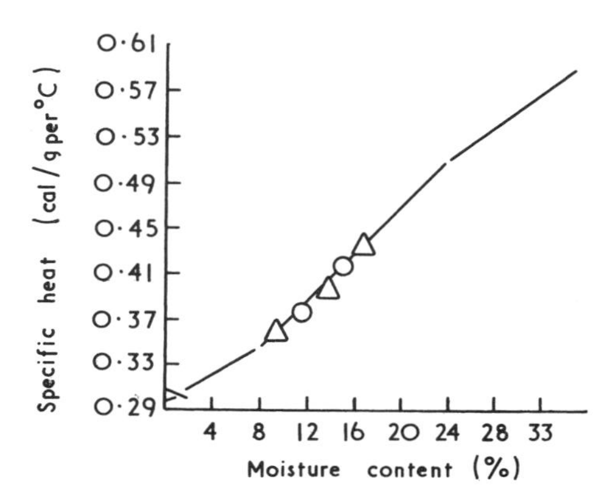

Fig. 10.5 The specific heats of barley (△), rough rice (○) and wheat (the regression lines) at different moisture contents (after Disney, 1954).

(Fig. 10.4a). However, if malts having different degrees of growth are dried at one temperature, the initial rates of drying differ but the 'equilibrium moisture contents' are the same (Fig. 10.4b), indicating that, despite views to the contrary, at any given temperature the equilibrium RH values of green malts differing in degrees of modification will be about the same and that water in the different malts of a given moisture content is approximately equally firmly bound. The enhanced rates of drying of more-extensively grown malts may result from the greater evaporative surface provided by the rootlets and/or greater ease of movement of moisture from the interiors to the surfaces of the grains.

As the grain dries it shrinks (Table 10.3). This reduces the distance the residual moisture must travel from the interior to the surface, so this effect tends to hasten drying. However, the concomitant reduction in the surface area from which evaporation occurs tends to slow the drying process. It is virtually certain that most evaporation from within grains occurs at the embryo end of the grain, through the micropyle in barley or where the testa is ruptured by the acrospire and rootlets in malt. It has been proposed that green malt might be lightly crushed before kilning to facilitate the loss of water vapour and so hasten drying.

10.4 Barley drying

Barley drying has been extensively reviewed (Hukill, 1954, 1974; Briggs, 1978; Foster, 1982; Bathgate, 1989; Brook, 1992). In the UK to allow storage for extended periods without ventilation, barley is usually held at 15 °C (59 °F) or less at a m.c. of 12% or less. However, for prolonged storage of over a year or for shorter periods at elevated temperatures barley should be dried to *ca.* 9% if germination vigour and viability are to be maintained (Chapter 7). The conditions of drying that may be used are limited by the use to which the grain is to be put. If the grain is to be used for animal-feed, then grain viability need not be retained and conditions for drying are not

Table 10.3 Estimates of the effect of moisture content on the surface area of malt

Moisture in malt (%)	Surface area of dry malt (m^2/100 kg)
45.0	360
40.0	330
30.0	250
20.0	200
10.0	178

Data of Schlenk cited by de Clerck (1952).

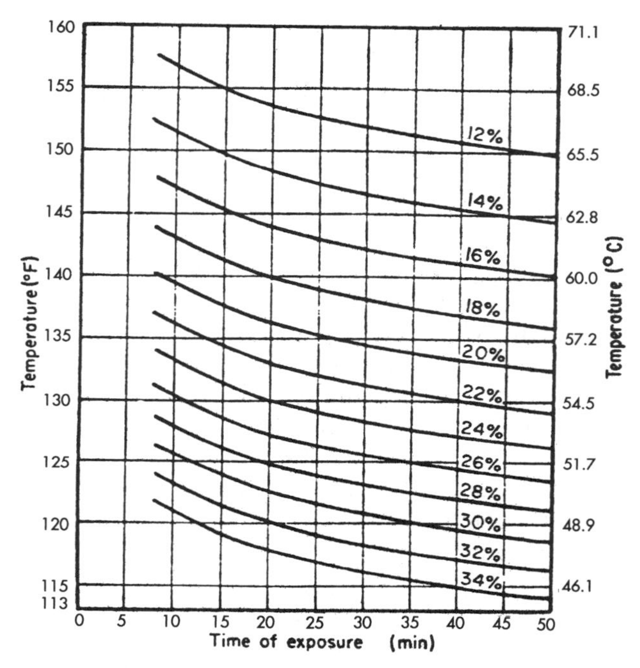

Fig. 10.6 Curves showing the combinations of temperatures and times of exposure that just cause a detectable delay in germination in barley or different moisture contents (after Oxley, 1959).

critical. However, with malting grain this is not the case. In Fig. 10.6 the combinations of moisture content, temperature and exposure time just causing detectable damage to barley are shown. Clearly a maltster drying barley wishes to avoid any detectable damage whatsoever, and so the conditions used should have a substantial 'safety margin'. The conditions used have been determined by trial and error. For instance, tower driers, in which forced evaporation (and hence grain cooling) occurs, operate at higher temperatures than batch driers, where the dwell time may be 8–24 h. In the latter, grain over 24% m.c. has a recommended maximum air-on drying temperature of 43 °C (110 °F), while for grain of less than 24% m.c. the recommended maximum temperature is 49 °C (120 °F). In a flow-through tower drier the dwell time may be 2 h and the air-on temperatures may rise from 54 °C (*ca.* 130 °F) to 66 °C (*ca.* 150 °F). The air-off temperatures, reduced below the air-on temperatures by the evaporation of water from the grain, may be 43 °C (110 °F) and 52 °C (125 °F), respectively. In the cooling section, the grain temperature may be reduced to 30 °C (86 °F) or less.

10.5 Conditions during germination

During grain germination, cooling is provided by the passage of humid air and by a certain amount of evaporation. For example, if during a 4-day germination period the respiration loss is 4%, this is equivalent to 40 kg starch being utilized for respiration (95% of the energy released being released as heat) for every tonne of dry barley. The heat output is 15.6 kJ/g and this heat must be removed (Cole, 1982). The air entering and leaving the bed of grain is almost saturated with moisture, and a 3 °C (5.4 °F) temperature rise may be accepted. Under these conditions, 1 kg air at an average of 18 °C (64.4 °F) removes 9 kJ of heat. The average rate of heat production is 1.81 kJ/s per t dry barley, so 0.2 kg air/s per t dry barley is needed for cooling **on average**. In fact, heat output peaks and declines as germination proceeds and allowance must be made for this. Economics dictate that, where the climate is cool enough to permit it, refrigeration plant should not be installed. However, where it is installed the 'waste heat' from the heat pump should be used, for example in warming steep water or in pre-warming air at the kiln air inlet. If the target air-on temperature is 16 °C (60.8 °F), for example, then clearly for much of the year this will differ from ambient conditions, and the air coming into the germination units may need to be warmed or cooled. The degree of warming or cooling needed may be reduced substantially by recirculation of a proportion of the air from the grain bed. This has the added advantage that this air is already saturated with water vapour. It has been reckoned that for each tonne of grain steeped 100 t ventilating air is used during the germination phase. Usually the air is supplied at a rate of 300–400 ft^3/min (8.5–11.3 m^3/min) for every tonne, giving a linear airflow through the grain of 10–15 ft/min (*ca.* 3–4.6 m/min). Air may be cooled with water sprays, water-cooled heat-exchange coils or a refrigeration plant. Water from the coils may be directed to the steep tank and the heat from the refrigeration unit to the kiln air inlet (Minch, 1978).

10.6 Kilning malt

While malt drying in a vacuum has been proposed, and proposals have been made for employing microwave and infrared drying, in practice malt is nearly always dried with currents of warm air. Many inter-related factors regulate the rate of drying including (i) the volume of air passing through the grain bed in a given time; (ii) the depth, uniformity and compactness of the grain bed; (iii) the moisture content of the grain and hence its hygroscopic state and the total amount of water to be removed; (iv) the temperature and moisture content of the air entering and leaving the grain bed; and (v) the specific heat of the grain or malt. Other related factors that are of importance are (vi) the temperature within the grains; (vii) the degree of grain shrinkage; and (viii) the quantities of roots present. Old accounts suggest that as malt is kilned the rates of drying fall into clearly divided stages. This is not strictly correct; however, it is true that, at a constant temperature, the decline in drying rate begins to become more noticeable as the moisture content falls below 30% and becomes progressively more apparent until it is obvious at values of 20% or less. In practice, this may be obscured by the use of successive small increases in the air-on temperatures and by deep beds of grain in modern kilns. Practical limitations on the temperatures used are set by the type of malt being made and by the type of kiln being employed.

Usually the temperature of the air is measured below and above the bed of grain on the kiln deck, using electrical thermometers. The data may be logged in a computer, displayed electronically or may be recorded on a chart. For some purposes, the outside air temperature and RH are also important, as are the RH values of the air in the kiln and the airflow rates. Accurate measurements of RH and airflow rates (with a sensitive anemometer) are not easy under kilning conditions and these measurements were not made routinely. However, the advent of robust electrically operated hygrometers is changing this. In addition grain-bed temperatures may be determined from thermometers thrust into the grain bed to a fixed depth. These values may be called in-grain temperatures, but in fact they are most likely to be temperatures of the air in the grain bed at that point. The values so obtained have their uses, for example in checking the uniformity of the temperature across the grain bed or the time of passage of the drying front (see below), but they are of no use in drying calculations.

In older, shallow-loaded kilns and a few more modern types, the grain may be loosened or even turned by mechanical devices to try and obtain even drying and uniform treatments. It is very difficult to obtain even drying in these kilns as uneven loading leads to local 'thin' areas where there is less resistance to airflow and drying is rapid, and 'thick' areas where it is slow. Therefore, drying malt in imperfect, thin layers must waste hot air. Heating air is costly in terms of the fuel used. Unless the heated air removes as much moisture as possible from the grain, that is unless the air emerges from the grain bed saturated with moisture (having an RH of 100%), heat is being wasted. In practice, the RH of the air must not exceed the equilibrium RH of the malt anywhere in the bed, as otherwise the upper layers will pick up moisture and become wetter. No evaporative cooling will occur and the malt will stew. For drying to continue, the RH of the air in each successive 'layer' of grain must be below the equilibrium RH of the grain. Furthermore, the air above the bed of malt must not be cooled appreciably, or saturation will be reached, the air in the upper part of the kiln will become foggy and 'reek', condensation may occur and water may drip back onto the bed of malt. In older, natural-draught kilns such occurrences were probably frequent but, as with modern kilns, the objective was to keep the RH of the air emerging from the grain bed as high as possible to maximize drying efficiency in terms of the fuel used. Other causes of heat waste are (i) imperfect combustion of fuel; (ii) heat 'wasted' in warming the structure of the kiln; (iii) heat lost from the kiln owing to poor insulation and air leaks; and (iv) the (inevitable) heat used in warming the wet malt and not in evaporating moisture from it. Furthermore, in older kilns, and in natural-draught kilns especially, the regulation of the airflow was inadequate. The draught was generated by the difference in density between the cooler outside air and the warmer, and less dense, air within the kiln. Therefore, in warm, humid weather the ambient air was warm, the density difference between it and the in-kiln air was small, the draught was poor and the drying power of the air was poor. The conditions in the kiln might be so damp and warm and so much condensation could occur that the top layers of malt sometimes continued to grow. Conversely, in cold and frosty weather, the draught was often excessive, air emerged from the grain bed in a comparatively dry state and so drying power, and hence fuel, was wasted. In some windy conditions, some older kilns exhibited down-draughts where there was no reliable flow of air through the malt. These problems were minimized by using (i) cowls to reduce the effects of winds; (ii) fans to obtain at least a minimum draught; (iii) 'dampers' to allow the airflow to be restricted or throttled; and (iv) thermostatically controlled furnaces to regulate the temperature of the air entering the hot air chamber.

Moisture removal from, for example, a pale-ale malt in shallow-loading kilns was divided into four or five merging phases. These divisions were to some extent arbitrary but were used to guide the kilning process. At the beginning of a kilning programme, there is a start-up phase in which the kiln structure and the bed of grain are warmed, the airflow is established through the grain bed and the rate of water evaporation builds up until the system is in dynamic equilibrium. Then, in the (mis-named) free-drying stage, with an air-on temperature of 50–60 °C (122–140 °F) for example, water is removed from the malt without much 'restriction' at an approximately linear rate. Because drying conditions are comparatively steady, the airflow can be adjusted so that the air-off has an RH approximating to 90–95%. When the malt has dried to a significant extent (from an initial value of 43–48% to about 25%, i.e. when about 60% of the moisture initially present has been removed) the slowing of the rate at which moisture reaches the grain surface and the falling equilibrium RH of the grain begin to restrict the drying rate to a serious extent and so the 'intermediate' stage of drying begins. The airflow is now reduced to allow for the longer time it takes for

Fig. 10.7 Kilning conditions used on shallow-loading, two-floor German kilns: (a) a pale malt being kilned on a 24 h cycle; (b) for a light malt being kilned on a 48 h cycle; and (c) a dark malt being kilned using a 48 h cycle (after Schuster (1962).

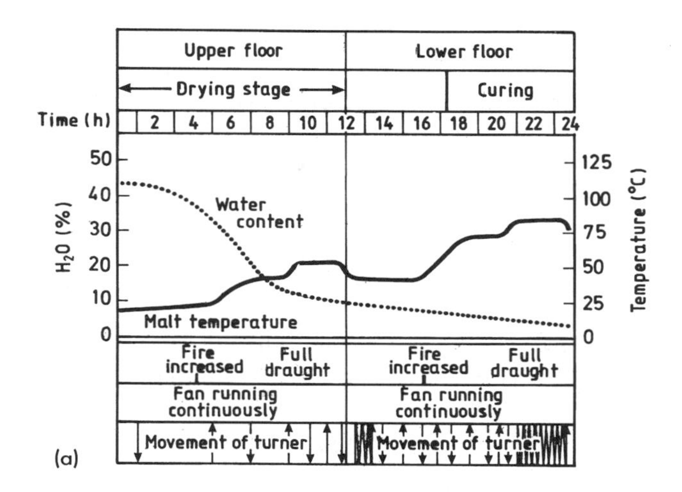
Upper floor
Lower floor
Drying stage
Curing
Time (h)
2 4 6 8 10 12 14 16 18 20 22 24
H2O (%)
Temperature (°C)
Water content
Malt temperature
Fire increased
Full draught
Fan running continuously
Movement of turner
(a)

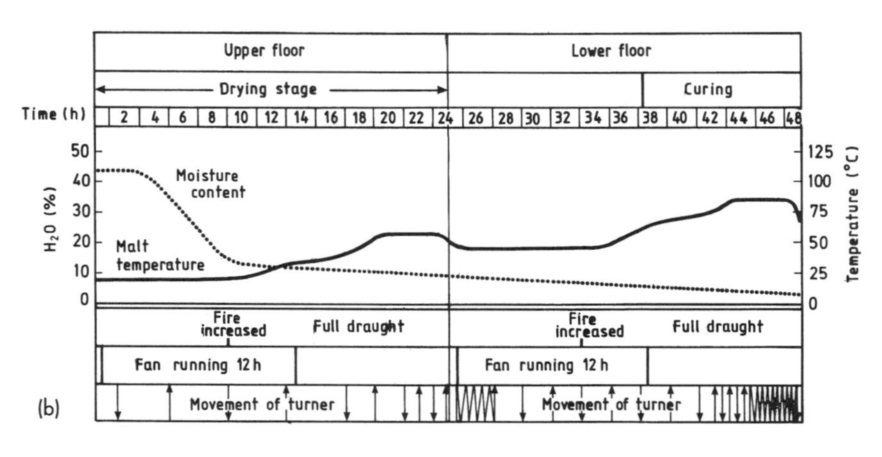
Upper floor
Lower floor
Drying stage
Curing
Time (h)
H2O (%)
Temperature (°C)
Moisture content
Malt temperature
Fire increased
Full draught
Fan running 12 h
Movement of turner
(b)

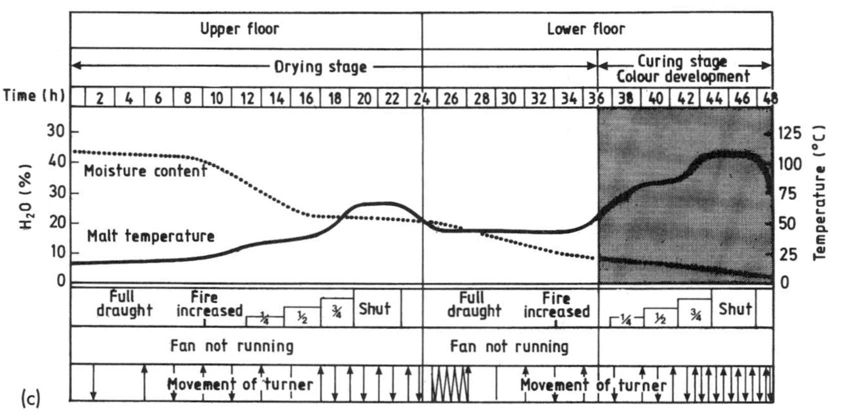
Upper floor
Lower floor
Drying stage
Curing stage
Colour development
Time (h)
H2O (%)
Temperature (°C)
Moisture content
Malt temperature
Full draught
Fire increased
¼ ½ ¾ Shut
Fan not running
Movement of turner
(c)

moisture to reach the surfaces of the grains and establish equilibrium between the grain and the surface layer of the air. The temperature of the air-on may now be increased to accelerate the rate of water removal. The bound-water phase was reckoned to start when the malt moisture content had fallen to about 12%. The temperature of the air-on would be raised to *ca.* 65–75 °C (149–167 °F) to permit continued drying, and the airflow might be reduced even more. If equipped with suitable ducts then some air recirculation may be started to save fuel. The airflow through the grain bed must not be reduced to too low a value, as with draughts of less than *ca.* 1.2 m/min (4 ft/min) it is unlikely that the bed of grain will be evenly dried. When the malt was judged to be 'hand-dry', that is when the moisture content had fallen to 5–8%, the curing stage was initiated. The airflow might be reduced more, and the air-on temperature was increased further to 'cook' the malt, to add to its character and to reduce the moisture content to its final value. In making traditional British malts, the curing temperatures would typically be in the range 80–105 °C (176–221 °F) for pale malts or 95–105 °C (203–221 °F) for darker malts. In making lager malts, the curing temperature may be as low as 65–70 °C (149–158 °F). During curing, particularly in deep-loaded kilns, up to 75% of air may be recirculated since very little drying is occurring at this stage. Interestingly, in traditional three-floor continental European kilns producing pale lager malts, the alterations in moisture content on the top, middle and bottom floors were about 43–20%, 20–10% and 10–5%, respectively, corresponding roughly to the 'free drying', 'intermediate' and 'bound-water and curing' phases. The temperatures, turning frequencies and times used for making pale and dark malts on traditional, shallow-loading two-floor kilns are shown in Fig. 10.7. In multiple-deck or linked kilns the 'waste' drying power of the air from the

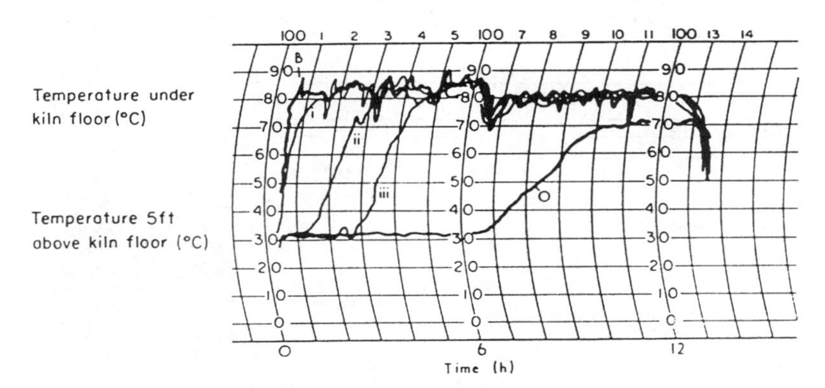

Fig. 10.8 Temperatures at different depths in the bed of malt (i, ii, iii) and above (O) and below (B) the grain bed in a deep-loading Winkler kiln on a 14 h kilning cycle. The curves illustrate the advance of the drying zone upwards through the bed of malt. The 'break' is between 6 and 7 h after the beginning of the cycle (after Fernbach, 1928).

last curing stages of kilning of one batch of malt is used in drying malt at an earlier stage (Chapter 9).

In older kilns, the malt on each floor is turned and is in such a thin layer that it may be regarded as receiving uniform treatment. In deep-loading kilns, in which the bed of malt is not disturbed, the situation is rather different. As the hot air, at a constant temperature, moves into the bed of green malt it begins to dry the lower layers and in the process is cooled as it evaporates moisture. It is then in equilibrium with the rest of the grain bed, which does not, therefore, become drier, nor does its temperature rise. The basal drying zone progressively moves up through the bed of malt, the temperature within the successive malt zones rising as it does so (Fig. 10.8). Until the drying zone reaches the surface of the malt bed the temperature of the air-off does not rise above the initial 'equilibrium' value, and the RH does not fall. In practice it is usual to raise the temperature periodically during kilning in deep-loaded kilns (Fig. 10.9). Air velocities are in the range 9–24 m/min (*ca.* 30–80 ft/min) and the main drying zone is about 20 cm (*ca.* 8 in) deep (Marsh, 1975). The fall in the RH of the air-off, leaving the grain bed, would be expected to occur as the temperature of the air-off begins to rise. This break or break-point is when the drying zone reaches the surface of the malt bed. In practice on a deep-loaded kiln, the RH of the air-off is falling rapidly by the time there is much of a rise in the air-off temperature (Abrahamson, 1978). This is because the temperature of the air is regulated not just by the surface layer of grain but also by the bulk of the grain below. The obvious break-point occurs when the malt in the surface layer has a moisture content of *ca.* 30%. The rise in temperature is usually used as a guide to initiate the curing stages and begin air recirculation, if this is appropriate. However, the decline in the RH of the air-off is a more sensitive indication of the break (Fig. 10.9); as robust, distant-reading hygrometers are becoming increasingly available these are being used to 'inform' the automatic controls of the kiln of the state of the kilning cycle. It should be noted

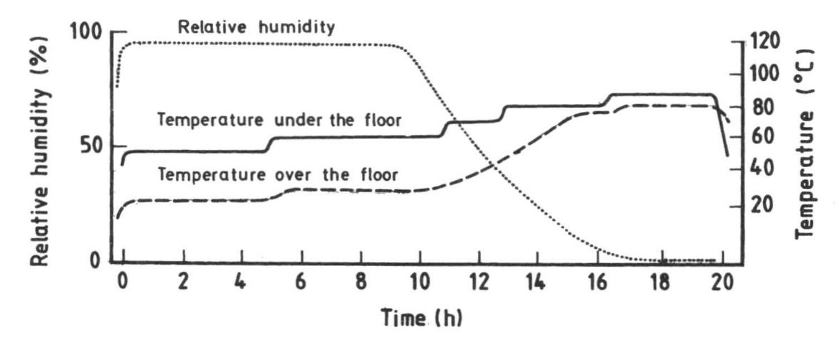

Fig. 10.9 The temperatures of the air under and over a bed of malt and the RH value of the air-off. The (idealized) kilning sequence is for a light malt being kilned on a single-deck, deep-loading kiln (after Scriban and Dancette, 1979).

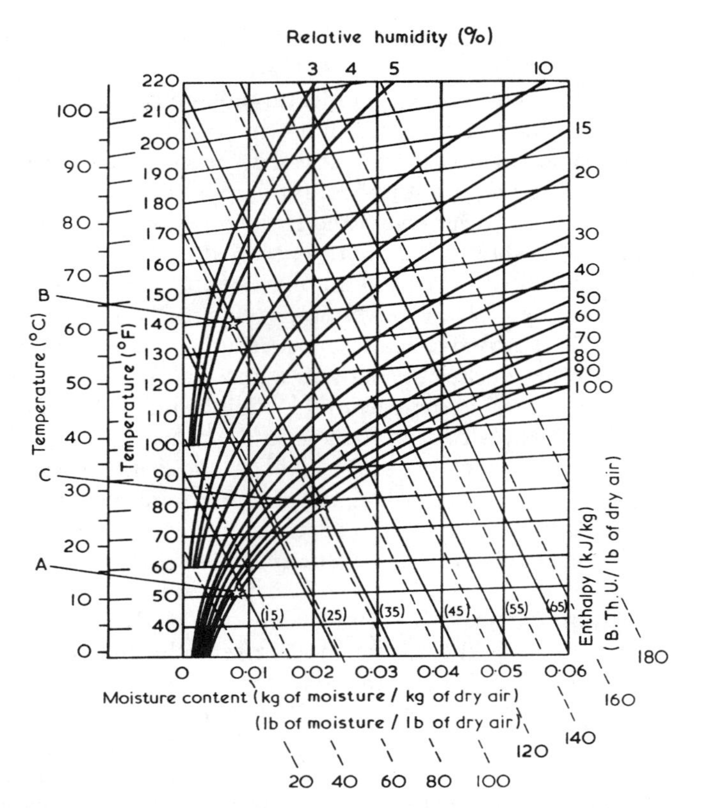

Fig. 10.10 Mollier drying diagram showing the relationships between air temperature, moisture content, RH and heat content (enthalpy; 1 Btu/lb of dry air equals 2.326 kJ/kg of dry air). The temperatures shown are dry-bulb values. To find the wet-bulb temperature for any set of conditions find the intersection of the equivalent enthalpy line with the '100% relative humidity' curve and read off the equivalent dry-bulb temperature. The points A, B and C refer to example 2 in the text. Modified from St Johnston (1954).

that shallow-loaded kilns were intended to treat all the grains of green malt in a batch equally. In a deep-loaded kiln this **cannot** occur, and the grain in the lower layers dries first and receives a longer exposure to the driest and hottest air than the grain in the upper layers (Chapter 14).

The type of malt being produced is controlled by: (i) the quality and degree of modification of the green malt; (ii) the temperatures and times used in drying and curing during kilning; and (iii) the moisture content of the malt at the different stages. In general, the better modified the green malt and the moister it is at the onset of curing and the higher the temperature of curing the greater will be the final colour, flavour and aroma and the smaller will be the final enzyme complement (Chapter 14). Agreed malt moisture contents have risen in recent years to reduce the amount of

Table 10.4 Weight of dry air (kg) corresponding to 1 m³ at the RH values indicated

Temperature °C (°F)	RH (%) 100	75	50	25
0 (32)	1.283	1.286	1.288	1.290
5 (41)	1.258	1.261	1.264	1.267
10 (50)	1.233	1.237	1.240	1.243
15 (59)	1.205	1.208	1.213	1.218
20 (68)	1.177	1.184	1.190	1.197
25 (77)	1.154	1.159	1.164	1.172
30 (86)	1.116	1.128	1.141	1.153
40 (104)	1.046	1.067	1.087	1.107
50 (122)	0.961	0.994	1.026	1.060
60 (140)	0.849	0.899	0.952	1.004
70 (158)	0.711	0.789	0.867	0.946
80 (176)	0.533	0.665	0.763	0.882

After de Clerck (1952).

water to be evaporated and hence the use of fuel and the costs. English traditional pale-ale malts had moisture contents of as low as 1.5%, with mild-ale malts about 2% m.c. and lager malts about 2.5% m.c. However, values of 4–5% m.c. are now common in the UK and values of 6% m.c. may be accepted in Europe. Some enzyme-rich malts may have moisture contents of as much as 8%.

10.7 The removal of moisture from green malt or barley

The amount of water that must be removed when malt is dried is very large indeed. For example, each 1 t 'finished' malt, with culms still present and a moisture content of 5%, contains 950 kg dry matter and 50 kg water. If the malt is loaded onto the kiln with a moisture content of 46%, then it will

Table 10.5 The theoretical air requirements to dry malt from 45% to 4% moisture content, at 60 °C (140 °F) and with air-off RH values of 95% under different ambient conditions

Air entering the kiln			Air leaving the kiln			Air needed (m³/t) for drying the malt
Temperature °C	(°F)	RH (%)	Temperature °C	(°F)	RH (%)	
−20	(−4)	50	21	(69.8)	95	35 421
+6	(42.8)	75	24	(75.2)	95	41 815
+18	(64.4)	85	28	(82.4)	95	49 894

From Narziss (1976).

carry 809 kg water, of which 809 – 50 = 759 kg must be evaporated. If allowance is made for the water removed in drying the original barley, then for each 1 t malt produced, in total, roughly an equal amount of water must be evaporated (Hudson, 1980a). It has been estimated that in the UK it takes a volume of air of *ca.* 10^6 ft^3/t barley steeped (28 317 m^3/t) to dry green malt to the break-point (Marsh, 1975).

If air enters a kiln at 10 °C and has an RH of 75%, its moisture content is 0.006 kg/kg dry air (Fig. 10.10). If the air is heated (without the addition of water) to 65 °C (149 °F) then its RH is reduced to 4%. When this air is fully saturated (RH = 100%) its moisture content is 0.02 kg/kg dry air and it has been cooled to *ca.* 25 °C (77 °F). Thus each kilogram of (dry) air can, at most, remove 0.022 – 0.006 = 0.016 kg water. To remove the 759 kg of moisture will require at least 759 ÷ 0.016 = 47438 kg dry air. The weight of 1 m^3 air at 10 °C (50 °F) and 75% RH is 1.237 kg (Table 10.4). Therefore each tonne of finished malt will have needed 47 438 ÷ 1.237 = 38 349 m^3 (*ca.* 13.5×10^5 ft^3) air to dry it.

Theoretical air requirements for drying malt have been presented (Table 10.5). It is not feasible to maintain an RH of 95% in the air-off with dry malt, but by linking kilns this value can be approached for the entire system. Ambient conditions, both air temperature and RH, have a very large effect on kilning operations. Bearing in mind the large batch sizes of green malt kilned, often equivalent to 200–300 t original barley, and the fact that kilning is not completely efficient, the enormous volumes of air needed are readily appreciated. The fans used must not only drive the large volumes of air needed but must also generate sufficient pressure to drive the air through the bed of grain. The bed depth used and the rate of drying have large effects on the power requirements of the fans (see below).

Estimates of the efficiency of kiln drying may be made with reference to the relationship between the amount of heat generated by burning a given amount of fuel, the amount of water that theoretically this heat can evaporate and a knowledge of the amount of water actually evaporated. In 1978, the thermal efficiency of kilns rarely exceeded 75% (Marsh, 1982), but since then there have been significant improvements. During the early stages of kilning, when the evaporation of water is relatively unrestricted because the equilibrium RH of the malt is high and moisture migrates rapidly to the surface of the grain, it is possible to use a knowledge of the temperatures and RH values of the air entering and leaving a thin bed of grain to calculate the mean in-grain temperature. However, allowance should be made for the true equilibrium RH values which, theoretically, define the conditions at the surface of each grain. Calculations have permitted economies in fuel to be achieved in kilning cycles and they allow an intelligent appreciation of the kilning process and the data available from the instruments used to monitor it.

10.7.1 Mollier drying diagram

Maltsters use a modified psychrometric chart, commonly called a 'Mollier drying diagram' in their drying calculations (Fig. 10.10). Various summaries of psychrometric data are available (Perry *et al.*, 1984; Quayle, 1987). The calculations used by maltsters make certain approximate assumptions. The moisture-containing state of the air is defined by any two of the following measurements: (i) the dry-bulb temperature, measured with an ordinary thermometer exposed to the air; (ii) the wet-bulb temperature, normally determined with a sling hygrometer in which the bulb of a thermometer is covered by a damp, water-saturated cloth from which evaporation is occurring freely; (iii) the absolute humidity, or dew point, of the air; and (iv) the relative humidity. The relationships between these values, together with the heat contents (enthalpies) of the air under different conditions are displayed in the Mollier diagram (Fig. 10.10). Many calculations are based on the assumption that the grain–air–moisture system being investigated in the grain bed is adiabatic, that is the system is in thermal equilibrium and no heat is being added to or subtracted from it. In some cases at least, as in the early stages of kilning when the green malt is generating heat, this assumption is not strictly true (Cole, 1982). The use of the Mollier diagram can most ready be understood by reference to specific examples.

Example 1: wet- and dry-bulb temperatures. If air has a dry-bulb temperature of 85 °F (29.4 °C) and a wet-bulb temperature of 72 °F (22.2 °C), what is the RH (%) and the moisture content (lb/lb dry air) of the air? At 72 °F (22.2 °C), the temperature line on the Mollier diagram intersects the 100% RH curve at a point that defines the enthalpy of the air. Following this line back to its intersection with the 85 °F (29.4 °C) line and reading off the values from the diagram shows that the RH of the air is 50% and the moisture content is 0.014 lb/lb dry air (0.014 kg/kg dry air).

Example 2: a stage in drying green malt. Air enters a kiln at 11 °C (52 °F) saturated with moisture (RH, 100%). From the chart, its moisture content is 0.0080 kg/kg dry air (point A, Fig. 10.10). It is indirectly heated (so its moisture content is unchanged as no moisture is added from the combustion of the fuel) to 60 °C (140 °F). Its enthalpy (heat content) is now 81.4 kJ/kg (35 Btu/lb dry air) and its RH is 6% (point B, Fig. 10.10). The air passes through the bed of green malt and, in so doing, it picks up moisture and is cooled by the latent heat of evaporation of the water. However the grain–air–moisture system is taken as being in thermal equilibrium so the system is adiabatic. Therefore, it is assumed that the enthalpy (heat content) remains unchanged so that as the moisture content and RH of the air rises its state is described by points along the line of equal enthalpy (heat content). This continues until it becomes equilibrated with the moisture in the malt, that is until its RH equals the equilibrium RH of the malt. If the (unrealistic) assumption is made that the vapour pressure of the malt is wholly unrestricted, so that the equilibrium RH is 100%, then the state of the saturated air leaving the bed of green malt is defined by the intersection of the enthalpy line with the 100% RH line (point C, Fig. 10.10). Under these conditions the air has a temperature of 25.5 °C (77.9 °F) and a moisture content of 0.0215 kg/kg dry air. In practice, the equilibrium RH of the green malt is unlikely to exceed 95% and rapidly falls below this figure as drying proceeds (Fig. 10.2). Therefore, at an RH of 80% the temperature of the air will be about 28 °C (82.4 °F) and will carry about 0.02 kg moisture/kg dry air. At an air-off temperature of 28 °C (82.4 °F), the air removes 0.020 – 0.008 = 0.012 kg moisture/kg dry air. As the equilibrium RH of the drying grain falls, and if the air and malt are truly in equilibrium, so the temperature of the air rises and its moisture content declines. In deeply loaded kilns, as the drying zone spreads upwards the successive layers of malt dry; break through and a significant temperature rise of the air-off do not occur until the drying zone reaches the uppermost layer of malt. The assumption that the system is truly adiabatic is incorrect. Under the warm, humid conditions that occur in the bed of green malt at the start of kilning, the grain respires strongly and generates heat. Under these conditions at low airflows, air-off wet-bulb temperatures 2 °C (3.6 °F) above those calculated have been observed (Cole, 1982). Therefore, for more precise calculations allowance should be made for the heat input from the malt. Furthermore, most Mollier diagrams are calculated for standard atmospheric pressure conditions. As atmospheric pressure varies to a substantial extent, allowances should be made for this. In direct-fired kilns or dryers, the moisture that is a product of the combustion of the fuel must be allowed for. At the start of kilning, the kiln structure and the bed of green malt must be warmed up to the initial steady-state temperatures. This heat too, which is not directly concerned with drying the malt, should be allowed for, as should heat losses to the surroundings.

Example 3: drying barley. Barley with a moisture content of 17% has an equilibrium RH of 80–85% (Figs 10.2 and 10.3). Air entering the dryer at 0 °C (32 °F) and saturated with moisture (RH 100%) has a moisture content of 0.004 kg/kg dry air. Radiators heat this air to 49 °C (120 °F), when it has an RH of *ca.* 6% and an enthalpy of about 64 kJ/kg (27.4 Btu/lb dry air). The air passes through the bed of barley and is in equilibrium with it. Thus at the surface of the grain the conditions are defined by the intersection of the enthalpy and 80% RH lines at 23 °C (73.4 °F). The moisture content of this air (23 °C; 80% RH) is 0.014 kg/kg dry air; therefore, 0.014 – 0.004 = 0.010 kg water is removed per kilogram dry air passing through the grain. This is a maximum value that assumes that the moving air comes into true equilibrium with the moisture in the grain. At an in-grain temperature of 23 °C (73.4 °F), barley viability is completely preserved for short periods, so these conditions are suitable for barley drying. If the airflow is increased to such an extent that equilibrium is not even approximately attained, the grain will be at a higher temperature. Under equilibrium conditions, as the moisture content of the barley falls, so does its equilibrium RH (Figs 10.2 and 10.3) and the grain temperature will rise along the line of equal enthalpy, being defined by the intersection of the appropriate RH line (given by the equilibrium RH of the barley at that moisture content) with the line of equal enthalpy. Therefore, if the grain is particularly well dried (to a moisture content of *ca.* 8%) the equilibrium RH will be 24–28% (Figs 10.2 and 10.3), and the in-grain temperature will be around 36 °C (96.8 °F) (Fig. 10.10). For such well-dried, sound grain, this temperature will not reduce the viability appreciably over a period of several weeks.

10.8 Drying on kiln

In drying barley and kilning malt, it is necessary for the air passing through the bed of grain to be at an RH appreciably below the equilibrium RH. To accelerate the drying process and to offset the fall in equilibrium RH that occurs as the malt or barley dries and the fall in the rate of moisture migration, the temperature of the air-on is raised and, when malt is being kilned, the airflow is reduced to allow a nearer approach to equilibrium. As curing takes over from drying, relatively little evaporation of water occurs. It is not possible to reduce the airflow through a bed of malt below a certain level because of the characteristics of the fans used to drive the air, the need to maintain combustion in the furnaces and because the safe, lower practical limit to airflow through a bed of grain to achieve uniform treatment is about 0.9–1.2 m/min (3–4 ft/min). To continually heat fresh air during this curing stage and to discharge it from the kiln hot and nearly dry is wasteful. This air may be used in supplying some of the drying power needed in a

Table 10.6 The effects on fuel consumption of draught control and air recirculation applied to older, direct-fired kilns burning solid fuel; the results are adjusted to a constant ambient temperature of 7 °C (*ca.* 45 °F)

Kiln operating conditions	Fuel used, as equivalent coal consumption +5 lb for loss in ash (lb fuel used/Qr (336 lb) malt kilned)
Kiln used in the traditional way, fuel consumption	125.8
Fuel saved when draught was regulated, being monitored with an anemometer	21.1
Saving by recirculating air during the curing stage	25.1
Average fuel consumption when kiln was used with both draught regulation and air recirculation	75.0

From St Johnston (1954).

Table 10.7 Energy consumption on kilns, based on 'average' fuel consumptions

Kiln type	Fuel consumption (therms/t malt)		
	Ale malt	Average	Lager malt
Traditional kiln, top fan, open furnace	65	–	55–60
Traditional kiln, top fan, Suxe furnace	50	–	45–50
Simple, natural draught kilns	–	40–65	–
Single, draught regulated kiln, exceptionally well run	–	40	–
Traditional European double-deck kiln, exceptionally well run	–	28–30	–
Converted pressure kiln, Suxe furnace	35–40	–	35
Müger pressure kiln, with air recirculation and volume control	28–30	–	25–28
Standard pressure kiln, with recirculation	–	35	–
GKV with air recirculation and volume control	30–32	–	25–30
Single kiln with recirculation and heat exchanger	–	23	–
Two pressure kilns, linked	–	28–30	–
Linked kilns with heat exchanger and heat pump (exceptional)	–	19	–

Hudson (1980a); P. A. Brookes, personal communication.

second, linked kiln or in an upper deck of a multiple-deck kiln. This approach is much more efficient in using fuel.

Alternatively, in single-deck, deep-loading kilns, a proportion of the air leaving the bed of malt during curing may be recirculated, so saving fuel. When the RH of the air-off has fallen to *ca.* 25%, a proportion of the air may be recirculated. As curing proceeds, the grain dries and little further drying is needed, so increasing proportions of air may be recirculated until 75–80% may be being recirculated at the end of kilning. This will allow fuel to be saved, equivalent to as much as 5 therm/t (0.5 GJ/t) (Tables 10.6–10.8); furnaces will be 'turned down' and one or more will be turned off when several serve one kiln. However, introducing air recirculation too soon can lead to an appreciable reduction in the rate of drying and the malt may, therefore, be overheated for its moisture content, leading to an increase in colour and a reduction in the content of active enzymes. Reducing and redirecting airflows efficiently requires well-controlled and maintained motorized dampers, louvres and fans. Very precise pre-programmed control systems, regulated by feedback from sensors in the kiln, save power equivalent to around 8 therm/t (0.8 GJ/t).

With older kilns, the heat wasted with incomplete combustion of the solid fuel was compounded by a lack of draught regulation, inadequate insulation and leaks from the structure. The savings that could be achieved by training staff to control the kiln draught as exactly as possible and by using recirculation were very substantial (Table 10.6). The problems of imperfect staff performance have largely been overcome by automation in modern kilns, with significant results (Tables 10.7 and 10.8). In some old kilns, air

Table 10.8 Estimates of fuel utilization and electrical power consumption by various types of kilns[a]

Type of plant	Electricity		Fuel	
	GJ/t	(therm/t)	GJ/t	(therm/t)
Kiln without recirculation or air-volume control	0.21	(2.0)	4.8	(45.5)
Kiln (16 t batch) with manual control of air recirculation; no volume control	0.38	(3.6)	3.86	(36.6)
GKV (160 t batch); automatic recirculation; radial vane air-volume control; fan fixed speed	0.35	(3.3)	3.10	(29.4)
Kiln attached to tower maltings (batch 206 t barley); heat exchanger, automatic recirculation, variable speed fan (DC thyristor), controlled motor	0.30	(2.8)	2.20	(20.9)
Semicontinuous kilns:				
Air-Fröhlich Triflex	–	–	1.25	(11.9)
Lausmann	0.08	(0.76)	1.80	(17.1)
Seeger	–	–	1.90	(18.0)
Bühler-Miag (double deck)	–	–	2.0	(19.0)

[a] For inter-conversion of heat/power units: 1 therm = 0.105506 GJ.
From Brookes (1988).

from the furnace was not used for direct drying at all but was admitted to the space above the malt to disperse reek, the 'fog' produced when condensation was occurring in the cool air above the malt and which was not removed because of an inadequate draught. This wasteful process has long been discontinued.

10.8.1 Minimizing kilning costs

Some aspects of efficient kilning have already been mentioned. These points will now be integrated with others that relate to the costs of kilning under modern conditions (Marsh, 1975, 1982; Minch, 1978; Ringrose, 1979; Hudson, 1980a,b; Griffiths, 1982; Malkin, 1982; Gluckman and Kidger, 1984; Anon., 1985a; Kidger, 1986a; Brookes, 1988). The main costs of kilning are the initial capital cost of the kiln, maintenance costs and the costs of the fuel and power. Building new, sophisticated and thermally efficient kilns can be expensive, and the savings in running costs must 'pay back' the original investment in a reasonable time (for example, three years). Similarly, modifications to existing kilns must be financially worthwhile. In practice, compromises have to be made between optimizing performance and minimizing cost, and the compromise reached will depend on many factors. The rapidly increasing prices of fuels and electricity have directed attention to reducing kilning costs by savings on these items, and very large capital expenditures have been made to install energy-saving equipment. The use of 'specialized kilns' to replace kilning in GKVs and SGKVs offers various chances for saving fuel as the need to repeatedly warm and dry the GKV or

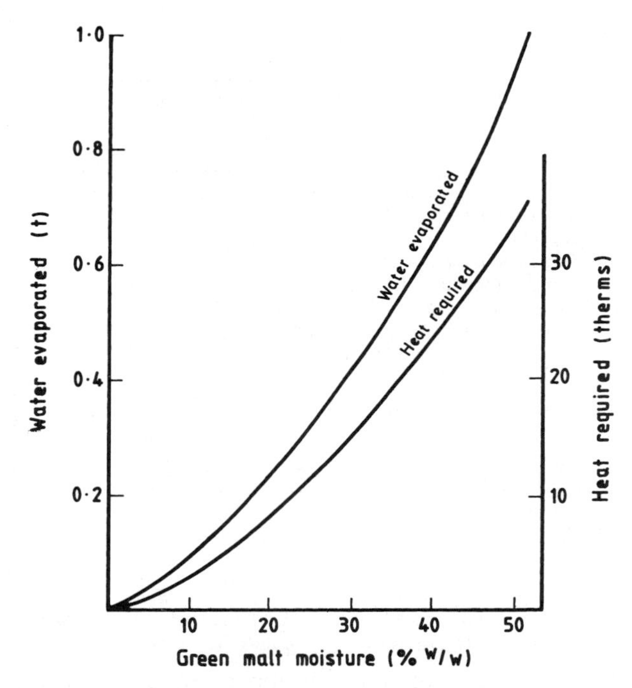

Fig. 10.11 Idealized graph showing the amounts of water to be evaporated and the amount of heat required (1 therm = 100 000 Btu = 105.5 MJ) to produce 1 t finished malt from green malts of different moisture contents. The data are for a 'perfect' kiln, lightly loaded and having no heat losses. (After notes; *Energy Conservation in the Malting Industry*.)

SGKV structure is avoided. This saving alone has been estimated to be as much as 5 therm/t malt kilned. The efficiency of heat recovery from the fuels burned has been optimized with heat recovery from the flue gases of gas-fired furnaces and the use of multiple furnaces that can be successively closed down as the requirement for heat declines with the lower volumes of air needed in the later, curing stages of the kilning process.

Sources of heat, other than the kiln furnaces, have been investigated or are in use. In particular trials in Colorado, more than 40% of the heat needed could be provided by the sun (Smith, 1980; Smith and Maupoux, 1981); 'recovered heat' from the air leaving the kiln is now widely used (see below). In addition, heat from the refrigeration units used to cool the air entering the germination compartments may be directed into the kiln as may waste heat from, for example, gas engines, which are used to generate electricity 'on site' (Stevens, 1993; Booth, 1995). Other suggestions, such as chemically pre-drying the air entering the kiln with a honeycomb-like structure packed with lithium chloride (to improve its drying power) have not been accepted. Fuel consumption can be reduced by reducing the amount of water to be evaporated from the green malt. This may be achieved by (i) reducing the moisture content of the green malt being loaded on the kiln (Fig. 10.11); and (ii) accepting finished malts at higher moisture contents. In recent years, there has been a tendency to germinate malt at higher moisture contents (45–47% or even more) than were once usual (42–45%) to ensure that the grain modifies rapidly. Preliminary drying (withering) can be achieved by turning off the humidification equipment before the end of the germination period. Claims have been made that using the squeeze-malting process (Chapter 14) adequately modified green malt can be obtained using 35–40% moisture contents. In many regions, customers now accept finished malts with higher moisture contents than they once would have done. For example, pre-1939, ale malts in the UK would have had 'target' moisture contents of about 1.5–2.0%. In the early 1970s, the values had increased to 2–2.5%, then to 3–3.5% in the 1980s and now in the late-1990s they are often 4–4.5%. For comparison, many North American malts have moisture contents of 3.8–4.1%; European lager malts are often at 5–6% and some high-DP malts have moisture contents of around 8%. Trials suggest that some distillers' malts could be used at 10% moisture and, in the limit, some distilleries use green (unkilned) malt in their mashes. Coupled with the rise in final malt moisture contents has been the acceptance of lower curing temperatures. At least in 'simple' kilns, curing is 'wasteful' in the sense that much of the drying power of the hot air is wasted. While in the past British ale malts were finished at temperatures of 105 °C (221 °F) or even more, in many instances 85–100 °C (185–212 °F) are now used. For standard and Pilsen lager malts, finishing temperatures may be about 75 °C (167 °F) and 60 °C (140 °F), respectively.

For high thermal efficiency, kilns should be compact, well insulated, leak-free and have a low thermal capacity to minimize the heat needed to warm the structure. Therefore, well-insulated, lightweight stainless steel or corrosion-resistant steel structures seem ideal and preferable to comparatively massive concrete or brick buildings. GKVs and SGKVs are disadvantaged in this sense, since with each production cycle the structure must be reheated and dried. A kiln should be operated as near-continuously as possible and with a rapid turn-around (rapid stripping and loading) to 'save' the heat in the structure. Other benefits also follow. It has been said that if a kiln working on a 24 h cycle has a turn-round time of 4–5 h, instead of 1 h, because of the shorter 'working period', the cost of the electricity driving the fan can double (Marsh, 1975). To understand this some fan characteristics need to be outlined.

The power used to drive the necessary amount of air through a bed of grain rises disproportionately as the loading of the grain bed (the depth) is increased and as the duration of the kilning cycle is reduced (Fig. 10.12 and 10.13). Furthermore, if the grain bed is compacted, for example by loading with a thrower instead of a carefully designed screw loader, which loads it

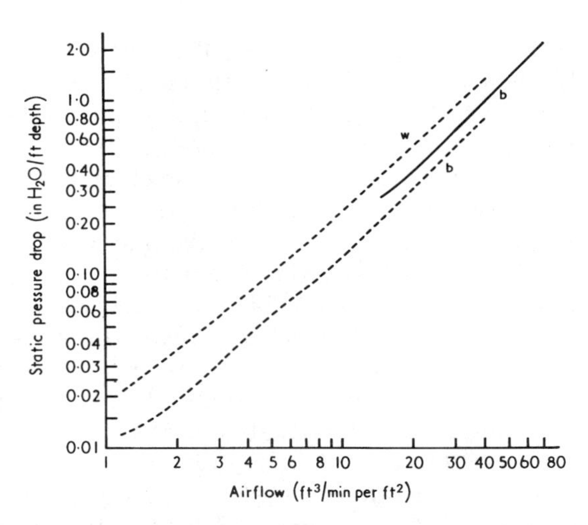

Fig. 10.12 The relationship between airflow and pressure drop through beds of barley (b) and wheat (w). The relationships are altered by compaction of the grain bed and differences in moisture content (after various sources).

very evenly and in an uncompacted state, the power needed to drive the air through the bed is substantially increased (Fig. 10.13). In energy equivalents, 'light' loading can save 2 therm/t malt kilned. Ideally, fuel consumption should depend mainly on the amount of water to be removed from the malt and should be independent of the kilning cycle. Clearly this is not the case with the electrical power used to drive the fan(s). Fan power usage is least for lightly loaded kilns with a shallow bed of malt (low floor loading) and an extended kilning cycle (Table 10.9). However operating constraints dictate the duration of kilning cycles (cycles of 18, 24, 36 and 48 h are common) and shallower beds mean larger kilns, which, therefore, are less compact and are more costly to build. Inevitably, various compromises are reached; recent choices have been to use kiln floor loadings of 350–500 kg (original barley)/m^2 giving bed depths of green malt of 0.85–1.2 m (2.79–3.94 ft) (Gibson, 1989). As the malt dries the bed shrinks and the resistance to airflow declines. To take advantage of this phenomenon, it has been proposed that kilns be loaded with successive layers of green malt, each new layer being added just before the drying zone breaks through the layer below, so that resistance to airflow – and hence fan power consumption – is minimized. To minimize the use of fan power, kiln structures should provide a minimum resistance to airflow. Ducts should be wide and short, with carefully designed low-resistance bends and should be heavily lagged (e.g. with mineral or glass wool) to reduce heat losses. Kiln decks should be of wedge-wire or other easily cleaned material having a large 'free space', and consequent low resistance to airflow, and the designs of heat exchangers should be carefully chosen to minimize resistance to airflow. Airflows

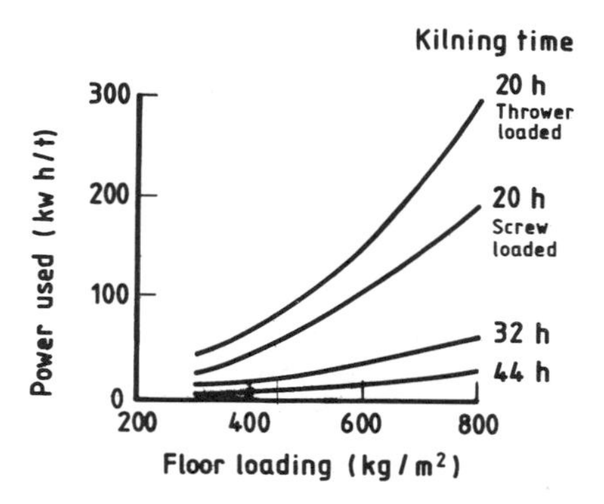

Fig. 10.13 The relationships between the specific floor loading of the green malt and the fan power used in entire kilning cycles lasting 20, 32 and 44 h. It can be seen that a lightly loaded bed (screw loaded) requires much less fan power than a compacted, thrower-loaded bed (after Hudson, 1980a).

delivered by kiln fans must be variable and ideally should be under automatic control. When not required, fans should be switched off. To reduce airflows, one of a set of fans may be turned off. Airflows will also need to be varied in steep and germination ventilation systems. The 'fan laws' are (i) that the airflow generated is proportional to the fan speed; (ii) that the pressure generated is proportional to the fan speed squared; and (iii) that the power used to drive the fan is proportional to the cube of the fan speed. For economy, fans must be of an adequate size and not be over-powered by an excessively large electric motor. Fans may be fixed speed and powered by an AC electric motor. With these, the air speed is varied by inlet guide-vanes or devices (dampers) that throttle the airflow or by controlling the pitch of the fan blades (with axial fans). Such systems are cheap to install but are inefficient. Variable-speed fans powered by variable-speed AC motors with inverter controls or variable-speed, DC electric motors controlled by thyristors are used.

A development of recent years has been the use of devices to 'recover' heat from the hot air leaving the kiln, (the exit air) and deliver it to the kiln's inlet air (Chapman and Walker, 1979; Ringrose, 1979; Scriban and

Table 10.9 Consumption of electricity by fans driving the airflow on malt kilns at two loading levels and two cycle times

Loading (kg/m²)	Kilning time (h)	Electrical power consumption (kWh/t)
500	24	70
500	48	20
800	24	200
800	48	50

From Brookes (1988).

Dancette, 1979; Hudson, 1980a; Malkin, 1982; Marsh, 1982; Watson, 1987). Before 1976, such devices were not used in the UK. By 1980, around 20% of UK kilns were equipped, saving around 7 therm/t. Now all new kilns and many older kilns are so equipped. The simpler devices have been spectacularly successful in saving fuel. The economics and reliability of more complex systems are disputed. Heat is most easily recovered from the exit air when it is nearly saturated with water vapour. This is the period before the break in single-deck kilns, in the period before the break on the drying deck in double-deck and linked kilns, and can be continuous in continuous kilns. With single-deck kilns equipped with heat-recovery units, it is usual to begin recirculating a proportion of the air after the humidity break. The greater proportion of the time the air is saturated, the more efficiently the drying power of the kiln is being utilized and the easier it is to recover heat from the exit air since its RH is high and, as it is cooled, the moisture will condense on the cooling surfaces of the heat-recovery unit and give up latent heat. Furthermore, the larger the proportion of available time that the heat-recovery unit is in use the more fuel is saved and the faster the rate of payback of the capital cost. Unfortunately, the inclusion of radiators in the kiln inlet airstream and cooling systems in the exit gases provide resistances to airflows and create the need for more fan power.

Various devices have been proposed for heat recovery. Rotating heat rings or regenerative heat exchangers have not found favour for use in kilns, nor have 'heat pipes': sealed tubes filled with refrigerant that evaporates in the warm exit air, so taking up heat, and condenses in the cool inlet air, releasing the heat. The devices used are: (i) passive air–air heat exchangers; (ii) air–liquid–air heat run-around exchangers; and (iii) active heat pump systems of varying degrees of complexity. The use of air-to-air heat exchangers in kilning was proposed by Galland in 1874 for use in his 'wonder-kiln' (Wahl, 1944) (Fig. 10.14). The first practical devices were made of sheets of plate glass, which separated the outgoing and incoming airstreams, but allowed the transfer of heat between them. Now most devices are made of very large numbers of borosilicate glass or stainless steel tubes. The tubes are mounted vertically and the incoming air moves through them and is pre-warmed on the way to the kiln inlet. The wet and warm kiln exit air moves across the tubes and is cooled. Condensation occurs and the condensed liquid runs down the tubes, helping to keep them clean. Even so, at intervals the tubes must be hosed down to remove dust and organic deposits that reduce their heat-exchange efficiency. The tubes are not equipped with fins, to enlarge the heat-exchange surfaces, because of the problems this would create for cleaning. Such heat exchangers are large. For example, a glass-tube heat exchanger for a comparatively small 40 t kiln measured 13.5 × 2.7 × 2.45 m (44.3 × 8.86 × 8.04 ft) and contained 25 000 tubes of *ca.* 1.9 cm (0.75 in) diameter (Hudson, 1980a). The tubes were of borosilicate glass in a stainless steel frame, sealed in with a silicone

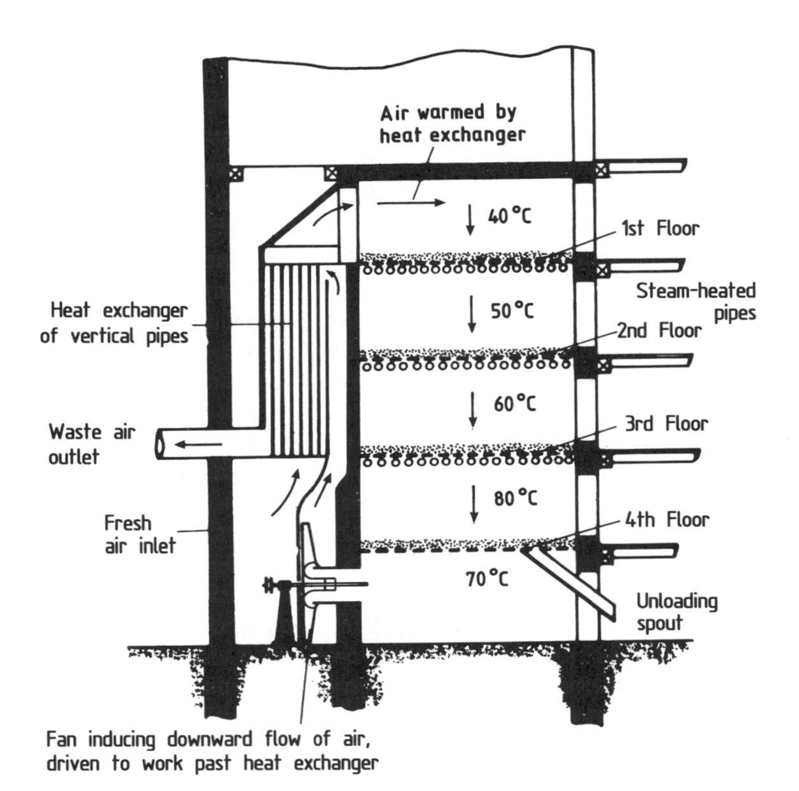

Fig. 10.14 A four-floor kiln with a fan-driven airflow and an air–air heat exchanger proposed by N. Galland in 1874. It was referred to as a 'wonderkiln' by some writers but may never have been built (after Wahl, 1944).

rubber. Such devices are simple, contain no working parts and require minimal maintenance. However, they require extensive ductwork to direct the inlet and exit gases to cross and so are not always easy to install retrospectively in old kilns. Air–liquid–air, or run-around heat exchangers are easier to install in older kilns, although they require more maintenance and use electrical power. In these devices, liquid (usually water containing glycol as an anti-freezing agent and pressurized from a header tank) is driven around by an electric pump through heavily insulated tubes. In the kiln air exit, the liquid is warmed as it passes through a finned heat exchanger. This must be made of corrosion-resistant material such as stainless steel. (Copper or aluminium heat exchangers, even though coated with PVC or an epoxide layer, are rapidly corroded in gases containing sulphur dioxide.) The warmed liquid is circulated and gives up its heat to the cold air entering the kiln via another heat exchanger/radiator. This second heat exchanger need not be made of such expensive, corrosion-resistant materials. Such 'passive' devices can reduce fuel costs by 25% or even 30% and they have attractive payback times. How a heat-recovery unit operates on a linked (effectively double-deck) kiln is illustrated by the following example (Hudson, 1980b). Ambient air enters the kiln at 11 °C (51.8 °F) and with a RH of 80%. On moving through the inlet heat exchanger it is warmed to 26 °C (78.8 °F) and its RH falls to 32%. Then, on moving through the direct-fired furnace, it is heated further to 80 °C (176 °F) and its RH declines to 3%. The air then moves through the bed of malt being cured. The limited evaporation that occurs cools the air to 75 °C (167 °F) and raises its RH to 5%. This air is now mixed with more pre-heated air (26 °C (78.8 °F), 32% RH) and this larger volume of air moves through a second, direct-fired furnace to be warmed to 70 °C (158 °F) and to achieve an RH of 5%. This air now passes through the drying bed of grain and is substantially cooled and wetted (30 °C (86 °F), approximately 100% RH). This warm, wet air now passes out over the heat exchanger where condensation occurs. The air emerges saturated with water vapour and cooled to 27 °C (80.6 °F). The heat liberated warms the incoming air, as originally noted, from 11 °C (51.8 °F) 80% RH to 26 °C (78.8 °F) 32% RH.

'Active' heat-recovery systems use heat pumps to transfer heat from the kiln exit gases to the inlet air (Marsh, 1975, 1982; Parsons and Marsh, 1981; Aznavorian, 1983; Chapman, 1983; Gluckman and Kidger, 1984; Watson, 1987). In the simplest cases, the refrigerant is driven by electrical pumps. In practice, the heat is usually conveyed to and from the heat-pump system by secondary water circuits (Fig. 10.15), but in principle these are not needed. A heat pump 'takes in' heat from the kiln exit air by evaporating refrigerant in a heat exchanger. The heat exchanger is cooled by the latent heat of evaporation. The refrigerant is pumped around and is condensed under pressure in a second heat exchanger or radiator, where the latent heat of condensation is given out into the air entering the kiln. Electrical power is consumed in substantial amounts, but the heat pump has the unique ability to 'collect' heat from a low-temperature source and release it at a higher temperature, so that, in this sense, the heat is upgraded. For electrically driven heat pumps, it is reported that 3.6–4.0 units of heat are recovered per unit of input (pumping) energy. It is only economic to install heat pumps in tandem with passive heat-recovery units. It is claimed that they reduce fuel usage by a further 25–30%. Simple heat pumps (electrically powered) need to be custom-designed and have a series or cascade of heat-exchange units (often four) each with its own refrigerant working in its optimal temperature range (Watson, 1987). Often each refrigerant circuit has two compressors. A schematic diagram of a heat-recovery system used on a double-deck kiln (based on a double-deck barley drier) is shown in Fig. 10.15. Electrically driven heat pumps have been applied to different patterns of kilns (e.g. Watson, 1987).

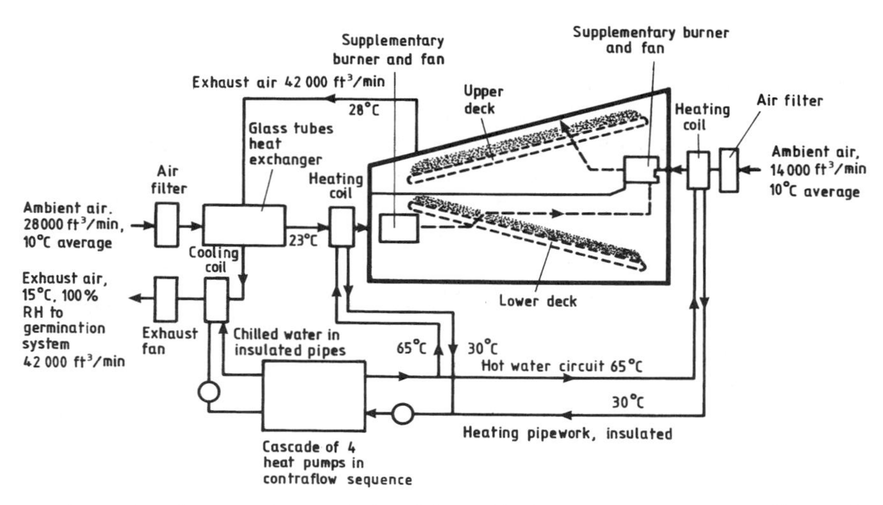

Fig. 10.15 Flow diagram of a two-decked kiln and its heat-recovery system working under average conditions. Ambient air 10 °C (50 °F) enters at 28 000 ft³/min and after filtration passes through a glass tube, cross-flow heat exchanger that warms it to 23 °C (73.4 °F). This air is further warmed by a water-heated heat exchanger. The water is heated by the heat pumps. This air passes, via a supplementary heater and fan, through the lower deck on which malt is being cured. It then mixes with a second inflow of air (14 000 ft³/min (approximately 396 m³/min)), which is also pre-heated, and the combined airflow passes through a fan and supplementary heater before moving through the upper bed of drying malt. The airflow (42 000 ft³/min (approximately 1189 m³/min)) emerges at 28 °C (82.4 °F) and is nearly saturated with moisture. It passes successively through the glass-tube heat exchanger and the cooling coils, chilled with water, connected to the heat pumps. The heat pumps, which run continually, provide water at 65 °C (149 °F) for warming the incoming airstreams and water at 12 °C (53.6 °F) for cooling the outgoing air. The exit air, exhausted through a fan, is saturated with water and is at 15 °C (59 °F). This conditioned air is directed to the germination units to cool the growing grain, resulting in savings on refrigeration costs for the germination units. This kiln produces 22.35 t of finished malt every 12 h. This kiln is based on an Alvan Blanch drier (Fig. 8.28). (After Watson, 1987).

The high cost of electricity has resulted in some maltsters installing their own generator units. It seemed an attractive idea to use reciprocating gas engines to drive kiln fans, to drive heat pumps and to generate electricity. The 'waste' heat from the cooling jacket of the engine and in the exhaust gases has been used to pre-warm the kiln air (Parsons and Marsh, 1981; Marsh, 1982; Suhr, 1984; Schu, 1984). At least in the UK, these complex but potentially highly efficient systems have fallen out of use. It seems that the reliability of the systems was low and this made them uneconomic. By comparison, the use of simpler combined heat and electrical power-generating units can be economic (West and Sleight, 1994), the electricity being available throughout the maltings, the heat being used to pre-heat the kiln inlet air.

10.8.2 The comparative thermal efficiencies of kilns

Estimated fuel and power consumptions of various types of kiln are summarized in Tables 10.7, 10.8 and 10.9. The data show impressive improvements in kiln efficiencies. Unfortunately data for truly continuous kilns (e.g. as used in the Saturne maltings) are not available. Extremely high efficiencies are claimed for new patterns of (semi) continuous kilns (Table 10.8). Apparently these novel and complex kilns have not been tested in the UK, but several installations have been made in mainland Europe (Anon., 1985a; Schlimme, 1985; Schuster, 1985; Grandegger, 1991; Grandegger and Traber, 1991). In one design, the air, heated by using it to cool the finished malt, is returned together with the air from the curing malt to an earlier temperature-increasing section. This is preceded by two low-temperature withering zones. Air leaving the unit is discharged via a heat exchanger and heat pump (Schlimme, 1985).

11 Experimental malting

11.1 Introduction

Experimental malting is an essential tool for maltsters and plant breeders (Isebaeart, 1959; Meredith, *et al.*, 1962). Breeders are concerned with deciding, using closely regulated comparative trials, the relative malting values of their various varieties and hybrids. In consequence, they must evaluate large numbers of grain samples in exactly the same way and the equipment they use reflects this. The samples may be quite small (as little as 5 g), particularly in the early stages of a breeding programme, and so small-scale and rapid analytical methods may be used which are regarded as being too imprecise for use by commercial maltsters. There is no uniformity in the conditions used in small-scale malting trials to compare different genotypes, so it is important to note that trials made in different ways generally 'rank' barleys in the same order in terms of their malting qualities.

Maltsters often wish to compare different batches of barley, differing in genotype, provenance, growth conditions, storage history and so on. For these purposes, uniform small-scale malting conditions are used. Usually larger amounts of grain are available to maltsters than are available to breeders. In addition to direct 'comparative trials' maltsters frequently use micromalting to test the feasibility or desirability of new production methods or programmes. To perform these investigations, it is essential to have equipment that is flexible, i.e. that may be used in different ways. It is necessary to be able to treat samples in a range of different ways in trials that are carried out in parallel. Some processing variables that have been investigated include steeping temperatures, steeping times, the use of water changes, aeration during immersion, and air-rests, the use of additives (including growth regulators and anti-microbial agents), spray-steeping, resteeping, variations in steeped-grain moisture content, variations in germination time and temperature, the ambient relative humidity, the frequency of turning, the extent of aeration and the influence of various gaseous atmospheres. Small-scale kilning trials have been less successful and have mostly been concerned with producing malts matching the production-scale materials.

Small-scale micromalting trials may be carried out on very small samples of grain (5–100 g) and the numbers of samples processed at any one time may be large, often up to 100. This approach has the advantage that well-designed experiments using replicated samples and various sampling times and ranges of other variables can be carried out. For example, the trials may be arranged in a 'factorial' way or in a 'latin square' design (Bishop and Spaul, 1959) and the results may be assessed using statistical techniques.

In addition, many trials are made using barley samples of about 1 kg (2.205 lb). These samples are more difficult to handle and often the equipment used is such that only 8–12 samples can be handled at any one time. Using larger samples reduces the chances of sample replication but it also reduces the uncertainties caused by sampling errors inherent in using small batches of grain and it provides more material for analysis. Larger-scale, 'pilot' malting plant, for example using 50 kg to 1 t grain, may be needed to provide conditions more nearly like those encountered in the production plant and to provide sufficient experimental material for particular purposes, such as small-scale brewing trials. The problems associated with scaling-up processes developed on the small scale to the production scale should be recognized. In particular in experimental malting, grain will be well mixed and uniform, each batch is quickly moved, (e.g. between the steeps and germination units, and between these and the kiln) and the temperature of the steep and the germinating grains is closely controlled. Also conditions for the grains in the kiln will be relatively uniform. These conditions are usually less closely controlled in production malting. The gradients of conditions that occur in the grain masses in production steeps, germination compartments and kilns are virtually absent in small-scale malting equipment.

11.2 General considerations

In some of the first reports of small-scale malting trials, it was recognized that the conditions used were superior to those that occurred in commercial floor-malting (Luff, 1898, 1899; Bleisch, 1899; Frew, 1901). This realization led to attempts to improve full-scale malting, for example by achieving better control of the temperature of germinating grain and having cooler germination temperatures (Luff, 1900).

The early experimenters recognized the problems that followed from the failure of small-scale malting to mimic exactly commercial malting conditions. They recognized that the differences should be minimized, but that for comparative trials (e.g. between different barley varieties) it was more important to have the experimental conditions exactly controlled and perfectly reproducible than to have them mimic the variable commercial conditions precisely.

It is still true that micromalting conditions never reproduce those found in industrial plant exactly, and it is informative to consider why this is so. The barley used in small-scale malting trials is usually well mixed, screened and sampled with sample dividers. Consequently, differences between the moisture contents and the malting performances of sub-samples are very small indeed when the experimental conditions are correctly controlled. On the small scale dry barley (12% m.c. or less) can be maintained for at least 2 years in, apparently, an unaltered physiological state (degree of maturity, level of dormancy) by storage in closed, airtight and moisture-proof containers at −18 °C (0 °F). Furthermore, it is easy to break dormancy by drying the grain to 7–8% m.c. in a rapid airflow at 40–43 °C (104–109.4 °F) and then storing, in a closed container, at 40 °C (104 °F) for a suitable period. In small-scale malting, handling should be carried out in such a way that there are no grain losses. As small amounts of grain may easily be weighed with an accuracy of better than ±0.1%, the malting losses associated with different sets of malting conditions may be accurately determined.

The control of process times and process temperatures and the uniformity of processing conditions used in small-scale malting are superior to those found in commercial malting. For example, on the small scale, adding grain to steep liquor and steep draining are very rapid procedures that can be timed closely. Each process will take a few seconds; both the grain and the water can be pre-attemperated and all the grains will be treated in nearly identical ways. In contrast, the steps of filling a steep, draining it and emptying the grain from it into a germination compartment on the commercial scale may take extended times of 1 h or more. Under these conditions and with grain bed depths of, for example, 2 m, timing is inexact and it is impossible to achieve equal treatments for all the grains in a batch. Similar conditions apply to commercially germinated and kilned malt. It is inevitable that grain that has been malted on the small scale will produce a more homogeneous malt than if it was malted on the large scale.

In many types of micromalting steeping equipment, the ratio of water to grain is unrealistically large (compared with production malting) and so the grain is unusually well washed and aerated. This is particularly clear in the equipment described by Römheld (1897) in which grain held in a perforated metal container was suspended in a large volume of attemperated water.

In modern micromaltings, it is desirable to be able to steep grain in water at an accurately controlled temperature, to be able to aerate it in the steep and to air-rest it with low ventilation. It is sometimes desirable to steep it under anaerobic conditions in 'degassed' water. Steeping periods are best chosen with reference to water-uptake curves, but the different ways in which these are determined influence the results obtained. In some methods in which samples are centrifuged or blotted, the weights found are in the absence of the surface film of water, while methods in which samples are only drained before weighing include this substantial amount of water. Furthermore, the aeration consequent on the repeated removal of the grain from water, to allow it to be weighed, can cause it to begin to metabolize rapidly, and even to chit, with consequent alterations in the water-uptake rate. While barley will withstand gentle centrifugation to remove the surface water, triticale, wheat and rye will not (Blanchflower and Briggs, 1991a,b,c).

It is, of course, possible to take known weights of barley (of a known moisture content), steep them for various periods, drain, blot or centrifuge, then determine the wet weight before discarding them. The moisture content can be calculated approximately from the increase in weight (assuming there is no appreciable loss in dry matter) or it can be measured by 'double-drying' to obtain the true dry weight. While accurate, this latter approach is laborious and uses large amounts of grain. We have routinely measured the increase in the fresh weight of well-screened grain by rapidly removing samples from the steep and returning them after weighing. At the end of the steeping period, the final, calculated moisture content may be checked by double-drying the sample. As the grain samples are held in nylon mesh bags or in woven-wire metal baskets, the weights after draining are corrected for the wet weights of the grain containers. When making comparative micromalting trials, it is essential that all grain samples have the same amount of surface moisture. We achieve this by draining under standard conditions for exactly 1 h before transferring to the germination unit.

Heat is readily lost from or gained by small samples of germinating grain and so in most micromalting equipment the grain temperature remains essentially the same as that of the surroundings, or rises (in some pieces of well-insulated equipment) to 1–2 °C (1.8–3.6 °F) above the ambient temperature. In some micromalting devices, maintaining a water-saturated, attemperated atmosphere in the germination compartment has been a problem, while in others it has been necessary to use baffles to prevent water-droplets from the humidification unit, which are entrained in the airstream, from reaching the grain. Sometimes the conditioned air circulates around or over the germinating grain, while in other micromaltings it is driven through the grain, as in large-scale pneumatic maltings. In this last case it is impossible to ensure that even airflows move through different batches of grain at the same rate when the air is supplied to the grain containers from one pressurized plenum chamber. Often, in pneumatic micromalting plant, the airflow through the grain is disproportionately high relative to production-scale units.

A major disadvantage with some micromalting systems is that the grain must be turned by hand, usually 2 or 3 times each day, throughout the germination period. While this can work well, it is extremely time consuming. In newer equipment, turning is usually controlled automatically, but it should be noted that excessively vigorous or prolonged periods of mixing can check rootlet growth and lead to abnormal germination (Sipi and Briggs, 1968; Ponnet, 1969).

In many respects, small-scale kilning, while it can be made extremely reliable and reproducible, does not closely resemble the commercial processes. Ovens or vacuum ovens are not suitable for drying experimental malts. The lack of forced evaporative cooling leads to stewing and enzyme destruction.

The green malt should be dried with a flow of air moving through it. Early small-scale kilns were based on the natural-draught principle and often only dried one sample of green malt at a time. However, it was realized around 1900 that while commercial kilning, in shallow-loaded grain beds, took up to 3–4 days the process could easily be completed in *ca.* 8 h in the laboratory. Usually, it is not possible in small experimental kilns to mimic the range of conditions experienced by a large batch of grain in a modern, deep-loaded kiln; in the latter, the grain in the base of the bed is rapidly dried while that in the upper layers is exposed for various periods to warm air saturated with water vapour and so, in the initial stages, modification can proceed in this area. A further complication is that if numerous samples are dried on one experimental kiln and these samples are added to the kiln at different times (e.g. after different periods of germination), it is impractical to vary the kiln temperature or airflow for each batch of grain. Our solution has been to run a kiln under constant conditions, supplying a rapid airflow at 43 °C (109.4 °F). Under these conditions, the green malt dries rapidly to a moisture content of *ca.* 7% and the activities of the amylase enzymes are largely preserved. The malt is stable in this state and may be stored to be cured later. On the experimental scale, green malt may be stored, frozen or may be freeze-dried, with maximal retention of many enzyme activities. Obviously this option is not available on the large scale. In our micromalting system, if a malt is to be cured it is kilned further in a second kiln in which the air can be recirculated and the air-on temperature increased (e.g. to 65 °C (149 °F)). Such malts have moisture contents of 4.5–5.0% and closely resemble British premium pale-lager malts (white malts). In attempts to make coloured malts, our limited experiments with a small, commercial coffee roaster were unsatisfactory in that, although coloured malts were obtained, we failed to obtain reproducible results.

Finished malts may be de-rooted by agitation and sieving. The weights of the malts and rootlets are then determined. The malt samples should be stored in carefully labelled, airtight and waterproof containers, ready for analysis.

There has been no standardization of small-scale malting equipment. Many designs have been described and others, which have not been described in writing, have been used. The most flexible and least costly small-scale units, which are capable of processing large numbers of samples (e.g. 50 or more), are those in which many of the operations are carried out by hand. The extreme demands this can make on labour has meant that increasingly micromalting plants are being automated and instrumented. If many hundreds of samples are to be processed in exactly the same way to assess their relative 'malting values', as plant breeders need to do, then automation can be relatively simple. Several devices have been described that allow about 100 samples to be processed in parallel. However, for maltsters who wish to carry out trials to optimize processing conditions, the situation is much more complex since they will need to vary the duration, water-change frequency, temperature, aeration and ventilation parameters in steeping, possibly alter or experiment with the use of additives, and alter the germination and/or kilning conditions. For such trials to be carried out simultaneously, in parallel, requires each 'process stream' to be 'programmed' individually. Such extreme flexibility can be provided by a skilled operator. It is, however, usually too expensive to automate independently numerous process streams; as a result, the 'automated', small-scale plants used by maltsters represent compromises in complexity. Doubtless, as automatic process control becomes more commonplace, maltsters will obtain increasingly sophisticated and flexible micromalting equipment.

11.3 Types of small-scale malting equipment

No doubt some maltsters have always carried out 'comparative trials' on their malting floors. However, it was widely recognized by the late 1800s that the control of the growth conditions in old-fashioned floor-maltings was inadequate and that, in consequence, comparative trials carried out on different occasions were rarely reliable. Beavan (1902) suggested that samples of trial barleys, 'contained in suitable receptacles', could be processed embedded in a larger piece of grain so that, in effect, the small samples would be conditioned and grow as a section of the large piece. Thus valid, comparative small-scale malting trials could be performed. Beavan noted that the analytical assessments of the malt made from the bulk of grain agreed well with those of small samples of the grain malted in contact with it in the receptacles. It seems that the receptacles initially used were of woven wire or perforated metal, but later containers of loosely woven cloth or netting containers were widely used and the technique came to be called bag – or stocking – malting in consequence. The technique is still sometimes used (often with nylon net sacks) in modern maltings, but the samples must be withdrawn from the piece when the steep is being loaded or stripped, and when the germinating grain is being turned. The technique was described in detail by Hind and Lancaster (1932), who used it in an extensive barley evaluation programme. Samples carefully drawn from a bulk of barley were dried to a moisture content of *ca.* 11% in a room held at 80 °F (26.7 °C) to reduce dormancy. Samples of barley (e.g. 1 kg (2.2 lb), exactly weighed) were malted at weekly intervals and the three finished samples were weighed and mixed and the mixture was analysed. Samples of about 0.45–1.0 kg (1.0–2.2 lb) were widely used by others in similar trials. All samples were loosely held in open-weave bags of the same pattern, with room to allow the grain to grow and be mixed. Mixing was carried out by hand morning and evening. All samples were steeped, with a bulk, in a cistern at about 50 °F (10 °C) and were floored together, buried in a piece that was thickened or thinned to control the temperature in the usual way. The temperatures achieved were usually 57–63 °F (13.9–17.2 °C). After kilning, embedded in the piece and not in contact with the kiln deck, the finished stocking-malts were screened to remove rootlets and each finished malt was weighed and stored in an airtight and moisture-proof container, ready for analysis.

Laboratory micromalting, by convention with samples often of between 5 g and 1 kg (0.176 oz and 2.2 lb), is now widely used. Sometimes the equipment in use has been described but many maltings run their own unique micromalting type of equipment. The methodologies used are very numerous. Only a few will be mentioned here.

Possibly the first micromalting unit (made before 1880 by a Dr Gruber) was a hexagonal wooden drum in which the germinating grain could be viewed through sliding glass windows. This device was turned by a hand crank (Aubrey, 1881). Ulsch (1895) germinated 500 g samples of barley in a short metal cylinder, fitted with an internal perforated deck and a cover containing a thermometer. Air gained access through the lower part. The cylinder was placed in a room at 12–14 °R (15–17.5 °C, 59–63.5 °F). Uniform kilning conditions were obtained using a small natural draught kiln in which air entered the hot air chamber via metal tubes running through a water bath heated by a bunsen burner regulated by a thermostat.

A small pneumatic system was described by Römheld (1897). Barley (2.5 kg (5.51 lb)) was steeped in a perforated metal container immersed in a large volume of water. Water uptake was followed by weighing the contained plus the grain. The grain was germinated in a vertical metal cylindrical container. It was turned with helical screws powered by a hand-crank, which rotated around the central axis. Forced ventilation was by sucking air down through the grain bed. The grain was dried on a gas-fired kiln. Luff (1898) attempted to achieve better control of the small-scale malting conditions, steeping at 10–11 °C (50–51.8 °F) and germinating in vessels with heavily insulated walls, held either on a malting floor surrounded by a piece or surrounded by a constant temperature, water-saturated atmosphere. The expected rise of in-grain temperature occurred but to a lesser extent and with an earlier decline than was found in the surrounding large-scale piece. Samples were kilned in air, heated by passage through metal tubes immersed in a gas-heated thermostatically controlled oil bath, which also heated the base of the hot air chamber. Interestingly Luff based his calculations on the initial dry weight of the barley used. Bleisch (1899) expressed doubts about the value of micromalting pointing out that, for some purposes at least, barleys should be equally 'storage-ripe' for valid comparisons to be made between them. He took the view that micromalting should mimic production conditions exactly and this it failed to do since, for instance, in production in-grain temperatures of 16–17 °R (20–21.3 °C, 68–70.3 °F) were reached while in the laboratory temperatures of only 12–13 °R (15–17.3 °C,

59–63.1 °F) were obtained. Malting losses were higher in samples made on the production scale. In reply Luff (1899) agreed with this finding but pointed out that micromalting was surely valid for comparative trials and that as superior malts were obtained in higher yields by micromalting the grain cool it made more sense to try and match the micromalting conditions on the production scale. This view was subsequently reinforced by many findings, including those of Frew (1901), who carefully carried out his experiments on at least 1 kg (2.2 lb) samples of grain in the 'German type of equipment'. He expressed the view that it was impossible to achieve reproducible conditions in commercial maltings 'as the ordinary maltster is a thoroughly unscientific person with a profound contempt for regularity in working', but that the comparative malting values of different barleys were correctly determined in small-scale trials. A different type of micromalting system, which seems to have achieved reproducible germination conditions (constant temperature, high humidity), was used by Blaber (Anon., 1906b; Blaber, 1907). Schjerning (1910) described a glass micromalting germination unit that was ventilated with a flow of air.

Bode and Westhofen (1910) described a small laboratory kiln in which green malt could be dried in 6 h and so be used to dry successive small samples of malt from a piece, permitting its progress to be followed. However the rapid drying and curing, wholly unlike that in use on the commercial scale at that time, underlined the difficulty of reproducing commercial kilning conditions on the small scale.

Much valuable micromalting work has been carried out using comparatively simple and inexpensive laboratory equipment. Such equipment usually allows great experimental flexibility but, because it is not automated, experiments with it can be laborious. Whitmore and Sparrow (1957) and Whitmore (1963) devised a system for evaluating different genotypes. Samples of barley (60 g), steeped in tap water at 12 °C (53.6 °F) to at least 42% moisture, were divided between three large borosilicate glass test-tubes (6 in × 1 in (15.2 cm × 2.54 cm)) that were closed with corks in which a 3/16 in (0.48 cm) hole had been bored. These were incubated upright, in the dark, at 17 °C (62.6 °F) for 3 days. The tubes were emptied; the grains were well mixed to separate the rootlets, then they were returned to the tubes. After 2 more days of incubation, the process was repeated. After 7 days, the green malt was dried on wire gauze trays in a forced draught oven for 24 h at 50 °C (122 °F), 8 h at 60 °C (140 °F) and 16 h at 72 °C (161.6 °F). The dried malt was de-rooted and weighed before analysis.

Small-scale malting has also been carried out in small tubes immersed in temperature-controlled water-baths; a slow stream of washed, attemperated air is passed through the tubes. An advantage of this arrangement is that the respiration of the grain (50–100 g) can be estimated by measuring the carbon dioxide in the outlet gas stream (Day, 1896; van Roey and Hupé, 1955; Meredith *et al.*, 1962). Others have chosen to germinate grain in beakers, sometimes covered with watch glasses (Banasik *et al.*, 1956; Bishop and Spaul, 1959). The differences between the reported results obtained with and without watch glasses covering the beakers are substantial and illustrate how closely micromalting conditions must be regulated (Banasik *et al.*, 1956). Failure to regulate and standardize all aspects of malting trials can yield highly variable results (Henry and McLean, 1984a). The need to evaluate closely the performance and reproducibility of a micromalting system before it is used experimentally is now generally appreciated (e.g. Mändl *et al.*, 1970).

Since the early 1960s, we have used and adapted a micromalting system originally described by Pollock (1960). The system has been used for teaching and research. It is flexible, but labour intensive. Barley samples are screened, well mixed before use and are stored in airtight and waterproof containers. For some experiments, the barley is dried to 7–8% moisture in a rapid airflow at 43 °C (109 °F). If dormancy is to be eliminated, the dried grain is stored in closed containers at 40 °C (104 °F) for weeks or months, as required. If the experiments are to be completed in a few weeks, the grain may be stored at 5–8 °C (41–46.4 °F), but if they are to take place over months or years then storage at −18 °C (0 °F) is preferred. Sub-samples, for analysis or malting, are taken using sample dividers. Usually they are screened to limit the range of corn sizes using a set of EBC-type shaking sieves. Samples of grain (exactly 50 g or 75 g fresh weight, within one grain) are weighed into glass bottles (nominally 500 ml capacity), which are then closed with nylon mesh held in place with elastic bands. The bottles are clearly labelled. The bottles of grain and the water to be used in steeping are pre-attemperated for at least 12 h, and usually 24 h, in a constant temperature, high-humidity cabinet. This cabinet, which is also used to hold germinating samples, is usually held at 16 °C (60.8 °F). An internal fan circulates the air over a refrigeration coil. High humidity is maintained by buckets of water that wet projecting 'wicks' of thick chromatography paper and by pumping air around and through the buckets of water, using aquarium pumps and air dispersers. The airstream is not allowed to blow directly onto the perforated metal shelves that support the bottles of grain. By staggering the operations, a practised worker can process 80 samples in one trial.

Before use, the attemperated steep water may receive additives or dissolved oxygen may be removed by 'degassing', that is exposing the water to a vacuum while the vessel is immersed in a sonicating bath or the contents are vigorously mixed with a magnetic stirrer. For most purposes, steeping is begun by the addition of measured volumes of steep liquor to the grain, at a noted time, through the mesh on the top of the bottles. When all the bottles have received their water, the contents are briefly 'swirled' and each bottle is struck to displace air bubbles and to 'sink' floating grains. At the end of the first 'wet' or immersion period, the water is poured out, the bottles are shaken so that the grain forms a compact bed in the neck, and water droplets are shaken down from the sides of the bottle and the bottle is placed in an inverted position in the constant temperature cabinet to drain. Normally an hour is allowed for draining, but if a longer air-rest is required, the bottle remains inverted for this period. If an air-rest is to be avoided, fresh steep water may be added immediately. For a second, or subsequent, immersion, the grain is shaken to the base of the bottle and attemperated water is added as before. For routine purposes, barley is steeped for two or three 24-h immersions. We have experimented with various additives including sodium or calcium hypochlorite, formaldehyde, lime, gibberellic acid, potassium bromate, hydrogen peroxide and mixtures of antibiotics. Steeping at other temperatures is easily carried out by holding the bottles in laboratory constant-temperature water-baths, but draining and air-resting requires the use of high-humidity incubators held at the appropriate temperatures. By using aquarium pumps and 'air-stones' (porous gas diffusers), it is possible to aerate grain being steeped in bottles. If anaerobic steeping is desired, it may be necessary to completely fill the bottles with degassed water and seal them, for example with a plastic film.

At the end of the steeping period the grain is drained. If additives are to be applied to the grain they are usually sprayed on in solution at the end of the drain period. A solution containing the correct weight of the additive(s) is sucked into a disposable plastic syringe that is fitted with a needle the end of which has been cut and crushed so that liquid emerges from it as a fine, fan-shaped spray. The bottle containing the grain to be sprayed is placed on its side and the needle is pushed through the nylon mesh closure. Then, as the bottle is rolled to and fro to continuously mix the grain, the appropriate volume of solution (usually 6 ml) is sprayed into the bottle from the syringe. After a final mix, the bottles are usually inverted and drained. The same technique may be used for spray-steeping grain or sprinkling during germination. For germination, the grain is shaken down in the bottle so it forms a compact bed on one side, with no corns in the 'neck'. Rows of bottles are placed on the shelves all facing in one direction. Germinating grain may be resteeped in the same bottles. Nearly always, steeping and germination have been carried out in the same bottles. However, with some extremely water-sensitive grain, it is necessary to transfer the steeped grain, after draining and blotting, to fresh, dry bottles to obtain even germination. To carry out comparative trials using different germination temperatures it is necessary to have several constant-temperature, high-humidity incubators. Keeping the bottles upright during the germination period leads to abnormal growth or even growth failure, presumably because the carbon dioxide generated by grain metabolism 'ponds', fills the bottle and smothers the grain. The grain is turned, usually twice a day at fixed times in the morning and evening, until the germination period is finished. Alternatively, the bottles may be turned, in our 'drum-malting' apparatus (see below). Measured amounts of gases have been applied to germinating grains and gaseous atmospheres have been regulated by closing the bottles with rubber bungs and introducing and removing gases through hypodermic needles that penetrated the bungs and could be closed with small plastic taps.

Grain has been 'stewed' held in closed containers, chosen to prevent water loss, in ovens held at various temperatures in attempts to mimic the production of caramel malts. As noted above, our 'standard' system is to dry malts and quickly 'fix' their composition in a rapid airflow at 43 °C (109.4 °F). If malts are to be cured this is usually done at a chosen temperature, about 65 °C (149 °F), in a second kiln. Each kiln contains 36 removable metal containers with woven-wire mesh bases and lids. The fan-driven

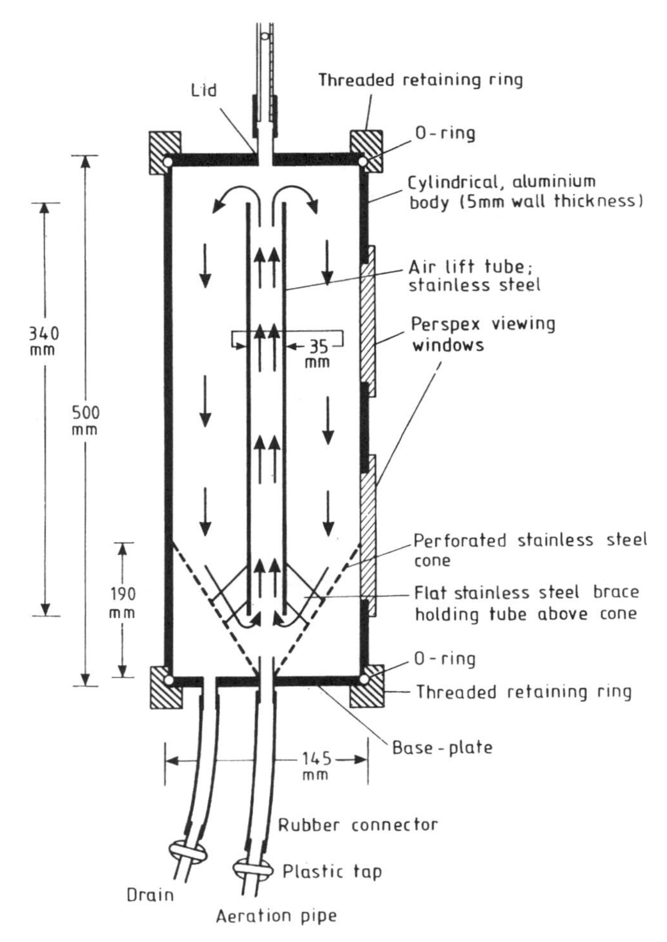

Fig. 11.1 A longitudinal section of a steeping vessel for 800–1000 g barley (Kelly and Briggs, 1992a).

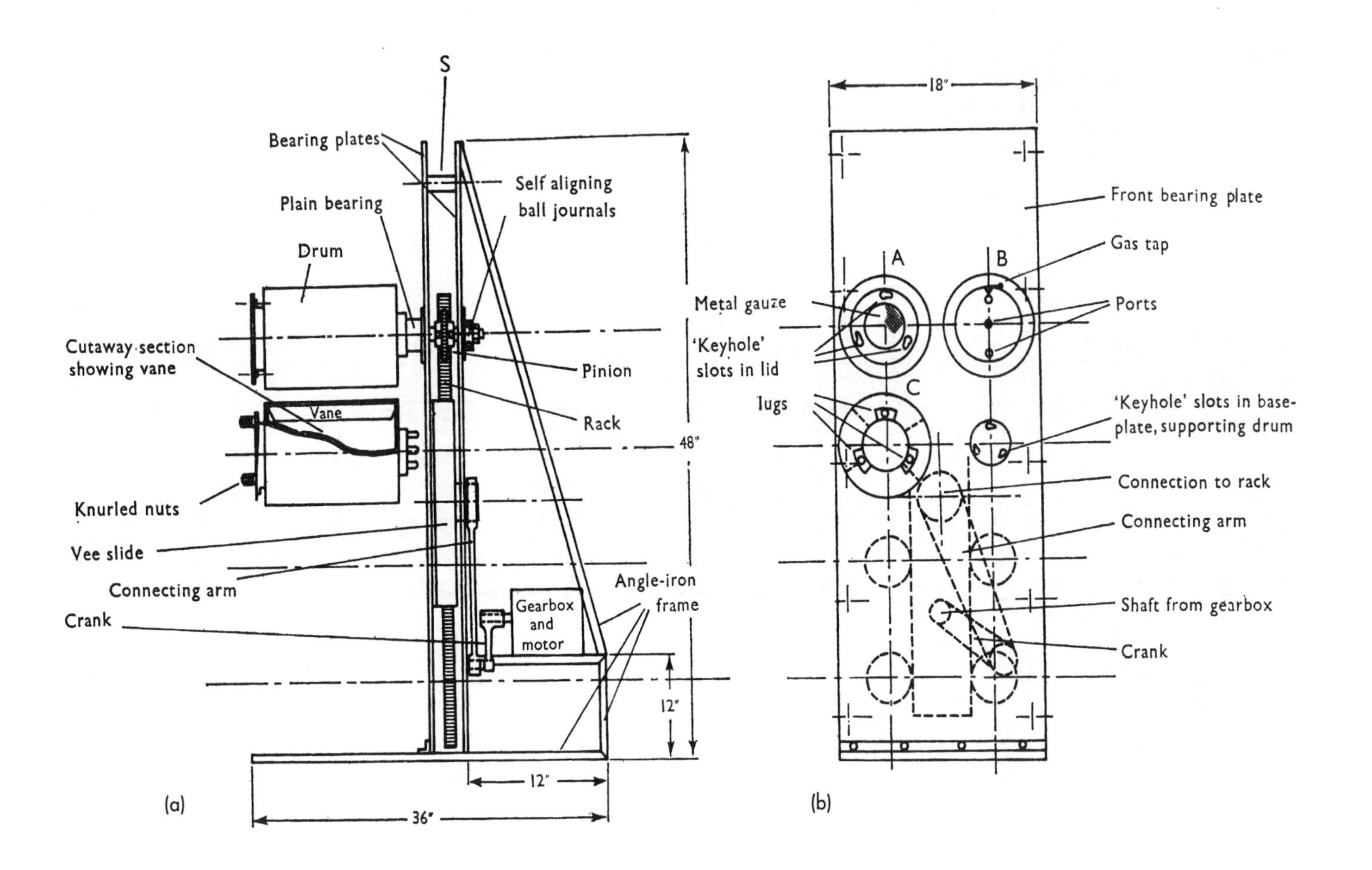
S
Bearing plates
Plain bearing
Self aligning
ball journals
Drum
Cutaway section
showing vane
Vane
Pinion
Rack
48"
Knurled nuts
Vee slide
Connecting arm
Crank
Gearbox
and
motor
Angle-iron
frame
12"
12"
36"
(a)
18"
Front bearing plate
Gas tap
A
B
C
Metal gauze
Ports
'Keyhole'
slots in lid
lugs
'Keyhole' slots in base-
plate, supporting drum
Connection to rack
Connecting arm
Shaft from gearbox
Crank
(b)

airflow is heated electrically, the temperature being regulated by a thermostat. In the drying kiln, the airflow can be limited using a butterfly valve in the 'chimney'. In the curing kiln, limited air recirculation is allowed. The removable containers fit either kiln. Because large numbers of samples must be kept separate during kilning, often each is held in a clearly labelled, nylon mesh bag. Several of these bags can be placed in each container. The kilned malts are de-rooted by sieving and the rootlets, as well as the malts, are collected. The fresh weight of the malt is measured prior to analysis and the dry weight of the rootlets is determined.

This type of malting is simple, reliable and repeatable. Indeed although we frequently malt samples in triplicate when sufficient grain is available, the third of each three has rarely needed to be analysed because the analyses of the first two agreed well. The weaknesses are that, because grain turning and all other operations are manual, an operator must be in attendance at least twice each day, often for 10 or 11 successive days. Furthermore, the amounts of malt produced are inconveniently small for some purposes, e.g. for small-scale brewing trials (Briggs *et al.*, 1985). To reduce these problems we developed a second, drum-malting system, designed to malt 0.7–1 kg (1.54–2.2 lb) samples of barley. Grain may be steeped in buckets or in the germination drums held 'on end' in a constant-temperature cabinet and closed with a perforated metal cover through which water may be added or drained (Briggs and Harcourt, 1968; Sipi and Briggs, 1968). Alternatively grain may be steeped in a more controlled way in purpose-built steeps (Fig. 11.1) (Kelly and Briggs, 1992a). Eight steeps are mounted in an angle-iron frame in a constant-temperature cabinet. For loading and emptying, they are dismounted and held on a hinged frame that allows one person to tip them in a controlled fashion and lock them in a tipped or an upright position. Usually exactly 1 kg (2.2. lb) barley is steeped in 5.6 l (1.23 gal, UK) of water. The grain may be drained and re-covered with water *in situ*. The grain-water mixture may be thoroughly mixed and aerated by discharging air (at about 5 l/min) into the base of the central air-lift tube. Aeration frequency and duration is automatically controlled. Carbon dioxide extraction is by pumping air downwards through a drained bed of grain.

Germination is carried out in eight steel drums horizontally mounted in two vertical rows of four. A timer actuates an electric motor that drives a reciprocating vertical rack (Fig. 11.2) (Briggs and Harcourt, 1968). The

Fig. 11.2 The side (a) and front (b) elevations of a small-scale drum malting equipment (Briggs and Harcourt, 1968). The dimensions are in inches (1 in = 2.54 cm). The side elevation shows the positions of two drums. The front elevation shows the positions of three drums. The drums illustrated are the older, mild-steel type equipped with internal, longitudinal vanes (see text). A, drum with a lid having a metal mesh insert; B, a gas-tight lid; C, showing the positions of the securing lugs; S, a spacer holding apart the structural parts that support the rack-and-pinion mechanism which actuates the drums and the bearings on which the drums are mounted.

movement of the rack, via cogs or pinions, causes the drums to oscillate to and fro twice a minute, through 320°. This turns the grain. Usually turning is carried out for 5 min every 6 h. The entire device is mounted in a constant-temperature, high-humidity cabinet. The original drums were of mild steel internally coated with enamel. Six longitudinal vanes ran down the inside of each drum to prevent the grain 'slipping' during turning. Corrosion problems with the mild steel and the unnecessarily vigorous turning given by the vanes lifting and dropping the grain caused us to replace the original drums with others of stainless steel. These are hexagonal in cross-section and 1 ft (0.305 m) in length. The width of each hexagonal side is 4 in (10.16 cm) and the distance between two parallel sides is 6 in (15.24 cm). The end carrying the mounting to lock onto the turning device is closed. The other end contains a circular hole 11.4 cm (4.5 in) in diameter. In use, this is covered by a stainless steel lid containing a rectangular hole 5.08 cm (2 in) square covered with a stainless steel, woven-wire mesh of about 1 mm (0.254 in) spacing. This 'vent' allows gas exchange between the interior of the drum and the atmosphere in the constant-temperature chamber. Most heat loss is through the drum walls. The oscillating rather than rotating method of turning the drums was chosen because, when using special gas-tight lids, it allows flexible tubing connections to be made to the drums. These may be used for regulating the atmosphere in the drums or to add 'sprinkling' solutions to the grain, even while it is turning. If steeping is being carried out in the drums, it is also feasible to aerate the grain by applying a vacuum to an airtight lid and admitting air to the bottom of the steep through a metal tube that extends from the lid. Steeping and grain washing have been carried out by immersing grain in the drums while these are oscillating continuously. By packing several micromalting bottles into each drum it is possible to reduce the amount of manual turning when 75 g samples of barley are being used. The kilns used to process the small-scale 'drum' malts are those used to dry bottle-malted samples.

'Families' of micromalting equipment were derived from that described by Harrison and Rowland (1932) (Meredith *et al.*, 1962). Grain was held in wire-mesh cylindrical containers for steeping. Steeping was started, at an appropriate time using a mechanical timer, by lowering the container into a temperature-controlled water bath. The steeping of different samples could be started at different times. The system produced two batches of malt samples each week, each batch from twelve 250 g samples of barley (Meredith *et al.*, 1962). The steep-tank could be filled with water or emptied according to a pre-set programme. Germination took place in 24 cylindrical cages rotated on a frame in a stream of humidified and attemperated air. Kilning was carried out in a similar device in which 12 sample cages were rotated in a stream of dry, electrically heated air.

This system was progressively refined. For example in the germination equipment, baffles were introduced into the airstream to prevent water droplets from the humidifying unit reaching the grain. Also it was noted that in wire-mesh cages, germinating grain tended to dry out, and so the cages were replaced by metal cylinders containing a limited number of perforations. It is believed that significantly different rates of water loss occur from grain germinated in different types of micromalting equipment.

Many other types of small-scale malting plant, with various degrees of automation, have been described (e.g. Davis and Pollock, 1956; Bettner *et al.*, 1962; Meredith *et al.*, 1962; de Batz, 1963; Reeves *et al.*, 1979) (Figure 11.3). In most micromalting plants, the temperature control of the germinating grain is 'unrealistically' good in that the temperature rises and differentials seen across grain beds in production plants do not occur. In an attempt to overcome this 'defect', Graff and Spitz (1951) constructed equipment in which heat loss from the germinating grain was minimized by using Dewar vacuum-walled flasks as the grain containers. The equipment appears to have been complex and not widely adopted.

Various commercial micromalting plants have been, or are, available. In the Seeger plant, originating in the 1960s, 1 kg (2.2 lb) samples of grain held in rectangular stainless steel containers were steeped, germinated in a flow

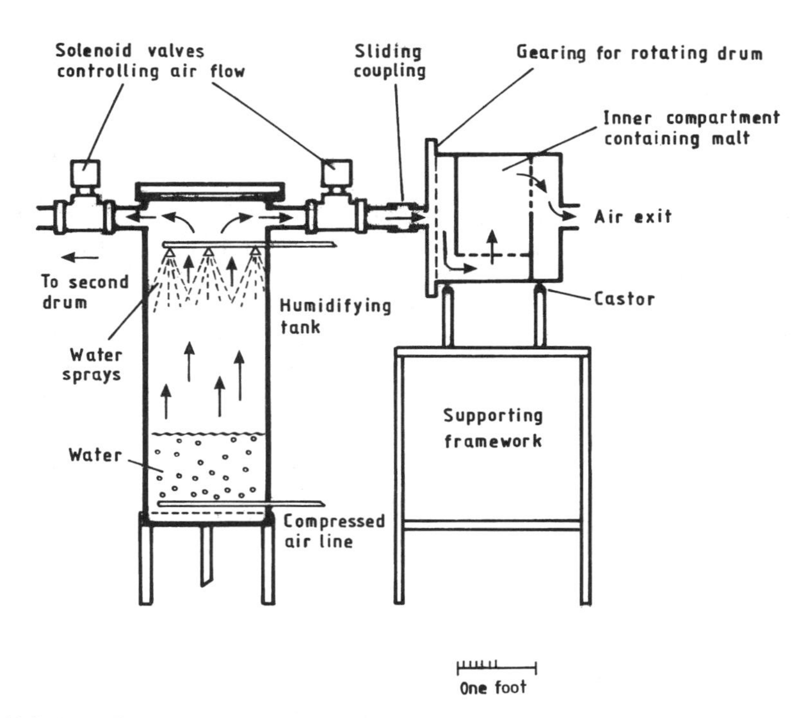

Fig. 11.3 A small-scale germination drum and the air-conditioning unit that supplies it with conditioned air. The drums are 20 in (50.8 cm) long and 20 in (50.8 cm) in diameter. The initial sample size is 15 lb (6.804 kg) barley. (After Bettner *et al.*, 1962.)

of attemperated and humidified air and kilned in separate units. It performed well (e.g. Mändl *et al.*, 1970) and was widely used but required continual attention, with manual grain turning and transfers between units. In contrast, the micromalting plant offered by Joe White Malting Systems (Australia) is almost completely automated and is constructed of stainless steel (Dagg *et al.*, 1981). The plant can be adjusted to hold four 2 kg (4.85 lb) samples of grain, 8 at 1 kg, 16 at 0.5 kg, 32 at 240 g, or even larger numbers of smaller-sized samples of grain. These are held in rectangular stainless steel boxes mounted side by side, having perforated bases and lids. The device, a single unit, is programmed through a computer that monitors the programme, runs it throughout and logs performance. Steeping, germination and kilning are carried out in the same unit. Temperatures, the durations of wet immersion periods and air-rests, steep aeration, germination temperatures (in a flow of attemperated and humidified air), turning frequency and the kilning cycle (with variable temperatures and airflow rates and various proportions of air recirculation) are all controlled. Turning is achieved by rotating the grain containers. It is so arranged that, as in box-drums, the surface of the grain bed is parallel to the perforated metal base of the container so the depth of the grain bed is uniform during ventilation. Because of the flexibility of the computer programming, it is possible to mimic the temperature changes occurring in the production-scale plant. A programmed malting sequence will run entirely without operator intervention. The saving in labour is so large that for most purposes it is clear that computer-controlled, automated micromalting will become standard equipment.

A somewhat similar device for handling 87 samples of 15 g or 129 samples of 5 g grain was described by Haslemore *et al.* (1985). In this device, the grain samples are held in cylindrical containers of polycarbonate plastic (with holes in the sides); the base of each container acts as the lid of the container below. The containers are held in stainless steel racks that are mounted longitudinally in a stainless steel insulated tank which is humidified and temperature controlled. During steeping, the containers are half-immersed. During germination, the water is below the rack but the bottom of the tank is still covered with a layer of water. Turning is by rotating the racks about their longitudinal axes. Kilning is carried out in the same containers but mounted in a separate, electrically heated enclosure with a forced air circulation. The steeping, germination and kilning programmes are automatically controlled. When replicate samples of barley were malted in this equipment the malts obtained were very uniform.

Several other devices have been described in which barley is held in cylinders of perforated metal or woven wire, which are supported on rollers and are turned when the rollers are actuated. All are microprocessor controlled and in all cases steeping and germination are carried out in one tank. Differences occur in whether or not kilning is performed in separate vessels and in details of construction (Gothard and Smith, 1986; Sparrow *et al.*, 1986; Glennie-Holmes *et al.*, 1990). At least one of these devices is commercially available. Because the cylindrical sample containers are supported on rollers, containers of different capacities can be accommodated. In the most recently described device, steeping, germination and kilning take place in the same vessel and the device can process 40 samples of 100 g or 70 of 60 g simultaneously (Glennie-Holmes *et al.*, 1990). In each case, twice the number of samples (of half the size) can be processed if the grain containers are divided with longitudinal partitions. In all these devices, variability between samples processed in parallel in the malting unit is acceptably small. Some highly versatile micromalting plants have been custom-built (Oliver, 1994).

Experimental malting devices with capacities of more than a few kilograms are often referred to as 'pilot' units. It is believed that numerous units have been employed, but often these have not been described. At least some of these have had capacities of around 50 kg (*ca.* 1 cwt) original barley. Meredith *et al.* (1957) and Bettner *et al.* (1962) described a box-drum capable of holding 6.8 kg (15 lb) barley (Fig. 11.3). Eight drums worked in pairs. The grain was held in a central compartment of the drum. Steeping and kilning were carried out in separate pieces of equipment. The devices were housed in constant-temperature rooms.

Baker and Burant (1956) described a pilot malting laboratory that had a unit which was a development of an earlier design, introduced in 1938. Half a bushel (US) of barley was held in a cylindrical, stainless steel steep/germination unit, 35.6 cm (14 in) diameter and 61 cm (24 in) high, equipped with a false bottom 5.1 cm (2 in) above the base. A helical ribbon turner, about half the width of the container, was suspended from the lid. As the turner rotated (at 9 r.p.m.) it also slowly revolved around the centre of the vessel at 0.6 r.p.m. The mixing system worked well. Six units were used, working in three pairs.

A sophisticated pilot malting is described by Kelly (1987) in which steeping, germination and kilning are fully automated. The equipment is thoroughly instrumented. Three decked drums, each with a capacity of 50 kg (*ca.* 110 lb) dry barley are served by two kilns. The drums are SGVs. Their mounting is novel, as they can be tilted upward or downward from the horizontal to facilitate loading and emptying. Because the drums are used for a range of experiments, it is possible to steep at any chosen point in the temperature range 10–30 °C (50–86 °F) and to germinate at 10–25 °C (50–77 °F). Steep aeration is at 80 m^3 air/t original barley, air (carbon dioxide) extraction is at 250 m^3/t and germination air-ventilation rates are up to 750 m^3/t. Solutions of additives may be sprayed onto the grain as a fine mist as a drum is rotating. Finished, green malt is transferred manually to one of the stainless steel kilns. The kiln air is heated electrically. The plant is regulated by an electronic programmable controller (PC) and the process can be followed by colour graphic representations of the state of the plant. The PC can store up to six programmes for steeping, four for germination and eight for kilning. In addition, the PC contains a diagnostic alarm programme. The output from this plant is sufficient for extensive analyses and small-scale brewing trials.

11.4 The assessment of micromalts

With micromalting experiments, only very small samples of malt are available and this has necessitated the development of small-scale methods of analysis. Broadly, maltsters attempt to use small-scale methods of analysis that give results agreeing with the standard macromethods, while plant breeders are often content, or at least are forced by circumstances, to rely on methods that give results which correlate more or less closely with those obtained using the macromethods when applied to reference samples of malts. The latter approaches may permit barley samples to be ranked in order of their 'malting value' but are inherently less satisfactory, and some few of the proposed methods seem to be of dubious validity.

Because all the barley and malt and rootlets can be weighed with high precision, and moisture contents can be accurately determined, malting losses and yields may be determined rather exactly, provided that no grains are lost at any stage in the micromalting programme. The relationships are:

$$\text{malt yield (\% d.b.)} = [\text{malt (d.b.)} \times 100] \div \text{barley (d.b.)}$$
$$\text{malting loss (\% d.b.)} = [\text{barley (d.b.)} - \text{malt (d.b.)}] \times 100 \div \text{barley (d.b.)}$$
$$\text{rootlets (\% d.b.)} = [\text{rootlets (d.b.)} \times 100] \div \text{barley (d.b.)}$$

Therefore,

$$\text{malting loss (\%)} - \text{rootlets (\%)} = \text{respiration} + \text{steeping losses (\% d.b.)}$$

Steeping losses may be determined by measuring the difference between the dry weights of barley samples before and after steeping.

Micromalts have been examined for many characteristics including hot water extracts, cold water extracts, total nitrogen, total soluble nitrogen, free amino nitrogen, wort fermentability (boiled or unboiled), diastatic power, α-amylase, β-glucanase, wort viscosity, wort β-glucans, 70 °C mash viscosity, extents of modification (sanded block techniques with Calcofluor, Methylene Blue or Oil Red O staining), friability, homogeneity and polyphenol or anthocyanogen contents. In general, the most frequent source of problems is that the small number of corns used for particular small-scale analyses can lead to large statistical errors.

Most studies have been concerned with the determination of hot water extracts on the small scale. Kirsop and Pollock (1958) and Pollock (1962c) derived calculations for determining the hot water and cold water extracts of small amounts of grists (2.5–7.5 g) of exactly known weights. Basing their method on the, then current, IoB isothermal mash in which 50g of grist was adjusted at the end of mashing at 150 °F (65.6 °C) to a mash volume of 515 ml, Whitmore and Sparrow (1957) mashed exactly 5 g grist and adjusted the final volume to 51.5 ml. We routinely used 10.0 g grist and adjusted the final volume to 103 ml. The larger scale was preferred because sampling errors were less and, malting 50 g or 75 g barley, this amount of malt was readily available for duplicate determinations. Also the larger volume of wort produced was more convenient for the specific gravity determination. Many workers have overcome the problem of small wort volumes by determining specific gravity indirectly using very precise measurements of refractive index made on a few drops or a few millilitres of wort (e.g. Essery, 1954; Essery *et al.*, 1954). When it was necessary to use specific gravity bottles or pycnometers for evaluating worts, those with 50 ml nominal capacity gave more accurate results than those of 25 ml capacity. However, the larger volume of wort was not always available. The availability of density meters requiring only a few millilitres of wort has overcome this problem. When the IoB modified its method for determining hot water extract so that a mash containing 50 g grist was adjusted to a final weight of 450 g (Chapter 13) we initially modified our micromash so that the 10 g grist gave rise to a 90 g mash. Adjusting to a fixed weight is more accurate than adjusting to a fixed volume. However, with many undermodified and experimental malts, filtration was so slow that a useable volume of wort could not be collected in the specified time. Rather than separate the solids by centrifugation, which is an acceptable technique, we adjusted final mash weights to 100 g. These more dilute mashes filtered more rapidly and the extracts obtained were identical to those obtained with the reduced-scale *Recommended Method* when the appropriate equation was used:

$$\text{Hot water extract (1°/kg)} = \frac{9.776 \times G}{SG}$$

where G (degrees saccharine) is the excess specific gravity of the extract at 20 °C over water, taken as 1000.00, also at 20 °C and SG is the specific gravity of the solution at 20 °C (Markham *et al.*, 1991). Furthermore, the wort characteristics more nearly resembled those of the older type of determination in which adjustments were to a final volume. An 'ultra-micro' method has been described for measuring extract on as little as 1.0 g ground malt, for plant-breeding purposes (Fox and Henry, 1993). Other recently described, very small-scale methods for determining hot water extracts are those of Glennie-Holmes (1990), Henry and McLean (1984b) and Slack *et al.* (1986). Many parameters besides hot water extract are used to evaluate malts (Chapter 13). Microversions of many of the 'standard' analytical methods have been developed. In particular, one may mention the use of micro- and pilot-scale breweries (e.g. Stollberg *et al.*, 1957; Groulx, 1970;

Briggs *et al.*, 1985). It may be noted that to maximize analytical differences between samples, e.g. between samples steeped in various ways or prepared from different barley cultivars, it is sometimes appropriate to germinate malts for a short time so that they are 'under-modified'. The hot water extracts of barley samples that are malting well rise more rapidly than those that are malting badly, but then the values attained tend to plateau. Differences between better and worse samples are greater before the plateau is reached.

12 Competitors for malt

12.1 Introduction

Malts serve as sources of soluble extract and/or enzymes and/or colour and character for the products made from them. Inevitably they compete with other materials that also serve as sources of extract or with extract having different characteristics. In addition, preparations of enzymes from other sources (usually microbes but also other plant materials) may be used to supplement or act as alternatives to malt enzymes in producing extract for mashes or to modify products derived from malts, for example by altering beer characteristics. Caramels are used in beer, in malt-vinegar brewing and whiskey manufacture, and in foodstuffs to adjust the colour of the final products.

In many respects, the distinction between special, coloured malts and the adjuncts that are based on raw or cooked cereal preparations which are used in mashing or in baking are artificial since coloured malts are defective or wholly lacking in a range of enzymes, as is the case with cereal adjuncts. However, the traditional distinctions between these products, which are based on the manufacturing processes, will be retained here. The brewer's system of analysis (EBC, IoB, ASBC; Chapter 13) include methods for the analysis of adjuncts that resemble those used for coloured malts. Those that require mashing are milled and are incubated in small-scale mashes mixed with milled, enzyme-rich malt. The extracts obtained are 'corrected' for the extract derived from the malt in the mixtures to obtain estimates of the yields obtained from the adjunct.

Adjuncts are classified with reference to where they are used in the production process (in the mash, dissolved in the copper in beer brewing or added to beer as 'priming sugar' or caramel), the raw material(s) they are made from, whether or not they must be cooked by the user, and the method of manufacture. Many substances were once used as adjuncts, additives or flavouring materials in brewing, including orange powder, coriander seeds, capsicum, caraway seeds, salts of tartar, oyster shells, alum, quasi-lignum, gentian ('grains of paradise'), some of which were 'of a warm nature and that they disperse wind and give warmth to the stomack' (Morrice, 1802 quoted by Imrie and Martin, 1969). Fortunately the use of these, and even less desirable additives, during the brewing process has long been discontinued.

In the UK, it has been legal to use sugars in brewing since 1847 and unmalted cereal adjuncts in mashing since the repeal of the malt tax in

1880. Throughout the world unmalted cereal preparations are commonly used in the mash during brewing, both of clear and opaque beers, and in the manufacture of vinegar, whiskey and extracts. However in certain regions, in parts of Germany and in Norway for example, malt is the only permitted source of extract in the manufacture of beer, as is the case in making malt whisky in Scotland. Large quantities of unmalted cereals are used in the manufacture of grain whisky and in making Irish and most other whiskies.

The classification of adjuncts adopted here divides them into: (i) those used in mashing, commonly called mash-tun adjuncts even when other mashing equipment is in use (these are subdivided into those that are used 'raw', such as wheat flour, those which are pre-cooked, such as maize flakes or torrefied grains, and those which must be cooked immediately before use, such as cereal grits); (ii) those that are dissolved in the wort, during the copper-boil stage of brewing (such as solid sugars or syrups) (called copper- or kettle-adjuncts); (iii) those that are added to beer to support secondary fermentations (priming sugars); or (iv) to add sweetness or other flavours to the product; and (v) those added to adjust the colour of the final product (caramels).

Adjuncts used during mashing include crushed, ground or milled preparations of entire raw grains (such as barley, wheat, rye and, possibly, triticale), roasted, micronized or torrefied entire grains, flaked preparations of entire grains (barley, wheat, oats), or flaked preparations of partly refined maize, sorghum or rice grits. Pre-cooked, entire grains of maize and rice have reportedly been flaked but these have probably not been in use recently. Grits (derived from maize, rice or sorghum) may be cooked by users as a part of the mashing process. Refined starches may also be employed directly in mashes (starches from wheat, potatoes or tapioca) or after pre-gelatinization and liquefaction (starches from maize, sorghum and rice). Historically, enzyme-rich highly diastatic malts or malt extracts were added to mashes to make good insufficient enzyme levels. Highly enzymic malts are still used when unmalted adjuncts are used in mashing but such mashes may be supplemented by enzymes derived from microbes, which may also be used to overcome difficulties encountered when using 'problem malts'. The use of enzymes during brewing, other than those derived from malt, is restricted or forbidden in several countries.

Sugars (solid or in solution), malt extracts and wort replacement syrups may be dissolved to yield extract. In brewing, such copper-adjuncts can be dissolved or dispersed in the wort during or before the hop-boil. These include preparations of sucrose, invert sugar (equimolecular mixture of glucose and fructose), comparatively pure mixtures of saccharides derived from the controlled hydrolysis of purified starch, and malt extracts, green malt extracts or extracts based on unmalted cereals. These last preparations contain significant quantities of other substances in addition to carbohydrates, including nitrogenous materials and minor yeast nutrients. Such cereal-based syrups may be regarded as wort replacements, while the comparatively pure carbohydrate preparations are wort extenders, the use of which reduces the ratio of nitrogenous and polyphenolic substances in the wort to the carbohydrate extract content. In some home brewing, and in some unusual commercial conditions, beer is brewed from concentrated, hopped wort syrups. Complete replacement of malts with malt extracts (hopped or otherwise) is usual in home brewing. 'Adjunct' fermentation is a proposed process in which concentrated syrups or sugar worts are fermented and the bland, alcoholic liquids so obtained are used to dilute 'all-malt beers', saving, *inter alia*, on beer-conditioning capacity (Moll and Duteurtre, 1976; Pollock and Weir, 1976). The preparations used in priming and sweetening are either starch-derived or sucrose-derived. The possibility exists that 'high-fructose' syrups, derived from high-conversion starch syrups, may be used for sweetening.

The proportions of malt used in the **average** brewer's grist in the UK has varied relatively little in recent times, being 78.5% malt, 9% mash-tun adjuncts and 12.5% copper-adjuncts in 1984 (Collier, 1984). Changing relative costs have resulted in starch-derived sugar preparations largely displacing sucrose-based materials. The rates of adjunct usage may be regulated by legislation, by product definition or by local commercial considerations. In beer-brewing, adjunct usage varies between zero and a maximum of 60% grits in some grists used in the USA. It is said to be feasible to produce an acceptable wort from 95% raw barley and 5% malt, provided microbial enzymes are also used in the mash and the mashing period is extended.

Adjuncts often provide extract more cheaply than malt although flavour and other aspects of product quality must also be considered (Lloyd, 1988a). However the economies of using adjuncts are not simple, since many require special plant to be installed, for cleaning, handling, storing and milling. Availability and costs may fluctuate widely. Because the plants needed to cook grits, or to handle flours, flakes, raw grains or syrups are specialized, it is not feasible to change easily from using one type of adjunct to another. Consequently, users must ensure that a continuing supply of good quality, competitively priced material is available before committing themselves to installing plant to allow it to be used. An advantage of special malts, roast barley and torrefied cereals is that they can be used in normal malt-handling machinery. There is a need for equipment that will mix solid adjuncts thoroughly with the malt either before or during mashing. Some adjuncts create problems, for example of wort separation during lautering run-off (because they elevate wort viscosity) or creating impermeable layers of fine particles in the mash. In addition, high viscosities can cause problems with beer filtration, undegraded β-glucans from barley adjuncts may cause hazes and wort fermentabilities may be lower. The prime attraction of using adjuncts is that they provide cheap extract. There are other advantages to their use. For example, they may yield little or no nitrogenous, polyphenolic, lipidic or coloured materials to the wort, so wort composition may be optimized for fermentation without excessive yeast growth; this produces a pale product with a reduced tendency to become hazy, an enhanced shelf life and improved flavour stability (pale beers, malt vinegar). Some adjuncts are used to provide colour and/or characteristic flavours or aromas, or to improve head retention in particular beers. Brewers may use syrups and sugars to increase the 'brewing capacity' of their plant in periods of high demand, since these, being dissolved in the copper, do not require the use of more mashing or wort-separation equipment. Sugars and syrups are also convenient for obtaining concentrated worts for 'high gravity brewing'.

'Industrial (non-malt) enzymes', where permitted, may be used to supplement malt enzymes during mashing to aid in the conversion of poor-quality malt and/or adjuncts, to aid wort run-off, to increase wort fermentability or to reduce the viscosity of the wort. They may be used during adjunct cooking (usually to liquefy the starch) or to convert raw cereal preparations or starches during the manufacture of some syrups or sugars. They are also added to worts and beers to break down dextrins to sugars, to adjust fermentabilities and to confer sweetness to the product. Sometimes proteases are added to beers to degrade polypeptides and so reduce haze potential. Originally, enzymes derived from malt were added to mashes as preparations of diastase or diastatic malt extract. There have been attempts to use enzymes from animals, and enzymes from ungerminated cereals and soya beans are used to a limited extent. Considerable quantities of the plant proteases papain, ficin and bromelain are used in beer-stabilization treatments. However most food-grade industrial enzymes are derived from microorganisms. Those of most interest include a wide range of dextrinogenic and saccharogenic amylases, starch and dextrin debranching enzymes, amyloglucosidase, β-glucanases, pentosanases, proteases and possibly other hydrolases. In addition, other enzymes, such as glucose oxidase, are used to a limited extent.

12.2 The common cereal grains

Most brewers' adjuncts are prepared from a limited range of cereal grains (Table 12.1), but in particular cases adjuncts may be prepared from other cereals (e.g. triticale, millet) or from other materials such as potato starch or cassava. Tables of proximate analyses of locally available cereals are of some use in judging their values as adjuncts. However, all cereals vary widely, both among varieties of one species and among samples of one variety. For example the usual limits of crude protein contents in barleys are

Table 12.1 Representative compositions of some common cereal grains on a percent dry weight basis

Cereal	Cellulose or fibre (% d.wt)	Starch and sugars; extract (% d.wt)	'Lipids' oil/fat (% d.wt)	Other non-N substances (% d.wt)	Crude protein (% dw.t)	Ash (% d.wt)
Barley	5.7	60–71	2.3; 3.5	4.0	8.0–14.0	3.1
Wheat	2.9	76	2.0; 3.0	2.8	14.5 (8–18)	2.2
Rye	2.4	74	2.0; 3.0	5.8	9.0–15.4	1.6–2.4
Oats	12.4	61	6.1	2.4	13.4	3.5
Maize	4.2; 9.5	70; 74.3	3.9; 5.8	7.6	9.5; 11.6	1.2; 1.4
Sorghum	0.4–7.3	65–81	1.4–6.2	(–)	8.1–16.8	1.2–7.1
Rice	0.7; 2.3	57; 81	0.5; 1.9	(–)	7.2; 9.0	0.4; 1.0

After de Clerck (1952); Morrison (1953) quoted in Stewart and Hahn (1965); Lorenze and Kulp (1991); various other sources.

probably 7.8–15.6% (d.b.) (1.25–2.50% N, (d.b.)) and the levels of starch, soluble sugars and β-glucans vary widely (Briggs, 1978). Different grades and types of cereal, notably wheat, maize, sorghum and rice, differ in their suitabilities for preparing adjuncts. The oil (fat) contents of cereals differ and, because high levels of fat are deleterious in brewing, oats are now only rarely used and maize, sorghum and rice grains are processed to remove the oil-rich germs and bran before being used to prepare adjuncts. Oats, like ungerminated barley, are rich in β-D-glucans (3–6%). In addition sorghum, like barley, contains proanthocyanidin tannins, sometimes at such high levels that they can substantially reduce enzyme activities in a mash. These are removed with the surface layers when grits are prepared from sorghum. Even so, lipids derived from adjuncts can be important in brewing.

Cereal starches in adjuncts differ in the difficulty with which they are degraded during mashing. In some adjuncts, such as grits, the starch grains are invested with enzyme-impermeable cell walls, which must be broken before the starch can be degraded, and whole grains must be milled, or otherwise broken open. However amylases only attack starch granules rapidly when these have started to swell and gelatinize, that is lose their 'semicrystalline' structure and start to disperse into solution: processes which occur when the granules are heated in water. So ease of degradation is associated with the ease with which starch grains swell and gelatinize as the temperature of an aqueous suspension is raised. Swelling is a progressive, time-dependent process, as is gelatinization (expressed as the percentage of grains gelatinized); because starch granules differ, gelatinization occurs over a range of temperatures (Radley, 1968). Consequently when malt is mashed over an extended period, more than 90% of its starch-derived extract can be recovered at temperatures **below** the notional gelatinization temperature of its starch. Estimates of gelatinization temperatures vary with the methods used to determine them. Reported gelatinization temperature ranges of various starches are: maize, 62–74 °C (143.5–165.2 °F); sorghum, 69–75 °C (156.2–167 °F); millets, 54–80 °C (129.2–176 °F); rice, 61–78 °C (141.8–172.4 °F); short- and medium-grain rice (USA), 65–68 °C (149–154.4 °F); long-grained rice (USA), 71–74 °C (159.8–165.2 °F); wheat, 52–64 °C (125.5–147.2 °F); barley, 60–62 °C (140–143.5 °F); malted barley, 64–67 °C (147.2–152.5°F); and potato, 56–69 °C (132.8–156.2 °F) (Maiden, 1971; Canales, 1979; Lorenz and Kulp, 1991). These values correlate with the fact that while refined starches or starch in flours of wheat, barley, malted barley, potato and triticale can be converted by malt enzymes during mashing, the starches and grits of maize, sorghum and rice all need to be gelatinized or pre-cooked before adequate conversion will occur. This is because, in this latter group, malt enzymes are rapidly inactivated at the elevated temperatures needed to gelatinize the starch granules and make them susceptible to enzymic attack. However, incomplete conversion of starch has been reported from wheat and barley adjuncts in infusion mashing, probably because the small granules are not sufficiently swollen or gelatinized. Cereal starches contain small amounts of polar lipids associated with the amylose. These amyloses–lipid complexes are more or less resistant to enzymic attack. Temperatures in excess of 100 °C (212 °F) are needed to dissociate some of the lipid–polysaccharide complexes. Considerations like these dictate how a raw material can be used as an adjunct.

12.3 Analyses of adjuncts

For brewing purposes, adjuncts (with the possible exception of some sugars and syrups) are analysed according to one of the major published sets of analytical methods (IoB, EBC, ASBC; Chapter 13). Distilleries may use other methods in addition. Sampling errors, which may be large, must be minimized and the samples are assessed for some or all of the following: physical characteristics and purity (odour, presence of foreign matter, moulds, insects, etc.), clarity and colour of a 10% solids solution (in the case of soluble materials), moisture content (%), solids content (%), extract yield (various units, %, 1°/kg) and sometimes extract fermentability, the spectrum of carbohydrates present, the levels of sugars present, the iodine reaction after mashing (residual starch), the oil/fat content, acidity/pH of an extract or solution, the nitrogen ('protein') content and/or the yield of soluble nitrogen ('protein'), the ash content, the enzymic complement/diastatic power, the refractive index and specific gravity (syrups) and levels of iron, copper and arsenic. In many instances, microbiological contamination is also assessed. Distillers estimate the 'spirit yield' (final alcohol yield of adjuncts).

In some respects, the analytical methods used for mash-tun adjuncts are unsatisfactory. For example, estimates of oil contents are often surprisingly lower than those found in the starting materials when modern methods of extraction and analysis are employed. Methods for determining extract yield are often inadequately standardized and in any case do not give true estimations of the (variable) extract recoveries attained in commercial practice. This is partly because unmalted adjuncts lack the enzymes necessary to solubilize their starch and degrade it to an acceptable spectrum of carbohydrates. Usually the extract of an adjunct is determined by mixing the adjunct (milled or not and then cooked or boiled with or without an addition of enzyme(s) as specified in the chosen method) with malt to give a 50:50 mix. The mixture is then mashed either isothermally at 65 °C (149 °F) or using a programmed temperature sequence. After cooling, the wort is separated and the extract of the adjunct is calculated from the specific gravity on the basis that the malt will have contributed half of what it would have done if it had made up the entire grist. Essentially similar methods are used with coloured malts (Chapter 13). Such methods are incapable of high degrees of accuracy. Other proposed methods, based on converting adjuncts with partially purified and standardized mixtures of enzymes (microbial or malt-derived), with or without pre-boiling or autoclaving, seem not to have gained acceptance (e.g. Rosendal and Madsen, 1982). Adding increasing quantities of a particular enzyme preparation to mashes of particular adjuncts led to progressively higher yields of extract, (Albini *et al.*, 1987) showing that, for analytical purposes, a closely defined amount of enzyme would have to be employed. In the IoB method the extract of a syrup is calculated as extract (l°/kg) = 10.0 × G (20 ± 0.5 °C) where G = SG – 1000, taking water at 20 °C (68 °F) as 1000, SG being determined on 25.00 g of syrup diluted to 250.0 ml also at 20 °C (68 °F).

Control in sugar factories and refineries, in which sucrose is processed, is based on measurements of specific gravity and optical rotation (Meade and Chen, 1977). Procedures originated by Balling and improved by Brix were used to graduate hydrometers on the basis of percentage by weight of pure sucrose solutions. Plato and his collaborators made still more accurate measurements in 1900, and these results provide the standard scale (Appendix, Table A5). Thus °Balling, °Brix or °Plato are identical except in the 5th and 6th decimal places, but while the sugar industry refers to °Brix, brewers employ °Balling and °Plato, as if their worts were solutions of pure sucrose. In continental European brewing practice, °Plato are normally used to express the strength of the wort obtained from cereal mashes. In the USA, results are usually calculated using Balling tables. In each case, the results are expressed as 'percentage extract'. Sugars are optically active; their sugars rotate the plane of polarized light. Polarimetry, therefore, provides a method for assaying pure sugar solutions (Meade and Chen, 1977). The specific rotation is defined from:

$$[\alpha]_D^{20} = \frac{100\alpha}{lc}$$

where $[\alpha]_D^{20}$ is the observed rotation in circular degrees (at 20 °C (68 °F)); c is the concentration of the solution (g/100 ml); and l is the length of the solution (in dm (10 cm)). Measurements are made using the D-line of the sodium emission spectrum (wavelength 589 nm); for pure sucrose $[\alpha]_D^{20}$ is +66.53°. In sugar factories and refineries, special polarimeters called saccharimeters, calibrated in International Sugar Degrees, are used. By definition a solution of 26.00 g pure sucrose dissolved in 100 cm³ of distilled water in a 2 dm (20 cm) tube rotates the plane of polarized light 100 International Sugar Degrees (34.62 circular degrees). Using a saccharimeter a direct reading of the strength of a sucrose solution (called the 'Pol') can be obtained. The purity of a sugar solution can then be defined as:

$$\text{purity} = \frac{\text{Pol}}{°\text{Brix}} \times 100\%$$

This expression assumes that the impurities present have no optical activity, which is usually untrue. Commonly other sugars, including glucose and fructose, are present. Sucrose is hydrolysed either by acids or the enzyme invertase into an equimolecular mixture of glucose and fructose:

$$\underset{\text{sucrose}}{C_{12}H_{22}O_{11}} + H_2O = \underset{\text{glucose}}{C_6H_{12}O_6} + \underset{\text{fructose}}{C_6H_{12}O_6}$$

$$[\alpha]_D^{20} + 66.53° \qquad \underbrace{+52.7° \qquad -92.4°}_{\substack{-39.7° \\ \text{invert sugar}}}$$

While glucose (dextrose) is dextrorotatory, fructose (laevulose) is strongly laevorotatory; as a result, invert sugar has a negative rotation. (The optical rotations given are those that pertain when mutarotation is complete.) Since invert sugar is a common 'impurity' in sucrose-based sugar and syrup preparations, single polarimetric measurements are inadequate for determining their concentrations and more refined analyses, based on measurements made before and after complete inversion, are required (Meade and Chen, 1977). Representative analyses of various sugar-rich and other adjuncts are given in Tables 12.2–12.5.

Sugars and syrups made from purified starches are also evaluated with reference to their densities and often their sugar spectra (section 12.8). Higher oligosaccharides can often be partly resolved by column chromatography (e.g. gel-exclusion chromatography) and lower saccharides are well resolved by HPLC or the gas–liquid chromatography (GLC) of volatile derivatives such as the trimethylsilyl ethers. However, routine monitoring is by the 'dextrinization equivalent' (DE) in which the reducing power of the sugar solution under test is determined as glucose (dextrose) and is

Table 12.2 Typical analyses of sugar-rich brewing adjuncts, priming sugars and caramels

Preparation	Hot water extract (f.wt) (l°/kg)[a]	Total nitrogen (f.wt)(5)	Colour, 10% w/v solution (EBC units)	Fermentability solids (%)	Specific gravity (20 °C)
Sugar preparations					
Raw cane sugar syrup	258	0.01	3	95+	1.33
Invert sugar liquid	318	0.01	3–12	95+	1.43
Sugar primings	310	0.01	3–12	95+	1.42
Refined maize-starch hydrolysates					
Brewing syrup	310	0.02	Colourless or adjusted[b]	77–78	1.42
Confectioners' glucose	318	<0.01	Colourless	30–50	1.43
Solid brewing sugar	310	0.02	Colourless or adjusted[b]	86–87	–
Glucose chips	318	0.01	20–50	82	–
Other materials					
Grain-based syrup	302	0.4–0.8	4	65–70	1.40
Malt extract	302	0.65–1.3	4	70	1.40
Caramel, 46 000 liquid	242	–	4600	–	1.29
Caramel, 32 000 liquid	284	–	3200	–	1.36

[a] Dry, solid sugar preparations, e.g. sucrose, have values of 382–386/l°/kg.
[b] FAN values 0.01–1.15%. The colour may be adjusted to specification by the addition of other sugar-based products.
Analyses IoB (1993).
After Lloyd (1986, 1988a).

expressed as a percentage of the value that would be attained if the starch had been completely hydrolysed to dextrose (theoretical DE = 100%). Preparations obtained by hydrolysing starch in different ways may have similar DE values but widely different sugar spectra. The importance of the sugar spectrum is a consequence of the relative fermentabilities (and, less usually, different degrees of sweetness) of the various sugars when they are presented as a mixture to brewers' or distillers' yeasts. Results depend on the yeasts being used, but generally sucrose, glucose and fructose are rapidly fermented, maltose is fermented more slowly, maltotriose is fermented very slowly and higher oligosaccharides, $\alpha(1\rightarrow6)$-linked gluco-oligosaccharides and some other di- and trisaccharides are not fermented at all. In the presence of initially high levels of glucose, maltose and maltotriose may be incompletely metabolized by brewing yeasts.

12.4 Whole cereal grains

12.4.1 Raw cereal grains

The cereals used as adjuncts are 'conventional'; unconventional naked barleys or barleys that contain enough α-amylase to make their starch 'self-liquefying' (Goering and Eslick, 1976), for example, are not used. In

Table 12.3 Typical analyses of some starch-rich adjuncts

Adjunct	Moisture (%)	Hot water extract			TN (% d.b.)	TSN (% d.b.)	Bulk density (kg/l)
		(l°/kg as-is)	(l°/kg d.b.)	(% d.b.)			
Maize grits	12	301	342	90	1.5	–	0.76
Maize flakes	9	313	344–355	–	1.5	0.04	0.46–0.66[a]
Starches, refined	10	347	380–390	102–105	<0.1	Negligible	0.60–0.70
Rice grits	11	316	355	93	1.0	–	0.85
Rice flakes	9	325	357	–	0.85	–	0.30
Wheat flour	11	304	342	–	1.5	0.36	0.51
Wheat, torrefied or micronized	4–9	291	310–315	–	1.6–2.0	0.15	0.55
Wheat flakes	5–8	287	302	–	1.8	0.12	0.37–0.4
Wheat, raw	12	260	295	–	1.6	–	0.77
Barley, torrefied	5–6	254	267	72	1.8–2.2	–	0.37–0.4
Barley, flaked	9	253	278	–	1.8	0.12	0.25–0.26
Barley, raw	12	250[b]	284[b]	–	1.8	variable[b]	0.65–0.66

Analyses IoB (1993) except HWE (%), determined by the ASBC method (ASBC, 1992).
[a] Depends on the degree to which they are crushed.
[b] Depends on added enzymes.
From Lloyd (1986, 1988a); Brookes and Philliskirk (1987).

Table 12.4 Typical analyses of some roasted products and special malts; figures for pale malts given for comparison

Product	Moisture (%)	Hot water extract (d.b.) (l°/kg)	Hot water extract (d.b.) (%)[a]	Colour (EBC units)	Cold water extract (f.wt, %)	Bulk density (kg/l)
Roasted barley	1.5–2.0	270	–	1400	60	0.51
Roasted malt	1.5	265	–	1300	60	0.47
Chocolate malt	1.5	268	–	1100	60	0.47
Amber malt	2.0	280	79	55–60	16	0.54
Crystal malt	4.0	280	–	140	50	0.52
Cara pils	7.0	280	79	27–30	50	0.62
Brown malt	2.5	275	–	130	–	–
Pale malt	3–6	300–310	80–83	2.5–5.0	18–21	0.55

[a]*Analytica-EBC* (EBC, 1987); all other analyses IoB (1993).
After Lloyd (1986).

practice, it seems that triticales are not widely used and neither is rye, except in rye whiskey manufacture. Unlike many raw cereal preparations, triticale adds appreciable amounts of soluble nitrogenous materials to wort, and the pentosans make wort viscous (Charalambous and Bruckner, 1977; Blanchflower and Briggs, 1991c). In addition to flavour problem in beers, it may be that the high levels of *n*-alkylresorcinols present in rye, which are regarded as 'anti-nutritional factors', are deleterious to yeasts. Raw cereals contain significant amounts of β-amylase as well as some α-glucosidase, phosphatase and minor protease activities. However, they also contain protein inhibitors of some proteases and α-amylases. The low gelatinization temperatures of the starches from barley, wheat, rye and triticale, combined with their low oil contents, indicate that, when suitably milled, these grains can be usefully used as mash-tun adjuncts without prior cooking (Wieg *et al.*, 1969; Koszyk and Lewis, 1977; Lloyd, 1978; Martin, 1978; Canales, 1979; Wieg, 1987).

The feasibility of using raw barley in brewing had been recognized by 1800 (Maiden, 1978) and this material was used by British brewers in wartime (Letters, 1990). Irish whiskey distillers always use high proportions (for example 68% of the grist) of 'low-dried' barley (i.e. dried to *ca.* 4% moisture) originally ground between stones or, more recently, in hammer mills (Briggs, 1978). Both wheat (1.8–2.1% N) and barley (1.5–1.8% N) have been used in infusion mashing in breweries after hammer milling to pass through about 1.6 mm (0.063 in) and 4.8 mm (0.19 in) screens, respectively. At high malt-replacement rates of 26–35% extract, yields were 261–268 and 208–234 l°/kg (20 °C, as-is) (87–90 and 68.6–78.0 lb/Qr), respectively (Scully and Lloyd, 1965, 1967). It is likely that these extracts were low because of incomplete gelatinization of the small starch granules at these mashing temperatures. It is doubtful if such high malt-replacement rates are now used in the UK. Raw, entire sorghum grains have been tested as

Table 12.5 Analyses (ASBC) of various adjuncts

Adjunct	Moisture (% as-is)	Extract		Protein (% as-is)	Fat/oil (% as-is)	Fibre (% as-is)	Ash (% as-is)	pH	Gelatinization temperature range	
		(% as-is)	(% on dry)						(°C)	(°F)
Corn (maize) grits	9.1–12.5	78.0–83.2	87.7–92.8	8.5; 9.5	0.1–1.1	0.7	0.3–0.5	5.8	61.6–73.9	143–165
Corn (maize) flakes	4.7–11.3	82.1–88.2	91.0–93.4	–	0.31–0.54	–	–	–	–	–
Refined maize grits (maize starch)	6.5–12.3	90.6–98.3	101.2–105.6	0.4	0.04	–	–	5.0	61.5–73.9	143–165
Rice grits	9.5–13.4	80.5–83.8	92.2–96.1	5.4; 7.5	0.2–1.1	0.3–0.6	0.5–0.8	6.4	61.1–77.8[a]	142–172[a]
Sorghum grits	10.8; 11.7	81.7; 81.3	91.4; 91.1	8.7; 10.4	0.5; 0.65	0.8	0.3–0.4	–	67.2–78.9	153–174
Wheat flour	11.5	80.1	90.7	11.4	0.7	–	0.8	–	–	–
Wheat starch	11.1; 11.4	86.5; 95.2	105.2; 97.5	0.2	0.4	–	–	5.7	51.7–63.9	125–147
Torrefied wheat	4.9	74.4	78.2	12.2	1.0	–	–	6.2	–	–
Torrefied barley	6.0	67.9	72.2	13.5	1.5	–	–	5.9	–	–

[a] Variation in US rice types: 65–68 °C (149–154.4 °F); 71–74 °C (*ca.* 160–165 °F) (long grain rice).
Data of Canales and Sierra (1976); Coors (1976); Bradee (1977); Canales (1979).

adjuncts in brewing clear beers. They tend to confer a darker colour and bitter flavour to the products, and so the use of sorghum grits is preferred (section 12.5.1). Most interest was shown in the use of whole barley in brewing (an interest which seems to have declined) and wheat flours (section 12.5.4). Whole grain is easy to handle, but milling may present difficulties.

Raw barley for brewing is selected for its cleanliness, its nitrogen and moisture contents, and for plump corn size so that it can be suitably broken up by the milling system in use. Viability and germination characteristics are irrelevant. Often the grain is washed before use. Major problems with raw, or cooked, barley adjuncts are caused by the β-glucans that are present. For adequate starch conversion, a mash with raw barley should contain a sufficient level of α-amylase; β-glucanase is necessary to achieve good run-off and the avoidance of filtration and other problems with the beer. The relatively recent availability of heat-stable microbial β-glucanases has increased the ease with which barley adjuncts can be used (section 12.10).

If significant contributions of soluble nitrogen need to be made to the wort by adjuncts during mashing then the level of proteolytic activity in the mash is also important. Comparatively low levels of barley in brewer's grist (10–20%) can usually be adequately converted by the enzymes from a good-quality pale malt, particularly if an extended, temperature-programmed mashing schedule is used. However in many instances, and particularly if higher levels of malt replacement are used, it is necessary to supplement mashes with a mixture of α-amylase(s), protease(s) and β-glucanase(s) of microbial origins (e.g. Brenner, 1972; Button and Palmer, 1974; Lisbjerg and Nielsen, 1991). Without enzyme additions, there is a risk of obtaining a low yield of extract and a poorly fermentable wort containing reduced levels of soluble nitrogenous substances and slightly lower levels of polyphenols. Such enzyme mixtures were used, often with a small addition of enzyme-rich malt, in making barley syrups and in barley-brewing (Wieg *et al.*, 1969; Scherrer *et al.*, 1973a,b; Wieg, 1987). If barley varieties with low levels of β-glucans become commercially available, they will probably be natural choices for use as adjuncts in brewing. If raw barley is ground too finely it can cause problems with wort separation and in most breweries it is advantageous to keep the husk as intact as possible. Besides hammer milling, barley has been successfully used in brewing after dry roller milling, after impact milling to loosen the husk followed by burr milling or roller milling, after steam conditioning followed by roller milling or after a pre-soak followed by wet milling (Martin, 1967; Imrie, 1968; Witt, 1970b; Pfenninger *et al.*, 1971; Roberts and Rainbow, 1971; Button and Palmer, 1974; Rao and Narasimham, 1976; Wieg, 1987). In the interests of cleanliness and product flavour, barley should be washed before it is wet milled and it has been suggested that an alkaline wash be used followed by a treatment with sulphur dioxide (Wieg, 1987) or a warm wash at 70 °C (158 °F) for 4 h even if a little extract is lost.

When using high levels of barley to replace malt in a brewery mash, it is desirable to use a decoction or a temperature-programmed mashing system and to add supplementary enzymes. Depending on the enzymes used and the details of the grist, temperature rests at about 50 °C (122 °F) for proteolysis and β-glucanolysis, 63–65 °C (145.4–149 °F) for starch liquefaction and saccharification and 70–75 °C (158–167 °F) for dextrinization of residual starch are usually suitable, with the durations of the rests adjusted to meet product requirements. The mashing period may need to be extended to several hours (Button and Palmer, 1974; Wieg, 1987). When barley-rich grists are supplemented with a balanced mixture of enzymes, the wort obtained can be very similar to that obtained from an all-malt grist.

Sorghum is being used as an adjunct in African breweries. It seems that extended mashing with microbial enzymes is the best means of obtaining worts for clear beers from this material (MacFadden, 1989; Bajomo and Young, 1992).

In the preparation of grain spirit, whole, raw cereal grains may be cooked under pressure and then blown into the mashing vessel when the pressure is released. The pressure drop causes the cereals to disintegrate. Spirit yields obtainable (litres of alcohol-tonne dry cereal) are 347 from barley, 439 from maize and 430 from wheat (Walker, 1986).

12.4.2 Cooked, intact cereal grains

Whole grains cooked in various ways have long been used in brewing. Torrefied and kibbled (chopped into short lengths) wheat, rye and oats have been used in some special breads. Such materials have the advantages that they can be stored, handled and milled in the same manner as malt and, in contrast to grits, they do not need to be pre-cooked before mashing. Their extract yields are greater than those of raw cereals (Tables 12.3, 12.4, 12.5). It has been claimed that extract recoveries would be greater if these materials were pre-mashed with enzyme additions and cooked before being mixed with a malt mash. Frequently used preparations are roasted barley and torrefied or micronized barley or wheat. These cooked materials contain no active enzymes but the heating they have received, which denatures the enzymes and other proteins, usually causes at least a partial expansion of the grain, or even 'popping', the endosperm structure is disrupted, some cell walls are broken and some of the starch granules are gelatinized (Reeve and Walker, 1969). Heating can partly degrade β-glucans and other hemicellulosic materials so that extracts from torrefied grains are less viscous than those from raw grains. In some preparations, where the temperature during cooking has reached about 140 °C (284 °F), starch is partly broken down to 'pyrodextrins' or 'torrefaction dextrins'. Also depending on the conditions of heating, and probably the moisture content of the grain, sugars may be caramelized, colours develop as melanoidins are formed from interactions between sugars and amino acids or other nitrogenous substances and polyphenols may be oxidized (to coloured 'phlobaphenes'). Many flavour and/or aroma substances are also formed, including sulphur-containing compounds, aldehydes, ketones, lactones, aromatic and heterocyclic substances, together with bitter diketopiperazines (Briggs, 1978). Therefore, cooking facilitates extract recovery during subsequent mashing and generates flavour, colour and aroma. Cooked grains release very little soluble nitrogen during mashing.

Cereal grains have been pre-cooked in a wide variety of ways and, experimentally, microwave and dielectric heating have been tested. In the traditional method, barley is roasted in slowly rotating steel drums using either direct heating or hot air (cf. special malts, Chapter 14). Heating must be carefully controlled to obtain the final desired colour and to avoid overheating, which can result in carbonization and even fire. Cooking grains at high temperatures always involves a real risk of fire, and plant is arranged with this in mind. Roasted barley has a characteristic sharp, 'dry' flavour and, in addition to colour and character, it is used to provide extract. Stouts usually owe their colour and character to this product. Typically, roasted barley is prepared from clean, plump, high-nitrogen feed-type barley. It usually has a hot water extract of about 267–277 l°/kg (90–93 lb/Qr), a moisture content of 1–2% and a colour of 1000–1550 EBC units.

Historically, grain was cooked or torrefied by passing it through a stream of red-hot sand where it popped, and gained a nutty flavour. At present, grain may be pre-wet to 14–18% moisture and be pre-heated to about 65.6 °C (150 °F) before being cooked, when it swells and pops, while being moved in a stream of hot air at 260 °C (500 °F) (Britnell, 1973; Coors, 1976). Alternatively grains may be micronized (Collier, 1970, 1975; Brookes and Philliskirk, 1987). Grain is metered onto the end of a moving wedge-wire belt or a vibrating feed tray 5 m (16.4 ft) long and 1.5 m (4.9 ft) wide. As it moves along, the grain passes below ceramic tiles heated to red heat with a double row of gas burners. The radiant heat (infrared radiation) heats the grain to an internal temperature of about 140 °C (284 °F) and cooks it. The vibrating feed causes the grain to move about relative to the heat source and ensures an even treatment. As the grain is heated, it softens, the rising interial water vapour pressure swells it and may cause it to burst. After cooling, the products have moisture contents of about 4–6% (Tables 12.3, 12.5). Sometimes the hot grain is passed immediately between heated rollers and is flaked (section 12.5.2). Wheat and barley are the cereals usually micronized. The products have low colours (e.g. about 2 EBC units) and yield extracts with bland flavours. When rapid wort separations are needed, the use of micronized wheat in the mash can cause problems. The slow wort separation has been associated with the formation of disulphide links between the grain proteins during the micronizing process. The problem is reduced by heating the grain less, to give products with higher final moisture contents (South, 1991, 1992). The inclusion of torrefied wheat in a mash can increase wort viscosity and this effect is even more apparent with torrefied barley, which is rich in β-glucans (Lloyd, 1986).

12.5 Grain preparations

12.5.1 Grits

Grits are uncooked, nearly pure fragments of starchy endosperm derived from cereal grains. The preparations most employed are from maize (corn), sorghum and rice but preparations from other cereals have been used. Grits may be used directly in breweries or distilleries, where they must be cooked during the mashing sequence, or they may be flaked (section 12.5.2). During the preparation of grits, the surface layers and embryos of the grains are removed, reducing the lipid, ash and fibre contents of the materials. A product has a higher final percentage starch content than the original grain.

Experimentally pearled or debranned barley, prepared by abrasive milling, has been used in brewing (Briggs, 1978; Munck and Lorenzen, 1977). The debranned material, prepared in 75% yield, had an extract of 86% (compared with 75% for the original barley). Its lipid, fibre, ash and polyphenol contents were less than those of the barley, but because the material was richer in β-glucans, it yielded more viscous worts. This viscosity could be reduced by additions of β-glucanase.

Rice grits or broken rice are usually fragments of endosperm produced while debranning and milling grain for human consumption. As the efficiency of milling has improved so the production of this by-product has declined (Coors, 1976; Canales, 1979; Lorenz and Kulp, 1991). Although the product is relatively poor in lipid (and fibre) it must be stored carefully to avoid the development of off-flavours owing to fat rancidity. The material used must be well cooked before it will yield its extract in mashing (Tables 12.3, 12.5). Sometimes the adjunct is cooked under pressure. It is advantageous to incorporate a thermostable α-amylase in the mixture to liquefy the starch. The selection of rice for grits to be used in brewing is important. Long-grained rice is unsuitable, but short and medium-grained rice is acceptable (Bradee, 1977). It has been proposed that gel-point detection be used as a selection technique for choosing rice grits that are suitable for brewing (Teng *et al.*, 1983). In general, the flavour imparted to beers is comparatively 'neutral', making this adjunct popular, but the flavour does seem to differ from that of beers made with maize grits. Light, clean-tasting beers are produced using rice grits.

Maize grits, prepared from yellow dent corn, are widely used as adjuncts in the USA (Bradee, 1977; Canales, 1979). Grits may be prepared by one of several milling processes (Kent, 1983; Lorenz and Kulp, 1991). In an integrated process, in which valuable by-products such as oil are separated, selected maize is screened and washed, then conditioned with steam to soften the hull and the germ. The conditioned grain is decorticated and degermed by abrasive milling and is successively passed through break rolls and screens to prepare a pure grits fraction, in *ca.* 55% yield. The grits are rich in starch and contain much less oil than the original grain. After cooking, maize grits yield a good extract during mashing (Tables 12.3, 12.5).

Originally the use of sorghum as a brewing adjunct gave poor-quality, bitter-tasting beers. Partly this was because the grits were contaminated with the surface layers (which are rich in lipids and polyphenols). These difficulties have been overcome by selecting improved varieties and by improving the efficiency of the milling operations (Bradee, 1977; Canales and Sierra, 1976; Canales, 1979; Kent, 1983). In a particular milling process, the grain yields 35% grits and 12% flour. The remainder is used for animal-feed. The technical feasibility of making grits from millets has been demonstrated.

For use in mashing, grits must be cooked to break open their structure and gelatinize their starch. This is necessary because the high gelatinization temperatures of the starches of these materials means that they are not attacked by malt enzymes at temperatures at which the latter survive. Afterwards, the cooked grits are normally mixed into a malt mash to complete the starch conversion and provide the other components needed in the wort. For cooking, a slurry of grits mixed with a small proportion of highly diastatic malt (e.g. 5%) or a microbial α-amylase is heated. Frequently, heat-stable bacterial α-amylases are now used. The pH of the slurry may be adjusted and calcium salts may be added to favour α-amylase activity. Sometimes the slurry is held for a period at the optimum temperature for the α-amylase before the mixture is boiled or cooked under pressure. It is likely that the high temperatures achieved are needed to dissociate amylose–lipid complexes in the starch and achieve total starch liquefaction. However, in some trials, grits were cooked at 85 °C (185 °F) with a bacterial α-amylase. During cooking, the cell walls are disrupted, the starch swells and is gelatinized and it is partially degraded or liquefied by the α-amylase before this is inactivated by the rising temperature. Consequently, the viscosity of the mixture does not rise too high, nor does the mixture set to a gel on cooling. It does not cling to the stirrers or vessel surfaces, it does not burn onto the heating surfaces and the slurry can be pumped to, and mixed with, the malt mash (Briggs *et al.*, 1981).

12.5.2 Flaked cereals

By flaking cereals, or cereal preparations, the starch is gelatinized, the endosperm structure is disrupted, the thin products are easily permeated by enzymes in a mash and the breakdown products of starch hydrolysis are readily leached into solution. Consequently, the starch of flaked cereals is easily converted and the extract produced is easily recovered. Flaked cereals, particularly rice and maize flakes, were widely used in UK brewing practice because they do not need to be cooked in the brewery (avoiding the need for a cooker), they readily yield their extract in infusion mashing and they are easily handled and mixed with malt. They are now being partly displaced by other adjuncts. When mashed, they yield relatively little soluble nitrogen so their extract serves as a 'wort nitrogen diluent'. Usually they contribute little to product flavour.

In principle, many materials could be flaked. In practice, flakes from whole grains of barley, wheat and oats, from pearl barley and from yellow or white maize grits or rice grits have been used (Tables 12.3, 12.5) (Collier, 1970, 1975, 1984; Bradee, 1977; Lloyd, 1986, 1988a). Flakes are made by two different processes. In the older process the moisture content of the raw material is adjusted to 15–20% and the raw material is held for 24 h before being cooked either with steam at 90–100 °C (194–212 °F) for 20 min or under pressure. This softens the internal structure and begins starch gelatinization. The pre-cooked material is delivered through feed rolls to flaking rolls held at about 85 °C (185 °F). The rolls squash and flatten the material, which is dried in a counter-current of warm air (105 °C (221 °F)), usually to 8–12% moisture before being cooled. The product, which has a low bulk density (i.e. it is light in weight for the volume it occupies; Table 12.3), may be slightly crushed to make it more compact (Maiden, 1971). In the second, newer process, the raw materials are micronized and, while still hot and soft, they are immediately passed between flaking rolls and are cooled. An advantage of this method is that the product has a moisture content of 7–8% and so no drying stage is needed (Collier, 1975).

In the past, oat flakes, which appeared rather soft and grey, were used in the preparation of certain stouts, to which they added a distinctive character. On a dry weight basis, their extract yield was low, about 254–282 l°/kg (20 °C (68 °F) (85–95 lb/Qr)); their protein content was often about 15% and they contained a high level of oil and lipid, about 9% (Hind, 1940). The high husk content of flaked oats 'opened' the mash and favoured run-off (wort separation). The high β-glucan content of this cereal gave rise to viscous worts and the worts were often poorly fermentable. Flaked rice, which is normally white and is prepared from rice grits, is not used at present for economic reasons, but technically it is a well-liked product with a good extract yield (Table 12.3). The extract has a 'neutral' flavour. Often lipid contents are about 0.4%. Flaked maize grits are widely used, but the wide range of lipid contents quoted (Tables 12.3, 12.5; up to 0.75%) shows that this value should be limited in specifications.

Whole flaked grains of wheat and barley are available (Table 12.3) and flaked, debranned or pearled barley used to be manufactured. In some pilot brewhouse trials, flaked micronized barley yielded higher extracts (73.5% d.b.) than some traditionally prepared flakes (72.0% d.b.), but micronized maize flakes gave lower extracts than the more traditional product (76.5 d.b. against 78.0% d.b.) (Collier, 1970). On occasion, barley flakes produce highly viscous extracts possibly in part resulting from uneven steam cooking before flaking. The problem was also noted with flaked pearl barley but is less with micronized, flaked barley. The problem is caused by β-glucans and can be overcome by spraying the flakes with a solution of bacterial enzymes that contains α-amylase and β-glucanase. These enzymes were said to be destroyed during drying of the flakes, but the flakes, which otherwise appeared normal, had a high cold water extract and yielded wort with a low viscosity (Collier, 1970).

12.5.3. Extruded cereals

Extruded preparations of cereals are widely used in snack foods, breakfast cereals and animal-feeds (Lorenz and Kulp, 1991). The extruded products can be varied in size, shape and density so that easily conveyed and milled material may be prepared. Depending on the cooking conditions, cereal starch may be gelatinized, or liquefied by added heat-stable α-amylase (Anderson *et al.*, 1970; Hakulin *et al.*, 1983; Østergard *et al.*, 1989). Experiments have shown that extruded cereal preparations can be successfully used in mashing (Briggs *et al.*, 1986; Dale *et al.*, 1989; Laws *et al.*, 1986) and can be converted by microbial enzymes (Albini *et al.*, 1987), but apparently they are not yet widely used commercially as adjuncts.

A diagram of a simple, single-screw extrusion cooker is given in Fig. 12.1. Such machines vary widely in whether they have one or two screws, the pitches of the screws, whether they have variable pitch screws, and so on (Kirby *et al.*, 1988; Lorenz and Kulp, 1991). In the simple machine illustrated, cleaned, ground ingredients are fed from a vibrating feed at the base of the bin via a delivery and pre-conditioning screw, where steam, water or other substances may be added, through a mixing cylinder to the extruder. During its passage through the extruder, driven by a worm screw, the material is worked into a dough that is continuously heated from the screw jacket and cooked and is subjected to pressure, shearing and forced mixing. The sections of the barrel may be individually heated or cooled. Temperatures may exceed 200 °C (392 °F). At the end of the barrel, the plastic material emerges from a die; it expands and is cut into sections by a rotating knife before being cooled and dried. The products are firm, relatively dust-free and may be produced in a wide range of shapes and sizes. The pellets are easily permeated by water or enzyme solutions. During extrusion cooking, starch is gelatinized and cellular structures are disrupted. Proteins are denatured and microorganisms are killed. Experiments have been carried out with extruded barley, wheat, maize and sorghum. During mashing, the gelatinized starch in extruded pellets is easily converted, but there may be problems if the mash is made without supplementary enzymes because 'fines' reduce the rate of

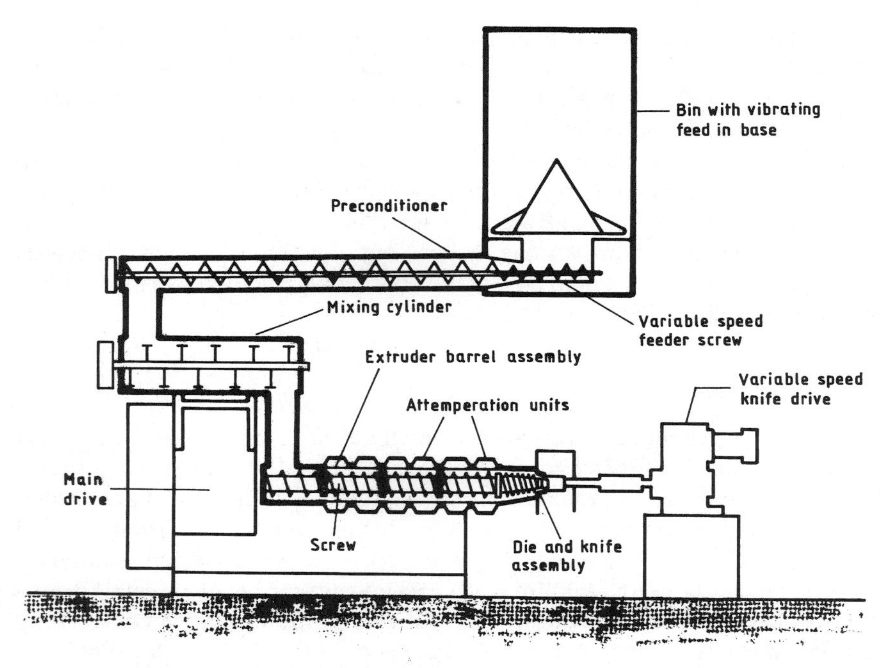

Fig. 12.1 A schematic view of a simple, single-screw extruder and ancillary equipment (Briggs *et al.*, 1986).

run-off from mashes and the worts may be viscous owing to extracted β-glucans and possibly pentosans. These problems may be overcome by adding suitable enzymes to the mash or possibly by the inclusion of heat-stable β-glucanase in the ground mixture being fed into the extruder. As with all mashes containing high percentages of adjuncts (20–30% malt replacement), the best results are obtained when supplementary enzymes are added and a temperature-programmed mashing sequence is used. In a series of trials using all lager malt or lager malt (70%) plus adjunct (30%), the calculated extract yields (l°/kg d.m.) after a temperature programmed mash were as follows (figures in parentheses, bacterial enzymes added): lager malt, 304.6 (310.2); wheat flour pellets, 348 (365); extruded barley, 292 (304); extruded maize, 311 (325); and extruded wheat, 315 (330). Different enzyme preparations gave different results (Briggs *et al.*, 1986).

12.5.4. Flours

Flours are prepared from raw cereals by milling procedures. Raw wheat and barley can be hammer milled on site to produce flours used in mashing, and sorghum flour is produced as a by-product of the manufacture of grits. However, flour milling, in the usual sense, is a complex process carried out

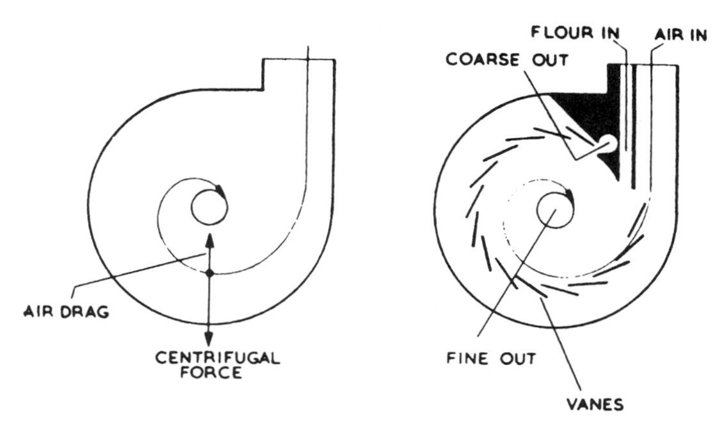

Fig. 12.2 The principle of one type of air classifier for fractionating flour. Air and flour are introduced into an air vortex in a modified cyclone, seen in plan-view in the diagrams. The particles sort into streams regulated by their sizes and densities and the air speed so that the centrifugal outward force balances the drag drawing them into the vortex. By altering the position of the guide vanes it is possible to regulate the size at which the 'cut' between the two fractions occurs (Rumpf and Kaiser, 1952).

in specialized mills; the finished products are purchased by the brewers or other users. Flours can be produced from all cereals or cereal preparations, usually by continuous processes, and the products are often separated into mill-streams, for example by sieving or air classification to give streams of different compositions (Kent, 1983). Flours are formed during the preparation of maize or sorghum grits, or from barley or pearled barley. Using air classification (Fig. 12.2), a barley flour fraction containing 72% starch and 10.2% protein was prepared from a material containing 57% starch and 15.2% protein (Briggs, 1978; Vose and Youngs, 1978; Canales, 1979; Kent, 1983). Maize and sorghum flours need to be cooked before mashing; in many mashing systems hammer-milled wheat and barley give wort-separation problems. However, when using mash filters, and particularly if supplementary enzymes are used, wort separation may be less of a problem.

Cereals break up in different ways when milled. Wheat milling is a highly developed process. Wheat flour preparations have been widely used in brewing. Wheat is a naked grain with a deep crease, or furrow, running along its ventral side, a feature that complicates cleaning and milling. Wheat selected for making 'brewing flour' is usually a soft variety, having a mealy or floury type of endosperm (Chapter 2) and a nitrogen content up to about 1.8%. Wheat milling is complex and many processing variations occur (Kent, 1983). The prime objective is to collect the starchy endosperm as flour, separated as cleanly as possibly both from the germ (embryo tissues) and the bran (pericarp, testa and aleurone layer). The grain is cleaned by various screening (sieving), dry-scouring and agitated-washing processes; other impurities are removed, for example by disc separators. The grain is conditioned to achieve an optimum moisture value, usually in the range 15.0–17.5%, and it is held to allow the moisture to redistribute itself and equilibrate throughout the grain. The conditioned wheat is fed into the 'break system' of a roller mill in which the grains are successively split open and the endosperm is scraped from the bran. Between each stage, fractions are separated by sieving ('scalping' in the early stages) before being processed further. Endosperm-rich streams are roller milled and reduced to flour, while germ and bran are usually collected separately. Care is taken to maximize the yield of the 'best-size' range of endosperm fragments that make up the flour. In principle, finer grinds lead to a range of smaller particle sizes and easier extract recoveries when the flour is mashed; however, 'fines' (small particle sizes) interfere with wort recovery and the more readily extracted cell-wall polysaccharides yield more viscous worts and beers. In the final flour, starch and protein fragments are never entirely enclosed in unbroken endosperm cell walls. Such wheat flour may be used directly in the mash tun. However, by using a series of air-classification treatments (Fig. 12.2), it is possible to obtain a fraction with particles 17–35 μm diameter that is enriched in starch and depleted in protein (e.g. having 7% crude protein as against 9% in the original flour) and so has a better extract and fewer fines than the original flour (Rumpf and Kaiser, 1952; Eggitt, 1964; Kent, 1983). The other grades remaining after this procedure are used in several ways, for example to enrich bread flour (<17 μm diameter) or in biscuit manufacture (>35 μm diameter). An alternative method for making brewing flours is to adopt a milling procedure in which some 45% of the wheat endosperm particles consists of 'clusters' of starch granules and protein of greater than 100 μm average diameter and in which particles of less than 20 μm make up only 3–4% of the whole (Hough *et al.*, 1976).

Flours provide handling problems with which brewers may be unfamiliar. Flours may be bought in sacks or in bulk; they have a high bulk density (Table 12.3) (Maule and Greenshields, 1970). They need to be held in steep-sided containers equipped with arch breakers or vibrating feeds to ensure that they flow freely. Horizontal movement is in pneumatic or vibrating conveyors. Allowance must be made for the abrasive characteristics of the flour. To minimize the risks of fires and flour–dust explosions, the risks of build-up of static electricity are reduced by avoiding the use of plastic chutes, etc. All electrical equipment should be spark-proof and the all-metal containers for storing and handling flour (and refined starches) is grounded or earthed. It requires skill and the correct equipment to mix flour evenly with ground malt to obtain a uniform grist during mashing. Large-particle flours flow more easily and are more readily handled than fine flours (Hough *et al.*, 1976). 'Agglomerated' wheat flours, bound together in pellets, are still more easily handled, although the run-off characteristics they contribute to a mash are comparable with those of a normal flour (Webster, 1978; Canales,

1979). There have been proposals to make wheat flours or pellets supplemented with microbial enzymes (β-glucanase, pentosanase, cellulase) to minimize any run-off problems that may occur.

Wheat flour starch need not be gelatinized and liquefied before the flour is mashed in with malt grist, although this treatment may give a slightly enhanced extract recovery. It was reported that pre-soaking flour before mashing in gives an elevated recovery of extract. In North American practice, wheat flour may be slurried and cooked, the slurry being initially made up at less than 52 °C (125 °F) to minimize clumping and then cooking takes place at less than 98 °C (*ca.* 208 °F) to minimize frothing (Canales, 1979). In infusion mashing, using a slow wort run-off, a 10% replacement of malt by wheat flour is usually well tolerated, but in fast wort-collection systems, using lauter tuns, fines may significantly slow wort recovery. Replacement levels of 25% and even 36% of malt with wheat flour have been successfully used when special precautions have been taken with mixing the grist into the mash tun and when using an enzyme-rich malt (Scully and Lloyd, 1965, 1967). However, the use of these exceptionally high replacement levels has probably been discontinued. Wheat flour contributes a limited amount of β-amylase to the grist. Run-off problems encountered when using wheat flour appear to be caused by the formation of impermeable layers of fines, largely hemicellulosic in nature, in the grist. Supplementation of grists with enzyme preparations containing hemicellulases, including pentosanases, can overcome this problem (Forrest *et al.*, 1985). As well as contributing little or nothing to the total soluble nitrogen and α-amino nitrogen levels of worts, the inclusion of wheat flours in mashes alters the spectrum of nitrogenous substances in wort to a significant extent. A high-molecular-weight fraction, probably containing glycoproteins, materially adds to head retention (foam stability) of beer (Anderson, 1966; Leach, 1968). For this reason, many brewers choose to use wheat flour, even with the risk of slowing wort separation. It has been shown that, on a weight basis, finely ground, raw or pre-gelatinized preparations of rye and millets are preferable to wheat flour in supporting head retention in beer, and their use has been proposed (Stowell, 1985). Beer brewed with wheat flour has an enhanced shelf life, i.e. a reduced tendency to form non-biological haze relative to all-malt beers, probably because of the overall reduction in high-molecular-weight polypeptide material and the reduction in tannin polyphenols.

Soya flours have been used in the USA. The β-amylase they contribute to the mash improves wort fermentability.

12.6 Refined starches

In principle, refined starches, which are relatively pure preparations of starch granules, can be prepared from any cereal grains and from other materials such as tapioca (manioc, cassava) and potatoes (Tables 12.3 and 12.5). Starch can be prepared from finely ground barley using an alkali-wash procedure but this is not used commercially (Goering and Imsande, 1960; Letters, 1990). The starches that are used are dictated by ease of preparation and local economic conditions. Refined wheat, maize, potato and tapioca (manioc) starches have been used. Refining must be thorough to remove unwanted flavouring and toxic substances. Maize starches are widely used in the preparation of syrups (section 12.7.2). Potato starch, or farina, is sometimes available at a low price and, because of its low gelatinization temperature, it is readily converted during either infusion or temperature-programmed mashing even at a 20% replacement rate (Klopper, 1973). Even so, brewers are generally unwilling to use it because of consumer resistance to 'potato beer'. Wheat starch has been employed in Canadian and Australian breweries, when local conditions made it economic to do so. Extruded starch preparations have been used to reduce dust (Geiger, 1972). Methods are available for preparing wheat starch by dry-milling or wet-processing procedures, the economics depending largely on the recovery of the gluten (Coors, 1976; Canales, 1979).

Maize (corn) is the most important source of refined starch. Known as corn flour, refined maize grits and wet-milled pearl starch, maize starch is made in around 67.5% yield from selected and clean grain by a continuous, wet-milling procedure (Swain, 1976; Smith, 1977; Canales, 1979; Kent, 1983; Lorenz and Kulp, 1991). Clean maize grain, from which broken corns have been removed, is steeped in warm water containing 0.1–0.2% sulphur dioxide for about 2 days at a temperature below the gelatinization range of the starch, 48–52 °C (*ca.* 118–126 °F). Among other changes, the proteins of the endosperm are weakened through cleavage of the interpeptide disulphide links by the sulphur dioxide. The grain is hydrated and softened. The corn-steep liquor (6–8% solids) is concentrated to approximately 50% solids and is sold for use in animal-feeds or as a nutrient for the industrial culture of microorganisms. The softened grain is coarsely ground and the detached embryos, which are rich in oil and so have a low density, are separated by flotation in hydrocyclones. After drying, they are pressed or are extracted with solvents to yield maize oil (*ca.* 50%) and a solid residue that is used for cattle food. The degermed material is ground more finely, then the hulls and fibre are collected on sieves while the starch and gluten pass through. The starch slurry, 35–40% solids, is purified by counter-current washing. The gluten, which is mainly in suspension in the washing liquor, is recovered by centrifugation and, together with the hulls and fibre, is used in cattle-food. The slurry of washed starch may be used in the manufacture of syrups (section 12.7.2). Alternatively the granular starch, which typically has an amylose:amylopectin ratio of 28:72, a crude protein content of 0.35% and a fat content of 0.04%, is collected in continuous centrifuges and dried to 1–12% moisture.

Refined starches are the 'purest' mash-tun adjuncts available. When dry they must be handled with all the precautions used with flours; as with flours, some handling and dust problems may be reduced by 'granulation' or pelletizing (Canales, 1979). Potato and wheat starches, having low gelatinization temperatures, can be added directly into a mash with the malt grist. Starches may be conveniently handled as a suspension or slurry in water, so avoiding the chance of explosions of starch dust–air mixtures, but it is essential to keep the slurry mixed, as settling ('setting') must be avoided. Maize starch must be cooked and liquefied before use and wheat starch may be. Cooking is carried out as for grits and, indeed, starch and grits may be mixed in the cooker. The slurry is mixed with a source of α-amylase (high-DP malt or bacterial α-amylase) to partly degrade or 'liquefy' the starch as it gelatinizes, so producing a runny product that is less prone to burning onto heating surfaces and that does not set on cooling. Refined starches contain very little nitrogenous material; on a dry weight basis they have extracts of 102–105%, or about 380 l°/kg (120–130 lb/Qr), because of hydrolysis gains (Tables 12.3, 12.5). Since starches are completely converted during mashing, they do not cause run-off problems. The products of hydrolysis of pure starches make no direct contribution to the flavour of beer or other products. However, many preparations of wheat and maize starches contain appreciable amounts of lipids (e.g. 0.5%) and for some uses this is undesirable.

12.7 Sugars

12.7.1 Sucrose and invert sugars

Sucrose (α-D-glucopyranosyl-(1→2)-β-D-fructofuranoside) and its hydrolysis product invert sugar were once widely used in the brewing industry, but they are now relatively expensive and their use is restricted. Historically, lactose (β-D-galactopyranosyl-(1→4)-D-glucopyranose) derived from milk whey was used, for example in the manufacture of milk stouts. Recently the use, as a brewing adjunct, of lactose from whey treated with a β-galactosidase to cleave the lactose to galactose and glucose has again been suggested. Probably more than 80 million tonnes of sucrose are produced annually of which *ca.* 20 million tonnes are traded (Birch and Parker, 1979). Sucrose is prepared from sugar cane, a giant tropical grass (*Saccharum officinarum* L. and related spp.), and from the swollen root of the sugar beet (*Beta vulgaris* L.), an annual that grows in temperature climates. Fully refined sucrose or invert sugar may be used in brewing. Impure grades of beet sugar have unpleasant flavours and cannot be used. In contrast incompletely refined preparations of cane sugar have flavours that may be desirable for some purposes.

Sugar cane is a perennial plant that is propagated by cuttings. When fully grown it is harvested, by machine or by hand, and the cut cane is taken to be processed immediately to minimize sucrose inversion and sugar losses (Meade and Chen, 1977; Birch and Parker, 1979). The cane occurs in long, jointed lengths 2–5 cm (1–2 in) in diameter. Details of processing differ. The cane is crushed between rollers then shredded and the juice is collected. The solid residue (bagasse) is washed and pressed to recover the last sugar. The bagasse may be burned as fuel or used to make hardboard or insulating board, or it may be used as cattle bedding. About 10–16% of the cane is soluble solids, and it contains 73–76% juice. Of the solids, 75–92% is sugars. Sucrose makes up 70–88% of the solids (Meade and Chen, 1977). Suspended solids are removed from the dark green, acidic (pH 5–5.5) and muddy-looking juice by screening and passing it through a cyclone. The pH is then raised to 7.5–8.0 by calculated additions of suspended lime and, perhaps, other substances, and the juice is heated under pressure. Protein coagulates and calcium phosphate precipitates; these solids separate with wax and other impurities. The 'mud' is separated by settling and filtration and is returned to the cane fields. The clear yellow juice contains sucrose, lesser amounts of invert sugar, traces of tri- and tetrasaccharide sugars, minerals and organic substances. It is concentrated from 12–15% solids to 60–65% solids by heating in multiple-effect vacuum evaporators. The syrup is concentrated further in vacuum pans until sugar crystallization begins. The impurities present prevent complete crystallization. The suspension of sugar crystals in the syrup (molasses) is called massecuite. Solid sugar is collected by centrifugation and, after washing, is despatched to a refinery. The residual syrup is concentrated further and two more collections of progressively less pure sugar crystals are taken. The final crop of sugar is dissolved and returned to the start of the process stream. The final molasses, from which no more sucrose can be crystallized, may be used to make rum or as a medium for propagating bakers' yeast. The raw sugar is checked for microbes. It is about 95–98% sucrose and contains minerals, some insoluble and some coloured impurities and residual syrup. This material is widely traded but is usually refined before use.

Sugar-beet roots are harvested and washed and the green leaves, which are separated, may be fed to cattle (McGinnis, 1971; Birch and Parker, 1979). In the factory, the washed and destoned beet is cut into cossettes (strips *ca.* 7.8 cm × 0.6 cm × 0.3 cm (*ca.* 3 in × 0.2 in × 0.1 in). The sugar, with impurities, is extracted by diffusion. The warm extracting liquid and the cossettes move counter-current so that fresh cossettes are extracted by the strongest sugar solution and those that are nearly exhausted are rinsed with fresh water. The spent sugar-beet pulp is dried and used in animal-feed. The pH and microbial purity of the juice, initially containing 10–15% sugar, is controlled. It is screened then treated with lime followed by carbon dioxide. The calcium carbonate formed, together with impurities, is separated as a sludge. This type of treatment is repeated, sometimes with other precipitating reagents. The product is a less-coloured, clear juice, which is concentrated by evaporation to 50–65% solids. This in turn is boiled in vacuum pans until crystallization occurs. The solid sugar is collected by centrifugation. This product requires considerable further refinement before it can be used.

Brewers usually use partially or completely refined cane sugar; the composition of the raw material is too variable. Residual impurities add to the character and the colour of some preparations of brewing sugars (Table 12.2). Sugar refineries are large, highly efficient units (Meade and Chen, 1977). In the first refining step (affination) molasses are removed from the raw sugar crystals by successively washing with warm syrup and water. The liquids and solids are separated in centrifuges. The impure affination syrups are processed separately. The washed sugar is dissolved, 'melted' in water. Insoluble impurities are removed and lime is added to the solution, followed by carbon dioxide. The precipitate of calcium carbonate carries down many impurities. Alternatively, liming followed by additions of phosphoric acid or sulphur dioxide are used to form adsorbent inorganic precipitates. The amber-coloured solutions may be decolorized with animal charcoal (bone char) and perhaps other adsorbents and may be treated with mixed-bed ion-exchange resins, before being concentrated under vacuum until crystallization occurs. The sugar is separated in centrifuges.

Partially purified cane sugar may be purchased as brewing grades of sugar, for example, brewing syrup with an extract of 258 l°/kg, a colour of 30°EBC units and a nitrogen content of *ca.* 0.01% (Table 12.2). The main sugar is sucrose, but the syrup is still 'impure' and only about 95% fermentable. Because its sugars content is low, often about 62% (w/w), it will not keep (Anon., 1980). If the concentration is increased, crystallization will occur at about 67% (w/w). Solid preparations of sugar, varying in purity, may be dissolved in the wort during the copper-boil, but precautions must be taken to prevent material sinking and burning onto the heaters. It is safer to dissolve the sugar in a separate, stirred container before it is added to the copper. Granular, fully refined sugar is better than 99.9% pure sucrose, is colourless and is fully fermentable. However it may partly decompose and interact with other wort constituents during the copper-boil, generating colour. Its extract is high and completely recovered in brewing (Table 12.2). Some brewers used candy sugar, crystals of sucrose formed into string or rods. Others prefer brown, less refined preparations.

Invert sugars are prepared by the hydrolysis ('inversion') of sucrose, often raw sugar. This can be achieved with the enzyme invertase, but it is usually carried out with dilute sulphuric or hydrochloric acid. The product is neutralized with calcium carbonate or soda ash and, after partial purification with a charcoal treatment, is concentrated under vacuum to a syrup. The syrup may contain up to 83% (w/w) solids and must be stored warm at

40–50 °C (104–122 °F) to prevent crystallization (Maiden, 1978) (Table 12.2). Alternatively, it may be seeded with solid invert sugar and cooled. The solidified material, which still contains *ca.* 17% water, is supplied to brewers in 50 kg (*ca.* 1 cwt) blocks. Extract yields are good, 321–326 l°/kg. Invert sugar preparations may contain 0.01–0.04% nitrogen and have colours in the range 30–100 EBC units (in 25 mm cells, 100% sample). The more deeply coloured samples contain more nitrogenous material and the solids are less fermentable, e.g. 91% against 95% for the paler types (Anon., 1980). As the main sugars glucose and fructose are completely fermentable, it must be that the unfermentable materials are impurities and substances formed during acid inversion. Many substances may be formed during the inversion process, including melanoidins from condensation reactions between the sugars and nitrogen-containing substances. It is the by-products, impurities and colours that give character to invert sugars and may lead to their being preferred to the cheaper sucrose, even though the pure sugars are equally well used by yeast. Typical older types of brewers' priming sugar (primings) contain sucrose and invert sugars, often in a ratio of 45:55. (Now starch-derived sugars are often used instead; section 12.7.2.) Sugars do not crystallize readily from these mixtures, which are supplied as concentrated syrups with extended storage lives even at room temperature and have extracts of 290–340 l°/kg. Primings can be provided with a variety of colours and nitrogen contents and with fermentabilities resembling those of invert sugars. In addition, they are chosen for their special flavours: bland, sweetness (various intensities), burnt note, raw sugar, and so on.

12.7.2 Sugars and syrups from purified starch

Sugars or syrups from the partial or essentially complete hydrolysis of refined maize starch contain no significant quantities of nitrogenous materials or other non-carbohydrate yeast nutrients. Therefore, like sucrose-based products, these materials are a source of extract (but of variable, specified fermentability) that acts as a wort extender and protein and polyphenol diluent rather than as a wort replacement. They are easy to handle and, by choosing the appropriate type, can be used to adjust the properties of worts. For economic reasons, these are now much more widely used than sucrose-based products. During their preparation contaminating lipids may need to be removed (Howling, 1979). In general, the starch, prepared with a wet-milling process (section 12.6), is used as a 35–40% slurry in water. The hydrolytic breakdown of the starch is carried out in two stages. In the first stage, the starch slurry is solubilized (pasted, cooked, liquefied or dextrinized) using either an α-amylase or a mineral acid. The granules are completely gelatinized and the polysaccharides are sufficiently degraded that the products of partial hydrolysis are soluble, have a low solution viscosity and do not set to a gel or retrograde (separate from solution) when the mixture is cooled. In the second, 'saccharification', stage the dextrins are converted into the desired mixture of sugars. The two hydrolytic stages may be completed using dilute mineral acid throughout as the catalyst (acid-acid procedures), or an acid-catalysed liquefaction first stage can be followed by an enzymic saccharification stage (acid/enzyme procedures) or both stages may be carried out using enzymes (enzyme/enzyme procedures). A wide range of products, with closely controlled sugar compositions, may be prepared (Howling, 1979; Anon., 1980; Hebeda, 1987; Palmer, 1987, 1992).

Acid/acid processes may be used to prepare 'confectioners glucoses' with a wide range of compositions and fermentabilities (Table 12.6). Starch is hydrolysed with hydrochloric acid. Subsequently the mixture is neutralized with soda ash, filtered and treated with charcoal. Residual ions may be removed with ion-exchange resins. This is desirable, since residual chloride ions can potentiate the corrosion of stainless steel vessels and pipework and the ions can confer undesirable flavours to the product. These preparations have good extracts (Table 12.2) and have fermentabilities of 30–50%. If acid hydrolysis is extended, coloured by-products, bitter-flavoured substances and reversion products, such as polymeric materials and gentiobiose, may be formed (Imrie and Martin, 1969). The product, termed 'chip-sugar', may be supplied as blocks or 'chips', or as a syrup blended with other materials. The extract may be 319 l°/kg (Table 12.2) and is about 82% fermentable. Colours are in the range 200–500 EBC units (25 mm cell, 100% sample). The bitter and coloured components may be partly removed by charcoal treatments. Chip sugars confer characteristic 'dry' flavours to beers.

Products prepared using acid/enzyme or enzyme/enzyme processes may be supplied as solids or syrups, with a wide range of closely defined compositions (Anon., 1980) (Table 12.6). Some syrups must be held warm at about 50 °C (122 °F) to prevent solidification and may be supplied at 60–71 °C (140–160 °F). Enzyme-converted materials contain fewer inorganic salts than those converted with acids. Fermentabilities of 75–85% are common, but lower and higher values (within the range 30–95%) can be achieved. The products are usually colourless but can be coloured if desired (e.g. by the addition of caramel) to 0–500 EBC units. In the UK, these materials are usually made from maize starch, but materials made from wheat starch are also available.

Enzymic liquefaction of a starch slurry is normally achieved with one of the bacterial α-amylases, stabilized by an added calcium salt and at a chosen pH. With the more labile enzymes, the starch slurry–enzyme mixture is progressively heated to 90 °C (194 °F). Using a particularly heat-stable α-amylase, the starch in a slurry can be continuously liquefied in a tube converter, at about 140 °C (294 °F). Alternatively the slurry may be pasted at 120 °C (248 °F), released from a jet and be flash-cooled to 105 °C (221 °F), when the heat-stable amylase is added to liquefy the gelatinized material. These high temperatures are used to ensure breakdown of the

Table 12.6 Composition of glucose syrups derived from starches by various routes

					Acid/enzyme conversion					
Syrup type	Acid conversion				High maltose		High conversion		Enzyme/enzyme conversion	
Dextrose equivalent	30	34–36	42–43	55	42	48	63	70	42	65
Dextrose (glucose)	10	13.5	19	31	6	9	37	43	2.5	34
Maltose	9	11.5	14	18	45	52	32	30	56	47
Maltotriose	10	10	12	13	12	15	11	7	16	3
Maltotetraose	8	9	10	10	3	2	4	5	0.7	2
Maltopentaose	7	8	8	7	2	2	4	3	0.4	1.5
Maltohexaose	6	6	7	5	2	2	3	2	0.7	1
Maltoheptaose	5	5.5	5	4	–	–	–	–	–	–
Higher sugars	45	36.5	25	12	–	–	–	–	–	–
Maltoheptaose plus higher sugars	–	–	–	–	30	18	9	10	23.7	11.5

Data from Howling (1979).

amylose–lipid complexes in the starch. The thermostable α-amylase is able to degrade the amylose before the complexes re-form and retrogradation occurs. After liquefaction, it may be necessary to remove oil and precipitated materials. Carbon adsorption and/or ion-exchange treatments may be used. After cooling and pH adjustment, saccharification may be carried out with fungal amylases, highly diastatic malt, or β-amylases from wheat, soya or bacteria (Table 12.6). Sometimes these are supplemented with debranching enzymes, such as pullulanase or isoamylase.

Highly fermentable preparations of nearly pure glucose (dextrose; 95–98% if an enzyme/enzyme process is used, 92–93% if an acid/enzyme process) are obtained using amyloglucosidase (glucamylase; with no or only low levels of transglucosidase activity), with or without α-amylase and preferably supplemented with a debranching enzyme to accelerate the breakdown of α(1→6)-linkages. Conversion may proceed for 2–3 days at 55–60 °C (131–140 °F). Syrups with a high glucose content must not be concentrated over about 35% solids or crystallization will occur. With many yeasts, high levels of glucose in the wort suppress the fermentation of higher saccharides, causing 'hanging fermentations' and incomplete attenuation. To overcome this problem 'high-maltose' syrups are used instead of 'high-glucose' syrups (Table 12.6). These are prepared by saccharifying liquefied starch preparations with plant β-amylases (malt, soya flour), fungal enzymes or *Bacillis polymixa* β-amylase, preferably supplemented with a debranching enzyme. Syrups with DE values as high as 70–80% can be prepared (Howling, 1979). Before use, or during manufacture, syrups must be refined to remove residual lipids (as much as 0.6% in some starches), mineral salts and other unwanted materials; this is achieved by centrifugation, filtration and treatments with ion-exchange materials and active carbon. In addition anti-foaming agents may be added and sulphur dioxide may be used as a preservative (Howling, 1979). Sulphur dioxide is commonly present in many syrups (2–40 p.p.m.) but the maximum acceptable levels must be specified by users since excessive amounts can cause problems, such as reduced foam stability in beers.

In recent years high-fructose content syrups have become available (Smith, 1977; Howling, 1979). Using α-amylase liquefaction and then saccharification with amyloglucosidase supplemented with debranching enzyme, a syrup is prepared with a DE of 98% and about 96% glucose. After purification, this is passed through columns packed with immobilized microbial 'ketol isomerase' enzyme, which catalyses the reversible isomerization of glucose to fructose. After carbon treatment, the syrup is concentrated to 70–72% solids; these are 99% fermentable. A typical composition (% d.b.) is fructose 42, glucose 54, maltose 2, maltotriose 1 and higher saccharides 1. Potentially this preparation can be used to prime beer to confer sweetness. Fructose is sweeter than sucrose, which, in turn, is sweeter than glucose. The syrup is as sweet as sucrose but is less expensive.

Experimentally other types of syrup have been prepared, but as yet they do not compete with malt or extract derived from it.

12.8 Malt extracts and wort replacement syrups

Malt extracts and other syrupy materials prepared by concentrating worts from mashes vary widely in composition, colour, flavour, aroma, sweetness, fermentability and enzyme content. They are used in home brewing, occasionally in commercial breweries (usually as wort extenders and priming sugars), in baking special breads, cakes and biscuits, as yeast foods, in malted drinks (often with dried milk), in confectionery and as carriers for vitamin-rich cod- or halibut-liver oils. In the past enzyme-rich, high-diastase preparations were used to remove starch size from textiles and in cloth-making but now heat-stable bacterial amylases are preferred for these purposes (Griffin *et al.*, 1968; Burbridge and Hough, 1970; Maule and Greenshields, 1970; Roberts and Rainbow, 1971; MacWilliam, 1971; Briggs, 1978). Extracts are prepared by mashing all-malt or other grists and then collecting the worts using one of the methods used in breweries. The wort is concentrated to a syrup. This used to be carried out by evaporation in heated vacuum pans, but now this is done in single-, double- or triple-effect evaporators. Sometimes the products are further dried to crusty, powdery materials (95–98% solids) by spray-drying or by heating on belts moving through vacuum chambers. The treatments used depend, in part, on the nature of the final product and whether or not it is desired to retain the enzymes from the mash in an active state.

In the past, diastatic extracts, rich in malt enzymes, might be added to brewers' mashes at rates of about 1 part per 36 parts of grist (w/w), to improve or supplement the enzymes there and to enhance extract recovery, wort fermentability and the supply of assimilable yeast nutrients, especially low-molecular-weight nitrogenous substances (FAN). In addition, such extracts have high buffering capacities. Alternatively, extracts were added to cooled wort, despite the risk of causing microbial infections, to adjust fermentability. Non-diastatic malt extracts, or other syrups, may be dissolved in the copper during hop-boiling. They represent a convenient source of extract, but they are rarely, if ever, used in UK commercial brewing at present. By comparison, a variety of other concentrated syrups made from various grists and converted in a variety of ways are sometimes used as wort replacements, particularly in high-gravity brewing (Table 12.7) Malt extracts and hopped malt extracts (wort boiled with hops before concentration) are usually used by home brewers to avoid the need for milling and mashing. These materials are made using a wide range of malt types and adjuncts and are diluted for use to give worts differing widely in colour,

Table 12.7 Compositions of some barley syrups compared with a pale-ale wort and a malt extract

	Carbohydrate fractions (as % of total carbohydrate)						Nitrogen fractions[b] (mg/100 ml at SG 1038.8)			Total Nitrogen (% solids)
	Fructose	Glucose	Sucrose	Maltose	Maltotriose	Dextrins[a]	Total N	α-Amino N	HMW-N[c]	
Pale ale, all-malt wort	1–4	8–11	1–5	42–50	13–16	21–30	65–75	14–19	15–23	–
Malt extract	12–34[d]		0–4	18–54	6–9	21–39	–	–	–	0.7–1.6
Barley syrup	7–27[d]		0–6	46–60	7–17	5–27	–	–	–	0.6–0.8
Green malt syrup	6–8[d]		5–8	48–55	9–19	23	–	–	–	0.6–0.7
Standard syrup (A)	1–5	7–19	1–7	41–51	7–15	19–27	62–97	12–23	14–33	–
Low-dextrin syrup (A)	2–3	9	2–4	54–56	13–14	17–18	92–93	22–23	27	–
Standard syrup (B)	2–5	9–14	1–2	36–58	7–17	17–31	52–81	13–19	16–24	–
Low-dextrin syrup (B)	2–3	9–10	2–6	47–57	10–15	16–22	62–75	11–18	20–24	–

[a] Dextrins, the unfermentable carbohydrate fraction.
[b] Nitrogen fractions, concentrations in extracts diluted to an SG of 1038.8, approximately 10% solids.
[c] Nitrogen in the high-molecular-weight material (>5000).
[d] Fructose plus glucose values.
From MacWilliam (1971); Roberts and Rainbow (1971).

aroma, etc. Hopped malt extracts have been used to produce beers in remote locations in temporary or permanent breweries. For example, in the 'Conbrew system', a hopped wort syrup (minimum of 75% solids) was produced in Newfoundland and shipped to the Bahamas, or other small and isolated markets, in 19 hl (500 US gal) collapsible rubber containers lined with food-grade rubber. On arrival, it was diluted and fermented to produce beer (Papadopoulos, 1964; Ramsay, 1964).

For diastatic malt extracts, processing is carried out to maximize enzyme survival (Briggs, 1978; Hall, 1978). A high-nitrogen (e.g. 1.7–2.0%), highly diastatic pale malt is ground and mashed at less than 49 °C (120 °F). After a stand, the first worts are collected. The residual solids are remashed at 68 °C (154 °F) to complete the conversion of the starch and recover the extract. Spargings and last-runnings may be used in the mashing-in liquor of the next mash to maintain a high extract concentration in the wort. This minimizes the amount of water to be evaporated and so saves heat. The combined worts are concentrated at the lowest possible temperatures in multiple-effect vacuum evaporators. Initial temperatures are usually 32–35 °C (90–95 °F), but the temperature rises to 43–46 °C (109–115 °F) as the extract concentration rises towards 80% solids. Typically these extracts are pale and highly fermentable. In contrast, malt extracts used in sweets and cakes may need to be totally devoid of α-amylase and other enzyme activities. To achieve this, extracts are heated to 80–90 °C (176–194 °F), sometimes with prior acidification. Many extracts used in foodstuffs are dark, have rich and luscious flavours and characteristic, attractive aromas.

Malt extracts and syrups provide compact sources of extract, as 3.7 volumes are approximately equivalent to the extract yield of 10 volumes of malt. An extract of *ca.* 302 l°/kg is usual (Table 12.2). Malt extracts contain 75–82% solids, with SG values of 1400–1450. The high concentrations are needed to prevent the growth of microbes, the more concentrated products being intended for use in the tropics. The most common contaminants are osmophilic yeasts. The extracts must be handled warm (*ca.* 49 °C; 120 °F) to reduce their viscosities and allow them to flow. However this causes changes in their flavours and colours and limits their lives. At ambient temperatures, they are almost indefinitely stable provided that local dilution does not occur, for example as the result of condensate from the upper parts of the storage containers running or dripping back and 'pooling' on the surface. To overcome condensation problems, the container space above the syrup may be ventilated with filtered, warm air. Alternatively, the tops of the containers may be closed with a fine, porous mesh on which condensation does not occur. Malt extracts and syrups may be delivered to users in bulk tankers or, in smaller quantities, in sealed metal drums. Malt extracts may be prepared to meet a wide range of specifications (Tables 12.2 and 12.7; colours, 3–520 EBC units (3–50 EBC units with high diastase extracts); enzyme activities range from zero to high levels (amylases) (DP

0–400°L) and proteases. Wide variations in sugar spectra can be achieved giving products with fermentabilities of 65–93%. Pale extracts may be extremely sweet but otherwise have bland flavours while some dark, melanoidin-rich products have fine, rich flavours and aromas. Specifications of malt extracts and syrups, in addition to stipulating colours, enzyme levels, fermentabilities, extracts and percentage solids and the degree of microbial purity, may also specify details of the sugar spectra, the nitrogen content of the extract, its pH, its arsenic content and its refractive index.

Malt extracts are expensive both because of the cost of the malt and because the equipment and operations for preparing the wort and concentrating it to a syrup are costly. Numerous syrups, intended to be less expensive, have been prepared from mashes made from malt mixed with adjuncts, from raw cereal grains converted with microbial enzymes (with or without malt), from green malt and, at least experimentally, with enzymes following preliminary acid hydrolysis. These last are rich in salts and the wort solids may contain 1.6–3.8% of ash. Some conversion times for producing syrups were very long: 6–12 h periods being quoted for making barley syrups. These various wort-replacement syrups are used much less in the UK than was once the case.

The various production processes can produce syrups with the wide range of sugar spectra similar to those available in products made from refined starches. However, these grain-based products are intended for use as wort replacements rather than merely extenders, so their nitrogen-containing and other components are also important. Syrups can be prepared that closely resemble all-malt worts, typically having fermentabilities of 60–75% (Tables 12.2 and 12.7) (Imrie and Martin, 1969; Roberts and Rainbow, 1971; MacWilliam, 1971; Lloyd 1986). Compared with worts from all-malt mashes, proline levels are low and the amino acid spectra of cereal syrups are not exactly the same. They do not lack essential yeast nutrients, but they may need to be supplemented with zinc salts to achieve maximum fermentation rates. Syrups are used as copper adjuncts to increase the effective capacities of breweries, to produce concentrated worts for high gravity brewing and to adjust wort composition. Probably it is technically feasible to wholly replace wort with diluted cereal syrup. Brewers buy syrups from outside suppliers and demand highly consistent products.

The technologies used to produce syrups from cereals are largely those used to convert adjuncts in breweries and distilleries. 'Liquid malt', legally a malt extract and so exempted from the restrictions applied to the use of raw grain extracts in some countries, was produced exclusively from green (i.e. unkilned) malt (Griffin *et al.*, 1968; Collier, 1970). The difficult milling and wort-separation problems were overcome by Collier, who used bowl centrifuges. The unpleasant green-grain flavour, characteristic of green malt mashes, was removed during the evaporative concentration steps. The syrup was low in anthocyanogens and so beers brewed with it had reduced haze-forming potentials. Because the green malt was unkilned, the considerable expense of kilning was avoided, which helped to offset the cost of concentrating the wort to a syrup. Many of the precautions used in making barley syrups are naturally the same as those required when using barley as an adjunct in a brewery. Selected, clean barley is milled wet or dry in a variety of ways (Griffin *et al.*, 1968; Imrie and Martin, 1969; Collier, 1970; Maule and Greenshields, 1970; Roberts and Rainbow, 1971; MacWilliam, 1971; Wickens, 1973; Wieg, 1987). In addition, barley can be dehusked by impact milling in a pin-mill and the dehusked fraction coarsely ground between fluted rolls before recombining this material with the husk (Wickens, 1973). A coarse grist is needed if wort is to be separated from residual solids in a lauter tun, but finer and more readily converted grists can be used if centrifuges, filter presses or continuous-belt vacuum filters are employed. By using weak last-runnings from one mash as the foundation liquor for the next mash, high-gravity worts can be produced to supply the evaporator. Typically a grist containing 90–95% of ground barley and 5–10% of high diastatic malt would be mashed with additions of a mixture of bacterial and possibly fungal enzymes and would be subjected to a rising temperature programme, which might last 6–12 h (Rao and Narasimham, 1976; Wieg, 1987). The malt enzymes act with the β-amylase from the barley and the β-glucanase, protease and α-amylase (and possibly other enzymes) from the mixture of microbial enzymes. Typically constant-temperature periods (stands or holds) in the rising temperature programme occur at 50–55 °C (122–131 °F) to allow β-glucanolysis and proteolysis, at 63–65 °C (145.4–149 °F) for the major part of amylolysis and saccharification, and at 75 °C (167 °F) to ensure complete starch solubilization. Alternatively an all-barley mash, supplemented with a mixture of microbial enzymes including an active α-amylase is temperature programmed to 85 °C (185 °F), and this temperature is held until starch liquefaction is complete. By this stage, adequate proteolysis and β-glucanolysis should have occurred but the mixture of starch degradation products is incorrect, giving a poorly fermentable wort. The mixture is then cooled to 62–65 °C (143.6–149 °F) and is mixed with malt and/or fungal enzymes to obtain adequate saccharification of the dextrins and hence wort fermentability. Worts are vacuum-concentrated to syrups having solids contents of 78–83%, specific gravity values of 1400–1430 and crude-protein contents of 3–5%. About 85% of the solids are carbohydrates (compared with *ca.* 98% for starch-based syrups). Colours may be adjusted to a wide range of values. Fermentabilities are normally in the range 65–75%, although these too can be adjusted. Clearly the potential range of cereal-derived adjuncts, sugars and syrups is very large, and only a small proportion of the possibilities is commercially exploited.

12.9 Caramels

Caramels are colouring materials, the use of which is permitted in foods and drinks. As such they may be regarded as competitors for coloured malts (Greenshields and Macgillivray, 1972; Greenshields, 1973; Buckee and Bailey, 1978; Hough *et al.*, 1982; Thornton, 1990). Colours are in the red–brown range (Table 12.2). Caramels are used in soft drinks, beers, spirits, vinegars, meats, bakery products, confectionery and pickles. Different types of caramel are required; they are not all suitable for use in every product. They are made using food-grade chemicals and, in each case, sugars are heated in mixtures in solution so that Maillard and other coloured reaction products are generated. For a time it was feared that caramels might contain toxic or mutagenic by-products, including 4-methylimidazole and 2-acetyl-4(5)-tetrahydroxy-butylimidazole (THI). The products now marketed are checked to ensure that the legal limits of these substances are not exceeded. Nevertheless, there have been discussions on the desirability of replacing caramels with coloured malts as colouring agents. One or more of a wide range of reagents, including acetic, citric, phosphoric, sulphuric and sulphurous acids, ammonium, sodium and potassium carbonates, phosphates, sulphates or sulphites, ammonium, sodium and potassium hydroxides, and gaseous ammonia, may be used. These are added to sucrose or dextrose (starch-derived) preparations and the mixtures are heated in carefully controlled ways. Two main types of colouring reaction appear to occur. Firstly, true Maillard reactions occur in which ammonium ions interact with the sugars and, in a complex series of reactions, amino-sugar adducts polymerize and coloured materials are formed. Carbon dioxide is released in the course of these reactions. Secondly, true caramelization occurs. Carbohydrates heated well above their melting points with one of several reagents but, in the absence of ammonium ions, become highly coloured and develop characteristic bitter or burnt flavours. The products, which exhibit a wide range of colours, viscosities, stabilities and flavours, consist of complex mixtures of polymeric and low-molecular-weight substances. Specifications may include colour intensity, microbiological purity, upper levels of iron, copper, ash, total or ammonium nitrogen, sulphur dioxide and phosphates and will demand a particular pH.

Caramels are divided into different classes. Class I, caustic or spirit caramels, which have colours of 17 000–25 000 EBC units, are made by cooking sugars with sodium or potassium hydroxides. Their main use is in colouring spirits; they are soluble in up to 80% (w/v) ethanol. Class II consists of 'caustic sulphite' caramels. These materials, which are little used in the UK, give stable, clear solutions in the presence of both alcohol and tannins. They are used to colour aperitifs, vermouth and brandies. The class III caramels are the electropositive ammonia caramels and they are widely used, in brewing for example. In their preparation, carbohydrates (usually glucose syrups) are heated with ammonia. The sugar–ammonia mixture may be allowed to stand for a week at ambient temperatures then be held at 90 °C (194 °F) overnight and finally be heated at 120 °C (248 °F) for 3 h. Considerable care is needed at this stage to reach the correct colour and viscosity. The product is cooled and diluted with water. Different batches are blended to meet specifications. Usually the products contain about 65–75% solids and 2.5–5.0% nitrogen. Because they have isoelectric points in the range pH 6.0–6.5 they normally carry a positive charge and so are termed electropositive. These caramels have the highest colour intensities, e.g. 32 000–500 000 EBC units. These materials are used in beers, some vinegars, bakery products and confectionery. Class IV products, the ammonium sulphite or electronegative caramels, are made by heating carbohydrates with ammonia and sulphites. The products, which have colours in the range 20 000–50 000 EBC units, are stable in the presence of acids and tannins and have isoionic points at around pH 1.0, hence they are normally negatively charged in solution. These products are used for colouring colas and soft drinks.

12.10 Industrial enzymes

Industrial enzymes are used for many purposes, for example in 'tenderizing' meat and in biological detergents. Bacterial α-amylases have displaced malt enzymes in desizing (removing a starch coating from the threads) textiles (Linko, 1987). In this section, interest is confined to non-malt enzymes that are used to replace or supplement malt enzymes for various purposes. These include enzymes added to dough in bread making; those added to mashes, particularly adjunct-containing mashes; those used in the preparation of sugars and syrups from starch and unmalted cereals (section 12.7.2 and 12.8); and also those added to worts, fermentations and beers to adjust the quality of the final products. Examples of the last are additions to increase the fermentability of the product by degrading residual dextrins or to improve its quality by degrading β-glucans, pentosans (to reduce product viscosity and the risk of gel hazes or precipitation) or high-molecular-weight peptides (to reduce the chances of haze-formation in beers) (Anderson, 1984).

In some countries, the use of 'industrial enzymes' is not allowed in preparing foods and beverages. However elsewhere, even when enzymes are not routinely used, a brewery may hold a supply for 'trouble shooting', for example to aid in the treatment of set mashes, of slow wort run-off, or incomplete starch degradation (Bamforth, 1986).

Industrial enzymes come from various sources. Limited amounts come from plants, such as the proteolytic enzymes papain (papaya latex), bromelain (pineapple) and ficin (fig) used to 'chill-proof' beer by causing the limited degradation of high-molecular-weight peptide haze precursors, or the β-amylases from soya bean flour, sweet potatoes, wheat flour or other cereals used to adjust fermentabilities by increasing maltose production from dextrins present in syrups and worts. Rarely small quantities of enzymes from animal sources, such as the proteolytic enzyme pepsin from pig gastric juice, have been employed. However, most industrial enzymes are prepared from specially cultured microorganisms, usually fungi or bacteria. Because most of the enzymes are secreted by the microorganisms and so are extracellular, it is relatively simple to obtain preparations that are microbe-free. The strains of microbes used must be non-pathogenic and must not produce mycotoxins or other undesirable materials. Often only 'agreed' microbes may be employed. Historically, the microbes were grown on solid media such as cooked bran, cereals or even malt flour, in the way that *Aspergillus oryzae* is grown on steamed rice to make the traditional 'koji' in Japan (Meyrath, 1966). At present most industrial enzymes are produced by microbes grown in deep, stirred liquid cultures (Fogarty, 1982; Sheppard, 1986; Beckerich and Denault, 1987; Linko, 1987; Byrne, 1991). It is thought that most fungal enzymes are derived from *Aspergillus* spp. (*A. oryzae* and *A. niger*) and from *Trichoderma viride*, *T. reesei*, *Rhizopus* spp. and *Penicillium emersonii*. Bacterial enzymes come mainly from *Bacillus* spp. such as *B. subtilis*, *B. stearothermophilus* and *B. licheniformis*. They may be supplied 'in the liquid form' (in solution) or as a dried powder and are often mixed with 'extenders' or carrier substances. The preparations used in bulk are not pure and, in addition to the 'named' enzyme and possibly an extender (e.g. starch), usually contain other enzymes, 'junk' proteins, residual microbial culture medium, diluents, in some instances some microbes or their spores, and preservatives (which may include sodium chloride, other inorganic salts, glycerine, potassium sorbate, corn syrup, propylene glycol, sodium benzoate, sodium metabisulphite and parabens) (Byrne, 1991). Products containing surviving microbes or their spores are not acceptable for many purposes. Although only enzymes from 'permitted' microbes prepared to food-grade standards under hygienic conditions may be used for many purposes, the acceptance and use of these enzymes is hampered by a lack of agreed standards of quality and purity and methods for assessing their activities. The preparations must be adequately stable, that is have acceptable shelf lives when stored under recommended conditions. Preparations usually contain a mixture of enzymes, many of which can be very simply detected with radial diffusion tests (Albini *et al.*, 1987). The laboratory methods used to assess enzyme activities have not been standardized, so different suppliers and users employ different assay methods and units of activity. Laboratory enzyme assays on starch-degrading enzymes differ widely from the conditions encountered when starch is liquefied for syrup manufacture or is converted in distillers' or brewers' mashes. Because of considerations like these users must often undertake extended, complex trials to decide which of the preparations on offer is the 'best' and most cost-effective. Because of the complexity of the situations, brewers test enzyme preparations under as realistic conditions as possible: in small-scale mashes or brews for example (Bamforth, 1986). The complexity of the situation may be illustrated with reference to pH and temperature optima and enzyme stability. These are not true constants but depend on the conditions under which a test is carried out and the duration of the test period. One key to understanding the results of trials is to appreciate how rapidly the enzyme is destroyed under test conditions. For example α-amylases are usually more resistant to high temperatures at higher substrate (starch) concentrations, with higher concentrations of calcium ions and at particular pH values. In mashes, the pH values used may not be optimal from the point of view of enzyme stability or activity and there are practical limits to the concentration of calcium salts. 'Contaminating' enzymes in a preparation may be insignificant, or they may have beneficial or deleterious effects. For example β-glucanase present in bacterial α-amylase preparations may be advantageous to brewers who have problems with β-glucans from barley adjuncts or incompletely modified malt. However, the protease that is present may be of use in increasing the levels of soluble nitrogenous substances, in making cereal syrups or in barley-brewing, but too high protease activities may give rise to excessive levels of wort nitrogen and hence excessive yeast growth. Also, too extreme degradation of the high-molecular-weight polypeptides in the wort can change the product so that the beer has a low head retention (poor foam stability). The product may have its character changed in more subtle ways. By selecting particular microbial strains, by culturing them under different conditions and by modifying the conditions used for partially purifying and concentrating the enzymes, manufacturers can vary the proportions of the different enzymes that occur in their products. It is difficult to ensure that the proportions of different enzymes present are always the same.

The preparations that compete more or less directly with malt are those which contain hydrolytic enzymes, particularly those that catalyse the degradation of starch, dextrins, β-glucans, pentosans, proteins and possibly lipids and complex phosphates (including phytate). For many purposes, nucleic acids should be completely degraded and the hydrolysis of lipids should be limited because excessive levels of fatty acids are undesirable. However data on nucleases, lipases and other 'minor' enzymes, such as lipoxygenases and phosphatases, are usually not available. In addition to enzymes used in mash conversions, brewers may use enzymes to regulate other beer properties. For example, glucose oxidase may be added to beer to 'scavenge' oxygen from the pack, and diacetyl reductase may be used to reduce vicinal diketones into diols, with an improvement in beer flavour. Carbohydrases may be added to worts to maximize dextrin degradation to produce fermentable sugars. This approach may be used to produce low-carbohydrate beers or to maximize ethanol production in distilleries making fuel alcohol or potable spirits. Brewers may also add enzymes to cause a limited breakdown of dextrins in beers to create 'priming' sugars to support secondary fermentations or to sweeten some yeast-free beers. As already noted, plant proteases may be used to reduce the likelihood of chill-haze formation. The wide and increasing range of enzymes known to be produced by microbes, and improved ways of producing them in increased yields, means that enzyme preparations with new and unfamiliar properties will continue to become available. Some may be accepted for use in the preparation of foods, beverages etc. (Imrie, 1968; Aschengreen, 1969; Nielsen, 1971; Fogarty *et al.*, 1974; Enevoldsen, 1975; Marshall, 1975; Rao and Narasimham, 1976; Willox *et al.*, 1977; Fogarty, 1982; Harada, 1984; Linko, 1987). Characteristics of some widely used enzyme preparations are outlined below.

Starch- and dextrin-degrading enzymes have been extensively studied, and differences in characteristics between enzymes from the same class but from different organisms are extremely important. For example, fungal α-amylases (usually from *Aspergillus* spp.) have convenient pH optima for use in mashes (about 5.5), but they are heat labile, under most conditions being inactivated at 60–65 °C (140–149 °F). Therefore, these enzymes are inconveniently thermolabile for use in cereal mashes. However, this thermolability is a positive advantage when these enzymes are used to 'prime' beer, by producing sugars from dextrins, since they are thermally inactivated during pasteurization. Similarly these enzymes may be added to dough to produce sugars to support yeast leavening and are destroyed when the bread is baked. In each case, thermal inactivation is used to stop the activity of an enzyme when its function is complete and further activity would be undesirable. Enzyme preparations from *Bacillus subtilis* have been widely used. The α-amylase has a pH optimum of 6–7 but has useful activity in the usual mashing range pH 5–6. Its temperature optimum is usually given as 65–70 °C (149–158 °F), but depending on the concentration of the starch, the pH and the concentration of calcium and chloride ions, it can act usefully at 75–80 °C (167–176 °F); it is destroyed on boiling. A bacterial strain *B. subtilis* (var. *amyloliquefaciens*) produces an α-amylase with a pH optimum of 5.7–5.9 (at 40 °C (104 °F)) that is more heat stable, having a temperature optimum of about 70 °C (158 °F) but which can be used to liquefy starch slurries (34–40%) at 85–90 °C (185–194 °F) if the calcium ion concentration is about 105 p.p.m. The main limit hydrolysis product is maltohexaose (Fogarty, 1982). The most readily available heat-stable bacterial α-amylase is produced by *B. licheniformis*. Quoted pH optima for this enzyme vary greatly: 5–8, 6–6.5 and 7–9. Usually it appears to be used at around pH 6. The enzyme is stable at comparatively low levels of calcium ions (e.g. 6 mg/l) and has a temperature optimum of about 90 °C (194 °F).

At higher concentrations of calcium ions (e.g. 50 mg/l) the enzyme will act at around 100 °C (212 °F) or, for short periods, at even higher temperatures (e.g. 115 °C (239 °F)) during starch liquefaction (section 12.7.2). The final products of limit hydrolysis are rich in maltose, maltotriose and maltopentaose (Fogarty, 1982). There are reports that 'α-amylases' from various organisms produce large proportions of maltose in the products of starch hydrolysis. Heat-stable α-amylases are useful in liquefying starch slurries but their resistance to high temperatures is such that conditions are chosen such that they do not survive into the finished product. Commercial preparations of bacterial α-amylases usually contain β-glucanases and proteases, although protease-free preparations are available. Selected preparations have been used to make cereal syrups and to supplement brewery and distillery mashes.

Starches usually contain high proportions of amylopectin (typically 70–80%) and to break the $\alpha(1\rightarrow6)$-links in this substance debranching enzymes are required. These enzymes have low activities in cereal malts, and they are inconveniently thermolabile. Therefore, to obtain high degrees of starch or branched-dextrin hydrolysis it is desirable to employ added debranching enzymes. The most easily available debranching enzymes are isoamylases (which debranch amylopectin but are without activity on pullulan) and pullulanases, which are useful in degrading both amylopectin and dextrins. These enzymes are actually or potentially available for preparing syrups or worts, in conjunction with other enzymes such as amyloglucosidase or β-amylase (Enevoldsen, 1970, 1975, 1986; Willox *et al.*, 1977; Howling, 1979; Fogarty, 1982; Harada, 1984). *Bacillus cereus* (var. *mycoides*) produces a pullulanase and a β-amylase. However the pullulanase that is widely available is produced by *Klebsiella pneumoniae* (syn. *Aerobacter aerogenes*). It is thermolabile and must be used at 45–55 °C (113–131 °F). The pH optimum is given as 5.5–6.0, but the enzyme has been used at pH 4 at 45 °C (113 °F) and at pH 5.0 at 50–55 °C (122–131 °F). Wort having a normal fermentability was produced from a mash of 90% raw barley and 10% malt using an unusual mashing cycle in which the temperature was reduced after liquefaction and pullulanase was added and allowed to act for 2 h at 45 °C (113 °F) (Enevoldsen, 1970). Pullulanase, or pullulanase with β-amylase, has been added to beer to degrade dextrins with the production of fermentable sugars, so replacing priming sugars. Other enzymes used for this purpose have been fungal α-amylase and amyloglucosidase, but as pullulanase is destroyed more certainly by pasteurization the enzyme is preferred. If enzymes survive pasteurization, the beer continues to become sweeter. Isoamylases are less widely used. They are variously reported to have pH optima in the range 3.5–4.5 and temperature optima in the range 45–52 °C (113–125.6 °F).

β-Amylases, extracted from raw cereals or in flour from soya bean or sweet potatoes, have been used as saccharifying agents either with or without other enzymes. The feasibility of using cereal β-amylase in mashes and during fermentation has been demonstrated (Norris and Lewis, 1985). Many of these preparations contain α-glucosidase, which hydrolyses maltose and dextrins to glucose; this enzyme is widely ignored. Some fungal preparations contain a saccharifying enzyme in addition to α-amylase and distinct from amyloglucosidase (which generates glucose) that preferentially produces maltose. These preparations are often used to make syrups with high-maltose contents. Some bacteria, e.g. *Bacillus polymyxa*, produce a β-amylase that, like the cereal enzyme, is exo-acting on the starch chains and so cannot, of itself, wholly degrade starch to maltose since it cannot pass the α(1→6) branch points. However, when used together with pullulanase conversion of starch can be much greater. For example, if starch is liquefied with α-amylase then incubated with a β-amylase, maltose yields of 55–60% are obtained, but if saccharification is with β-amylase and pullulanase, maltose yields of 80–95% are achieved (Fogarty, 1982). *Bacillus cereus* (var. *mycoides*) produces β-amylase and pullulanase.

Amyloglucosidase (AG, AMG, gluc(o)amylase) is normally prepared from cultures of *Rhizopus* spp., *Aspergillus niger*, or *A. oryzae*. Preparations differ in their properties and in particular in the presence of additional enzymes. The presence of α-amylase may be desirable, but the presence of transglucosidases is not. Amyloglucosidases are exo-acting hydrolytic enzymes attacking starch chains from the non-reducing ends, with the liberation of glucose. They hydrolyse α(1→4)-linkages comparatively rapidly but attack α(1→6)- (and α(1→3)- linkages only slowly. These enzymes are usefully active at mash pH values (e.g. 5.3–5.5) but they are optimally active at pH 4.4. They are sufficiently thermostable to be used at 60–65 °C (140–149 °F). Although impure preparations will eventually degrade liquefied starch, chiefly to glucose, the process is slow and is accelerated by the addition of pullulanase when glucose yields of 96–98% may be achieved. Many preparations contain a transglucosidase, which is disadvantageous since it catalyses the formation of unwanted and unfermentable sugars such as isomaltose, panose and perhaps other disaccharides. In addition, some preparations contain protease, lipase, β-glucanase and cellulase enzyme activities. Amyloglucosidase is invariably used in the preparation of glucose (dextrose) from starch (section 12.7.2). It has also been used to a limited extent as a saccharifying agent in mashing and, either alone or with pullulanase, added to wort to obtain highly fermented (lite, low-carbohydrate) beers and to maximize ethanol production in distilleries. As already noted its thermostability makes it an unsuitable enzyme for replacing priming sugars in brewing, as it can survive pasteurization. It has been suggested repeatedly that beer be treated by using columns of amyloglucosidase or other enzymes immobilized by attachment to insoluble materials. However, problems with the loss of enzyme activity, the 'leakage' of enzyme into the beer, the need for extended residence times and the difficulty of maintaining sterility have prevented the acceptance of this approach.

Preparations of β-glucanase have been widely used in brewing to break down β-glucans, particularly in under modified malts and barley adjuncts, to increase extract recovery, speed wort run-off and simplify beer filtration by reducing its viscosity, and, in extreme cases, preventing the formation of hazes and even of gelatinous precipitates (Macey *et al.*, 1967; Crabb and Bathgate, 1973). Having enzymes in the mash that break down both soluble and insoluble β-glucans may assist wort recovery (Webster, 1978). The malt enzyme, which has a pH optimum of 4.8–5, is comparatively heat labile and is destroyed at temperatures over 40 °C (104 °F). Consequently it is readily destroyed during kilning and if it survives into the malt it is of limited use in mashing unless the temperature programme has an extended, cool rest. The various microbial enzymes are more heat stable. *Bacillus subtilis* produces a β-glucanase with a pH optimum of around 7, and a temperature optimum of 50–60 °C (122–140 °F). Despite the high pH optimum, this enzyme, which is usually present in *B. subtillis* α-amylase preparations, can act to a useful extent in cereal mashes (pH usually *ca.* 5.3) in breweries and is briefly active even at 70–75 °C (158–167 °F) (Sorensen, 1970; Erdal and Gjertsen, 1971; Enkenlund, 1972; Enari and Markkanen, 1975; Albini *et al.*, 1987). *Aspergillus* spp. also produce β-glucanases, with optimal temperature and pH ranges variously reported as 55–60 °C (131–140 °F) at pH 5.0 and 45–50 °C (113–122 °F) at pH 4–6. Cellulase enzyme preparations from *Trichoderma* spp. (e.g. *T. reesei* and *T. viride*) contain β-glucanases with temperature optima of 50–55 °C (122–131 °F) and pH optima in the range 3.5–5.5, but these enzymes are effective in brewery mashes (Stentebjerg-Olesen, 1980; Oksanen *et al.*, 1985; Pajunen, 1986). Enzymes from *T. reesei* have been added to malting barleys but the benefits observed result, in the writer's view, from a carry-over of enzyme into the mash tun. Recently there has been considerable interest in the more heat-stable β-glucanase enzymes produced by *Penicillium* spp. such as *P. funiculosum* and *P. emersonii.* With pH optima around 4.3–5 and temperature optima of 80 °C (176 °F) they seem to be ideal for use in mashes. Indeed there have been reports that in countries without controlling regulations enzyme preparations containing heat-stable β-glucanases have been sprayed onto undermodified malts, presumably so that when they are mashed their low qualities will be obscured. These enzyme preparations usually contain starch- and protein-degrading activities and some also have pentosanase activity.

The use of wheat-based and the possible use of triticale-based adjuncts in mashing has caused difficulty with wort separations owing, at least in part, to the presence of pentosans, which can be highly viscous in solution, and to occur in fine particles (Blanchflower and Briggs, 1991c). Several fungal 'pentosanase' preparations are now on offer: from *Disporotrichum* spp. and -

Trichoderma spp. (Bamforth, 1986; Byrne, 1991). Some enzyme preparations from *Aspergillus* spp. also contain pentosanase activities. The activities of the effective enzymes are not adequately characterized in biochemical terms, but it seems likely that the most effective enzymes in the mixtures (which often contain acetyl esterase, feruloyl esterase, xylanase, arabinosidase, xylobiase, cellulase and amylase activities) are endoxylanases, with low temperature optima, of *ca.* 50 °C (122 °F). These enzymes are evidently active in mashes and worts. It has been suggested that useful extract might be recovered by remashing spent grains with these enzymes.

Proteases occur frequently in preparations of bacterial and fungal enzymes (Albini *et al.*, 1987). Often one species may produce a mixture of enzymes (Macey *et al.*, 1967; Sorensen and Fullbrook, 1972). *Bacillus* spp. produce proteinases with neutral and alkaline pH optima. The former can solubilize barley proteins during mashing and making syrups. The optima are often given as 45–60 °C (113–140 °F), pH 6–7. *B. subtilis* protease is more heat labile than the α-amylase but, provided there is a sufficiently long 'stand' at about 50 °C (122 °F) in the mashing programme, sufficient α-amino nitrogen can be solubilized in barley-enzyme mashes by this enzyme to support yeast growth.

12.11 The future

Undoubtedly adjuncts and enzymes will be used in the future and so will continue to compete with malt. In recent years, and despite a great deal of interest, their use has not increased as much as some predicted and they have not displaced malt. New enzyme preparations are not easily introduced since regulatory authorities must be convinced that they are safe and contain no toxins or pathogenic organisms, and potential users have to be persuaded that their use is beneficial. Factors that have limited the use of some enzyme preparations are patent restrictions, food laws, high costs, poor stability in store, variable composition of the enzyme mixtures, the need for particular metal ions as cofactors, the need for excessively high concentration of calcium ions (to stabilize some α-amylases) and inconvenient temperature and pH optima. Then the introduction of new raw materials and enzymes into existing production processes may not be easy. Not only may this be a matter of 'sentiment', in that customers object to alterations in the materials used in the manufacture of traditional products (indeed this is forbidden in some instances), but new plant may need to be installed and the revised process may not be easily controlled or able to produce a consistent product. Also the character of a product may change in subtle ways that customers may come to prefer or to dislike.

The choice of adjuncts and whether or not to use industrial enzymes will also depend on local supplies and economies. The restrictions imposed on the importation of barley malts into Nigeria has given brewers of European-style beers some major problems. If sorghum adjuncts are widely used, this will almost certainly dictate that industrial enzymes will have to be used in quantity to produce worts and beers of acceptable quality. Sorghum malts used alone may not be suitable for making clear beers (Chapter 15).

13 Malt analysis

13.1 Introduction

Malt analyses are carried out for several purposes: (i) to provide data for the maltster to use for quality control and to guide process adjustments; (ii) to provide a basis for product valuation, used in buying and selling the malt; and (iii) to indicate the potential value to the customer, allowing the specification of a consistent raw material, the prediction (for example) of the extract recovery that will be achieved and whether or not a particular malt is likely to give production difficulties. However, when a malt is processed in different ways, different yields of extract of differing qualities are obtained. For simplicity it has been necessary to adopt agreed, standardized methods of analysis for commercial use. Malt users generally establish correlations between the analytical values of various malts and their performances in the production processes. For experimental purposes, the composition of grains and malts have been investigated by a very large number of chemical, biochemical and physical techniques. Estimates of the chemical composition of malts are generally of little direct commercial value. The analyses of most value are those designed to mimic some aspects of the processes of milling and/or mashing. Three sets of 'standard' or recommended methods of analysis are widely used.

- *Recommended Methods of Analysis of the IoB, London* (IoB, 1993). The first versions of these methods were recommended, against vociferous opposition, in 1906; they were essentially the methods proposed by Heron in 1888 and were intended for use by ale-brewers using infusion mashing systems with mash tunes.
- *Analytica–EBC* (EBC, 1987). These methods are also still called the *Congress Methods of Analysis*, after the agreement reached by Continental European brewing scientists at an International Congress of Applied Chemistry, in 1898. These methods are intended for use by lager-brewers.
- *Methods of Analysis of the ASBC* (ASBC, 1992). These methods are chiefly used in North America. They closely resemble the Congress methods.

All these methods have frequently been refined and tested in collaborative trials, and continuing efforts are being made to agree methods that are generally acceptable. Introductions of revised methods are noted in the technical literature. Other 'unofficial' methods of analysis are used because of their simplicity or rapidity or because of a particular user's requirements.

The various reasons these analytical methods are not 'approved' include (i) the fact that some are 'secondary' methods and so each must be regularly calibrated against the appropriate primary standard method; (ii) that the errors in the determinations are considered to be too large; and (iii) the methods are only considered to be of value by one, or very few, companies.

Most of the analytical methods relate to 'white' brewing malts, although some methods for coloured malts, dark malts and adjuncts are specified. Distillers and other malt users use many of the brewers' methods of analysis. Here some of the principles of malt analysis will be outlined. The chapter will not provide a catalogue of methods or describe any in detail. Methods of malt analysis have frequently been reviewed (de Clerck, 1952; Hudson, 1960; Pollock, 1962b; Wainwright and Buckee, 1977; Moll, 1979a; Burger and La Berge, 1985; Briggs, 1987b,c). In addition the systems usually specify analytical methods for malt competitors such as sugars, syrups, caramels and unmalted adjuncts. However, standardized methods for analysing non-malt enzyme preparations have not been specified. With the passage of time, the methods of analysis used have been added to and improved. In comparatively recent years, accelerated methods of malt production have been adopted while brewers have adopted shorter mashing sequences and more rapid methods of wort collection. Problems, particularly of wort separation and beer filtration, were traced to variations in malt quality. However, the analytical methods then in use often could not distinguish between satisfactory and problem malts. Newer analyses have been introduced to determine heterogeneity among the grains within a malt sample and to detect incomplete modification, for example by detecting residual β-glucan.

Modification is an 'omnibus word' variously taken to mean either all the physical and chemical changes that occur when barley is converted into malt or just the physical changes that occur, particularly the physical weakening caused by the degradation of the cell walls of the starchy endosperm (Chapter 2). The many attempts made to find one simple measure of modification have failed and will continue to fail, because the various components of modification that alter during malting, such as the accumulation of hydrolytic enzymes, the degradation of the endosperm cell walls, the changes in various soluble nitrogen fractions and alterations in the ease of extract recovery, do not change 'in proportion' with time or with alterations in malting conditions. Therefore, malts have to be described using the results of a range of analyses. It is in both the maltsters' and the malt users' interests to use the minimum number of analyses needed to adequately describe the malt. A century ago characterization of malt was made by inspection of whole and cut grains, by chewing and by the determination of the diastatic power, the hot water extract, the colour and the specific optical rotatory power of the wort; now malt users may require many other analytical values. In addition, they may require statements regarding the variety(ies) of barley used to make the malt, whether or not additives, resteeping or abrasion have been employed, and so on.

13.2 Sampling malts

Single batches of malt are commonly prepared from 200–300 t lots of barley. Malts from different batches are usually blended by vendors and users. It necessarily follows that, like the barleys from which they are made, malts are heterogeneous in two senses: (i) individual malt corns differ in quality; and (ii) the compositions of small samples from a bulk of malt will differ to some extent. It follows that sub-samples for analysis must be drawn from all parts of a bulk of malt and then these 'primary samples' should be combined, mixed to give a 'bulk sample' and sub-divided in order to obtain 'laboratory samples' that are truly representative of the bulk. However precisely and carefully analyses are carried out, if they are performed on inadequately prepared samples they are most unlikely to reflect the true, 'average' composition of the bulk of grain and so are likely to be misleading. Every set of analytical methods specifies how primary samples from different lots of malt (whether in sacks, in bulks, in lorries, in railway wagons or moving in conveyors) should be drawn and bulked. These guidelines minimize the chances of statistical errors. In general, the procedures follow those used for barley and should avoid safety hazards, such as exposing operators to dusty atmospheres (Chapter 7). The bulked samples, which must be of adequate size (often several kilograms in weight), must be well mixed and divided, using sample splitters to avoid accidental bias or grain sorting, to obtain sufficiently small laboratory samples for analysis. Because malt is highly hygroscopic, samples must be placed immediately into sealed, air- and waterproof containers that are dated and fully labelled. Samples must also be handled with care to avoid physically damaging the malt kernels. Sometimes vendors take samples, retain half and supply the other half to the purchaser, who may analyse it before accepting delivery of the bulk.

13.3 Statistics of analyses

The analytical methods used are, to some extent, empirical. In addition to the real differences (hopefully small) that occur between sub-samples drawn from one lot of malt, there are inevitable small differences between the same analyses when they are replicated, made on different occasions or made in different laboratories. The results of analyses made by different operators, particularly if they work in different laboratories, usually vary appreciably more than those obtained by one operator using one piece of equipment. Because of the inevitable discrepancies that occur when malts are analysed by different operators, great efforts have been and are being made to refine the analytical methods that are in use, to train analysts and to determine the likely ranges of variation of the results of different analyses. Frequent collaborative trials are made in which samples of malt are analysed independently in different laboratories, and the results are collected and subjected to statistical analyses. Analysts whose results deviate appreciably from the mean values can revise and check their techniques, while from the bulked results the ranges of variability of analyses within and between laboratories can be determined. These exercises maintain the high standards of the analytical laboratories. The data obtained are also of value when malt specifications are being set. From time to time, specifications have been proposed that made no allowances for the tolerances needed when interpreting analytical results. Excessively tight specifications are impossible to meet, both because of the variability inherent within malts and because of inevitable small differences between the results of replicated analyses. Currently the recommended methods of the IoB include experimentally determined estimates of repeatability, $r(95)$, and reproducibility, $R(95)$, for the results of barley and malt analyses (Baker, 1989; Bernard, 1992) (Table 13.1). The statistics are complex, but broadly each repeatability value is the maximum acceptable difference at the 95% confidence level between repeat determinations carried out in one laboratory, within a short time interval by one operator. The reproducibility value is the maximum acceptable difference between single determinations at the 95% confidence level made in two different laboratories by two different operators using two different sets of analytical equipment.

Frequently, laboratories retain samples of well-characterized, standard malts that they reanalyse at intervals, often with replicates, to check that their analytical methods are not changing with time (drifting). In addition, routine results of analyses of successive batches of particular grades of malt are analysed numerically to monitor malthouse performance by determining if there are any trends indicating progressive changes in malt quality (Lloyd, 1989, 1994; Seward, 1992). These analyses may be carried out conveniently using computer packages. The results can be shown graphically on, for example, Shewhart, Manhattan, Pamto or Cusom charts. These allow deviations of analytical values from those within the acceptable range to be quickly identified so that the appropriate remedial action can be taken to adjust the production process.

13.4 The sensory evaluation of malts

It is traditional to group malt analyses under sensory, physical and chemical methods. In fact the distinctions are blurred. Undoubtedly simple

Table 13.1 The repeatability values, *r*(95), and reproducibility values, *R*(95), for various analytical methods applied to barley and malt

Analytical method (units)	Mean value (range)	*r*(95)[a]	*R*(95)
Methods for barley			
Moisture (% m/m)	(12.50–15.70)	0.352	0.867
	(13.10–16.19)	0.244	0.791
TCW (g, d.m.)	(31.2–35.8)	1.17	1.44
	(40.4–42.3)	1.16	2.01
TN (% m/m)	(1.37–1.58)	–	0.084
	(1.60–1.86)	–	0.096
Pre-germinated grains (%)			
Fluorescein dibutyrate method	1.8	1.33	1.75
	7.0	2.84	4.15
Methylene blue method	2.3	1.61	2.36
	7.3	2.73	6.38
Sieving test (% m/m)			
<2.2 mm	6.1 (1.0–9.5)	1.70	2.34
2.2–2.5 mm	12.6 (3.0–24.4)	1.75	2.84
2.5–2.8 mm	30.3 (15.9–45.3)	2.53	8.79
>2.8 mm	60.0 (22.3–79.7)	3.49	9.12
Methods for pale malts			
Moisture (% m/m)	3.94	0.433	0.606
Hot water extract (l°/kg)			
Miag 0.7 mm	295.3	2.14	3.48
Miag 0.2 mm	298.2	1.93	2.90
Colour (EBC units)	2.38	0.225	0.512
Cold water extract (% m/m)	17.5	1.19	1.29
Diastatic power (°IoB)			
Fehling's method	69.2	6.07	14.50
Ferricyanide method	31.5	3.00	5.29
	45.8	3.24	6.81
	86.4	3.94	14.67
α-			
Amylase (international method)			
DU, d.b.)	30.7	3.51	14.34
TN (% m/m, d.b.)	1.64	0.049	0.085
TSN (% m/m, d.b.)	0.594	0.028	0.061
FAN (international method)			
(% m/m, d.b.)	0.114	0.0088	0.0308
Wort viscosity (mPa s)	(1.40–5.10)	0.020	–
Wort fermentability (real attenuation) (%)			
Boiled, brewing worts	72.91	1.66	5.26
Unboiled, distilling worts	85.74	2.72	4.20
Friabilimeter analyses (% m/m; f.wt.)			
Friability	93	1.6	2.7
	88	1.5	2.7
	80	1.5	6.7
	65	4.8	8.0
Homogeneity	98	0.6	1.1
	97.6	0.7	1.3
	93.7	1.5	6.5
	79.4	3.9	9.9

[a] Repeatability values for analyses of β-glucanase determinations, coloured malts, dark malts and roasted barley are not available.
Data from the Recommended Methods of the *IoB* (1991 edition).

sensory inspections of malts were the original methods of evaluation, and many were well worked out at least by 1830. The corns should be full and plump but not 'mellowed' or artificially swollen by sprinkling water onto the malt at the end of kilning, a practice used when malts were sold by volume and before malt-moisture determinations were routine. The grain should be of uniform size and colour, of a 'bright' appearance, with no local scorching, indicating local overheating or the presence of 'magpie' malt. A dull appearance might be caused by a heavy microbial infestation. The malt should be made of grain of a known variety and not a mixture, unless this has been agreed. The variety of each grain may sometimes be determined from its physical characteristics (lodicules, rachilla form, aleurone colour, etc.) but because of the changes undergone in malting this is even more difficult to do than to distinguish the varieties of ungerminated grains (Chapters 2 and 6). Sometimes the banding patterns obtained when hordeins are extracted from individual grains and are separated by polyacrylamide gel electrophoresis assist in the identification of malt grains, but the partial degradation of the proteins that occurs in malting can make this difficult. The corns should be undamaged, without broken or half-grains or loose husks and without holes, which show that insect attack has occurred. If insect infestation is suspected, corns should be split open to look for grubs or hollowed-out grains. There should be no more than a specified minimal contamination with foreign grains, weed seeds, extraneous materials, dust or rootlets. The grains should 'snap' freely and the endosperm should be completely modified (by British standards) and be floury from end to end. Under-modified or slack (damp) malt corns do not break freely. In older terms, the endosperm should be cretaceous (chalky) and, like chalk, should make a white mark when drawn across a black surface. When bitten, the kernels of white malt (correctly dried) should be brittle and floury. The undesirable presence of unmodified, hard ends is detected by biting along the grains. These tests are analogous to 'rubbing out' the endosperms of grains of green malts, the test used to check the state of modification during germination. The odour of the malt should be clean and normal for the type, with no trace of mustiness, sulphury odours or 'cooked vegetable' smell. It should have the correct flavour before and after being bitten, with no mouldy, bitter or other off-flavours. When split transversely or (better) longtudinally in a farinator, or when groups of kernels are glued to a flat support and half the thickness is ground away with an electric sander, all the endosperms should be of a uniform colour and of the correct consistency, glassy in the case of caramel malts but mealy or floury in the case of white malts (Chapter 15). For some years, malt vitreosity was assessed using the Diaphanoscope, or a similar device, in which corns were held in a perforated rack and inspected by transmitted light. It was supposed to be possible to distinguish between mealy and steely grains, but in fact the test was too subjective to be of value. Steeliness in white malts may be caused by under-modification or by faulty kilning. Having the temperature too high too soon, while the green malt is still damp, can cause local endosperm liquefaction (as is deliberately produced in making crystal malts) and the liquid sets hard when the malt cools, causing vitreosity or local 'case-hardening'. As progressively less mealy, less well-modified malts are analysed, it is found that the hot water extract declines, wort filtration time and viscosity increase, and the wort colour and soluble nitrogen content decline, while wort turbidity, the grain nitrogen ('protein') content and residual starch increase (Ullman *et al.*, 1987). Various 'scales' of vitreosity have been proposed; for example, malt with an average corn vitreosity of less than 2.5% is regarded as very well modified, 5–7.5% is satisfactory, while over 7.5% is bad. At least 200, and better 500, kernels should be tested to obtain a reasonably reliable estimate. This test has been largely replaced by staining sanded grains or by using the Friabilimeter.

The acrospire grows as germination proceeds and, within one batch of malt, there is a rough relationship between acrospire length (expressed as a fraction of the length of the grain) and modification. To facilitate this assessment a malt sample is usually boiled with a 2% solution of copper sulphate until the greenish acrospires can be clearly distinguished through the husks, or the corns may be cut longitudinally through the acrospire. Alternatively, malt corns may be peeled to expose the acrospires. The acrospires of about 200 corns are 'scored' for length and are grouped into classes, for example 0, 0–1/4 kernel length, 1/4–1/2, 1/2–3/4, 3/4–1, and 1+ or overgrown (called huzzars, cockspurs or bolters), in which the acrospires have emerged from the ends of the grains. Various other classifications have been used. A range of devices have been used as measuring aids having magnifying lenses and ruled scales with diverging lines that are arranged over the grain with the base and tip at the '0' and '1' lines and then divide the grain length into 1/4, 1/2, 2/3, 3/4, 1. The results may be expressed as an average value or, more usefully, the percentages of the grains with acrospires falling in the different length categories are recorded. Inspection of the data gives an immediate indication of how uniformly the malted grain germinated and hence the homogeneity of the sample and the proportion of lie-backs (dead or dormant, ungerminated grains). It has been proposed that, for a malt to be regarded as homogeneous, at least four-fifths of the corns should be in two adjacent acrospire-length classes. It has long been realized that obtaining extra acrospire growth usually results in malt with enhanced diastatic power, while over-grown malt (acrospire lengths exceeding 1, bolters) yields less extract and produces 'thin' beers. For traditional, well-modified ale malts, acrospire lengths of 2/3–3/4 or 7/8 were preferred. If grains from malts are separated into acrospire-length classes and are analysed separately, those with shorter acrospires have lower extract yields, colours and soluble nitrogen values, while wort filtration times, viscosity, turbidity and total nitrogen values and the amount of residual starch are higher (Ullmann *et al.*, 1987).

While within malts made from one grade of barley, using regular malting conditions, there is a loose correlation between acrospire length and the extent of modification, this relationship is not necessarily the same for all barleys and certainly breaks down when barleys are malted in different ways. For example, if a barley is malted at an elevated temperature or is sprinkled with water during germination, the acrospire growth can be stimulated relatively more than modification, while if carbon dioxide accumulates around the grain the growth of the acrospire is retarded relatively more than is modification.

Irregular acrospire growth, like other aspects of irregular grain modification, can be caused by irregular steeping, germination or even kilning conditions, the use of partly pre-germinated, damaged, dead or dormant barleys, or the use of mixtures of different barleys that malt at different rates. Corns of differing sizes from one lot of barley hydrate and malt at different rates, so uniformity is favoured by screening barley into particular size fractions and malting them separately.

13.5 Corn size

The size of malt corns in a sample may be estimated by determining their TCW or by screening (sieving). TCW (g, d.m.) of malt (m) is laborious to determine and is of limited value unless the corresponding value for the original barley (b) is known, when the data can be used to calculate the malting loss:

$$\text{malting loss (\% m/m)} = \frac{\text{TCW}_b - \text{TCW}_m}{\text{TCW}_b} \times 100$$

However, as TCW determinations are not highly accurate, it follows that estimates of malting losses determined in this way are likely to be significantly in error. The measurement of corn size (width) distribution is more useful; this is conveniently determined by grading the malt on a set of mechanically shaken slotted sieves, as is done with barley. In the IoB and EBC methods, the slots in the standard sieves are 2.8 mm, 2.5 mm and 2.2 mm wide, while in the ASBC method they are different, at 7/64 in (2.778 mm), 6/64 in (2.381 mm) and 5/64 in (1.984 mm). Corns are divided into four classes. The boldest are retained on the topmost screen, which has the widest slots. The rest pass through this screen and are either retained on the screens with the successively narrower slots or are passed through into the catch tray. Malt users who break up the malt with roller mills are concerned to set the gaps between the mill rollers to an optimum width and so frequently specify a maximum proportion of thin corns, i.e. those that pass through the narrowest test screen. This is because narrower corns are inadequately crushed in roller mills, so their extract is not wholly recovered in a mash.

To a degree, the distribution of kernel sizes is typical of a particular type of malt; aberrant size distributions may indicate that a mixture of malts is being investigated. As is well known, malt properties alter with corn size (Chapter 14; Ullman *et al*, 1987). Generally as corns become bolder (more plump) the extract increases, malt colour declines, wort turbidity declines, the total nitrogen content ('protein') declines and the amount of soluble nitrogen decreases.

13.6 Grain density

The term grain density is, confusingly, used to refer to the bulk density of grain lots and to the specific gravities of grains. Bulk density refers to the weight of grain or malt per unit bulk volume and is usually expressed as kg/hl, t/m^3 or lb/bu. To determine bulk density grain is run into a vessel of known volume and is levelled by 'striking', that is by running a straight-edged implement (the striker) across the top of the contained, sweeping off excess grain. The weight of the vessel contents is then determined. Traditionally this determination was made with a chondrometer, a device in which the vessel for the grain was attached to a steelyard. Knowledge of the weight per unit volume of grain or malt is essential in calculating the capacities of lorries, stores or other containers. Historically, around 1730, bulk densities of British malts were in the range 27.5–39.5 lb/bu (Mathias, 1959). In 1907, values of 38–44 lb/bu were quoted. The official, and arbitrary, British values of 336 lb and 448 lb are for quarters of malt and barley, respectively (i.e. 42 lb/bu and 56 lb/bu, respectively). Elsewhere these values are different (Appendix, Table A4). More recent values for Continental European malts are 45–58 kg/hl. The determination of bulk density was of value when malt, and grain, was traded by volume, and a 25% loss in bulk density was expected when barley was converted into malt (weights being determined on an as-is basis). However, the practices of deliberately causing the malt to swell by suddenly raising the temperature on the kiln (to obtain 'blown malt') or by sprinkling it with water undermined the value of the bulk-density determination. Furthermore, the values obtained are highly dependent on corn shapes and sizes. The older literature specifically warns against 'bold malts' and indicates that the bulk density of malt should be as high as possible, provided the malt is fully modified. In modern usage, bulk density has little or no value in evaluating malt quality.

Densities or, more properly, the specific gravities of malt corns (dried) decrease with increasing modification. This is the basis of one of the oldest tests (described in 1757 and probably in use before): the 'sinker test'. In the traditional test 200–300 dry malt kernels are thrown into a glass container of cold water and are rapidly stirred. Fully modified malt kernels are expected to float horizontally on the water, slightly under-modified corn float vertically, apical (least modified) ends downwards, fully under-modified grains sink and lie flat on the bottom, while grains that have started to modify rest on the bottom, but in a vertical position embryo-ends up. While this test certainly gives an impression of malt uniformity, in detail it has proved to be unreliable. The dictum that all good grains of malt are 'swimmers' or 'floaters' and only poorly modified grains are 'sinkers' and that any malt in which more than 5% of the grains are sinkers is bad quality is not true. A good malt may yield 20%, or even 40%, of sinkers, depending on the criteria employed. Complications arise because the floaters depend, in part, on air trapped beneath the husk and if this is dislodged or escapes because the grains are cracked and broken, then the corns are likely to sink. Even a proportion of dry barley grains float initially. With the passing of time more and more grains sink as they hydrate and air is lost. Many efforts have been made to make this simple test more reliable. Mechanical, standardized discharge of the grain into water followed by mechanical stirring has been tried; standardized conditions of temperature and time have been proposed and water has been replaced as the buoyancy medium by salt solutions of various densities (e.g. SG 1.150) or by mixtures of benzene and carbon tetrachloride. The best 'improvement' seems to be to collect the sinker corns and test their individual degrees of modification (e.g. by cutting or chewing). This gives a comparatively reliable final estimate of the number of unmodified corns in a sample.

Some success has been achieved in evaluating the specific gravity of small quantities of malt, using toluene as the liquid in the specific-gravity bottle. This solvent is said not to penetrate malt grains and it dislodges any trapped air. The specific gravity of corns declines as malting proceeds, but the values finally attained alter with changes in moisture content, the sample of barley used and, possibly, the malting methods employed. Following the decline in specific gravity is a useful way of determining the progress of modification with malting sorghum. It is said that rapid malting systems for barley malts give low malting losses and some pneumatic-transfer systems give rise to malts with high specific gravities; with traditional malting methods, the decline in grain specific gravity is complete before chemical modification is adequate. Malts may be separated, non-destructively, into different density classes, e.g. by floating on a solution of ammonium sulphate or by using a vibrating gravity table. When this is done, and the separated malt fractions are analysed, it is found that as the specific gravities increase there are declines in the Kolbach index, the extent of modification and homogeneity, diastatic power, the lauter tun wort-separation rate, the beer filtration rate, the extract yield and the attenuation of the wort, while there are increases in gel-protein content, β-glucan content and the TCW (Drost *et al.*, 1987). Therefore, within one lot of malt, increasing density is related to lower modification. However, the differences between different malts were so large that density determinations did not give a reliable, absolute measure of quality.

13.7 Permeability, porosity and compressability

Using the moisture content and the specific gravity (toluene) of barleys or malts it is possible to calculate figures for the 'dead-space' or 'porosity number' of the corns. It has been claimed that the increase in porosity number in converting barley (porosity numbers 3–5%) into malt (porosity number 15–25%) is a good measure of malt quality. The porosity of sorghum malts is also said to be a good indication of modification, but a better indication is given by a compression test (Daiber *et al.*, 1973; Wainwright and Buckee, 1977). In this test, a fixed volume of sorghum grain or malt is compressed by a piston that advances at a constant rate until the grain column resists a load of 90 kg/cm^2. The decrease in the height of the column of grain is measured. The values give an indication of two stages of resistance to compression, owing to the porosity of the grain and to its hardness or cohesiveness.

13.8 Malt germinability

Pale malts that have been carefully dried before the kilning temperature is allowed to rise retain some ability to germinate. This test is said to give highly variable results and to give higher values if certain precautions are taken: to wash the malt corns with a mild disinfectant and limit the microbial population on the germination dishes. Nevertheless, it was used by some brewers as a check to determine whether or not their pale malts had been carefully kilned.

13.9 Patterns of modification

Several methods may be considered together because they allow regions of endosperm modification to be identified and roughly quantified. Thin (10 μm) sections of frozen grains may be stained with Congo Red or Trypan Blue or (less conveniently) fluorescent Calcofluor. Each stain, being selective for β-glucans, stains undegraded cell walls in the starchy endosperm, but not the modified regions in which the walls have been degraded. Inspection of the sections reveals the extent of endosperm modification. Similarly the exposed surfaces of cut malt grains can be inspected using the scanning electron microscope and modification can be assessed by noting where the endosperm cell walls have been degraded. These methods are not suitable for routine use (Briggs, 1987b).

However, several 'sanded grain' techniques are suitable, and two of them are specified in *Analytica-EBC*. In the older Kringstad test, the tips and bases are cut away from malted grains, which, if undried, are dried by soaking in alcohol. The dry grains are soaked in an alcoholic solution of Methylene Blue, which only penetrates the starchy endosperm in regions where the cell walls have been degraded. After about 1 h the grains are dried and are split longitudinally . The areas of endosperm remaining white indicate unmodified zones. In the Heinecken test, groups of dry, malted grains are attached, ventral or crease side downwards, to a glass or wooden supporting layer, using a strong glue. About 0.25% of the grain's thickness is ground away with a sander, exposing the interiors. The sanded grains are soaked in an alcoholic solution of Methylene Blue. After draining and drying, the grains are sanded further, so that half of each grain has been removed. Modified regions of the starchy endosperm, which are permeated by the dye solution, are stained blue. Each grain can be scored for the extent of modification, and so the average modification can be calculated and the degree of heterogeneity can be assessed. A similar result can be obtained by exposing sanded grains to clear polyurethane varnish, coloured with Oil Red 0. In this case, the need to resand the grain is avoided since the excess of coloured varnish is simply wiped from the surface.

Sanded grains can be used to decide on the extent of steeliness of malts and, by gently scratching the exposed surfaces, a rough idea can be obtained of the extent of endosperm modification since the modified regions tend to break up. Many enzymes, including esterase, can only penetrate the modified regions of the starchy endosperm. If sanded grains, coated with a solution of fluorescein dibutyrate (FDB), an esterase substrate, are incubated in a humid atmosphere and are subsequently viewed in ultraviolet light, the brilliant fluorescence of the fluorescein liberated from FDB by esterase marks the modified region of the starchy endosperm.

In the Carlsberg Calcofluor method, groups of dry malt grains, fastened ventral-side down and sanded away to about half their thickness, are successively stained with Calcofluor and Fast Green. The Calcofluor binds to undegraded β-glucan to give an adduct that fluoresces in ultraviolet light. The Fast Green quenches the autofluorescence of the cell walls substituted with ferulic acid. When viewed under ultraviolet light of the correct wavelength, undegraded regions of the starchy endosperm fluoresce strongly, while the degraded regions do not. The percentage modification of each grain is estimated either subjectively by eye or by using a digital, computer-based video-image analysis system. It is sometimes assumed that sample sizes of 50 grains are sufficient, but if lots of 50 corns from a malt sample containing 10% of raw barley are tested there is a high probability that a fair proportion will contain no raw grains. The approved method specifies the use of 100 corns (*Analytica–EBC*, EBC, 1987). It is likely that for quantitative purposes not less than 300 grains should be investigated, and this is tedious. Formulae are given for calculating average modification and homogeneity.

13.10 Mechanical assessments of malt modification

In view of the undoubted value of the 'bite test' it is unsurprising that many tests designed to provide mechanical, and more objective, methods for assessing physical modification have been proposed. The quest for better methods has revealed many complications.

An early suggestion was that 100–200 malt corns should be gently broken up, using a pestle and mortar, the ground material should be inspected and sieved, and the number of hard fragments retained on the sieve (the unmodified material) should be noted. Other early approaches involved milling malt using 'pre-set' mills and then separating the floury, modified material from the unmodified material by sieving. The extent of modification was determined by weighing the fractions. Others proposed that the grist should be characterized by the sedimentation pattern it gave after shaking in water or aqueous alcohol. Problems arose with obtaining reproducible mill settings; variations were also caused by alterations in moisture content (and probably corn size). The differences between good- and poor-quality malts were small. More recently, similar approaches have been made with other devices, including pearling machines. In the determination of 'mechanical mellowness', malt was processed in a barley dehuller. The broken malt material was fractionated using a seed blower. The grits, representing unmodified materials, were collected and weighed. Malt 'crushability' increases with increasing modification and has been evaluated by using a dynamometer to measure the work needed to crush a given amount of malt in a particular mill. The result was a measure of 'average' malt hardness. Bishop (1950) extended this approach and estimated malt homogeneity by passing individual malt grains between a pair of rollers. One roller was fixed, while the other was moveable and spring-loaded. Hard corns, or unmodified fragments, displaced the moveable roller more than the soft, well-modified corns did, and the displacements were recorded on a chart. This promising approach was not pursued, apparently because of the problems of interpretation caused by variations in displacement caused by grains of different sizes.

Malt hardness, determined using the Brabender hardness tester, was widely used. In this device 130 g malt, or a coarse grist, is fed into a pre-set cone mill and the time taken to grind the material to a pre-set extent and the work done in grinding (determined by a dynamometer) are recorded on a chart. The results, recorded in arbitrary 'Brabender hardness units', may be analysed in terms of the maximum hardness, the total energy used and the shape of the recorded energy-time curve. Therefore, interpretation of objective results may, in part, be subjective. Because the shape and size of the malt corns, or even rough grain handling or threshing, alter the results when whole corns are used, it appears desirable to rough grind the malt before it is fed to the fine-grind cone mill. As malting proceeds, hardness declines. However, differences in moisture content alter the results (even the relative humidity of the atmosphere can have an effect) and the hardness of malts, and of the barleys from which they are prepared, varies with the variety, the growing season and the location. However, with a limited range of malts prepared from similar barleys, there do appear to be useful correlations between declining malt hardness and some other malt analyses, such as declining coarse–fine extract differences, wort viscosities and increasing levels of α-amylase. The equipment needs to be carefully standardized. The results give 'average' values for malt hardness. The shape of the recorded trace gives some estimate of malt homogeneity.

Two devices, which were used for a number of years, gave estimates of malt hardness and homogeneity. The Murbimeter, in its final form, determined the mechanical resistance of individual corns to penetration by two blunt needles 1 mm (0.0394 in) in diameter and set 5 mm (0.197 in) apart. The device could automatically test 400 corns in about 30 min. The results were recorded in six hardness classes, The results gave average hardness values and estimates of homogeneity. The Sclerometer, also developed in the Brewing School at Nancy, automatically measured the resistance of individual corns to cutting, or 'bursting', along their lengths. The blunt 'knives' (12), or teeth, were arranged around the perimeter of a steel disc and 400 kernels could be evaluated in 30–40 min. The results, divided into five classes of toughness, were recorded automatically. They were little affected by the ambient temperature, by changes in malt moisture content in the range 3.8–5.9% or by corn size. The results obtained with these machines are said not to parallel each other. Using these devices, it was demonstrated that there is no absolute relationship between hardness and other malt qualities. Therefore, when English, Scotch and Canadian distilling malts were investigated, loose correlations occurred between extract yields and Sclerometer values, but the correlations were quite different for each of the different groups of malts. Measurements of hardness and homogeneity are of value, but the 'norms' vary among malt types, and among malts made from different barley varieties, barleys grown at different locations or in different years. The machines need to be calibrated and carefully maintained; the malt samples must be undamaged and their moisture contents must be standardized. Irregular modification in malt may be caused by malting mixtures of barleys or barley containing some damaged or dormant grains, by uneven water uptake during steeping, by using barley with a range of corn sizes, or by uneven conditions during germination or kilning. Even 5% of under-modified corns in malt can adversely influence some wort preparation systems' giving viscous worts that separate slowly, fine particle formation in the mash bed and viscous, slow-filtering beers rich in β-glucans. Under-modified malt in a grist may cause more problems than an equal proportion of raw barley. Over-modified malts also give rise to problems. They tend to break up during handling and transportation, and on milling they give rise to too high proportions of flour in grists and hence problems of wort separation in mash tuns or lauter vessels. Beers from over-modified malts often lack the correct character and flavour and may have reduced shelf lives. Steeping begins to mellow barley, and its mellowness or friability increases with germination and continues to increase after the malt is adequately modified. Homogeneity increases with increasing germination time.

The most successful and, now, officially accepted machine for assessing physical modification is the Friabilimeter (friability meter), which was also devised at Nancy. The machine is simple to use, when properly maintained it is reliable and adequately reproducible and it gives a result in a short time. The machine successfully combines earlier milling and grist-screening methods. Samples (exactly 50 g) of malt, recommended to have moisture contents of 3.5–5%, are loaded into a cylindrical, stainless steel sieve. The sieve rotates for 8 min, while the malt is rubbed and pressed against the inner face of the sieve by a rubber-faced spring-loaded roller (Fig. 13.1). The husks and the well-modified, friable endosperm material break up to fine fragments, which drop through the sieve. Unmodified fragments and entire endosperms of ungerminated grains are retained on the screen. These are weighed and friability (%) is calculated as 2 × weight (g) of the material that has passed through the screen (determined by difference). In the EBC method, the material retained by the screen is sorted and the 'whole grains',

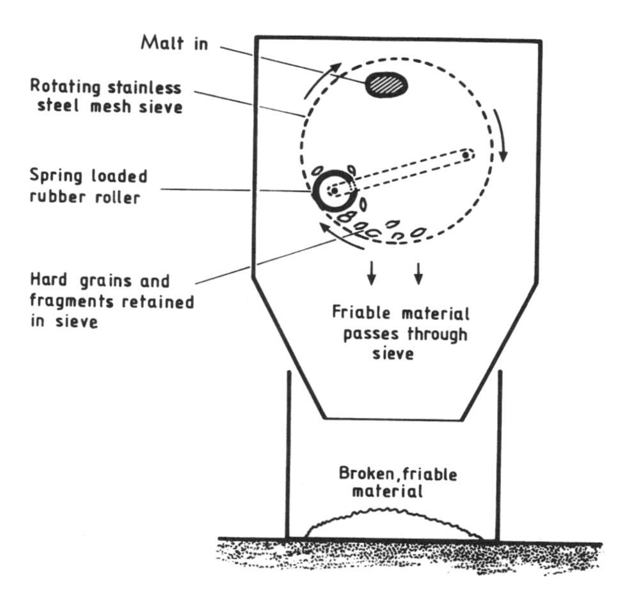

Fig. 13.1 A Friabilimeter. A load of 50 g malt is used. Each test has a running time of 8 min.

retaining three-quarters of the endosperm or more, are collected and weighed. This can be very time consuming. The result (2 × weight) is expressed as the whole grain percentage. For the 'partly unmodified' grain (or homogeneity) determinations, all the material from the Friabilimeter sieve that is retained on a 2.2 mm width shaken, slotted sieve is weighed. 'Partly under-modified grain' (%) is 2 × weight of material retained on the 2.2 mm sieve (*Analytica–EBC*). 'Homogeneity' (%) is 100 − (2 × weight of material retained on 2.2 mm sieve) (IoB). The determination of 'partly under-modified grain' as 'homogeneity' is much more convenient than the determination of whole grain, but it does have the disadvantage that thin, unmodified malt corns can be reduced to less than 2.2 mm in the Friabilimeter and so are not retained by the 2.2 mm sieve. If the barley from which the malt was prepared has been screened over, for example, a 2.5 mm screen, then the sieving method is more satisfactory. The material retained in the Friabilimeter drum sieve should be carefully inspected to ensure that falsely low estimates of modification are not being obtained because, for example, some corns have been toughened and hardened by partial caramelization (case-hardening) in the sub-aleurone layer caused during kilning or because particular barleys give tough malts or because malt has been in a humid atmosphere and has become slack, causing the husks to lose their brittleness, toughen and so fail to break up. Case-hardened corns may be recognized because they have grooves along the dorsal side, where the growing acrospire pressed into the grain. The precision of Friabilimeter analyses varies with the degree of malt modification (Table 13.1). The 50.0 g samples used in the Friabilimeter usually contain about 1400 corns, a comparatively satisfactory sample size. Even so, for critical analyses, samples should be evaluated in triplicate. Measurements of friability are useful for quality-control purposes and to inform a user that the malt will break up in a satisfactory way when milled. However, there is no absolute relationship between friability and other malt analyses; for example, malts having the same friability value may differ in other analytical values, and malts prepared to one specification, using one variety of barley grown in different years, may have widely different friability values. During malting, friability continues to increase after the hot water extract and soluble nitrogen : total nitrogen ratio values have plateaued. With a series of malts, as friability increases so does wort filterability, while the viscosities of 70 °C (158 °F) mash worts and EBC–Congress worts decline and the β-glucan contents of the malts decline. There is no clear relationship between friability and fine–coarse extract difference. The unmodified material retained on the Friabilimeter screen is poorer in diastase and richer in β-glucan and total nitrogen than the friable material and, when mashed, it gives a wort with lower tannoid levels, lower extract, a lower Kolbach index, more hazy and less well-fermentable but more viscous worts. In general, the Friabilimeter gives estimates of modification that are consistent with those of the Methylene

Blue (Heinecken) and (Carlsberg)/Calcofluor techniques. It is not clear, however, that this device can detect 'problem malts' when their analyses are otherwise 'within specification'.

13.11 Half-grain mashing

In the half-grain mashing procedure, as originally described, malt grains are cut longitudinally along the ventral furrow and the half-grains are 'mashed' by holding them in water at 65 °C (149 °F) (Palmer, 1975b). The ventral, sheaf cells remain intact but the starchy endosperm adjacent to the aleurone layer liquefies and washes away. If the grains are cut transversely and mashed, the liquefaction is seen to have occurred around the endosperm adjacent to the aleurone layer and stopping at the ventral furrow by the sheaf cells. This method appears to detect the presence of endosperm-degrading, heat labile enzymes from the aleurone layer, since not all highly kilned but friable malt corns liquefy when mashed, although experimental malts dried at low temperatures do so. The pattern of endosperm liquefaction that occurs during half-grain mashing does not coincide with the pattern of cell-wall degradation, as revealed by Calcofluor-staining for example (Briggs and MacDonald, 1983a; Briggs, 1987b,c).

13.12 Moisture content

As with barley, the primary methods for determining the moisture contents of malts or solid adjuncts are based on the loss in weight that occurs when a ground sample is dried in an oven (e.g. at 105–106 °C (221–223 °F)) under rigidly standardized conditions. The moisture of green malt is determined using a double-drying procedure. In the first stage, the entire malt corns are dried in a rapid airflow, while in the second stage the pre-dried malt is ground and oven dried in the method used for kilned malt. These methods do not give 'absolute' values for moisture content and so the details specified in the standard methods must be adhered to exactly to obtain reproducible results. Moisture contents may be determined more rapidly and conveniently using NIR absorbance in machines that variously operate in reflectance, transflectance or transmittance modes. Some require samples to be ground; others operate with whole grains. Infrared methods need to be carefully standardized and frequently checked, using samples that have also been analysed using a 'primary' method. Malt moisture contents, like those of grains, are expressed on a percentage fresh weight basis. Other malt analyses are variously expressed either on an as-is or fresh weight (fr. wt) basis, on a dry weight (d. wt) or dry basis (d.b.). Most malts now have moisture contents in the range 1.5–6.5%. The user needs to know the moisture content to know that the malt meets the specification and he/she does not wish to purchase water. The more moisture malt contains the heavier it weighs for a given yield of extract, so the 'potential extract' is more expensive to transport. The keeping quality of the malt can be impaired if the moisture content is too high, that is the malt is slack. Slack malt can lose its aroma on storage and will not break up normally when milled.

13.13 Cold water extract

The determination of malt cold water extracts is mostly used in Britain. Ground malt is mashed under conditions that prevent enzyme activity. Originally ice-cold water was used for this determination, but ground malt is now extracted at 20 °C (68 °F) with a dilute solution of ammonia (IoB, 1993). The alkaline conditions prevent any enzyme action. From the specific gravity of the extract, the cold water extract is calculated and is expressed as percentage malt dry matter. Cold water extract, also termed 'matters soluble' or 'pre-formed sugars' is a rough measure of the water-soluble materials that are present in malt. The values obtained are characteristic for particular malts. In the past, for pale-ale malts made from British barleys, values of less than 18% were regarded as a sign of under-modification, 18–21% were a sign of good modification, while values of more than 22% were taken to indicate either over-modification or that the malt had been stewed, that is had been kilned with an insufficient air-flow. In fact the cold water extract is not a good indicator of modification and malts with values of over 21% have produced excellent beers. The use of this method of analysis is declining. However, it is of value in guiding maltsters in that malting methods which favour high cold water extract values also favour colour – and flavour – development in kilning since high concentrations of sugars and amino acids favour melanoidin formation.

13.14 Hot water extract

When malt is ground and mashed and the liquid wort is separated, the materials dissolved in the solution make up the extract. In brewers' wort, prepared from an all-malt mash, the extract consists of 90–92% carbohydrate materials together with a range of nitrogeneous substances, polyphenols, salts and many other substances. The ways in which the milling and mashing operations are carried out influence the yield of extract (the amount of material which dissolves) and the quality of the extract (the proportions of the different components). Rigidly standardized methods for determining the extracts of malts, adjuncts and sugars are now used on the laboratory scale. The methods of the ASBC and EBC (Congress) systems give closely similar results, which are expressed using the same units, while the method for determining the hot water extract used by the IoB is distinct. Small samples of malt, or other grist materials, are milled to a closely specified extent and fixed weights are mashed using specified volumes of distilled water, temperatures, times and degrees of agitation. At the end of the mashing period, the mash is cooled and is adjusted to a fixed volume (or weight) and is then filtered, The specific gravity of the liquid wort is measured at a fixed temperature and from it the extract derived from a known weight of malt is calculated. These analytical mashes differ from brewery and distillery mashes in important ways. (i) Analytical mashes are very dilute ('have a high liquor : grist ratio') when compared with those of commercial practice; for example, the collected laboratory wort may have a specific gravity of about 1029 (that for water is 1000.00) while first brewery worts may have a value of 1085. (ii) Laboratory mashes are made with distilled water, but commercial liquors contain various salts. (iii) In commercial mashes, the mash pH is usually adjusted, while in laboratory mashes it is not. (iv) The spent grains from laboratory mashes are not sparged. (v) The mills used in laboratory mashes differ from any of the wide range of mills used on the manufacturing scale. (vi) The temperature/time patterns used on the small scale do not closely follow any of those used on the commercial scale.

The extract value of a malt is the most valuable single analysis for evaluating its quality. However, it is not a sufficient analysis on its own, since it cannot predict how the malt will perform mixed in a grist with other malts or adjuncts and the extract value (yield) gives no direct measure of the quality of the wort. To achieve this, other analyses are needed. An analytical wort is frequently assessed for its colour, its odour, its flavour, its appearance, the nitrogeneous materials it contains, the fermentability of its carbohydrates, its viscosity, its β-glucan content and its pH. To obtain results of more direct value in breweries and distilleries, some companies also carry out other, different laboratory mashes that more nearly resemble the conditions used in their production plants. They may also carry out trial mashes on the pilot-scale. The worts so obtained may then be boiled and hopped (for brewing trials) and can be fermented to evaluate them in terms of beer quality or spirit (alcohol) yield.

The dry weight of extract obtained from malt exceeds the decline in dry weight of residual malt solids that occurs during mashing by about 4%, because of the water of hydrolysis that is incorporated into the dissolved material, most importantly in carbohydrates, as the hydrolytic processes proceed. However, the extract obtained is inversely related to the dry matter remaining in the draff or spent grains, the insoluble residue from the malt that remains at the end of mashing. Historically, the extract in the wort was obtained by drying samples of wort and weighing the residues. This approach was unsatisfactory since the wort solids were difficult to obtain dry and tended to decompose if heated to accelerate drying. At present, the extract content of wort is nearly always calculated from the specific gravity of the liquid determined at a controlled temperature (e.g. at 20 °C (68 °F)). One of the first to use a hydrometer to determine the gravities of brewing worts was Reddington in 1760 (Corran, 1975). Quite soon at least seven makes of saccharometers or hydrometers were in use and the results they gave disagreed. As Tizard (1843) remarked, 'The visible diversity, as a curiosity, is entertaining, but shows great instability and inaccuracy as a matter of business'. For precise results, the specific gravity of wort is determined by weighing, using a specific gravity bottle, a pycnometer or a vibrating U-tube density meter. More rapid, but less accurate, determinates have been made using various methods including: (i) by successively weighing objects (sinkers) in air, suspended in water and suspended in wort, as in a Mohr–Westphal balance; (ii) with hydrometers; (iii) by floating drops of liquid worts in density gradients made of organic liquids that are immisible with water; (iv) by Doemen's float, a glass float of known density that was used to guide the dilution of wort with water until the mixture had a density equal to that of the float; (v) indirect determination by NIR; and (vi) indirectly by refractometry. Both of the indirect methods must be calibrated using worts of known and similar specific gravities prepared in similar ways from closely similar grists. Refractive index measurements require the use of very precise immersion refractometers at a temperature controlled to within 0.1 °C (0.18 °F) (Essery, 1954; Essery *et al.*, 1954).

In the pre-1977 IoB system of analysis, the hot water extract was expressed in terms of brewers' pounds per quarter of malt (lb/Qr). The brewer's pound unit derived from the method of determining the weight of a barrel (36 Imp. gal; 163.66 l) of wort in pounds (1 lb = 0.4536 kg) at 60 °F (15.6 °C)) and subtracting from it the weight of an equal volume of water (360 lb; 163.30 kg). Thus the extract in the wort, in brewer's pounds, was the excess weight of a barrel of wort above 360 lb at 60 °F (15.6 °C) and the specific gravity of a wort is the (excess brewer's weight (lb) + 360) ÷ 360. In UK brewing usage, the specific gravity is multiplied by 1000, so that the SG of water is 1000.0 (in contrast to sg = 1.0000). Therefore:

$$\begin{aligned}\text{extract (brewer's lb)} &= [\text{SG wort} - 1000] \times 0.36\\ &= \text{excess SG} \times 0.36\end{aligned}$$

In the pre-1977 method (IoB), malt grist (50 g) prepared using a laboratory two-roll Seck mill was mixed with a fixed volume of water at 150 °F (65.6 °C) in a special glass flask or a beaker and was agitated every 10 min. After exactly 1 h the mixture was cooled to 68 °F (20 °C), the volume was adjusted to exactly 515 ml, using water at the same temperature and the flask's contents were mixed and filtered. Within 1 h the specific gravity of the wort was determined at 60 °F (15.6 °C). When the mashing flasks, calibrated to 515 ml, were originally introduced, it was assumed that the volume of draff from 50 g malt was 15 ml, so that 50 g malt gave rise to 500 ml of wort, that is a

'10% wort'. Thus the extract of the malt, in brewer's lb/Qr (336) lb of malt was taken as:

$$\text{apparent HWE (lb/Qr)} = \frac{\text{Excess SG of 10\% wort (at 15.6 °C)} \times 10 \times 0.36 \times 336}{360}$$
$$= (\text{SG} - 1000.0) \times 3.36$$

However, even when the formula was in use it was widely recognized that the average volume of draff from 50 g malt was significantly less than 15 ml, that the volume of draff varied and that the volume occupied was inversely proportional to the quantity of solids dissolved as extract. Therefore, 50 g of a well-modified malt yields 8–9 ml draff while 50 g of oat adjuncts yields about 16 ml of draff. When it was accepted that the specific gravity of the wort altered (i) with the quantities of materials extracted, and (ii) with the volume occupied by the draff, the original, approximate formula was corrected. The use of 515 ml mashing flasks was retained, for convenience. The volume of the draff (ml) = washed, dried weight (g) ÷ 1.48. It was shown, by trial, that:

$$\text{true HWE (lb/Qr, as-is)} = (1.061 \times 3.36 \times \text{excess SG}) - 4.8$$
$$= (3.565 \times \text{excess SG}) - 4.8$$

Malts contain varying amounts of water and so, for absolute comparisons, the hot water extract as-is (that is on the fresh weight basis, the value of most use to malt users) is converted into hot water extract on-dry (that is to the dry weight basis). The results of analyses were calculated either from the formula or from tables. To obtain sufficient accuracy, the specific gravity of the hot water extract wort must be determined to at least five or, better, six significant figures, since the extract is calculated from the excess specific gravity (wort SG − 1000.00) and the result is intended to be accurate to 0.1%.

In 1977, the IoB recommended method was altered in several ways and has been revised since. (i) The malt is now ground in a Bühler–Miag DLFU disc mill (instead of the old two-roll Boby or Seck mill) set to a clearance of 0.7 mm (revised from 0.5 mm) for the standard coarse grind and to 0.2 mm for the fine grind. (ii) The units of extract are now litre-degrees per kilogram (l°/kg), i.e. extract in wort is expressed as excess SG × wort volume (l) derived from 1 kg malt. (iii) Since 1979, wort specific gravities have been measured at 20 °C (68 °F), instead of at 60 °F (15.6 °C), reducing the reported extracts by 1.6 l°/kg. The units brewers' lb/Qr are now obsolete. It is now assumed that the volume of draff from 50.0 g of malt is 8.5 ml, the 515 ml flasks may still be used and so a fixed volume of wort is also assumed, and the formula used to calculate the extract (l°/kg) is 10.13G, where G is the excess SG at 20 °C (68 °F). For the approximate interconversion of the old and new units see the Appendix (Table A6). In 1985, the practice of mashing in stirred beakers and adjusting the mash to a fixed weight (450 g) instead of a fixed volume was introduced. Both methods are in use and give the same results. The formula for calculating the extract from a weight-adjusted mash is:

$$\text{HWE as-is (l°/kg)} = \frac{G \times 8.773}{SG}$$

where G and SG have the meanings indicated previously. For some purposes, it is now acceptable to express extract values as percentage soluble extract (% SE), using the appropriate formulae for 515 ml or 450 g mashes. For small-scale methods, used to measure the extracts of micromalts for example, it is convenient to work on the 0.2 scale and to mash 10 g grist and adjust the mash + wort volume to 103 ml, or the mash weight to 90 g. In practice, adjusting the micromash weight to 90 g yields too small a volume of wort for convenience. It is better to adjust the mash to 100 g and calculate the results using an adjusted formula (Markham *et al.*, 1991). To determine the extracts of coloured malts, which lack enzymes, it is necessary to make mixed mashes in which the ground coloured malt (25 g) is mashed with ground, pale-ale malt (25 g) of a specified enzyme (DP) content. The extracts are calculated from the gravity of the wort obtained with the mixed mash, and from the gravity of a wort from a mash made with 50 g of the pale-ale malt alone. The pale-ale malt provides the enzymes that convert the starch of the coloured malt. With dark malts (colours exceeding 500 EBC units) and roasted barleys, the IoB methods specify making mashes with boiling water. Flours, cereal flakes and torrefied products are analysed for the HWE values using mixed mashes, with no pre-treatment of the adjuncts. Cereal grits (rice, maize, sorghum), however, must first be gelatinized before being analysed using a mixed mash. The extracts of sugars, syrups and caramel preparations, which vary in their moisture contents, are determined by making up solutions of 100 g/l. Then the extract of a preparation is given as E (l°/kg) = 10 × G, where G is the excess gravity at 20 °C (68 °F).

The methods adopted by the EBC (Congress analyses, *Analytica–EBC*) and the ASBC are very similar and were designed with traditional lager-brewing practice in mind. This involved decoction mashing with changing temperatures, rather than the ale-brewing practices, in which infusion mashes were more nearly isothermal, that guided the development of the methods of the IoB. The methods have been revised and refined over the years. The EBC and ASBC methods differ very significantly from those of the IoB. In particular, the mashing conditions are different, and different assumptions are made in calculating extracts, so there are no valid factors for interconverting extract units, although approximate equivalents may be established (see the Appendix, Table A4).

In the EBC, 'Congress' analytical mash, 200 ml water at 45 °C (113 °F) is added to finely ground grist (50 g; 0.2 mm clearance on the Bühler–Miag DLFU mill) (also at 45 °C). The mash is continually stirred. After 30 min at

45 °C, the temperature is increased at the rate of 1 °C/min, for 25 min, to 70 °C (158 °F). More water (100 ml at 70 °C) is added. Saccharification rate is measured from this time. The temperature is maintained at 70 °C (158 °F) for 1 h. The mash is then cooled and the stirrers are rinsed, the rinsings going into the mash. The mash weight is adjusted to 450 g by the addition of water. The mash is filtered and the specific gravity of the wort is determined at 20 °C (68 °F). The ASBC mashing procedure is very similar. Both methods use tables relating specific gravity to the concentration of sucrose solutions to determine the concentrations of the worts. While the EBC uses tables that were originated by Plato, the ASBC calculations are based on the, slightly less accurate, tables of Balling (see the Appendix, Table A5). The calculated results are expressed as the weight of extract produced as a percentage of the original weight of malt. The extract (E) is given by:

$$E\ (\%,\ \text{as-is}) = \frac{P\ (M \times 800)}{100 - P}$$

Where P is the weight of extract (g)/100 g wort (from the Plato table) and M (%) is the moisture content of the malt. The extract on dry weight is given by:

$$\text{Extract}\ (\%)\ \text{d.b.}) = \frac{E \times 100}{100 - M}$$

These calculations rely on assumptions that are invalid. Thus no account is taken of the fact that, because of the 'water of hydrolysis' that is incorporated into sugars and dextrins when starch is hydrolysed, and into amino acids and peptides when proteins are hydrolysed, the dry weight of extract solids is about 4% greater than the dry weight loss of the grist. Also, because water is 'bound' the wort occupies a smaller volume than is assumed in the calculations. Furthermore, the relationship between the concentration of sucrose solutions and their densities is slightly different to that given by starch conversion products and other substances present in the extract. The calculations also ignore the non-carbohydrate components of the extract. Therefore, despite the apparent simplicity of 'percentage extract' values of malt, they are not exact and are less satisfactory than results expressed as litre-degrees/kilogram, since in calculating the latter several invalid assumptions are avoided. Both the EBC and ASBC systems specify methods for determining the extracts of unmalted adjuncts.

The methods outlined do not recover near-maximum extracts from sorghum malts or malts made from other 'tropical' cereals in which the starches have high gelatinization temperatures (Chapter 15). Several similar methods have been proposed designed to overcome this problem. For example, ground sorghum malt may be mashed with water containing calcium sulphate at 50 °C (122 °F) for 30 min. After settling the solids, the supernatant liquid (about 20% of the mash) is collected. The residual, thick suspension of solids is heated to 100 °C (212 °F), at the rate of 1.5 °C/min. and is held at boiling for 15 min. During this time, the starch is gelatinized and partly liquefied and all enzymes are inactivated. The 'thick mash' is cooled to 60 °C (140 °F); the unboiled supernatant, containing extracted enzymes, is returned and mixed in. After 1.5 h at 60 °C (140 °F), the whole mash is warmed to 78 °C (172.4 °F) and is then filtered (Byrne *et al.*, 1993; Chapter 15).

Finely ground (f.g.) malt gives a higher extract in laboratory mashes than coarsely ground (c.g.) malt. The difference in extract obtained between such mashes (the f.–c. extract difference) is an indirect measure of the modification of the malt. The official analytical methods specify the milling and mashing conditions to be used. The finely ground malt has a large proportion of its cell walls disrupted and so its starch granules are exposed. Consequently the starch is more accessible to the diastatic enzymes in the mash and substances can readily diffuse into and out of the small particles. Therefore, it is sometimes supposed that the fine-grind mash will yield the total recoverable extract from the malt consistent with the amounts of starch, sugars and diastatic enzymes that are present. The extract recovered when a coarsely ground sample of malt is mashed is regulated by the degree of modification of the malt, since under-modified pieces of endosperm will not break up into fine particles when malt is coarsely ground and so will remain impervious to amylases and will not yield their potential extract. With increasing fineness of grinding, the extract recovered in a laboratory mashing system rises and plateaus; a good-quality malt will yield a high maximal extract while a less good malt will yield a lower maximal extract. As malting proceeds, and modification improves, so the extract difference obtained between coarsely and finely ground samples at first decreases but then becomes approximately constant. The idea of the f.–c. extract difference has been familiar for many years, but the reliable determination of this value presents many problems. In particular, the older types of mill were not capable of closely reproducing the two types of grist needed. Then, for most malts, the difference between the extracts obtained with the f.g. and c.g. mashes are small, relative to the extract values themselves and the variabilities associated with them. Therefore, to obtain the necessary degree of reliability, each mash should be replicated four to six times, and the results subjected to statistical analysis. With EBC analyses, f.–c. extract differences of about 2–2.6% are obtained with normal malts, greater values being given by less well-modified malts and values of less than 1.3% being given by very fully modified or over-modified malts. Many attempts have been made to improve this analysis. For example, Bourne and Wheeler (1982) showed that by replacing the IoB coarse-grind grist (0.7 mm) with a coarser grist (1.0 mm mill clearance) and by making the mash with a lower liquor : grist ration giving a 'coarse-concentrated mash' (c. conc. mash), the difference in extract obtained, the f.–c. conc. extract difference, was substantially greater than the standard f.–c. extract difference and was inversely related to extract recovery in the brewhouse. Finely ground grists do not necessarily yield maximum extracts, particularly if the malt is deficient in enzymes. To overcome this, mashes have, as with the old Graf method, sometimes had enzyme preparations added to them, with the result that extracts are enhanced, particularly when under-modified malts are being investigated. The 'limit' of this approach is when extracts are determined on unmalted barley or other cereals. Problems arise from the lack of standardized enzyme preparations.

The Tepral method was designed to evaluate malts using thicker mashes, with salts in the mashing liquor, controlled sparging and selected wort collection times after temperature-programmed mashing, using conditions more closely resembling those used in breweries. In 1936 Hartong first proposed his mashing system for evaluating malts. The original method was modified subsequently by Hartong and Kretschmer. The malt was mashed using the standard Congress (EBC) temperature-programmed method and in isothermal mashes made at 25 (later 20), 45, 65 (later 60) and 85 (later 80) °C (77 or 68, 113, 149 or 140 and 185 or 176 °F). The Hartong evaluation number, dissolution number or index was calculated by expressing the extract yields obtained using the isothermal mashes as percentages of the Congress extract, taking the average and subtracting an arbitrary constant (60 or, later, 58). Higher numbers were given by better-modified malts. Values of 5 were regarded as average. Different values have been suggested as being ideal for malts being used for different purposes. In addition it is claimed that the individual values obtained from the isothermal mashes gave additional indications of malt quality. Thus the 20 °C (68 °F) (or 25 °C (77 °F)) extract supposedly indicated the pre-existing extract in the malt, although actually some enzymic degradation of malt substance does occur in these mashes. The 45 °C (113 °F) value varied with the activities of key heat-labile enzymes in the malt, supposedly proteases and β-glucanases. The extract recovered from the 60 °C (140 °F) (or 65 °C (149 °F)) mashes reflected α- and β-amylase activities, while that recovered at 80 or 85 °C (176 or 185 °F) was attributed to α-amylase activity. Many attempts have been made to simplify this highly laborious system, which was once used quite widely. Probably the most frequently reported value now is the extract obtained at 45 °C (113 °F), the VZ-45 °C in German reports. This value is influenced by the friability of the malt, the level of pre-formed extract, the activities of heat-labile enzymes, the variety of barley used to make the malt and the year of its harvest.

During mashing, by the EBC method, the mash is smelled to test if it has a normal odour for the particular type of malt. Any abnormal odour must be noted. When doubts arise, the wort may be tasted. Using the EBC or ASBC systems, it is usual to note the 'conversion time' or 'saccharification time' of the mash. As soon as the mash reaches 70 °C (158 °F) the time is noted and small samples are taken, cooled and mixed with an iodine reagent on a porcelain or gypsum plate. This is repeated at intervals until the blue-black colour given by residual starch fails to appear and the mixture remains clear yellow. The time is noted. The destruction of the starch is mainly through the activity of α-amylase. Excessively long conversion times indicate that the malt has a low enzyme complement.

13.15 Some determinations made on laboratory worts

Sometimes the time taken to collect a given volume of wort, or the volume of wort collected in a fixed time, is noted at the end of a laboratory mash when the mash is being filtered. This test is not highly reproducible, but slower wort separation occurs with under-modified barley malts and with triticale malts, for example, and is likely to be reflected in slower wort separations in the brewery. A number of laboratory devices have been tested, with limited success, to see if they are better able to predict brewhouse wort run-off rates.

The viscosities of laboratory worts are often determined. With the EBC analytical system, the value of wort viscosity determinations has long been recognized. However, viscosity declines to an approximately constant level as modification improves. For a long period, with IoB analyses of ale malts, the determination was regarded as valueless. However, as more rapid malting and mashing procedures have come into use, so has the regular determination of the viscosity of 'HWE worts'. Elevated values indicate that a malt is under-modified or is unevenly modified. However, differences between malts are small. Various workers have mashed green or kilned malts at 70 or 80 °C (158 or 176 °F), and determined the viscosities of the resulting worts. In a standardized method, fine grist at 70 °C (158 °F) is mashed with a solution of diastase also at 70 °C, which supplies α-amylase and so ensures starch hydrolysis (Bourne *et al.*, 1977; Bourne and Wheeler, 1982). Pre-heating the diastase solution to 70 °C (158 °F) ensures that any β-glucanase present is destroyed. After 1 h, the wort is collected and the viscosity is determined. At this temperature, β-glucanases are rapidly inactivated and gums, including β-glucans, are largely dissolved. The viscosity values obtained are substantially higher than those obtained with standard laboratory worts. High values in the range give an indication that malt is irregularly modified or under-modified.

The appearance of laboratory worts is frequently noted. Well-modified malts give clear worts, while less well-modified malts may give slightly opalescent or hazy worts. Such malts do not necessarily give hazy beers.

In the past, worts were often titrated to obtain measures of acidity or buffering power. Now the pH values of worts are measured, using a glass electrode system. Departure from the normal range indicates some problem. For example, a low pH value may indicate that a heavy infestation with acid-forming bacteria was present during the production process.

13.15.1 Colour

Wort colours, or tints, are regularly determined, as are those of sugar preparations and caramels used in brewing. This determination is not easy to standardize in a satisfactory way because worts vary in the intensity of their colours and in the nature of these colours; that is different worts can have different absorbtion spectra. Originally, wort colour intensities were matched with iodine solutions, but this method was not satisfactory since, at least with worts from darker malts, the colour qualities were different to those of the iodine solutions. Mixtures of synthetic dyes gave better colour matches. However, the best results were obtained using 'colour comparators'. Lovibond introduced his coloured-glass reference slides in 1883. In a comparator, the colour of a wort, held in a parallel-sided glass cell of known width and illuminated from behind, is matched with the colour of a glass disc, both being viewed simultaneously using one light source. By varying the cells to give different thicknesses of wort for the light to traverse, by diluting worts and by choosing the appropriate sets of colour discs, operators with good eyesight obtain reproducible results. For analytical purposes, hot water extract worts are usually used, although sometimes these are first boiled to give the boiled wort colour (Kockfarbe), to allow brewers to make some allowance for the wort darkening that occurs during the copper-boil. The results give the malt colour. The *Analytica–EBC* system has recently adopted a spectrophotometric measurement of wort colour intensity at a wavelength of 430 nm. Such measurements are very precise but inevitably cannot match a range of colours in which the absorption spectra differ significantly. Suggestions are being made that values based on combining the results of measurements made in the red, green and blue spectral regions be adopted. This proposed revival of the 'tristimulus' approach is theoretically sound but has not been formally adopted.

13.15.2 Wort fermentability and attenuation limits

The decline in the specific gravity of wort that occurs as it is fermented by yeast is referred to as attenuation, and the lower limit, which is reached when fermentation ceases, is the attenuation limit. The reduction in gravity results from the removal of the fermentable sugar and the accumulation of ethanol, which has a density below that of water. In the IoB system, the apparent attenuation, F(app. %) is

$$F(\text{app }\%) = \frac{\text{OG} - \text{PG}}{\text{OG}} \times 100$$

where OG, the original gravity, is that of the wort before fermentation and PG, the present gravity, is the value when fermentation is complete. The real attenuation, F(real, %), is the calculated decline in gravity owing to the removal of carbohydrates and is given by F(app. %) × 0.819 for boiled, brewers' worts or F(app. %) × 0.814 for unboiled, distillers' worts. Because brewers boil their worts, they destroy any residual malt enzymes, and so the ratio of fermentable : unfermentable carbohydrates is fixed. By comparison, in unboiled, distillers' worts enzymes survive and continue to degrade dextrins into sugars. From these figures, distillers calculate the predicted spirit (alcohol) yield (PSY) of a malt (Dolan, 1976, 1983; Dolan *et al.*, 1981). Wort attenuation depends on the availability of fermentable sugars and on the yeast remaining in contact with the wort. Most brewing yeasts ferment glucose, fructose, sucrose and maltose. Some ferment maltotriose, but only slowly. A few will ferment the minor sugars isomaltose, panose and isopanose. Therefore, wort fermentability can differ when tested with different yeasts. In analytical 'forced fermentation' tests, the worts are shaken with a pure sample of a specified yeast in a container closed with a non-return, fermentation gas-lock and incubation is continued until fermentation is complete. However, it has long been known that in a fermenter the premature separation of yeast, either into the head (froth) or by flocculation and sedimentation, can limit fermentation even while fermentable sugars remain. Different malts give worts that cause yeasts to separate at different stages of fermentation. Several proposals have been made to carry out fermentation tests in unstirred, thermostatically controlled tall tubes to allow a more complete characterization of malts by determining the fermentabilities of worts prepared from them in a standard way (Kruger *et al.*, 1982; Axcell *et al.*, 1986; Inagaki *et al.*, 1994). In these tubes, the yeast can either remain in suspension, sediment or rise into the head. These tests have not been widely adopted.

13.16 Mash viscosity

Various attempts have been made to characterize malts by measuring the changes in viscosity that occur when a stirred mash is processed through a heating cycle and is then held at a final temperature. Most experiments have been carried out with Brabender Amylographs or Viscographs, devices originally designed for use in the baking industry (Yoshida and Yamada, 1970; Bourne *et al.*, 1977; Lloyd, 1979; Moll, 1979a; Woodward and Oliver, 1989). While it is not a standard method, Amylograph analysis is of value for investigative purposes. Generally a fixed amount of finely ground malt is mashed under standard conditions and is continually stirred while being subjected to a rising temperature programme. As the temperature rises, the mash viscosity initially declines slightly, but at around 55 °C (131 °F) it begins to rise rapidly, reaching a peak value at about 62–66 °C (143.6–150.8 °F) and then declines to an approximately constant value. The details vary with the mashing conditions and the malt being investigated. The mash may be supplemented with enzymes, salts or enzyme inhibitors as the experiment requires. The viscosities and temperatures just before the onset of the rise in viscosity, the temperature at the onset of the rapid rise, the peak value and the temperatures at which it occurs, and the value at the end of the decline are noted, as is the area of the viscosity–time peak. The changes observed vary with the barley being malted, the degree of modification and the malting conditions being employed. The results appear to depend chiefly on the gelatinization of the starch, which causes the viscosity to rise, the levels of enzyme present (particularly α-amylase), the rate at which the enzymes present degrade the starch (causing a fall in viscosity) and the accessibility of the starch to enzymic attack. Among many interesting correlations is the fact that high peak mash viscosity values correlated with poor brewhouse performances in a range of malts.

13.17 Nitrogen fractions of malt

The determination of nitrogen in malt, in extracts prepared from malt, in wort and in various wort fractions provides a variety of important results for assessing malt qualities. In the EBC and ASBC methods, nitrogen values are usually multiplied by the arbitrary factor 6.25 to give estimates of 'protein'. Malts, like unmalted grains, contain a wide range of substances that contain nitrogen and are not proteins; in addition, proteins themselves contain a range of proportions of nitrogen and indeed may be glycosylated and/or acylated, so that they carry nitrogen-free substituents (section 5.3). As a result, the convention of using nitrogen values × 6.25 for protein values is incorrect. The IoB system of reporting results in terms of weights, or proportions of nitrogen, is to be preferred.

For many years, the total nitrogen (TN) value ('protein') content of a malt was determined using the Kjeldahl technique as the reference method. With this technique, introduced in 1883, the organic material is destroyed in a wet digestion and most organic nitrogen is converted into ammonium ions. A finely ground malt sample is strongly heated with strong sulphuric acid and highly toxic catalysts, variously containing potassium sulphate, copper, selenium, mercury and titanium. The ammonium ions are deprotonated by the addition of sodium hydroxide to the digestion mixture and the ammonia released is steam distilled into a trapping solution, to be determined by titration. Alternatively, the ammonia may be determined using a colorimetric technique.

Nitrogen in worts or various fractions was determined in a similar way. Automated and semiautomated Kjeldahl methods recover slightly more nitrogen than the manual method. This method is sound but slow and laborious and uses dangerous reagents; as a result, various 'secondary' methods are used, which have to be calibrated against the primary method. For example, by heating finely ground malt directly with caustic soda a fixed proportion of the nitrogen in a sample was supposed to be released as ammonia, which was estimated. A more usual method in recent years is the use of NIR measurements, in the reflectance, transflectance or transmittance mode. These methods require continual recalibration and checking using malts also analysed by a primary method. More recently the pressure to use less dangerous laboratory techniques and the availability of suitable equipment has led to the use of the Dumas method as a primary method (Buckee, 1995). In the Dumas technique, introduced in 1826, samples are combusted in pure oxygen at about 1000 °C (1832 °F). The water and carbon dioxide that are generated are removed. The oxides of nitrogen that are also produced in the combustion are reduced to nitrogen gas and this is measured, by thermal conductivity for example. The Dumas method recovers organic nitrogen more completely than the Kjeldahl methods. It works well on grain samples but at present a method using it for analysing liquids, such as wort, has not been perfected.

Since there are inverse relationships between malt extract yields and their total nitrogen contents, the latter value is of interest. Of greater interest is the amount of nitrogen that dissolves when a malt is mashed. The yield of soluble nitrogen is influenced by the mashing conditions used. Thick brewery mashes produce more soluble nitrogen than thin laboratory mashes. More soluble nitrogen is produced using ASBC or EBC temperature-programmed, laboratory mashes than IoB isothermal mashes. For many years, the yield of soluble nitrogen obtained with a laboratory mash (IoB) was recorded as permanently soluble nitrogen (PSN), the nitrogen remaining in the laboratory wort after this had been boiled for a standard period (e.g. for 5 h, under reflux, in a vessel immersed in a boiling, saturated brine-bath at 108 °C (226.4 °F)) and the coagulated material (coag. N) had been separated by filtration. This boiling process was tedious and poorly reproducible and so the convention was adopted that PSN = TSN × 0.94, where TSN is the total soluble nitrogen in the unboiled laboratory wort. The ratio (PSN/TN) × 100 was called the nitrogen index of modification. However, the proportion of soluble nitrogen that is coagulated when wort is boiled is variable and the ratio PSN/TN can be varied independently of other aspects of modification. The ratio now employed in IoB analyses is the soluble nitrogen ratio (SNR), which is given by (TSN/TN) × 100 (%). The same ratio, but determined using temperature-programmed mashes and so yielding higher absolute values, is called the S/T (soluble/total) nitrogen ratio in ASBC analyses and the Kolbach Index (KI, or N_K) in Congress (EBC) analyses. In practice, brewers find that they require malts having SNR or equivalent values in particular ranges.

The soluble nitrogenous materials in solution serve various purposes. The true proteins are partly precipitated in the copper-boil and those (or at least the very large polypeptides) that survive into beer contribute to head (foam) stability, the mouth-feel of the beer (body) and, sometimes, to the formation of undesirable hazes. The low-molecular-weight materials, in particular ammonium ions and most of the amino acids (but usually not the imino acid proline) serve as the sources of nitrogen that support yeast growth. The proportions of the different nitrogen fractions differ among worts.

There are no 'official' methods for analysing proteins in worts. Historically these were coagulated by boiling and were determined by measuring the nitrogen content of the coagulum or they were precipitated by various agents such as tannins (in the Lundin method), magnesium sulphate (Myrbäck method) or trichloracetic acid, and the nitrogen present in the precipitate was determined. At present 'proteins' may also be determined by measuring the spectral change that occurs when a Coomassie Brilliant Blue dye binds to proteins in solution or when the high-molecular-weight nitrogen-containing substances (HMW-N) are separated from smaller molecules by gel-filtration chromatography. In the saturated ammonium sulphate precipitation limit test (SASPL), successive small additions of saturated ammonium sulphate solution are added to a sample of wort, and the turbidity of the mixture is determined after each addition. A sudden rise in turbidity occurs at a particular concentration of ammonium sulphate. This SASPL value is related to the quantity of complex, potentially haze-forming protein in the wort. Lundin, the Myrbäcks and others presented fractionation schemes that distinguished between high-molecular-weight ('protein'), moderate-molecular-weight and low-molecular-weight nitrogen-containing materials. For experimental purposes, the ammonium ion content and the levels of the individual amino acids are determined using ion-exchange chromatography. For quality-control purposes, the smaller nitrogenous substances were determined by formol titrations. They are now determined by colorimetric methods. Attempts to determine 'assimilable nitrogen' or 'yeast utilizable nitrogen' directly by incubating wort with yeast and measuring the uptake of nitrogen by the yeasts failed; the tests did not give reproducible results.

In formol titrations, wort is adjusted to a pH of about 9, then formalin (formaldehyde solution) is added. The formaldehyde reacts with amino groups, displacing protons and causing a fall in the pH (acidification) of the mixture. The amount of sodium hydroxide solution needed to restore the pH to the initial value is, therefore, an indirect measure of the number of amino groups present in solution and gives the formol-N value. The method detects ammonium ions, amino acids and simple peptides. Because this method detects peptides, which yeasts cannot utilize, it can indicate apparently adequate values of yeast nutrients when free amino acids are in short supply. Consequently, determinations of formol-N have been largely replaced with estimates of free amino nitrogen (FAN). Two colorimetric methods for determining FAN are in use, one employing ninhydrin (the method of choice), the other using trinitrobenzene sulphonic acid (TNBS). The methods are calibrated with amino acid standards. Neither of these methods is truly specific and they do not give identical results and so the method used must be specified. The results of the various nitrogen ('protein') analyses can be recorded in different ways. For example, values for total nitrogen or 'protein' (N × 6.25) of malt are usually expressed as % m/m (d.b.) as are values for the total soluble nitrogen. Free amino nitrogen is variously reported as % m.m (d.b.) or as mg/1 wort.

13.18 Enzymes in malt

Many enzymes have been determined in malt; however, for quality-control purposes the methods used are for diastatic power (DP), α-amylase and β-glucanase. For various reasons these determinations are not fully satisfactory. The IoB has been unable to recommend the use of any assay for β-glucanase.

The assay of diastatic power extends back to 1833 when Payen and Persoz precipitated 'diastase' from malt extracts. In principle, malt extracts containing a mixture of carbohydrases catalyse the hydrolysis of soluble starch to sugar and dextrins. As each intersugar bond is broken, a reducing group is liberated. The enzymes chiefly involved are α-amylase and β-amylase, but α-glucosidase and limit dextrinase activities may also be significant. After a period of incubation, the reaction is stopped and the increase in reducing groups is determined by colorimetry or by titration. The various methods do not give highly reproducible results. The values are of most importance when the malt user requires high enzyme activities because the malt is being used in a mash rich in unmalted adjuncts, which must be converted by enzymes from the malt. Using the IoB methods, the results may be expressed as °IoB or °Lintner where DP(°L) = DP (°IoB) × 1.1. In the EBC Congress method, diastatic power measurements are expressed in Windisch–Kolbach units (°W–K).

Two methods for determining α-amylase are in use. In the first, the 'International Method', soluble starch is incubated with β-amylase to yield a β-limit dextrin, to prepare the substrate mixture. Malt extract, which contains α- and β-amylases among other enzymes, is added to the β-limit dextrin–β–amylase mixture. Samples are taken at intervals and are mixed with iodine solution. When the colour given matches that of a standard the time is noted. The amount of α-amylase is inversely proportional to the time taken to reach the reference colour. The results are expressed in dextrinizing units (DU). The principle is that whenever the α-amylase hydrolyses an interglucose bond within the β-limit dextrin chain, the large amount of β-amylase that is present rapidly degrades the exposed, unbranched chain until another new branch point is reached or the molecule is wholly hydrolysed. Because of the large amount of β-amylase already in the substrate mixture, the addition of more β-amylase in the extract containing the α-amylase is without effect on the degradation rate of the β-limit dextrin. This method is not truly specific for a-amylase since limit dextrinase and possibly α-glucosidase can hydrolyse the α-(1→6)-branch points in the β-limit dextrin, exposing more linear -α(1→4)-linked glucan chain for hydrolysis by the β-amylase. Therefore, this method actually determines α-amylase plus limit dextrinase and (probably) plus α-glucosidase. In the second method for determining α-amylase, the substrate is a commercially prepared, cross-linked starch to which blue dye is covalently linked. During digestion with α-amylase, soluble blue substances are released into solution and the colour is determined spectrophotometrically. Enzyme activity is calculated from the amount of colour released in a 15 min incubation period.

Endopeptidase is assayed by following the release of colour from blue-dyed hide-powder (collagen) in the *Analytica–EBC* method. This assay may be of value in determining protease activities in malts or in microbial or other plant preparations. Protease activites are not determined routinely.

The determination of endo-β-glucanase activity can be important, particularly when malts are to be used in mashes containing barley-based adjuncts. Several methods have been proposed and two are frequently employed, but neither has proved sufficiently reproducible to be recommended by the IoB (Buckee and Baker, 1988). In one method, the rate of decline in the viscosity of a solution of β-glucan, when incubated with an extract of malt containing β-glucanase, is followed. The time taken for the viscosity to fall by a defined extent is inversely proportional to the enzyme activity. In another method the substrate is a β-glucan preparation covalently linked to a coloured dye. As the substrate is progressively degraded by β-glucanase in a test incubation, so it is less readily precipitated by the reagent used to stop the reaction and more colour remains in solution. The increase in soluble coloured material that occurs in a particular time can be measured spectrophotometrically and is a measure of enzyme activity.

13.19 Gums and hemicelluloses

Worts rich in β-glucans and/or pentosans are viscous, but wort viscosity also depends on many other dissolved substances and viscosity is not an adequate, specific measure of particular dissolved polysaccharides. In the past 'soluble pentosan' measurements were tried as a measure of modification, but this test was inadequate. Several methods for determining β-glucan have been devised, since the presence of substantial amounts in barley malts is an indication of inadequate endosperm modification. A specific, precise, simple and reproducible assay of β-glucan is difficult to achieve. A commonly used method is to determine the glucose generated when samples containing β-glucan are incubated with a partly purified, fungal cellulase preparation. Unfortunately, it is difficult to destroy the α-amylase and amyloglucosidase enzymes that occur in the cellulase preparations. If these enzymes survive, then starch and dextrins in the samples being examined are converted, in whole or in part, into glucose; as a result, the quantities of β-glucan actually present are over-estimated. In addition, these cellulase preparations may attack polysaccharides other than mixed-link β-glucans.

In one of the two EBC-approved methods, samples containing β-glucan are incubated with a highly purified bacterial endo-β-glucanase ('licheninase'). The oligosaccharides liberated by the specific hydrolysis of β-glucan are degraded to glucose by incubation with purified β-glucosidase, and the glucose is measured. The weight of the original β-glucan is given by 0.9 × weight of glucose. The other EBC-approved method is based on the fact that when Calcofluor binds to β-glucan a highly fluorescent complex is obtained. Solubilized β-glucan samples are introduced into a flow-injection analysis system (FIA). They are mixed with Calcofluor reagent and then passed through a recording fluorimeter. By comparing the fluorescence intensities, the amounts of β-glucan in test samples are found by comparison with reference samples of β-glucan. The variations encountered with this method are caused, at least in part, by the variations in fluorescence yield that occurs with equal amounts of β-glucans of different chain lengths, the strong influence of alterations in the ionic strength of the medium on fluorescence intensity and the absence of recognized β-glucan standards.

13.20 Phenolic materials

Occasionally the levels of phenols or polyphenols are determined in extracts of malts, or in the worts produced in laboratory mashes. None of the methods used is 'official'. Because of the complexity of the mixtures, HPLC is needed to determine individual substances such as the catechins or the proanthocyanidin (anthocyanogen) dimers and trimers. Colorimetric methods, such as that based on the reaction between polyphenols and ferric ions, give an estimate of 'total polyphenols'. Proanthocyanidins may be extracted from malt using 60% ethanol containing ascorbic acid. By adsorbing proanthocyanidins onto insoluble PVPP, they are considerably purified and can be eluted with N-methylpyrrolidone and then degraded to give measurable colours by heating with butanol/hydrochloric acid. Flavanoids have also been quantified using a colorimetric *p*-dimethylamino-cinnamaldehyde reagent. Polyphenols with tanning properties, which are likely to take part in haze formation in beer, can be precipitated with soluble polyvinylpyrrolidone and so be estimated.

Routine analysis of some phenols is carried out on Scotch distillers' malts to determine the degree to which malt has been 'peated' during kilning. While a large number of substances are taken up from peat-smoke, those that are used as 'indicators' are the simple phenols, phenol, the cresols, the xylenols and guaiacol. Colorimetric and GLC methods have been used, but they are not fully satisfactory (Bathgate and Taylor, 1977; Taylor, 1979).

13.21 Some other analyses

A variety of other analyses is carried out on particular malts. Dimethyl sulphide (DMS) and its precursor substance *S*-methylmethionine (DMS-P, SMM) are determined by GLC of DMS before and after the precursor is broken down to DMS by a treatment with alkali. Brewers of lighter, pale lager-type beers often specify limits for DMS-P. Hydroxymethyl furfural (HMF) increases in amount in step with the development of colour and the accumulation of heterocyclic, nitrogen-containing compounds that occur during kilning as the Maillard reactions proceed. Sometimes HMF is determined, by colorimetry or GLC, as a measure of the progress of the Maillard reactions.

Arsenic levels are determined, using a modified Gutzeit test, in malts made using direct-fired kilning. Because of its toxic nature, strict upper limits on arsenic content have been set for many years. Upper limits are also often set on the 'total residual sulphur dioxide' content (sulphur dioxide plus sulphite). Determination is by distilling the malt, collecting the gases that are liberated, oxidizing them and determining the sulphate ions that are formed. Sulphur dioxide, which is readily oxidized to sulphate, is taken up from furnace gases in direct-fired kilns, and from the sulphur dioxide used to treat malts to minimize the formation of NDMA. 'Fatty substances', materials extracted from grists using specified organic solvents, are rarely, if ever, now measured in malts. However, they are routinely measured in unmalted adjuncts. Measurable cyanide (MC), derived from the cyanogenic glycosides of the malt, is measured by some distillers, who set upper limits to the amounts malts may contain. Glucose on the surfaces of malt corns may be determined (e.g. with indicator test-papers) to decide whether the malt has been sprayed with a glucose solution (as in the Belmalt process).

A malt may be analysed for bromide ions to determine whether bromates have been used in its manufacture, or for high levels of gibberellic acid, in an attempt to determine if this has been applied. Since some maltsters provided malts that had been loaded with heat-stable microbial enzymes, particularly α-amylases, β-glucanases and cellulases, no doubt these will have been analysed at intervals. *N*-Nitroso-compounds are usually determined in specialist laboratories. Frequently NDMA is measured using a thermal energy analyser connected to a gas–liquid chromatograph (GLC-TEA). Similarly, residues of pesticides, fumigants, fungicides, herbicides and growth regulators (variously applied in the field or in store) are determined by specialists.

13.22 Microbes and microbial metabolites

Increasingly malts are being tested to assess their microbiological populations using a range of methods of preparation and selective culture media (Flannigan, 1996). Chiefly this is because certain moulds, particularly *Fusarium* spp., form substances that potentiate gushing (overfoaming) in beers. Donhauser *et al.* (1991) developed a test in which boiled, unhopped worts are highly carbonated and bottled. After 2 days in storage, the bottles are opened and the extent of the gushing that occurs is assessed. Malts that give rise to gushing beers also give 'gushing worts' when tested in this way. Fungi are capable of making a chemically very diverse range of mycotoxins, and some mycotoxins can be formed by fungi during malting. The situation is highly complex, but tests for individual mycotoxins are available in a few instances and other tests are being developed. Usually mycotoxins cannot be detected on malt and it is probable that small amounts of mycotoxins on malt are more of a risk to cattle that eat brewery spent grains than to those that drink beer (Flannigan, 1996). Recently proposals have been made that barley should be inoculated with chosen, 'acceptable' microorganisms to suppress the growth of others. Adoption of this approach on the production scale, if it occurs, will necessitate the reliable measurement of these organisms on the 'finished' malts.

14 Malting conditions and their influences in malting

14.1 Introduction

The type of malt produced is regulated by the species of cereal used, the quality of the grain sample, the details of the hydration processes (immersion- and/or spray-steeping) and whether or not additives are used, details of the germination process (including grain moisture content, the temperature profile and the duration of the germination period, and the frequency and vigour of turning) and, finally, details of the drying and cooling processes (withering and kilning or roasting). The product needs to be deculmed in most instances, although roots and shoots are not removed from sorghum or millet malts used for making African fermented 'porridges' and opaque beers.

Different qualities and types of malt are made for distillers, brewers of clear beers and stouts, brewers of opaque beers, bakers, confectionery manufacturers, and for use in preparing weaning foods (Chapter 1). Malts are predominantly made from barley although they are also prepared from other cereals, and various leguminous seeds are sprouted for use in foodstuffs. The preparation of barley malts will be considered in most detail. The details of sprouting legume seeds will not be considered here.

The requirements of barley malts for making European-style beers (ales, lagers, stouts), distilled products (whiskeys) and malt vinegars are fairly clear – in particular maximal extracts of the correct quality and, in some instances at least, a minimum spirit yield – and are reasonably well understood. A battery of analytical methods is available for barley brewing malts (Chapter 13). These analytical methods are often applied to malts intended for other purposes, or made from different cereals, and it is doubtful if the results are then reliable or if they are necessarily relevant (e.g. for the production of opaque beers).

The malting processes, originally developed by trial and error, have been greatly refined in the light of systematic investigations carried out over the last 200 years. The result is that better-quality malts can now be produced on a routine basis and increasingly on a large scale; batches of 200–500 t barley are regularly processed, and yet some small maltings are still in operation, making high-quality specialist malts. Maltings varying widely in size and sophistication are operating in parallel. Advances in malting knowledge and technology have resulted, among other advantages, in: (i) more evenly modified, more homogeneous malts as a result of better grain selection and more uniform processing; (ii) reduced processing times, giving greater throughputs in plants (the shorter the processing time of a batch the more batches can be produced in a year); (iii) reduced malting losses, so a greater quantity of malt is made from a given amount of barley and the amounts of by-products produced (culms) are proportionately less; (iv) production of better-quality malts, having higher hot water extracts at a given nitrogen content; (v) production of malts meeting a wide range of analytical specifications; (vi) an enhanced ability, to some degree, to malt less high-quality barleys, and other cereals; and (vii) less requirements for personnel, water, power and fuel in malt production and, as a result, less effluent.

It is convenient, initially, to consider floor-malting and then pneumatic-malting processes and the range of changes that occur in malting grain. This is followed by a consideration of how changes in the barley used and the operating conditions influence the quality of the malt.

14.2 Floor-malting

There have been many reports over 400 years discussing floor-malting (Harrison, 1587; Markham, 1615; de Coetlogon, 1745; Thomson *et al.*, 1806; Ford, 1849; Ure, 1860; White, 1860; Thausing *et al.*, 1882; Stopes, 1885; Lévy, 1899; Petit, 1899; Lancaster, 1908; Wahl and Henius, 1908; Leberle, 1930; Schönfeld, 1932; de Clerck, 1952; Northam, 1965a; Narziss, 1976). Traditionally making malt was regarded as a job for women but, as the scale of working became larger and the work heavier, it became usual for men to carry it out. Markham (1615) though that, while men turned malt using shovels, women should turn it using their hands! Originally malting was a winter occupation, because low temperatures were needed to check and give good control over the germination process. Farm labourers were available to 'work' the malthouses from the autumn, post-harvest period to the time of spring ploughing. Air conditioning now allows floor-malting to be carried out around the year and under more consistent conditions. Some maltsters did risk malting in the summer, usually producing poor-quality products. More recently, the use of sodium or potassium bromates (with or without gibberellic acid) has allowed heat output to be controlled, with a consequent lesser requirement for cool ambient conditions. The use of bromates also encouraged the use of air-rest steeping, which, like elevated temperatures, favours rapid and vigorous germination.

Cleaned barley was wetted by 'swimming'; it was poured into clean water in the cistern or steep and was stirred. Dust, light corns (floaters) and other light debris were skimmed from the surface of the liquid, which covered the grain to a depth of about 15–20 cm (*ca.* 6–8 in). In the 'ditch' steeps, the preferred depth of grain was about 60 cm (2 ft), but it was more in the later conical-bottomed, self-emptying steeps. The desirability of washing grain before steeping and frequently changing the steep liquor has long been recognized. Many 'antiseptics' were added to steeps at various times to check mould growth on the grains during subsequent flooring.

Clean water from wells at 10–15 °C (50–59 °F) was preferred and steeping would normally last for 50–80 h, with several changes of water. In cold weather, steeping would be continued longer, for a week or more, until the maltster judged that the grain was steep-ripe (judged by the appearance and feel when cut or bitten, the ease of penetration of a needle and whether the husk separated when the ends were squeezed). White (1860) noted that if the grain was steeped warm and the water was not changed, and especially if the water came from a stagnant pond or a canal, the steep water became dark and had an offensive smell, it acquired a 'peculiar viscid character' and might require 2–3 h to drain from the grain! Therefore, there were and are absolute requirements to use clean water, under most circumstances to steep cool (less than 16 °C (60.8 °F) and 11 °C (52 °F) was preferred), and to change the steep water as often as necessary to keep it 'sweet'. The conditions actually used depended on the barley samples and the ambient conditions. In modern floor-maltings in the UK, ducts leading to fans are often connected to the drainage channels in the older types of steep. This permits carbon dioxide extraction (downward ventilation) to take place during air-rests, with subsequent more rapid, even germination of the steeped grain. When the grain was judged steep-ripe (probably with a moisture content of 42–45%), the water was drained. Sometimes the grain was rinsed with fresh water in warm weather and, after 6–10 h, was moved onto the floor and was heaped or placed in a couch-frame. Until 1880, this last was a legal requirement in the UK, where tax was levied on the volume of the steeped grain. Afterwards couching was often omitted but, particularly in cold weather, the grain was sometimes couched (heaped or rucked up) to a depth of 23–76 cm (0.75–2.5 ft) or even to 91 cm (3 ft) so that heat loss was minimized. Excess water drained from the grain in the couch. The rise in temperature that occurred favoured the faster uptake of the surface water on the grains, so the surfaces became dry and chitting occurred. As soon as chitting was detected in the warm, middle of the heap, the couch was 'broken down' and remade so that the grain initially on the outside was turned into the middle in an attempt to give the grains a more even treatment. Couching lasted about 24 h, or longer if the conditions were cold. Sometimes the grain appeared to 'sweat'; that is it became bedewed with condensed moisture if the centre of the heap was relatively warm.

At the end of this period, the grains should have been uniformly chitted. Depending on ambient conditions, and whether the grain was to be 'worked' warm or cool, the heap was broken down (if couching had been used) and the grain was spread on the floor to a depth of 8–30 cm (*ca.* 3–12 in). The piece was spread thinner to facilitate heat loss if the air temperature in the malthouse was warm or if the piece was to be worked cool. Clearly the floor area covered by each piece depended on how thinly it was spread. Ford (1849) suggested that 200 ft^2 (18.58 m^2) of floor be allowed for each quarter (448 lb, 203.2 kg) of barley steeped (91.4 m^2/t). For green malt spread 10 cm (3.94 in) deep, Narziss (1976) allowed 32–36 m^2 floor/t barley steeped, or about 30 kg green malt/m^2 of floor.

Preferred air temperatures varied between about 12 and 15 °C (53.6–59 °F). The piece was typically turned and inverted two to four times a day to ventilate it, to reduce the temperature, to equalize the growing conditions of the grains and to prevent the rootlets becoming entangled and matting together. Between turns, the malt might be forked or ploughed to disturb the grain, to ventilate it and to separate the roots. In traditional malting, the maximum heat output occurred about 4 days after steep-out. With modern air-rest steepings and/or if gibberellic acid is used, this now occurs sooner. The major period of heat output coincides with the period of maximum respiration and accumulation of carbon dioxide in the piece, which develops a characteristic smell, described as apple-like. As malting progresses, the smell of the grain alters. However, only if the green malt is attacked by microbes does the smell become unpleasant.

Flooring temperatures (determined by thermometers in the bed of grain) of 13–17 °C (55.4–62.6 °F) are often aimed at, but at the end of the germination period they may be allowed to rise to 21 °C (69.8 °F). Floor-maltings normally have thermometers suspended from the ceilings to determine the air temperature and embedded in the piece to indicate the grain temperature. These are read regularly and the temperatures are recorded. For a piece 7.6 cm (3 in) deep and with an air temperature of 9–11 °C (*ca.* 48–52 °F), the temperature in the grain bed may be 14–16 °C (*ca.* 57–61 °F) with a temperature differential of around 0.5 °C (1 °F) between the top and bottom of the piece. For a piece 15 cm (*ca.* 6 in) deep and with an ambient air temperature of 13 °C (about 56 °F) the mean temperature of the piece will be about 20 °C (68 °F), with a temperature differential of about 1.5 °C (2.7 °F). These values are not constant, since the heat output of the grain varies with the stage of germination. Turning the piece ventilates it and reduces the temperatures by about 1 °C (1.8 °F). There are wide variations between the systems used by maltsters and these are varied depending on the type of malt being produced. For many years, cool malting with the grain remaining on the floor for 11–14 days was regarded as normal. Now 5–6 days on the floor are regarded as reasonable, much time being saved by inducing rapid germination with air-rest steeping and sometimes by an application of gibberellic acid. Sometimes soluble bromate salts are also used. Historically, some maltsters regularly watered or sprinkled their pieces (using fine sprays from home hoses or watering cans) to increase the vigour of grain growth. Others only sprinkled if the grain appeared to be drying too soon and the roots were becoming flaccid. It was suggested that water should be applied at the rate of 1–5 gal/Qr (0.022–0.11 l/kg), in small increments with turning in-between applications to try and ensure even wetting of all the grains. The progress of malting was judged by the smell of the grain, the extent of root growth, the growth of the acrospire under the husk and the rub-out of the corns, which gives an indication of endosperm modification. It has been known for centuries that it is undesirable when making normal malts to allow the acrospire to grow beyond the husk, giving bolters, since if this occurs malting losses increase rapidly and the extract given by the malt declines sharply. Flooring conditions were varied, depending on local customs, the ambient conditions, the maturity and grade of the barley, the judgement of the maltster and the type of malt being produced. Malt might be kilned early if it showed signs of becoming mouldy. In the past, with long germination times (3 weeks in good weather, 4–5 weeks in the cold (de Coetlogon, 1745)), mouldiness was an ever-present risk. Whole pieces might appear greeny-blue, owing to *Penicillium* or other moulds; Harrison (1587) speaks disparagingly of the 'musty malthouse'. Up to 1880, in England, the laws regulating malting were so stringent that they discouraged or prevented rational working (Champion, 1832; Tizard, 1843; White, 1860). Modern practices, with much shorter germination times, are greatly superior since 'mouldy floors' are never seen. Authors speak of contrasting 'cold sweat' and 'warm sweat' systems of working, but differ in the conditions described. Using the cold-sweat system for pale malts the objective was to keep the malt temperature below 18 °C (64.8 °F), or 15–18.7 °C (59–63.7 °F), with turning every 12 h until the grain chitted, then every 5–8 h. At the finish, the roots would be 1–1.5% the length of the grains, the acrospires 0.66–1. The grain might remain 8–10 and even up to 14 days on the floor. In the warm-sweat system, the pieces might be 30–50 cm (11.8–19.7 in) deep but were usually less, depending on the ambient temperature. The pieces were turned three times a day. Thausing *et al.* (1882) regarded a temperature profile of 25–31.2–37.5 °C (77.0–88.2–99.5 °F) over a 4–5 day flooring period as normal. Others stated that temperatures should be kept below 30 °C (86 °F) until just before kilning, when they could rise to 30–35 °C (86–95 °F). At the end of the 5-day flooring period, the roots would be matted together and acrospires would be about half the length of the corns. The warm-sweat method of working produced darker malts in shorter processing times but the malts were often poorly and unevenly modified. In English floor-maltings (1985–90), flooring took 6–7 days, less if gibberellic acid or gibberellic acid with bromates were used. Such malts are of high quality and are often preferred by specialist or traditional brewers.

Often the last stage of malting was withering, preparing the green malt for the kiln. Some maltsters spread the green malt on special, perforated withering floors to hasten drying. Withered malt was 'dry to the touch' and the roots were partly shrivelled. Some claimed that this process 'mellowed' the malt, while others claimed that it was a waste of time. Withering allowed the malt to become air-dry and it helped if the kiln had a poor draught. If kilns with good draughts were available, withering, as a separate stage, was not needed (Stopes, 1885). With modern, fan-driven kilns, withering before kilning is not needed. However, with older natural-draught kilns, with their uncertain performances, withering would have reduced the chance of the temperature of the malt rising unduly while it was still wet, with consequent stewing and reduction in malt quality.

14.3 Pneumatic malting

Floor-maltings are no longer being built. New plant is constructed to carry out malting on the pneumatic principle, that is relatively deep beds of grain are ventilated and cooled with forced flows of water-saturated air adjusted to a selected temperature. For economic reasons, large batches of grain need to be malted in relatively compact buildings, under closely regulated conditions, on a routine basis according to preset schedules, using highly automated equipment and employing minimum personnel. These requirements can only be met in pneumatic plant by using relatively inflexible steeping, germination and kilning schedules.

Floor-maltings are large for the amount of malt produced in them and have relatively high workforce requirements. In the relatively shallow beds of grain on a malting floor, the air within the grain is stagnant, is fully saturated with moisture and contains variously reduced levels of oxygen and elevated levels of carbon dioxide (values of up to 7.5% CO_2 were quoted by Leberle, 1930). These levels are sufficient to reduce the respiration intensity of the grain. In some European floor-maltings, fresh air was admitted through the ceilings and 'foul' air was removed by ventilators situated at floor level to carry away the carbon dioxide-rich layer. Heat is lost from the grain bed by conduction into the malting floor and into the atmosphere. When moisture condensation occurs on the floor or on the cooler surface layer of grain (sweating), it is available for reabsorbtion by the grains, and drying is relatively slow.

In contrast, in pneumatic maltings where the grain beds are deeper (1 m (39.4 in) or more) and are ventilated with a forced airflow, the conditions are different. Carbon dioxide levels rarely rise significantly. Isebaert (1965) noted that the carbon dioxide content of the air (%, v/v) rose to 3.5% on the 5th day of floor malting, but in a pneumatic malting it never exceeded 0.2% if fresh air was employed or 0.5% if the air was recycled using a refrigerator (and not water-sprays) for cooling. Water sprays dissolve carbon dioxide and so remove it from the airstream. The beds of grain tend to dry. In most pneumatic maltings, the rates of airflow are relatively constant (rising after a piece is lightened by turning). The air entering the grain bed is as nearly saturated with water as possible (RH target 100%) but the temperature of the air-on (entering the grain bed) is varied. Usually the airflow is maintained throughout the germination period but sometimes it is interrupted to avoid a power overload or because it is desirable to allow the grain temperature to rise, for example to couch, that is to accelerate the uptake of the surface moisture after wet-transfer from the steep. The target temperature differential across the grain bed is often 2 °C (3.6 °F), or less, but it often reaches 3 °C (5.4 °F) and in some plants it is even higher. Since, in most plants, turning does not invert the grain bed, it follows that grain growing at the air-on side will malt under significantly cooler conditions than that on the air-off side. This must contribute to heterogeneity in the finished malt. The temperature regimes used in germination will be discussed later. For most purposes they occur in the range 13–18 °C (55.4–64.4 °F), sometimes being allowed to exceed 20 °C (68 °F) towards the end of the germination period. Under ideal conditions, the air enters and leaves the grain bed saturated with water vapour, i.e. with a RH of 100%. However, the grain leaving the grain bed is warmer than that entering it and so carries more water. Thus, water is removed from the grain, which, inevitably, dries. Water may be sprinkled onto the grain during turning to make good these losses, but this is not wholly satisfactory and so, to allow for evaporative losses, grain to be malted in a pneumatic malting plant is often steeped to a higher moisture content than would be used in a floor-maltings. In a particular maltings, the air-on was normally saturated with moisture (RH 100%) and, on average, the moisture content of the green malt fell at a rate of about 0.5% per day. A fault in the humidification system allowed the RH of the air-on to fall to about 95%. The rate of grain drying increased to about 2.5% per day and in 2 days the green malt was so dry that modification effectively ceased. Humidification efficiency is not easily checked since high RH values (90% and above) are not easily measured with precision. Another problem is when moisture droplets are entrained in the ventilating airstream. If these are not separated before the air reaches the grain bed the grain on the air-on side will be continuously wetted and will bolt and grow too vigorously. Sometimes it is necessary to heat or cool the ventilating air. The cost of these processes is minimized by recirculating a proportion of the airflow through the grain bed. Refrigeration plant is expensive, so in some temperate climates air cooling relies on the temperature of the water in the humidification compartments and evaporative cooling. At some times of the year, these will not provide adequate cooling; this is a recognized risk.

Grain beds are turned: (i) to prevent rootlets matting together; (ii) to try and ensure that all the grains in a batch are treated as nearly evenly as possible; (iii) to lighten the piece (i.e. reduce its bulk density), encouraging a higher and more even airflow through the grain and cooling hot-spots. The degree of grain turning and mixing varies depending upon the equipment used. It is probably excellent in drums and equipment using Wanderhaufen-type turners but is less good in equipment using helical screw turners. On the one hand, too little turning allows rootlets to mat and the malt to clump, leading to the formation of unventilated hot-spots in the grain mass. On the other hand, excessive or too vigorous turning can damage green malt, damaging or breaking the rootlets (root pruning), checking modification and providing foci for fungal growth.

14.4 Malting losses

Malting losses are defined in different ways. Here the convention is adopted that malting losses are the reduction in dry weight, expressed as a percentage, that occurs when a batch of barley, or other cereal grain, entering a steep is converted into finished, de-rooted malt. These measurements are used to determine the efficiency with which the barley is turned into malt. These losses are conventionally divided into steeping losses, respiration losses and culms (sprouts or rootlets). These losses, which are not exactly specified by their names, are made up in various ways.

Steeping losses Steeping losses, which occur during the steeping period, are made up of displaced dust and materials dissolved from the grain by leaching and the products of metabolism released by the grain (carbon dioxide and ethanol). Probably microbes assist in 'solubilizing' husk materials. Both fermentative processes and respiration lead to losses of dry matter during steeping, respiration being most intense during air-rest periods. In addition, fragments of husks and floaters may be lost during steeping. Steeping losses may or may not include 'skimmings' (floating corns, awns and trash), which, together with materials dissolved from barley, usually give a steeping loss of 0.5–1.5% of the initial dry weight of the grain. In more modern steeping systems, where chitting is initiated in the steeps, an extra 0.5–1.5% losses of dry matter may occur owing to vigorous respiration during air-rest periods.

Respiration losses Respiration losses are usually calculated as the loss of dry matter that occurs during germination (total losses **less** the weight of the culms). These are attributed to the formation of carbon dioxide (and small amounts of other volatile metabolites) and water that occur during germination and kilning. The intensity of respiration, and heat output, varies with the stage of germination and how germination and kilning are carried out.

Rootlet losses The rootlet losses are given by dry weight of the rootlets collected after kilning. With barley malts, these are indeed mainly rootlets, with a little grain dust, but with malts made from naked cereals this material contains a proportion of coleoptiles or acrospires (wheat, rye, triticale), or even coleoptiles and leaves (sorghum, maize, millets). For some purposes malts are not deculmed (e.g. green barley malt used in distilleries; sorghum malt used for making Bantu beers).

Plant out-turn In production malting, different calculations are used to discover the 'out-turn' of the plant. In this case, the malt yield is calculated as the weight of malt as-is (typically 3–6% moisture content) leaving the maltings as a percentage of the weight of the barley (typically 8–18% moisture content) entering it. The decline in weight includes loss of water, loss of screenings, losses of impurities and losses of broken corns as well as the malting losses incurred in the production process itself (often 6–12%).

White (1860) remarked, 'The principal aim of the maltster is to attain as great an amount as possible of saccharine matter [we now say extract] with the smallest amount of loss of substance, together with the best outlay in the shape of labour'. Therefore, the need to know malting losses and out-turns and to minimize them, while producing desirable malts, has long been recognized. Malting losses can be accurately determined in small-scale trials in which all the barley, malt and rootlets can be weighed to better than 0.1%, and without sampling errors. In production malting, it is unusual for weighing to be better than 'within 0.5%' and 'within 1%' has been considered realistic. Barley arriving at a maltings in the UK will typically have a moisture content of 12–18% or more before drying. Until recently, in the UK, the fresh weight was expressed in units of barley quarters (203.21 kg, 448 lb). After malting, kilning and deculming, the malt (fresh weight, 1–5% moisture) was weighed in malt quarters (152.41 kg, 336 lb). It was expected that 1 Qr barley would yield about 1 Qr malt, an out-turn of about 75%. Because of the different moisture contents of the barley and malts, and the two units used, values of malting losses and gains, calculated this way, meant very little. At a still earlier time, barley and malt were sold by volume (bushels and quarters; see Appendix). Inevitably in that period maltsters attempted to swell malt and 'puff it up' on the kiln to have a larger volume to sell. Undoubtedly the best measure of malting loss (ML) is the percentage reduction in dry weight occurring in the conversion of prepared barley to the finished malts. Similarly, the malt yield (MY) is expressed as the yield (percentage) dry malt from the dry barley, as MY (%) = 100 – ML (%). Malting losses may be followed by determining the TCWs of samples during the process. Unfortunately TCW values are tedious to determine and are not very accurate.

Some average malting losses are summarized in Table (14.1). More examples are given in Chapter 15. The bulk of respiration losses and all the rootlets originate from the metabolism and growth of the embryo. Numerous techniques have been tried for minimizing malting losses. These are indicated in subsequent sections. Careful choice, and control, of the grain

Table 14.1 Losses in dry weight (%) incurred in the conversion of barley to malt (other data are quoted in the text)

	German data for various malt types					British data probable current range[c]
	Pale[a]	Dark[a]	Floor[b]	Pneumatic drum[b]	Kropff[b]	
Steep losses	1.0	1.0	1.5	1.5	1.5	0.5–1.5
Rootlets	3.7	4.5	4.5	3.8	2.4	2.5–4.0
Respiration losses	5.8	7.5	5.4	4.9	3.1	3.5–5.0
Total malting losses	10.5	13.0	11.4	10.2	7.0	6.5–10.5

[a] Data of Kolbach cited by de Clerck (1952)
[b] Data of Lüers (1922)
[c] Excluding malts made with resteeping or application of bromate and some experimental malts (see text).

quality, the steeping regime, the moisture content of the steeped grain and the green malt, the temperature programmes used in germination and kilning, and germination time are all important. Malting losses are higher when dark malts are being made. In the 1890s, malting losses for pale malts were probably highly variable and may have averaged 13–14%. By using applications of gibberellic acid and potassium bromate, and air-rest steeping schedules, pale malts can now be prepared with malting losses of as little as 6.5%. With well-controlled malting, with no use of additives, using carefully regulated steeping conditions and cool germination temperatures, total malting losses of around 10% are probably usual.

Potentially, there are important differences between the objectives of user-maltsters (brewers or distillers who make malt for their own use) and sales-maltsters. A sales-maltster must produce particular malts of the correct type, characteristics and analyses (to meet buyers' specifications) regardless of the malting losses incurred. Having a variety of customers means that a wide range of different malts will be made. However a user-maltster may aim to produce as great a quantity of extract from a given amount of original grain as is feasible, subject to the extract being of the correct quality. This may be achieved by producing malt in greater yield but with less than the maximum achievable extract per unit weight of malt. In addition a user-maltster will make only a limited range of malts. However, malt users now set such close specifications that often sales-maltsters and user-maltsters have to meet the same criteria.

14.5 Changes that occur during malting

When the grain is steeped it swells and substances are lost to the steep liquor and the atmosphere. In traditional malting (UK), steeps were not ventilated and signs of germination did not appear until 1–2 days after casting. Now,

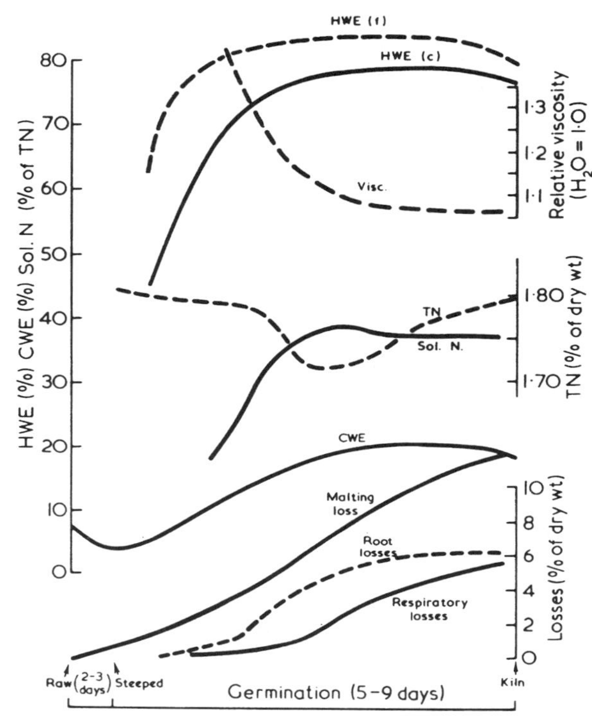

Fig. 14.1 Generalized diagram of the accumulating losses and changing analyses of barley under traditional floor-malting conditions. Usually the process is terminated by kilning before the hot water extract begins to decline. HWE, hot water extract (% dry matter); (f), determined using a fine grind; (c), determined using a coarse grind; Visc., viscosity of wort, relative to water; TN, total nitrogen of derooted malt, as % dry weight; Sol N, soluble nitrogen found in an analytical wort as a % of TN; CWE, cold water extract of malt, expressed as % dry weight; malting loss, loss in dry matter, as % of initial dry weight of the barley; made up of weight of roots, losses owing to the conversion of grain solids to carbon dioxide and water (curves shown) and losses owing to dust washing away and grain substances being dissolved during steeping.

using ventilated steeping, grain is chitting when it is cast (Figs 14.1 and 14.2; table 14.2). During germination the embryo grows, and the endosperm is progressively degraded by enzymes from the scutellum and aleurone layer. Maltsters use the growth of the rootlets and acrospires as indications of the progress of malting. The progressive breakdown of the endosperms can be followed by 'rubbing out' the green malt, and by the decreasing hardness, increasing friability and increasing homogeneity of dried samples of malt. Dry matter is lost as ethanol, carbon dioxide and water are formed by fermentation and respiration. Materials migrate to the growing acrospire, which is retained in barley malts, and to the rootlets, which are normally removed. While there is a net breakdown of complex materials in the

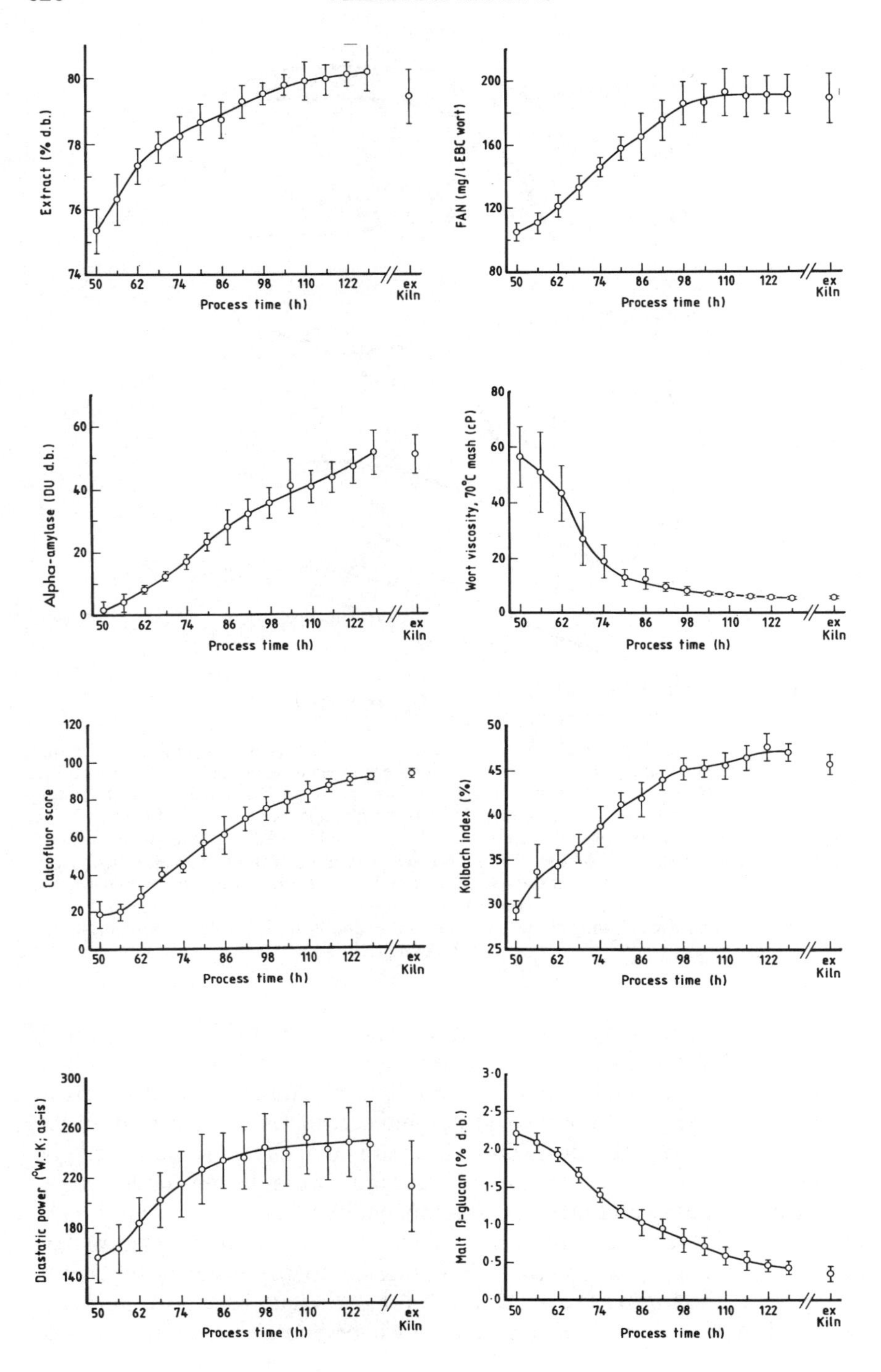
Extract (% d.b.)
74
76
78
80
50
62
74
86
98
110
122
ex Kiln
Process time (h)
FAN (mg/l EBC wort)
80
120
160
200
Alpha-amylase (DU d.b.)
0
20
40
60
Wort viscosity, 70°C mash (cP)
80
Calcofluor score
100
120
Kolbach index (%)
25
30
35
40
45
50
Diastatic power (°W.-K; as-is)
140
180
220
260
300
Malt ß-glucan (% d.b.)
0·0
0·5
1·0
1·5
2·0
2·5
3·0

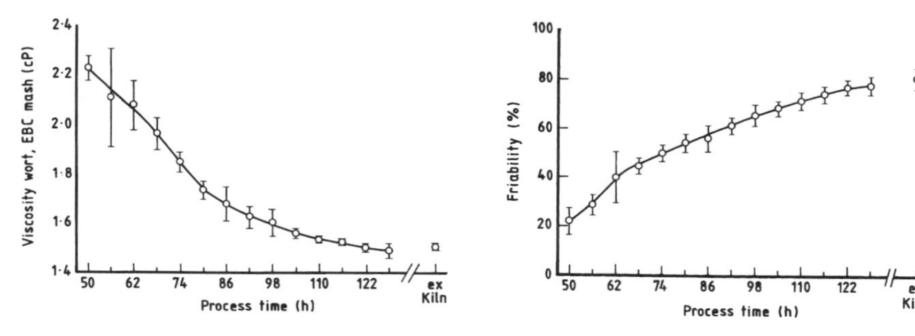

Fig. 14.2 The changes occurring in pneumatically malted barley (after Morall and Basson, 1989). Clipper barley was malted in 320 t batches. The mean analytical values and the 95% confidence limits are given for the replicated samples. The steeping regimen, carried out in flat-bed steeps, was 7 h wet, 18–20 h dry, 10 h wet and 8 h wet, a total of about 44 h. Temperatures during immersions were *ca.* 15 °C (59 °F) and during air-rest periods, with carbon dioxide extraction, were *ca.* 17 °C (62.6 °F). Steep-out moistures were about 44%. Gibberellic acid (0.04 mg/kg barley) was applied during the transfer to the germination compartment. Germination occurred at a temperature of *ca.* 18 °C (64.4 °F).

starchy endosperm and aleurone layers, there are rises in all the components of the growing embryo. These movements of materials within malting grains, and the eventual loss of some of them (by leaching, by respiration and as rootlets) complicates the interpretation of analyses relating to the malting process. For example, if the 'potential extract' of barley is 80.8%, the extract of the malt is 78.0%, and the malting loss is 6.8%, then the loss of the potential extract of the barley is: $80.8 - (78.0 \times 0.932) = 8.1$ or $8.1 \times 100 \div 80.8 = 10.0\%$ of the potential extract of the barley.

Altering process conditions alters the type of malt produced. During malting, the malting losses increase, but at very varying rates (Figs 14.1 and 14.2; Table 14.2). Rootlet production rises and the rate of production declines faster than the respiratory losses (Figs 14.1 and 14.3). Since nitrogen (or 'protein'; N × 6.25) is not lost from the entire grain (including roots) during the germination period but does migrate into the rootlets, it follows that the TN of de-rooted malt depends on the initial nitrogen content of the barley (less that lost in steeping), the proportion that has migrated into the rootlets, and the decline in weight of the root-free grain. Therefore, TN varies with germination time (Fig. 14.3) but, under most malting conditions, is nearly the same in the finished malt as in the original barley, sometimes a little more, probably more often a little less (Ballesteros and Piendl, 1977; Briggs, 1978). The level of TSN obtained by mashing malt or total soluble protein (TSN × 6.25%) rises and tends to plateau during germination. However the different fractions within this material (salt-soluble N, PSN, formol-N, FAN) do not necessarily occur in the same proportions in worts prepared from malts germinated for different periods or under different conditions. The ratio of soluble nitrogen ('protein') to total nitrogen ('protein'), either the SNR or the Kolbach Index, also increases with germination, usually to a maximum value.

Table 14.2 Changes in analyses (*Analytica-EBC*) of Wisa barley during micromalting. Grain was wet-steeped and air-rested at 13 °C (55.4 °F) then, after moisture adjustment to 45%, was germinated at 15 °C (59 °F) for the times shown. Analyses were carried out after kilning for 1 h at 50 °C (122 °F), 2 h at 60 °C (140 °F), 1 h at 70 °C (158 °F) and 5.5 h at 84 °C (183 °F)[a]

Germination time (days)	0	1	2	3	4	5	6	7	8	9
Extract (f.g. % d.b.)	72.6	75.6	80.6	81.5	82.6	82.4	82.7	82.7	81.6	81.8
Extract (c.g. % d.b.)	56.2	64.2	71.8	76.5	78.0	80.3	81.1	81.5	81.3	81.7
Fine–coarse extract difference (%)	16.4	11.4	8.8	5.0	4.6	2.1	1.6	1.2	0.3	0.1
Saccharification time (min)	>60	>60	20–25	10–15	5–10	5–10	5–10	5–10	5–10	5–10
Colour (EBC units)	2.0	2.2	2.7	3.0	3.4	3.9	4.1	4.3	4.6	4.1
Protein (% d.b.)	10.1	9.8	9.6	9.7	9.6	9.6	9.7	9.5	9.8	9.5
Soluble protein (mg/100 ml)	35.6	45.5	63.0	73.9	81.2	79.5	79.8	80.2	80.2	78.4
Kolbach index (%)	19.4	25.8	36.3	42.2	43.8	45.8	45.5	46.9	45.7	45.5
pH	6.03	5.85	5.80	5.80	5.81	5.70	5.82	5.77	5.84	5.77
Viscosity (cP)	2.79	2.74	2.24	1.84	1.80	1.58	1.53	1.50	1.47	1.47
VZ 45 °C (%)	16.2	18.6	26.8	32.9	38.4	39.7	44.3	45.3	45.4	44.4
Brabender hardness (relative)	665	605	560	430	330	295	250	212	210	180
Total malting loss (%)	1.5	3.7	5.6	7.2	7.9	8.7	9.4	10.1	10.4	11.2

[a] The multiple air-rest steeping schedule ensured the rapid onset of modification and so it was possible to analyse 'germination day 0' material.
Data of Piendl (1971).

As modification proceeds, the endosperm of the green malt is more easily 'rubbed out' and is more friable in the dried product. Hot water extract increases to a plateau value then declines; the fine–coarse extract difference decreases, as does the wort viscosity. Friability and modification continue to increase after the hot water extract has plateaued (Figs 14.2, 14.4 and 14.5). The β-glucan content of the green malt progressively declines and, as a result of this, the viscosities of extracts (both of the standard analytical worts and the 70 °C mash worts) also progressively decline (Figs 14.2 and 14.4; Table 14.2). Close correlations were noted between the β-glucan in EBC wort and in the malt, and between malt β-glucan and the Calcofluor modification index or friability. There was a positive correlation between the β-glucan content of malts and the viscosity of the worts from 70 °C (158 °F) mashes.

Barley has a cold water extract of 5–10%. These 'pre-formed soluble substances' decline during classical steeping but subsequently rise to an elevated level as sugars accumulate during malting (Fig. 14.1). Eventually the value plateaus with increasing germination time as a balance is reached between the generation and the utilization of these low-molecular-weight substances. Cold water extract was used by British brewers as a criterion of malt quality. It is of most value to the maltster, indicating how his malting procedures are working (Hind, 1940; Hopkins and Krause, 1947).

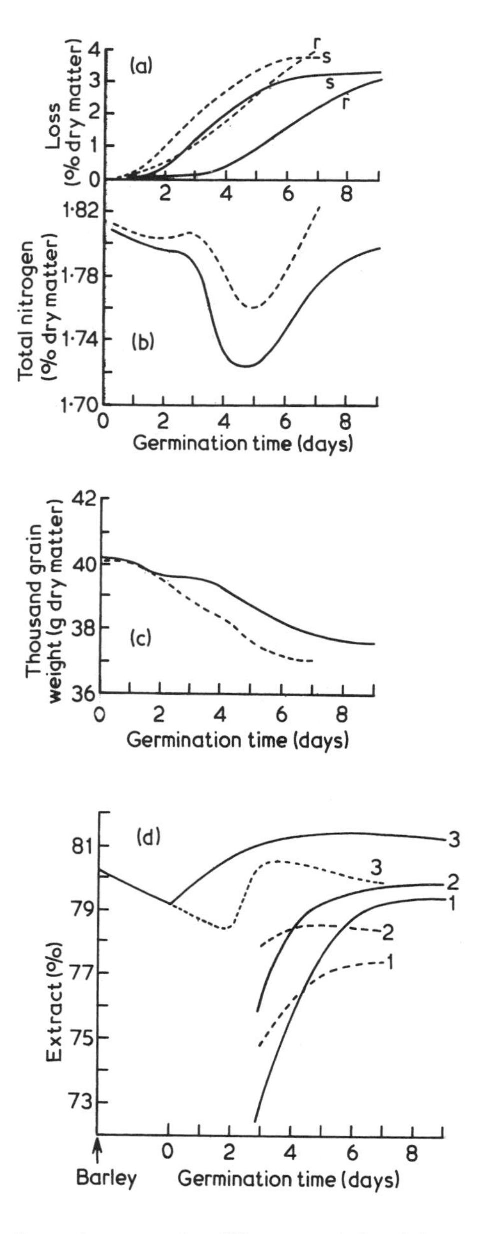

Fig. 14.3 Analyses for malts grown for different periods of time at low temperatures of 13–17 °C (*ca.* 55–63 °F) (—) and high temperatures of 19–22 °C (*ca.* 66–72 °F) (---). (a) Malting losses, r = respiration losses; s = rootlets; (b) total nitrogen content; (c) TCW; (d) extracts obtained with different types of mash, (1) made with coarsely ground malt; (2) made with finely ground malt; (3) 'Graf' mashes, made with additions of enzymes to supplement those provided by the finely ground malt. (After Piratzky and Rehberg, 1955.)

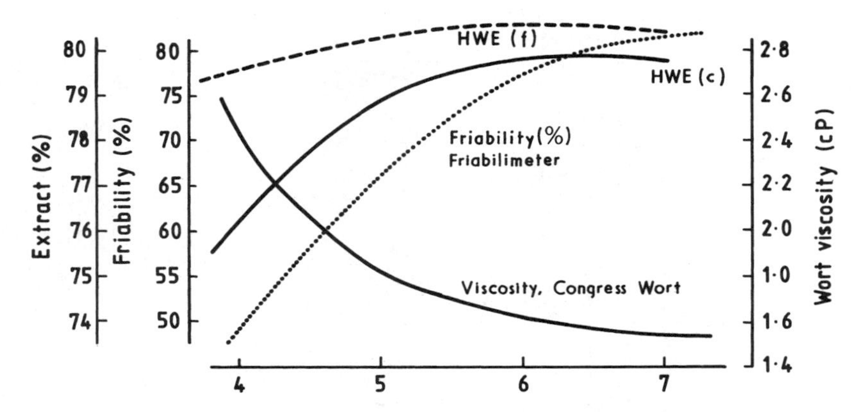

Fig. 14.4 Changes in the hot water extract of finely ground (HWE (f)) and coarsely ground (HWE (c)) samples of malt grown for different periods of time as well as friability and wort viscosity values (after Bathgate, 1983).

Similarly the hot water extract of malt rises with germination time and tends to plateau (Figs 14.1–14.5). If germination time is prolonged, and particularly if malt is allowed to overgrow, both the hot water extract and cold water extract decline. The degree to which the malt is ground influences the recovery of the hot water extract. With increasing germination time the fine–coarse extract difference values decline. The composition of the hot water extract changes with germination time so that, for example, fermentability increases. The 45 °C (113 °F) extract value (VZ 45 °C) also tends to plateau during germination (Table 14.2).

During the germination period the diastatic power of the malt initially rises rapidly and then more slowly or remains relatively constant (Figs 14.2 and 14.5; Table 14.2). In contrast the values for α-amylase continue rising with time under most germination conditions. The ratio of α-amylase to other starch-degrading enzymes increases with germination time. As α-amylase values increase so mash saccharification time, which depends largely on this enzyme, declines to a limiting value (Table 14.2). Many of the analytical values of finished malts are strongly influenced by kilning. In particular, the higher the kilning temperatures the less well enzymes survive and the higher the colour of the final malts. The effects of mild kilning are illustrated in Fig. 14.2. The effects of more 'stringent' kilning conditions are discussed later. The changes that occur during malting are most easily understood if the 'intermediate' samples taken for analysis during processing are quickly dried at a low temperature to fix the composition of the samples rapidly while minimizing changes caused by heating.

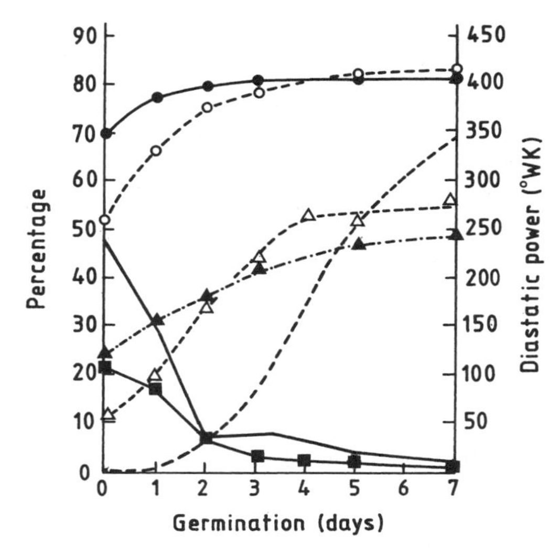

Fig. 14.5 Changes in Congress wort analytical values with germination: unmodified corns (—); fully modified corns (---); final wort attenuation attenuation (○); extract (d.b.) (●) ; diastatic power (°W–K) (△); Kolbach index (▲); fine–coarse extract difference (■). (After Van Eerde, 1983).

14.6 Barley characteristics and malting

The evaluation and purchase of malting barley is outlined in Chapter 6. Barley for malting must meet a number of criteria. The grain must be of the correct variety and have an acceptable nitrogen or protein content and range of corn sizes. It must be undamaged, smell sweet and be unstained, showing minimal or no signs of fungal contamination, pre-germination, peeling, chipping, splitting, breakage or husk shedding. The grain must be free of some contaminants (ergot sclerotia, storage insects) and must not contain more than specified amounts of other contaminants. Its viability (germinative capacity) must equal or exceed a specified value and, at the time of malting, germinative energy (4 ml and 8 ml tests (section 3.7.6), or others) must be sufficiently high, that is residual dormancy and water sensitivity must be low or absent. As the grain matures during storage, so its malting characteristics improve (Chapter 3). For example, Witt (1970a) noted that during storage from August to October the germination of a batch of grain increased from 97 to 99%, the hot water extract of malt prepared from it increased from 72.5 to 74.3%, the ratio of soluble nitrogen to total nitrogen increased from 35.5 to 39.2% and the clarity of the worts improved as the malts were made from longer stored barley samples. Seward (1986) suggests that barley is ready to malt when germination values exceed 80% and 95% on the second and third days of the 4 ml germination test and exceed 75% on day 3 of the 8 ml germination test. Steeping schedules have to be altered as barley matures. As far as possible, barleys of different grades and varieties should be stored and malted separately. The proportion of thin grains should not exceed a specified value. Thin grains are defined in different ways depending on the season, the variety of barley and the location. In England, thin grains are often defined as those of 2.25 mm (0.089 in) in width, or less (since 1992), while in Continental Europe and Scotland the value is often 2.5 mm (0.098 in) or less. In addition, the proportion of various grain sizes, for example those with widths between 2.2 and 2.5 mm, between 2.5 and 2.8 mm and exceeding 2.8 mm may be considered. Often grains in the 2.5–2.8 mm size range, but sometimes those of > 2.8 mm, have the lowest nitrogen content and malts made from them have the highest SNR or Kolbach index values. Thinner barley grains take up water faster (Chapter 3) and have higher nitrogen contents than bolder grains (Table 14.3) and they are relatively richer in phenolic substances, including anthocyanogens (Sommer, 1977a). Malts made from thin corns have lower extracts and are less well modified in terms of fine–coarse extract difference and SNR or Kolbach Index (Table 14.3). However, malts from thin corns usually have high diastatic powers (making them useful for grain-distilling purposes, or as a source of diastase) and are richer in β-glucanase. Older varieties of six-rowed barleys usually gave malts with lower extracts than those from two-rowed barleys, but with modern cultivars there is a wide overlap in the ranges of extracts encountered. Malts giving low extracts do not necessarily have large fine–coarse extract difference values. Usually six-rowed barley malts have higher Kolbach index values than two-rowed barley malts (Sommer, 1977a).

Table 14.3 Size of malt corns (separated by screening) and quality

Analysis	Original sample	Fractions (mm)			
		<2.2	2.2–2.5	2.5–2.8	>2.8
Sieve analysis (%)	100	11.7	42.0	25.3	21.0
Extracts (1°/kg d.m.) at Miag mill setting					
1	294	267	290	296	302
5	284	255	281	289	296
10	270	230	263	275	285
Fine–coarse difference (1–10; 1°/kg)	24	37	27	21	17
Moisture (%)	2.2	2.4	2.3	2.3	2.4
TN (% d.m.)	2.02	2.40	2.06	1.92	1.78
TSN (% d.m.)	0.54	0.59	0.52	0.52	0.52
SNR (%)	27	25	25	27	29
β-Glucanase (IRV units; dry)	457	583	455	479	426

From Hyde and Brookes (1978) and personal communication.

Grain varieties are classified as of malting quality primarily because they yield a high hot water extract in a reasonable malting period and because the quality of the wort is acceptable. In a good harvest year, a plump, two-rowed barley variety might contain less than 10% of grains thinner than 2.5 mm, but in a bad year the value might be as much as 25%. Barleys may be chosen for special reasons, for example because they lack anthocyanogens (commercial quantities of these have been available in some areas; malts from these barleys yield beers with high haze stabilities) or because they yield malts with very high diastatic power values, as is the case with some Canadian and Finnish barleys, such as the variety Pokko. These types of grain are desirable for making malts for grain distillers and brewers who use high proportions of unmalted adjuncts in their mashes. Distillers who use pot stills choose barleys that give malts with high spirit yields and contain only low levels of cyanogenic glucosides. There are wide differences between the levels of enzymes produced by different malting barleys. For example, in six varieties, optimally malted, the ranges of diastatic powers were 33–94 (grains 1.75% TN) and 62–102 (grains 1.90% TN). The equivalent values for α-amylase (DU) values were 32–47 and 34–52, respectively (Seward, 1986). It is no disadvantage to have malting barleys with blue aleurone layers. For example, Halcyon, a malting-quality winter barley, has a blue aleurone layer and was widely used in the UK, while in Canada all six-row malting barleys have had to have blue aleurone layers since 1929.

Malting barley should not have a high proportion of steely corns as these tend to malt slowly, take up water slowly and be high in nitrogen. However the nitrogen content range of mealy, semisteely and steel grains overlap (Ehrich and Kneip, 1931). Mealy and steely corns differ in density and so can be separated by flotation in salt solutions and be investigated separately (Munck, 1993; Table 14.4). Differences between the grains can be considerable, the mealy grains taking up water more rapidly in steeping, having higher starch and lower protein contents and modifying more readily during malting. Mutants low in β-glucans have thin endosperm cell walls, tend to

Table 14.4 Analyses of fractions of 'Triumph' barley separated by flotation in sodium nitrate solution

	Barley fraction		
	Mealy	Intermediate	Steeley
Hectoliter weight (kg)	64.9	69.4	70.7
TCW (g)	40.1	41.9	43.1
Protein (% d.m.)	9.5	10.5	11.7
Starch (% d.m.)	65.7	61.9	60.1
Total β-glucan (% d.m.)	4.04	4.34	4.36
Soluble β-glucan (% d.m.)	0.43	0.63	0.74

From Munck (1993).

be mealy and malt very rapidly (Aastrup *et al.*, 1985). However, at present no varieties with this character are commercially available.

Non-malting, feed-grade barleys are cheaper to purchase than malting grades and many efforts have been made to malt them (e.g. Narziss, 1975; Ballesteros and Piendl, 1977; Baxter, 1988). Actions taken to maximize the chances of success have included (i) thorough screening to remove thin grains; (ii) the use of grain samples with low nitrogen contents; (iii) the use of a period of warm storage to ensure full maturity; (iv) applications of gibberellic acid with or without the use of abrasion and/or potassium bromate; (v) the use of air-rest steeping schedules; (vi) extended periods of germination at low temperatures; (vii) malting with the grain steeped to high moisture contents; and (viii) the use of low-temperature kilning schedules to maximize enzyme survival. Adequate malts may be made from some poor feed-quality, 'difficult' barleys if several stimulating techniques are employed (Baxter, 1988). Often considerable extra effort and expense is involved in preparing an adequate malt from a poor-quality barley. In one example, the malt prepared with 9 days of germination and a malting loss of 9.4% was not as good as that prepared in 6 days, with a 7.7% malting loss, from a good barley (Narziss, 1975).

Inferior malting barleys probably owe their poor qualities to various independent factors, including low contents of starch and sugars (which give rise to poor extracts), structural characteristics that cause slow water uptake and slow modification and inadequate development of one or more hydrolytic enzymes (giving malts with low SNR or Kolbach index values), low levels of TSN and FAN and/or worts with poor fermentabilities. In some instances, malts from poor malting-grade barleys give worts with poor flavours, or high levels of polyphenolic substances. Only comparative micromalting trials accurately determine a barley's malting quality. Predictions based simply on the electrophoretic pattern of alcohol-soluble proteins (hordeins) or β-glucan contents, sedimentation tests, grain hardness or behaviour when milled, or the behaviour of autolysing mashes of raw grains are not reliable.

Plant breeders frequently release new varieties, some of which have improved malting performances. In Canada the extract yields from two-rowed malts were about 78% at the beginning of the period 1955–90, but at the end had risen to 80.5%. For malts made from six-rowed barleys, extracts were about 75.3% in 1945, 77.5% in 1950 and 78.7% in the period 1975–90 (Pitz, 1990). When six-rowed North American barleys originating from 1910 onwards were grown together for 3 years (1990–2) and then malted, the oldest three gave extracts of 73–75% and the most recent three gave values of 77–79% (Schwarz and Horsley, 1995). There has been sporadic interest in malting naked (huskless) barleys for more than 100 years (e.g. Ballesteros and Piendl, 1977; Rennecke and Sommer, 1979; Singh and Sosulski, 1985). Naked barley, like all naked grains, is easily damaged during malting, although the acrospire is often retained within the pericarp, which is **not** what occurs when wheat or rye are malted. In general, naked grains take up water rapidly and give high extracts, as expected since the husk limits grain swelling (and hence water uptake) and does not contribute to extract. From the brewing point of view, the malts are not attractive both because the absence of husk gives wort separation problems and because of other undesirable properties such as high grain nitrogen contents, high levels of polyphenols, low α-amylase levels and attenuation limits, excessive levels of soluble nitrogenous substances, high wort viscosities and high malting losses. However these malts would be very suitable for use in human food-stuffs. Naked barleys have not been bred to maximize malting quality and very little naked barley is commercially available in western countries. When the barleys Harrington (hulled) and Scout (naked) were malted without (and with) additions of gibberellic acid, the extracts obtained were 80.5% (82.3%) and 83.9% (85.8%), respectively (Singh and Sosulski, 1985). The effects of varying doses of gibberellic acid and grain moisture on malting a husked and a naked barley are illustrated in Table 14.5

The conditions under which grain grows and matures greatly influence its quality. The variety Golden Promise rarely gave malting-quality grain when grown in southern England but regularly produced excellent material when

Table 14.5 Comparison of characteristics of malts made from husked and naked barleys. In all cases the saccharification time was less than 10 min

	Gibberellic acid (mg/kg barley)											
	0				0.03				0.10			
Moisture after steeping (%)	42	42	46	46	42	42	46	46	42	42	46	46
Naked (n) or husked (h)	h	n	h	n	h	n	h	n	h	n	h	n
Steeping time (h)	55	37	83	63	55	37	83	63	55	37	83	63
Extract (f.g. % d.b.)	81.5	86.1	81.9	87.6	82.0	87.0	82.6	89.0	82.6	89.1	82.4	90.4
Extract (c.g. % d.b.)	78.5	77.9	80.3	83.8	79.0	80.6	81.6	86.5	79.9	83.2	81.8	89.0
Extract (f.-c. difference, %)	3.0	8.2	1.6	3.8	3.0	6.4	1.0	2.5	2.7	5.9	0.6	1.4
Colour (Brand)	0.21	0.19	0.21	0.23	0.22	0.20	0.23	0.27	0.26	0.28	0.26	0.47
Kolbach index (%)	41	35	47	43	46	42	49	48	48	49	50	57
Viscosity (10%, cP)	1.69	2.19	1.59	1.68	1.76	2.11	1.59	1.71	1.85	1.96	1.58	1.59
Protein, total (% d.b.)	10.6	11.6	10.6	11.4	10.6	11.4	10.6	11.3	10.6	11.4	10.5	11.3
Malt yield (% d.b.)	93.4	93.4	91.1	89.8	95.0	94.6	92.6	92.0	95.0	94.0	92.8	92.0

Data of Rennecke and Sommer (1979).

grown further north, in Scotland. The growing season can also have dramatic effects. In a drought year, UK maltsters, who would normally purchase barley with TN values of about 1.65%, were compelled to buy grain with TN values of 1.8–2.3% because the hot dry weather had caused premature drying before the grains had finished filling. Unexpectedly the beers made from these malts were of good quality. The further north grain is harvested in the UK, the more likely it is to be dormant and water sensitive. Dormancy is common after a cool, damp and cloudy growing and ripening period but is rare after a hot, dry and bright ripening period. Changes in the growing season can alter the malting characteristics of different barleys to different extents and even in different senses (Kieninger and Rumelin, 1984). In some seasons, particular barleys have modified so easily during malting that it has been necessary to restrict modification (e.g. by reducing the moisture content of the grains or shortening the germination time) to restrict the levels of soluble nitrogen formed. The effects of environmental differences on malts made from one variety of barley grown in one year can also be considerable (Pitz, 1990).

The way barley is handled and stored can dramatically alter its malting performance. Reportedly, in 1805, a Mr Gilpin discovered that if he slowly dried barley at low temperatures on the kiln its malting quality was greatly improved. Later 'sweating barley on the kiln' became common practice. Grain germinability can be greatly enhanced by storing dried grains for a limited period at an elevated temperature (Briggs *et al.*, 1994; Chapter 3). Grain can also deteriorate in store and such deterioration should be minimized or avoided (Chapter 7).

14.7 Correlations between barley analyses and malt quality

For many years, attempts have been made to predict what quality of malt can be made from particular lots of barley, based on analyses of the barleys. Usually the assumption has been that pale malts would be made using a 'conventional' malting procedure. Of course very different types of malt can be prepared from one barley, using different malting programmes (Bishop and Day, 1933; Bishop and Marx, 1934; Bishop 1948; Hudson, 1960; Hoggan *et al.*, 1962, 1963; Meredith *et al.*, 1962; Reiner, 1973; Sommer, 1975, 1980; Schildbach, 1977; Briggs, 1978; Ulmer *et al.*, 1985). In fact, a knowledge of an unfamiliar barley's composition has only a limited predictive value, because of the wide varietal differences in enzyme development etc. that appear during the malting procedure. Barleys must be screened in breeding trials, but the first reliable evaluation of malting quality depends on micromalting trials. However, a knowledge of the composition of a sample of a familiar barley is very valuable in guiding which

Table 14.6 The brewing characteristics of malts made from barley samples of different protein contents (TN × 6.25) all steeped to 43% moisture and germinated for 7 days

	Malt analyses			
Barley protein (% d.m.)	F–c extract difference (%)	Kolbach Index (%)	Lauter time (min/2 l)	Brewhouse extract (%)
9.5	0.6	45.7	13	75.0
11.8	1.6	38.7	22	72.5
13.0	1.8	39.0	28	71.9
14.7	2.3	35.3	37	70.5
15.5	2.5	33.9	51	69.6

From Schildbach (1977).

lot to purchase, in predicting the type of malt that may be made from it and in checking if batches of malt are of as good qualities (or better) as would be expected from the barleys used to make them. By 1857, Lawes and Gilbert were aware that the higher the nitrogen content of a barley the worse the quality of malt made from it was likely to be. By the 1880s, it was established that nitrogen content and malt extract were inversely related (Meredith *et al.*, 1962). The explanation is that as the nitrogen content of the grain (protein) increases, there is a corresponding decrease in the proportion of the carbohydrates, particularly the starch and sugars, that provide most of the extract when the malt is mashed. The regularity principle of barley composition was noted in Chapter 4. The next advance was the clear separation of barley varieties and the realization that varieties differ and that, for a given variety, when sound, non-dormant grain samples of a known range of nitrogen contents are processed in a standard manner, the extracts of the malts vary in a regular way. For example, for a set of malts having crude protein contents of 9, 11, 13 and 15%, the average extracts obtained were 81.8, 79.3, 77.3 and 75.6% of dry matter, respectively (Piendl and Oloff, 1980). Similar relationships apply when other cereals are malted. In addition, higher extracts tend to be given by malts of a particular nitrogen content, prepared from bold, plump grain, from grain with a high TCW value and, possibly, although this has been discounted, from grains having higher bulk densities (e.g. kg/hl or lb/bu).

It is difficult to compare barleys with different nitrogen contents, and malts made from them having 'equal degrees of modification' when it is frequently found that it becomes progressively more difficult to induce barleys to modify as the nitrogen content of the grain samples increases (Table 14.6). Bishop and his collaborators devised various extract prediction equations and these were revised and extended by others (Bishop and Day, 1933; Bishop, 1934, 1948; Bishop and Marx, 1934). Equations were applicable to the British barley varieties then available and using the method of analysis then current (IoB methods) gave:

$$TE = A - 11.0N + 0.22G \quad (14.1)$$
$$TE = A_1 - 10.0N - 0.3Z - 0.2X \quad (14.2)$$
$$TE = 138.2 - 9.5N - 3.0I \quad (14.3)$$

where TE is the hot water extract (dry basis: lb/malt Qr of 336 lb); A, A_1 are varietal 'constants' for each type of barley (for Proctor A = 112.5; the range of values for two-rowed varieties was 109.5–114.5; in fact these constants were found to vary a little between seasons); G is the dry weight of 1000 corns (TCW, g); N the nitrogen content of the dry barley (%); Z is 1000/G, that is the reciprocal of the mean dry weight of 1 corn (g); X is the percentage of dead or dormant grain found in the sample by a specified method; and I is the percentage of 'insoluble carbohydrate' remaining after a specified series of hydrolyses and extractions. For reasons that are explained below, the IoB no longer publishes prediction equations, but in *Analytica–EBC* (3rd edn) an equation is given as:

$$E(\%) = A - 4.7N + 0.1G \quad (14.4)$$

where E(%) is the extract (% d.m.) determined by the *Congress Method* on finely ground malt and N and G are as defined above. If 'protein' (P, % d.m.) is used instead of total nitrogen, the term –4.7N is replaced by –0.75P. For most older six-rowed barleys, the varietal constant is 80.0, while for two-rowed varieties the values vary within the range 84.0–86.8.

Equation 14.1 uses readily available analytical data for the prediction but takes no account of the dead and dormant corns, which are allowed for in equation 14.2. Equation 14.3 does not suppose a knowledge of the barley variety but relies on a new chemical measurement, that of I, the insoluble carbohydrate. This empirical measurement probably describes the quantity of carbohydrate material in the grain that does **not** contribute to extract when the malt is mashed. Clearly such equations only accurately predict malt extract for malts made using the given type of malting procedure that was used to determine the constants. They do, however, indicate the trends in the relationships between barley analyses and malt extract when any malting process is used. The TCW value G in equations 14.1 and 14.4, is a tedious and often imprecise determination and can detract from the precision of the prediction. With Proctor barley, it was shown by regression analysis, that using a particular malting programme each rise of 0.1% in the total nitrogen content of the barley led to a decline in coarse-grind extract of 1.2 lb/Qr in the final malt, using the units of extract then current (Hoggan *et al.*, 1962, 1963). The revised equation for the 1960 and 1961 crops of Proctor was:

$$TE\ (lb/Qr) = 119.9 - 11.5N \pm 1.1 \quad (14.5)$$

Several factors have emerged that complicate this approach. It seems from analyses of 2600 samples that the inverse relationship between barley grain

Table 14.7 Relationship between barley protein content and decline in malt extract; the relationship is not linear

Barley protein (N × 6.25, %)	Malt extract decline (%)
10.5–11.5	0.2
11.5–12.5	0.5
12.5–13.5	0.9
13.5–14.5	1.3

From Reiner (1973).
[a] Relationship established by analysing 2600 samples using *Analytica-EBC* Congress methods (EBC, 1987).

protein (N × 6.25) and malt extract is not truly linear (Table 14.7). Furthermore, when the prediction equations were originally derived the inverse relationship between barley nitrogen and malt extract was the same for all the barleys in use, as indicated by the single nitrogen constant used (–11.0 in equation 14.1). This is certainly untrue for modern barleys. At least one company determines the inverse relationships between nitrogen and extract values using linear correlations (Fig. 14.6) (Lloyd, 1982; Hudson, 1986b). With poorer quality barley, as the nitrogen content increases the extract obtained from the malts declines faster than it does from malts prepared from good quality barley (Fig. 14.6). The best-fit regressions between hot water extract and total nitrogen differ depending on how well the malts are modified as judged by the SNR or TCW values of the grain samples (Hudson, 1986b). Thus malts made with additions of gibberellic acid have hot water extracts greater than those made without, using barley samples having the same nitrogen contents.

Barley extracts can be determined by mashing finely ground grain with carefully standardized mixtures of enzymes that contain α-amylase. Linear

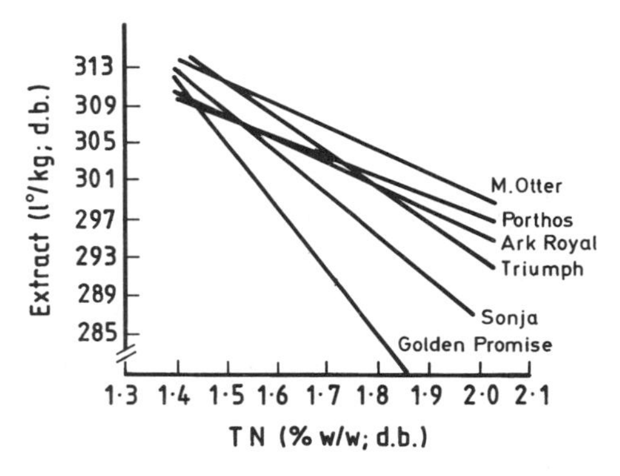

Fig. 14.6 The linear regressions between malt extract and total nitrogen for six barley varieties, all malted for a total time of 7 days (Hudson, 1986b).

Table 14.8 Mean analyses of 12 Canadian barley varieties grown at different sites, arranged in order of nitrogen contents, and data on some derived malts[a]

Variety no.	TN (%)	Salt-soluble proteins			Alcohol-soluble protein nitrogen (%)	Insoluble protein nitrogen (%)	Starch (%)	Crude cellulose-lignin (%)	Barley extract (%)
		Non-protein nitrogen (%)	Protein nitrogen (%)	Total (%)					
1	1.540	0.292	0.305	0.598	0.480	0.462	59.2	10.0	80.0
2	1.738	0.342	0.320	0.662	0.582	0.494	57.7	10.8	78.2
3	1.932	0.333	0.312	0.646	0.699	0.587	56.1	11.0	77.5
4	2.278	0.338	0.348	0.686	0.893	0.699	54.2	10.9	76.8
5	2.294	0.349	0.462	0.812	0.893	0.589	55.1	9.6	77.9
6	2.333	0.359	0.398	0.757	0.921	0.655	53.6	10.2	74.7
7	2.357	0.342	0.399	0.741	0.962	0.654	54.8	9.8	76.5
8	2.381	0.352	0.445	0.797	0.932	0.652	52.4	10.6	74.6
9	2.526	0.361	0.411	0.772	1.030	0.724	53.3	9.9	75.5
10	2.668	0.395	0.396	0.791	1.182	0.695	52.7	10.0	74.4
11	2.674	0.349	0.393	0.742	1.208	0.723	51.7	10.4	73.6
12	2.687	0.394	0.412	0.807	1.065	0.815	52.7	9.9	74.2

Table 14.8 (Contd)

Variety no.	Barley			Malting data			Malt		
	TN (%)	1000-kernel weight (g)	Plump barley (%)	Steeping time (h)	Malting loss (%)	Sprouts (%)	Extract (%)	Wort nitrogen (%)	Diastric power (°L)
1	1.540	34.8	92.2	99	6.8	1.6	78.0	0.61	63
2	1.738	34.8	84.7	83	6.5	1.7	76.3	0.65	85
3	1.932	32.9	79.3	74	7.0	1.6	75.5	0.78	100
4	2.278	33.5	89.5	77	6.9	2.0	74.3	0.82	121
5	2.294	33.5	86.6	84	7.2	2.0	74.7	0.73	116
6	2.333	30.6	65.0	75	7.2	2.3	73.0	0.75	105
7	2.357	28.0	84.0	77	7.0	2.0	73.9	0.78	117
8	2.381	32.5	50.3	65	6.6	1.8	72.8	0.87	122
9	2.526	28.0	69.7	81	7.4	1.9	73.3	0.87	133
10	2.668	32.5	82.9	75	7.1	2.0	71.9	0.79	140
11	2.674	31.8	78.0	74	8.0	2.4	70.5	0.78	149
12	2.687	30.8	67.3	76	7.4	1.7	71.6	0.81	133

From Anderson and Ayre (1938); Meredith and Anderson (1938); Ayre *et al.* (1940) by permission of the National Research Council of Canada.
Results as dry matter basis

correlations are often found between barley extracts and the extracts of malts prepared from them (Table 14.8) (Meredith *et al.*, 1962; Briggs, 1978). Barley extracts are not routinely determined, although they seem to be a reliable guide for predicting potential malt extracts.

Many other correlations have been investigated in attempts to predict malt qualities, other than hot water extracts, from analyses of barley. Some correlations seem to break down in particular years, or in particular instances, and many seem to hold only for samples of one variety of barley. Some additional correlations (positive (+) or negative (–); not necessarily linear) occur between malt extract and barley starch (+), barley nitrogen (–), barley salt-soluble nitrogen (–) and barley gel-protein content (–). There are correlations between barley nitrogen and various nitrogen fractions (+), malt saccharifying activity (+) and malt proteolytic activity (+), and between wort, fermentability (–) and fine–coarse extract difference (+); between extract and the SNR or Kolbach index (+), malting loss (+), hardness (–), friability (+) and colour (+); and between corn size and hot water extract (+), fine–coarse extract differente (–), total nitrogen (–), diastatic power (–), fermentability (+), predicted spirit yield (+) and friability (+) (Meredith *et al.*, 1962; Reiner, 1973; Sommer, 1975, 1980; Schildbach, 1977; Smith and Lister, 1983; Ulmer *et al.*, 1985; Hudson, 1986b). These relationships differ among varieties. For example, the increase in malt diastatic power increases roughly linearly with increasing barley nitrogen content, but the extent of the increase varies greatly among varieties (Meredith *et al.*, 1962; Hudson, 1986b).

In commercial practice, malts of different specifications must be produced, as required, from the barleys available. For this to be achieved maltsters must know the potentialities of their barleys and must know how to alter malting conditions to obtain the desired products. In practice, conditions are chosen with regard to experience, the outcome of micromalting trials and a knowledge of published experimental malting trials. Various factors will be discussed in the following sections.

14.8 Steeping

14.8.1 Steep liquor

The water used for steeping should be clean and of potable quality, free of pathogens and free of dissolved organic matter. Toxic elements, such as heavy metals or significant residues of herbicides, fungicides, etc., must be absent. The water should be clear, colourless and odour free and not contain appreciable amounts of ammonium ions, nitrates, chlorine or any oils or lubricants. Depending on the malting plant, the water is either used at the temperature at which it arrives in the malthouse or, in more modern plants, it is heated or cooled as necessary to achieve a chosen temperature. Traditionally, temperatures of about 12 °C (53.6 °F) were preferred, with as many as five or six water changes. However, temperatures of anywhere between 5 and 20 °C (41 and 68 °F) were used if conditions dictated it. Longer, cool steeps gave fewer problems with irregular germination. There were 'rules of thumb' that indicated to what extent steeping times should be altered as the temperature changed. However, better knowledge of the influence of temperatures and water uptake now allows steeping times to be adjusted exactly (Briggs, 1967; Chapter 3). With modern steeping schedules using air-rests, higher steeping temperatures of 16–18 °C (60.8–64.4 °F) are common, and even higher temperatures may be used for particular purposes (see below).

The quantities of salts dissolved in potable waters are too small to have significant effects on the rates of water uptake. However, the substances dissolved from the grain by the steep liquor can have a significant effect on the germination of the barley and the eventual quality of the malt. Chalky, alkaline water seems to check the growth of microbes in the steep and favours the solution of a mixture of acidic, proteinaceous and phenolic materials (collectively called 'testinic acid') from the grain. The first steep liquor may be made alkaline by additions of lime, to produce lime water (a clear, saturated solution of calcium hydroxide), which may be used at full strength (saturated) or at various dilutions (e.g. 50%; 0.05–0.1%), or by additions of sodium hydroxide, which is a more efficient extractant (Stevens, 1958). Milk of lime, a suspension of calcium hydroxide in water, is not used because of the undesirable chalky and dusty deposits that remain on the malt. To avoid damage to the grain, barley must only be exposed to these alkaline conditions for short periods (1–3 h) and then the alkaline liquor is replaced by fresh water. It is widely, but not universally, believed that malts from alkali-washed barleys (0.05–0.1% sodium or potassium hydroxide) produce beers with improved, non-biological colloidal stabilities and less astringent, 'cleaner' flavours (Dadic *et al.*, 1976). Around 1900, it was often reported that steeping in lime water reduced dormancy, but under modern conditions where grain is harvested and stored dry in bulk, this no longer seems to be true. Short alkaline treatments, rarely used in England, also 'clean' grain and remove musty smells. Many additions have been made to steep liquor. Some, such as gibberellic acid, soluble bromate salts and hydrogen peroxide, will be discussed later. Steep liquor has been made alkaline with sodium carbonate or sodium hydrogen carbonate (bicarbonate). These are safe to use, being less damaging to grain, but are less effective than caustic soda in removing testinic acid. A wide range of antiseptics and other agents has been employed in steeps, including chlorine, solutions of alkaline hypochlorites such as sodium hypochlorite and calcium hypochlorite (bleaching powder), potassium permanganate, sulphur dioxide, hydrogen peroxide, sodium bisulphite, salicylic acid, sodium hydrogen fluoride, ammonium hydrogen fluoride, fluorosilicic acid, various mineral acids, chloramine-T and formaldehyde (Lévy, 1899; Weichherz, 1928; Leberle, 1930; Witt, 1970a; Narziss, 1976). These were variously employed to overcome dormancy and/or to limit the multiplication of microbes in the steep or subsequently during germination. Many of these treatment agents are no longer acceptable. Chlorinated water and hypochlorites are objectionable as they 'taint' malt, giving it a 'chlorophenol', disinfectant-like flavour that is carried into beer. Sulphur dioxide (sulphurous acid; 0.02–0.05 g/l; Lévy, 1899) is objectionable to use and, like bisulphites, seems to be of only limited value. Strong fluoride solutions are toxic and mineral acids, and while effective, are dangerous and corrosive. Spraying grain with sulphuric acid (0.1%) before drying and storage subsequently reduces dormancy, probably by controlling microbes, and brief steeps in solutions of dilute mineral acids have similar beneficial effects (Doran and Briggs, 1993a). The reuse of steep water, without prior extensive purification, delays the germination of grains and gives unsatisfactory malts (Sommer, 1977b). Formaldehyde (e.g. 0.02%; Lévy 1899) is not widely used. It is such an effective antiseptic that its incorporation into steep liquor has allowed this to be reused (Piratzky, 1958). It kills *Fusaria* and other fungi that can generate mycotoxins and induce gushing in beer. Short exposures to it effectively reduce water sensitivity in grains and, applied at 500–1000 mg/l liquor in the last steep, it reduces extractable anthocyanogen levels in malts, which, in turn, give beers less prone to develop non-biological haze. Its use in resteeping liquor controls microbes very effectively (Gjertsen *et al.*, 1963, 1965; Withey and Briggs, 1966; Whatling *et al.*, 1968). Formaldehyde has been used as a fungicide when steeping wheat for malting (Fleming *et al.*, 1961). It has also been used to reduce levels of extractable tannins in birdproof sorghums, when these are malted (Daiber, 1978). Waters containing iron salts (ferrous waters) are avoided because the iron interacts with phenols in the surface layers of the grains, forming blue-black substances that give the grain a dull, greyish, unattractive appearance (Heron, 1912).

While short, initial steeps in alkaline liquors (0.05–0.1% sodium hydroxide) are widely used, sometimes more complex washes have been employed. For example, dormant barley was treated with lime water then, after a rinse, the pH of the steep was reduced to 2.5 by additions of sulphuric acid, mixing being achieved by aeration. After 3–5 h the steeps were drained and steeping was continued using ordinary water.

The use of surface water, particularly if it contains any organic matter and is from stagnant ponds, canals or dirty rivers, is to be avoided. Pure well or borehole water is preferred and this usually has a relatively constant temperature. The temperature of mains water fluctuates and it may be contaminated with residual chlorine. Ozone has been proposed for use as a sterilant in steeps and for overcoming dormancy, but no detailed reports of the use of this agent are available and the gas is certainly able to damage and kill dry grain. Experimentally, used steep liquor has been sterilized with chlorine dioxide.

Obtaining steep liquor and disposing of effluent, the waste and dirty steep liquor, are both costly. It is, therefore, less costly to change steep liquor as few times as possible. However, with older steeping schedules it was found preferable to change the liquor every 8–12 h to obtain 'clean' grain that germinated evenly. In Norway and some other areas, it was traditional to steep small amounts of grain in running water. Split and damaged barley kernels readily release nutrients into the steep, resulting in excessive microbial growth and poor germination unless the steep water is changed more frequently. White (1860) noted that in warm weather material from the husk dissolved to such an extent that 'if the water be not changed the day after steeping it will take two or three hours to drain. The steep water at such times is of a peculiar viscid character'! Muspratt (1855) remarked that after the steep had drained it was desirable to rinse the grain to remove the slime. Of course, such working conditions are now wholly unacceptable. If the steep water is becoming unduly dirty it should be changed.

At present, there is no widely accepted substance that may be used as an antiseptic in steeps. Hydrogen peroxide, which is an antiseptic, has been added to steep water to hasten the subsequent rate of germination of dormant barley (e.g. Green and Sanger, 1956). In view of the increasing concern about the influences of undesirable microbes, it has been proposed that these should be controlled by 'seeding' grain with cultures of 'unobjectional' bacteria. These, it is hoped, will outnumber and 'swamp' the undesirable organisms, preventing their growth. Additions of lactic acid bacteria check the activities of at least some undesirable microbes (Haikara and Home, 1991; Boivin and Malanda, 1993; Haikara *et al.*, 1993; Stars *et al.*, 1993). The lactic acid bacteria are not without effect on the malting grain. No doubt they will compete with the grain tissues for oxygen, for example. It is uncertain if this approach will be widely adopted.

14.8.2 Steeping: some general considerations

Steeping has been widely discussed in the literature (White, 1860; de Clerck, 1952; Witt, 1970a; Brookes *et al.*, 1976; Narziss, 1976; Adamic, 1977). Newer maltings often have facilities for washing barley before it enters the steep. Nearly all maltings use at least some immersion-steeping and this is desirable (Maule, 1996). Spray-steeping is most often used at the end of the steeping schedule to adjust final moisture contents. Barley is usually loaded into steeps from above, normally into water. Historically, the steeps were stirred by hand. At present, the steeps are often aerated as the grain is run into water to agitate it and help clean the grain. Traditionally water was run into the base of the steep and, in rising through the grain, it brought dust, light impurities and thin, 'unfilled' grains to the surface. The water was allowed to weir over and run to waste. Suspended material, removed by sieving the washing water, was dried and sold with other by-products as cattle-food. Alternatively, the floaters were collected by skimming. In many modern maltings, using well-cleaned grain, weiring over is not employed since this uses water and produces effluent, so increasing costs.

As the grain takes up the water it swells to a variable extent, eventually occupying around 130–140% of the space it initially occupied. In the early stages, the oxygen requirements of the grain are relatively low, but steeps are aerated to loosen the grain, to help clean it and to prevent it packing tightly as it swells. Gradually, materials dissolve from the grain, the water becomes yellow-brown and begins to froth as carbon dioxide is liberated; microbes multiply on the grain surfaces and in the liquor. Materials appearing in the steep liquor include uncharacterized pigments, amino acids, hexose and pentose sugars, betaine, phenolic and carboxylic acids, phosphate (said to be as much as 10% of the grain's phosphate), ethanol and acetic acid. Under anaerobic conditions, grain glycolyses, generating ethanol and carbon dioxide. Steep aeration checks ethanol production, and ethanol levels can be used as an index of aeration efficiency (Smith and Briggs, 1979; Cantrell, 1987). In traditional immersion-steeping, grain lost 0.5–1.5% of its dry weight as dust, skimmings and dissolved materials. With modern, air-rest steeping schedules, in which germination and vigorous respiration begin in the steep, the losses may be higher, perhaps 2% or even 3%. The dissolved materials support the growth of microbes that compete with the grain for oxygen and possibly produce off-flavoured materials, substances which induce gushing in beer, mycotoxins and phytotoxins. While consuming oxygen, the microbes produce acidic metabolites, such as acetic acid, and the steep pH falls. Undoubtedly microbes interferer with germination and modification in malting and with the quality of the final malt (Kelly and Briggs, 1992a; Stars *et al.*, 1993). There are no generally accepted antiseptics with which to control the population of microbes, so it is necessary to change the steep liquor at least once and probably more than once to remove microbes and microbial nutrients. The large volumes of effluent liquor, with its high levels of suspended solids and dissolved materials and consequent high BOD, makes it costly to purify or discharge to public sewers (Chapter 9).

For many years, hydrogen peroxide (0.1–1.0%) has sometimes been incorporated into steep liquor to accelerate the germination of dormant grain (Becker, 1926; Enders *et al.*, 1940; Green and Sanger, 1956). Another suggestion was that the steep should contain hydrogen peroxide (0.1%) and lactic acid (0.1%). The optimum concentration of hydrogen peroxide needed to induce satisfactory germination varies between grain samples. Hydrogen peroxide is well known to check the growth of microbes. Since it breaks down to oxygen and water, the pure substance leaves no undesirable residues. However, it is expensive and consequently it is not often used. We have proposed that it be sprayed onto the grain (e.g. 0.75% solution) thereby reducing the amount of material used (Briggs, 1987a). In many plants, spraying grain in the steep is not practical. To overcome this, 'plug rinsing' has been proposed as an economical way to wash the grain at the end of a steeping period and apply hydrogen peroxide or other dissolved materials (Briggs, 1995). It is proposed that the steep be partly drained and then fresh liquor (containing hydrogen peroxide or other substance, if desired) would be added from above until this just covered the grain. Draining would then be resumed. The clean liquor, following the dirty down through the grain, would replace the film of dirty liquor (containing microbes and microbial nutrients) from around the grains with clean liquor and with hydrogen peroxide if this had been added. The plug-rinsing technique could be used to apply other additives such as gibberellic acid or potassium bromate at the start of an air-rest period or before casting. This approach appears to be successful (Maule, 1996).

No practical use has been found for used steep liquor. It has been used as an irrigant and weak fertilizer sprayed onto farmland, and it has been proposed as a supplementary source of yeast nutrients by adding it to mashing liquor in breweries. However, because of its uncertain composition, its large load of microbes, its large volume and the distances between most maltings and breweries, this last proposal is impractical. Maltsters increasingly try to minimize the use of water to reduce water abstraction and effluent-disposal charges. Different plants were said to use 27–90 hl water/t malt produced (90–300 gal/Qr of 336 lb).

The objectives of steeping are to clean the grain, to hydrate it to the correct extent and to prepare it so that it grows steadily and modifies uniformly in the germination stage. The division between the steeping and germination stages is no longer clear since, in contrast to traditional programmes in which the grain might not chit until 1–3 days after steep-out, grain leaves modern steeps fully chitted and beginning to grow. From the production aspect, problems arise because barleys need to be cast at different moisture contents and because barleys from different batches take up water at different rates.

In traditional British floor-malting, mellow, two-rowed barley was transferred to the germination floor with a moisture content of about 43%. Increasingly, higher steep-out moistures are used so that casting at 45–46% moisture is common and poor quality, steely barleys may be cast at 47–48% moisture, or they may be sprayed in the germination compartment to increase their moisture contents to within this range. Malting grain at different moisture contents substantially alters the progress and pattern of modification. In the past after a 40–84 h immersion, for example, barley was judged to be steep-ripe when the grain appeared fully swollen by the appearance of the embryo and the uniform wetting of the endosperm when it was cut, by the lack of a 'pricking' feeling and the separation of the husk when the ends of a corn were squeezed between finger and thumb, if the root sheath (chit) appeared when the sides were squeezed, by whether the grain was easily pierced with a needle, by whether the grain bent (instead of snapping) when pressed with the thumb-nail, and by its softness when bitten (Muspratt, 1855; White, 1860; Graham, 1874; Hunt and Rudler, 1878; Thausing *et al.*, 1882; Lévy, 1899; Wahl and Henius, 1908). Now water-uptake times are usually determined on a small scale before the bulk of the barley is committed to the steep. This is not quite straightforward as handling the grain in the steep alters its behaviour, and surface water is easily dislodged. Furthermore, during air-rests the temperature of the grain in bulk is liable to rise more than in small samples, accelerating both evaporation and water uptake. Usually the moisture content of the grain is determined a fixed time after draining (to allow some surface moisture to be absorbed) or the moisture may be determined after the removal of the surface film of water. A double-drying technique must be employed to obtain accurate estimates. NIR may be used as a simple and rapid 'secondary' method of determination.

Grain may be under-steeped or over-steeped. Under-steeping means that the grain is cast at too low a moisture content to permit adequate modification to occur. Initially germination is rapid but it slows, root growth is poor and modification is incomplete (Reynolds *et al.*, 1964). Sometimes grains are deliberately cast under-steeped, and then the moisture content is adjusted by spraying water onto the grain in the germination compartment. Strongly divergent views are expressed about the desirability of spray-steeping. Over-steeping can cause grain death, but usually irregular germination is the consequence. Problems are particularly likely with immature grains or grains that are damaged so that microbes rapidly multiply on the materials leached from the broken corns. The grain may reach high moisture contents and germination is irregular so that while some grains do not germinate others grow strongly, modify rapidly and sustain high malting losses, so giving an inhomogeneous malt. With mature grain, the use of steep aeration and air-rests reduces the dangers of over-steeping. However, with immature and, particularly, undried grain, the situation is less clear, and with air-sensitive grain aeration of the steep **reduces** the extent to which it subsequently germinates, presumably because of an excessive proliferation of oxygen-requiring microbes (Scriban, 1963; Kelly and Briggs, 1992a). Many microbes increase on barley and in the liquor during steeping (Petters *et al.*, 1988). Whether they are harmful seems to depend on the maturity of the barley, the mixture of microbial species present and the steeping programme employed. Some depress germination, the 'vigour' of malting and the rates of modification and enzyme production. Infestations may reduce grain responsiveness to added gibberellic acid. Others can strongly influence malt pH by altering the organic acids in the malt and hence alter malt quality (Stars *et al.*, 1993). Some microbial infections result in variously altered malt analyses (including over-modification in some instances). Yet others produce factors on the malt that induce beers to gush (overfoam), and there is the concern that mycotoxins may be produced. Experimentally, surface-sterilizing grain greatly improves its germination characteristics (Kelly and Briggs, 1992b). Undoubtedly, the discovery of an acceptable, inexpensive and effective microbicide for use in steeps would be a great advance. The use of such an agent might alter the nature of malts since various enzymes (e.g. β(1→4)-glucanase) originate from the microflora.

14.8.3 Commercial steeping schedules

In traditional floor-malting practice, early in the 20th century, steeping temperatures were not controlled and so steeping times were very variable. However, temperatures of 9–13 °C (48.2–55.4 °F) were usually preferred and steeping was often for about 3 days (59–75 h for British barleys, 100–120 h for imported barley; Ross-Mackenzie, 1927). Good quality grain was cast with a moisture content of about 43%. The water was changed three times or more in warm weather or if 'dirty' barley was being used. Usually the only aeration would have been as air was drawn down into the grain as water was drained away and if the grain was sprinkled from above, to wash it, before the next steep-water was added. Aeration or other treatments that allowed the grain to chit in the steep or caused it to grow too vigorously and uncontrollably on the malting floor were avoided. Consequently this near-anaerobic steeping system was regarded as desirable. However, elsewhere many other steeping techniques were in use or had been tried, often with very variable results (Graham, 1874; Briggs, 1987a). In Bavaria, as on Norwegian farms, steeping was sometimes carried out in running water. In Bohemia, steeping was variously reported to be carried out with air-rests by immersing the grain for 24 h (or 12 h) and then draining it and moving it to the floor for a period of 24 h (or 12 h). At the end of this dry, air-rest period it was returned to the cistern for a further period of immersion (Graham, 1874; Schönfeld, 1932). This system was used with older, shallow steeps and was very laborious. Increasingly, steeping was carried out in deeper, conical steeps; in the base of these the carbon dioxide content was higher and the oxygen content was lower than further up in the grain-water mass. Therefore, it seemed desirable to aerate and mix the grain mass and many steep tanks were designed to achieve this by blowing compressed air into the base. Aeration accelerated germination but, with immature grain, germination could be irregular. Aeration was never sufficiently effective at oxygenating the steep liquor to induce the grain to chit in the steep. Grain–water mixtures were often moved in air-lift tubes using compressed air, but sometimes propellers were employed (Chapter 9). In addition, grain and water might be pumped around in a steep or transferred between steeps. As early as 1887, the firm Steinecker patented a device for washing grain during transfers between steep tanks.

Windisch (1897, 1901, 1902) championed the use of air-rests in which the grain rested in air between immersions. These hastened germination and reduced the risks of over-steeping. During air-rests, the grain became warm and reimmersion served to cool it, as well as providing more moisture. The durations of the air-rests were adjusted to avoid inducing the grain to chit in the steep, a notable difference to modern practice. Some of these older steeping programmes employed an amazing three to four water changes each day. Grain air-resting in a simple, conical steep tank soon became anaerobic as the air was displaced by carbon dioxide generated by the grain. The Söding-Winde reform steep (section 9.8.2) was equipped with a series of tubes and inverted funnels to allow carbon dioxide extraction or downward ventilation, that is carbon dioxide was sucked out of the grain mass to be replaced with fresh air from above. At this period, the use of hydrogen peroxide in the steep liquor was also known and the use of oxygen was discussed.

The first systematic experiments in warm water steeping were reported in 1907, by Somló (Pollock, 1962a). The expected benefits included shorter steeping times (because of accelerated water uptake), more effectively cleaned grain and improved germination of some immature grain. Moufang reported that optional warm-water steeps, used as the first steep, were 90 min at 40 °C (104 °F), 60 min at 45 °C (113 °F) and 30 min at 50 °C (122 °F) (Pollock, 1962a). The grain then had to be quickly cooled by mixing with cold water and should have been vigorously aerated during the rest of the steep. These requirements, which are easily met in the laboratory, present many technical problems on the production scale when masses of cool grain must be mixed with warm water and rapidly mixed (to avoid over-heating); later they must be rapidly drained and cooled. Warm-steeped grain often showed reduced root growth but normal modification and enhanced acrospire growth, suggesting, probably for the first time, that root-free malt might be prepared, with reduced malting losses. However many problems were encountered. The responses of different batches of grain varied so much that the process was unreliable and could not be used routinely. Warm-water steeping was reinvestigated more recently in connection with both the use of gibberellic acid and in multiple-steeping/resteeping programmes (see below). Spraying-steeping, yet another novel steeping system, was tried in 1875 but failed because the grain at the top of the grain bed chitted while that at the base did not (Vermeylen, 1962).

In the UK, a limited amount of aeration was used in steeps, but its intensity was such that it chiefly served to mix the grain–water mixture and help clean the grain, although it did accelerate subsequent germination. Chambers and Lambie (1960) and Enari *et al.* (1961) experimentally sparged carbon dioxide, nitrogen or oxygen into steeps and investigated the malts produced after fixed germination periods. If the steeps were sparged with nitrogen or carbon dioxide then germination was delayed, but the malts were obtained in 1% and 2.5% greater yields, respectively. It was noted that the grains chitted more easily if air-rested between immersions in water saturated with carbon dioxide. However, if the steeps were sparged with oxygen the grains chitted rapidly, subsequently they bolted, growing uncontrollably and producing malts at the expense of unacceptably high malting losses (*ca.* 20%). Unfortunately no analyses were reported for malts grown for less than 7 days, to determine whether good quality malt had been produced in a reasonable yield in a shorter time.

With older maltings, both floor-maltings (whether air-conditioned or not) and pneumatic maltings, it was not possible to cool vigorously growing grain sufficiently to prevent its over-heating and bolting, whether this was induced by a highly aerobic steeping programme, casting at a high moisture content or by the use of gibberellic acid. Consequently, slow chitting followed by slow, cool malting was preferred, or the grain was germinated at a lower moisture content than usual. More recently, soluble bromates have been employed (when permitted) to check heat production and excessive growth (see below).

In modern practice, where large batches of grain are malted as quickly as possible, steeping conditions are chosen so that the grain is uniformly chitted (but rootlet production has not started) at the end of the steeping period. Conditions are chosen to suit the particular barley being malted and the type of malt being made. The temperature of the steep liquor is now often controlled, but in many (but not all) cases other aspects of temperature control (the steep room, the air used to ventilate the grain during air-rests) are inadequate.

Originally grain was not malted until its 'profound dormancy' had nearly gone and its water sensitivity had reached a low figure (e.g. >85% germination in the 8 ml test). Unexpectedly, it was found that grain giving low values in germination tests could often be malted on the pilot scale. It was found that small-scale tests, varying steeping and germination conditions for small samples held in test tubes, gave better estimates of grains' abilities to germinate in commercial maltings. Still better estimates were obtained with bag tests in which samples (0.5–1 kg) of grain, held in nylon net bags, were steeped and germinated in contact with large bulks of malting grain. However, with all these test systems, discrepancies were found between the results of these tests and full-scale malting trials. The lack of fully reliable, small-scale tests is unsatisfactory (Macey and Stowell, 1959). However small-scale tests are essential for screening barley lots and they usually err on the safe side in that they give pessimistic estimates of a barley's readiness for malting. Consequently,. although some few 'maltable' samples are held back until further maturation has occurred, unmaltable, immature barleys are recognized and are not steeped.

The recognition that water-sensitive grain germinates if steeped to a moisture content of only 35% (Chapter 3), but that this moisture content is insufficient to allow adequate modification led to the reintroduction of air-rest steeping. Steeping to about 35% moisture and then holding the grain in air for a period of 8–24 h overcomes water sensitivity. Then the grain can be steeped to the correct final moisture content either by immersion or, less usually, by spray-steeping (Macey and Stowell, 1959; Sims, 1959; Reynolds *et al.*, 1966). Theoretically, incremental spray-steeping is very attractive, since nearly all the water applied to the grain is taken up and so water usage and effluent production are reduced. However, the effluent that is produced contains more suspended matter and has an elevated COD. At intervals, water is sprayed evenly onto the grain until it is all wet and some minimal run-off occurs. Then the grain is ventilated downwards. When the film of surface moisture has been taken up the grain is sprayed again, and this is repeated at intervals until the correct final moisture content is reached. This attractive approach saves water and effluent charges and, if correctly carried out, causes grains to germinate vigorously and evenly, reducing malting time. However, spraying a bed of grain uniformly is technically difficult in some malting plants, although not in others (Domalt continuous maltings for example, section 9.10.4). In many parts of Europe, adjusting grain moisture content (e.g. from 43 to 46%) following 'conventional steeping' by spraying grain in the germination compartments is a routine practice. Undoubtedly good malts can be prepared using spray-steeping with great savings in water use and effluent production (Sommer, 1977b). In some processes, a comparatively short immersion-steep, or wash, is supplemented by extensive spray-steeping. Only rarely is all the steep water applied by spraying or sprinkling. Many spray-steeping trials have failed, probably because the plant used did not allow grains to be washed or treated evenly. Recirculating water through conical-bottomed steeps, withdrawing it from the base and spraying it back onto the top of the grain bed did not work well. Perhaps large temperature gradients developed, water already dirtied by contact with grain accumulated in particular zones of the grain bed, the water was only aerated in the upper level of the grain bed or, more simply, 'channelling' occurred and not all grains were adequately wetted. It seems that spray-steeping is best applied to grain being thoroughly turned and mixed. Late applications of water to germinating grain can rehydrate the embryo fully and cause excessive embryo growth without appreciably wetting the endosperm or accelerating modification. The dryer regions of the endosperm may not modify (Reynolds *et al.*, 1966; Axcell *et al.*, 1983). Judgements on the value of spray-steeping vary widely (Dahlstrom, 1965; Northam, 1965b; van den Beken *et al.*, 1975; Sommer, 1977b; Willmer, 1977; Doncheck and Sfat, 1979; Narziss, 1987). In one North American system, the grain is first steeped by immersion for 8 h. Then it is ventilated (carbon dioxide extraction) and intermittently, when the grain temperature begins to rise, it is rewetted using high-volume sprays delivering water at 15 °C (59 °F). Spraying is continued until temperature probes indicate that the temperature in the tank is uniform. The water is recirculated but is replaced with fresh water three times in the 48 h spraying period. The key to the system, which reduces water usage, is the use of a chiller to adjust the water temperature and a high-volume spraying system (Haley and Stokes, 1987). In a German programme, taking a total of 7 days for steeping and germination, a preferred routine is to steep for about 4 h at 12 °C (53.6 °F) to a moisture content of about 30%. The grain is then air-rested for a period of 24 h **minus** the duration of the first steep. During this period, the grain is ventilated with air at a RH 90–100% and at 15 °C (59 °F). The grain temperature rises towards 18 °C (64.4 °F). The second steep, using water adjusted to the grain temperature to avoid a thermal shock, lasts for 2–3 h at 18 °C (64.4 °F). This steep is vigorously aerated. The grain is cast, wet or dry depending on the plant, with a moisture content of 37–39%. During the initial germination stages, the moisture content is adjusted to the required value by spraying. In an alternative programme, the second steep is followed by another 20 h air-rest and then a 2 h immersion and wet transfer. Less spraying is subsequently needed to adjust the grain to the desired moisture content (Narziss, 1987, 1991a). Malts produced by immersion-steeping can have different qualities to those produced by spray-steeping. Usually spray-steeping favours a more rapid onset of germination and a reduced malting time but gives higher malting losses, less good fine–coarse extract differences, higher Kolbach indices and higher enzyme levels (Narziss, 1969). Others have claimed that malting losses are reduced while malt quality is maintained or improved by spray-steeping (Donchek and Sfat, 1979). It seems that spray-steeping schedules are successful only if carried out in correctly designed, instrumented plant (Kitamura *et al.*, 1988).

'Multiple steeping' is now commonly used in the UK. Typically, grain is immersed three times. During immersions it is aerated at intervals, while during the 'dry' periods it is ventilated with air sucked downwards (carbon dioxide-extraction). The objectives are to obtain uniform grain hydration and rapid, uniform and controllable germination and growth. There are wide variations in this approach, dictated by the type of plant used, whether or not additives can be applied to the grain, the barley growing season, the type of malt being made and so on. Examples are given below. In general, it is expected that the grain will have chitted by the end of the steeping period.

Flush-steeping was developed for use in floor-maltings with small steeps, relatively simple facilities and lacking equipment for controlling steep-water temperatures (Macey and Stowell, 1959; Northam, 1965a,b). The objectives were to hasten water uptake (so shortening steeping times to about 40 h) and to cast the grain chitted so that germination times could be reduced by about 2 days. In one particular schedule grain was steeped for about 6 h at 11–12 °C (53 °F) then drained. During the subsequent air-rest (*ca.* 12 h) the moisture film around the grains was taken up and, as respiration started to increase, the temperature of the grain rose a little. The grain was then 'flushed', that is it was covered with water for about 5 min and aerated, so cooling it and moving it in the steep, which reduced its 'packing' and renewed the film of surface water. In the next air-rest (*ca.* 8 h), the grain was ventilated, the surface moisture was taken up by the grains, respiration and heat output increased (causing the temperature to rise, e.g. to 14 °C (57.2 °F) and germination began. The grain was flushed again, reducing the temperature to about 11 °C (51.8 °F). In the air-rest which followed (*ca.* 6 h), despite downward aeration, the temperature rose rapidly to 21–27 °C (69.8–80.6 °F). At the end of this period, the grain had chitted. Then **either** the grain was flushed again and cast, when the rootlets had often grown to 2–3 mm, **or** it was steeped again (e.g. for 8 h) in a solution of potassium bromate (and perhaps gibberellic acid), which subsequently checked rootlet growth, protein modification and heat output. At the end of this steeping programme, the grain had a moisture content of 45–47%, was growing strongly and was liable to bolt on the floors if potassium bromate had not been used. To employ this technique, it was necessary to be able to fill and empty the steeps quickly and to spread the grain on the floors thinly and turn it frequently. A problem with this steeping programme is the high water usage (47 hl/t (210 gal/Qr)) and the high volume of effluent produced (40.3 hl/t (180 gal/Qr)) (Northam, 1965b).

In older plant, downward ventilation was inadequate. The grain was not cooled uniformly and uneven germination could occur. A later steeping system, sometimes called kangaroo steeping because the grain was moved between steeping vessels (Macey, 1977, 1980; Stowell, 1976), utilized 'conventional' steeps as well as steeps with 'fully perforated cones' designed to achieve more uniform and rapid downward ventilation (Chapter 9). The steeping programmes described are very flexible and were intended to be varied so that different batches of grain could be processed in a fixed time to simplify production schedules. The first immersion lasted for 5–10 h, in the temperature range 15–25 °C (59–77 °F) until the moisture content was 36–38%. Using too high temperatures resulted in uneven chitting, growth and modification and favoured the growth of microbes. The steep was vigorously aerated to mix the contents and obtain a uniform temperature. After draining, the grain was transferred to the ventilated tank in which strong downward ventilation could be applied, but this was probably unable to cool all the grain. Sometimes gibberellic acid (e.g. 0.25 mg/kg grain) might be applied during the transfer. The grain was air-rested in the second vessel for 12–24 h with ventilation at the rate of 2–5 m^3/t barley steeped. The temperature of the grain slowly rose to 20–30 °C (68–86 °F), higher temperatures being used for poorer quality barleys. Germination and modification began but proteolysis and rootlet growth were checked. This was ascribed to the 'stunning' effects of the unusually high temperatures. The grain was evenly chitted by the end of this air-rest. Finally, the grain was again covered with water and remained immersed for 5–16 h at 15–25 °C (59–77 °F) until the moisture content was 46% for good quality barleys but was as high as 48% for less-good quality grain. For average barleys, this steep would last for 12 h at 20 °C (68 °F). The grain was then cast, with or without applications of gibberellic acid and/or potassium bromate, into a box or drum and germinated for 2–4 days at 15–20 °C (59–68 °F).

With increased batch sizes, it has become less practical to transfer grain between steep vessels, whether steeping is carried out in large flat-bed steeps or in sets of smaller, hopper-bottomed vessels. Consequently immersions, aeration and ventilation during air-rests are commonly carried out in the same vessels. Downward ventilation is most uniform in flat-bed steeps. Air-rests, by allowing temperature rises and grain chitting, accelerate the rate at which grain can be steeped and germinated. As the season advances and grain stocks mature, the lengths of air-rests are shortened to avoid over-vigorous growth. In the UK, it is a common practice to use two or three immersions separated by one or two ventilated air-rests. In Continental Europe, two-water systems are popular, with spraying being used to adjust final moisture contents. One type of schedule is steeping for 6–8 h at 15–16 °C (*ca.* 60 °F) to a moisture content of 32 ± 1%, then air-resting for 10–14 h with carbon dioxide extraction. Then a second steep of variable duration (*ca.* 8 h) is used to increase the moisture content to 42%. In older plant, the grain may be transferred between steeps permitting mixing and washing. During the next (second) air-rest, lasting 8–10 h, the grains chit uniformly. Water enters chitted grains very quickly, so subsequent steeping times are shortened. The final moisture content is reached in the third steep of 6–8 h. This is not extended for too long to avoid checking germination. After a final ventilated rest of 6 h in the steep tank the grain is cast dry. The final moisture content chosen would depend on the degree of modification required, being 43–45% for less well-modified malts made with additives, 45–46% for better modified malts made with additives and 48% for over-modified malts, malts made from 'stubborn' barleys or made without additives (Seward, 1986). The duration of the air-rests may have to be reduced if the ventilating air cannot be cooled or the airflow is insufficient to hold the grain temperature down to a chosen level. As steeping progresses, so the heat output of the grain rises and the required cooling capacity increases. So, for example, in the first air-rest airflows of 0.85–1.42 m^3/min per t (30–50 ft^3/min per t) are adequate while in the second air-rest airflows of 2.83–5.66 m^3/min per t (100–200 ft^3/min per t) are needed (Seward, 1986). With downward ventilation, there are risks of excessive grain drying and grain being at the wrong temperature if ambient air is used. The use of humidified, temperature-controlled air for downward ventilation is to be preferred (Chapter 9). In steeping–germination (resteeping)/kilning vessels, with full air-conditioning facilities, it has been possible to extend air-rests to 24–36 h (Seward, 1986). A steeping regime for use with abraded barley, processed in flat-bottomed steeps without the facility to condition the ventilating air, is 8 h wet, 8 h dry, 8 h wet, 6 h dry, 8 h wet and 8 h dry, all carried out at about 18 °C (64.4 °F) (Brookes, 1981).

In practice, grain is usually aerated at intervals while it is immersed. At the start of steeping, grain metabolism is slow and its requirement for oxygen is minimal; therefore, except for aeration to 'unpack' or lighten the grain bed as it swells, vigorous aeration can be undesirable at this stage as it encourages the growth of microorganisms. In addition, it encourages fouling in the plenum chamber of flat-bottomed steeps (Cantrell, 1987). The oxygen dissolved in the steep water, is rapidly utilized (Cantrell *et al.*, 1981; Cantrell, 1987). It is not entirely clear how beneficial aeration can be when applied during immersions within an air-rest steeping programme. It has been reported that, although aeration accelerated the rate of chitting, the final malts were exactly the same under the conditions tested regardless of whether aeration was applied (French and McRuer, 1990). These authors also pointed out that grain at the bottom of steep tanks is immersed for longer than grain at the top, causing it to have a higher moisture content. Excessive aeration, used without air-rests, can lead to uneven chitting of the grain in the steep, with consequent uneven moisture contents and germination and an inhomogeneous malt (Ekström *et al.*, 1959; Kelly and Briggs, 1992a). Steeping programmes are not invariable and must be modified to suit the grain, the malt to be made and the season (Brookes *et al.*, 1976; Narziss, 1966, 1987, 1991a; Narziss and Kieninger, 1967).

As previously noted, steeping with warm or hot water was extensively investigated around 1900. The advantages of cleaning the grain and shortening steeping times are offset by the occurrence of 'induced water sensitivity' (Lubert and Pool, 1964), followed by irregular germination and grain death. Furthermore, steeping programmes that succeed with some grain samples fail with others. Despite the known difficulties, the potential advantages of warm steeping are such that the use of higher temperatures has often been reinvestigated, using aeration, air-rests and applications of gibberellic acid in attempts to overcome the problems (Witt, 1959; Pool, 1964a,b; Dahlstrom, 1965; Brookes *et al.*, 1976; Home and Linko, 1977; Baxter and O'Farrell, 1980; Reeves *et al.*, 1980; Sfat, 1966). Sometimes steeps at 20 °C (68 °F) may be used throughout the process, but steeps at 25 °C (77 °F) are generally unsatisfactory. By comparison, briefly prewashing barley in a spiral in a countercurrent of warm water (e.g. 25–30 °C; 77–86 °F) seems to be acceptable and was used in Domalt plants (Chapter 9). The second approach is to steep for 8 h at 40 °C (104 °F), for example, deliberately damaging the root initials of the grain and then inducing grain modification with applications of gibberellic acid (Pool, 1964a,b, 1967a,b). Good malts have been prepared in this way with malting

Table 14.9 The effects of the temperatures used during steeping and germination on indices of modification. Values are the means of results obtained by malting four different barleys at five different moisture levels

Indices of modification	Steeping and Germination			Steeping 20 °C (68 °F); germination 12 °C (53.6 °F)
	12 °C (53.6 °F)	16 °C (60.8 °F)	20 °C (68 °F)	
Friability (%)	72	85	84	81
Partially under-modified grains (%)	11.7	4.3	5.9	6.2
Fine–coarse extract difference (%)	2.4	1.5	1.3	1.5
Viscosity of laboratory wort (cP)	1.61	1.50	1.47	1.48
Kolbach index (%)	46.6	45.4	40.7	45.0
Malt yield (% d.m.)	92.0	88.7	87.9	–

Data of Smart *et al.* (1993) using *Analytica-EBC* methods of analysis.

losses of only 3.5%, on a semi-industrial scale. However, the difficulties caused by variations in the behaviour of different batches of grain and the problems of achieving prompt mixing and maintaining exact and uniform temperatures in large amounts of grain and water meant that the development of this process was discontinued. However warm-water steeping was used successfully for resteeping to reduce malting losses (see below). The impression remains that for steeping 16 °C (60.8 °F) is about the most uniformly successful temperature (Shands *et al.*, 1941; Dahlstrom, 1965; Narziss, 1969). Processing at different temperatures can significantly alter malt quality and yield (Table 14.9).

14.9 Casting the grain

The original distinction between steeping and germination vessels is blurred, since it is now usual to have the grain chitted at the end of the steeping period and in SGVs and SGKVs steeping and germination occur in the same unit. Traditionally, ungerminated grain was cast wet or dry. These days, the grain has normally chitted before casting. In wet casting, the grain, slurried with water, is pumped to the germination unit where it drains and begins to grow. This is a simple and flexible procedure. However, the film of surface water initially remaining on the grain delays chitting in traditional malting practice. Furthermore, care has to be exercised to ensure that the grain is not physically damaged in the pumps or by high hydrostatic pressures in the system. The alternative is to drain the grain and cast it 'dry', conveying it mechanically or (more cheaply and with less risk of damage) transferring it by gravity from the steep to the germination compartment.

Table 14.10 Extracts of small samples of malt separated from commercial batches (A–D) when they chitted then allowed to complete their germination period

Time at which grains began to chit (days from steep)	Hot water extract (1°/kg)			
	A	B	C	D
0–1	301.2	309.5	310.5	297.2
1–2	298.3	307.3	307.9	293.8
2–7	296.0	–	–	–

After Kirsop (1964).

In traditional practice, the absence of a film of surface moisture when the grain arrived at the germination unit ensured that it began to chit faster. Indeed an old patented process used warm air to remove the surface moisture from the grain as it was conveyed from the steep. Another advantage of dry casting is that the grain may be sprayed to coat it with a solution of one or more additives (e.g. gibberellic acid and/or potassium bromate) as it is transferred. Damage caused by pumping during wet casting can delay germination for a day (Bathgate and Cook, 1989). As well as during pumping, pressure damage can occur in grain situated at the bases of deep steeps. This is caused by the pressure of the grain bed and the water layer (Anon., 1900b; Eyben and van Droogenbroeck, 1969; Gomez, 1971; Yoshida *et al.*, 1979). Variable results have been recorded from application of high gas pressures. Damage caused during steeping is related to the pressure (mechanical and hydraulic), the duration of exposure to the pressure and the frequency of exposure. Some pumps generate pressures of over 4 kg/cm^2. Chitted grains are particularly easily damaged. Exposure to a pressure of 1.5 kg/cm^2 for 10 s causes detectable damage. Pressures of 1, 2, 3, 4 and 5 kg/cm^2 are progressively more damaging, as are exposures of 1, 15 and 30 s. The pressure at the base of a 5 m column of grain and water is sufficient to cause damage. One reason for using flat-bed steeps is to avoid the high pressures that can occur in the bases of large, conical bottomed steeps. Water is forced into the grains, embryos are damaged, respiration is depressed and later root growth is slower, as is acrospire growth. If the damage is not too severe, it can be offset by lengthening germination times or, where allowed, by applying gibberellic acid. Deliberate pressure damage used in conjunction with gibberellic acid has been used to produce malt with reduced malting losses (see below).

Newer steeping schedules can save 2 days processing time, relative to older schedules, by achieving uniformly chitted grains at steep-out. Even chitting is important since without it inhomogeneous malt is made, which will 'take colour' unevenly on the kiln, have an irregular appearance and will perform badly in the brewhouse. Grains that are slow to germinate do not 'catch up', and so produce less extract after malting (Table 14.10).

14.10 The effects of varying germination conditions in malting

The changes undergone by germinating grain are influenced by the steeping procedure, the quality of the grain, how it is handled, the duration of the germination period, whether or not additives are used, the temperatures used and the moisture content of the grain. In this section, the effects of varying temperatures and moisture contents will be discussed. The effects of using carbon dioxide rests, gibberellic acid, potassium bromate, resteeping and other less usual procedures will be considered below.

In the UK in the 19th century, and earlier, grain might remain on the malting floor for up to 3 weeks. By the 1930s the flooring period was typically 8–10 days at 12–16 °C (53.6–60.8 °F) when well-modified, pale malts were being prepared, or 6–8 days when less well-modified, traditional lager malts were being made. In the period 1980–95, using 'broken' (air-rest) steeping schedules, which hastened germination without the use of additives, germination periods in floor-maltings were typically 6 days. Schuster (1962) describing German practice noted germination times of 7–9 days for making pale malts of two-rowed barleys, with temperatures rising from 12 °C to 17–18 °C (53.6 °F to *ca.* 64 °F). Enzyme-rich, six-rowed barleys could be germinated in 5–6 days. Darker malts are often made using higher germination temperatures. In North America, germination temperatures of 16–25 °C (*ca.* 61–77 °F have been used with germination times of 4–6 days (Witt, 1970a; Adamic, 1977). In pneumatic maltings, the temperatures used depend in part upon the cooling capacity of the equipment, since the higher the temperature, and the higher the moisture content of the grain, the faster heat is generated. Unless the plant has adequate cooling capacity and this heat can be removed, the temperature will rise out of control. Varying the

Table 14.11 The influence of germination moisture, using immersion steeping with or without additional spray-steeping, and process time on malt composition[a]

	Germination moisture (%)					
	48	48	48	48	46	43
Steep + germination time (h)	153	177	201	226	225	224
Extract (% d.m.)	79.7	81.1	81.3	81.9	81.6	80.5
Fine–coarse extract difference (%)	5.6	1.9	1.5	1.0	1.0	1.6
Kolbach index (%)	35.9	43.4	45.4	47.8	44.5	39.1
Formol-N (mg/100 ml)	25.4	31.5	33.3	35.0	23.3	19.7
VZ 45 °C (%)	28.1	39.3	42.8	45.1	43.3	36.1
α-Amylase (DU)	34.9	55.8	68.0	76.1	70.1	57.6
Diastatic power (°W-K)	209	247	279	338	307	234
Malting loss (%)	5.1	8.2	10.0	12.8	10.2	7.8

Data of Narziss and Hellich from Narziss (1969).
[a] Moisture after steeping 43% in all cases.

moisture content and the germination temperatures of the grain alters the analyses of the malts obtained. Confusingly, these parameters interact.

Grain steeped to 37% moisture germinates rapidly, but with steep-out moisture contents of less than about 43%, modification is unlikely to be complete however extended the germination period. Conversely, over-steeping grain, particularly grain with residual dormancy, leads to uneven germination. The influences of malting at different moisture contents have been investigated (Shands *et al.*, 1941; Narziss, 1966, 1969, 1987; Weith, 1967; Piendl, 1976; Ballesteros and Piendl, 1977; Seward, 1986; Smart *et al.*, 1993). The longer the steeping period and the higher the moisture content of the barley (in the normal range, 43–48%), the more difficult it is to obtain even germination. This problem is now overcome by using relatively mature barleys and broken or air-rest steeping schedules, with or without final spray-steeping to adjust the moisture content. Once germinated, grain with a high-moisture content grows vigorously, with strong respiration and rootlet growth, relative to grain germinating with a lower moisture content. The malting equipment must be able to cool the grain and dissipate the heat produced. Malting times tend to be shortened; malting 'wet' (at a high moisture content) can be helpful in processing 'difficult' barleys that are slow to modify. The cold water extracts, 45 °C (113 °F) extracts and the levels of soluble nitrogen are elevated by wet malting; the final values for total nitrogen ('protein') are little altered; diastatic power and α-amylase levels are usually increased; fine–coarse extract differences, β-glucan contents, and wort viscosity values are reduced. Malt friability, colour and fermentability are increased. Hot water extract values of 'wet-grown' malts are sometimes lower than those germinated with lower moisture contents, and malting losses are higher, owing to elevated respiration and extra rootlet growth (Table 14.11; Fig. 14.7). In comparative trials, as the moisture content increases and acrospire growth is increased, so various 'indices of modification' improve, but not 'in parallel' (Table 14.12), and DMS formation is

Table 14.12 The effects of germination moisture content on different indices of modification. Values are the means of results obtained using four barley varieties using four different temperature regimes. The germination period was 4 days

Indices of modification	Moisture content (%)				
	41	43	45	47	49
Friability (%)	69	78	83	85	88
Partially under-modified grains (%)	15.3	8.6	5.7	2.9	2.6
Fine–coarse extract different (%)	2.9	1.9	1.3	1.0	1.2
Viscosity of laboratory wort (cP)	1.61	1.52	1.48	1.48	1.48
Kolbach index (%)	39.1	41.9	44.2	47.5	49.7
Malt yield (% d.m.)	91.5	90.6	89.7	88.7	87.1

Data of Smart *et al.* (1993) using *Analytica-EBC* methods of analysis.

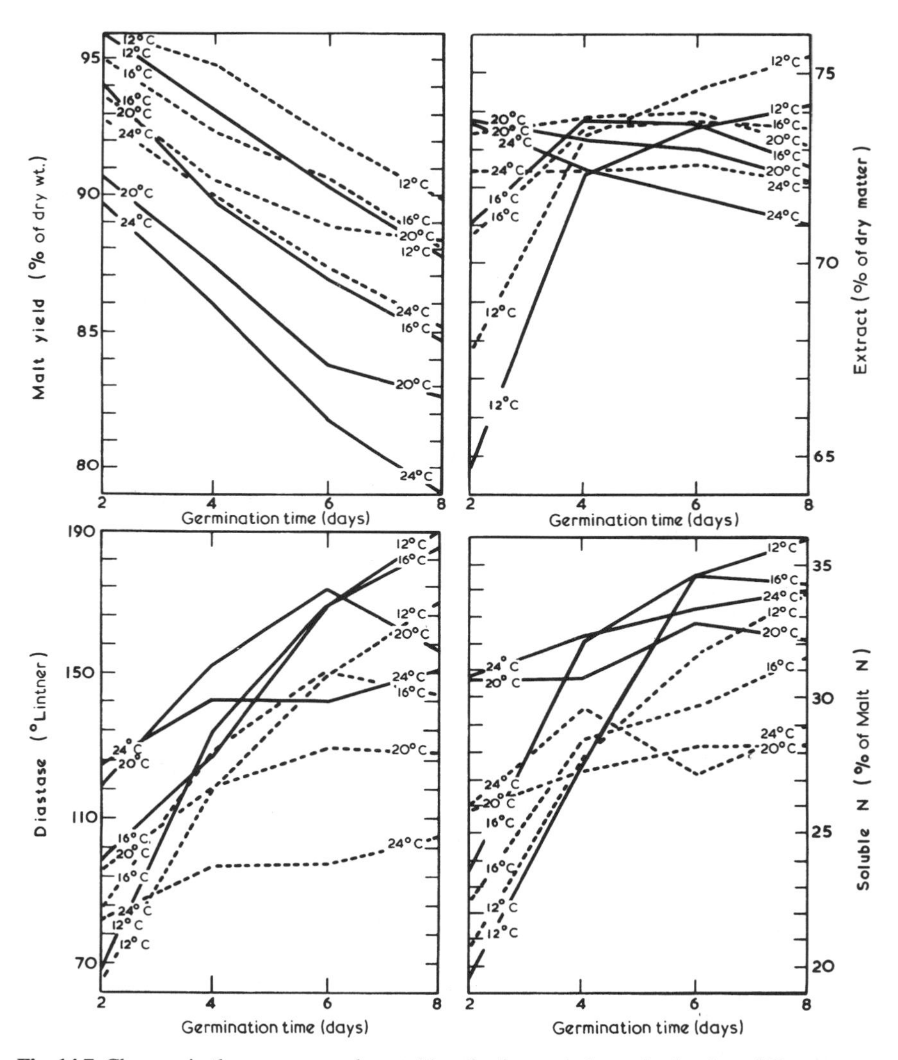

Fig. 14.7 Changes in the average analyses of four barley varieties malted at four different temperatures (12, 16, 20 24 °C; 53.6, 60.8, 68.0, 75.2 °F) and at two different moisture contents, 43% (---) and 49% (—) (After Shands *et al.*, 1941).

increased. Different barley varieties respond differently to some degree when malted using a range of moisture contents.

When trials involving germination at different temperatures are considered, germination times and grain moisture contents must also be taken into account (Fig. 14.7). The influence of temperature on malting has been thoroughly investigated (e.g. Piratzky and Rehberg, 1935; Piratzky, 1936; Shands *et al.*, 1941, 1942; Witt, 1959, 1970a; Weith, 1960, 1967; Pollock, 1962a;

Table 14.13 Influence of germination temperature on the soluble nitrogen fractions of malt

	Germination temperature		Levels of soluble nitrogen (mg N/100 g dm) at germination times (days)			
	°C	°F	0	2	5	9
TSN	13	55.4	188	359	515	526
	16	60.8	188	380	462	459
	19	66.2	188	453	443	449
PSN	13	55.4	113	234	330	314
	16	60.8	113	253	280	270
	19	66.2	113	294	257	262
Formol-N	13	55.4	37	80	138	128
	16	60.8	37	93	112	107
	19	66.2	37	111	105	93

Data of Kolbach and Schild (1939) from Sommer (1980).

Dahlstrom, 1965; Narziss, 1976, 1991a; Ballesteros and Piendl, 1977; Briggs, 1978; Seward, 1986; Oliver, 1996). Since, within limits, grains grow faster at elevated temperatures it seems attractive to malt at the highest possible temperature that the grain will tolerate and so finish the process in the shortest possible time. In practice, this does not work out well. As the temperature increases, there is an increasing tendency for grain to germinate unevenly, heat-output increases and respiration and grain growth are accelerated. In floor-maltings and older pneumatic plant, the grain is liable to grow 'wild' and the temperature rises uncontrollably. There is a greater risk of fungal growth at higher temperatures. Initially enzyme levels, the accumulation of malting losses, extract values and the levels of soluble nitrogen increase faster at elevated temperatures. However, as the germination period goes on, malting losses increase steadily, while the rates of increase in the other characteristics (hot water extract, total soluble nitrogen, etc.) slow down, and their values may plateau and even decline (Fig. 14.7). This is illustrated for some soluble nitrogen fractions in Table 14.13. As the grain germinates so the endosperm progressively modifies and the derived sugars and nitrogenous substances migrate to the embryo. Some are incorporated into the structure of the growing tissues, while some of the sugars are 'lost' as they are converted, by respiration, to carbon dioxide and water. Thus, the dry weight of the grain declines. In contrast, the total amount of nitrogen in the grain **plus** rootlets remains unaltered. Respiratory losses and rootlet growth do not increase in parallel (Fig. 14.3). The total nitrogen (%) in the de-rooted malt depends on the amount that has migrated to the roots and the loss in dry weight. The levels of total nitrogen in the barley and the final malt are often very similar (Figs 14.1 and 14.3). Similarly, the maximum recoverable extract (%) is regulated by the removal of starch and sugars from the endosperm, which is, to an extent, offset by the loss in weight. These 'balances' are different when grains are grown at different temperatures and so

Table 14.14 The average results of malting four barley varieties (Oderbrucker, Wisconsin Barbless, Peatland and Chevron) using a range of different temperatures

Malting quality factor	Malting times, in days, at various temperatures (°C)														
	6 at 12	6 at 16	6 at 20	4 at 12 2 at 16	2 at 12 4 at 16	4 at 12 2 at 20	2 at 12 4 at 20	4 at 16 2 at 12	2 at 16 4 at 12	4 at 16 2 at 20	2 at 16 4 at 20	4 at 20 2 at 12	2 at 20 4 at 12	4 at 20 2 at 16	2 at 20 4 at 16
Steep and respiration loss (%)	3.9	5.2	6.7	3.6	4.3	5.0	6.3	4.6	4.2	5.7	5.9	5.7	5.8	6.6	5.5
Malt yield (%)	93.2	90.5	88.2	92.8	91.4	91.0	88.8	92.3	92.6	89.4	89.2	89.9	90.0	89.6	90.7
Extract (dry basis %)	74.3	73.8	72.9	74.1	73.9	73.9	72.7	74.3	74.5	73.5	73.2	73.8	74.3	73.4	73.5
Colour (Lovibond)	1.3	1.5	1.6	1.3	1.4	1.4	1.7	1.6	1.5	1.3	1.5	1.4	1.4	1.6	1.6
Diastase (°L)	162	172	164	159	163	165	163	182	179	165	159	154	173	166	169
Conversion time (min)	8.3	6.4	7.5	7.6	8.0	8.5	5.5	5.8	5.0	7.9	8.5	10.9	8.9	5.8	4.8
Ratio of wort N to malt N (%)	34.6	34.4	34.7	33.9	33.3	33.3	31.9	35.8	36.3	32.5	31.5	33.1	36.4	32.1	35.3
Ratio of formol-N to malt N (%)	13.4	13.4	11.6	13.4	13.6	12.4	11.5	14.2	14.4	12.0	11.9	12.8	14.5	12.1	13.1

After Shands *et al.* (1942).

Table 14.15 The effects of malting at two fixed temperatures and reducing the temperature during germination in barley samples germinated for 7 days at two moisture contents

	Max. moisture content 48%			Max. moisture conent 46%		
	17 °C	13 °C	17/13 °C	17 °C	13 °C	17/13 °C
Extract (% d.m.)	81.2	81.8	82.1	80.9	82.0	81.6
Fine–coarse extract difference (%)	0.8	1.0	0.9	0.8	1.6	0.8
Kolbach index (%)	44.1	46.3	46.8	41.9	44.9	44.8
Formol-N (mg/100 ml)	25.4	23.1	23.6	22.8	25.4	21.9
VZ 45 °C (%)	40.1	46.0	43.0	38.1	38.4	38.6
α-Amylase (DU)	73.1	69.2	74.2	68.5	68.5	76.5
Diastatic power (°W-K)	309	319	303	307	333	340
Fermentability (%)	84.4	82.1	84.7	81.7	80.5	81.0
Malting loss (% d.m.)	13.1	10.2	10.4	11.2	8.0	9.1

Data from Narziss (1976).

malts prepared at different temperatures from one lot of grain have different compositions. Generalizations are dangerous (e.g. because of varietal differences or the use of different moisture contents) but usually as malting temperatures increase the extracts of the final malts decline, the total nitrogen is altered relatively little, the 45 °C extract, soluble nitrogen levels, free amino nitrogen levels and wort viscosities are reduced; the saccharification time is increased, α-amylase levels and diastatic power tend to be reduced and malting losses are increased, so TCW values are reduced (Tables 14.14 and 14.15). Wort viscosity is least and endosperm modification is best in grain malted at 19 °C (66.2 °F) (Oliver, 1996).

In practice, malting is not carried out at a single temperature. Attempts to malt at temperatures consistently over 20 °C (68 °F) have usually failed. Under production conditions, temperatures are altered in attempts to regulate malt composition. The results of systematic trials to investigate the effects of various temperature combinations are given in Tables 14.14 and 14.15. Table 14.14 shows that maximum malt yield (minimum malting loss) occurred with 6 days of germination at 12 °C (53.6 °F), which also gave the the highest soluble nitrogen/total nitrogen ratio. The lowest soluble nitrogen/total nitrogen ratios were obtained by germinating for 2 days at 16 °C (60.8 °F) and 4 days at 20 °C (68 °F). α-Amylase accumulates slowly but steadily at lower temperatures (e.g. 13–14 °C (55.4–57.2 °F)) so that prolonged germination at low temperatures gives an enzyme-rich malt. The enzyme level initially rises faster at higher temperatures (e.g. 17–18 °C (62.6–64.4 °F)) but then the rate declines so that finally the enzyme content is below that obtained by using the lower temperature. By beginning germination warm and then reducing the grain temperature, malt rich in α-amylase can be obtained with a shortened germination period and in good yield (Dickson *et al.*, 1947; Narziss, 1975). In the example given (Table 14.15), the effect is more striking in the grain malted at 46% rather than

48% moisture content. However, altering the temperatures of large masses of grain in a short time may be difficult in commercial maltings, and it is considered that malting may be checked if a large, sudden change in temperature gives the grain a 'thermal shock'.

Sometimes temperatures are allowed to rise towards the end of the germination period to reduce the levels of soluble nitrogen (Table 14.14). Holding steeped grain at 20, 25 or 30 °C (68, 77 or 86 °F) for the first 16 h of a 5-day germination period dramatically reduced the levels of soluble nitrogen in the final malt (Seward, 1986). However, increases to 20 or 25 °C (68 or 77 °F) or even more (from 16 °C (60.8 °F)) at various stages of germination generally had undesirable effects on the qualities of malts (Baxter and O'Farrell, 1980; O'Farrell *et al.*, 1981). By altering grain germination temperatures, moisture contents and malting times, it is possible to vary malt characteristics and alter the malting losses sustained (Narziss, 1975) (Table 14.16). In a particular instance, when floor-malts and pneumatic malts were compared, in each case using gibberellic acid with a 4-day germination period and with applications of sodium bromate in the floor-maltings, it was found that the pneumatic malts contained substantially more β-glucan than the floor-malts. In the floor-malting, β-glucanase activity increased faster and β-glucan degradation was more rapid than during pneumatic malting. Whereas in the pneumatic malting the grain-bed temperature was 15–17.5 °C (59–63.5 °F) on the floors it increased from 15 °C (59 °F) to 25 °C (77 °F) over the 4-day germination period. By increasing

Table 14.16 Comparison of malts produced by different germination schemes (barley) (Carina 1974)

	Malt analyses				
Germination schedules					
Temperatures (°C)[a]	12/15/17	17/13	17/13	18/12	18/12[b]
Moisture (%)[c]	44	38/43/47	38/43/48.5	38/43/48.5	38/50 Resteep
Steep + germination (days)	9	7	6	6	6
Extract (% dry matter)	82.5	82.8	82.4	82.7	82.8
Fine–coarse difference (%)	1.5	1.3	1.5	1.5	1.0
Viscosity (cP)	1.55	1.53	1.57	1.57	1.60
Protein (% dry matter)	10.2	10.2	10.3	10.3	10.4
Kolbach index (%)	40.2	44.7	42.4	44.0	44.2
α-Amylase (ASBC)[d]	55	64.2	57.1	60.9	60.3
Colour (EBC units)	3.0	3.0	2.8	2.8	2.5
Colour of boiled wort (EBC units)	5.0	5.5	4.8	5.2	5.0
Malting loss (%)	9.2	8.5	7.7	7.7	6.3

[a] Temperatures at successive stages of germination.
[b] Resteeping process.
[c] Moisture levels at successive stages of germination.
[d] Method of the ASBC (1992) (DU).
Data of Narziss (1975).

the processing temperature in the pneumatic plant, the Kolbach index of the malt was depressed, β-glucanase activity was greatly enhanced and the β-glucan contents of the malts were reduced (Anderson *et al.*, 1989). There are other ways in which the course of modification and the malting losses can be regulated. Many experimental approaches have been tried, for example physical treatments, the use of additives or carbon dioxide rests. For various reasons, many of these are not used and others are used to only a limited extent. These techniques will be discussed next. The use of antiseptics (anti-microbial agents) in steeping to control microbes has already been mentioned.

14.11 The use of gibberellic acid in malting

A range of additives and special techniques has been used in malting to reduce dormancy (warm storage of grain, air-rest steeping, hydrogen peroxide, gibberellic acid), to reduce the microbial population (anti-microbial agents in the steep liquor), to hasten germination and modification (gibberellic acid, hydrogen peroxide), to reduce the levels of soluble polyphenols (formaldehyde), to augment the extract (glucose), to reduce malting losses (various chemical additives, carbon dioxide rests; resteeping, 'battering') and to alter the composition of the malt. Common objectives are to enhance the levels of enzymes (e.g. using air-rests, temperature changes and gibberellic acid) and to regulate the soluble nitrogen ratio (e.g. potassium bromate, salts of octanoic acid, ammonium persulphate). The special treatments have sometimes only been used experimentally, to a limited extent or for a short period. Others have been, or are, used more widely.

Probably the most widely used additive (where local legislation and/or customers allow) is gibberellic acid. Mixtures of gibberellins were shown to accelerate malting in the 1940s in Japan, but the outstanding effects of gibberellic acid were not widely recognized until the publications of Sandegren and Beling (1958, 1959). Gibberellic acid (GA_3) and the closely related GA_1 are natural plant hormones that occur in barley (Chapter 4). Pure gibberellic acid is prepared on the industrial scale from culture filtrates of the fungus *Gibberella fujikuroi* (syn. *Fusarium moniliforme*). Gibberellic acid added to malting grain augments the grain's own supplies and enhances the grain's metabolism, shoot (but not necessarily root) growth and the rate of endosperm modification, so malting is accelerated. Gibberellic acid enters the grain in the region of the micropyle and spreads along the grain, inducing enzyme formation and release from the aleurone layer (Briggs, 1978; Briggs and MacDonald, 1983a,b). The production of enzymes that catalyse modification, and the rate of modification itself, are enhanced relatively more than embryo growth, so under the correct conditions malting is completed in a shortened time, with smaller malting losses. Malting barley samples respond differently to a given dose of gibberellic acid, depending on the variety and quality of the grain, the temperature and the grain moisture content. Gibberellic acid is particularly beneficial when malting 'difficult', poor quality barleys or samples in which the grains have been damaged or are slightly dormant. Gibberellic acid is, or has been, used in combination with many other treatments, as will be outlined below. Additions of gibberellic acid (i) tend to overcome dormancy; (ii) accelerate the malting process so that a given degree of modification is achieved in a shortened time, so accelerating malt production; (iii) enhance the respiration rate and heat output; (iv) have variable effects on rootlet growth, which may be little altered or slightly enhanced or reduced; (v) stimulate acrospire growth; (vi) stimulate the production of some enzymes including some proteases, peptidases, α-amylase (hence diastatic power also), some other carbohydrases, nucleases and phosphatases, but not others (β-glucanase activity, for example, being little enhanced); (vii) in a fixed, short malting time enhance the cold water extract, the hot water extract, the soluble nitrogen level, the 45 °C Hartong extract, the fermentability of the wort, and malt friability and reduce the wort viscosity, the 70 °C mash wort viscosity and the fine–coarse extract difference. It should be noted that enzymes do not all increase to the same proportional extent in response to a dose of gibberellic acid and nor do all other malt parameters alter 'in parallel'. Because large proportions of the enzymes involved in modification are produced in the aleurone layer and are formed and/or are released in response to gibberellins and because most malting losses are consequent upon embryo growth and metabolism, many malting processes have been proposed in which the embryo is damaged or killed, to reduce malting losses, while modification is induced by added gibberellic acid.

Table 14.17 The malting performance of two different barley cultivars, with different nitrogen contents, malted with different applications of gibberellic acid. The results shown are the hot water extracts (1°/kg) with SNR (%) in parentheses

Total nitrogen (% d.b.)	Gibberellic acid (mg/kg) to Triumph (good malting quality)			Gibberellic acid (mg/kg) to Atem (moderate malting quality)			
	0	0.1	0.2	0	0.1	0.2	0.5
1.45	307	308	313	300	306	–	–
	(40)	(43)	(44)	(35)	(38)	–	–
1.60	305	306	308	296	302	303	306
	(38)	(39)	(41)	(32)	(36)	(37)	(40)
1.75	302	305	307	288	294	297	302
	(35)	(40)	(40)	(27)	(31)	(34)	(37)
1.90	298	299	303	284	291	–	208
	(32)	(35)	(37)	(25)	(28)	–	(34)

Data of Seward (1990).

Responses to added gibberellic acid are highly dose dependent and vary between barley varieties and different samples of one variety (Table 14.17; Fig. 14.8). Within limits, some responses are proportional to the logarithm of the dose applied (Briggs, 1963b; Palmer and Bathgate, 1976) (Fig. 14.8). The manner in which gibberellic acid (which is always used in aqueous solution) is applied to the grain is important. The hormone may be applied in the steep, in which case application in the last immersion is the most efficient, even so only 14–26% of the gibberellic acid is taken up by the grain, the rest going to waste with the effluent steep liquor (Ault, 1961). It is much more efficient to spray the grain with a solution of gibberellic acid as it is conveyed from the dry-cast steep to the germination unit, particularly if the grain has already chitted. Gibberellic acid may also be applied to grain at the onset of an air-rest period between immersion- or spray-steeps. Hydrogen peroxide can be incorporated into the solution of gibberellic acid sprayed onto grain, often with a useful increase in the rate of chitting (Briggs, 1987a). With suitable doses of gibberellic acid maximal extracts are obtained 1–3 process days sooner than would otherwise be the case and in greater yield than is obtained from undosed controls. If germination is terminated (by kilning) in a shorter time, at the correct stage of modification, malting losses are not enhanced and may even be reduced by up to 4%. This is because the hormone accelerates the rate of modification relatively more than the rate of increase in malting losses. With a carefully chosen dose of the hormone applied as a sprayed solution (usually in the range 0.005–0.25 mg GA_3/kg 'original' barley) all the advantages mentioned

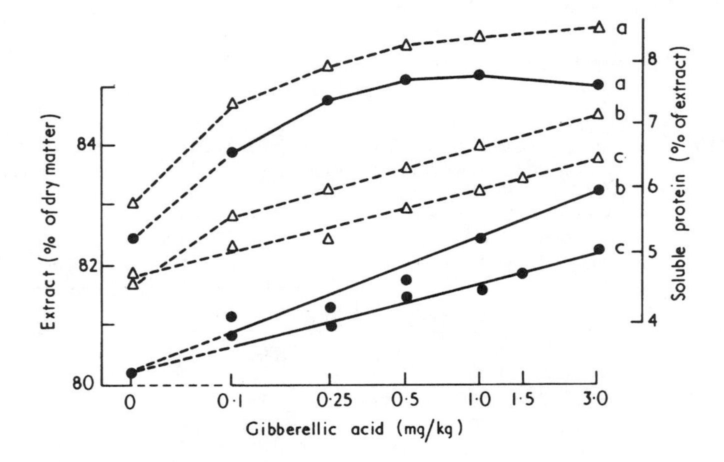

Fig. 14.8 Increased values of extract (●) and soluble nitrogen (△) found after malting three varieties of barley for a fixed period after dosing with various amounts of gibberellic acid. Varieties were (a) Beka, (b) Carlsberg and (c) Aurore. The gibberellin dosage axis is on a logarithmic scale. (Data of Secobrah, cited by Briggs, 1963b).

will be obtained, with the production of a malt that is at least equal, and in many ways may be superior, to traditional malts; indeed the malts cannot be reliably distinguished from malts made without gibberellic acid that have been germinated for longer periods. In many of the early trials excessive doses of gibberellic acid were employed (e.g. 2–3 mg GA_3/kg barley) (Sandegren and Beling, 1959). The results were unsatisfactory. Besides the benefits of shortened germination times and enhanced extracts, heat output was so increased that the grain temperature was hard to control, over-modification occurred with excessively high levels of the soluble nitrogen fractions (TSN, FAN) and highly fermentable worts were obtained. The levels of pre-formed reducing sugars and amino acids were highly elevated in the green malt, leading to excessive colour development from melanoidin formation during kilning. The worts also were very dark and, as well as being unusually fermentable (having high sugar:dextrin ratios), had greatly increased amino acid:protein ratios leading to excessive yeast growth. None of these adverse effects need occur if lower doses of gibberellic acid are used and other malting conditions are carefully selected.

14.12 Physical treatments of grains

Various techniques have been proposed to alter the responsiveness of barley to applications of gibberellic acid. For example, it was suggested that barley steeped to a very low moisture content (e.g. 40%) be induced to modify by a substantial application of gibberellic acid (0.25 mg GA_3/kg barley). This technique is not used, probably because modification is irregular and incomplete. However, barley has been malted at a lower moisture content following a physical 'squeezing' treatment. The selectively permeable layers of the grain, specifically the testa, limit or prevent the penetration of water and gibberellic acid, except at the micropyle (Smith and Briggs, 1979, 1980b; Briggs and MacDonald, 1983b; Briggs, 1987b). Both substances enter chitted grain faster than unchitted grain. By lightly crushing dry grain by passage through a roller mill so that in 95% of the grains small splits occurred through the pericarp and testa and into the endosperm, the impermeable layers were breached and the uptake of water or a solution of gibberellic acid (0.1–1 mg GA_3/kg barley) was greatly accelerated (Sparrow, 1965). The time course of enzyme development in such split, GA_3-dosed grain was substantially different to that occurring in whole grain, but modification was unusually rapid and was complete in about two-thirds of the time taken by intact grain that had received an equal dose of GA_3. Split grains could not be used on a commercial scale because of the risk of grains breaking up, causing handling problems, and because leakage of nutrients from the splits supported heavy microbial infestations. Other processes were tested, including 'beating' grain during steeping with an impeller rotating in a mesh 'cage' (Block and Morgan, 1967). The agitation evidently bruised the embryo and reduced grain growth. Heavily dosed with gibberellic acid (2 mg GA_3/kg barley) such barley gave malt in 96% yield. The process was not adopted for commercial use. Trials in which grains were squeezed with the intention of 'loosening' the structure of the starchy endosperm, without deliberately breaking the surface layers, gave variable results (Githua and Barrell, 1974; Powell and Barrell, 1974). Subsequent studies showed the way to malt squeezed grain with applications of gibberellic acid, and this process has been in limited industrial use (Pool, 1978a,b; Pollock and Pool, 1979, 1987; Northam, 1985; Collier, 1986; Maule *et al.*, 1987). The conditions used in squeeze-malting could be varied widely, depending on whether GA_3 was used, on the barley sample and on the product required. Typically the grain was steeped to a chosen low moisture content in the range 34–37% (e.g. using a 14 h immersion at 16 °C (60.8 °F)) and after a 6 h drain with downward ventilation, during which time the surface moisture was taken up, the grain was squeezed by passage through a roller mill. It was desirable to change the steep liquor at least once to clean the grain. The chilled, lightly spiral-patterned steel rolls (460 mm (18.1 in) diameter, 1525 mm (60.04 in) long, 180 r.p.m.), which were scraped to dislodge adherent grain, were typically set with a gap of 1.8 mm (0.071 in), (1.7–2.0 mm) (0.067–0.787 in). Grain was processed at 20 t/h. Squeezing pressed water from beneath the husk and slightly flattened the grains. These resumed their original shape after about 24 h. The squeezing may have bruised the grain, loosening the internal structure and creating small splits in the pericarp and testa. After a sprayed application of a solution of gibberellic acid (0.25–0.75 mg GA_3/kg barley) the grain was germinated. The low moisture content of the grain (about 30% less than in conventional malting) limited heat generation and the production of rootlets. Because of this, the germination compartments could be more heavily loaded, increasing their capacities, and the grain required less turning. Usually the germination period was short (4 days), extract was slightly enhanced and the levels of soluble nitrogen (TSN and FAN) were elevated unless a bromate was used. The cost of kilning this squeeze-malt was low because there was less water to evaporate. The splits in the surface layers may have facilitated drying. Thus the products, which had acceptable analyses, were made more cheaply, the capacity of the plant was effectively increased, the processing time was reduced, less NDMA was produced on kilning, less DMS precursor was formed (Chapter 4) and there were less rootlets to dispose of. The process could be used with or without applications of gibberellic acid or other additives (Table 14.18). However, if gibberellic acid was not employed, the grain had to be steeped to a higher moisture content to ensure adequate modification, although at 41% this was still less than is used in conventional malting. Squeezing improved the malting performance of dormant grains and tended to enhance the levels of soluble nitrogen

Table 14.18 Influence of squeezing (rolling) on the composition of malts made with barley steeped to a moisture content of 41%, with and without applications of gibberellic acid (GA_3, 0.33 mg/kg)

	Not rolled		Rolled	
	$-GA_3$	$+GA_3$	$-GA_3$	$+GA_3$
Hot water extract (1°/kg)	294	308	301	315
SNR (%)	33	45	34	54
Friability (%)	59	79	70	85
Homogeneity (%)	78	94	87	97

and some enzymes, including those contributing to diastatic power and β-glucanase.

Barley experimentally decorticated (husk and pericarp removed) with sulphuric acid malts extremely rapidly (Sparrow, 1964; Smith and Briggs, 1979). Chemical decortication (dehusking) is not practicable on a commercial scale, but malting mechanically dehusked barley has been evaluated (Macey *et al.*, 1969; Collier, 1973). Grain was mechanically dehusked in machines that relied on mechanical impacts and grain-to-grain impacts and abrasion. Conditions usually chosen were such that 7–9% of the corn's weight was separated as 'hulls' (mainly husk), with minimal reductions in germination or corn breakage. Drier grain was easier to treat. The dehusked grain germinated rapidly and, because of bruising, less rootlets were formed. Dormancy was eliminated and the rate of water uptake in steeping was enhanced. The grain could be malted at lower moisture contents, leading to reduced kilning costs. With up to 3% husk removal, the grain could be malted without additions of gibberellic acid. However, the process was not adopted for commercial use for a number of reasons. It was difficult to treat the grains evenly and husk removal could be regarded as a weight loss. The dehusked grain 'packed down' in the steep, during germination and in the kiln, obstructing airflows. The grains became sticky, adding to the handling problems. Despite enhanced extracts (expected because of the removal of non-extract yielding husk) the de-hulled malts could only be used as a small percentage of the mash-tun grist, unless the separated husk was mixed back with the product, eliminating the advantage of elevated extract per unit weight. Husk removal was too vigorous a treatment, the damaged grains tended to become mouldy. The process proved to be uneconomic (Macey, 1974; Witt, 1970a).

A much milder process, which has achieved some commercial success, is abrasion. In the original laboratory investigations, grains were abraded by being 'brushed' with rotating wire-brushes working in cylindrical metal containers (Palmer, 1969, 1974; Palmer *et al.*, 1972). Commercial abraders rely on grain impaction or friction to alter the grain's malting characteristics (Chapter 9; Northam and Button, 1973; Halford *et al.*, 1976; Brookes, 1980,

1981b). Originally it was proposed that abrasion to the apex of the grain permitted the entry of externally supplied gibberellic acid and water so that two-way modification occurred, spreading from both the base and the apex of each grain. However this is clearly not the mechanism involved, at least in grains treated in commercial abraders and some laboratory devices, since two-way modification can rarely be detected (Smith and Briggs, 1979; Briggs and MacDonald, 1983a,b; Briggs, 1987b). It seems that gibberellic acid still enters treated grains predominantly in the region of the embryo, and abrasion causes limited damage to embryos and creates faults in the surface layers that permit more atmospheric oxygen to reach the living tissues. Lightly abraded grains malt slightly faster and enzyme levels may increase more than in untreated controls in the absence of added gibberellic acid. Heavy abrasion evidently damages grain embryos and the malting performance of heavily treated grain is depressed unless gibberellic acid is applied. Typically 0.5–1% of the grain is lost as dust when barley is treated in particular commercial abraders. When gibberellic acid is applied, enzyme production and malting occur rapidly in abraded grains. Modification extends from the base of the grain in the usual way but occurs more rapidly than in unabraded control grains. Maximum hot water extract values are reached sooner and the levels of soluble nitrogen-containing materials are elevated. To offset this, gibberellic acid is often used in conjunction with sodium or potassium bromate (see below and Tables 14.19 and 14.20). Typically, germination time is reduced by about 1 day, from 4 to 3 days in some cases (Brookes, 1980, 1981), by using abraded grain. Abrasion sometimes accelerates the rate of water uptake when air-rest steeping is used. Up to 25% reduction in steeping times have been noted. Presumably this is because the rate of chitting is accelerated since in continuous immersions, in which chitting does not occur, the rate of water uptake is not altered by prior abrasion. Abrasion sometimes reduces dormancy in barley. It was

Table 14.19 The influence of abrasion on malting Maris Otter barley (1978 crop) having a nitrogen content of 1.55% and a TCW of 34.1 g. All samples were treater with gibberellic acid (0.25 mg/kg barley)

Steeping + germination time (days)	Potassium bromate (mg/kg)	Abrasion (%)	Extract (l°/kg d.m.)	Colour (EBC units)	TSN (%)	TN (%)	SNR (%)	Fine–coarse extract difference (l°/kg)
7	0	Control[a]	304	5.0	0.57	1.52	38	6
7	25	90	311	8.0	0.64	1.54	42	2
6	25	84	307	5.7	0.60	1.54	39	3
6	50	75	305	4.6	0.56	1.53	37	3
6	50	72	309	5.0	0.58	1.54	38	2

[a] Only 'natural' abrasion; or 2%.
After Brookes (1980).

Table 14.20 Hot water extracts of malts prepared in various ways from three varieties of barley (Proctor, Abacus and Julia).

	Hot water extract		
	Proctor	Abacus	Julia
Total nitrogen of barley (%)	1.72	1.75	1.78
Malt preparation			
Control (6 days of germination)	100.2 (297)	98.5 (292)	98.2 (291)
Abraded (6 days of germination)	100.9 (299)	99.9 (296)	99.5 (295)
Plus GA_3 + bromate (4 days of germination)[b]	101.6 (301)	101.6 (301)	100.8 (299)
Abraded + GA_3 + bromate (3.5 days germination)[b]	101.8 (301)	101.4 (300)	101.1 (299.5)

From Macey *et al.* (1975).
[a] Hot water extract as lb/Qr (approximate l°/kg at 20 °C).
[b] GA_3 (0.25 mg/kg barley) and potassium bromate (100 mg/kg barley) applied in solution at steep-out.

noted many years ago that 'running barley round' in conveyors, when 'natural abrasion' occurs, sometimes reduces dormancy.

There are varietal differences in responses to abrasion and the moisture content of the grain influences the ease of abrasion; as as result, in practice, treatments must be optimized for each class of grain and the intensity of the treatments must be controlled to limit dehusking and grain breakage. Abrasion seems particularly beneficial when applied to poor quality barleys. Grains are often unevenly treated and the final malts may appear a little variable, which purchasers may not like. This is possibly the reason why abrasion seems mostly to be used by brewer-maltsters. Passing grain through two abrading machines, working in tandem, gives more even results (Northam and Button, 1973). There is a tendency for abraded grain to be fragile and over-modified. Malting and kilning conditions need to be adjusted. Abraded grains can be malted at lower moisture contents than untreated grains, allowing the use of shorter steeping times and/or lower steeping temperatures and reducing the costs of kilning the green malt, since there is less water to evaporate.

14.13 The use of bromates in malting

After gibberellic acid, the soluble salts sodium bromate and potassium bromate have probably been the most widely used additives in malting, where customers and local regulations permit their use (Macey and Stowell,

Table 14.21 Analysis of malts prepared with additions of potassium bromate in the second steep

	Potassium bromate (mg/kg barley)			
	0	500	1000	2000
Hot water extract				
l°/kg at 20 °C	286.1	287.1	285.6	283.6
lb/Qr	96.2	96.9	96.4	95.6
Cold water extract (%)	17.3	17.8	16.8	16.6
PSN (%)	0.48	0.52	0.46	0.40
Malting loss (%)	9.1	6.6	5.0	4.1

Macey and Stowell (1961).

1957a, 1961). Their use seems to have declined in recent years. The two salts seem to be equally effective. Originally used as 'flour improvers' in baking, these salts reduce respiration (and hence heat output) and rootlet growth when they are applied to malting grains and so overall malting losses are reduced. The lower heat output during germination is beneficial in older maltings, since the tendency of the germinating grains to warm uncontrollably and bolt is reduced. The necessary refrigeration capacity in pneumatic maltings is also reduced by the use of bromates. The rootlets of treated grains are characteristically shortened and twisted and are thicker relative to untreated controls. Because rootlet production is reduced by bromates, the green malt from a given amount of barley occupies a smaller volume and so the capacity of germination plant is effectively increased. Also, at particular doses and at stages of germination, bromates reduce the levels of soluble nitrogen produced by malts. Bromate ions inhibit many of the proteolytic enzymes of green malt. The salts penetrate grains within about 4 h, gaining entry mainly at the embryo end, but also over the whole surface (Macey and Stowell, 1961; Brookes and Martin, 1974). During germination, the bromate ions within the grains are progressively reduced to bromide, and little or no bromate survives in the finished malt (Brookes and Martin, 1974). None reaches finished beer. Varying the doses of bromates induces markedly different effects on malt extract yields, malting losses and levels of soluble nitrogen, whether or not gibberellic acid is also used (Tables 14.21 and 14.22; Fig. 14.9). Doses of potassium bromate applied in the last steep water (e.g. 500 mg of $KBrO_3$) need to be about five times higher to achieve the same result as an application (e.g. of 100 mg $KBrO_3$/kg barley) sprayed onto the grain in solution at steep-out. Therefore, as with gibberellic acid, it is most economical to apply bromates at the end of steeping and a bromate and gibberellic acid may be applied together in solution. Applying bromates during the germination period reduces levels of soluble nitrogen in the malts produced (Kerslake, 1960), but it is most effective to apply them at the start of the germination period (Dudley and Briggs, 1977). The ability of bromate to reduce the heat output and levels of soluble nitrogen in

Table 14.22 Analyses of malts prepared with various additions of gibberellic acid and potassium bromate sprinkled onto the grain, in solution, at steep-out

	No additions	Gibberellic acid (0.25 mg/kg barley) plus potassium bromate (mg/kg barley)			
	0	125	188	250	375
Hot water extract					
l°/kg at 20 °C	302.9	307.1	307.4	306.1	308.1
lb/Qr	102.4	103.9	104.0	103.5	104.2
Cold water extract (%)	17.0	19.9	19.1	18.7	18.1
Diastatic power (°L)	58	58	58	52	54
TN (%)	1.55	1.54	1.52	1.50	1.51
PSN (%)	0.56	0.61	0.58	0.53	0.50
Index of modification:					
(PSN/TN) × 100	36	40	38	35	33
Malting loss (% d. wt)	7.2	5.3	5.3	4.6	4.5

Macey and Stowell (1961).

germinating grain make it attractive to use in combination with gibberellic acid to offset the possible adverse effects of this latter additive. Indeed, by varying the gibberellic acid:potassium bromate ratio, a variety of malt types with varying hot water extracts and soluble nitrogen ratios can be produced at the expense of a range of malting losses (Macey and Stowell, 1961). There are no immutable relationships between these analyses and other malt characteristics. Consequently, the various 'indices of modification', established using traditional malting methods, frequently disagree when used to assess the value of newer types of malt. Levels of soluble nitrogen in germinating grain vary with germination time, the dose of bromate used, the germination conditions and whether or not gibberellic acid has also been applied (Tables 14.19–14.22; Fig. 14.9). Thus, in a trial made without the use of gibberellic acid and for a fixed germination period, increasing doses of potassium bromate increasingly depressed malting losses, caused an increase and then a decline in the hot water extract, the cold water extract and the level of soluble nitrogen (Table 14.21). The different levels of cold water extract represent a balance between the pre-formed sugars 'saved' by the depression of malting losses caused by depressing embryo growth and respiration and a slowing down in their rate of formation owing to a reduced rate of endosperm degradation. At very high doses, applications of bromates may reduce the diastatic power of malts as well as their proteolytic activity. During mashing, bromate-treated malts release reduced quantities of proline into the wort. As yeasts do not use proline under anaerobic conditions, less nitrogenous material remains in the finished beer. In commercial malting, but not in micromalting in which grain temperatures are completely controlled (and hence the respiration rate is limited), a correct dose of a bromate enhances the hot water extract (Fig. 14.9). Increasing doses of potassium bromate, applied with a fixed dose of gibberellic acid,

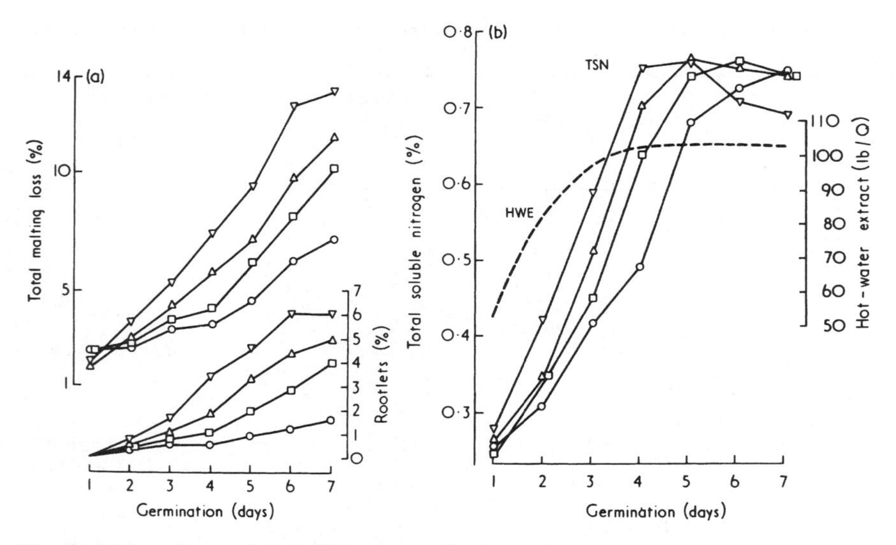

Fig. 14.9 The effects of four different applications of potassium bromate (▽, none; △, 75 mg/kg barley; □, 125 mg/kg; ○, 250 mg/kg) applied at steep-out with gibberellic acid (0.25 mg/kg fresh grain) during subsequent micromalting. (a) Rootlet production and total malting loss; (b) HWE and TSN. The applications of potassium bromate had no detectable effects on the HWE in these micromalting experiments (Ducley and Briggs, 1977).

depress the build up of malting losses and rootlet production to relatively different extents, delay the rises in soluble nitrogen and delay the attainment of the maximum level of soluble nitrogen (Fig. 14.9). Bromates depress the formation of SMM during malting, and so their use is avoided in the preparation of high-DMS content lager malts. The depression of rootlet formation by bromates probably leads to the formation of lesser amounts of hordenine and other amines. This explains the reduced formation of unwanted nitrosamines, e.g. NDMA, during kilning.

14.14 The use of other additives during malting

Many chemicals have been tested for their effects on malting grain. It is probable that few, if any, of these substances are used now, but the findings are of considerable interest.

Ammonium persulphate has been tested, at dose rates of about 1 g/kg barley, as a possible replacement for bromates (Wainwright *et al.*, 1986). This substance is used as a flour improver in baking. In water it decomposes to ammonium hydrogen sulphate and ozonized oxygen. Ammonium persulphate applied to malting grain at the start of germination reduced malting losses, rootlet production and sometimes malt friability. Sometimes it enhanced extract yields. However, in contrast to bromates, ammonium persulphate was without effect on the soluble nitrogen ratio.

Soluble salts of short-chain, saturated fatty acids (C_6–C_{10}) have also been tested for their abilities to replace bromates in malting (Anderson *et al.*, 1979). These substances, at appropriate doses, control some of the unwanted effects of gibberellic acid, depressing malting losses, TSN and FAN and colour formation on the kiln. They do not significantly depress DMS formation. The C_8 and C_9 acids are the most effective at reducing TSN, and the C_9 and C_{10} acids are best at reducing malting losses. Using applications of sodium octanoate at steep out, doses of 50–100 mg/kg barley began to reduce the levels of TSN and FAN found in the final malt. Doses of 150–200 mg/kg barley reduced levels of soluble nitrogenous substances and began to reduce malting losses. At a dose of 350 mg/kg, the fine–coarse extract difference increased, while at doses of 400–500 mg/kg extracts and malt modification were reduced. The rootlets of green malt, treated with octanoic acid, were long and thin, making the malt difficult to turn. Beers prepared from octanoate-treated malts were normal.

Other chemical and physical treatments have been sought for causing selective, limited damage to grain embryos and so reducing malting losses. Often excessive damage reduces or prevents the ability of the grain to malt. If gibberellic acid is not applied then embryo-damaging treatments must not be applied until after the embryos have germinated and released the endogenous gibberellins that induce enzyme formation and release of enzymes from the aleurone layer. In older floor-malting systems, this occurred after 3–4 days of germination, but in modern malting systems, using air-rest steeping, it must occur much sooner. However, if gibberellic acid is used the survival of the hormone-generating capacity of the embryo is largely irrelevant and growth-regulating treatments may be applied in the steep or at steep-out. The doses used, the timing of the applications of chemicals and the timing and intensity of physical treatments are critical.

Chemicals or preparations that have been tested for their abilities to reduce malting losses, usually on the experimental scale only, include concentrated steep liquor, 2(3)-benzoxazalone, phthalide, indole-3-acetic acid, naphthalene acetic acid (NAA), abscisic acid (ABA), lactic acid, acetic acid, salts of various fatty acids, 2,4-dichlorophenoxyacetic acid (2,4-D), nitric acid, urea nitrate, sulphuric acid, phosphoric acid and acid phosphate salts, perchloric acid, sulphur dioxide (as a gas or in solution), coumarin, copper sulphate, formaldehyde, many organic solvents (including methanol, acetone and various other aldehydes, ketones and esters), ammonia (as a gas or in solution), sodium carbonate, sodium bicarbonate and ethylene (ethene) gas (Paula, 1932; Lambert, 1953; Luchsinger and Fleckenstein, 1960; Pollock, 1962a; Pool and O'Connor, 1963; Prentice *et al.*, 1963; Beckord and Fleckenstein, 1964; Kneen *et al.*, 1965; Spillane and Briggs, 1966; Withey and Briggs, 1966; Griffiths and MacWilliam, 1967; Ponton and Briggs, 1969;

Tahara, 1969, 1973; Cole and Hemmens, 1977; Yamada, 1985; Kosar *et al.*, 1989). As well as reducing malting losses, treatments often increase the hot water extract and soluble nitrogen ratios of finished malts. The use of some of these substances requires comment. Some are inherently dangerous to handle, others impart unpleasant flavours to malts or they may leave (or be suspected of leaving) undesirable residues on the malt.

Steep liquor is dirty and rich in microbes. Its reapplication to green malt as a growth inhibitor is highly undesirable and has not been adopted. At more than trace levels, copper ions are toxic to yeast, hasten haze formation in beer and inhibit many enzymes, and so the proposed use of copper sulphate to check malting losses was surprising (Lambert, 1953). Applying a 0.8% solution to barley on the 4th or 5th days of germination on the floors reduced the malting losses by about 1%. The diastatic power of the malt was reduced, but the extract yield and Kolbach index were slightly increased. Little or no extra copper reached the final beer, presumably because the 'extra' copper, above the levels always present, was retained in the mash, by the trub or by the yeast. Acetic acid, a 'natural product' sprayed onto grain on the 4th day of germination during micromalting, or sooner if gibberellic acid was used, reduced the malting losses while enhancing the hot water extract, the cold water extract and the soluble nitrogen ratio but, in contrast to some mineral acids, did not depress wort fermentability (Spillane and Briggs, 1966; Ponton and Briggs, 1969). However, on the pilot-plant scale, using a pneumatic malting system, doses of acetic acid had to be increased to uneconomic levels to be effective, presumably because the acid was evaporating and being removed in the airstream (J. Adams and D. E. Briggs, unpublished data). The non-volatile 'natural' substance lactic acid has also been used to check malting losses while producing acid or 'lactic' malts (Chapter 15). In Germany, around 1930, it was proposed to 'feed' malting barley by applying nitrogenous plant foods, such as ammonium nitrate, nitric acid and various mixtures of substances. These processes are grouped under the Wulfrum patent. Experiments showed that nitric acid alone, or mixed with urea, reduced malting losses, enhanced levels of soluble nitrogen and extract, but slowed saccharification in the mash and reduced wort fermentability (Paula, 1932; Hausmann, 1937; Hajek, 1938). Applications of other mineral acids, at the correct time, reduce malting losses and give malts with enhanced extracts and otherwise acceptable analyses except that the fermentability of the extract is frequently reduced, whether or not gibberellic acid is used (Paula, 1932; Pool and O'Connor, 1963). In contrast, it has been reported that 'acidulation' with sulphuric acid can be used to reduce malting losses when malts are made using gibberellic acid and abrasion and that worts from these malts have normal fermentabilities (Crabb *et al.*, 1972).

Applications of sulphur dioxide (as a highly soluble gas or in solution) to germinating grain 3 days after casting have been shown to reduce malting losses and enhance extract yields. The gas is rapidly taken up by the grain in a closed container (Spillane and Briggs, 1966; Ponton and Briggs, 1969) (Fig. 14.10). Rootlet growth ceased and heat output by the grain was reduced. Worts had reduced pH values, enhanced levels of soluble nitrogen and almost normal fermentabilities. Sulphur dioxide is corrosive and toxic but, in view of its recent widespread use in kilning to bleach pale malts and prevent the formation of nitrosamines, its use to check the growth of microbes on green malt (Maule, 1989) and the increasing use of corrosion-resistant materials in making malting equipment, the use of this substance as a growth-inhibitor might now be reconsidered. The plant hormone abscisic acid (0.1 mg/kg), used in conjunction with gibberellic acid (0.1 mg/kg) and sprayed onto the grain in solution at steep-out, reduced rootlet growth and increased malt yield. At higher doses (1 mg/kg) it also reduced the Kolbach index of malts, but the fermentabilities of worts from these malts was enhanced (Yamada, 1985; Kosar *et al.*, 1989).

In addition to the use of alkaline washes to clean grain, applications of alkaline substances have been proposed to reduce malting losses (Beckord and Fleckenstein, 1964). In particular, the very soluble and toxic gas

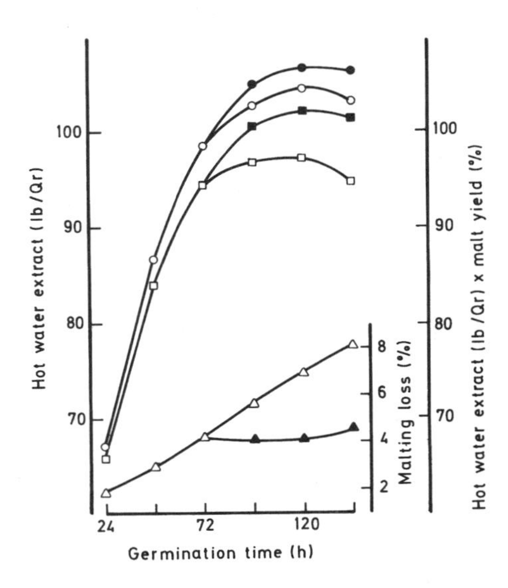

Fig. 14.10 The influence of an application of sulphur dioxide on malting grain (Ponton and Briggs, 1969). Proctor barley was micromalted with an application of gibberellic acid (0.25 mg/kg barley). After 3 days of germination, the grains were sprayed with water and 30 min later the experimental samples were exposed to sulphur dioxide gas (467 ml/kg barley). Controls: HWE (lb/Qr) (○); malting loss (%) (△); HWE (lb/Qr) × malt yield (%) (□). Experimental sulphur dioxide-treated samples: HWE (lb/Qr) (●); malting loss (%); HWE (lb/Qr) × malt yield (%) (■).

ammonia has been used as a gas, or in solution, to check malting losses. It was first proposed that ammonia was best applied to germinating grain (Beckord and Fleckenstein, 1964), but more recently grain has been steeped in ammonia solutions (0.02–0.5 g/100 ml for 2–24 h) to kill the embryos. Then, after a wash, modification has been induced by applications of gibberellic acid. Some of the doses of gibberellic acid used are very high (0.02–5 mg GA_3/kg barley). Good malts were obtained with malting losses of only 4–5% (Tahara, 1969, 1973; Narziss, 1976).

Formaldehyde was used in first steep liquors at the rate of 300–500 mg/l to sterilize the liquor and allow it to be reused (Piratzky, 1958). Brief exposure of barley to solutions of formaldehyde (diluted formalin) reduces grain water sensitivity, kills *Fusaria* and many other fungi and reduces the chances of malt made from the grain giving rise to gushing in beers (Gjertsen *et al.*, 1963, 1965). It is probable that formaldebhyde treatments reduce the possibility of mycotoxin formation on malting grain. Formaldehyde added to brewery mashes reduces anthocyanogen levels in worts and beers and the tendency of the beers to become hazy. Formaldehyde (500–1000 mg/l) applied in the final steep also reduces the levels of anthocyanogens in malt and these malts given more stable beers (Withey and Briggs, 1966; Whatling *et al.*, 1968). Formaldehyde added to the liquor used for resteeping makes the process more reliable, giving cleaner malts having reduced levels of anthocyanogens (see below). Spraying formaldehyde solution onto malting barley also usefully reduces levels of anthocyanogens and, to some extent, levels of soluble nitrogenous materials (Narziss, 1976). Despite these advantages, there is a view that the use of formaldehyde is undesirable, although evidence for this seems to be lacking. Many reports indicate that altering malting conditions change the phenol and anthocyanogen levels of malts (Narziss, 1976), yet some direct measurements indicate that malting conditions do not influence these values (Jerumanis, 1972). This conflict may be apparent rather than real, since altering malting conditions may not alter the levels of **total** phenols and polyphenols in the grain but may well alter the proportions of these fractions that are extracted during mashing.

Ethylene gas, which is a natural plant hormone, reduces root growth in many varieties of barley. However different barleys respond differently to this gas, and it has not come into commercial use (Cole and Hemmens, 1977). The gas is highly inflammable and can form explosive mixtures with air, facts which militate against its use.

In a recently described process, grain is steeped in water saturated with oxygen and is air-rested in oxygen. The onset of germination is extremely rapid and dormancy is overcome. Modification is also very rapid in grain germinated in oxygen (Wainwright, 1995), although this technique is not yet in use.

14.15 Substances applied to green malt for purposes other than controlling malting losses

In the Belmalt process, glucose syrup solutions (e.g. 3.5 kg/100 kg original barley) are applied to green malt some time (e.g. 16 h) before kilning. The malts obtained are said to be 'acid' and, on mashing, yield higher extracts and levels of soluble nitrogen and worts with higher attenuation limits (de Clerck and Isebaert, 1959; Scriban and Vazart, 1960; Kolbach and Zastrow, 1961; Urion and Barthel, 1961). Claims were made that treated malts were better modified, the acrospires were better grown and root yield was enhanced. However, the analytical results can be explained by the adherence of glucose to the malt and rootlets, increasing their 'apparent yields' and extract yield. Increased acidity is caused by some of the glucose being metabolized to acids by microbes on the grains. The consequent pH reduction gives an increased yield of soluble nitrogen in the mash (Vermeylen, 1962).

Gum arabic dissolved in water has been sprayed onto malt, where it is retained. Beers prepared from such malts have better foam retention characteristics and colloidal stabilities (Narziss, 1976). It would almost certainly be more efficient for the brewer to add the gum arabic directly to the beer, if he wished. There have also been reports that additions of heat-stable preparations of microbial enzymes, particularly cellulases and β-glucanases, to green malt leads to the production of better-modified malts, giving worts with lower viscosities and higher extracts. It is most unlikely that these enzymes penetrate the interiors of grains and reach the starchy endosperm. Since it is probable that they survive on the finished malt and are active in the mash, it is likely that their beneficial effects result from their activities in the mash. Unless this approach is used as a device to circumvent the addition of enzymes directly to the mash in the brewery, it appears to be pointless, since, if the use of enzymes is permitted, it is more efficient for the brewer to add them directly to the mash.

14.16 Kropff malting

Grain respiration, heat output and growth can be checked by restricting the supply of oxygen to the grain and/or by allowing carbon dioxide to accumulate. The realization that this was so gave rise to one of the earliest processes for controlling malting losses, the Kropff system (Table 14.1). This process was operated in various ways, with varying degrees of success (Bleisch, 1911, 1912; Lüers, 1922; Leberle, 1930; Schönfeld, 1932; de Clerck, 1952; Vermeylen, 1962). It appears to have only been successful in the manufacture of dark malts. In this system, green malt was initially steeped to a low moisture content of about 41% and was then transferred after a germination period of 4–5 days to a Kropff box (Chapter 9). The grain was formed into a bed about 1 m (3.28 ft) deep on a perforated deck. No arrangements were made for turning the bed of grain since rootlet growth was checked. The box was well-insulated and completely enclosed and could be sealed. The box might be aerated for 1–2 h in every 24 h. This periodic ventilation was necessary to maintain the viability of the grain. Between aerations the boxes were sealed, the carbon dioxide levels rose, for example to 25–30% of the atmosphere in 12–13 h, and the oxygen levels fell. Respiration was halted, glycolysis occurred and the rate of grain metabolism was reduced. The heat output of the grain was checked and a temperature rise of not more than 5 °C (9 °F) occurred. There were no arrangements to recirculate the gases and, therefore, temperatures and carbon dioxide levels were uneven throughout the grain bed (Leberle, 1930). Because of the largely anaerobic metabolism, ethanol and possibly organic acids accumulated, damaging the embryo and checking rootlet growth. Malting losses were typically reduced by 2–3%. The mass of grain did not 'felt' because the rootlets did not grow and bind it together. The extract of the malt was elevated by 0.5–0.8% dry matter, but the enzyme content was reduced. The results of Kropff malting were apparently very variable, possibly owing to the use of grain of variable quality, the lack of exact control of the conditions in the box and variations in the technique used. Kropff malts had high cold water extracts and high levels of soluble nitrogen, which reacted during kilning to form melanoidins, giving dark malts that also tended to be acid. Therefore, Kropff-made malts tended to resemble Munich and lactic malts in their properties (Chapter 15). The Kropff process was useful when long germination periods were usual, the control of malting conditions was less exact than is now usual and malting losses of 11–13% were common. However, in more recent, small-scale trials, using gibberellic acid and comparing with efficiently malted controls having malting losses of 6–8%, carbon dioxide treatments or periods of anoxia merely reduced the quality of malts under conditions where malting losses were reduced (Ponton and Briggs, 1969). In general, it appears that 'carbon dioxide rests' lead to reduced enzyme production and reduced modification. Nevertheless, schedules for giving green malt rests in moist atmospheres enriched with carbon dioxide have been described (Sommer and Antelmann, 1967; Narziss, 1976).

14.17 Undermodified malts

In areas where the use of unmalted adjuncts in the mash is forbidden, the regulations may be circumvented by using chit malts or short-grown malts (Chapter 15). These have many of the characteristics of raw barley yet are cheaper to produce and are produced more quickly and in substantially

Table 14.23 Analyses of chit- and short-grown malts compared with normal malt. The chit malt was steeped to a moisture content of 39–40% and had a total steep + germination time of 2 days. The short-grown malt was steeped to a moisture content of 43% then germinated for a further 2 days (4 days in all). The normal malt had a steeping plus germination time of 7 days

	Chit malt	Short-grown malt	Normal malt
Respiration losses (% d.m.)	1.0	2.6	5.2
Rootlets (% d.m.)	0.7	2.2	3.7
Total malting losses (% d.m.)	1.7	4.8	8.9
Extract (% d.m.)	76.9	80.4	81.7
Fine–coarse extract difference (% d.m.)	12.2	5.6	1.8
Viscosity, Congress wort (cP)	2.34	1.85	1.54
Protein (%)	11.1	10.9	10.8
Kolbach index (%)	22.2	33.1	40.2
VZ 45 °C (%)	19.7	28.8	39.2
Fermentability (%)	62.3	75.7	81.4
α-Amylase (DU, d.b.)	6.9	50	74
β-Amylase (°W-K, d.b.)	124.2	158	197

Data of Narziss (1976).

better yield, (much lower malting losses) than fully modified malts. Germination times are very short. Because the barley is steeped to comparatively low moisture contents, kilning these under-modified malts is relatively inexpensive (Chapter 15). Examples of analytical values of chit and short-grow malts are given in Table 14.23.

14.18 Physical methods of checking malting losses

Several physical techniques have been tested for their abilities to selectively damage embryos of germinating barley and so reduce malting losses. In treated grains, modification is normally triggered by applications of gibberellic acid. One impetus for these investigations has been the desire to avoid the use of chemical growth regulators.

Doses of gamma-irradiation, from ^{60}Co sources, have been used to kill insects in grain; at high doses, grain germinability is depressed. Very high doses of gibberellic acid (e.g. 3 mg GA_3/kg barley) are needed to induce modification in irradiated grain, presumably because the aleurone layer is damaged as well as the embryo, and malting losses are reduced (Fujii and Lewis, 1965; Lewis and Fujii, 1966). This process does not seem to be used.

Another proposed method for reducing malting losses was double-malting (Linko, 1961). In a version of this process, grain was steeped to a moisture content of 43–44%. Then, after a short initial germination period of, for example, 1 day at 12 °C (53.6 °F), it was dried, during a day, in an airflow at 30 °C (86 °F) until it reached a moisture content of about 10%. The rootlets were then physically removed. The de-rooted grains were then rehydrated to 44–44% moisture by steeping for about 1.5 days at 12 °C

(53.6 °F). 'Germination' was then continued for a further 2–3 days at 14–16 °C (57.2–60.8 °F) when modification and acrospire growth continued. Malt yields were enhanced, as were the diastatic power and α-amylase levels, and extracts were increased by 0.5–2%. In view of the extra handling and drying costs involved, this process is never likely to be adopted.

Various attempts have been made on the experimental scale to damage embryos selectively, leaving the aleurone layers functional, by freezing wetted grain and then inducing modification by additions of gibberellic acid (Smith *et al.*, 1964; Powell and Barrell, 1974). In this approach, grain was steeped to 28–40% moisture then frozen at –15 to –27 °C (5 to –16.6 °F). Freezing was carried out in a short (1 h) period. The grain was then thawed, steeped to 45% moisture and dosed with gibberellic acid (0.5 mg/kg barley). The yields of 'frozen' malts were 94–97% (d.b.) compared with 85–92% (d.b.) for the unfrozen controls. Virtually no rootlets were produced by the frozen grain, respiration was reduced and extracts were as good as, or better than, those of the controls. The 'frozen' malts yielded high levels of soluble nitrogen and coloured readily when kilned. These effects could be offset by additions of bromate. Results were best when the grain was frozen at higher moisture contents, possibly because the structure of the endosperm was disrupted. Apparently, this technique cannot be applied economically in commercial malting.

As previously noted, when barley is exposed to high pressures under water the embryo is damaged. If gibberellic acid is applied, pressure-damaged grain can be caused to modify and, since the embryo is damaged, malting losses are reduced. Malt yields as high as 96% have been quoted (Pool, 1978b) but it is not clear if this process is in use. Mechanically bruising grain, by turning it for 40% of the germination period in small, experimental drums that lifted and dropped the grain, prevented growth and modification. However, if the grain was dosed with gibberellic acid, modification occurred, rootlet production was checked and malt yields of up to 96.5% were obtained. The malts had higher extracts than the controls (Sipi and Briggs, 1968). It has been confirmed that vigorous turning will also reduce root growth in malting (Ponnet, 1969). In some maltings, turning may be sufficiently damaging to reduce malting losses, but it does not seem that any plant has been designed specifically to exploit this effect.

Hot-water steeping for short periods can be adjusted so that it rapidly hydrates grain while selectively damaging embryos. Correctly treated grains can be induced to modify by applications of gibberellic acid and, on the experimental scale, good malts can be produced in high yields (Pool, 1964a,b) (Table 14.24). This process has been tested successfully on the pilot scale, but it has not been adopted commercially, probably because of the difficulty of mixing many tons of grain and hot water rapidly to obtain the chosen steeping temperature of 40 °C (104 °F) sufficiently quickly to obtain even treatment of all the grain. In addition, enzyme development has

Table 14.24 Analyses of malts prepared by steeping barley for 8 h at 40 °C (104 °F) then treating with different quantities of gibberellic acid and germinating at 15.5 °C (59.9 °F)

	Germination time (days) with GA_3 0.125 mg/kg[a]			Germination time (days) with GA_3 0.25 mg/kg[a]			Germination time (days) with GA_3 0.5 mg/kg[a]		
	4	5	6	4	5	6	4	5	6
Hot water extract									
l°/kg at 20 °C	279.1	294.7	297.4	279.7	299.6	307.7	290.4	302.6	311.1
lb/Qr	94.0	99.5	100.4	94.2	101.3	104.1	98.0	102.3	105.3
Cold water extract (%)	9.9	13.6	15.2	11.4	14.3	17.9	11.8	15.1	19.1
TSN (%)	0.39	0.50	0.54	0.39	0.51	0.65	0.40	0.52	0.63
Malting loss (% dry matter)	2.3	2.9	3.1	2.3	2.9	3.6	2.4	2.7	3.5
HWE (lb/Qr) × malt yield (%)	92.3	96.9	98.0	92.4	98.6	100.7	96.0	99.1	102.0

Data from Pool (1964a).
[a] Gibberellic acid (GA_3) added per kilogram barley.

sometimes been inadequate and different batches of grain respond differently to hot-water steeps. More prolonged steeps at lower temperatures (30 °C (86 °F)) combined with applications of gibberellic acid (0.5 mg/kg barley) have been proposed as an alternative more easily achieved method of selectively damaging grain (Home and Linko, 1977).

Resteeping, a malting process that accelerates malt production and reduces malting losses, has had considerable success although, owing to changing economic conditions, its use is declining. Special plant must be used for the process to be successful (Chapter 9). In the original, simplest resteeping programme, barley was steeped to a low moisture content of about 40% and was then germinated for 3 days. Then the roots and root initials were killed by resteeping, that is by reimmersing the grain for 24 h in water that was allowed to become anaerobic, at 18 °C (64.5 °F) (Pollock, 1960, 1962a; Kirsop and Pollock, 1961; Pool and Pollock, 1962; Pool, 1964b, 1967a,b). At the end of the resteep period, the liquor was drained and the grain was rinsed (flushed) with fresh water to remove the materials dissolved from the killed roots and the microbes living on them. The grain, now containing 49–52% moisture, was returned to the germination compartment for a further autolysis period of 1–3 days prior to kilning. During the autolysis period, the grain was kept cool. If the resteep had been successful, rootlet growth did not resume and the acrospires did not bolt. Malting losses were reduced to values as low as 4% and the high moisture content of the grain ensured that modification in the autolysis period was extremely rapid, saving considerable amounts of 'germination' time (Table 14.25). Because rootlet growth was inhibited, the grain lay compactly in the germination compartment, effectively increasing its capacity. In some early resteeping equipment (SGKVs), no provision was made for turning the grain, so the process was sometimes referred to as 'static malting' (Chapter 9). However in later plant, turning equipment was installed, if only to lighten the grain mass after it swelled from taking up water during steeping. Particular care was needed to ensure an adequate level of killing of roots and root initials and embryo damage by resteeping, otherwise subsequently new roots grew, the acrospires of the grain tended to bolt and the malting losses could actually be increased. Furthermore, inadequate flushing allowed the growth of microbes on the dead roots and materials leached from them, giving rise to greyish, discoloured, smelly or offensively flavoured malts. Correctly carried out, resteeping produces excellent malts in shortened times and in better yields, having extracts equal to or better than those of conventionally prepared controls, with reduced fine–coarse extract differences, elevated soluble nitrogen ratios, α-amylase levels and diastatic powers. The simple resteeping process was soon displaced by more sophisticated procedures as the details came to be better understood. For example, it was discovered that during the resteep soft water was more efficient in root killing than hard water, the calcium and magnesium ions in the latter protecting the roots from damage. Furthermore, as the temperature at which the resteep was carried out was elevated, so the duration necessary to achieve root killing could be reduced. A resteep of 24 h at 18 °C (64.4 °F) was roughly equivalent to a 3 h resteep at 30 °C (86 °F) or a 1–1.5 h resteep at 40 °C (104 °F). Resteeping at the last temperature prevented the development of unpleasant odours and eliminated the need to flush the grain. The resteeping process had to be adjusted for different batches of grain. In small-scale experiments, the inclusion of formaldehyde (0.1%) in the resteep liquor prevented the growth of microbes, achieved a more certain root-kill and led to clean malts that released fewer anthocyanogens and less soluble nitrogen into the wort during mashing and so gave beers that were less likely to become hazy (Withey and Briggs, 1966; Whatling *et al.*, 1968). It is improbable that formaldehyde was ever used on the production scale. Malting with resteeeping was hastened and complicated by using aerated and broken steeps (steeps with air rests), giving 'multiple steeping processes', and by using gibberellic acid. Warm- and hot-water steeps were employed in some instances. One resteeping programme that was used, probably with applications of gibberellic acid, was (i) steeping for 1 h at 28 °C (82.4 °F); (ii) draining then air-resting the grain for 22 h at 15 °C (59 °F); (iii) steeping the grain again for 1 h at 35 °C (95 °F); (iv) a second period of ventilation in which the grain was germinating; (v) a root-killing resteep at 35 °C (95 °F) lasting for 1.5°h; (vi) a further period for autolysis while the grain was ventilated; and, finally, (vii) the green malt was kilned (Ringrose, 1979). Malt yields from this process were about 96%. In German practice, in which gibberellic acid was not employed, malt yields of about 94% were evidently usual (Narziss, 1976). Despite its successes, resteeping

Table 14.25 Analyses of commercial malts prepared by conventional and resteeping processes. Resteeping reduced the malting loss by 4% and the malting time by 20%

	Hot water extract		Diastatic power (°L)	Cold water extract (%)
	(l°/kg at 20 °C)	(lb/Qr)		
Conventional malt	299.8	101.4	34.5	17.0
Resteep malt	300.5	101.6	38.0	20.0

From Riviere (1961).

is falling out of use and indeed may have been discontinued completely. The process used large quantities of water and produced correspondingly large volumes of strong effluents (the resteep liquors having exceptionally high BOD values) and kilning green malts with moisture contents of around 50% requires large amounts of heat. Therefore, the high water charges and effluent disposal charges and the high cost of fuel to provide heat made the process uneconomic.

14.19 Kilning

The chemical and biochemical changes which occur when green malt is kilned are outlined in Chapter 4 (de Clerck, 1952; Pollock, 1962a; Schuster, 1962; Vermeylen, 1962; Narziss, 1976; Briggs, 1978; Seward 1986). The physical principles involved and some other features of kilning are summarized in Chapter 10, while the types of kilning or roasting conditions for making particular types of malt are indicated in Chapter 15. Kilning schedules are very variable and are chosen to take account of the type and performance of the equipment, the time available (e.g. if an 18 h, 24 h or a 48 h turn-round is needed), the minimal level of fuel costs and power use, the degree to which the green malt has been modified and the type of product being made. Malt is kilned to give a dry and stable product, of fixed composition, from which the brittle roots can be removed. The dry product is friable, that is it is easily crushed. Its stability facilitates storage and transport. In the initial stages of kilning, the development of some enzymes can be encouraged and mellowing (i.e. continuing malt modification) can be achieved. In the later stages of kilning, some enzyme destruction occurs. As kilning continues, unwanted 'raw grain' flavours are removed, DMS precursor is destroyed and, to various extents, the malt develops character, that is it develops colour, a malty flavour and a characteristic aroma. The development of these latter characteristics is largely dependent upon Maillard reactions between amino acids and reducing sugars. Therefore, colour development occurs more readily in well-modified malts having high cold water extracts because these are rich in reducing sugars and pre-formed soluble nitrogenous substances. It occurs more readily at higher temperatures and at higher moisture contents. Enzyme survival occurs best when malt is kilned at the lowest possible temperature and the temperature is not allowed to rise until the malt is comparatively dry. These facts dictate the nature of the kilning schedule needed to make each type of malt. Thus pale malts are dried at comparatively low temperatures that rise slowly; enzyme survival in these products is relatively high. In coloured malts, which are heated to high temperatures, while still moist, extensive or complete enzyme destruction can occur. In the highly coloured special chocolate and black malts, prepared in roasting drums, heating is so intense that enzyme destruction is virtually complete.

Kilning is conceptually divided into the drying and curing stages. In older, shallow-loaded kilns making pale malts, low-temperature drying continued in a rapid airflow until the malt was judged to be hand-dry (typically 8–10% moisture), then the temperature was raised and curing began. With the oldest kilns, draughts were irregular and so were the kilning periods. Until the malt was hand dry the temperature of the air-on was kept at about 30 °C (86 °F) and certainly below 40 °C (104 °F). However, because of a lack of draught, the grain temperature could rise through a lack of evaporative cooling. Such stewing is undesirable in the preparation of most malts. The exceptions are caramel malts in which the temperature is raised sufficiently while the malt is moist to mass and liquefy the contents of the endosperm, which set to a hard, semicrystalline mash on cooling (Chapter 15). To decrease the risks of stewing, green malt was sometimes withered, that is it was partly dried by spreading it thinly on floors or on special, slatted withering floors that allowed free passage of cool air, and the grain was frequently turned (e.g. six times per day (Lévy, 1899; Northam, 1965a). When fans were installed, ensuring that kilns could maintain steady draughts, withering was generally discontinued. However, in multi-decked kilns, the drying that occurs on the upper deck is still sometimes called withering, although in fact drying is occurring in a stream of heated air. For traditional pale-ale malts the curing temperature was raised very slowly to the final value, which would normally be in the range 75–105 °C (167–221 °F) or even up to 110 °C (230 °F), depending on the product being made. There has been a tendency for curing temperatures to be reduced, even to 65 °C (149 °F) for some pale-lager malts. To minimize fuel costs, lower curing temperatures are preferred.

In simple, modern, deep-loading kilns, as operated by one company, the air-on temperature may initially be 50–70 °C (122–158 °F). During the free-drying stage, before the drying front reaches the surface of the grain and breakthrough occurs, the temperature of the water-saturated air leaving the bed of grain is around 30 °C (86 °F). After the break, during the falling drying-rate period, the air-on temperature may be increased (e.g. to 65–75 °C (149–167 °F)) and the airflow is progressively reduced to about

25% of its initial value. Air recirculation may be started when the relative humidity of the air leaving the grain bed is about 40%. The proportion of air recirculated is gradually increased until, at the end of curing, it may be 75–80% of the air passing through the malt. In the final, curing stage, the temperature of the air-on might be increased in stages to 80–85 °C (176–185 °F) for lager malts and to 100 °C (212 °F) for ale malts (Seward, 1986).

The moisture contents of malts vary. At the upper limit, green malts, for example of 45% moisture content as used in some distilleries, are undried. Hand-dried malts, at 13–15% moisture content, have been used in distilleries and were evidently more convenient to handle than green malts (Wahl and Henius, 1908). Air-dried wind-malts are used in the preparation of Old Louvain beers (de Clerck, 1952). Traditional English pale-ale malts had moisture contents of about 1.5%, but values of 2.0–5.0% are now common and the malts are less costly to kiln in consequence. The moisture contents of various lager-type malts are frequently in the range 4–6%.

For the preparation of pale malts high levels of reducing sugars and free amino acids, either in the green malt or developed on the kiln, can be a problem, since these are the precursors of coloured melanoidins. To minimize the Maillard reactions the malt should be dried in a high airflow and at a low temperature. In contrast, for coloured malts, it is desirable that the levels of pre-formed soluble substances be elevated in the green malt and that these be induced to interact by raising the temperatures of the kiln while the malt is comparatively moist. However, for most malts, it is not desirable that endosperm liquefaction occurs, as is required for the preparation of caramel malts (Chapter 15). Because of the number of variables involved, the controlled preparation of darker malts on kilns is difficult. Broadly, down to a moisture content of about 25% and at temperatures below 40 °C (104 °F), the drying malt is still respiring, the acrospires continue to grow, some enzymes continue to increase in amount, modification continues, β-glucan levels continue to decline and the levels of soluble nitrogen and cold water extract increase (Narziss, 1976). If these processes are to be encouraged, as when preparing some dark malts, the grain must be warmed and not dried too quickly. The flow rate of the warm air through the grain may be reduced to limit evaporative cooling, and the air may be recirculated so that it still transfers heat to the bed of grain but, since its relative humidity is increased, evaporation from the grain and the associated cooling are reduced. The changes going on in the grain depend on both the temperature and the moisture content. The drier the grain the higher the temperature must be for amylolysis and proteolysis to occur, increasing the levels of reducing sugars, peptides and amino acids (Runkel, 1975). By holding green malt at 40 °C (104 °F) for 20 h the α-amylase content may be increased by as much as 30% (Linko and Enari, 1966).

Curing is carried out to alter the characteristics of the malt, and again the changes that occur depend both on the temperature achieved within

Table 14.26 Influence on malt analyses of curing at different temperatures on the bottom deck of a two-floored kiln

	At dropping from upper to lower floor	Curing temperature on the lower floor				
		70 °C (158 °F)	80 °C (176 °F)	90 °C (194 °F)	100 °C (212 °F)	110 °C (230 °F)
Moisture (%)	6.93	4.77	3.79	2.57	2.01	1.36
Extract (% d.m.)	80.7	80.8	80.8	80.4	79.7	78.4
Wort appearance	Clear	Clear	Clear	Slightly turbid	Slightly turbid	Turbid
Wort colour (cc$^{N}/_{10}$ I_2)	0.21	0.21	0.22	0.26	0.32	0.80
TSN (% extract)	0.76	0.77	0.80	0.79	0.75	0.72
PSN (% extract)	0.68	0.68	0.70	0.70	0.68	0.60
Diastatic power (°W-K)	248	215	169	111	75	34

Data of Hopkins and Krause (1947).

individual grains and on the moisture contents of the grains. As the curing temperature increases, the moisture content of the malt is reduced, enzyme activities are reduced to variable extents, the fine–coarse extract difference increases but the hot water extract is reduced to only a small extent, or not at all, under moderate conditions (Table 14.26). The colour of the malt increases with curing temperature, as do flavour and aroma; the levels of total soluble nitrogen and free amino acids decline (depressed by protein denaturation and melanoidin formation); the malt total nitrogen content and total malting losses are little altered and wort viscosity is little altered (or increases or decreases, according to various reports). By comparison, the wort attenuation limit (fermentability) is reduced as is wort pH and the diastatic power of the malt, while the saccharification time of the malt is increased. Wort filtration time increases, as does wort turbidity. Beer foam may be improved or remain unchanged by curing the malt at higher temperatures. One judgement is that beers are too 'soft' and insipid if made from under-cured malts. Lager malts cured at 65–70 °C (149–158 °F) often lack character. Curing at 75–80 °C (167–176 °F) is ideal for making Pilsener malts while curing temperatures of 85 °C (185 °F) and upwards give malts with increasingly harsh and caramel-like flavours (Runkel, 1975). In contrast, ale malts should be cured at the higher temperatures to produce the correct character. However, for maximal survival of DMS precursor, well-dried malt should be cured at less than 65 °C (149 °F). Malts made with higher curing temperatures give beers with different flavours, which are less prone to haze and flavour instability and which contain little DMS. Various effects occur as the duration of curing is increased (Shands *et al.*, 1942; Hopkins and Krause, 1947; Isebaert, 1965; Barrett *et al.*, 1967; Kirsop *et al.*, 1967a; Runkel, 1975; Narziss, 1976; Ballesteros and Piendl, 1977). How large the observed changes are depends, in part, on the moisture content of the malt when curing begins. If the malt is already very dry (8% moisture or less) then at moderate curing temperatures (less than 95 °C (203 °F)) and for moderate periods (e.g. 4 h) many changes will be undetectable. In general α-amylase activity is reduced comparatively little, for example 0–15% by moderate curing, but diastatic power values can easily be reduced by 30–70%. To maintain any endo-β-glucanase activity, malts must be dried well at low temperatures and then be cured as gently as possible. Besides altering the levels of enzymes, curing reduces the susceptibility of malt starch to enzymic degradation (MacWilliam, 1972).

In traditional, shallow-loaded kilns in which the malt is turned, all the grains may be regarded as being equally treated, although, in fact, as the grain bed contracts channelling, can occur, which leads to high local airflows and variations in heating and drying. In deep-loaded, unturned kilns, the situation is different. In these, before the break, the air-on may be at 50–60 °C (122–140 °F) and the air-off at 25–30 °C (77–86 °F). Therefore, at the base of the grain bed the grain will dry rapidly and will be at a higher temperature than more moist grain; the latter is at a lower temperature because of the evaporative cooling, which reduces the temperature of the airstream. Conversely, grain at the top of bed will be moist and warm and will become hot last of all when the drying front reaches it (Chapter 10; Lloyd, 1988b; Piglas *et al.*, 1988) (Table 14.27). At each level in the grain bed, the malt will be subjected to a different temperature/moisture and time regime. Thus, kilning malt in an unturned, deep bed adds to the inhomogeneity of the final malt. Enzyme increases, and subsequent declines, are different at different levels in the bed, as are the durations of grain growth and modification and all the chemical changes which accompany these processes. By many criteria (low 70 °C mash wort viscosity, low β-glucan content, high hot water extract, low fine–coarse extract difference, high friability, high SNR and level of FAN, high diastatic power, high α-amylase activity and wort fermentability), the malt from the top of the bed of grain on the kiln is the best modified. However, β-glucanase survival is best at the base of the grain bed, where drying is most rapid. Sulphur dioxide uptake and nitrosamine formation vary through the bed of the grain. A comparison between malts made on two kilns showed that both β-amylase and endo-β-glucanase (but **not** α-amylase) were inactivated more in malts held for longer periods at high moisture contents and that inactivation was reduced by increasing the airflow (Bourne *et al.*, 1990). Evidently a slower kilning schedule (e.g. 48 h) will give different results to a more rapid one (e.g. 18–24 h). The use of sulphur-rich fuels in direct-fired kilns and applications of sulphur dioxide to green malt to minimize nitrosamine formation have been discussed (Chapter 4), as has the need to prevent arsenic-rich fumes from solid fuels reaching grain. Brewers sometimes set maximum levels of residual sulphur dioxide or sulphites on malts because these may influence the mashing process and the amount of break formed in the copper-boil. In addition, during fermentation, the yeast may convert residual sulphite to undesirable flavour and aroma compounds, and the head retention of the beer may be altered. Recently small amounts of sulphur dioxide (100 p.p.m.) have been applied to green malt 20 h before kilning to reduce the level of microbes on the grain's surface (Maule, 1989) Table 14.28). A problem that was sometimes encountered when using direct-fired kilns was the occurrence of magpie malt. In this malt, the husks of some grains were speckled and appeared to be scorched (Anon., 1898; Macey and Stowell, 1957b; Macey, 1958). Only the husk was discoloured, the endosperms of the corns were white. The black speckles were acidic to test papers. The phenomenon was not caused by over-heating. It was most likely to occur in grains in contact with the kiln deck in kilns in which sulphur-rich oils or other fuels were used. Under some conditions, sulphur in the fuel is oxidized to sulphur trioxide, rather than dioxide, and this, or kiln dust containing it, was taken up by the husk of the damp malt in contact with it, turned to sulphuric acid, which marked the husk. Magpie malt is satisfactory to use but appears unsightly. A means of preventing its formation is to inject small amounts of ammonia gas into the kiln airstream. This basic gas interacts with the acidic sulphur trioxide forming unobjectionable ammonium sulphate.

Table 14.27 Composition of barley malt from different levels in a deep-loaded (0.8 meter) pressure kiln

	Place in grain bed		
	Bottom	Middle	Top
Time to reach temperatures shown (approximate, h)			
35 °C (95 °F)	0.5	5.3	9.5
50 °C (122 °F)	3.2	8.1	12.7
65 °C (149 °F)	7.5	10.0	15.1
78 °C (172 °F)	13.6	16.5	18.6
Moisture (%)	3.8	3.9	4.2
Friability (%)	76	80	77
Intact endosperms (%)[a]	3.1	2.5	2.4
α-Glucan (%)	0.83	0.54	0.49
β-Glucanase (U/kg)	170	158	139
Hot water extracts (l°/kg)	301.2	302.7	304.3
Fine–coarse extract difference (l°/kg)	6.5	6.0	4.4
TSN (%)	0.67	0.68	0.70
SNR (%)	37.5	38.0	39.1
FAN (mg/1)	130	135	140
Colour (EBC units)	3.7	3.5	3.6
Wort β-glucan (mg/1)	135	90	82
Diastatic power (IoB)	70	70	73
Fermentability (%)	82	82	83

[a] Whole endosperms, retained by 2.2 mm screen, from the Friabilimeter.
Data from Piglas *et al.* (1988).

Table 14.28 Part of results of a micromalting trial, varying applications of bromate at cast, squeezing steeped grain at casting and variations in aeration or anaerobiosis in the last 20 h of the germination period and germination time[a]

	Trial samples with varying treatments														
Treatment															
Sodium bromate (400 p.p.m.)	+	+	+	+	+	–	–	–	–	–	+	+	+	+	+
Squeeze (1.8 mm gap)	+	+	+	+	+	+	+	+	+	+	–	–	–	–	–
Aeration last 20 h germination	anox.	anox.	anox.	air	air	anox.	anox.	anox.	air	air	anox.	anox.	anox.	air	air
Temperature (°C) last 20 h germination	36	36	18	18	36	36	36	18	18	36	36	36	18	18	36
Steaming (100 °C/20 min) before kiln	–	+	–	–	–	–	+	–	–	–	–	+	–	–	–
Total germination time (days)	5	5	6	6	5	5	5	6	6	5	5	5	6	6	5
Malt analyses															
Extract (l°/kg IoB coarse; d.b.)	302	294	303	302	301	318	311	318	312	315	309	295	306	310	308
Extract (%; EBC, fine, d.b.)	79.7	78.9	80.6	80.5	80.0	84.7	82.5	83.8	82.8	83.9	81.4	78.5	80.4	81.2	81.6
Colour (EBC units)	2.5	2.5	2.8	2.8	2.5	12	26	6.9	5.3	4.7	2.8	3.4	4.7	3.1	3.1
TN (%)	1.68	1.68	1.66	1.63	1.71	1.63	1.62	1.61	1.59	1.70	1.68	1.62	1.64	1.62	1.65
SNR (%; IoB)	38	31	35	36	35	71	66	68	59	61	35	29	34	35	35
Kolbach index (%; EBC)	41	34	39	40	40	72	70	72	66	69	38	32	36	39	39
Boiled wort colour (EBC units)	5.3	6.9	7.5	7.5	5.3	25	658	16.8	11.6	12	7.5	6.8	10.6	6.1	6.9
Wort viscosity (mPa, EBC)	1.47	1.58	1.47	1.47	1.49	1.50	1.55	1.48	1.46	1.48	1.47	1.51	1.46	1.46	1.47
Titratable acidity (ml)	16.8	14.0	18.0	15.2	16.6	29.2	32.8	26.0	22.8	21.8	11.8	10.4	28.0	11.6	12.0
Diastatic power (IoB)	72	20	60	62	62	56	20	78	88	96	84	22	60	68	82
α-Amylase (DU)	20	10	20	27	21	34	16	40	71	43	38	<10	35	63	36
β-Glucanase (IRVU)	150	0	450	300	200	220	50	610	850	250	200	50	200	430	240
Wort fermentability (%; IoB)	79.7	65.6	75.6	80.9	79.2	65.2	57.8	70.9	77.7	75.2	83.8	66.0	80.0	85.3	83.9
Friability (%)	77.9	59.5	81.6	88.8	78.0	90.8	78.0	96.7	98.0	88.6	54.9	54.3	58.5	74	59.0
Homogeneity (%)	88.9	93.7	92.9	98.0	89.7	99.2	98.1	98.8	99.5	98.8	60.4	89.4	75.7	90.5	78.7
Acrospires															
0–0.25	5	9	5	3	11	52	5	7	6	45	1	7	4	2	2
0.25–0.5	87	89	91	88	84	23	59	63	18	20	98	88	95	15	94
0.5–0.75	7	2	4	8	3	19	33	27	59	26	1	5	1	51	2
0.75–1.0	1	0	0	1	2	6	3	3	16	9	0	0	0	32	2
Over 1	0	0	0	0	0	0	0	0	1	0	0	0	0	0	0

[a] The barley variety was Pipkin. Steeped barley was drawn from a production steep at casting then treated as indicated. Steaming was carried out at 100 °C (212°F) for 20 min. All samples were treated with sulphur dioxide (100 p.p.m.) for 20 h before kilning.

[b] Aeration: air, with aeration; anox., anaerobic.

Data selected from Maule (1989).

In making peated malts, used for making whiskeys and in making other smoked malts used for manufacturing some special beers, smoke from slowly burning peat or hardwood is directed into the hot-air chamber of the kiln to give flavour to the malt. These smoking processes are difficult to regulate (Chapter 15).

Historically, malt that had become slack, that is its moisture content had increased above a specified level, was recovered by redrying. However, these redried, clung- or thane-malts did not regain their original qualities. Provided that the slack malt has not deteriorated detectably in any other way, it is probably better used for roasting.

14.20 Controlling wort fermentability

Altering almost any aspect of the malting process will alter the quality and nature of the malt produced and, quite probably, the yield with which it is obtained from the barley. Comparatively recently, the many factors that influence wort fermentability have received attention (Axcell *et al.*, 1986; Healy and Armitt, 1986; Maule, 1989; Pollock, 1989). The fermentability of wort prepared from a malt in a standard way is loosely referred to as malt fermentability. Distillers require maximum wort fermentability to maximize spirit yield and brewers require high fermentabilities when making low-carbohydrate beers. For most beers, brewers require moderate, but reproducible fermentability values unless low-alcohol beers are to be produced, when low fermentabilities are advantageous. It is possible to prepare malts with good extracts but poor fermentabilities. Fermentability depends not just on the fermentable sugars:dextrin ratio but also on the supply of other yeast nutrients, such as zinc. Fermentabilities that are too low can be improved by increasing the modification of the green malt and by maximizing enzyme survival by reducing kilning temperatures. Conversely, fermentability can be depressed by stewing the malt on the kiln and by increasing the curing temperature. The brewer can decrease the fermentabilities of his worts by increasing the mashing temperature. Other proposed means of reducing fermentability include (i) steaming the green malt at 100 °C (212 °F) for 20 min before kilning, when fermentabilities as low as 58% are achieved; (ii) applying a period of anoxia during the last 20 h of the germination period; and (iii) making malt from barley varieties or mutants inherently poor in β-amylase (Maule, 1989) (Table 14.28). The table gives a good idea of the complexities involved when malting conditions are altered.

14.21 Dressing malt

After kilning or roasting, malt must be cooled, preferably to 20 °C (68 °F) or less. If this is not done it will remain hot in store, colour development and enzyme destruction will continue and the nature of the malt will change. With special malts or roasted barley from roasting drums it is usual to place the product immediately into a strongly ventilated cooler to check further colour development and reduce the risk of fire (Chapter 9). After some preliminary cooling malts are dressed, that is to say the dried and brittle rootlets and dust are broken up and removed and broken corns are separated (de Clerck, 1952; Schuster, 1962; Vermeylen, 1962). Culms must be removed (by processes of agitation, sieving and aspiration) since they are extremely hygroscopic, rich in soluble nitrogenous substances, contain poorly flavoured and bitter substances and can be rich in sulphur dioxide and/or nitrosamines. Deculming should be carried out soon after the malt is stripped from the kiln to help cool it and before the rootlets pick up moisture from the air, become slack and pliable (less brittle) and, therefore, more difficult to break and separate. The rootlets and dust are collected from the aspirating airstream. Before dispatch from the malt store, malt may be screened again and aspirated to remove broken corns, separated husk, dust and any contaminants. Rarely it may also be passed through a destoner or be polished by passage between vegetable-fibre brushes to remove dust and adherent mould spores (essentially a cosmetic procedure) (de Clerck, 1952). Technically, it is feasible to remove the most dense, under-modified malt corns on a shaking table (Leberle, 1930) but it is unlikely this slow process is ever used in practice.

14.22 Malt storage and blending

Each batch of freshly dressed malt is stored in a holding or analysis bin or store (de Clerck, 1952; Vermeylen, 1962; Schuster, 1962). Only when the analyses have been checked will it be transferred to a bulk bin, or silo, to be mixed with other malt having the same or closely similar analytical values. To some degree, each batch of malt is heterogeneous since the barley corns in a batch are not of uniform maturity and composition, and inevitably they are subjected to ranges of steeping, germination and kilning conditions owing to their differing positions in the production equipment. Later, the malts from different stores will be blended, in carefully chosen proportions, to give a mixture that matches the user's specification. To supply large quantities of malt of uniform quality and to minimize the effects of the unavoidable minor variations in quality that occur in successive production runs, blending is both inevitable and desirable. However in the UK, batches that are blended are normally prepared from one barley variety (or agreed group of varieties) within a particular range of nitrogen contents and degrees of maturity, all processed to produce one particular grade of malt. Unless the customer permits it, it is not considered desirable to blend malts made from different types of barley or to blend different types of malt. In contrast, in

North America, malt deliveries may even contain mixtures of malts made from two- or six-rowed barleys. In the UK, in most instances, users blend their chosen malts when preparing their grists. Blending malts to obtain the right mixture of characters is **not** a simple matter (Yamada and Yoshida, 1976; Moll *et al.*, 1982). For example, if two malts differing in character are blended, then some analytical values, such as the hot water extract and FAN, change in proportion to the proportions in which the malts are mixed. However wort fermentabilities tend to be above those predicted by simple proportions, probably because enzymes from the malt giving the more highly fermentable extract break down dextrins derived from the extract from the poorer malt during mashing. It has been concluded that account must be taken of seven analyses if three malts are to be blended successfully and of 15 analyses if four malts are to be blended (Moll *et al.*, 1982).

Malt should be stored cool in water-tight and air-tight containers to minimize the uptake of moisture from water or from humid air. Some moisture uptake is inevitable, and so malts are initially dried to a value below that of the customer's specification. Historically, malt was stored in hessian sacks, often covered with layers of culms to absorb moisture from the air, or, even worse, in heaps on open floors, sometimes covered with a tarpaulin. Now, and particularly for transport across the world, malt may be placed in several-layered, waterproof sacks. Most often malt is handled and stored in bulk. Bins or silos are normally used and these are best filled to capacity and then closed. It is absolutely essential that malt should not become contaminated with unmalted grain or malt of another type. Contamination usually results from poorly cleaned or badly operated conveying equipment and storage facilities.

Insect infestations must be prevented. Insecticides are not applied directly to malts. Malt handling, movements and transfers should be minimized to reduce the inevitable corn breakage and losses that occur. Malt stocks should be used in sequence and not be allowed to remain in store for extended periods. Malt stores and conveyors must be emptied completely and cleaned, and possibly fumigated, before fresh malt is loaded into them. There is an old belief, doubted by some, that white malts should be stored for at least 2 weeks, and preferably 3 to 6 weeks, before use. It is believed that beer from freshly made pale malts produce worts that clarify with difficulty and ferment abnormally, giving beers with poor head retentions, which tend to become hazy and which have poor organoleptic qualities. There is some support for this belief, and it seems that different malts behave differently during storage (Rennie and Ball, 1979). The reasons for the changes that occur in white malt quality during storage are obscure. In contrast, coloured special malts should be used fresh as aroma and flavour are lost during storage.

Before dispatch, each batch of malt (normally blended) is screened and aspirated to remove broken corns, dust and loose husk and to ensure the absence of contaminants. The analysis of the batch is checked. Excessively

Table 14.29 Kilned malt rootlets: approximate ranges of composition; compositions vary very widely

Fraction	Proportion (%)
Moisture	Less than 7.0
Crude protein (N × 6.25)	25–34
Fat (oil)	1.6–2.2
Minerals	6–8
Nitrogen-free extractives	35–44 (or even 50)
Pentosan	15.6–18.9
Cellulose	6–10

From Briggs (1978).

dry or over-modified malts easily break up when handled or transported, causing waste and generating extract-rich dust. In addition malts from some barleys readily shed their husks. Where possible, maltsters avoid using these varieties.

14.23 Culms

Culms (rootlets, coombes, cummins, malt, sprouts) are usually 5–8 mm (*ca.* 0.2–0.31 in) long and 0.3–0.4 mm (*ca.* 0.10–0.16 in) thick; they are produced during malting at the rate of about 30–50 kg/t (*ca.* 88/lb ton) barley steeped (Briggs, 1978). The yields, and qualities, vary depending on the barley being used, the details of the malting and kilning processes and whether light or dark malts are beings produced. Malts from naked huskless grains, such as wheat, rye, triticale and sorghum, also produce sprouts, acrospires and, in the cases of the 'tropical' cereals, leaves. Culms are recovered in sweepings from the kilns, from the transporting equipment and from the deculming equipment. In addition, malting generates barley screenings (including thin corns), recovered floating grains from the steeps, barley dust, malt dust and other malt screenings. This 'offal' is usually sold for use as animal-feed. In the past, the components were sold separately, but increasingly maltsters blend and pelletize them. The pellets are convenient to handle and are comparatively easily stored, in contrast to culms, which occupy a large volume, having a low bulk density (Appendix). The pelleted materials have a cash value, but it is less than that of an equal weight of malt.

Some ranges in the compositions of rootlets are given in Table 14.29. The variations in composition can be large. Moisture contents of 5–20%, crude protein contents of 10–35%, fibre contents of 8.6–15% and ash contents of 6–8% have been quoted. Of the crude protein (N × 6.25) some 70–80% is true protein. Rootlets from pale malts are rich in various enzymes, a fact that does not seem to be exploited. The rootlets are comparatively rich in lysine and contain many basic substances, including choline, betaine, tyramine, hordenine and candicine, as well as allantoin, purine bases and nucleotides. The lipid fraction contains free fatty acids and is comparatively rich in tocopherols (vitamin E). The vitamin C that is present in the rootlets of green malt is destroyed on kilning, but the culms are rich in B vitamins including pantothenic acid. Under some conditions, the rootlets are liable to accumulate undesirable levels of sulphur dioxide and nitrosamines formed by interactions between their basic nitrogenous components and oxides of nitrogen present in the kiln gases. In addition to their use in compound animal feedstuffs, rootlets have been used as a nitrogen-rich organic fertilizer, in compost for orchid growing and extracted to provide nutrients for baker's yeast, the mould *Eremothecium ashbyii* (which is used to produce riboflavin) and some microbes that produce antibiotics.

15 Types of malt

15.1 Introduction

Maltsters aim to make malts that meet users' requirements as closely as possible. It is apparent that an almost infinite number of different malts might be made, but it is not possible, for example, to produce a malt with a low total nitrogen (crude protein) content when only barley with a high nitrogen content is available, nor is it possible to make malts that are both highly diastatic and highly coloured. In short, malt users need to discuss their requirements with maltsters so that they do not draw up specifications that are impossible to meet (Gromus, 1988; Brookes, 1989).

Malts are classified into particular groups by tradition and with reference to the ways in which they are made. These divisions are not wholly logical and since different types of malt have ranges of specifications that differ between localities and customers and have altered with time, no tidy or universally accepted classification is possible. The descriptions that follow must be viewed in this light.

Malts have been made from a wide range of cereal grains as well as peas, beans and vetches. Of the malts made from sprouted pulse seeds, only some aspects of their food qualities are known (Chapter 1) and they are not considered further. Barley is by far the most important cereal for making malts, and so barley malts are considered in most detail. From the technological point of view, malts made from other 'temperate' cereals (wheat, rye and oats) resemble barley in many respects, while malts made from 'tropical' cereals (sorghum, maize, the various millets and rice) differ significantly from barley malts, both in the ways they are made and in their final properties. There is increasing interest in the production of malts from 'tropical' cereals. Barley malts may be divided into relatively pale, white, enzyme-rich malts (such as Pilsener, pale-ale, mild, distillers' and vinegar-brewers' malts) coloured malts (such as Vienna, Munich and brumalts) and special malts (such as caramel, amber, brown and black malts and, by tradition, roasted barley). Special malts are now prepared in roasting drums.

In the last 150 years the scale of malting (batch sizes) has increased, the processes have become more mechanized and automated and control of the processing conditions has steadily improved. Pneumatic malting has mostly replaced floor-malting and the better choice of barleys, the use of steep aeration, air-rests and, where permitted, additives such as gibberellic acid and potassium bromate has allowed the production of malts in shorter times (Table 15.1). The figures in the table are not absolute since, for example, in

Table 15.1 Relative times taken to make pale malts in the UK. These figures are approximate. With old floor-maltings, in very cold weather, steeping might take 5 days. Flooring could be between 9 and 21 days. With old-fashioned kilns, working under adverse circumstances, kilning might exceptionally take as long as 6 days

Year	Floor-maltings					Pneumatic-maltings				
	Steep (days)	Germ.[a] (days)	Kiln (days)	Total time (days)	Malting loss (%)	Steep (days)	Germ.[a] (days)	Kiln (days)	Total time (days)	Malting loss (%)
1850	3–4	13–14	3–4	20–22	–[b]	–	–	–	–	–
1900	3	11–12	4	19–20	–[b]	2–3	10	3–4	15–17	–[b]
1930	3	10–12	4	17–19	11–13	–	–	–	–	–
1950	3	8–10	3–4	14–17	10–12	3	8	2	13	9
1970[c]	2–2.5	4–6	2	8–10	7–8	2	4–5	1–1.5	7–8	5–6
1990[d]	2–2.5	4–6	2	8–10	7–8	1.8–2.5	3–4.5	0.75–1.5	5.55–8.5	*ca.* 6

[a] Germination period.
[b] No reliable data available.
[c] Using gibberellic acid and potassium bromate.
[d] Using abrasion, steep aeration and air-rests, plus gibberellic acid and potassium bromate.
From Macey (1971) and other sources.

old, unheated floor-maltings in cold weather, steeping might take nearly a week and flooring could last 3 weeks. The duration of kilning in natural-draught kilns depended upon the weather and the skill of the kilnman and was very variable. In very cold weather, the draught might be excessive and the kilning period was shortened, while, conversely, in warm, humid weather it could be extended (Table 15.1). Under modern conditions, malt is made on a production-line basis using fixed-time, pre-selected schedules. At present, steeping times vary, usually in the range 1.75–2.5 days, germination times range between 3.5 and 6 days and kilning times between 0.75 and 2 days. As a result of improved understanding of the process, better control of the process, better analytical techniques and more stringent specifications by malt users, barley malts are now at least as good and, in many instances, very much better than they were in the past.

15.2 Barley chit malts and short-grown malts

Where the use of raw grains as mash-tun adjuncts is forbidden by law, brewers may take advantage of the relatively cheaply produced chit and short-grown malts, both to reduce their grist costs and to have the advantages of better head formation and retention and the creation of a 'dry' flavour in their products. Problems caused by using these under-modified malts include higher wort viscosities and slower wort separation rates (Schuster, 1962; Narziss, 1976). The products are used at rates not exceeding 10% of the grist. In the past, barley was steeped normally and after 2–3 days cool germination (below 18 °C (64.4 °F)) was gently kilned to maximize enzyme survival or, more rarely, was used green (i.e. unkilned) or after being flaked. Under more modern conditions, with aerated and air-rest steeping, under-steeped grain chits during the second day of steeping and may be kilned directly to give chit malt with a malting loss of only 1–2%. Alternatively, to produce short-grown malts, the grain may be under-steeped to a moisture content of around 40%, so that steeping time is reduced and the grain's moisture content is low, then, after 2–4 days germination, or sometimes more, the product is lightly kilned. Kilning is cheaper than usual because there is less moisture to evaporate from the green malt. This short-grown malt is better modified than chit malt and is produced with malting losses of 2–4% (Table 14.23). Chit and short-grown malts, while containing some enzymes and being partly modified, retain many raw grain characteristics. Green chit malts have been used after wet-milling. This approach has the advantage of maximizing enzyme survival but it risks green-grain flavours in the product. Green chit malt has been used after flaking, which, since the enzymes are destroyed, is very close to producing barley flakes by an expensive route (Chapter 12).

15.3 Green malts and lightly kilned malts

Most malts are dried to low moisture contents, either in a kiln or in a drum. This is an expensive operation and, to a greater or lesser extent, enzymes are destroyed. By using conventionally prepared but undried green malt, grain distillers can convert grists containing 90% cooked, unmalted cereal (wheat, maize or barley) or starch from various sources, such as potatoes. Petit, 1899; de Clerck, 1952; MacWilliam *et al.*, 1963; Scully and Lloyd, 1965; Narziss, 1976; Macartney, 1988) Green malt mashed alone has a high extract and produces highly fermentable worts, rich in soluble nitrogen and amino nitrogen and containing very low levels of anthocyanogens. The worts have very low viscosities. Green malt appears to be ideal for grain distilling provided that it can be used on site as it is produced, since it will not keep for long and is awkward to transport because of its tendency to heat and its low bulk density owing to the rootlets. It has also been used in the production of cereal syrups. It is not used in brewing because of the unpleasant green-grain, harsh flavour and after-taste, which is often apparent in the beers, and because the rootlets cannot be removed and the high levels of amino acids and sugars in the wort react during hop-boil to generate melanoidins, which darken the wort.

Experimentally, green malts have been freeze-dried to give pale, enzyme-rich products that can be deculmed, stored and easily milled. This process is too expensive for production purposes (Duff, 1963; MacWilliam *et al.*, 1963; Brudzynski and Roginski, 1969).

Air-dried malts (wind malts) were obtained by allowing green malt to dry spread on perforated screens in thin layers in large, airy lofts (Petit, 1899; de Clerck, 1952). Wind-malts were used in making some traditional Belgian, Old Louvain beers. In Africa, homemade sorghum and millet malts may be air dried in the sun. This process is not practical for large-scale malt production. To reduce kilning costs and maximize enzyme survival, many distillers' malts are lightly kilned. There have been various trials reported in which hand-dry malts have been used. These are less expensive to kiln, since the final moisture contents are 8–12%, enzyme survival is good, raw-grain flavours are removed, the rootlets can be separated, the product is sufficiently stable to be transported and stored and it can be broken up using conventional dry mills. A more radical proposal is that the green malt be comminuted and slurried at the maltings and the slurry, held at less than 5 °C (41 °F) and possibly treated with a preservative such as potassium metabisulphite or formaldehyde, be transferred to the distillery (Macartney, 1988).

15.4 Providence malt

Until 1880 malt was taxed in the UK. However, if grain in the field sprouted in the ear, which sometimes happened if the weather was wet and warm,

Table 15.2 Some 'mid-range' analyses of conventional German malts

Malt type	Pale lager	Pilsener lager	Vienna-type lager	Munich, dark lager	Diastatic malt	Wheat malt	Proteolytic malt
Extract (fine grind, % d.wt)	80.7	80.7	80.9	79.5	78.8	83.7	82.5
Extract (coarse grind, % d.wt)	79.1	78.9	79.3	77.5	77.3	82.2	78.3
Fine–coarse extract difference (%)	1.6	1.8	1.6	2.0	1.5	1.5	4.2
Moisture (%)	4.4	4.6	4.5	3.8	7.6	5.7	5.4
Conversion time (min)	10	10	10	20	5	15	10
Colour (EBC units)	3.4	3.0	7.1	17	2.6	4.1	4.5
Colour boiled wort (EBC units)	5.0	4.3	8.6	17.0	4.3	5.7	5.1
Total protein (g/100g d.wt)	11.0	11.4	11.0	11.5	12.1	13.3	10.9
Soluble protein (mg/100 g d.wt)	734	716	783	714	811	808	867
Kolbach index (%)	41.9	39.4	44.4	39.0	42.1	38.1	49.5
pH (EBC)	5.76	5.74	5.63	5.54	5.82	5.94	5.01
Viscosity (cP)	1.57	1.60	1.57	1.61	1.54	1.80	1.67
Extract (% at 45 °C; VZ 45 °C)	39.1	38.1	42.1	39.7	43.7	36.0	47.0
Hardness (Brabender)	347	358	371	365	510	694	548
Diastatic power (°W-K/100 g d.wt)	289	307	215	145	433	317	187
α-Amylase (ASBC, DU/100 g d.wt)	44.0	46.0	40.0	30.5	64.0	47.0	25.5
Fermentability of wort (%)	83.4	81.2	81.3	75.9	83.9	79.2	73.1

From Oloff and Piendl (1978) using *Analytica-EBC* methods of analysis.

and particularly if the plants were lodged, and if the sprouted grain was dried, then it was regarded as providence malt and was not subject to tax (White, 1860). The product was highly irregular in appearance. White (1860), an Englishman, noted that much advantage of providence malt was taken in Wales.

15.5 Pilsener and other pale lager malts

The very pale Pilsener and lager malts, which may be equated with Bohemian malts, used to be withered on the floor while still under-modified and were kiln-dried at a low temperature in a fast airflow (Lévy, 1899; Petit, 1899; Wahl and Henius, 1908; Hind, 1940; Schuster, 1962; Narziss, 1976, 1991b; Blenkinsop, 1991). The malt grains had hard ends. Pilsener malts are now made from plump, two-rowed barleys with TN contents of 1.52–1.84% (crude protein 9.5–11.5%) steeped to a moisture content of about 43%. After a long, cool period of germination, at less than 17 °C (62.6 °F), the fully modified malt is well dried to about 8% in a cool, fast airflow before being cured with slowly rising temperatures up to 70–85 °C (158–185 °F). The very pale products, which have no trace of caramel or melanoidin coloration (Table 15.2) have very weak aromas.

In the UK lager malt is now taken to mean a very pale, fully modified malt made from a two-rowed barley of moderate nitrogen content (1.65–1.80%), that has been dried lightly and is close to a modern European Pilsener malt or other pale-lager malt in its characteristics (Tables 15.2 and 15.3). In the past (e.g. in the 1930s) lager malts were under-modified and, with infusion mashes, yielded less extract than pale-ale malts. This is no longer true and because the light curing used for pale-lager malts allows more enzymes to survive they often yield slightly more extract than a more highly cured pale-ale malt. The many darker lager malts (Viennese, Munich; sections 15.10 and 15.11) used in mainland Europe are virtually unknown in the UK. British lager malts are sometimes made from barleys with moderate nitrogen contents, sometimes using gibberellic acid and potassium bromate, although the latter is avoided if high DMS levels are required. Kilning temperatures (air-on) are often in the range 50–70 °C (122–158 °F). The malt may retain some 'grain flavour'. Moisture contents are typically 4–6%, colours are usually below 3 EBC units and with TSN values of 0.5–0.7%; SNR values of 31–41% are common. A wide range of malts may be produced, having different flavour profiles. Many users require the malts to be rich in DMS precursor (S-methyl methionine, SMM) but others do not.

15.6 Pale-ale, mild-ale and standard malts

Pale-ale, mild-ale and standard malts are mostly used to brew traditional British top-fermented beers, or 'ales' (Lévy, 1899; Petit, 1899; Hind, 1940;

Table 15.3 Representative analyses (IoB analyses) of high-quality floor-malts

Type of malt	Best pale ale	Mild ale	Standard	Lager
Hot water extract				
1°/kg at 20 °C (d.b.)	307.0	296.0–304.9	301.0–306.2	307.0
lb/Qr (d.b.)	103.7	100.0–103.2	101.8–103.6	103.9
Moisture content (%)	1.7; 3.0	1.8–3.5	1.6–2.0	2.4
Colour (EBC units)	4–5	6–7	5–7	2–3
Cold water extract (%)	18.0–20.0	17.0–19.0	18.0–20.0	18.0–19.0
Diastatic power (°L)	35	33–48	39–47	63
TN (%)	1.35–1.50	1.44–1.65	1.46–1.70	1.60–1.75
TSN (%)	0.49–0.54	0.50–0.63	0.55–0.65	0.66–0.68

From Northam (1965) and more recent analyses.

Northam, 1965a; Brookes, 1989; Blenkinsop, 1991). The qualities required have varied a little over the years (cf. Hind, 1940) (Tables 15.3 and 15.4). Pale-ale malts, regarded as the finest malts, are particularly in demand for cask-conditioned beers. Some brewers still try their utmost to obtain pale-ale malts made in floor-maltings. The best, low-nitrogen, two-rowed barleys of preferred varieties are used for making these malts. Pre-1940, the barleys used had, for choice, total nitrogen contents of around 1.35%, but over the years these values have moved upwards so that, although values of 1.5% or less are preferred, in some seasons or by particular brewers, values of 1.60–1.65% may be accepted. Newer varieties of barley tend to give malts with better extracts than the older varieties did, even at higher total nitrogen content. The malt must be evenly and well modified. Neither under-modification nor over-modification is acceptable. Head retention is favoured by using slightly less well-modified malt. Under-modified malts give poor extract recoveries, slow wort recovery rates and problems with beer filtration and instability. Over-modified malts break up too readily on handling, causing losses. Often they also give problems with wort separation, the beers tend to have a thin character, lacking body and having poor head retention.

The green malts are first carefully dried in a rapid, cool airflow but after the break has occurred, the air-on temperatures are raised in the range 60–90 °C (140–194 °F) until curing is finished at a high temperature, up to 105 °C (221 °F). Hot water extracts are regularly in excess of 300 l°/kg (d.b.) and sometimes reach values over 310 °l/kg (d.b.). Cold water extracts of around 19% are sometimes specified, to show that the malt has not been 'forced'. Traditionally pale-ale malts were removed from the kiln with a moisture content of about 1% and if it was delivered to the brewery with a moisture content of over 2.5% it might have been rejected as slack. Depending on the customer, moisture contents (delivered) of 2.5–4.5% are more usual now. The acceptance of higher moisture levels has resulted in significant savings in kilning costs. Under modern handling conditions

Table 15.4 'Average' analyses of various British malts

	Best pale ale	Pale ale	Best mild	Mild	Pot-distilling malt
Hot water extract					
(as-is; 0.7 mm, 1°/kg)	301	300	298	296	296
(as-is; % soluble extract)	–	–	–	–	76.8
(d.b.; 0.7 mm; 1°/kg)	311	310	307	305	309
Fine–coarse extract difference					
(0.2–0.7 mm, 1°/kg)	4	4	5	5	–
Moisture (%)	3.2	3.3	3.0	3.0	4.3
Colour (25 mm; EBC units)	4.7	5.0	5.8	6.0	–
Diastatic power (°IoB)	40	45	55	58	65
Cold water extract (%)	18.0	18.2	18.4	18.5	–
TN (%)	1.48	1.53	1.62	1.72	1.60
TSN (%)	0.58	0.60	0.64	0.68	0.62
SNR (%)	39.2	39.2	39.5	39.5	38.5
Unboiled wort fermentability (%)	–	–	–	–	87.0
Fermentable extract (%)	–	–	–	–	66.8
Predicted spirit yield (l ethanol/tonne)	–	–	–	–	405
Peatiness (p.p.m.)	–	–	–	–	3–12+

From *Pauls Brewing Room Book, 1995–7*, pp. 246–7; IoB methods of analysis.

there is much less chance of the malt picking up moisture than was once the case.

Finished pale-ale malts have high extracts and moderate diastatic powers. TSN values of 0.5–0.7% are common, with TSN : TN ratios of about 40%. Colours are low (4–6.5 EBC units) but still higher than those of lager malts, and the flavours of the products are more malty. No raw-grain flavour is detectable.

Before 1940, English maltsters made malts from a wide range of barleys imported from around the world. Since that time, most British malts, with the exception of some distillery malts, have been made almost exclusively from home-grown barleys. Consequently, some older types of malts, such as those made from husky, six-rowed Californian barleys (Hind, 1940) that gave low extracts but were helpful in giving open, husky grists from which worts were easily separated, are no longer produced. The range of qualities of pale-ale malts merge with those of mild-ale malts (Tables 15.3 and 15.4). Mild-ale malts are made from barleys with higher nitrogen contents (1.5–1.75%, or more in some seasons). Typically, they are slightly less well modified than pale-ale malts. Kilning, with curing temperatures up to 105 °C (221 °F), with a shorter initial drying period, gives colours of 6–8 EBC units. Extracts and diastatic powers were usually less than those of pale-ale malts but now diastatic power values are often higher (Tables 15.3 and 15.4). Traditional moisture contents were about 1.7–2.6% but values of 3–4% are now more usual.

Pale-ale and mild-ale malts are usually made to meet a brewer's specification. To simplify production processes, maltsters may offer standard malts, often made to mild-ale or pale-ale specifications. Since these are made in large bulks to one specification they are simpler to produce. Examples of analyses are given in Tables 15.3 and 15.4.

15.7 North American malts

Although small quantities of a wide range of malts are used in North America, particularly in microbreweries and boutique breweries, the greatest amounts of malts are standard, pale-malt types (Table 15.5). The production of low-enzyme malts from Californian six-row barleys has virtually ceased, being replaced by malts from two-rowed barleys giving better extracts (Briggs, 1978; Schuster, 1962; Kneen and Dickson, 1967) (Table 15.5). Probably most malts are made from six-rowed barleys and, to the surprise of most Europeans, some customers accept loads of malt in which malts from six-rowed and two-rowed barleys are mixed. The six-rowed barleys tend to have thinner corns than the two-rowed barleys (Tables 15.5 and 15.6). Total protein contents of 10.4–14.3% (TN 1.66–2.29%) may be encountered. A wide range of production processes are used, including extensive spray-steeping in some instances. Grain may be steeped to a moisture content of 46% and germinated for 4–6 days at 12–18 °C (53.6–64.4 °F). The green malt is dried in a rapid airflow at 35–50 °C (95–122 °F) and is cured at temperatures of 70–85 °C (158–185 °F). In general, the malts are well modified, with most acrospires grown to 3/4–1 of the grain length and with low fine–coarse extract differences. Because they are very lightly kilned, they have low colours and moisture contents of 3.7–5.3%. Diastatic power values are relatively high by UK standards and the TSN : TN ratios are often high (e.g. 43–48%). In some areas, it seems that 9–12% of the grains may be overgrown, that is they have bolted, but this is not universally true (Table 15.6). α-Amylase values are often high (up to 50 DU) and mash conversion times are usually low (5–7 min). Levels of FAN are also often high; these are needed for yeast nutrients when making beers with large amounts of cooked, unmalted adjuncts, which produce little soluble nitrogen, in the grists. Extracts (77–82% d.b.; ASBC, fine grind) are below those of many European pale malts. This is of minor importance when much of the brewers' extract comes from unmalted adjuncts. Worts have low viscosities, around 1.5 cP or less, and are clear.

15.8 Diastatic malts, grain-distillers' malts and pot-distillers' malts

Highly diastatic malts are made as sources for diastatic extracts and diastase and for processes in which they must convert large amounts of adjuncts

Table 15.5 Representative analyses (ASBC methods) of some North American malts[a]

	USA					Canada	
	Klages (two-rowed)	Piroline (two-rowed)	Karl (six-rowed)	Californian (six-rowed)[b]	Mid-western Larker (six-rowed)	Six-rowed	Two-rowed
Moisture (%)	3.9 ± 0.2	3.9 ± 0.2	4.0	4.4	4.1	3.8 ± 0.2	3.8 ± 0.2
Extract (%, d.wt, fine grind)	80–81	78.5–79.5	81.7	77.0	77.0	78–79	79.5–80.5
Fine–coarse extract difference (%)	1.5–2.1	1.5–2.1	1.7	2.0	1.7	1.3–2.2	1.8–2.5
α-Amylase (DU)	35–41	33–38	33	25	40	35–45	30–40
Diastatic power (°)	110–120	100–122	102	65	156	120–145	90–120
Total protein (%)	11.5–12.5	11.3–12.3	10.4	11.0	13.3	11.5–12.5	11.0–12.0
TSN : TN ratio	39–44	38–42	45	35	39	38–42	38–42
Colour (°Lovibond)	1.45–1.75	1.55–1.85	1.68	1.4	1.74	1.40–1.60	1.20–1.40
Corn sizes							
Over 7/64 in (2.78 mm)	64.2	61.1	43.8	68	44.0	30–40	62–67
Through 5/64 in (1.98 mm)	0.5	0.6	1.0	0	0.8	1–3	1–2

[a] Courtesy of R. Vogel and R. Zytniak, Great Western Malting Co., Vancouver, USA; W. W. Sisler, Brewing and Malting Barley Research Insitute, Winnipeg, Canada.
[b] From Kneen and Dickson (1967).

Table 15.6 Standard and special malt analyses of Canadian two- and six-row malts[a]

	Typical values	
	Two-row malts	Six-row malts
Standard analyses[b]		
Hectoliter weight (kg/hl)	52–54	51–53
Acrospire		
0–0.25	0–2	0–2
0.25–0.5	0–4	0–4
0.5–0.75	0–8	0–8
0.75–1	87–94	87–94
Over 1	0–5	0–5
Mealiness (%)		
Mealy	94–98	94–98
Half mealy	2–6	2–6
Glassy	0–4	0–4
Assortment (%)		
on 7/64 in	62–67	30–40
on 6.64 in	26–30	40–55
on 5/64 in	4–8	15–20
<5/64 in	1.2–1.6	1–3
Moisture (%)	3.8–4.2	3.8–4.2
Extract (fine, % d.b.)	79.9–80.9	78.0–79.0
Extract (coarse, % d.b.)	78.1–79.1	76.8–77.8
Fine–coarse extract difference (%)	1.5–2.1	1.0–1.5
Wort colour (°SRM)	1.4–1.7	1.4–1.6
Diastatic power (°ASBC)	120–145	120–145
α-Amylase (DU)	39–49	37–47
Malt protein (%)	10.8–12.3	11.5–12.5
Wort protein (%)	4.9–5.6	4.5–5.25
S/T (%)	40–48	38–42
Special analyses		
FAN (mg/l)	180–220	160–210
Wort viscosity (cP)[c]	1.38–1.48	1.38–1.48
Wort β-glucan (mg/l)	25–150	25–150
Friabilimeter value (%)	70–86	60–76
Fermentability (AAL)[c,d]	78–82	78–82
SO_2, residual (mg/kg)	0–40	0–40
NDMA (μg/kg)	<5.0	<5.0

[a]Values for conventional malt analyses as per ASBC *Methods of Analysis* (ASBC, 1992).
[b]All standard analyses test methods have origins that date back at least 50 years.
[c]Special analysis tests in which the origin of the test dates back at least 50 years.
[d]AAL, apparent attenuation limit.
From Pitz (1990).

during mashing, as in many North American breweries and in grain distilleries (Lévy, 1899; Petit, 1899; Schuster, 1962; Kneen and Dickson, 1967; Bathgate *et al.*, 1978; Bathgate and Cook, 1989; Walker, 1990; Blenkinsop, 1991). Distilleries may use 85–90% cooked, unmalted adjuncts. Breweries use rather less, probably 30–60% in North America, with smaller proportions, down to none, being used in Europe. For these purposes, extract yield is of less importance in judging malt quality than the production of high levels of diastatic enzymes and soluble nitrogenous substances

Table 15.7 Analyses of two Finnish high-amylase content malts. The varieties are spring barley six-rowed types of the 1994 crop (courtesy M. Sipi, Oy Lahden Polttimo AB)

Barley variety	Pokko	Kilta
Moisture (max.; %)	6.5	6.5
Extract (%, fine grind, d.m.)	79.5	79.0
Extract difference (%)	1.0	1.3
Wort viscosity (cP)	1.45	1.45
Protein content (%)	12.5	13.0
Kolbach index (%)	50	55
α-Amylase		
DU (EBC)	108	103
DS (IoB)	100	95
Diastatic power		
°W-K	700	750
°IoB	217	232
Measurable cyanide potential (g/t)	1.2	2.5

that serve as yeast nutrients. For distilling purposes, the levels of diastatic enzymes must be such that not only is the adjunct starch completely dextrinized but it is also degraded as far as possible to relatively simple, fermentable sugars. The yeast then converts these sugars to ethanol in the greatest practical yield. Often thin barley (<2.5 mm) or six-rowed barleys of selected varieties, having high nitrogen contents (1.8–2.3%), are used to produce highly diastatic malts; the barleys are steeped (with aeration and air-rests) and sprayed to a high moisture content (e.g. 47–50%) and are germinated cool, with regular turning, for 6–8 days to allow the maximum levels of enzymes to develop. If permitted, applications of gibberellic acid are likely to be made (Linko and Enari, 1966). For traditional reasons, the use of gibberellic acid-treated malts in making Scotch whiskys is not permitted. The grains are well grown and a proportion may be overshot. The malt is dried with a low air-on temperature (e.g. 33 °C finishing at 55 °C, about 91–131 °F) to maximize enzyme survival. In Britain, most high-diastatic grain distilling malts have α-amylase levels of 50 DU or more, and diastatic power values of at least 200 °L. It is hard to prepare such malts from the barleys currently available, so suitable barleys or finished malts may be imported. By British standards, many North American brewing malts have high diastatic power values (Tables 15.5 and 15.6). Analyses of two malts with exceptionally high α-amylase contents, as used by grain distillers, are given in Table 15.7. It will be remembered that some grain distillers use green malt (see above).

Malt or pot distillers, who distil their products from fermented worts made using all-malt mashes, desire the malt to yield the greatest possible yield of alcohol (the spirit yield) in mixtures that on distillation will have the correct positive characteristics, such as peatiness in Scotch malt whiskys, and without undesirable characteristics, such as off-flavours or substances such as ethyl carbamate (urethane). Such malts should not be 'difficult' to process. Spirit yields depend on the yield of extract, the proportion of the extract that can be converted into alcohol (the percentage fermentability) and the ability of the extract to support adequate yeast growth, but not overgrowth, which is wasteful since when it occurs fermentable extract is diverted into superfluous yeast. Extract yield and fermentability increase and then decrease with malting time, while malting losses continue to increase (Fig. 15.1). Part of the maltster's skill is to know when to stop germination, when the spirit yield is maximal and malting losses are not unacceptably high. It is possible to produce malts that have high extracts but poor spirit yields because the worts are poorly fermentable. The enzymology of starch conversion during mashing, which regulates fermentability, is incompletely understood (Bathgate *et al.*, 1978). On the large scale, spirit yields frequently exceed those predicted from small-scale experiments. Malts for pot distillers are typically made from plump, two-rowed barleys with nitrogen contents of 1.5–1.7%. A wide range of malting schedules are in use, resembling those used to make Pilsener malts. Final kilning is at low temperatures to permit adequate enzyme survival. Melanoidin formation is avoided. The temperature of the air-on may initially be 50–65 °C (122–149 °F), rising to 70 °C (158 °F) after the break. The malts are very pale (their colour is irrelevant for distilled products) and have a moisture content of about 5%.

Originally Scotch malts were kilned using peat as the fuel. Kilning removed some unwanted, green-grain flavours from the malt while the smoke or reek from the peat flavoured the malt and gave characteristic flavours to whiskys made from them. Later peat, mixed with anthracite, was

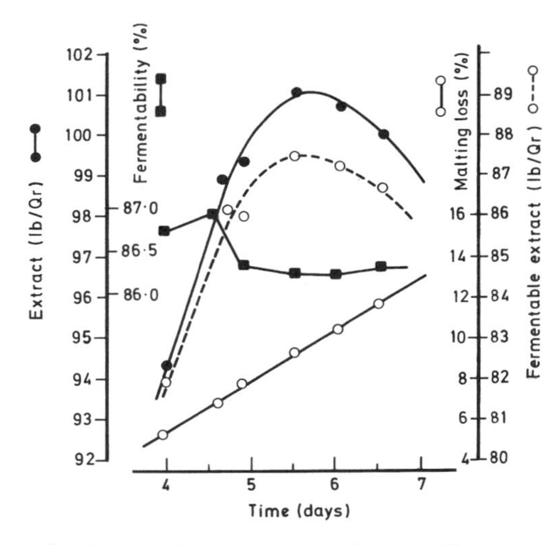

Fig. 15.1 The relationship between hot water extract, fermentable extract, wort fermentability, total malting loss and malting time: Barley, Golden Promise (after Bathgate *et al.*, 1978).

used as fuel. At present, peat is slowly burned, with a deficit of air, in a secondary furnace and the smoke, caused by partial pyrolysis, is conducted into the hot air chamber of the kiln and then in the fan-driven airstream into the malt. Peat is very variable in quality and tends to burn unevenly. To reduce this variability, the peat may be pre-dried before use to a constant moisture content. Peating is still an 'art' and it is important that the correct amount and mixture of phenols and other substances be taken up from the reek by the malts to give the product the correct flavour. Uptake varies with the depth of the grain bed and the stage of kilning. The best uptake occurs when the reek is applied when the green malt has a moisture content of more than 20%. Uptake increases with increasing exposure times. Once taken up, the phenols are not readily desorbed during the later stages of kilning. Because peat burns with a wet flame, peating increases the relative humidity of the air-on and so slows the drying process. Peating can reduce the fermentability of wort from a malt. It is usually necessary to mix and blend peated malts to meet specifications. In Scotland, malts for lowland whiskys are lightly peated (e.g. 1–5 p.p.m. total phenols). Other malts have different values; for example, malts for highland whiskys are heavily peated (having, for example, 15–50 p.p.m. total phenols). Distillery worts are not boiled and so microbes, other than yeast, occur in the mashes and worts and can cause flavour problems, which may be carried forward to the whisky and lower spirit yields. Consequently, distillers may set limits to the microbial populations on their malts. No practical way is known of eliminating the microbes. Another concern is the level of measurable cyanide (MC) in the finished malts. The cyanogenic glucosides in the malts, such as epiheterodendrin, are converted during mashing and fermentation into hydrogen cyanide (prussic acid). During distillation, the volatile cyanide is converted, in contact with copper surfaces, into urethane (ethyl carbamate, a carcinogen (Fig. 4.19)) and this can reach the whisky. The cyanide level varies with distillery design and distillery practice. To overcome this problem, barleys with low cyanide-forming potentials (such as Pipkin, Grit and Derkado) are used to make malts. Measurable cyanide levels of malts vary from 29 to 1200 p.p.b. Derkado makes less than 0.4 g cyanide/t malt. To minimize cyanide production, the most plump grains are used, germination temperature is minimized, as is germination time, moisture content and acrospire growth.

15.9 Other smoked malts

Of the malts flavoured by smoke, peated whisky malts are the only ones now commonly in use. However, this was not always so. When kilns were fired with woods or straw, the selection of the fuel and the way it was burned (with a bright or a smoky fire) influenced the flavour of the product and indeed the smoke contributed a necessary part of the flavour of Porter and amber malts, as they were originally made. Smoked malts are now rarely used for brewing in the UK. In Germany, small amounts of smoked malt (rauch-malz) are produced and this gives its distinctive flavour to Bamberger rauch-bier. The green malt of the chosen degree of modification is kilned while smoke from burning beech logs is allowed to mix with the kiln airstream.

15.10 Vienna-type malts

Vienna-type lager malts (Wiener malz) are darker than those usual in the UK and have colours of about 5–10 EBC units, with values between those of Pilsener and Münchener-type malts (Table 15.2) (Petit, 1899; Schuster, 1962; Narziss, 1976, 1991b; Bemment, 1985). These malts seem to be less popular than was once the case. They are used in the manufacture of some European golden brown lagers. Two-rowed barley, sometimes with higher protein contents than those used in making Pilsener malts, is well (but not over-) modified, sometimes using an elevated temperature on the last day of germination. Initially drying is slow, using air recirculation on the kiln to encourage the formation of soluble sugars and amino acids, and from them melanoidins and hence colour and aroma. In contrast to making caramel malts, endosperm liquefaction must be avoided. When the malt is hand-dry, the temperature is raised and curing is completed at about 90 °C (194 °F). This kilning regime develops the appropriate colour, flavour and aroma while depressing the enzyme content of the malt.

15.11 Munich-type malts

Munich malts (Münchener malz) and some other melanoidin-rich malts are dark, aroma-rich and are used for making dark, rich, aromatic, full-bodied lagers (Lévy, 1899; Petit, 1899; Wahl and Henius, 1908; Schuster, 1962; Narziss, 1976, 1991b; Blenkinsop, 1991). Nitrogen-rich barley (1.84% N; 11.5% protein) of moisture content over 45% is allowed a long, warm period of germination. Often, in the past, the rootlets were allowed to mat together. In an older process, using two-deck kilns, the malt was slowly dried in a slow airstream in a relatively deep bed (24 cm (9.45 in)) at 38 °C (*ca.* 100 °F) to allow a limited amount of stewing, when reducing sugars and amino acids accumulated (Chapter 4). Later, the draught was increased. After 24 h, when the moisture content had fallen to 15–25%, the malt was dropped to the lower floor and the temperature was raised to 75 °C (167 °F) for 12 h, then to 87–100 °C (*ca.* 189–212 °F), or even to 105 °C (221 °F), for 2 h. Using modern deep-loading, single-deck kilns, the green malt is held in warm recirculating air for an extended period to slow evaporation and allow extensive further modification to occur. A quite complex series of stages may be used, with drying periods being followed by periods of recirculation. The highly coloured products (15–30 EBC units) have characteristic flavours and aromas. Being relatively poor in enzymes, the malts have slow conversion times and yield slightly reduced extracts (relative to pale malts) that are not highly fermentable (Table 15.2).

Similar dark malts are dignified by various names. For example, rH malts and melanoidin malts are so called because they slow oxidation-induced flavour changes, promote haze stability in beers made from them and they are dark in colour. Brumalt (Brühmalz) was made by transferring highly steeped, 6-day germinated green malt into special compartments (Bruhaufen) where it was spread in layers about 50 cm (19.7 in) deep. The grain layer was covered and left for 30–40 h. Initially a steep rise in temperature, often to about 50 °C (122 °F) occurred in 24 h and then, as the oxygen was consumed and carbon dioxide accumulated, respiration with its associated heat output was checked and then repressed, allowing the temperature to fall, as occurred in Kropff box malting. Under these conditions, rootlet growth ceased, the starchy endosperm became pulpy and sugars and amino acids accumulated, with a consequent rise in the cold water extract. On kilning, drying at about 65 °C (149 °F) and finishing at less than 100 °C (212 °F), a very dark product (15–30 or even 40 EBC units) was obtained, yielding a sugar and melanoidin-rich wort when mashed. The high buffering and reducing power of the extracts from these malts favoured beer stability. Often they imparted an acidic flavour to beers. These malts were used in proportions of up to 15% of the grist.

15.12 Proteolytic, enzymic and acid malts

A considerable number of special proteolytic, enzymic and acid malts, which are variants of white malts, have been prepared in a variety of ways (Hind, 1940; Schuster, 1962; Vermeylen, 1962; Narziss, 1976, 1991b; Bemment, 1985; Cantrell, 1992). All the malts are enriched in lactic acid so that, when added to a brewhouse grist, they reduce the pH of the mash, ideally adjusting it to the pH optimum for amylolysis and towards the pH optimum of proteolysis. The malts themselves have high cold water extracts (e.g. 29%) and are rich in soluble nitrogenous materials, including FAN. Consequently, they readily form melanoidins and become coloured during kilning. Originally these acid malts were used to offset the unwanted alkalinizing effects of carbonate-rich waters when they were used in mashing, but the worts are also enriched in yeast nutrients and phosphates. Such malts are still used to adjust the pH of mashes, usually being added at the rate of 3–5% of the grist. Even up to 10% may be used, particularly where the direct addition of mineral acid or lactic acid (or acidified wort to the mashing liquor or mash) either is considered undesirable or is forbidden. In addition to improving enzymolysis during mashing, the use of acid malts increases extract recovery (by as much as 0.9% EBC); it is said to improve break formation during the copper-boil and gives a cleaner flavour to pale beers. Standard pale malts have Congress (Analytica-EBC) hot water extract worts with pH values of 5.8–6.1 while acid malts give worts with pH values of 3.8–4.4. Some acid malts carry 2–4% lactic acid. A representative analysis (IoB) of an acid malt is moisture 5–6%, hot water extract >297 l°/kg (d.b.), colour 10–25 EBC units, TSN 0.8–1.2% and lactic acid 2.1–2.5%.

Acid malts are made in various ways. For example germinating barley is held under anaerobic conditions for 24 h or more until it becomes acid. This works better if the grain is first macerated, to give the lactic acid bacteria access to the sugars within the grains. Alternatively, green malt may be sprayed with a suspension of the lactic acid bacterium *Lactobacillus delbrückii* and then be held at about 50 °C (122 °F) for 24–36 h before kilning. Yet another process is to steep kilned malt in water in a well-insulated steep tank held at 47 °C (116.6 °F). Sugars leach from the malt and the lactic acid bacteria multiply and convert the sugars to lactic acid, which accumulates to a concentration of 0.7–1.2% in the liquid in 24–30 h. The malt is removed and kiln-dried at 60–65 °C (140–149 °F) to a moisture content of about 5.5%, so the non-volatile lactic acid becomes concentrated to 2–4% on the malt. The liquor is saved and may be used once or twice more. With reuse, the desired acid concentration is reached in a shorter time. An older method of making 'Dixon's enzymic malt', an acid malt, was to spray the germinating barley twice or more with solutions of lactic acid (10%) and to steep it in a lactic acid solution before kilning. The application of lactic acid checks root growth. Most often lactic acid is prepared by the biological acidification, or 'souring', of mashes or worts. Occasionally, where allowed, chemically prepared lactic acid or biologically soured skimmed milk has been used.

15.13 Food and vinegar malts

Food and vinegar malts are a diverse mixture of products tailored to meet particular requirements. They are not all white malts. Vinegar-brewers require high spirit yields to give ethanol, which will be converted into acetic acid during the acetification stage, and so their malts resemble those needed by distillers. Generally, if unmalted adjuncts are used in a vinegar-brewer's grist, the nitrogen content and the diastatic power of the malt will be higher, resembling the malts used by grain distillers. 'Neutral spirits' are sometimes derived from mashes in which various starchy adjuncts, or starches, are converted with highly enzymic malts before fermentation and rectification.

Spirit vinegars are made from neutral spirits, but these need not have been derived from malt mashes.

Food malts are of many different types, ranging from the pale, enzyme-rich white malts to highly coloured products such as the enzyme-poor caramel malts. They have a wide range of uses (Chapter 1). Food malts are, or have been, variously made from barley, wheat, triticale, rye and oats in the UK; and elsewhere various millets, sorghum and maize malts are used in foods. The malts may be used entire, cracked, ground as flours, or as syrupy or dry extracts. They may be enzyme rich or virtually enzyme free depending on their intended use. In general, food malts made from barley are manufactured in the same ways as brewing malts, but they tend to be prepared from protein-rich, high-nitrogen barleys (e.g. TN 1.8–2.2%).

Crystal and caramel malts (see below) and preparations made from them are often used as sources of flavour. The darker, special malts are increasingly used as alternatives for caramels, as sources of colouring for foodstuffs and confectionery. In general, food malts may also be used as sources of enzymes, of sweetness, of texture, of mass (especially in some sweets) and for nutritional purposes. Extracts may be used as carriers, e.g. of vitamin-rich fish oils for dosing children.

Roasted malts and roasted barleys are used as the bases of various hot drinks approximating to tea or coffee (e.g. Mugi-cha in Japan; Malzkaffee in Germany (Briggs, 1978)). Extracts from these materials may be dried in drums or spray-driers (Chapter 12) and are used in 'instant' drinks that are reconstituted by the addition of hot water or hot milk. Malt extracts are used in the preparation of various non-alcoholic beverages and, less often, in some commercial breweries. They are widely used in home brewing.

15.14 Special barley malts

Special malts are usually regarded as being those products that, in modern practice, are made or are finished in roasting cylinders. They were not always made in this fashion (Tizard, 1843; White, 1860; Briant, 1907; Wahl and Henius, 1908; Hind, 1940; de Clerck, 1952; Schuster, 1962; Vermeylen, 1962; Narziss, 1976, 1991b; Bemment, 1978, 1985; Kieninger and Wiest, 1982; Blenkinsop, 1991; Cantrell, 1992; Jupp, 1994). Special malts include the light and dark caramel and crystal malts, amber, diamber, brown, chocolate and black malts and, by tradition in the UK, roasted barley. Some of these malts (e.g. carapils) are extremely pale (Table 15.8) while some kiln-made malts (e.g. Munich malt) are relatively dark. While most special malts are made from barley, some small amounts are made from other cereals, such as wheat.

In the UK, roasting was first permitted under the 'Roasted Maltsters' Act' (*sic*) of 1842. Because the process was patented, roasted malts were

Table 15.8 Analyses of some German caramel and dark malts, compared with a pale, standard EBC malt (Congress analyses)

Malt type	Pale (EBC) Malt	Carapils	Light caramel (Carahell)	Dark caramel (Caramünch)	Coloured malt[a] (Farbmalz)
Moisture content (%)	5.2	8.4(7–8)	8.4(5–8)	6.7(5–7)	7.7(1–2)
Extract (% d.m.)	80.8	77.9(77–80)	77.3(76–77)	75.9(76–77)	70.4(60–65)
Protein content (% d.m.)	10.2	12.4	12.5	12.1	11.9
Soluble N (mg/100g d.m.)	651	665	643	609	551
Kolbach index (%)	39.8	33.5	32.2	31.5	28.9
Wort viscosity (cP)	1.51	1.57	1.60	1.57	–
pH	5.94	5.75	5.48	5.30	5.07
Titrable acidity (ml N NaOH)	12.8	13.6	15.6	16.5	17.2
Colour (EBC units)	3.3	4.4(2–5)	22.5(20–35)	126(50–300)	1450(1300–1600)

[a]Unusually high extract and moisture content.
From Narziss (1976) with additions from various other sources.

sometimes referred to as ‘patent malts’. To produce high-quality products, plump, even-sized grains must be used and malts must be evenly grown and modified and evenly coloured. The objective is to obtain a uniform, homogeneous product. Absolutely no burning or charring must occur. Although drums are now well instrumented and may be controlled by a pre-set programme, it is still necessary to have operations overseen by a skilled person who continually takes samples and assesses them for texture and/or colour, who can quickly change the conditions in the drums or stop the process. Modern drums usually take charges of 0.5–5 t (Chapter 9). The contents may be directly heated, when the hot gases move through the grain and cause drying. Alternatively, the drums may be closed and indirect heating applied to the outside. Under these conditions, evaporation is largely prevented and stewing can occur. Because of the high dry-weight losses incurred, many of the special malts are relatively expensive to make.

15.15 Crystal and caramel malts

Crystal and caramel malts are distinct because, in contrast to all other malts, the grains are deliberately stewed: held wet and warm at such a temperature that the contents of the endosperm mashes and liquefies, becoming replaced by sugary liquid. When the product is cooled and dried, the liquid solidifies and the endosperm is replaced by a semicrystalline or glassy barley-sugar-like mass. Crystal and caramel malts have characteristic and delicious flavours.

In the past, green malt, sometimes given a preliminary sprinkle or wetting with water, was loaded onto a kiln in a thick layer, trampled down and covered with a tarpaulin to reduce evaporation, and the kiln fan (if present) was shut off. The temperature of the malt bed was then increased to about 60–75 °C (140–167 °F) and held for 30–45 min, or even for up to 2 h. During this period the grain endosperms liquefied. The tarpaulin was then removed, the temperature was raised and the grain bed was ventilated to dry the malt, caramelize the sugars and form melanoidins. If the temperature was raised very little, caramelization and melanoidin formation were minimal and very pale carapils malt was produced, a very sweet product with little caramel flavour. If, however, the temperature was elevated to 120–150 °C (248–302 °F) and held for 1–2 h, relatively dark products were obtained.

At present crystal malts are finished in drums. Fully modified (not over-modified) green malt or, less economically, very pale malt rewetted by steeping (to up to 50% moisture) is loaded into a roasting cylinder. For 5–10 min the cylinder is fired directly (at about 50 °C (122 °F)) to remove the surface moisture from the corns. The drum is then closed to prevent further evaporation, and the temperature is slowly increased to 65–70 °C (149–158 °F) with external heating, to obtain the maximum production of reducing sugars and until the endosperm has been replaced by a clear, sweet liquid that emerges if the corns are squeezed. The temperature may then be increased suddenly. Alternatively, the temperature may be slowly and continuously increased, over 40 min, to 100 °C (212 °F). Changing the moisture content and temperature alters the nature of the product, for example the levels of reducing sugars. When the corn contents have liquefied, direct heating is resumed to dry them and to develop the desired colour, flavour and aroma. For carapils drying is at a low temperature (55–60 °C (131–140 °F)). The final temperature is higher for more highly coloured products, being in the range 120–160 °C (248–320 °F), higher temperatures giving darker coloured products. The malt grains are smooth, round, swollen, bright and evenly coloured without any dark spots. When cut across, more than 90% should appear hard and glassy, not floury or mealy. These malts are rich in reductones and reducing sugars. Their addition to grists gives beers with characteristic flavours, having greater body, with less tendency to become hazy or to have flavour deterioration. Caramel malts favour palate-fullness, and head (foam) formation and retention. They also confer a golden colour to beers. Some analyses of German caramel malts are given in Table 15.8. The data illustrate the fact that the darker the malts the lower the extract yields and the lower the pH values of the worts. British caramel malts come in a wide range of colours from the lightest (20 EBC units) to quite dark colours (300–500 EBC units). Colours of about 140 EBC units are the most popular. Extracts of these malts are in the range 260–285 l°/kg, or 76–80% (Analytica-EBC, d.b.). Moisture contents are in the range 3–7.5%. The very pale caramel malts are sweet to taste. As progressively darker malts are investigated, their flavours become more caramel- and toffee-like, more malty, aromatic, honey-like and luscious. In the very dark malts, newer flavour-notes (burnt flavours) are detectable. These malts should be used soon after manufacture, as their characteristic flavours and aromas are lost during storage.

15.16 Amber malts

The manner in which amber malt is made and, no doubt, its character have altered a great deal (White, 1860; Hind, 1940; Bemment, 1978, 1985; Jupp, 1994). Historically, amber malts were made by carefully drying well-modified, green malt on a coke-fired kiln having a metal or woven wire deck, and then suddenly raising the temperature to 85 °C (185 °F) by adding billets of oak or other hardwoods to the furnace. It is believed that this process was used to some slight extent in the UK as recently as 1959. The product had a 'fine amber colour' when ground up and had a smoky 'empyreumatic' flavour, imparted by the wood smoke. While most amber malts were made from barley, some were made from wheat. The modern product most nearly resembling the original amber malt, because of its wood-smoke flavour, is German raüchmalz.

At present amber malts are made in roasting drums. They have no smoked flavour. Normally finished pale-ale malt or, more usually, mild-ale malt is heated indirectly in a roasting drum with a temperature profile rising from 49 °C (*ca.* 120 °F) to as much as 170 °C (338 °F). Lower finishing temperatures may be used, the value chosen depending on the specification of the product. The malts, which are amber-coloured, have a pleasant, dry, baked or biscuit-like flavour. They are used in special ales that are golden coloured and have dry palates. These malts lack enzymes. They give extracts of 270–285 l°/kg and colours of 40–85 EBC units, with values of about 60 EBC units being most common. Their moisture contents are low: typically less than 3.5%.

15.17 Porter, brown, 'snapped' or 'blown' malts

Traditional blown, porter and brown malts are no longer made (Tizard, 1843; White, 1860; Stopes, 1885; Bemment, 1978). In use, porter malts probably gave low extracts with poor and variable fermentabilities. They were used in making rich, dark acidic beers called porters. They were displaced in brewing use by mixtures of other dark malts and pale malts' which together gave superior extract recoveries; the newer 'porter' beers undoubtedly differ in character from the old. Porter malts were made on special kilns with long, tapering chimneys and swinging cowls. The kiln deck was of woven wire or thin perforated iron plates, and the fire was well below the deck. Water was always available to quench the malt and to put it out if/when the malt (apparently not infrequently) caught fire. Describing a porter-kiln fire, White (1860) writes, 'When an accident of this kind happens, the flame shoots out from the top of the kiln several yards in height

. . . accompanied by a humming noise resembling a deep-toned organ'. Uniformly germinated and well-withered (air-dried) green malt was spread in the kiln to a maximum depth of 3.8 cm (1.5 in). Sometimes out-of-season summer malts, blue with moulds, were said to have been used in making some porter malts, because the mouldiness of the green malt could not be detected in the finished product. The kilns were fuelled with wood. Initially, moderate heat was used to partly dry the malt, the fire being dampened down at intervals. Then the fire was allowed to go out and, after cooling the kiln, the malt was carefully turned. Occasionally, it was sprinkled at this stage 'to toughen the husk'. The fire was then renewed and the heat was intense as the hardwood billets and faggots (oak, elm, beech, birch, hazel, ash and hornbeam were used, with a preference for beechwood, 4–8 in (10.2–20.3 cm) in diameter) blazed up, sometimes supplemented with straw. The blaze was regulated by sprinkling with water. Estimates vary, but the first kilning stage might be carried out at 104–121 °C (*ca.* 220–250 °F) over 1.75 h. The final stage might be at 160–170 °C (*ca.* 320–340 °F). In the intense heat, the grains became swollen (by 20–25%) and burst with loud snapping noises. Hence the terms 'snapped' and 'blown' malt.

The product was a dark, deep rich brown and had a caramel, bitter flavour with a harsh smoky characteristic imparted by the wood smoke. Malt colours were judged on the powdered product, which had a biscuity (not caramelized) appearance when cut. The product must have been deficient in enzymes and produced poorly fermentable worts. Old analyses indicate that these malts had extracts of 88–90 lb/Qr (*ca.* 262–267 l°/kg), colours of 125° Lovibond, and moisture contents of around 1%. Despite the important place of porter in the history of brewing in the UK (Mathias, 1959) true porter malts are no longer made and indeed Stopes (1885) referred to their use as 'an absurdity fast dying out'.

The nearest modern equivalents to porter malts are brown or drum-brown malts. They lack any wood-smoke flavour and are not snapped or blown. The brown malts are prepared from green malts or pale malts in roasting drums using direct heating, so evaporation is not limited. Temperatures are gradually increased to about 130 °C (266 °F). These brown malts are used in making some specialist bottled beers, brown ales and sweet stouts. They have a dryer, less sweet character than crystal malts, having the same colours, which are normally in the range 90–180 EBC units, most being about 130 EBC units. These malts have extracts of 265–275 l°/kg and moisture contents below 3.5%.

15.18 Roasted barley

In the UK roasted barley is regarded as a 'special malt' (Hind, 1940; Bemment, 1978, 1985; Blenkinsop, 1991; Jupp, 1994). In some countries, the use of unmalted grain is forbidden and so chit malt may be roasted.

Well-dressed, bold and sound barley with a moisture content of 12–16% is used as starting material. Its malting quality is irrelevant. The roasting process, which is carried out in drums using direct firing, takes about 2.5 h. The temperature is raised over the first 2 h from 80 °C (176 °F) to 180 °C (356 °F). Subsequently, it is raised to 230 °C (446 °F). The very rapid colour development that occurs is monitored by inspecting grain samples, which are taken every 2–3 min. The amount of heat applied in the final stages is continually reduced until, when the temperature reaches about 215 °C (419 °F), the burners are turned off. However the grain is near combustion and the temperature continues to rise spontaneously for the last 5 min or so. When the operator judges the colour is correct, and before the load catches fire, the barley is cooled with quenching sprays of water, variously given as 45 or 200 l/t. The risk of fire is great and roasthouses are generally isolated and constructed with this risk in mind. The roasted grains appear reddish black, shiny and swollen, some 50% being split and having doubled in size. The product should have no charred or carbonized grains. The flavour is characteristic, being variously described as sharp, dry, astringent, acidic, hard and burnt with no hint of sweetness. The flavour is distinct from the flavours of roasted malts. Roasted barley is used in making particular stouts, to which it imparts pleasant, characteristic flavours. Roasted barleys usually have colours of 1200–1400 EBC units, although values of 1500 EBC units have been noted. The products lack any enzymic activity. Hot water extracts are in the range 260–275 l°/kg. Moisture contents are commonly 2% or less.

It may be noted that washed raw barley has been used as an adjunct in some breweries and that 'high-dried barley', dried to a moisture content of about 5%, has been used in some Irish whiskey mashes after being ground between stones or in hammer mills. These materials are not regarded as special malts.

15.19 Chocolate and black, roasted malts

Chocolate and black-roasted malts are highly coloured 'patent malt' products made in a manner closely resembling that used for roasting barley. They have no active enzyme complements, and their characters are quite distinct from those of less highly coloured malts. Typically they are used at rates of 3–5% of a brewery grist to give character (particularly colour and particular flavours) to the final product, usually a stout. Plump barley, with a moderate nitrogen content (1.5–1.7%), is malted to a slightly under-modified lager specification. After 4–6 days of germination, the green malt is lightly kiln-dried to a moisture content of 4–6% without curing. This material, which can be stored, is transferred in batches to a roasting drum that is used about 0.5–0.75 full and rotates at 15–20 r.p.m. The roasting process begins at about 75 °C (167 °F) and the temperature is increased over about 30 min to 160–175 °C (320–347 °F), when the malt emits white fumes. These are extremely unpleasant and must be destroyed in the drum's fume-destruction system (Chapter 9). Gradually the temperature is increased further for a period and to a degree dictated by how dark a product is being produced, for example to 215 °C (419 °F) over 30 min. At this stage, the vapours being emitted become bluish and have an acrid smell. Chocolate malts may be finished at this temperature, but with black malts the temperature is increased further to about 220–225 °C (428–437 °F) (below that used for roasting barley), and frequent colour checks are made. When the operator judges the colour to be correct, the heat is turned off and the malt is sprinkled. The water instantly turns to steam and the malt swells. In all, about 15% of the dry weight is lost during roasting as dust and fumes. In a German roasting programme, perfectly modified pale malts, with moisture contents of 4–5%, are then mixed in a closed drum heated at 70 °C (158 °F) for 2 h, then the roasting process at 173–200 °C (343–392 °F) is carried out for 1.5 h (Narziss, 1991b). Depending on the grade being made, the colour should be chocolate brown to black. The husk should appear shiny and polished. When cut, the corns must appear of the right colour and floury, and the endosperm must be friable, not steely, glassy or charred. Grains should not have been turned to charcoal, which readily occurs at temperatures above 250 °C (482 °F). Pale chocolate malts have colours of about 500–600 EBC units, while more standard chocolate malts have colours of 900–1100 EBC units and roasted, or black malts of 1200–1400 EBC units. Different companies use different colour ranges to define these malts. Hot water extracts are usually in the range 255–275 l°/kg and moisture contents are 2% or less. The flavours of these two malts are variously described as dry, burnt, acid and astringent but they are clearly distinct from the flavour of roast barley and often a hint of a rich, sweet taste can be detected.

15.20 Malts made from cereals other than barley

Malts made from cereal grains other than barley, are produced in relatively small amounts. Some are used in brewing European-style beers, others are used in opaque beer manufacture, in making distilled products, in foodstuffs and in confectionery. Much less information is available for non-barley malts. Broadly the malts may be divided into those made from temperate-zone cereals and those made from 'tropical' cereals. The former are taken to be wheat, rye, triticale and oats. The 'tropical' cereals are sorghum, maize, various millets and rice. No attempt is made to discuss the properties of germinated or malted peas, beans, lupin seeds, vetch seeds or buckwheat, although, historically, peas and beans were malted in England while mixed with barley, and germinated legume seeds are valuable foods in some areas of the world (Chapter 1). Of the temperate cereals commonly cultivated, only barley and oat grains are husked. The naked (huskless) nature of wheat, rye and triticale gives rise to practical problems in the malthouse. However, in many respects, these cereals resemble barley in their malting characteristics and many of the methods of analysis applicable to barley and barley malt can be applied to these cereals with a fair degree of confidence. With the possible exception of wheat, none of these cereals has been selected by plant breeders to improve their malting quality. With the 'tropical' cereals the situation is less satisfactory, although current studies are filling many gaps in our knowledge. The conditions under which the 'tropical' cereals can be malted in optimal ways are slowly being defined and are not the same as those applicable to barley. Work by plant breeders on selecting 'superior malting lines' of 'tropical' cereals is only just beginning. The indiscriminate application to malts made from 'tropical' cereals of methods of analysis developed for use with barley malts is often unsound and has given rise to misconceptions and errors.

Again, while clear-beer brewers and distillers generally know what analyses they require for their barley and wheat malts, many other malt users are less clear. Where malts are made at home or in small village communities, quality control, if any, is rudimentary in the extreme.

15.20.1 Wheat malts

Wheat malts are prepared on the industrial scale, and have been for many years (Lévy, 1899; Wahl and Henius, 1908; Leberle, 1930; Hind, 1940; de Clerck, 1952; Fleming *et al.*, 1960, 1961; Narziss, 1976; Bemment, 1985; Blenkinsop, 1991; Lersrutaiyotin *et al.*, 1991; Byrne *et al.*, 1993). Most wheat malts used in brewing beers are pale and lightly kilned. Many of those used in foodstuffs are dark and give sweet and luscious flavours to the products containing them, as well as colour, texture and fibre. Wheat malts may be used in foodstuffs entire, lightly crushed, kibbled, milled to a flour or as flakes. The absence of an abrasive husk is a substantial advantage over barley malt in foodstuffs. In the UK, wheat malts may be used in the grist for beers and stouts (at the rate of 3–10%) to improve the head formation and head retention of the beer and possibly to enhance the diastatic activity of the grist and the supply of yeast nutrients. Wheat malt lacks husk and this contributes to the way in which its presence in the grist slows wort separation. In contrast to barley malts, wheat malts contain very little polyphenolic materials that are tannins. In Europe, much larger proportions of wheat malt may be used in particular beers. For example, in the traditional wheat beers of Germany (Weizenbier or Weissbier) 75–80% of the grist may be wheat malt. Wheats are not widely developed and selected by breeders for their malting qualities, but regular reports appear on the comparative malting characteristics of German wheats (e.g. Sacher and Narziss, 1991, 1992).

Wheat for malting should be undamaged, fully viable and free from dormancy and should have the minimum contamination level of microbes, particularly *Fusarium* spp. As with barley, there is an inverse correlation between total nitrogen ('protein') content and extract yield. Wheats have TN values in the range of 1.5–3.1%. For brewing purposes, wheats with TN values of less than 1.6%, and certainly less than 1.9%, are preferred. For food malts, grain lots with higher nitrogen contents may be used with advantage. Because the grain lacks a protective husk it must, like all naked grains, be handled with care to minimize physical damage. The lack of husk allows wheat to pack down tightly in the steep, in the germination compartment and on the kiln. Because the grain is readily crushed, it is preferable to load steeps, germination compartments and kilns less deeply than would be usual with barley. Furthermore, the coleoptile (acrospire), in contrast to that of barley, grows free and is easily damaged, impairing malting and giving a risk of mould growth on the damaged tissue. To avoid this, grain should be turned gently. Water uptake is more rapid than with barley (steeping times of 38–55 h being common), probably because there is no husk to resist grain swelling. Hard wheats take up water more slowly than soft wheats. Often antisepctic agents have been incorporated in the steep water to minimize problems with moulds. Anti-microbial agents used have included lime water and calcium hypochlorite and numerous others have been tested. It is necessary to change the steep water from time to time and it is desirable to aerate the grain–water mixture to loosen the grain and minimize packing. Traditionally, wheat was under-steeped to a moisture content of about 42% and was germinated cool (at less than 16 °C (60.8 °F)) on floors in thin layers, with minimal turning carried out gently, for a short period of about 5 days before kilning. The product was under-modified but was produced in this way to minimize the risk of the grain becoming mouldy. However, for adequate modification the grain should be steeped to a moisture content of 45% and be germinated for not less than 5 days at 12–15 °C (53.6–59 °F) (Piendl, 1975). The changes undergone by the grain are similar to those occurring in malting barley (Table 15.9).

Wheat malting is hastened and α-amylase levels are enhanced by applications of gibberellic acid (Fleming and Johnson, 1961; Sethi and Bains, 1978; Singh *et al.*, 1983). In wheat malts produced by using a 5-day germination, applications of gibberellic acid increased malt extracts from 82.5 to 87.6% (Singh and Sosulski, 1985). The absence of extract-poor husk material explains why the yield of extract from wheat malts can substantially exceed those from barley malts with comparable nitrogen contents (e.g. Table 15.10). For maximum survival of diastatic power wheat malts should be kilned with air-on temperatures not exceeding 43 °C (109.4 °F). Temperatures of 35–40 °C (95–104 °F) are often used to finish kilning wheat malts (Fleming *et al.*, 1960; Narziss, 1976).

Figure 15.9 The influence of germination time on analyses of wheat malt

	Germination time (days)									
	0	1	2	3	4	5	6	7	8	9
Moisture content (%)	4.7	4.1	4.3	5.1	4.9	6.3	6.1	6.6	5.2	4.9
Extract (fine grind, % d.b.)	69.2	80.6	82.8	83.9	83.4	84.0	84.3	84.3	84.0	84.0
Extract (coarse grind, % d.b.)	62.2	76.2	79.0	81.8	81.6	81.7	82.7	82.4	82.5	82.6
Fine–coarse extract difference (%)	7.0	4.4	3.8	2.1	1.8	2.3	1.6	1.9	1.5	1.4
'Saccharification time' (min)	60	60	55	15–20	15–20	15–20	10–15	5–10	5–10	5–10
Colour (EBC units)	2.0	3.1	3.9	4.1	4.3	4.6	4.1	4.2	7.0	6.6
Protein (N × 6.25, % d.b.)	11.6	11.4	11.6	11.6	11.8	11.7	11.6	11.7	11.7	11.6
Soluble nitrogen (mg/100 g)	317	417	544	545	560	595	626	670	673	711
Kolbach index (%)	17.0	21.6	32.1	29.5	29.8	31.8	33.6	35.8	35.8	38.2
pH	6.03	6.12	6.09	6.07	6.05	6.14	6.05	6.06	6.14	6.02
Wort viscosity (cP)	2.587	2.031	2.045	1.916	1.852	1.876	1.801	1.810	1.834	1.778
VZ 45 °C (%)	17.8	18.4	23.5	29.2	32.9	34.8	39.1	41.5	38.7	37.6
Brabender hardness (relative)	700	650	640	600	600	612	585	600	610	610
Total malting loss (%)	1.0	2.6	4.1	5.2	5.7	5.4	5.4	6.5	7.7	9.9

After Piendl (1975).

Table 15.10 Comparative analyses of various malts

	Barley	Wheat	Triticale	Sorghum[a]
Extract				
% d.m.	81.0	85.7	88.5	80.8
1°/kg d.m.	312	328	335	–
Moisture (%)	4.3	8.1	4.5	6.5
Colour (EBC units)	3.5	8.0	16.0	–
TN (% d.m.)	1.65	2.18	2.24	1.63
TSN (% d.m.)	0.61	1.12	1.50	0.50
Kolbach index (%)	37	51	67	33
α-Amylase (DU, d.m.)	35	55	126	40
Diastatic power (IoB)	66	106	120	23
FAN (mg % d.m.)	84	143	201	70

[a] Sorghum malt prepared on the small scale, using 0.1% formaldehyde in the steep. A special mashing procedure was used with the sorghum malt (see text).
After Byrne *et al.* (1993).

In addition to having high extract values, wheat malts are often rich in FAN, necessary for yeast nutrition. However, when mashed, wheat malts tend to give slow wort separations, possibly because the pentosans that are present sometimes give rise to viscous worts. The 70 °C mash wort viscosity values are often high. Wort pH values tend to be higher than those from barley malts. A wide range of malts, including roasted malts, can be prepared from wheats. It is believed that smoked wheat malts are being made in eastern Europe. In comparative trials making pale malts from a range of wheat varieties, Sacher and Narziss (1992) noted that samples with higher nitrogen contents malted less well than those with lower nitrogen contents, and that Kolbach indices were generally lower than those obtained with barley malts. The ranges of some of the analytical values obtained were hot water extract 81–87% d.m. (EBC), fine–coarse extract difference 0.8–2.5%, α-amylase 30–120 DU, diastatic power 260–440 °W–K, colour of Congress worts 2.7–8.5 EBC units and malting losses 7.8–10% (Tables 15.9, 15.10, 15.12).

15.20.2 Rye malts

Rye grains are naked, so malting them has all the problems (risk of tight packing and crushing, ease of damage) and all advantages (rapid water uptake, high potential extract values) of wheat. Therefore, rye must be used with all the precautions appropriate for making wheat malts. However, the products are substantially different. The grains of rye are relatively small and thin and, because this cereal is wind pollinated, it is generatically inhomogeneous (Chapter 2). Little of this grain is malted in the UK. Elsewhere, considerable amounts of rye may be malted, for use in preparing rye whiskey for example. In the past rye malt was used to a limited extent in brewing but evidently it caused slow wort separation and favoured haze instability in beer (Wahl and Henius, 1908). Rye malts can have very high levels of starch-degarding enzymes, which makes them attractive to distillers using adjunct-rich mashes. Ryes differ widely in their malting characteristics (Narziss *et al.*, 1965; Narziss and Hunkel, 1968; Lersrutaiyotin *et al.*, 1991) (Table 15.11). Extract values may be extraordinarily high; values over 91% have been recorded. Pomeranz (1974) reported that a rye malt with a TN value of 2.73% had an extract of 82% and α- and β-amylase activities comparable to those of barley malt. The proportion of grain nitrogen becoming soluble is very high. It is believed that most rye malts are pale and are kilned at low temperatures to maximize enzyme survival. However, caramel malts have been prepared from rye.

Steeping rye is a rapid process but 7 days' germination time may be needed to complete modification, unless it is accelerated by applications of gibberellic acid. In the past, rye was sometimes malted deliberately mixed with barley, apparently because it was believed that the barley protected the rye acrospires from damage during turning (Lévy, 1899; Petit, 1899).

15.20.3 Malts prepared from triticale

The high grain yields obtained with some newer varieties of the 'synthetic' cereal triticale (*Triticosecale*) has led to a search for markets for the grain. Inevitably trials of malting quality have been carried out. Wide differences in malting behaviour are found between different varieties (Pomeranz *et al.*, 1970; Pomeranz, 1971, 1974; Charalambous and Bruckner, 1977; Tombros and Briggs, 1984; Gupta *et al.*, 1985; Antkiewicz, 1987; Perez-Escalmilla *et al.*, 1988; Sobral, 1989; Blanchflower and Briggs, 1991a,b,c; Lersrutaiyotin *et al.*, 1991; Byrne *et al.*, 1993). No attempts to breed triticale for malting quality seem to have been made.

Table 15.11 Ranges of Congress analyses of rye malts prepared from different varieties and germinated for 5 or 7 days

Moisture (%)	5.1–7.5
pH	5.8–6.1
Colour (EBC units)	5.25–7.25
Extract (% d.b.)	85.0–90.3
Fine–coarse extract difference (%)	1.1–3.0
VZ 45 °C (%)	54.6–70.4
Diastatic power (°W-K)	300–651
α-Amylase (DU)	61–137
Dextrinization time (min)	0–10–15
Protein (%)	7.3–14.1
TSN (mg/100 ml)	707–1259
Kolbach index (%)	52–69
Formol-N (mg/kg d.m.)	168–294

Data of Narziss and Hunkel (1968).

Plump grains of sound samples of triticale behave much like wheat and take up water quickly and malt well. Several reports stress a need to use anti-microbial agents in the steeping liquor. Gibberellic acid hastens malting and increases malt extract and the levels of soluble nitrogen but potassium bromate fails to reduce the level of soluble nitrogen and sometimes even enhances it. The final pale malts often have excellent extracts (Table 15.10), values in excess of 90% and up to 370 l°/kg have been reported, as with rye malts (Freeman, 1986; Blanchflower and Briggs, 1991a; Lersrutaiyotin *et al.*, 1991). Friability values may be low. Like rye, the nitrogen contents of the triticale malt grains tend to be high and the levels of TSN and FAN are very high indeed. The malts are rich in enzymes. Germination times of 5 days seem to be adequate and, despite the fact that the acrospire (coleoptile and first leaf) is removed with the rootlets, malting losses are not excessive, being around 11%. Worts are slow to separate from small-scale mashes, both because of their high viscosities (owing to pentosans) and because of the way in which the malted grains break up when milled. The 70 °C mash wort viscosities are very high (e.g. 10–27 cSt). In addition, worts develop intractable turbidities owing to finely divided, suspended protein (Blanchflower and Briggs, 1991c). Malts readily darken on kilning. The characteristics of pale triticale malts make them unsuitable for use in beer brewing. They do appear to be ideally suited for use, perhaps when more highly kilned, to develop flavour and aroma in foodstuffs, and their high enzymic activities and high levels of soluble nitrogen make pale triticale malts potentially attractive to grain distillers (Freeman, 1986).

15.20.4 Malted oats

Malted oats were used sporadically by European brewers for many centuries, but their use is now rare. Sometimes oats were malted mixed with barley. Limited quantities of oat malts are used in some breakfast cereals and possibly some bakery products. Because oat malts are now unusual, there are few data on them (Markham, 1615; Lévy, 1899; Petit, 1899; Wahl and Henius, 1908; Weichherz, 1928; Hind, 1940; Potter, 1953; Larsson and Sandberg, 1995). Oats have large proportions of extract-poor husk, about 30% compared with barley's 10%, and so inevitably oat malts have relatively low extract values, often only about 70–75% of those of barley malts. However, the high husk content favoured an 'open' mash structure and rapid wort separation under older infusion-mashing conditions. Indeed, oat husks were sometimes added to barley-malt mashes to 'aid drainage'. Oat malts are poor in both α-amylase and β-amylase and, since the raw grain is rich in β-glucans, it is necessary to modify the grain well to avoid producing highly viscous, slow-draining worts. Oats reportedly take up water rapidly during steeping and malt rapidly. However, oats are rich in oils that are supposed to suppress beer foam and certainly could go rancid during storage, leading to off-flavours in beer. Kilning was reckoned to be slow, and this has been attributed to the high proportion of husk. Some brewers used about 5% of oat malt in grists for particular stouts, but there was no clear agreement why this was done. There was a widespread belief, however, that oat malts could cause slow wort separation, flavour problems, turbid worts and abnormal 'stormy' fermentations in breweries. Hind (1940) gave the following analyses: hot water extract 71.5 lb/Qr (roughly equivalent to 210 l°/kg), total nitrogen 2.05%, PSN : TN ratio 24.1%, TCW 28.6 g, cold water extract 14%, diastatic power 21 °L, and moisture content 2.3%.

15.20.5 Malts from 'tropical' cereals

Until recently the 'tropical' cereals were only malted in the home to make porridges, opaque beers and possibly other foodstuffs. However, the industrialization of the manufacture of Bantu (Kaffir) opaque beer in South Africa (Chapter 1) initiated an ongoing programme of research and development related to sorghum malting. More recently, the restriction or banning of the importation of barley malts into various African countries is resulting in strenuous efforts to make clear, lager-type beers from locally grown malted cereals. Major problems have arisen in analysing these malts to assess their brewing value. These problems will be discussed in connection with sorghum malts.

Malts of tropical cereals are also being made to incorporate into high-nutritive-density baby weaning foods (Chapter 1). Some considerable amounts of data are available regarding this use for malts. It is unclear if these malts for weaning foods are made on a commercial scale.

15.20.6 Sorghum malts

Sorghum has long been malted for the production of opaque Bantu beers and related products. The relative nutritional values of sorghum and maize are increased by malting, and malted sorghum has been used in making low viscosity grists for feeding children at weaning (Chapter 1). The introduction of high-yielding American forage sorghums to South Africa created problems because the grains had poor malting qualities and were unsuitable for making Bantu beer. Efforts are being made in several African countries to produce new varieties with grain having better malting qualities. Varieties of sorghum are known to vary very widely in their malting characteristics (Okon and Uwaifo, 1985; Aisien and Muts, 1987; Palmer *et al.*, 1989; Evans and Taylor, 1990; Ratnavathi and Bala-Ravi, 1991; Wenzel and Pretorius, 1994). Jayatissa *et al.* (1980) reported that malts made from Sri Lankan sorghums had malting losses of 4.2–19.4% after 4 days of germination and 8.4–30.3% after 6 days of germination. Malting losses approaching 20% were noted by Nout and Davies (1982). Extracts, determined by mashing for 1 h at 45 °C (113 °F) and then 3 h at 70 °C (158 °F) were in the ranges 17.5–69.8% for 4-day grown malts and 21.4–76.4% for 6-day grown malts. Pathirana *et al.* (1983), using the same mashing system, obtained extracts of 82–88% for 4–5 day germinated malts. The best conditions for malting sorghum differ substantially from those appropriate for barley.

The problems of interpreting experimental results are confused above all by the use of methods of analysis that may not be appropriate for sorghum malts or malts made from other 'tropical' cereals. This is true for the methods adopted for analysing the amylases and for determining the hot water extracts. The necessity to modify hot water extract determinations and use a different technique to that employed for barley malts was recently recognized. With barley malt, the extract is determined by mashing ground malt samples at 65 °C (149 °F) (IoB) or with a rising temperature programme to 70 °C (158 °F) (Analytica-EBC). Barley starch gelatinizes and is completely solubilized in well-made malts during mashing for determinations of hot water extract. However, the starch of sorghum and other 'tropical' cereals does not begin to gelatinize until substantially higher temperatures are reached. If the temperatures of mashes are raised much over 65 °C (149 °F), enzyme destruction is rapid and so starch conversion is incomplete and recovered extracts are low. It is rare that starch conversion is complete with sorghum malts in 'standard' barley malt-type mashes. For the manufacture of Bantu beers, incomplete starch conversion is desired. In contrast, in the manufacture of clear beers, total starch conversion is highly desirable. In one modified hot water extract determination, sorghum malt is mashed in at relatively low temperatures (e.g. 50 °C (122 °F)) with a liquor containing calcium sulphate, the calcium ions serving to stabilize α-amylase. After 30 min, the mash is allowed to settle and supernatant liquor (*ca.* 20% of the mash volume) is withdrawn and retained. The rest of the mash, containing most of the solids, is progressively heated to 100 °C (212 °F) and is held at this temperature for 30 min, when all the starch is gelatinized and all enzymes are destroyed. This 'main mash' is then cooled to 60 °C (140 °F) and the reserved liquid, containing amylolytic enzymes, is mixed in. The entire reconstituted mash is held at 60 °C (140 °F) for 90 min, during which time the gelatinized starch is hydrolysed and conversion is completed. The temperature is then increased to 78 °C (172.4 °F) for wort separation. Extracts obtained by this method (Tables 15.10, 15.12 and 15.13) are much higher than those obtained using 'conventional' procedures designed for use with barley malts (Palmer *et al.*, 1989; Seidl, 1992; Byrne *et al.*, 1993; Little, 1994). Clearly, for large-scale brewing with sorghum malts, mashing procedures different to those usually used with barley malts must be used to recover extract, unless highly enzymic barley malts or heat-stable microbial enzymes are incorporated in mashes.

Another problem is that when sorghum germinates, the seedlings accumulate substantial amounts of the cyanogenic glycoside dhurrin

Table 15.12 A comparison between analyses of a barley malt, a wheat malt and a sorghum malt

	Barley malt	Wheat malt	Sorghum malt
Hot water extract			
65 °C l°/kg[a]	295	–	112
45 °C/80 °C/65 °C l°/kg[b]	290	–	268[b]
(% d.m.)	81	85.7	81.0[b]
FAN (mg/100 g)	84.4	142.6	70
α-Amylase (DU)	35	55	40; 50
Diastatic power (°L)	66; 80	106	20; 23
α-Amino nitrogen (mg/l)	160	–	100
TN (%)	1.65	2.18	1.63
TSN (%)	0.61(0.7)	1.12	0.54(0.7)
SNR (%)	37	51	33
βD-Glucan (%)	0.5	–	0.5
Pentosan (%)	10	–	1.5
Malting loss (%)	7	–	15–20
70 °C mash wort viscosity (mPa)	6.79	10.92	2.78

[a] Standard IoB mash.
[b] Modified mashing procedure. Initial mash was at 45 °C for 30 min. After settling the supernatant was decasted. Starch in solids in deposit was gelatinized by heating to 80–100 °C. After cooling, the supernatant liquid containing enzymes was added back and the mixture was incubated further at 65 °C to convert the starch. Wort fermentabilities are low but their viscosities are acceptable.
From Palmer *et al.* (1989); Byrne *et al.* (1993).

(Aisien and Muts, 1987). When the seedling is broken up, the glycoside meets the enzymes β-glucosidase and nitrilase, which degrade it, ultimately to glucose, *p*-hydroxybenzaldehyde and highly toxic hydrogen cyanide (prussic acid, HCN). In the manufacture of malts for use in foods or for brewing clear beers the seedlings should be removed. However, this is not done for brewing opaque, Bantu beers, when the seedling is incorporated into the grist with the rest of the grain. There are wide variations in the amounts of cyanogenic glycoside formed by different varieties of sorghum during malting. Osuntogun *et al.* (1989) reported that cyanide contents of 15 varieties of sorghum increased during malting to values in the range 78–150 μg HCN/g dry matter. The prussic acid (precursor) was concentrated in the seedling, reaching values of up to 1400 μg HCN/g (d.m.) in the shoots and up to 860 μg HCN/g (d.m.) in the roots. It is estimated that the lethal dose for a 70 kg person is 50 mg HCN. Some sorghum cultivars are regarded as being too dangerous to use.

Sorghum grains have TCW values of 25–40 g. Grains should be thoroughly cleaned before malting. Typically they are infested with microbes. During steeping, grains are often treated with antiseptics such as lime water or dilute formaldehyde, and the steeping liquor is changed repeatedly. Sorghum malts are often poor in enzymes (Table 15.10, 15.12 and 15.13). Both α-amylase and β-amylase increase during germination (Dyer and Novellie, 1966). Some 18–50% of the observed diastatic power

Table 15.13 Ranges of analyses of malts prepared from sorghum coming from different regions of Africa

Extract (f.g., %)[a]	65–83.7
Fine–coarse extract difference (%)[a]	0.5–18.2
Wort viscosity (mPa; 8.6%)	1.29–2.10
Friability (%)[b]	29–99
Attenuation limit (%)	43–96
Wort colour (EBC units)[b]	2.2–16.8
Boiled wort colour (EBC units)[b]	2.2–5.6
pH[b]	5.43–6.07
Protein (%)	7.2–12.9
Kolbach index (%)	15.3–41.0
Diastatic power (°W-K)	11–151
α-Amylase (DU)	11–91

[a] Extract determined by an optimized mashing procedure.
[b] Only part of the range of samples.
From Seidl (1992).

results from β-amylase and maltase. The ratio is altered by altering the germination temperature but not by altering the degree of watering. Much of the maltase is insoluble. In tannin-rich sorghums, like the bird-proof and other coloured varieties, enzymes can only be extracted in the active state if special precautions are taken, for example by including 2% peptone in the extraction medium (Watson and Novellie, 1976). When high-tannin sorghum malts are mashed, amylolysis is inhibited by the tannins and so are the lactobacilli during the acidification stage in making Bantu beer. The tannins are concentrated at the surface of the grain, in the testa (McGrath *et al.*, 1982; Morrall *et al.*, 1982). Various attempts have been made to reduce the tannin levels without depressing the malting quality of the grain (Reichert *et al.*, 1980; Butler, 1982; Daiber and Taylor, 1982; Glennie, 1983). These treatments also reduce the levels of microbes on the grains. Grains are either briefly steeped in a dilute solution of sodium hydroxide or in formaldehyde (0.03–0.08% in the first 4 h steep) (Daiber, 1975, 1978). Higher levels of formaldehyde, or longer exposures to it, are damaging. It has been proposed that a mixture of boracic acid and borax might be a better sterilizing agent (Ajerio *et al.*, 1993). The criteria used for following modification in barley are not easily applied to sorgum, where the cell walls of the starchy endosperm are not wholly degraded and in the steely regions may remain unmodified (Aisien and Muts, 1987). As malting proceeds, grain hardness declines and its porosity increases. The process may be followed using a compression test or by measuring the specific gravities of grains in toluene (Daiber *et al.*, 1973).

Malting sorghum for making Bantu beers is carried out on the small scale in the traditional way and on the commercial scale (Dyer and Novellie, 1966; Daiber and Novellie, 1968; Novellie, 1968; Kamalavalli and Ramananda Rao, 1973; Novellie and de Schaepdrijver, 1986; Aisien and Muts, 1987;

Palmer *et al.*, 1989; Agu *et al.*, 1993; Byrne *et al.*, 1993). Bantu women steep grain in clay pots, gourds or calabashes or in woven rush bags or sacks immersed in streams. The grain is germinated in woven grass, cane or reed baskets (which may be lined with leaves) or it may be grown on the ground covered with mats or sacks. The grain is watered as often as is considered necessary. When growth is judged sufficient, the 'green malt' is air-dried, usually spread in a thin layer in the sun. Commercial malting is carried out in modified Saladin boxes or Wanderhaufen streets. Frequently turning was between boxes, by hand, because mechanical turners caused too much damage to the seedlings. Previously, commercial malting was carried out on concrete floors in open-sided buildings, the grain being covered with damp sacks. Steeping takes between 8 and 24 h, 16–18 h being usual times, and is best carried out in running water. The grain is well cleaned and preferably is washed before steeping. Air-rests are helpful in obtaining even chitting, especially if warm-water steeping (40 °C (104 °F)) is used (Ajerio *et al.*, 1993; Ezeogu and Okolo, 1994). By 'barley' standards, malting is carried out warm. Germination lasts for 5–7 days and maximum diastatic power values are obtained in 4–5 days at 30 °C (86 °F), 6 days at 28 °C (82.4 °F) and 7 days at 25 °C (77 °F). It has been concluded that in-bed grain temperatures of 25–30 °C (77–86 °F) are probably the best. Often too low malting temperatures have been used (e.g. Pal *et al.*, 1976; Aisien and Muts, 1987; Taylor and Robbins, 1993). In the tropics, Nigeria for example, it is difficult to malt at temperatures as low as 24–26 °C (75.2–78.8 °F) (Little, 1994). The grains are watered regularly, at least daily. The less watering that is used, the lower the diastatic power of the malt and the lower the malting losses. The sorghum seedlings develops a strong shoot and a single powerful root (2.5–5 cm (1–2 in)); as a result, the volume of the grain bed is increased markedly during germination. During malting, there is a massive migration of nitrogen into the seedling. The final moisture content of the green malt may be as high as 84%. Attempts are made to adjust the watering to maximize diastic power for the least feasible malting losses. Malting losses may be very variable and high, for example 9.5–35% with roots and shoots and 18.5–60.5% if seedling parts are removed. In Bantu beer brewing, the roots and shoots are generally 'milled in' with the 'berry'. Most, but not all, reports indicate that gibberellic acid does not accelerate malting in sorghums and that additions of potassium bromate are valueless. Recently it was reported that applications of gibberellic acid enhanced the diastatic power in one sorghum malt but depressed it in another (Nzelibe and Nwasika, 1995). Attempts have been made to reduce the high (compared with barley) losses incurred when sorghum is malted, but without evident success. For example, attempts to reduce malting losses by checking embryo growth with ammonia severely reduced malt quality (Ilori and Adewusi, 1991).

A major systematic investigation into the best conditions for malting sorghum involved steeping the grain to 33% moisture and then watering it according to three different regimes so that after 6 days the moisture contents were about 42%, 60% and 77%, respectively. Germination times were varied and the germination temperatures used were in the range 24–36 °C (75.2–96.8 °F). Germination temperatures of 24–28 °C (75.2–82.4 °F) were generally the best (Morrall *et al.*, 1986). All the variables had effects on the diastatic power, FAN, extract and malting losses. All these values increased with increasing moisture content, but diastatic power began to decline towards the end of the germination period. For practical commercial malting, compromises have to be made between germination times and moisture levels to obtain malt of adequate quality without intolerably high malting losses. Some particular peak values obtained were: for extract (determined in a 2 h mash at 60 °C (140 °F), 75.5% after 6 days of germination at 24 °C (75.2 °F) with a medium moisture content; for FAN, 180 mg/100 g dry matter in 6 days of germination at 32 °C (89.6 °F) under very wet growing conditions; for diastatic power, 46 SDU/g whole malt dry, 5 days of germination, well wetted, at 24 °C (75.2 °F). The highest malting loss, 23.7%, occurred in grain grown very wet for 6 days at 35–38 °C (95–100 °F).

Because of their low enzyme complements, sorghum malts should be kilned using low air-on temperatures and with high airflows to maximize enzyme survival. It is not clear what analyses should be carried out on sorghum malts to assess their values for making opaque beers. However, much attention is given to their diastatic power values and their development of FAN and sugars during mashing (Taylor, 1989, 1991).

For a long time it appeared that extra enzymes would be needed in mashes of sorghum malts to obtain worts suitable for making clear beers. Extract recoveries and FAN values were low, and wort run-off times were slow. However, the use of better analytical techniques, the choice of high quality varieties, the use of optimized malting conditions and 'divided mash' mashing regimens makes it more likely that all-sorghum malt beers can be produced economically (Seidl, 1992). The best sorgum malts, analysed using optimized techniques, stand comparison with other malts (Tables 15.10, 15.12 and 15.13). However, high (relative to barley) malting losses and high losses of starch appear to be inevitable when this grain is malted.

15.20.7 Malted maize

Maize is generally considered to malt less well than sorgum and most of the reported trials have given unsatisfactory results from the clear-beer brewers' standpoint. The malting characteristics of this cereal have been relatively little studied. Germinating maize can, in some respects, enhance its nutritive value (Wang and Fields, 1978). However, as the gelatinization temperatures of maize starches are high it is probable that, as with sorghum malts, much better extracts would be obtained if modified mashing schedules were used. The naked grains of maize are comparatively large and generate large seedlings during germination (Chapter 2).

Winthorp (1678) recorded that European settlers in North America made beers from maize, but that maize could not be malted successfully using the regimes chosen for barley, since the roots and shoots had to be well grown before modification was adequate. The preferred method for malting maize was to cover a layer of grain with 2–3 in (*ca.* 5.1–7.6 cm) of topsoil and to allow it to grow for 10–14 days. When the whole area was covered with green shoots, the plants were taken up. This was comparatively easy to do because the roots were matted together. After shaking off the soil, the seedlings were well washed and then dried in the sun or in a kiln. The endosperms of well-grown grains were floury and very sweet-tasting. The beer made from maize malt was judged to be acceptable and preferable to beer made from bread.

Lévy (1899) and Petit (1899) recorded that maize malts had low diastatic powers and were used in warm countries and in Hungary. Steeping was prolonged; at 25–30 °C (77–86 °F) it lasted for 46–51 h and if the temperature was as low as 20 °C (68 °F), it could last 120 h. Weichherz (1928) agreed that steeping maize was a slow process. Two methods of malting maize were outlined (Lévy, 1899). In the first, the germination compartment was held at 20–21 °C (68–69.8 °F). In floor-malting, the layer of germinating grain was covered with wet sacks, and after the first 52 h during which the temperature was held at 20–24 °C (68–75.2 °F), the temperature was allowed to rise until after 60 h it reached 24–26 °C (75.2–78.8 °F) and finished at about 30 °C (86 °F). The grain was sprinkled regularly, using warm water in winter, and germination continued for 7–8 days, by which time the roots were about three times the length of the grain and shoots were about 5 mm (0.2 in) long. In the second method, the maize grain was pre-dried at 36 °C (96.8 °F) for 12 h before steeping for 36 h at 10 °C (50 °F), to increase its weight by 45%. Salt, sulphuric acid and ammonium phosphate were added to the steep liquor, or other anti-microbial agents such as hydrofluoric acid, salicyclic acid or calcium bisulphite were used. Recently, washes with dilute sodium hydroxide have been suggested (Singh and Bains, 1984). Problems with microbial contamination can be reduced by frequently changing the steep water or, better, by steeping in running water (Singh and Bains, 1984). During flooring, the grain must be regularly watered and turned (every 12 h, or when the temperature of the piece had risen by 2 °C (3.6 °F)). Heavy mould growth was frequently a problem and the sacks used to cover the germinating grain had to be well washed and damped with a dilute solution of an antiseptic. Weichherz (1928) recommended drying maize malt slowly and at a low temperature (e.g. 50 °C (122 °F) over 48 h). Maize varieties differ in their malting qualities (Aniche, 1989; Aderinola, 1992). Extract values, enzyme levels, Kolbach indices and malting losses all increase with malting time (Singh and Bains, 1984; Malleshi and Desikachar, 1986b). A heavy application of gibberellic acid (2 mg/kg maize) elevated α-amylase, protease, extract and Kolbach index values (Singh and Bains, 1984). In malts dried at 45 °C (113 °F) for 24 h, extracts measured using a 'barley' analytical method were 57.7–71.4% after 5 days of germination and 62.7–73.7% after 7 days of germination. The malt yields were 90.9–93.6% after 5 days and 86.3–90.4% after 7 days of germination (Singh and Bains, 1984).

Therefore, as with sorghum, maize needs to be malted 'wet and warm'. Maize is extremely likely to develop mould infestations and green malts should be dried at low temperatures to favour enzyme survival. There is no apparent effort being made to select maize varieties for superior malting characteristics. Its low enzyme content and low extract levels, as determined by conventional techniques, means that clear-beer brewers regard maize malts with disfavour. For example, maize malts malted for 6 days at 28 °C (82.4 °F) would not saccharify adequately and, for adequate conversion to occur, needed to be mixed with 20% barley malt when mashed (Aderinola, 1992). Maize malts would probably be valued more highly if they were analysed and mashed using optimized mashing regimens, as is being done with sorghum.

15.20.8 Malted millets

The millets are a heterogeneous group of small-grained ''tropical' cereals (Chapter 2). They have been malted for use in foodstuffs (Table 15.14), in opaque beers and, at least experimentally, clear beers (Table 15.15). There are considerable differences in malting qualities between the different species of millet and between individual varieties of one species. The small size of millet grains (one variety having a TCW of 0.65 g; Nzelibe and Nwasike, 1995) makes them inconvenient to handle using 'conventional' machinery. Millets have been malted mixed with sorghum to prevent their blocking or falling through slots in the deck of a Saladin box. In general, the grains need to be steeped with an anti-microbial agent in the steep liquor. They need to be germinated warm (25–30 °C (77–86 °F)) and to be regularly wetted during germination. Steep out moisture contents are generally low, 30–35% (Malleshi and Desikachar, 1986b). When six millets were malted with a view to determining their values in weaning foods, finger millet malt had a high enzyme activity and was suitable for reducing starch-paste viscosities, but the enzyme activities in barnyard millet malt and Kodo millet malt were very low (Table 15.14) (Malleshi and Desikachar, 1986b). The α-amylase levels rise and decline in germinating millet, with time courses that strongly depend on the growth conditions (Malleshi and Desikachar, 1986a) (Fig. 15.2), as is the case in barley grown 'warm and wet' (Briggs, 1968), so the choice of germination time and temperature is critical for optimum enzyme yield. There are marked flavour differences

Table 15.14 The characteristics of a variety of cereal grains and of some malts prepared from them. In these selected data the malts, germinated for 96 h, were dried at 45 °C (113 °F), de-vegetated and then lightly cured at 70 °C (158 °F)

	Ungerminated control grains					Germinated samples (96 h)					
	Amylase activity[a]	Viscosity (cP)	Starch (%)	Free sugars (%)	Protein (N × 6.25) (%)	Starch (%)	Free sugars (%)	Protein (N × 6.25) (%)	Malting loss (%)	Amylase activity[a]	Viscosity (cP)
Rice	–	2460	77.9	1.2	8.2	70.8	5.4	7.5	7.0	105	84
Wheat	2.0	1700	71.0	2.3	12.4	59.6	8.7	11.4	11.5	230	36
Maize	–	2200	66.3	1.3	10.4	59.8	5.9	9.5	6.8	150	125
Sorghum	0.5	1980	68.7	1.7	9.6	58.9	6.3	9.0	8.6	170	72
Triticale	2.5	1540	70.4	2.6	12.6	57.4	8.5	11.8	9.2	260	25
Finger millet	1.0	2100	65.5	1.1	8.2	53.6	5.9	6.8	16.2	200	28
Pearl millet	4.6	1810	69.0	1.4	12.6	57.5	7.3	10.6	9.5	154	22
Proso millet	1.0	1650	60.5	1.8	12.0	53.0	5.0	10.9	9.1	145	52
Foxtail millet	–	1620	60.3	1.3	11.4	52.6	4.9	10.2	9.5	132	60
Barnyard millet	–	1500	57.0	1.0	8.9	52.0	3.7	8.1	7.4	70	125
Kodo millet	–	1420	54.9	1.2	8.5	50.7	3.9	8.0	7.4	55	620

[a] Amylase activity: equivalent to mg maltose released by the extract from 1 g malt acting on soluble starch at 37 °C and pH 5.8.
From Malleshi and Desikacher (1986b).

Table 15.15 Influences of germination time, temperature and gibberellic acid on the extracts of two varieties (1 and 2) of malting *Eleusine coracana* (finger millet; ragi)[a]

Germination temperature °C	°F	Extract in control at 48 h germination 1	2	Extract in control at 96 h germination 1	2	Extract at 48 h germination with GA_3 1	2	Extract at 96 h germination with FA_3 1	2
20	68	42.7	49.8	58.8	60.9	41.2	50.2	63.0	63.9
25	77	57.1	58.9	69.2	67.1	57.8	58.0	66.3	69.0
30	86	60.8	64.5	68.1	69.0	64.6	66.9	69.9	71.9
35	95	60.4	70.3	67.8	–	65.5	73.5	65.8	–

[a] Extract as AOAC method (%); GA_3 added, O.I mg/kg finger millet.
Data of Singh *et al.* (1988).

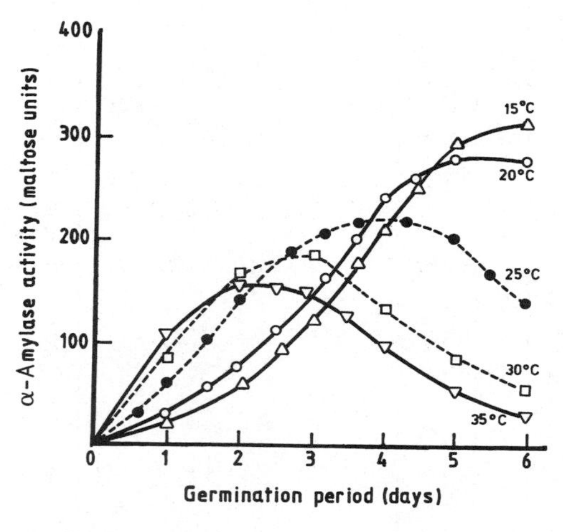

Fig. 15.2 The influence of temperature and germination time on the activity of α-amylase in germinating finger millet (*Eleusine coracana*) (after Malleshi and Desikachar, 1986a).

between different malted millets. The nutritive values of flours prepared from malted millet have been investigated (Malleshi and Desikachar, 1986c). Additions of potassium bromate depress the qualities of malted millets but, at least in some instances, additions of gibberellic acid enhance extract yields (Table 15.15) and cause enzyme activities to peak sooner, as is the case with wet-grown barley (Briggs, 1968). Often recovered extracts are low, but, as with sorghum malts, this might be remedied by using optimized mashing regimes.

Most studies have been made with the more important millets. Malting *Eleusine coracana* (finger millet, ragi) has been investigated (Malleshi and Desikachar, 1979, 1986a; Nout and Davies, 1982; Singh *et al.*, 1988). Water uptake occurs faster at higher temperatures, approaching a limiting value of about 34%. Generally lime or another anti-microbial agent is needed in the steep. Malting losses increase dramatically with increasing germination temperatures (15–35 °C (59–95 °F)), in one trial increasing to between 4.3 and 23.1% after 2 days of germination and to between 7.7 and 46.3% after 4 days. Levels of α-amylase are strongly influenced by germination temperature (Fig. 15.2). The malt is very suitable for use in baby weaning foods, having a high α-amylase content, being easily prepared with 1–2 days of germination and having an agreeable flavour. The malt has a high calcium content, which should favour the survival of α-amylase during mashing or cooking. In Kenya, finger millet malt is used, mixed with maize, to make the traditional opaque beer busaa. After 24 h steeping, the grain is germinated for 2–3 days in layers 5–10 cm (*ca.* 2–3.9 in) deep, covered with wet sacks or gunney bags. All operations are at 25–30 °C (77–86 °F). The malt is then dried for 1–2 days in the sun. With trials using the variety Imele, maximum starch liquefying power was attained after 6 days of germination, with a malting loss of 11.9% (or 9.5% if all the roots and shoots were retained) (Nout and Davies, 1982). Modification progressively reduced the specific gravity of the grains. Extracts determined using a barley-malt method were 47–52%. Additions of gibberellic acid accelerated the increase in hot water extract with germination time. Others have noted that, as with maize and sorghum malts, diastatic activities are low and that in conventional mashes starch conversion is incomplete. The small grain sizes (TCW 2.3–3.3 g) are inconvenient. Gibberellic acid accelerated α-amylase formation (Table 15.15), particularly at higher germination temperatures (Singh *et al.*, 1988). Extracts determined using an AOAC procedure ranged from 42.7 to 73.5% (d.b.). Germinating at a constant 35 °C (95 °F) temperature, malting losses were 12.4 and 24.9% after 48 and 96 h, respectively, in the absence of gibberellic acid and 19.8 and 44.7%, respectively, if gibberellic acid had been used. Wort filtration from mashes made with these malts was slow and the soluble nitrogen ratios were low. Other reports do not indicate that finger-millet malts yield viscous worts (Table 15.14).

Pennisetum spp. (*P. typhoides*, bajra, pearl millet; *P. maiwa*) have been investigated several times (Sheorain and Wagle, 1973; Pal *et al.*, 1976; Singh and Tauro, 1977; Opoku *et al.*, 1981; Malleshi and Desikachar, 1986b; Okeke, 1991, 1992; Agu *et al.*, 1993; Agu and Ezeanolue, 1993). During germination, some vitamin and enzyme levels increase while amounts of lipids, phytate and oxalate decline. α-Amylase activities rise, peak, then decline during germination. Analyses carried out using non-optimized techniques suggest that extracts of about 75% can be obtained. Bajra malts have amylase and protease activities comparable to those of barley malts. The malts have good physical properties and are used in making some traditional Nigerian opaque beers. However, during even short periods of storage, these malts develop bitter flavours. Investigating *P. typhoides* (TCW 8.7–9.9 g) it appeared that gibberellic acid stimulated α-amylase production but depressed protease activity (Nzelibe and Nwasika, 1995).

While the α-amylase was important in *P. typhodies* malt, β-amylase predominated in malt made from 'acha' (*Digitaria exilis*) (Nzelibe and Nwasika, 1995). This tiny grained cereal (TCW 0.65 g), when malted gave low extracts (max. 120 l°/kg, barley-malt analysis method, but the extracts reported for sorghum were even lower) and slow-filtering worts, which had low pH values of around 4.9. Malting losses ranged from 7 to 50%. It was proposed that malt of acha would usefully complement pearl millet and sorghum malts in brewery mashes.

15.20.9 Malted rice

Various attempts have been made to malt rice, but the earlier ones at least produced unsatisfactory products for brewing clear beers and the processes were discontinued (Lévy, 1899; Petit, 1899; Dax, 1932; Zimmerman, 1938; Malleshi and Desikachar, 1986b; Aniche and Okafor, 1989; Okafor and Iwouno, 1990; Aniche and Palmer, 1992). Dax (1932) described a rotatable drum with perforated sides that dipped into water, and in which rice was grown. About that period it was stated that beers made from malted rice had harsh flavours, but the high proportion of husk simplified wort separation from mashes made with rice malt. In later trials, before use the malts were milled, being dehulled and so removing the oil-rich germs and aleurone layers (Zimmerman, 1938). Rice starch has a high gelatinization temperature (about 72 °C (161.6 °F)), so rice malts should best be analysed using modified soghum-type analytical mashes. In early trials, Californian short-kernel paddy rice (TCW 29–30 g; germination energy over 95%) was preferred. The grain was steeped for 48–60 h, with water changes every 12–24 h to give a relatively low (by barley standards) steep-out moisture of 30–35%. Then, after germinating in drums for 72 h at 17.8–18.9 °C (64–66 °F), the malt was carefully dried over a 48 h period, with temperatures rising from 32.2 °C (90 °F) to 65.6 °C (150 °F). These low temperatures were used to prevent the vitrification of the grain. After the removal of the roots and shoots and pearling to remove the husks, embryos and aleurone layers, the yield of the residual material was about 64%. These preparations had moisture contents of about 5–5.5%, oil contents of less than 0.8% and extracts of 92–94%. The product was expensive to produce and was poor in enzymes. The endosperm material was poorly modified, as it had to withstand the processing, but, in contrast to unmalted rice, liquefaction in the cooker was rapid and occurred at a comparatively low temperature. The high extract yield was an advantage and the flavours and aromas of products made using this material were excellent.

The more recent interest in malting rice has been triggered by its possible use in foodstuffs (Table 15.14) (Malleshi and Desikachar, 1986b) and by the need in some parts of Africa to find alternatives to imported barley malts. Some trials have been carried out at 20 °C (68 °F), but temperatures

Table 15.16 Comparative analyses of barley, sorghum and rice malts

	Barley malt	Sorghum malt	Rice malt (range)[a]	
Moisture content (%)	6.4	5.0	4.0	
Cold water extract (%)	18.3	22.4	23.7	(22.4–33.7)
Hot water extract				
IoB (l°/kg)	302	186	265	(250–291)
EBC (%)	76.4	53.5	54.7	(51.0–55.1)
Diastatic power (°L)	90	41	42.5	(39.2–42.7)
α-Amylase (DU)	44	31	34	(31.5–39.5)
Kolback index (%)	37	35	36	(30.2–38.6)
Fat (%)	1.9	2.1	0.6	
TSN (%)	0.5	0.4	0.4	
TN (%)	1.4	1.3	1.1	
Malting loss (%)	–	–	22.5	(20.8–23.8)
Wort limiting attenuation (%)	74.4	78.0	79.4	

Rice: average results of six varieties germinated for 7 days.
From Okafor and Iwouno (1990).

of 28–30 °C (82.4–86 °F) are preferred. The rice grain sizes are moderate (TCW 25.6–31.5 g). Steep water is changed frequently (e.g. every 6 h). Steep-out moisture contents are 32–37%. Germination periods are usually 6–7 days and spraying and turning are regularly carried out. Drying should be at a low temperature. The roots and shoots (but not the hulls) were removed in recent trials. Malting losses are often high (e.g. 22%) (Table 15.16). When the malts are finely ground and mashed in the conventional way, saccharification is incomplete and wort run-off is slow. It is probable that optimizing the mashing regimes used would improve the extract recoveries.

Appendix
Units of measurement and some data on barley and malt

Over the years, the units used in trading barley and malt, and used in the malting and related industries, have been so diverse that the topic has a nightmare quality. In 1899 (Dec., *Brewers' J.*, p. 717) a writer complained that over 100 different units were then used as barley measures in the UK alone. Historical accounts are given by (i) Skinner, F. G. (1967) *Weights and Measures: their Ancient Origins and their Development in Great Britain up to AD 1855*, 117pp., HMSO, London; (ii) Zapko, R. E. (1977) *British Weights and Measures. A History from Antiquity to the Seventeenth Century*. 248pp., University of Wisconsin Press; and (iii) Connor, R. D. (1987) *The Weights and Measures of England.* 422 pp. HMSO, London. The situation was improved by the widespread adoption of the metric system in 1972 by the maltsters and malt users of the UK. Various definitive publications are available giving equivalents between metric (SI), non-standard and imperial measures. Of these, the booklet by Anderton, P. and Bigg, P. H. (1972) *Changing to the Metric System: Conversion Factors, Symbols and Definitions*, National Physical Laboratory HMSO, London is very helpful. Useful data are also in Weast (1977).

Nevertheless, the metric system is not universally employed worldwide and the units that have been, or are, used can be confusing. For example, in trading grain, quarters (Qr) and bushels (bu) were originally units of volume. However, in time these were given weight values so that in the UK 1 Qr (= 8 bu = 64 gall, UK) barley was taken to be 448 lb and 1 Qr malt was 336 lb. Oats were often traded at 320 lb/Qr and maize and wheat at 480 lb/Qr. Worse, the equivalent values varied between countries (Table A4). All these were distinct from the quarter (avoirdupois; qr) of 28 lb.

The results obtained by different systems of analysis may not be precisely inter-convertable. Therefore, the different mashing systems used by the IoB and the EBC can give truly different extracts and yields of soluble nitrogen and there are no valid factors for converting hot water extract (°l/kg, IoB) to or from E (%, EBC) or the soluble nitrogen ratio (%, IoB) and the Kolbach index (%, EBC). Indeed there are limited data which suggest that the relationship between the two extract values, determined on the same malts, varies between malts made from different barley varieties. In addition, the British (imperial) and US units have diverged so that, for example, the US gallon of water weighs 8 lb and is made up of 8 pints each of 1 lb, while in the UK a gallon of water weighs 10 lb and each pint 1.25 lb. In the UK, temperature is now usually measured in degrees Centigrade (°C) or Kelvin (K) (SI) but in the USA the Fahrenheit scale (°F) persists as does, it is said, the Réaumer scale (°R). Equivalents are given in Table A3. Table A5 contains a specific gravity and extract table. Table A6 gives the approximate equivalence values of extracts determined by the old and new analytical methods of the IoB. Table A7 gives some solution factors for several sugars; Tables A8 and A9 give some useful data on the properties of water, and Tables A10 and A11 give some representative data for barley. Table A12 gives a summary of the types of change in moisture content, weight and volume that occur when barley is malted.

Table A1 SI derived units

Physical quantity	Name	Symbol	Definition
Energy	joule	J	kg m^2/s^2
Force	newton	N	kg m/s^2 = J/m
Power	watt	W	kg m^2/s^3 = J/s
Electrical charge	coulomb	C	A s
Electrical potential difference	volt	V	kg m^2/s^3 per A = J/A per s
Electrical resistance	ohm	Ω	kg m^2/s^3 per A^2 = V/A
Inductance	henry	H	kg m^2/s^2 per A = V/A per s
Luminous flux	lumen	lm	cd sr
Illumination	lux	lx	cd sr/m^2
Frequency	hertz	Hz	per s

where m is the metre; kg is the kilogram; s is the second; A is the ampere; cd is the candela (luminous intensity); sr is the steradian (solid angle).
Other basic units are: kelvin (K), thermodynamic temperature and temperature interval; mole (mol) molecular (or atomic) mass.

Table A2 Prefixes for SI units

Fraction	Prefix	Symbol
10^{-15}	femto	b
10^{-12}	pico	p
10^{-9}	nano	n
10^{-6}	micro	μ
10^{-3}	milli	m
10^{-2}	centi	c
10^{-1}	deci	d
10^{1}	deka	da
10^{2}	hecto	h
10^{3}	kilo	k
10^{6}	mega	M
10^{9}	giga	G
10^{12}	tera	T

Table A3 Comparison of thermometers showing the relative indications of the Fahrenheit, centigrade and Réaumur scales

Scale	Boiling point	Freezing point
Fahrenheit (°F)	212	32
Centrigrade (°C)	100	0
Réaumur (°R)	80	0

Conversion of thermometer degrees

1 °C = 1.8 °F = 0.8 °R. °C to °R, multiply by 4 and divide by 5. °C to °F, multiply by 9, divide by 5, then add 32. °R to °C, multiply by 5 and divide by 4. °R to °F, multiply by 9, divide by 4, then add 32. °F to °R, first subtract 32, then multiply by 4, and divide by 9. °F to °C, first subtract 32, then multiply by 5, and divide by 9.

F	C	R	F	C	R
230	**110**	**88**	120.2	49	39.2
221	**105**	**84**	118.4	48	38.4
212	**100**	**80**	116.6	47	37.6
210.2	99	79.2	114.8	46	36.8
208.4	98	78.4	**113**	**45**	**36**
206.6	97	77.6	111.2	44	35.2
204.8	96	76.8	109.4	43	34.4
203	**95**	**76**	107.6	42	33.6
201.2	94	75.2	105.8	41	32.8
199.4	93	74.4	**104**	**40**	**32**
197.6	92	73.6	102.2	39	31.2
195.8	91	72.8	100.4	38	30.4
194	**90**	**72**	98.6	37	29.6
192.2	89	71.2	96.8	36	28.8
190.4	88	70.4	**95**	**35**	**28**
188.6	87	69.6	93.2	34	27.2
186.8	86	68.8	91.4	33	26.4
185	**85**	**68**	89.6	32	25.6
183.2	84	67.2	87.8	31	24.8
181.4	83	66.4	**86**	**30**	**24**
179.6	82	65.6	84.2	29	23.2
177.8	81	64.8	82.4	28	22.4
176	**80**	**64**	80.6	27	21.6
174.2	79	63.2	78.8	26	20.8
172.4	78	62.4	**77**	**25**	**20**
170.6	77	61.6	75.2	24	19.2
168.8	76	60.8	73.4	23	18.4
167	**75**	**60**	71.6	22	17.6
165.2	74	59.2	69.8	21	16.8
163.4	73	58.4	**68**	**20**	**16**
161.6	72	57.6	66.2	19	15.2
159.8	71	56.8	64.4	18	14.4
158	**70**	**56**	62.6	17	13.6
156.2	69	55.2	60.8	16	12.8
154.4	68	54.4	**59**	**15**	**12**
152.6	67	53.6	57.2	14	11.2
150.8	66	52.8	55.4	13	10.4
149	**65**	**52**	53.6	12	9.6
147.2	64	51.2	51.8	11	8.8
145.4	63	50.4	**50**	**10**	**8**
143.6	62	49.6	48.2	9	7.2

Table A3 *(continued)*

F	C	R	F	C	R
141.8	61	48.8	46.4	8	6.4
140	**60**	**48**	44.6	7	5.6
138.2	59	47.2	42.8	6	4.8
136.4	58	46.4	**41**	**5**	**4**
134.6	57	45.6	39.2	4	3.2
132.8	56	44.8	37.4	3	2.4
131	**55**	**44**	35.6	2	1.6
129.2	54	43.2	33.8	1	0.8
127.4	53	42.4	**32**	**0**	**0**
125.6	52	41.6	30.2	–1	–0.8
123.8	51	40.8	**23**	**–5**	**–4**
122	**50**	**40**	**14**	**–10**	**–8**

Table A4
Conversions
Metric system

m = 1.0936 yard = 3.280 ft; cm = 0.39370 in;
hectare = 2.471 acre; m^2 = 10.764 ft^2; cm^2 = 0.1550 in^2;
m^3 = 1000 dm^3 (or litre) = 33.315 ft^3 = 61024 in^3 = 1.30795 yd^3;
hl = 100 dm^3 (or litre) = 21.998 gal (British) = 26.418 (US) = 0.6111 barrel (British) = 0.8387 barrel (US) = 0.8522 *beer* barrel (US) = 0.1 m^3;
l = 35.196 fl. oz (British) = 33.815 fl. oz (US) = 0.21998 gal (British) = 0.26418 gal (US) = 0.035315 ft^3
tonne = 1000 kg = short ton = 0.984207 long ton = 2204.62 lb = 10 doppelzentner = 20 zentner; zenter = 50 kg = 0.984 cwt = 110.231 lb; 1 kg = 2.20462 lb
g = 0.03527 oz = 15.432 grain; kg/m^3 = g/dm^3 = 0.062428 lb/ft^3

British measures

yard = 3 ft = 36 in = 0.9144 m; ft = 0.3048 m lb/ft^2 = 4.88243 kg/m^2
in = 2.540 cm = 1000 thou (thousandth of an inch); lb/ft^3 = 16.0185 kg/m^3
thou = 25.4 micron or micrometre (μm);
acre = 4840 yd^2 = 0.4047 hectare (ha);
yd^2 = 0.8361 m^2; ft^2 = 9.290 dm^2; in^2 = 6.4516 cm^2;
yd^3 = 0.7646 m^3; ft^3 = 28.317 dm^3; in^3 = 16.3871 cm^3;
ton (long) = 20 cwt = 2240 lb = 1016 kg = 1.01605 tonne (metric);
lb = 16 oz = 256 dram = 7000 gram = 0.45359 kg;
oz = 28.35 g; grain = 64.80 mg;
gal = 160 fl oz = 8 pints = 1.201 gal (US) = 4.546 litre = 0.1605 ft^3 = 0.125 bu (British);
pint = 0.5682 litre; fl. oz = 28.412 ml;
butt = 2 hogshead = 3 barrel = 108 gal = 4.9096 hl;
brl = 2 kilderkin = 4 firkin = 36 gal = 1.6365 hl = 1.4 brl beer (US) = 4.5 bv (British);
bushel (bu, British) = 8 gal (UK) = 36.3687 dm^3 = 0.9690 bu (US); 8 by = 1 Qr.
100% proof spirit (British) = 57.10% ethyl alcohol (v/v) = 49.28% ethyl alcohol (m/m), relative density 0.91976 at 60 °F.

US measures

beer brl = 31 gal (US) = 25.81 gal (British) = 1.734 hl = 0.717 brl British;
standard brl = 31.5 gal (US) = 26.23 gal (British) = 1.1924 hl = 0.729 brl British;
gal = 8 pint = 128 fl. oz = 3.7853 litre = 0.8327 gal British = 231.0 in^3.
bushel (bu, US) = 35.23907 litre = 1.03203 bu (British).

Barley and malt measures

Britain and South Africa:	barley bushel = 56 lb = 75.401 kg
	barley quarter = 448 lb = 4 cwt = 0.2 ton = 203.209 kg
	malt bushel = 42 lb = 19.051 kg
	malt quarter = 336 lb = 3 cwt = 0.15 ton = 152.407 kg
Australia and New Zealand:	barley bushel = 50 lb; malt bushel = 40 lb
US and Canada:	barley bushel = 48 lb; malt bushel = 34 lb

Useful data (some equivalents are approximate)

kcal = 4.186 kJ = 3.968 btu = 1.1628 Wh = 3088 ft lb
btu= 1.055 kJ = 0.252 kcal = 0.2931 Wh = 778.2 ft lb
Wh = 3.6 kJ = 0.860 kcal = 3.412 btu = 2655 ft lb
therm = 105.506 MJ = 29.307 kWh;
standard ton refrigeration per 24 h = 12000 btu/h = 3024 kcal/h
atm = bar = 14.70 lb/in^2 = 760 mmHg = 10^5 N/m^2
lb/in^2 = 6896.76 N/m^2 = 0.06895 bar = 703 kg/m^3 = 27.7 inches water
lb/gal (British) = 99.76 g/l; lb/gal (US) = 119.8 g/l
lb/brl (British) = 3.336 g/l; lb/brl (US) = 3.865 g/l
grain/gal (British) = 14.25 mg/l; grain/gal (US) = 17.12 mg/l
CO_2 in beer: g/100 ml = 5.06 v/v beer;
v/v beer = 0.198 g/100 ml

Table A5 Specific gravity and extract table

The following table is based on those compiled by Dr Plato for the German Imperial Commission (*Normal-Eichungskommission*) and refers to apparent specific gravities, as determined in the usual manner by weighing in a specific gravity bottle in air or by means of a saccharometer. Cane sugar % w/v and % w/w represent g per 100 ml and g per 100 g of solution, respectively. The percentages by weight in column 6, corresponding with the specific gravities at 60 °F given in column 1, were computed by interpolation from Plato's table for true specific gravities at 15°/15 °C and 16°/15 °C corrected to 60°/60 °F and then brought to 60°/60° in air by adding (SG – 1) × 0.00121. The cane sugar weight percentages were converted to volume percentages and the solution divisors calculated. The column headed Plato gives the specific gravities in air at 20°/20 °C related to the cane sugar weight percentages and, with the latter, corresponds with the Plato Table commonly used in breweries and laboratories where 20 °C is the standard temperature. The column headed Balling similarly gives the specific gravities at 17.5°/17.5 °C from the Balling Table corresponding with the same sugar percentages. These specific gravities cannot accurately correspond with those at 60°/60 °F and 20°/20 °C on account of the errors in Balling's table. The following densities were used in the calculations:

Water at 15 °C/4 °C	0.999126
60 °F/4 °C	0.999035
20 °C/4 °C	0.998234

Specific gravity conversion table for cane sugar solutions

British units				Plato		Balling SG 17.5 °C	Baumé Modulus 145
SG 60 °F	Brewers' pounds	Cane sugar % w/v	Solution divisor	SG 20 °C	Cane sugar % w/w (°Brix)		
1002.5	0.9	0.643	3.888	1.00250	0.641	1.00256	0.36
1005.0	1.8	1.287	3.885	1.00499	1.281	1.00513	0.72
1007.5	2.7	1.932	3.882	1.00748	1.918	1.00767	1.08
1010.0	3.6	2.578	3.879	1.00998	2.552	1.01021	1.43
1012.5	4.5	3.225	3.876	1.01247	3.185	1.01274	1.78
1015.0	5.4	3.871	3.875	1.01496	3.814	1.01528	2.14
1017.5	6.3	4.517	3.874	1.01745	4.439	1.01776	2.48
1020.0	7.2	5.164	3.873	1.01993	5.063	1.02025	2.83
1022.5	8.1	5.810	3.872	1.02242	5.682	1.02273	3.17
1025.0	9.0	6.458	3.871	1.02490	6.300	1.02523	3.52
1027.5	9.9	7.107	3.869	1.02740	6.917	1.02776	3.86
1030.0	10.8	7.755	3.868	1.02989	7.529	1.03027	4.20
1032.5	11.7	8.405	3.867	1.03238	8.140	1.03277	4.54
1035.0	12.6	9.054	3.866	1.03486	8.748	1.03527	4.88
1037.5	13.5	9.703	3.865	1.03736	9.352	1.03775	5.22
1040.0	14.4	10.354	3.863	1.03985	9.956	1.04024	5.55
1042.5	15.3	11.003	3.862	1.04234	10.554	1.04273	5.88
1045.0	16.2	11.652	3.862	1.04481	11.150	1.04523	6.21
1047.5	17.1	12.303	3.861	1.04731	11.745	1.04773	6.54
1050.0	18.0	12.953	3.860	1.04979	12.336	1.05022	6.87
1052.5	18.9	13.604	3.859	1.05227	12.925	1.05269	7.20
1055.0	19.8	14.255	3.858	1.05476	13.512	1.05515	7.52
1057.5	20.7	14.907	3.857	1.05726	14.097	1.05760	7.84
1060.0	21.6	15.560	3.856	1.05975	14.679	1.06005	8.16
1062.5	22.5	16.213	3.855	1.06224	15.259	1.06252	8.48
1065.0	23.4	16.866	3.854	1.06472	15.837	1.06500	8.80
1067.5	24.3	17.519	3.853	1.06720	16.411	1.06747	9.12
1070.0	25.2	18.173	3.852	1.06970	16.984	1.06995	9.44

Table A5 *(Continued)*

British units				Plato		Balling	Baumé
SG 60 °F	Brewers' pounds	Cane sugar % w/v	Solution divisor	SG 20 °C	Cane sugar % w/w (°Brix)	SG 17.5 °C	Modulus 145
1072.5	26.1	18.827	3.851	1.07218	17.554	1.07244	9.75
1075.0	27.0	19.482	3.850	1.07467	18.122	1.07494	10.06
1077.5	27.9	20.135	3.849	1.07717	18.687	1.07743	10.37
1080.0	28.8	20.791	3.848	1.07965	19.251	1.07990	10.69
1082.5	29.7	21.446	3.847	1.08213	19.812	1.08237	11.00
1085.0	30.6	22.101	3.846	1.08462	20.370	1.08486	11.30
1087.5	31.5	22.758	3.845	1.08712	20.927	1.08737	11.61
1090.0	32.4	23.414	3.844	1.08960	21.481	1.08986	11.91
1092.5	33.3	24.071	3.843	1.09209	22.033	1.09235	12.21
1095.0	34.2	24.726	3.842	1.09457	22.581	1.09481	12.51
1097.5	35.1	25.384	3.841	1.09707	23.129	1.09730	12.81
1100.0	36.0	26.041	3.840	1.09956	23.674	1.09980	13.11
1102.5	36.9	26.700	3.839	1.10204	24.218	1.10230	13.41
1105.0	37.8	27.360	3.838	1.10454	24.760	1.10480	13.71
1107.5	38.7	28.019	3.837	1.10703	25.299	1.10730	14.00
1110.0	39.6	28.679	3.836	1.10952	25.837	1.10983	14.30
1112.5	40.5	29.339	3.834	1.11200	26.372	1.11235	14.59
1115.0	41.4	30.000	3.833	1.11450	26.906	1.11486	14.88
1117.5	42.3	30.660	3.832	1.11698	27.436	1.11735	15.17
1120.0	43.2	31.321	3.831	1.11947	27.965	1.11984	15.46
1122.5	44.1	31.981	3.830	1.12195	28.491	1.12231	15.74
1125.0	45.0	32.643	3.829	1.12445	29.016	1.12478	16.03
1127.5	45.9	33.305	3.828	1.12694	29.539	1.12729	16.31
1130.0	46.8	33.970	3.827	1.12944	30.062	1.12980	16.60
1132.5	47.7	34.632	3.826	1.13191	30.580	1.13228	16.88
1135.0	48.6	35.295	3.825	1.13441	31.097	1.13477	17.16
1137.5	49.5	35.958	3.824	1.13689	31.611	1.13723	17.44
1140.0	50.4	36.621	3.823	1.13938	32.124	1.13971	17.72
1142.5	51.3	37.285	3.822	1.14186	32.635	1.14221	18.00
1145.0	52.2	37.951	3.821	1.14435	33.145	1.14473	18.27
1147.5	53.1	38.617	3.820	1.14685	33.653	1.14727	18.55
1150.0	54.0	39.284	3.818	1.14934	34.160	1.14980	18.82

Table A6 Equivalence (approximate) between Institute of Brewing (UK) units of hot water extract

Litre °/kg (20 °C) against lb/Qr (15.5 °C, 60 °F) in the body of the table. Calculated from the Institute of Brewing Recommended Methods of Analysis (1993), on the basis that for l°/kg (20 °C) the equivalent values in lb/Qr will be 0.5 greater than those for l°/kg (15.5 °C). Example: 251 l°/kg (20 °C) is equivalent to 84.1 lb/Qr (15.5 °C); 251 l°/kg (15.5 °C) is equivalent to 83.6 lb/Qr (15.5 °C)

	240	250	260	270	280	290	300	310	320	330
0	80.2	83.7	87.2	90.8	94.2	97.8	101.3	104.8	108.3	111.9
1	80.5	84.1	87.5	91.1	94.6	98.1	101.7	105.2	108.6	112.2
2	80.8	84.4	87.9	91.4	94.9	98.5	102.0	105.5	109.0	112.6
3	81.2	84.8	88.3	91.8	95.3	98.8	102.4	105.9	109.3	112.9
4	81.5	85.1	88.6	92.1	95.7	99.2	102.7	106.2	109.7	113.2
5	81.9	85.5	89.0	92.5	96.0	99.5	103.1	106.6	110.1	113.6
6	82.3	85.8	89.4	92.8	96.4	99.9	103.4	106.9	110.4	113.9
7	82.7	86.1	89.7	93.2	96.8	100.2	103.8	107.3	110.8	114.3
8	83.0	86.5	90.1	93.5	97.1	100.6	104.1	107.6	111.2	114.6
9	83.4	86.8	90.4	93.9	97.5	100.9	104.5	107.9	111.5	115.0

Table A7 Solution divisors of some sugars at 20 °C (*Pauls' Malt Brewing Room Book* 1995–7, p. 231). The apparent solids content of a solution (g/100 cm^3) = G/D where G is the excess specific gravity and D is the solution divisor

SG (approx.)	D (sucrose)	D (invert sugar)	D (fructose)	D (glucose)	Wort solids
1016.0	3.872	3.875	3.907	3.805	–
1020.0	3.869	3.872	3.904	3.805	–
1028.0	3.864	3.866	3.900	3.80	3.92
1036.0	3.860	3.861	3.896	3.872	–
1061.0	3.849	3.845	3.882	3.794	3.90
1083.0	3.840	3.832	3.869	3.784	–
1106.0	3.831	3.819	3.853	3.772	–

Table A8 Some properties of water at various temperatures (various sources)

Temperature		–log Kw (pKw)	Oxygen content of air-saturated water	
°C	°F		(mg/l)	(cm^3 at NTP/kg)
0	32	14.9435	–	10.19
5	41	–	–	8.9
10	50	14.5346	–	7.9
15	59	–	9.8	7.0
20	68	14.1669	8.8	6.4
24	75.2	14.0000	–	–
25	77	13.9965	8.1	5.8
30	86	13.8330	7.5	5.3
35	95	–	7.0	–
50	122	13.2617	–	–
60	140	13.0171	–	–

Table A9 The density of pure, air-free water at various temperatures[a]

Temperature		
°C	°F	Density (g/ml)
0	32	0.99987
3.98	39.16	1.00000
5	41.00	0.99999
10	50.00	0.99973
15	59.00	0.99913
20	68.00	0.99823
30	86.00	0.99567
50	122.00	0.98807
60	140.00	0.98324
65	149.00	0.98059
80	176.00	0.97183
100	212.00	0.95838

[a] Data of Weast (1977).

Table A10 Recommended minimum slope and valley angles for chutes and hoppers for the materials indicated (after Forster, 1985)

Material	Angle (degrees)
Undried, unscreened barley	35
Dressed, dried barley	33
Fully steeped barley	45
Green malt (germinated barley)	45–50
Malt in culm	40
Dressed malt	30
Malt culms (rootlets)	50
Malt and barley dust, and light screenings	60

Table A11 Data for barley and wheat

Barley grain dimensions
Dimensions of barley grains are very variable (Briggs, 1978). Usual ranges (moisture contents 6–18%; quoted by Kustermann *et al.*, 1986, de Clerck, 1987; Hoseney and Faubion, 1992) are:

Length 7–11 mm
Width 3.1–3.8 mm
Thickness 2.4–3.1 mm
Volume 23–30 mm^3
Weights 30–54 mg/kernel
Bulk density 520–660, 650–750, 605 kg/m^3 (1.33–1.65 m^3/t)

Data of bulk grain
Angles of repose (from the horizontal) quoted vary depending on the variety, how clean the sample is and the moisture content; the higher the moisture content the higher the angle of repose. Air spaces make up 35–45% of the bulk.

Malt 26°
Barley 23–35°

Bulk densities
Malts 450–550 kg/m^3 (1.22–1.82 m^3/t) (de Clerck, 1952)
Chit malt 600 kg/m^3 (1.67 m^3/t) (Narziss, 1987)

Grain swelling
The 'swell' of barley in steeping is very variable; the volume is often estimated to increase by 1.3–1.4. Witt (1959) noted that barley, which occupied about 1.55 l/kg(dry), occupied about 10% more space when it was gently loaded into a steep tank. If the tank was periodically aerated then, at the end of steeping, when the grain was swollen, it occupied 30% or more space than the original barley.

Typical data for barley and wheat (MAFF, 1983)

Cereal grains	Typical bulk density		Equivalent weight		Storage space	
	(kg/m^3)	(lb/ft^3)	(kg/hl)	(lb/bu)[a]	m^3/t	ft^3/t
Wheat	780	48.6	78	62.4	1.28	45.3
Barley	700	43.6	70	56.0	1.43	50.5

[a] bu (British).

Table A12 Summary of representative changes occurring when a notional 100 t of barley is malted; data are calculated from widely accepted values or are derived from Forster (1985) or Gibson (1989)[a]

Sample	Moisture content (%)	Dry weight (t)	Fresh weight (t)	Volume original barley (m^3/100 t)
At start of malting				
Barley	12	88	100	142
Steeping loss 1% d.m., 0.88 t d.m.				
Steeped barley	45	87.12	158.4	200; 208
Germination, respiration losses 6% original barley (d.m. = 5.28 t d. wt)				
Green malt with rootlets	43	81.84	143.6	240; 246
Kilned malt in culm (undisturbed)	3	81.84	84.4	220
Dressing, rootlets removed, 4% d.m. barley = 3.52 t d.m.				
Dressed malt	3	78.32	80.7	181
Culms[b]	3	3.52	3.63	16.5[b]

[a] Values from other sources are given in Table A11.

[b] Density of pelletized material 0.56–0.64 t/m^3 as-is; density of culms 0.224 t/m^3 as-is (courtesy of D. Griggs).

References

Aalbers, V. J., Drost, B. W. and Pesman, L. (1983) *MBAA Tech. Quart.*, **20**, 74.
Aastrup, S. (1985) *Carlsberg Res. Commun.*, **50**, 37.
Aastrup, S., Erdal, K. and Munck, L. (1985) *Proc. 20th Congr. EBC*, Helsinki, p. 387.
Aastrup, S. and Outtrup, H. (1985) *Proc. 20th Congr. EBC*, Helsinki, p. 667.
Abrahamsohn, B. (1911) *Z. ges. Brau.*, **34**, 179.
Abrahamson, M. (1978) *Proc. 15th Congr. IoB* (Australia and New Zealand Sect.), p. 247.
Adamic, E. B. (1977) in *The Practical Brewer, a Manual for the Brewing Industry*, 2nd edn (Broderick, H. M. ed.) p. 21. MBAA, Madison, WI.
ADAS (1985) Booklet 2497; *Low-rate Aeration of Grain. Grain Drying and Storage No. 5*, pp. 13. MAFF, HMSO, London.
Aderinola, A. V. (1992) *Discovery and Innovation*, **4**(2), 103.
Adlam, E. G. (1907) Br. Patent 8800, April 16, 1907.
Agu, R. C. and Ezeanolue, J. C. (1993) *Process Biochem.*, **28**, 475.
Agu, R. C. and Okeke, B. C. (1991) *Process Biochem.*, **26**(2), 89.
Agu, R. C. and Okeke, B. C. (1992) *Process Biochem.*, **27**, 335.
Agu, R. C., Okeke, B. C., Nwufo, S. C., Ude, C. M. and Onwumelu, A. H. (1993) *Process Biochem.*, **28**, 105.
Ahluwalia, B. and Fry, S. C. (1986) *J. Cereal Sci.*, **4**, 287.
Ahrenst-Larsen, B. and Erdal, K. (1979) *Proc. 17th Congr. EBC*, Berlin, p. 631.
Aisien, A. O. (1982a) *J. Sci. Food Agric.*, **33**, 754.
Aisien, A. O. (1982b) *J. Inst. Brewing*, **88**, 164.
Aisien, A. O. and Muts, G. C. J. (1987) *J. Inst. Brewing*, **93**, 328.
Aisien, A. O. and Palmer, G. H. (1983) *J. Sci. Food Agric.*, **34**, 113.
Ajerio, K. O., Booer, C. D. and Proudlove, M. O. (1993) *Ferment.*, **6**(5), 339.
Albini, P. A., Briggs, D. E. and Wadeson, A. (1987) *J. Inst. Brewing*, **93**, 97.
Alexander, H. P. and Fish, J. (1984) *J. Inst. Brewing*, **90**, 65.
Alleman, R. (1981) *Brew. Dis. Int.*, **11**(7), 44.
Amaha, M., Kitabatake, K., Nakagawa, A., Yoshida, J. and Harada, T. (1973) *Proc. 14th Congr. EBC*, Salzburg, p. 381.
Åman, P., Hesselman, K. and Tilly, A. C. (1985) *J. Cereal. Sci.*, **3**, 73.
Åman, P. and Newman, C. W. (1986) *J. Cereal Sci.*, **4**, 133.
Anderson, F. B. (1966) *J. Inst. Brewing*, **72**, 384.
Anderson, G. K. and Saw, C. B. (1986) in *Paul's Brewing Room Book, 1986–8*, 81st edn, p. 37. Pauls Malt, Ipswich.
Anderson, I. W., Dickenson, C. J. and Anderson, R. G. (1989) *Proc. 22nd Congr. EBC*, Zurich, p. 213.
Anderson, J. A. and Alcock, A. W. (eds) (1954) *The Storage of Cereal Grains and their Products. Am. Assoc. Cereal Chem.*, St Paul, MN.
Anderson, J. A. and Ayre, C. A. (1938) *Can. J. Res.*, **16C**, 377.
Anderson, J. D. (1970) *Crop. Sci.*, **10**, 36.
Anderson, M. K. and Reinbergs, E. (1985) in *Barley* (Rasmusson, D. C. ed.), p. 231. *Am. Soc. Agron.*, Madison, WI.
Anderson, R. A., Conway, H. F. and Peplinski, A. J. (1970) *Stärke*, **22**(4), 130.
Anderson, R. G. (1984) *Brewers' Guard*, Oct., p. 15.
Anderson, R. G., Brookes, P. A. and Martin, P. A. (1979) *Proc. 17th Congr. EBC*, Berlin (West), p. 689.
Aniche, G. N. (1989) *Process Biochem.*, **24**(5), 183.
Aniche, G. N. and Okafor, N. (1989) *J. Inst. Brewing*, **95**, 165.
Aniche, G. N. and Palmer, G. H. (1992) *Process Biochem.*, **27**, 291.
Anness, B. J. (1984) *J. Inst. Brewing*, **90**, 315.
Anon. (*ca.* 1877) *Chemistry, Theoretical, Practical and Analytical, as Applied to the Arts and Manufactures*, Vol. I, p. 68, Vol. II, p.274. William Mackenzie, London.
Anon. (1881) *Brewers' J.*, April, 125.

Anon. (1898) *Brewers' J.*, 502.
Anon. (1900a) *J. Inst. Brewing*, **6**,140.
Anon. (1900b) *Brewers' J.*, 201.
Anon. (1903) *One Hundred Years of Brewing*. H. S. Rich and Co., Chicago and New York (Reprinted 1974).
Anon. (1904) *Brewers' J.*, **40**, 486.
Anon. (1906a) *Brewers' J.*, 554.
Anon. (1906b) *Brewers' J.*, 43.
Anon. (1907a) *Brewers' J.*, 513.
Anon. (1907b) *Woch. Brau.*, **24**, 289.
Anon. (1953) *Brewers' Guard.*, **82**(11), 19.
Anon. (1959) *Brewers' Guard.*, **88**(8), 23.
Anon. (1963a) *Brewers J.*, **99**, 14.
Anon. (1963b) *Brewers' Guard.*, **92**, 65.
Anon. (1963c) *Brewers' Guard.*, **92**, 21.
Anon. (1963d) *Brewers' J.*, **99**, 526.
Anon. (1964) *Brewers' J.*, **100**, 223.
Anon. (1965a) *Brewers' Guard.*, **94**, 31.
Anon. (1965b) *Brewers' Guard.*, **94**, 63.
Anon. (1967a) *Brewers' J.*, **103**(Aug), 41.
Anon. (1967b) *Brewers' J.*, **103**(Sept), 53.
Anon. (1967c) *Brewers' J.*, **103**(Sept), 60.
Anon. (1968) *Int. Brewers' J.*, **104**(May), 74.
Anon. (1969a, Jan.) *Int. Brewers' J.*, **105**, 77.
Anon. (1969b) *Int. Brewers' J.*, **75**, 47.
Anon. (1973) *British Pharmaceutical Codex*, p. 281, 684. The Pharmaceutical Press, London.
Anon. (1980) *Pauls and Whites Brewing Room Book*, 1980–1, 76th edn, p. 110. Pauls and Whites, Ipswich.
Anon. (1985a) *Brew. Dist. Int.*, Feb., 27.
Anon. (1985b) *Brewers' Dig.*, May, 36.
Anon. (1986) *Brew. Dist. Int.*, **16**(2), 21.
Anon. (1990a) *Adv. Malting Barley*, **5**, 4.
Anon. (1990b) *Brauerei, J.*, **108**, 496.
Anon. (1991) *Brauindustrie*, **76**, 1039.
Anon. (1994) *Brewers' Dig.*, **69**(12), 20.
Anon. (1996) *Brewers' Dig.*, **71**(3), 20.
Antkiewicz, P. (1987) *Acta Aliment. Pol.*, **13**, No. 2, 91.
Arnold, J. P. (1911) *Origin and History of Beer and Brewing*. Alumni Assoc. Wahl-Henius Inst. Fermentol., Chicago, IL.
Arnold, R. E. (1959) *J. Agric. Eng. Res.*, **4**, 24.
Arnould, M. (1971) *Agronomie Tropicale*, **26**, 865.
Arter, E. and Olsen, A. (1986) *Brygmesteren* (3), 16.
ASBC (1992) *Methods of Analysis of the American Society of Brewing Chemists*, 8th edn, Am. Soc. Brewing Chemists, St Paul, MN.
Aschengreen, N. H. (1969) *Process Biochem.*, **4**(8), 23.
Aubry, L. (1881) *Z. ges. Brau.*, **16**, 441.
Ault, R. G. (1961) *J. Inst. Brewing*, **67**, 405.
Axcell, B., Jankovsky, D. and Morrall, P. (1983) *Brewers' Dig.*, **58**(Aug), 20.
Axcell, B. C., Tulej, R. and Mulder, C. J. (1986) *Proc. 19th Conv. IoB* (Australia and New Zealand Sect.), Hobart, p. 63.
Ayre, C. A., Sallans, H. R. and Anderson, J. A. (1940) *Can. J. Res.*, **18C**, 169.
Aznavorian, A. (1983) *Bios*, **14**(1/2), 27.
Bacic, A. and Stone, B. A. (1981) *Aust. J. Plant Physiol.*, **8**, 453, 475.
Bajomo, M. F. and Young, T. W. (1992) *J. Inst. Brewing*, **98**, 515.
Baker, C. D. (1989) *J. Inst. Brewing*, **95**, 263.
Baker, D. L. and Burant, L. J. (1956) *Proc. Annu. Meet. Am. Soc. Brew. Chem.*, p. 109.
Bakhuizen, R., van der Veen, S. W., Sinjorgo, K. M. C. and Doderer, A. (1995) *Proc. 25th Congr. EBC*, Brussels, p. 169.
Ballesteros, G. and Piendl, A. (1977) *Brewers' Dig.*, Oct., 36.

Bamforth, C. W. (1981) *Proc. 18th Congr. EBC*, Copenhagen, p. 335.
Bamforth, C. W. (1982) *Brewers' Dig.*, June, 22.
Bamforth, C. W. (1983) *J. Inst. Brewing*, **89**, 420.
Bamforth, C. W. (1986) in *EBC Monograph-XI. EBC Symp. on Wort Production*. Maffliers, France, p. 149.
Bamforth, C. W. (1989) *J. Inst. Brewing*, **95**, 314.
Bamforth, C. W. and Martin, H. L. (1981) *J. Inst. Brewing*, **87**, 81.
Banasik, O. J., Myhre, D. and Harris, R. K. (1956) *Brewers' Dig.*, Jan., 50; Feb., 63.
Barrett, J., Bathgate, G. N. and Clapperton, J. F. (1975) *J. Inst. Brewing*, **81**, 31.
Barrett, J., Clapperton, J. F., Divere, D. M. and Rennie, H. (1973) *J. Inst. Brewing*, **79**, 407.
Barrett, J., Griffiths, C. M. and Kirsop, B. H. (1967) *J. Inst. Brewing.*, **73**, 445.
Bärwald, G., Niefind, H.-J. and ten Hompel, U. (1969) *Monatss. Brau.*, **22**(May), 114.
Bathgate, G. N. (1973) *Brewers' Dig.*, April, 60.
Bathgate, G. N. (1977) *J. Inst. Brewing*, **83**, 322.
Bathgate, G. N. (1983) *J. Inst. Brewing*, **89**, 416.
Bathgate, G. N. (1989) in *Cereal Science and Technology* (Palmer, G. H. ed.), p. 243. Aberdeen University Press, Aberdeen.
Bathgate, G. N. and Brennan, H. (1971) *Proc. Annu. Meet. Am. Soc. Brew. Chem.*, p. 22.
Bathgate, G. N. and Cook, R. (1989) in *The Science and Technology of Whiskies* (Piggott, J. R., Sharp, R. and Duncan, R. E. B. eds) p. 19. Longman, Harlow, UK.
Bathgate, G. N., Martinez-Frias, J. and Stark, J. R. (1978) *J. Inst. Brewing*, **84**, 22.
Bathgate, G. N. and Palmer, G. H. (1972) *Stärke*, **24**(10), 336.
Bathgate, G. N. and Taylor, A. G. (1977) *J. Inst. Brewing*, **83**, 163.
Battaglia, R., Conacher, H. B. S. and Page, B. D. (1990) *Food Additives and Contaminants*, **7**(4), 477.
Baum, B. R. (1977) *Oats: Wild and Cultivated. A monograph on the genus Avena L. (Poaceae)* Supply Services of Canada.
Baur, F. J. (ed.) (1984) *Insect Management for Food Storage and Processing*. Am. Assoc. Cereal Chemists, St Paul, MN.
Baxter, E. D. (1988) *Ferment.*, **1**(4), 37.
Baxter, E. D. and O'Farrell, D. D. (1980) *J. Inst. Brewing*, **86**, 291.
Beaven, E. S. (1902) *J. Inst. Brewing*, **8**, 542.
Beaven, E. S. (1904) *J. Inst. Brewing*, **10**, 520.
Beaven, E. S. (1947) *Barley – Fifty Years of Observation and Experiment*. Duckworth, London.
Becker, C. (1926) *Z. ges. Brau.*, **49**, 65.
Beckerich, R. P. and Denault, L. J. (1987) in *Enzymes and their Role in Cereal Technology* (Kruger, J. E., Lineback, D. and Stauffer, C. E. eds), p. 335. Am. Assoc. Cereal Chemists, St Pauls, MN.
Beckord, L. D. and Fleckenstein, J. G. (1964) US Patent 3,134,724.
Belderok, B. (1961) *Proc. Int. Seed Test. Assoc.*, **26**, 697.
Belderok, B. (1968) *Field Crop Abstr.*, **21**(3), 203.
Bell, G. D. H. and Kirby, E. J. M. (1966) in *The Growth of Cereals and Grasses* (Milthorpe, F. L. and Ivins, J. D. eds), p. 315. Butterworths, London.
Bell, G. D. H. and Lupton, F. G. H. (1962) in *Barley and Malt: Biology, Biochemistry, Technology* (Cook, A. H. ed.), p. 45. Academic Press, London.
Bemment, D. W. (1978) in *Pauls and Whites Brewing Room Book*, 1978–9, 77th edn, p. 82. Pauls and Whites Ltd, Ipswich, UK.
Bemment, D. W. (1985) *Brewer*, Dec., 457.
Bergal, P. and Clemencet, P. (1962) in *Barley and Malt: Biology, Biochemistry, Technology* (Cook, A. H., ed) p. 1. Academic Press, London.
Bergal, P. and Friedberg, L. (1940) *Ann. Epiphyt.*, **6**, 157.
Bernard, M. (1992) *J. Inst. Brewing*, **98**, 81.
Bettner, R. E. and Meredith, W. O. S. (1970) *Proc. Annu. Meet. Am. Soc. Brew. Chem.*, p. 118.
Bettner, R. E., Meredith, W. O. S. and Anderson, J. A. (1962) *Proc. Annu. Meet. Am. Soc. Brew. Chem.*, p. 5.
Bills, G. T. (1989) *Pauls Brewing Room Book, 1989–91*, 82nd edn, p. 28. Pauls Malt Ltd, Ipswich, UK.
Birch, G. G. and Parker, K. J. (eds) (1979) *Sugar: Science and Technology*. Applied Science, London.

Bishop, L. R. (1928) *J. Inst. Brewing*, **34**, 101.
Bishop, L. R. (1929a) *J. Inst. Brewing*, **35**, 316.
Bishop, L. R. (1929b) *J. Inst. Brewing*, **35**, 323.
Bishop, L. R. (1930) *J. Inst. Brewing*, **36**, 336.
Bishop, L. R. (1934) *J. Inst. Brewing*, **40**, 75.
Bishop, L. R. (1936) *J. Inst. Brewing*, **42**, 103.
Bishop, L. R. (1944) *J. Inst. Brewing*, **50**, 166.
Bishop, L. R. (1945) *J. Inst. Brewing*, **51**, 215.
Bishop, L. R. (1946) *J. Inst. Brewing*, **52**, 273.
Bishop, L. R. (1948) *J. Inst. Brewing*, **54**, 330.
Bishop, L. R. (1950) *Brewer's Guild J.*, **56**, 180.
Bishop, L. R. and Day, F. E. (1933) *J. Inst. Brewing*, **39**, 545.
Bishop, L. R. and Marx, D. (1934) *J. Inst. Brewing*, **40**, 62.
Bishop, L. R. and Spaul, D. M. (1959) *J. Inst. Brewing*, **65**, 504.
Blaber, S. (1907) *Brewers' J.*, June, 359.
Blanchflower, A. J. and Briggs, D. E. (1991a) *J. Sci. Food Agric.*, **56**, 103.
Blanchflower, A. J. and Briggs, D. E. (1991b) *J. Sci. Food Agric.*, **56**, 117.
Blanchflower, A. J. and Briggs, D. E. (1991c) *J. Sci. Food Agric.*, **56**, 129.
Bleisch, C. (1899) *Z. ges. Brau.*, **22**, 327.
Bleisch, C. (1900) *Z. ges. Brau.*, **23**, 309.
Bleisch, C. (1911) *J. Inst. Brewing*, **17**, 525.
Bleisch, C. (1912) *J. Inst. Brewing*, **18**, 702.
Blenkinsop, P. (1991) *MBAA Tech. Quart.*, **28**(4), 145.
Block, F. and Morgan, A. I. (1967) *Cereal Chem.*, **44**, 61.
Bode, G. and Westhofen, L. (1910) *Woch. Brau.*, **27**, 481.
Boivin, P. and Malanda, M. (1993) *Proc. 24th Congr. EBC*, Oslo, p. 95.
Booth, S. (1995) *Brewer*, **81**(2), 61.
Bourne, D. T., Auger, A. and Sturgeon, P. A. (1990) *Proc. 3rd Aviemore Comf., Malting, Brewing and Distilling* (Campbell, I. ed.), p. 309. IoB, London.
Bourne, D. T., Powlesland, T. and Wheeler, R. E. (1982) *J. Inst. Brewing*, **88**, 371.
Bourne, D. T. and Wheeler, R. E. (1982) *J. Inst. Brewing*, **88**, 324.
Bourne, D. T. and Wheeler, R. E. (1984) *J. Inst. Brewing*, **90**, 306.
Bourne, D. T., Wheeler, R. E. and Jones, M. (1977) *Proc. 16th Congr. EBC*, Amsterdam, p. 139.
Bowden, B. N. (1965) *Outlook Agric.*, **4**(5), 243.
Bradee, L. H. (1977) in *The Practical Brewer: a Manual for the Brewing Industry*, 2nd edn, (Broderick, H. M. ed.), p. 40. MBAA, Madison, WI.
Brain, G. F. (1983) in *Pauls and Whites Brewing Room Book, 1984–5*, 80th edn, p. 71. Pauls and Whites Ltd, Ipswich, UK.
Brandtzaeg, B., Malleshi, N. G., Svanberg, U., Desikachar, H. S. and Mellander, O. (1981) *J. Trop. Ped.*, **27**, 184.
Brenner, M. W. (1972) *MBAA Tech. Quart.*, **9**(1), 12.
Briant, L. (1907) *J. Inst. Brewing*, **13**, 486.
Briggs, D. E. (1963a) *J. Inst. Brewing*, **69**, 13.
Briggs, D. E. (1963b) *J. Inst. Brewing*, **69**, 244.
Briggs, D. E. (1967) *J. Inst. Brewing*, **73**, 33.
Briggs, D. E. (1968) *Phytochemistry*, **7**, 513.
Briggs, D. E. (1973) in *Biosynthesis and its Control in Plants* (Milborrow, B. V. ed.), p. 219. Academic Press, London.
Briggs, D. E. (1978) *Barley*. Chapman & Hall, London.
Briggs, D. E. (1987a) *J. Am. Soc. Brew. Chem.*, **45**, 1.
Briggs, D. E. (1987b) in *Brewing Science 3* (Pollock, J. R. A. ed.), p. 441. Academic Press, London.
Briggs, D. E. (1987c) in *Cereals in a European Context* (Morton, I. D. ed.), p. 119. Ellis Horwood, Chichester.
Briggs, D. E. (1992) in *Barley: Genetics, Biochemistry, Molecular Biology and Biotechnology* (Shewry, P. R. ed.), p. 369. CAB International, Wallingford, UK.
Briggs, D. E. (1995) in *Pauls Brewing Room Book, 1995–7*, 84th edn, p. 31. Pauls Malt, Ipswich, UK.
Briggs, D. E. and Clutterbuck, V. J. (1973) *Phytochemistry*, **12**, 1047.

Briggs, D. E. and Harcourt, J. (1968) *J. Inst. Brewing*, **74**, 440.
Briggs, D. E., Hough, J. S., Stevens, R. and Young, T. W. (1981) *Malting and Brewing Science*, Vol. 1, 2nd edn, Chapman & Hall, London.
Briggs, D. E. and MacDonald, J. (1983a) *J. Inst. Brewing*, **89**, 260.
Briggs, D. E. and MacDonald, J. (1983b) *J. Inst. Brewing*, **89**, 324.
Briggs, D. E. and McGuinness, G. (1993) *J. Inst. Brewing*, **99**, 249.
Briggs, D. E. and Morrall, P. (1984) *J. Inst. Brewing*, **90**, 363.
Briggs, D. E., Wadeson, A., Statham, R. and Taylor, J. F. (1986) *J. Inst. Brewing*, **92**, 468.
Briggs, D. E. and Woods, J. L. (1993) *Dormancy in Malting Barley: Studies on Drying, Storage, Biochemistry and Physiology. Report No. 84*. Home-Grown Cereals Authority, London.
Briggs, D. E., Woods, J. L. and Favier, J. F. (1994) *J. Inst. Brewing*, **100**, 271.
Briggs, D. E., Young, T. W. and Harcourt, J. (1985) *J. Inst. Brewing*, **91**, 257.
Britnell, J. (1973) *MBAA Tech. Quart.*, **10**(4), 176.
Broderick, H. M. (ed.) (1977) *The Practical Brewer: a Manual for the Brewing Industry*, 2nd edn, p. 475. MBAA, Madison, WI.
Brook, R. C. (1992) in *Storage of Cereal Grains and Their Products*, 4th edn (Sauer D. B. ed.), p. 183. Am. Assoc. Cereal Chem., St Paul, MN.
Brookes, P. A. (1980) *Brewer*, **66**, 8.
Brookes, P. A. (1981) *Brewers' Guard.*, **110**(4), 21.
Brookes, P. A. (1988) *Brew. Dist. Int.*, Nov., 16.
Brookes, P. A. (1989) *Brewer*, **75**, 579.
Brookes, P. A. and Dickenson, C. J. (1983) *Proc. 19th Congr. EBC*, London, p. 393.
Brookes, P. A., Lovett, D. A. and MacWilliam, I. (1976) *J. Inst. Brewing*, **82**, 14.
Brookes, P. A. and Martin, P. A. (1974) *J. Inst. Brewing*, **80**, 294.
Brookes, P. A. and Martin, P. A. (1975) *J. Inst. Brewing*, **81**, 357.
Brookes, P. A. and Philliskirk, G. (1987) *Proc. 21st Congr. EBC*, Madrid, p. 337.
Brown, A. J. (1907a) *Ann. Bot.*, **21**, 79.
Brown, A. J. (1907b) *J. Fed. Inst. Brewing*, **13**, 658.
Brown, H. T. (1895) *Brewing Trade Rev.*, **9**, 160.
Brown, H. T. (ed.) (1903) *Trans. Guinness Res. Lab.*, **1**(1), 96.
Brown, H. T. (1909) *J. Inst. Brewing*, **15**, 184.
Brown, H. T. and Morris, G. H. (1890) *J. Chem. Soc.*, **57**, 458.
Brown, J. (ed.) (1982) *The Master Bakers' Book of Breadmaking*. Turret Press, London.
Brown, J. (1983) *Steeped in Tradition. The Malting Industry in England since the Railway Age*. Inst. Agric. History, Museum English Rural Life. University of Reading, Reading, UK.
Brudzynski, A. and Roginski, H. (1969) *J. Inst. Brewing*, **75**, 472.
Brunner, H. (1989) *Brew. Bev. Ind. Int.*, (2), 54.
Brunswick, P., Manners, D. J. and Stark, J. R. (1987) *J. Inst. Brewing*, **93**, 181.
Bryan-Jones, D. G. (1986) *Proc. 2nd Aviemore Comb. on Malting, Brewing and Distilling* (Campbell, I. and Priest, F. G. eds), p. 247. IoB, London.
Buckee, G. K. (1995) *Ferment.*, **8**(6), 357.
Buckee, G. K. and Bailey, T. P. (1978) *J. Inst. Brewing*, **84**, 158.
Buckee, G. K. and Baker, C. D. (1988) *J. Inst. Brewing*, **96**, 387.
Bunker, H. J. (1972) in *Quality Control in the Food Industry* (Herschdoerfer, S. M. ed.), Vol. 3, p. 81. Academic Press, London.
Burbidge, M. (1989) *Brauw. Int.*, (11), 121.
Burbridge, E. and Hough, J. S. (1970) *Process Biochem.*, **5**(4), 19.
Burger, W. C. and La Berge, D. E. (1985) in *Barley* (Rasmusson, D. C. ed.), p. 367. Am. Soc. Agronomy, Madison, WI.
Burger, W. C. and Schroeder, R. L. (1976) *J. Am. Soc. Brew. Chem.*, **34**(3), 133, 138.
Burges, H. D. and Burrell, N. J. (1964) *J. Sci. Food Agric.*, **15**, 32.
Burges, H. D., Edwards, D. M., Burrell, N. J. and Cammell, M. E. (1963) *J. Sci. Food Agric.*, **14**, 580.
Burgess, H. J. L. (1962) *E. African Med. J.*, **39**(7), 437.
Bushuk, W. (ed.) (1976) *Rye: Production, Chemistry and Technology*. Am. Assoc. Cereal Chem. St Paul, MN.
Butler, L. G. (1982) *J. Agric. Food Chem.*, **30**, 1090.
Button, A. H. and Palmer, J. R. (1974) *J. Inst. Brewing*, **80**, 206.

Byrne, H. (1991) *Brew Distil. Int.*, **22**(11), 24.
Byrne, H., Donnelly, M. F. and Carroll, M. B. (1993) *Proc. 4th Sci. Tech. Conv.*, IoB (Central and S. African Sect.), p. 13.
Caldwell, F. (1957) *J. Inst. Brewing*, **63**, 340.
Canales, A. M. (1979) in *Brewing Science*, Vol. 1 (Pollock, J. R. A. ed.), p. 233. Academic Press, London.
Canales, A. M. and Sierra, J. A. (1976) *MBAA Tech. Quart.*, **13**, 114.
Cantrell, I. C. (1987) *J. Inst. Brewing*, **93**, 487.
Cantrell, I. C. (1992) *Brewer*, **78**, 103.
Cantrell, I. C., Anderson, R. G. and Martin, P. A. (1981) *Proc. 18th Congr. EBC*, Copenhagen, p. 39.
Carson, G. P. and Horne, F. R. (1962) in *Barley and Malt: Biology, Biochemistry, Technology* (Cook, A. H. ed.), p. 101. Academic Press, London.
Causton, J. W. F. (1965) *Brewers' J., Centenary number*, p.131.
Cauwe, Y. (1965) *Brewers' Guard.*, Mar., 31.
Chambers, A. R. and Lambie, A. D. B. (1960) *J. Inst. Brewing*, **66**, 159.
Champion, W. (1832) *The Maltsters' Guide, being a Statement from Long Experience of the Best Means of Making Good Malt, Both on the Sprinkling and Non-sprinkling Process*. 204pp. Redford and Robins, London.
Chapman, A. C. and Baker, F. G. S. (1907) *J. Inst. Brewing*, **13**, 638.
Chapman, J. (1983) *Brewers' Guard.*, **112**(7), 11.
Chapman, J. and Walker, R. (1979) *Brewers' Guard.*, Nov., 35.
Chapon, L. (1959) *Brasserie*, **14**, 222.
Chapon, L. (1960) *Brasserie*, **15**, 214.
Chapon, L. (1963) *Brasserie*, **18**, 327.
Chapon, L. and Chemardin, M. (1964) *Proc. Annu. Meet. Am. Soc. Brew. Chem.*, p. 244.
Charalambous, G. and Bruckner, K. J. (1977) *Proc. 16th Congr. EBC*, Amsterdam, p. 179.
Chavan, J. K. and Kadam, S. S. (1989) *Curr. Rev. Food Sci. Nutrit.*, **28**(5), 401.
Cherry-Downes, H. A. D. (1948) *J. Inst. Brewing*, **54**, 208.
Chevassus-Agnes, S., Favier, J. C. and Orstom, A. J. (1976) *Cahiers Nutr. Diet.*, **11**(2), 89.
Christensen, C. M. (ed.) (1982) *Storage of Cereal Grains and Their Products*, 3rd edn. Am. Assoc. Cereal Chem., St Paul, MN.
Christensen, C. M. and Kaufmann, H. H. (1969) *Grain Storage – the Role of Fungi in Quality Loss*. University of Minnesota Press, Minneapolis, MN.
Christensen, C. M. and Kaufmann, H. H. (1974) in *Storage of Cereal Grains and Their Products*, 2nd edn, (Christensen, C. M. ed.) p. 158. Am. Assoc. Cereal Chem., St Paul, MN.
Chubb, A. R. (1954) *J. Inst. Brewing*, **60**, 205.
Chubb, H. M. (1911) *J. Inst. Brewing*, **17**, 452.
Chumbley, J. D., Stenniken, M., Turnbull, T. and Horgan, R. (1986) *Proc. 19th Conv. IoB* (Australia and New Zealand Sect.), p. 101.
Clutterbuck, V. J. and Briggs, D. E. (1974) *Phytochemistry*, **13**, 45.
Cole, G. J. (1982) *Proc. Conv. IoB* (Australia and New Zealand Sect.), p. 41.
Cole, W. J. and Hemmens, W. F. (1977) *J. Inst. Brewing*, **83**, 93.
Collier, J. A. (1970) *Brewers' Guild J.*, **56**(May), 242.
Collier, J. A. (1973) *Brewer*, **59**, 507.
Collier, J. A. (1975) *Brewer*, **61**, 350.
Collier, J. A. (1984) *Pauls and Whites Brewing Room Book, 1984–5*, 80th edn, p. 76. Pauls and Whites Ltd, Ipswich, UK.
Collier, J. (1986) *Brew. Dist. Int.*, **16**, 19.
Collins, E. J. (1918) *Ann. Bot.*, **32**, 381.
Conner, H. A. and Allgeier, R. J. (1976) *Adv. Appl. Microbiol.*, **20**, 81.
Cook, A H. (ed.) (1962) *Barley and Malt; Biology, Biochemistry, Technology*, 740pp. Academic Press, London.
Cook, R., McCraig, N., McMillan, J. M. B. and Lumsden, W. B. (1990) *J. Inst. Brewing*, **96**, 233.
Cooke, L. A. and La Berge, D. E. (1988) *MBAA Tech. Quart.*, **25**, 36.
Cooke, R. J. and Morgan, A. G. (1986) *J. Nat. Inst. Agric. Bot.*, **17**, 169.
Coors, J. (1976) *MBAA Tech. Quart.*, **13**, 117.
Corbineau, F. and Côme, D. (1980) *C. R. Acad. Sci. Ser D.*, **290**, 547.

Corbineau, F., Lecat, S. and Côme, D. (1986) *Seed Sci. Technol.*, **14**, 725.
Corran, H. S. (1975) *A History of Brewing*. David and Charles, Newton Abbot, UK.
Crabb, D. (1970) *J. Inst. Brewing*, **76**, 469.
Crabb, D. (1971) *J. Inst. Brewing* **77**, 522.
Crabb, D. and Bathgate, G. N. (1973) *J. Inst. Brewing*, **79**, 519.
Crabb, D. and Kirsop, B. H. (1969) *J. Inst. Brewing*, **75**, 254.
Crabb, D., Kirsop, B. H. and Palmer, G. H. (1972) *Proc. Annu. Meet. Am. Soc. Brew. Chem.*, p. 36.
Crone, (1941) *J. S. African Chem. Inst.*, **24**, 35.
Crisp, J. W. M. (1967) *Brewers' Guard.*, May, 42.
Culpin, C. (1969) *Farm Machinery*, 8th edn, pp. 460, 489. Crosby, Lockwood and Co., London.
Dadic, M., van Ghelwe, J. E. A. and Valyi, Z. (1976) *J. Inst. Bewing*, **82**, 273.
Dagg, A. H. S., Tarr, A. W. and Portman, P. A. (1991) *Proc. 3rd Sci. Tech. Conv. IoB* (Central and S. African Sect.), p. 30.
Dahlstrom, R. V. (1965) *Cereal Sci. Today*, **10**(8), 466.
Dahlstrom, R. V., Morton, B. J. and Sfat, M. R. (1963) *Proc. Annu. Meet. Soc. Brew. Chem.*, p. 64.
Daiber, K. H. (1975) *J. Sci. Food Agric.*, **26**, 1399.
Daiber, K. H. (1978) *Manual on the Treatment and Malting of Birdproof Sorghum*. Special report (restr.) BB114 of the Sorghum Beer Unit of the CSIR, Pretoria, S. Africa. National Food Res. Inst., Pretoria.
Daiber, K. H., Malherbe, L. and Novellie, L. (1973) *Brauwissenschaft*, **26**(7), 220; 248.
Daiber, K. H. and Novellie, L. (1968) *J. Sci. Food Agric.*, **19**, 87.
Daiber, K. H. and Taylor, J. R. N. (1982) *J. Agric. Food Chem.*, **30**, 70.
Dale, C. J., Young, T. W. and Makinde, A. (1989) *J. Inst. Brewing*, **95**, 157.
Darby, W. J., Ghalioungui, P. and Grivetti, L. (1977) *Food: The Gift of Osiris* (2 vols). Academic Press, London.
David, A. (1975) *Farmers' Weekly*, Aug. 29th, 48.
Davies, N. L. (1989) *Proc. 22nd Cong. EBC*, Zurich, p. 221.
Davis, A. D. and Pollock, J. R. A. (1956) *J. Inst. Brewing*, **62**, 383.
Dax, P. J. (1932) *US Patent* 1,865,680 (via 1933, *J. Inst. Brewing*, **39**, 661).
Day, T. C. (1891) *J. Chem. Soc.*, **59**, 664.
Day, T. C. (1896) *Trans. Proc. Bot. Soc., Edinburgh*, **20**, 492.
Dean, R. F. A. (1953) *Plant Proteins in Child Feeding*. Medical Research Council Special Report Series, p. 279. MRC, London.
de Batz, B. (1963) *Wallerstein Lab. Commun.*, **26**, 101.
de Clerck, E. (1972) *Fermentatio*, **68**(1), 16.
de Clerck, J. (1938) *Bull. L'Assoc. Anc. Etud. Ecole Supér. Brass. Univ. Louvain*, **38**(3), 85.
de Clerck, J. (1952) *A Textbook of Brewing* (Barton-/Wright, K. Transl.) (2 vols). Chapman & Hall, London.
de Clerck, J. and Cloetens, J. (1940) *Bull. Assoc. Anc. Etud. Ecol Supér. Brass. Univ. Louvain*, **40**, 41.
de Clerck, J. and Isebaert, L. (1959) *Petit. J. Brass.*, **67**, 300; 318.
de Coetlogon, D. (1745) *Universal History of Arts and Sciences*, **1**, 385 (London).
Delbrück, M. (1910) *Illustriertes Brauerei-Lexikon*. Paul Parey, Berlin.
Derdelinckx, G. and Jerumanis, J. (1987) *Proc. 21st Congr. EBC*, Madrid, p. 577.
Derman, D. P., Bothwell, T. H., Torrance, J. D., Bezwoda, W. R., MacPhail, A. P., Kew, M.-C., Sayers, M. H., Disler, P. B. and Charlton, R. W. (1980) *Br. J. Nutrit.*, **43**, 271.
Desikachar, H. S. R. (1980) *Food Nutr. Bull.*, **2**, 21.
Desikachar, H. S. R. (1982) *Food Nutr. Bull.*, **4**(4), 57.
Dickenson, C. J. and Anderson, R. G. (1981) *Proc. 18th Congr. EBC*, Copenhagen, p. 413.
Dickson, A. D. and Burkhart, B. A. (1942) *Cereal Chem.*, **19**, 251.
Dickson, A. D., Olson, W. J. and Shands, H. L. (1947) *Cereal Chem.*, **24**, 325.
Dickson, A. D. and Shands, H. L. (1942) *Cereal Chem.*, **19**, 411.
Dickson, J. G. (1962) in *Barley and Malt; Biology, Biochemistry, Technology* (Cook, A. H. ed.), p. 161. Academic Press, London.
Disney, R. W. (1954) *Cereal Chem.*, **31**, 229.
Doggett, H. (1988) *Sorghum*, 2nd edn. Longman, Harlow, UK.
Dolan, T. C. S. (1976) *J. Inst. Brewing*, **82**, 177.

Dolan, T. C. S. (1979) *Brewer*, **65**, 60.
Dolan, T. C. S. (1983) *J. Inst. Brewing* **89**, 84.
Dolan, T. C. S., Dewar, E. T. and Gray, J. D. (1981) *J. Inst. Brewing*, **87**, 352.
Doncheck, J. and Sfat, M. R. (1979) *Brewers' Dig.*, Feb., 34.
Donhauser, S., Weideneder, A., Winnewisser, W. and Geiger, E. (1991) *Brau. Int.*, **4**, 294.
Doran, P. J. and Briggs, D. E. (1992) *J. Inst. Brewing*, **98**, 193.
Doran, P. J. and Briggs, D. E. (1993a) *J. Inst. Brewing*, **99**, 85.
Doran, P. J. and Briggs, D. E. (1993b) *J. Inst. Brewing*, **99**, 165.
Drost, B. W., Centen, A. H. J., Pesman, L. and de Vries, H. (1987) *Proc. 21st Congr. EBC*, Madrid, p. 167.
Dudley, M. J. and Briggs, D. E. (1977) *J. Inst. Brewing*, **83**, 305.
Duff, S. R. (1963) *J. Inst. Brewing*, **69**, 249.
Dyer, T. A. and Novellie, L. (1966) *J. Sci. Food Agric.*, **17**, 449.
EBC (1976) *Barley Varieties*. Barley Committee, EBC.
EBC (1987) *Analysis Committee of the EBC, Analytica EBC*, 4th edn and Supplements. Brauerei- und Getränke-Rundschau, Zurich.
Edwards, T. I. (1932) *Quart. Rev. Biol.*, **7**, 428.
Eggitt, P. W. R. (1964) *Brewers' Guild J.*, **50**(Nov), 533.
Ehrich, E. and Kneip, E. (1931) *Z. ges. Brau.*, **59**, 1, 98, 105, 118, 123.
Ekström, D., Cederqvist, B. and Sandegren, E. (1959) *Proc. 7th Congr. EBC*,. Rome, p. 11.
Ellis, R. P., Swanston, J. S., Taylor, K. and Bruce, F. M. (1989) *Ann. Appl. Biol.*, **114**, 349.
Enari, T.-M. (1974) *EBC Monograph-I. Wort Symp., Zeist*. p. 73.
Enari, T.-M. (1981) *EBC Monograph-VI. Symp. Relationship between Malt and Beer*, Helsinki, p. 88.
Enari, T.-M. (1986) *Cerevisia*, **11**(1), 19.
Enari, T.-M., Linnahalme, T. and Linko, M. (1961) *J. Inst. Brewing*, **67**, 358.
Enari, T.-M. and Markkanen, P. H. (1975) *Proc. Am. Soc. Brew. Chem.*, **33**, 13.
Enari, T.-M. and Mikola, J. (1967) *Proc. 11th Congr. EBC*, Madrid, p. 9.
Enari, T.-M. and Sopanen, T. (1986) *J. Inst. Brewing*, **92**, 25.
Enders, C., Nowak, G., Schneebauer, F. and Pfahlen, A. (1940) *Woch. Brau.*, **57**, 81; 89.
Enevoldsen, B. S. (1970) *J. Inst. Brewing*, **76**, 546.
Enevoldsen, B. S. (1975) *Proc. 15th Congr. EBC*, Nice, p. 683.
Enevoldsen, B. S. (1986) *EBC Monograph-XI. Symp. Wort Production*, Maffliers, p. 163.
Enkenlund, J. (1972) *Process Biochem.*, **7**(8), 27.
Erdal, K. and Gjersten, P. (1971) *Proc. 13th Congr. EBC*, Estoril, p. 49.
Erdman, J. A. and Moul, R. C. (1982) *J. Agric. Food Chem.*, **30**, 169.
Essery, R. E. (1954) *J. Inst. Brewing*, **60**, 21, 303.
Essery, R. E., Kirsop, B. H. and Pollock, J. R. A. (1954) *J. Inst. Brewing*, **60**, 473.
Essery, R. E., Kirsop, B. H. and Pollock, J. R. A. (1955) *J. Inst. Brewing*, **61**, 25.
Essery, R. E., Kirsop, B. H. and Pollock, J. R. A. (1956) *J. Inst. Brewing*, **62**, 150.
Evans, D. J. and Taylor, J. R. N. (1990) *J. Inst. Brewing*, **96**, 399.
Eyben, D. and van Droogenbroeck, L. (1969) *Proc. 12th Congr. EBC*, Interlaken, p. 107.
Eyben, D. and van Droogenbroeck, L. (1971) *Bios*, **11**, 3.
Eyben, D., Vriens, L., Franco, P. and Verachtert, H. (1985) *Proc. 20th Congr. EBC*, Helsinki, p. 555.
Ezeogu, L. I. and Okolo, B. N. (1994) *J. Inst. Brewing*, **100**, 335.
Faparusi, S. I. (1970) *J. Sci. Food Agric.*, **21**, 79.
Faparusi, S. I., Olofinboba, M. O. and Ekundayo, J. A. (1973) *Z. Allg. Mikrobiol.*, **13**, 563.
Farley, D. R. and Nursten, H. E. (1980) *J. Sci. Food Agric.*, **31**, 386.
Fernbach, A. (1928) *J. Inst. Brewing*, **34**, 119.
Fincher, G. B. (1989) *Annu. Rev. Plant Physiol. Plant. Mol. Biol.*, **40**, 305.
Fincher, G. B. (1992) in *Barley: Genetics, Biochemistry, Molecular Biology and Biotechnology* (Shewry, P. R. ed.) p. 413. CAB International, Wallingford.
Fincher, G. B. and Stone, B. A. (1986) in *Advances in Cereal Science and Technology* Vol. 8 (Pomeranz, Y. ed.), p. 207, Am. Assoc. Cereal Chem., St Paul, MI.
Finney, P. L. (1982) in *Recent Advances in Phytochemistry*, Vol. 17, *Mobilization of Reserves in Germination* (Nozzolillo, C., Lea, P. J. and Loewus, F. A. eds), p. 229. Plenum Press, New York.
Flannigan, B. (1983) *J. Inst. Brewing*, **89**, 364.

Flannigan, B. (1996) in *Brewing Microbiology*, 2nd edn (Priest, F. G. and Campbell, I. eds), p. 83. Chapman & Hall, London.
Flannigan, B., Okagbue, R. N., Khalid, R. and Teoh, C. K. (1982) *Brew. Dist. Int.*, June, 31.
Fleming, J. R. and Johnson, J. A. (1961) *J. Agric. Food Chem.,*, **9**, 152.
Fleming, J. R., Johnson, J. A. and Miller, B. S. (1960) *Cereal Chem.*, **37**, 363, 371.
Fleming, J. R., Johnson, J. A. and Miller B. S. (1961) *Cereal Chem.*, **38**, 170.
Fogarty, W. (1982) *Proc. 1st Aviemore Conf. Curr. Devel. Malting, Brewing and Distilling* (Priest, F. G. and Campbell, I. eds), p. 83. IoB, London.
Fogarty, W. M., Griffin, P. J. and Joyce, A. M. (1974) *Process Biochem.*, **9**(6), 11; and **9**(7), 29.
Foldager, L. and Jørgensen, K. G. (1984) *Carlsberg Res. Commun.*, **49**, 525.
Ford, W. (1849) *A Practical Treatise on Malting and Brewing, with an Historical Account of the Malt Trade and Laws, Deduced from Forty Years' Experience*. London.
Forrest, I. S., Dickson, J. E. and Seaton, J. C. (1985) *Proc. 20th Congr. EBC*, Helsinki, p. 363.
Forster, T. F. G. (1985) *Brewers' Guard.*, Jan., 8.
Foster, G. H. (1982) in *Storage of Cereal Grains and Their Products*, 3rd edn (Christensen, C. M. ed.), p. 79. Am. Assoc. Cereal Chem., St Paul, MN.
Foster, G. H. and Tuite, J. (1982) in *Storage of Cereal Grains and Their Products*, 3rd edn (Christensen, C. M. ed.), p. 117. Am. Assoc. Cereal Chem., St Paul, MN.
Fox, F. W. (1938) *J. S. African Chem. Inst.*, **21**, 39.
Fox, F. W. and Stone, W. (1938) *S. African J. Med. Sci.*, **3**, 7.
Fox, G. P. and Henry, R. J. (1993) *J. Inst. Brewing*, **99**, 73.
Frauenheim, E. E. (1959) *Proc. 7th Congr. EBC*, Rome, p. 100.
Freeman, P. L. (1986) *Proc. 2nd Aviemore Conf. on Malting, Brewing and Distilling* (Campbell, I. and Priest, F. G. eds), p. 258. University Press, Aberdeen.
French, B. J. and McRuer, G. R. (1990) *MBAA Tech. Quart.*, **27**(1), 10.
French, R. C. (1959) *Plant Physiol., Lancaster*, **34**, 500.
Frew, W. (1901) *J. Soc. Chem. Ind.*, **20**, 221.
Fröst, S., Harborne, J. B. and King, L. (1977) *Hereditas*, **85**, 163.
Fujii, T. and Lewis, M. J. (1965) *Proc. Annu. Meet. Am. Soc. Brew, Chem.*, p. 11.
Fulcher, R. G., Setterfield, G., McCully, M. E. and Wood, P. J. (1977) *Aust. J. Plant Physiol.*, **4**, 917.
Galland, N. (1874) *Faits et Observations sur la Brasseries. Suivis de la description d'un nouveau procédé de fabrication*. Berger-Leuvault et Cie, Paris et Nancy.
Galliard, T. (1982) in *Recent Advances in Phytochemistry*, Vol. 17, *Mobilization of Reserves in Germination* (Nozzolillo, C., Lea, P. J. and Loewus, F. B. eds), p. 111. Plenum Press, New York.
Galliard, T. (ed.) (1987) *Starch: Properties and Potential*. Wiley, New York.
Geiger, K. (1972) *MBAA Tech. Quart.*, **9**, 195.
Gent, J. F. (1889) *US Patent* 409,956.
Gibbons, G. C. (1980) *Carlsberg, Res. Commun.*, **45**, 177.
Gibson, G. (1989) in *Cereal Science and Technology* (Palmer, G. H. ed.), p. 279. University Press, Aberdeen.
Gibson, G. (1992) in *Pauls Brewing Room, 1992–4*, 83rd edn, p. 39. Pauls Malt Ltd, Ipswich, UK.
Gill, N. T., Vear, K. C. and Barnard, D. J. (1980) *Agricultural Botany*, 3rd edn, Vol. 2, *Monocotyledenous Plants*. Duckworth, London.
Githua, P. M. and Barrell, G. W. (1974) *MBAA Tech. Quart.*, **11**(3), 174.
Gjertsen, P., Myken, F., Krogh, P. and Hald, B. (1973) *Proc. 14th Congr. EBC*, Salzburg, p. 373.
Gjertsen, P., Trolle, B. and Andersen, K. (1963) *Proc. 9th Congr. EBC*, Brussels, p. 320.
Gjertsen, P., Trolle, B. and Andersen, K. (1965) *Proc. 10th Congr. EBC*, Stockholm, p. 428.
Glennie, C. W. (1983) *J. Agric. Food Chem.*, **31**, 1295.
Glennie, C. W. (1984) *Cereal Chem.*, **61**, 285.
Glennie, C. W., Harris, J. and Liebenberg, N. V. D. (1983) *Cereal Chem.*, **60**, 27.
Glennie-Holmes, M. (1990) *J. Inst. Brewing*, **96**, 17.
Glennie-Holmes, M., Moon, R. L. and Cornish, P. B. (1990) *J. Inst. Brewing*, **96**, 11.
Gluckman, R. and Kidger, P. B. (1984)*Pauls and Whites Brewing Room Book, 1984–5*, 80th edn, p. 59. Pauls and Whites, Ltd, Ipswich, UK.
Gocar, M. (1983) *Petit J. Brass.*, **91**, 92.
Goering, K. J. and Eslick, R. F. (1976) *Cereal Chem.*, **53**, 174.

Goering, K. J. and Imsande, J. D. (1960) *J. Agric. Food Chem.*, **8**, 368.
Gola, G. (1905a) *Ann. Botanica*, **3**, 59.
Gola, G. (1905b) *Mem. R. Acc. Sci. Torino, II*, **55**, 237.
Gomez, R. (1971) *Proc. 13th Congr. EBC*, Estoril, p. 29.
Gopaldas, T., Inamdar, F. and Patel, J. (1982) *Ind. J. Nutr. Dietet.*, **19**, 327.
Gordon, A. G. (1968) *J. Inst. Brewing*, **74**, 355.
Gothard, P. G. and Smith, D. B. (1986) *J. Cereal Sci.*, **4**, 71.
Graff, A. R. and Spitz, D. (1951) *Proc. Annu. Meet. Am. Soc. Brew. Chem.*, p. 122.
Graham, C. (1874) *J. Soc. Arts*, p. 188, 221, 245, 278, 297, 333, 366.
Gram, N. H. (1982) *Carlsberg Res. Commun.*, **47**, 143, 173.
Grandegger, K. (1991) *Brauindustrie*, **76**, 1264.
Grandegger, K. and Traber, U. (1991) *Brauwelt*, **131**, 2373.
Green, A. D. (1965) *Brewers' Guild J.*, **51**(June), 280.
Green, J. W. and Sanger, M. J. (1956) *J. Inst. Brewing*, **62**, 170.
Greenshields, R. N. (1973) *Process Biochem.*, **8**(4), 17.
Greenshields, R. N. and Macgillivray, A. W. (1972) *Process Biochem.*, **7**(12), 11.
Griffin, O. T. (1982) *MBAA Tech. Quart.*, **19**, 41.
Griffin, O. T., Collier, J. A. and Shields, P. D. (1968) *J. Inst. Brewing*, **74**, 154.
Griffin, O. T. and Pinner, B. C. (1965) *J. Inst. Brewing*, **71**, 324.
Griffiths, C. M. (1982)n *Proc. 1st Aviemore Conf. Curr. Devel. Malting, Brewing and Distilling* (Campbell, I. and Priest, F. G. eds), p. 235. IoB, London.
Griffiths, C. M. and MacWilliam, I. C. (1967) *J. Inst. Brewing*, **73**, 172.
Grime, K. H. and Briggs, D. E. (1995) *J. Inst. Brewing*, **101**, 337.
Grime, K. H. and Briggs, D. E. (1996) *J. Inst. Brewing*, **102**, 261.
Grist, D. H. (1986) *Rice*, 6th edn. Longman, London.
Groat, J. I. and Briggs, D. E. (1969) *Phytochemistry*, **8**, 1615.
Gromus, J. (1988) *Brau. Int.*, **II**, 150.
Groulx, R. A. (1970) *MBAA Tech. Quart.*, **7**(12), 101.
Gupta, N. K., Singh, T. and Bains, G. S. (1985) *Brewers' Dig.*, **60**(3), 24.
Haikara, A. and Home, S. (1991) *Proc. 23rd Congr. EBC*, Lisbon, p. 537.
Haikara, A. and Laitila, A. (1995) *Proc. 25th Congr. EBC*, Brussels, p. 249.
Haikara, A., Mäkinen, V. and Hakulinen, R. (1977) *Proc. 16th Congr. EBC*, Amsterdam, p. 35.
Haikara, A., Uljas, H. and Suurnäkki, A. (1993) *Proc. 24th Congr. EBC*, Oslo, p. 163.
Hajek, T. (1938) *Böhm, Bierbrau.*, **65**(6), 59.
Hakulin, S., Linko, Y.-Y., Linko, P., Seiler, K. and Seibel, W. (1983) *Stärke*, **35**, 411.
Haley, P. E. and Stokes, C. H. (1987) *MBAA Tech. Quart.*, **24**, 33.
Halford, M. H., Cordiner, B. G. and Clarke, B. J. (1976) *Proc. 14th Conv. IoB* (Australia and New Zealand Section), p. 45.
Hall, J. A. D. (1978) *Process Biochem.*, **13**(7), 20.
Hall, R. D., Harris, G. and MacWilliam, I. C. (1956) *J. Inst. Brewing*, **62**, 232.
Haman, R. W. and Hartzell, T. H. (1972) *MBAA Tech. Quart.*, **9**(4), 161.
Harada, T. (1984) in *Biotechnology and Genetic Engineering Reviews*, Vol. 1 (Russell, G. E. ed.), p. 39. Intercept, Newcastle-upon-Tyne, UK.
Harding, R. J., Wren, J. J. and Nursten, H. E. (1978) *J. Inst. Brewing*, **84**, 41.
Hardwick, W. A., Hickman, D. H., Jangaard, N. O., Ladish, W. J. and Meilgaard, M. C. (1982) *Regul. Toxicol. Pharmacol.*, **2**, 38.
Hardwick, W. A., Ladish, W. J., Meilgaard, M. C. and Jangaard, N. O. (1981) *MBAA Tech. Quart.*, **17**(2), 92.
Harrington, G. T. (1923) *J. Agric. Res.*, **23**, 79.
Harris, G. (1962a) in *Barley and Malt – Biology, Biochemistry, Technology* (Cook, A. H. ed.), p. 431. Academic Press, London.
Harris, G. (1962b) in *Barley and Malt – Biology, Biochemistry, Technology* (Cook, A. H. ed.), p. 583. Academic Press, London.
Harris, G., Hall, R. D. and MacWilliams, I. C. (1955) *Proc. 5th Congr. EBC*, Baden-Baden, p. 26.
Harrison, T. J. and Rowlands, H. (1932) *J. Inst. Brewing*, **38**, 502.
Harrison, W. (1587; reprinted 1968) *The Description of England* (Edelen, G. ed.). Cornell University Press, New York.
Hart, J. R. (1964) *Cereal Chem.*, **41**, 340.
Hartman, L. F. and Oppenheim, A. L. (1950) *J. Am. Oriental Soc.* Suppl., 10.

Hartong, B. D. and Kretschmer, K. F. (1961) *Proc. 8th Congr. EBC*, Vienna, p. 69.
Hartong, B. D. and Kretschmer, K. F. (1968) *Brauwelt*, **108**, 517.
Haslemore, R. M., Tunnicliffe, C. G., Slack, C. R. and Shaw, S. (1985) *J. Inst. Brewing*, **91**, 101.
Hausmann, E. (1937) *Böhm, Bierbrau.*, **64**(45), 489.
Healy, P. J. and Armitt, J. D. G. (1986) *Proc. 19th Conv. IoB* (Australia and New Zealand Section), Hobart, p. 71.
Hebeda, R. E. (1987) in *Corn: Chemistry and Technology* (Watson, S. A. and Ramstad, P. E. eds), p. 501. Am. Assoc. Cereal Chem., St Paul, MN.
Hector, J. M. (1936) *Introduction to the Botany of Field Crops*, Vol. 1, *Cereals*, p. 234. Central News Agency, Johannesburg, S. Africa.
Heinzelmann, R. (1907) *Woch. Brau.*, **24**, 13, 25, 33, 50, 61, 77, 93, 105.
Hejgaard, J. (1977) *J. Inst. Brewing*, **83**, 94.
Hejgaard, J. (1978) *J. Inst. Brewing*, **84**, 43.
Henry, R. J. (1985) *J. Sci. Food Agric.*, **36**, 1243.
Henry, R. J. (1988) *J. Inst. Brewing*, **94**, 71.
Henry, R. J. and McLean, B. T. (1984a) *J. Sci. Food Agric.*, **35**, 767.
Henry, R. J. and McLean, B. T. (1984b) *J. Inst. Brewing*, **90**, 371.
Heron, J. (1912) in *A Dictionary of Applied Chemistry*, revised edn, Vol. 1 (Thorpe, E. ed.), p. 526. Longmans, Green and Co., London.
Heslop, A. J. (1911) *J. Inst. Brewing*, **17**, 480.
Hesseltine, C. W. (1979) *J. Am. Oil Chemists' Soc.*, **56**, 367.
Hill, R. D. and MacGregor, A. W. (1988) in *Advances in Cereal Science and Technology*, Vol. 9 (Pomeranz, Y. ed.), p. 217. Am. Assoc. Cereal Chem., St Paul, MN.
Hills, S. (1986) *Brew. Dist. Int.*, **18**(2), 17.
Hind, H. L. (1940) *Brewing Science and Practice*. Chapman & Hall, London.
Hind, H. L. and Lancaster, H. M. (1932) *J. Inst. Brewing*, **38**, 290.
Hinton, J. J. C. (1955) *Cereal Chem.*, **32**, 296.
Hinz, H. (1954) *Z. Volkekunde*, **51**, 88.
Hoggan, J., Compson, D. G., Spencer, E. and Jennings, P. (1962) *J. Inst. Brewing*, **68**, 39.
Hoggan, J., Spencer, E. and Jennings, P. (1963) *J. Inst. Brewing*, **69**, 28.
Holbein, G. (1969) *Int. Brewers' J.*, **105**(Feb), 47.
Home, S. and Linko, M. (1977) *Proc. 16th Congr. EBC*, Amsterdam, p. 91.
Hopkins, R. H. and Carter, W. A. (1933) *J. Inst. Brewing*, **39**, 59.
Hopkins, R. H. and Krause, B. (1947) *Biochemistry Applied to Malting and Brewing*. Allen and Unwin, London.
Hopulele, T. and Piendl, A. (1973) *Proc. Annu. Meet. Am. Soc. Brew. Chem.*, p. 75.
Hoseney, R. C. (1986) *Principles of Cereal Science and Technology*. Am. Assoc. Cereal Chem., St Paul, MN.
Hoseney, R. C. and Faubion, J. M. (1992) in *Storage of Cereal Grains and Their Products*, 4th edn (Sauer, D. B. ed.), p. 1. Am. Assoc. Cereal Chem., St Paul, MN.
Hough, J. S., Briggs, D. E., Stevens, R. and Young, T. W. (1982) *Malting and Brewing Science*, 2nd edn, Vol. 2, *Hopped Wort and Beer*. Chapman & Hall, London.
Hough, J. S., Wadeson, A. and Daniels, N. W. R. (1976) *Brewers' Guard.*, **105**(3), 38.
Hough, J. S., Young, T. W. and Lewis, M. J. (1975) *Materials and Technology*, Vol. VII, *Vegetable Food Products*, p. 787. Longman de Bussy, the Netherlands.
Houtart, J. and Delatte, J.-L. (1979) *Bios*, **10**(4), 24.
Howard, K. A., Gayler, K. R., Eagles, H. A. and Halloran, G. M. (1996) *J. Cereal Sci.*, **24**, 47.
Howe, R. W. (1965a) *Nutr. Abstr. Rev.*, **35**, 285.
Howe, R. W. (1965b) *Stored Prod. Res.*, **1**, 177.
Howling, D. (1979) *Sugar: Science and Technology* (Birch, G. G. and Parker, K. J. eds), p. 259. Applied Science, London.
Hudson, J. R. (1960) *Development of Brewing Analysis: a Historical Review*. IoB, London.
Hudson, L. E. (1959) *J. Inst. Brewing*, **65**, 473.
Hudson, O. P. (1980a) *Pauls and Whites Brewing Room Book, 1980–1*, 78th edn, p. 55. Pauls and Whites Ltd, Ipswich, UK.
Hudson, O. P. (1980b) *Proc. 16th Conv. IoB* (Australia and New Zealand Section), p. 33.
Hudson, O. P. (1986a) *J. Inst. Brewing*, **92**, 115.
Hudson, O. P. (1986b) *Proc. 19th Conv. IoB* (Australia and New Zealand Section), Hobart, p. 109.

Hukill, W. V. (1954) in *The Storage of Cereal Grains and Their Products* (Anderson, J. A. and Alcock, A. W. eds), p 402. Am. Assoc. Cereal Chem., St Paul, MN.
Hukill, W. V.(1974) in *The Storage of Cereal Grains and Their Products*, 2nd edn (Christensen, C. M. ed.), p. 481. Am. Assoc. Cereal Chem., St Paul, MN.
Hulse, J. H., Laing, E. M. and Pearson, O. (1980) *Sorghum and the Millets: Their Composition and Nutritive Values*. Academic Press, London.
Hummel, B. C. W., Cuendet, L. S., Christensen, C. M. and Geddes, W. F. (1954) *Cereal Chem.*, **31**, 143.
Hunt, R. and Rudler, F. W. (1878) in *Ure's Dictionary of Arts, Manufacturers and Mines*, 7th edn, Vol. 1, p. 300; Vol. 3, p. 183. Longmans, Green and Co., London.
Hunter, H. (1952) *The Barley Crop*. Crosby, Lockwood and Son, London.
Hunter, H. (1962) in *Barley and Malt: Biology, Biochemistry, Technology* (Cook, A. H. ed.), p. 25. Academic Press, London.
Hunter, H. and Hartley, H. O. (1938) *J. Agric. Sci. Camb.*, **28**, 472.
Hutt, W., Meiering, A., Oelschlager, W. and Winkler, E. (1978) *Can. Agric. Eng.*, **20**(2), 103.
Hyde, W. R. and Brookes, P. A. (1978) *J. Inst. Brewing*, **84**, 167.
Hynard, W. (1947) *J. Inst. Brewing*, **53**, 187.
Ilori, M. O. and Adewusi, S. R. A. (1991) *J. Inst. Brewing*, **97**, 111.
Imrie, F. K. (1968) *Process Biochem.*, **3**(4), 21.
Imrie, F. K. and Martin, F. J. (1969) *Brewers' Guild J.*, **55**, 252.
Inagaki, H., Yamazumi, K., Uehara, H. and Mochzuki, K. (1994) *EBC Monograph-XXIII, EBC Symp. Malting Technol.*, p. 110. Andernach.
IoB (1993) *Analysis Committee of the Institute of Brewing: Recommended Methods of Analsyis*, IoB, London.
Isaac, P. G. (1969) *Effluent and Water Treatment J.*, **9**(11), 597.
Isaac, P. G. (1976) *Process Biochem.*, **11**, 17.
Isaac, P. G. and Anderson, G. K.(1973) *J. Inst. Brewing*, **79**, 154.
Isebaert, L. (1959) *Brauwelt*, **99**, 805.
Isebaert, L. (1965) *Proc. 10th Congr. EBC*, Stockholm, p. 36.
Izquierdo-Pulido, M., Marine-Font, A. and Vidal-Caron, M. C. (1994) *J. Food Sci.*, **59**(5), 1104.
Jacobs, M. B. (1944) *The Chemistry and Technology of Food and Food Products*, Vols 1 and 2. Interscience, New York.
Jacobsen, T. (1986) *EBC Monograph-XI, EBC Symp. Wort Production*, Maffliers, p. 196.
Jakob, G. (1913) *Kontrolle der Mälzerei- und Brauerei – Arbeitsmaschinen*. R. Oldebourg, Munich.
Jaques, G. and Mungham, H. H. (1964) *Proc. Irish Maltsters' Tech. Meet.*, p. 40.
Jayatissa, P. M., Pathirana, R. A. and Sivayogasunderam, K. (1980) *J. Inst. Brewing*, **86**, 18.
Jerumanis, J. (1972) *Brauwissenschaft*, **25**(10), 313.
Jones, B. L., Wrobel, R., Marinac, L. and Zhang, N. (1993) *Proc. 24th Congr. EBC*, Oslo, p. 53.
Jones, M. and Pierce, J. S. (1963) *Proc. 9th Congr. EBC*, Brussels, p. 101.
Jonsson, A. E. (1937) *J. Inst. Brewing*, **43**, 205 (abstr.).
Jupp, D. H. (1994) *Brewer*, **80**, 466.
Kahn, J. R. (1979) in *Fermented Food Beverages in Nutrition* (Gastrineau, C. F. ed.), p. 505. Academic Press, New York.
Kamalavalli, D. and Ramananda Rao, C. (1973) *J. Majaraja Sayajitao Univ., Baroda*, **22**, 9.
Kano, Y. (1977) *Bull. Brewing Sci.*, **23**, 1, 9.
Kano, Y., Kunitake, N., Karakawa, T., Taniguchi, H. and Nakamura, M. (1981) *Agric. Biol. Chem.*, **45**, 1969.
Kassenaar, J. D. C. (1974) *MBAA Tech. Quart.*, **11**(3), 179.
Kauert, G. (1995) *Fermentatio*, p. 229.
Kelly, G. (1987) *Brew. Dist. Int.*, **17**(12), 40.
Kelly, L. and Briggs, D. E. (1992a) *J. Inst. Brewing*, **98**, 329.
Kelly, L. and Briggs, D. E. (1992b) *J. Inst. Brewing*, **98**, 395.
Kelly, L. and Briggs, D. E. (1993) *J. Inst. Brewing*, **99**, 57.
Kent, N. L. (1983) *Technology of Cereals*, 3rd edn. Pergamon Press, Oxford.
Kerslake, R. T. (1960) *J. Inst. Brewing*, **66**, 373.
Kidger, P. (1986a) *Brewers' Guard.*, June, 17.
Kidger, P. (1986b) in *Pauls Brewing Room Book, 1986–8*, 81st edn, p. 29. Pauls Malt, Ipswich, UK.

Kieninger, H. and Rumelin, G. (1984) *Brau. Int.*, **11**, 140.
Kieninger, H. and Wiest, A. (1982) *MBAA Tech. Quart.*, **19**(4), 170.
King, H. J. H. and Hopper, W. S. (1936) *J. Inst. Brewing*, **42**, 403 (abst.).
Kirby, A. R., Ollett, A.-L., Parker, R. and Smith, A. C. (1988) *J. Food Eng.*, **8**, 247.
Kirsop, B. H. (1964) *Proc. Irish Maltsters' Tech. Meet.*, p. 52.
Kirsop, B. H. (1966) *Proc. Irish Maltsters' Tech. Meet.*, p. 38.
Kirsop, B. H., Griffiths, C. M. and Barrett, J. (1967a) *Proc. 11th Congr. EBC*, Madrid, p. 219.
Kirsop, B. H. and Pollock, J. R. A. (1958) *J. Inst. Brewing*, **64**, 227, 374.
Kirsop, B. H. and Pollock, J. R. A. (1961) *J. Inst. Brewing*, **67**, 43.
Kirsop, B. H., Reynolds, T. and Griffiths, C. M. (1967b) *J. Inst. Brewing*, **73**, 182.
Kitamura, Y., Yumoto, T., Yamada, K. and Noshiro, A. (1988) *Proc. 29th Conv. IoB* (Australia and New Zealand Sect.), Brisbane, p. 73.
Klopper, W. J. (1973) *Brewer*, **59**, 395.
Kneen, E. and Dickson, A. D. (1967) in *Kirk-Othmer's Encyclopaedia of Chemical Technology*, 2nd edn, Vol. 12, p. 861.
Kneen, E., Fleckenstein, J. C. and Beckord, L. D. (1965) US Patent 3,188,279.
Knorr, F. (1952) *Brauwissenschaft*, p. 70.
Koch, A. (1911) *Z. ges. Brau.*, **34**, 659, 669.
Kolbach, P. and Zastrow, K. (1961) *Monatss. Brau.*, **14**, 95.
Kosar, K., Psota, V., Vitkova, H. and Klicova, S. (1989) *Kvasny Prum.*, **35**(6), 163.
Koszyk, P. F. and Lewis, M. J. (1977) *J. Am. Soc. Brew. Chem.*, **35**(2), 77.
Kramer, S. N. (1963) *The Sumerians, Their History, Culture and Character*. University of Chicago Press, Chicago, IL.
Krause, G. A. (1892) US Patent 481,958 (Sept. 6th, 1890).
Kreis, M., Williamson, M., Buxton, B., Pywell, J., Hejgaard, J. and Svendsen, I. (1987) *Eur. J. Biochem.*, **169**, 517.
Kretscher, K. F. (1981) *Monatss. Brau.*, **34**, 41.
Kreyger, J. (1958) *Petit J. Brass.*, **66**, 811, 832.
Kreyger, J. (1959) *Petit J. Brass.*, **67**, 7.
Kringstad, H. and Kilhovd, J. (1957) *Proc. 6th Congr. EBC*, Copenhagen, p. 67.
Kruger, L., Ryder, D. S., Alcock, C. and Murray, J. P. (1982) *MBAA Tech. Quart.*, **19**(1), 45.
Kühtreiber, F. (1995) *Brau. Int.*, **13**(1), 45.
Kunze, W. (1994) *Technologie Brauer and Mälzer* (7 Auflage). VLB, Berlin.
Kustermann, M., Kutzbach, H. D. and Dörrer-Ibrahim, D. (1986) *Int. Agrophys.*, **2**(2), 161.
Laatsch, H. U. and Sattelberg, K. (1968) *Process Biochem.*, **10**(3), 28.
Laing, P. V. (1986) *Proc. 2nd Aviemore Conf. Malting, Brewing and Distilling* (Campbell, I. and Priest, F. G. eds), p. 255. IoB, London.
Lambert, J. G. (1953) *J. Inst. Brewing*, **59**, 324 (abstr.).
Lancaster, H. (1908) *Practical Floor Malting*. The Brewing Trade Review, London.
Lancaster, H. M. (1936a) *The Maltster's Materials and Methods*. IoB, London.
Lancaster, H. M. (1936b) *J. Inst. Brewing*, **42**, 496.
Landau, J., Chanda, S. and Proudlove, M. (1996) *Proc. 24th Conv. IoB* (Asia – Pacific Section), p. 139.
Larsson, M. and Sandberg, A.-S. (1955) *J. Cereal Sci.*, **21**, 87.
Laws, D. R. J., Baxter, E. D. and Crescenzi, A. M. (1986) *EBC Monograph-XI, EBC Symp. Wort Production*, Maffliers, p. 14.
Leach, A. A. (1968) *J. Inst. Brewing*, **74**, 183.
Learner, S. (1956) *J. Inst. Brewing*, **62**, 322.
Leaton, K. W. (1963) *Brewers' Guild J.*, **49**, 288, 337.
Leberle, H. (1930) *Die Bierbrauerei*, Vol. 1, *Die Technologie der Malzbereitung*. Ferdinand Enke, Stuttgart.
Lee, W. J. (1990) *J. Am. Soc. Brew. Chem.*, **48**(2), 62.
Lehtonen, M. and Aikasalo, R. (1987) *Cereal Chem.*, **64**, 133.
Lenoir, C., Corbineau, F. and Côme, D. (1986) *Physiol. Plant.*, **68**, 301.
Lermer, Dr and Holzner, G. (1888) *Beiträge zur Kenntnis der Gerste*. R. Oldenbourg, Munich.
Lersrutaiyotin, R., Shigenaga, S. and Utsunomiya, N. (1991) *Jap. J. Crop Sci.*, **60**(2), 291.
Letters, R. (1990) *Louvain Brewing Lett.*, **4**(3/4), 12.
Lévy, L. (1899) *La Pratique du Maltage*. G. Carré et C. Naud, Paris.

Lewis, M. J. and Fujii, T. (1966) *Proc. Annu. Meet. Am. Soc. Brew. Chem.*, p. 158.
Lewis, M. J. and Oh, S. S. (1985) *MBAA Tech. Quart.*, **22**, 108.
Lewis, M. J. and Wahnon, N. N. (1984) *J. Am. Soc. Brew. Chem.*, **42**, 159.
Lewis, M. J. and Young, T. W. (1995) *Brewing*. Chapman & Hall, London.
Linko, M. (1961) *Proc. 8th Congr. EBC*, Vienna, p. 99.
Linko, M. and Enari, T.-M. (1966) *Int. Brewer and Distiller*, **1**(1), 1.
Linko, P. (1960) *Ann. Acad. Sci. Fenn.*, Ser. A(II-Chemica), **98**, 1.
Linko, P. (1987) in *Enzymes and their Role in Cereal Technology* (Kruger, J. E., Lineback, D. and Stauffer, C. E. (eds), p. 357. Am. Assoc. Cereal Chem., St Paul, MN.
Lisbjerg, A. and Nielsen, H. (1991) *Proc. 3rd Sci. Tech. Conv. IoB* (Central and S. African Sect.), Victoria Falls, p. 157.
Little, B. T. (1994) *Ferment.*, **7**(3), 163.
Liu, D. J. and Pomeranz, Y. (1975) *Cereal Chem.*, **52**, 620.
Liu, D. J., Pomeranz, Y. and Robbins, G. S. (1975) *Cereal Chem.*, **52**, 678.
Lloyd, W. J. W. (1978) *Brewer*, **64**(Mar), 84.
Lloyd, W. J. W. (1979) *MBAA Tech. Quart.*, **16**(2), 60.
Lloyd, W. J. W. (1982) *Proc. 1st Aviemore Conf. Curr. Devel. Malting, Brewing and Distilling* (Priest, F. G. and Campbell, I. eds), p. 157. IoB, London.
Lloyd, W. J. W. (1986) *J. Inst. Brewing*, **92**, 336.
Lloyd, W. J. W. (1988a) *Brewer*, **74**(April), 147.
Lloyd, W. J. W. (1988b) *J. Am. Soc. Brew. Chem.*, **46**, 8.
Lloyd, W. J. W. (1989) in *Pauls Brewing Room Book, 1989–91*, 82nd edn, p. 40. Pauls Malt Ltd, Ipswich, UK.
Lloyd, W. J. W. (1990) *Proc. 3rd Aviemore Conf. Malting, Brewing and Distilling* (Campbell, I. ed.), p. 48. IoB, London.
Lloyd, W. J. W. (1994) *J. Inst. Brewing*, **100**, 339.
Lockwood, J. F. (1945) *Flour Milling*. Northern Publishing, Liverpool, UK.
Loi, L., Barton, P. A. and Fincher, G. B. (1987) *J. Cereal Sci.*, **5**, 45.
Lolas, G. M., Palamidis, N. and Markakis, P. (1976) *Cereal Chem.*, **53**, 867.
Longstaff, M. A. and Bryce, J. H. (1993) *Plant Physiol.*, **101**, 881.
Lorenz, K. J. and Kulp, K. (eds) (1991) *Handbook of Cereal Science and Technology*. Marcel Dekker, New York.
Lubert, D. J. and Pool, A. A. (1964) *J. Inst. Brewing*, **70**, 145.
Lubert, D. J. and Thompson, D. L. (1978) *MBAA Tech. Quart.*, **15**(4), 236.
Lucas, A. (1962) *Ancient Egyptian Materials and Industries*, 4th edn, (revised Harris, J. R.), p. 20. Edward Arnold, London.
Luchsinger, W. W. and Fleckenstein, J. G. (1960) Int. Patent 605158.
Lüers, H. (1922) *J. Inst. Brewing*, **28**, 192 (abstr.).
Luff, G. (1898) *Z. ges. Brau.*, **21**, 6, 16, 27.
Luff, G. (1899) *Z. ges. Brau.*, **22**, 354.
Luff, G. (1900) *Z. ges. Brau.*, **23**, 381, 397.
Lundin, H. (1936) *J. Inst. Brewing*, **42**, 273.
Lutz, H. L. F. (1922) *Viticulture and Brewing in the Ancient Orient*. Hinrichsäsche Buchhanduing, Leipzig.
Lyons, T. P. (1983) *Bevrages*, Apr./May, 14.
Macartney, I. (1988) UK Patent Application GB2194795 A.
MacDonald, J., Reeve, P. T. V., Ruddlesden, J. D. and White, F. H. (1984) in *Progress in Industrial Microbiology*, Vol. 19 (Bushell, M. E. ed), p. 47. Elsevier, Amsterdam.
Macey, A. (1957) *J. Inst. Brewing*, **63**, 306.
Macey, A. (1958) *J. Inst. Brewing*, **64**, 222.
Macey, A. (1963) *Brewers' Guard.*, May, 41.
Macey, A. (1971) *Brewers' Guard.*, **100**(Sept) (Centenary issue), 89.
Macey, A. (1974) *EBC Monograph-I. EBC Wort Symp.* Zeist. p. 339.
Macey, A. (1977) *Brewers' Guard.*, **106**(9), 81.
Macey, A. (1980) *Brewers' Dig.*, July, p. 37.
Macey, A., Marsh, J. B. and Stowell, K. C. (1975) *Proc. 15th Congr., EBC*, Nice, p. 85.
Macey, A., Sole, S. M. and Stowell, K. C. (1969) *Proc. 12th Congr. EBC*, Interlaken, p. 121.
Macey, A. and Stowell, K. C. (1957a) *J. Inst. Brewing*, **63**, 391.

Macey, A. and Stowell, K. C. (1957b) *Proc. 6th Congr. EBC*, Copenhagen, p. 104.
Macey, A. and Stowell, K. C. (1959) *Proc. 7th Congr. EBC*, Rome, p. 105.
Macey, A. and Stowell, K. C. (1961) *Proc. 8th Congr. EBC*, Vienna, p. 85.
Macey, A., Stowell, K. C. and White, H. B. (1966) *J. Inst. Brewing*, **72**, 29.
Macey, A., Stowell, K. C. and White, H. B. (1967) *Proc. 11th Congr. EBC*, Madrid, p. 283.
MacFadden, D. P. (1989) *Proc. 2nd Sci. Tech. Conv. IoB* (Central and S. African Sect.), Johannesburg, p. 306.
MacFarlane, C., Lee, J. B. and Evans, M. B. (1973) *J. Inst. Brewing*, **79**, 202.
MacGregor, A. W., Ballance, G. M. and Dushnicky, L. (1989) *Food Microstructure*, **8**, 235.
MacGregor, A. W. and Bhatty, R. S. (eds) (1993) *Barley: Chemistry and Technology*. Am. Assoc. Cereal Chem., St Paul, MN.
MacGregor, A. W., Macri, L. J., Bazin, S. L. and Sadler, G. W. (1995) *Proc. 25th Congr. EBC*, Brussels, p. 185.
MacGregor, A. W. and Yin, X. S. (1990) *J. Am. Soc. Brew. Chem.*, **48**(2), 82.
MacLeod, A. M. (1953) *J. Inst. Brewing*, **59**, 462.
MacLeod, A. M. (1967) *J. Inst. Brewing*, **73**, 146.
MacLeod, A. M., Travis, D. C. and Wreay, D. G. (1953) *J. Inst. Brewing*, **59**, 154.
MacMasters, M. M., Hinton, J. J. C. and Bradbury, D. (1971) in *Wheat Chemistry and Technology*, 2nd edn (Pomeranz, Y. ed.), p. 51. Am. Assoc. Cereal Chem., St Paul, MN.
MacWilliam, I. C. (1963) *Brewers' J.*, **99**(Feb, Mar), 68, 127.
MacWilliam, I. C. (1965) *Brewers' Guard.*, **94**(7), 20.
MacWilliam, I. C. (1971) *J. Inst. Brewing*, **77**, 295.
MacWilliam, I. C. (1972) *J. Inst. Brewing*, **78**, 76.
MacWilliam, I. C., Hudson, J. R. and Whitear, A. L. (1963) *J. Inst. Brewing*, **69**, 303, 467.
McFarlane, W. D. and Sword, P. T. (1962) *J. Inst. Brewing*, **68**, 344.
McGinnis, R. A. (1971) *Sugar Beet Technology*, 2nd edn. Sugar Beet Development Foundation, Fort Collins, CO.
McGrath, R. M., Kaluza, W. Z., Daiber, K. H., van der Riet, W. B. and Glennie, C. W. (1982) *J. Agric. Food Chem.*, **30**, 450.
Maeda, I., Jimi, N., Taniguchi, H. and Nakamura, M. (1979) *Carbohydr. Res.*, **61**, 309.
MAFF (1966) *Farm Grain Drying and Storage*. Bulletin no. 149. HMSO, London.
MAFF (1983) *Grain Drying and Storage Booklets* No. 1 (2415), No. 2 (2416), No 3 (2417). HMSO, London.
Maiden, A. M. (1971) *Brewer*, **57**, 76.
Maiden, A. M. (1978) in *Pauls and Whites Brewing Room Book, 1978–9*, 77th edn. p. 64 Pauls and Whites, Ipswich, UK.
Malcolm, A. (1990) *Brew. Dist. Int.*, Feb., 22.
Malkin, L. S. (1982) *Proc. 1st Aviemore Conf. Curr. Devel. Malting, Brewing and Distilling* (Priest, F. G. and Campbell, I. eds), p. 171. IoB, London.
Malleshi, N. G. and Desikachar, H. S. R. (1979) *J. Food Sci. Technol. (India)*, **16**, 149.
Malleshi, N. G. and Desikachar, H. S. R. (1982) *J. Food Sci. Technol. (India)*, **19**, 193.
Malleshi, N. G. and Desikachar, H. S. R. (1986a) *J. Inst. Brewing*, **92**, 81.
Malleshi, N. G. and Desikachar, H. S. R. (1986b) *J. Inst. Brewing*, **92**, 174.
Malleshi, N. G. and Desikachar, H. S. R. (1986c) *Qual. Plant. Plant Foods Hum. Nutr.*, **36**, 191.
Malleshi, N. G., Desikachar, H. S. R. and Tharanathan, R. H. (1986a) *Stärke*, **38**(6), 202.
Malleshi, N. G., Desikachar, H. S. R. and Venkat Rao, S. (1986b) *Qual. Plant. Plant Foods Hum. Nutr.*, **36**, 223.
Mändl, B., Wullinger, F., Wagner, D. and Piendl, A. (1970) *Brauwissenschaft*, **23**(10), 372.
Mann, A. and Harlan, H. V. (1915) *Bull. 138*, US Department of Agriculture (reprinted 1916, *J. Inst. Brewing*, **22**, 73).
Margiloff, I. B., Reid, A. J. and O'Sullivan, T. J. (1981) *Am. Chem. Soc. Symp. Ser.*, **159**, 47.
Markham, Gervase (1615, reprinted 1986) *The English Housewife: The English Huswife, contayning, The inward and outward vertues which ought to be in a compleat woman* (Best, M. R. ed.), McGill-Queens University Press, Kingston.
Markham, W. A., Blanchflower, A. J., Wadeson, A. and Briggs, D. E. (1991) *J. Sci. Food Agric.*, **56**(2), 125.
Marsh, J. B. (1975) *Brewers' Guard.*, July, 21.
Marsh, J. B. (1982) *Brew. Dist. Int.*, July, 24.

Marshall, J. J. (1975) *Stäke*, **27**(11), 377.
Marth, E. H., Hussong, R. V., Cremers, L. F., Goth, J. H., Hilker, L. D., Jackson, H. W., Krett, O. J., Stimpson, E. G., Sullivan, R. A. and Tumerma, L. (1967) in *Kirk-Othmer's Encyclopedia of Chemical Technology*, Vol. 13, p. 564. Interscience, New York.
Martin, F. J. (1967) *Brewers' Guild J.*, **53**, 354.
Martin, F. W. (1984) *CRC Handbook of Tropical Food Crops*. CRC Press, Boca Raton, FL.
Martin, P. A. (1978) *Brewers' Guard.*, **107**, 29.
Mathias, P. (1959) *The Brewing Industry in England, 1700–1830*. Cambridge University Press, Cambridge, UK.
Matz, S. A. (1970) in *Cereal Technology* (Matz, S. A. ed.), p. 221. Avi Publishing, Westport, ME.
Matz, S. A. and Milner, M. (1951) *Cereal Chem.*, **28**, 196.
Mauclaire, D. (1977) *Proc. 16th Congr. EBC*, Amsterdam, p. 105.
Maule, A. P. (1989) *Proc. 2nd Conv. IoB* (Central and S. African Sect.), Johannesburg, p. 16.
Maule, A. P. (1995) *Fermant.*, **8**(4), 244.
Maule, A. P. (1996) *Proc. 24th Conv. IoB* (Asia – Pacific Sect.), p. 181.
Maule, A. P. and Greenshields, R. N. (1970) *Process Biochem.*, **5**(2), 39.
Maule, A. P., Northam, P. C., Pollock, J. R. A. and Pool, A. A. (1987) *Proc. 21st Congr. EBC*, Madrid, p. 257.
Maule, A. P., Trumpess, C. R. and Thurgood, S. D. (1996) *Proc. 6th Int. Brewing Technol. Conf.*, Harrogate, p. 408. Brewing Technology Services Ltd, London.
Meade, G. P. and Chen, J. C. P. (1977) *Sugar Cane Handbook*, 10th edn. John Wiley, New York.
Meredith, W. O. S. and Anderson, J. A. (1938) *Can. J. Res.*, **16C**, 497.
Meredith, W. O. S., Anderson, J. A. and Hudson, L. E. (1962) in *Barley and Malt; Biology, Biochemistry, Technology* (Cook, A. H. ed.), p. 206. Academic Press, London.
Meredith, W. O. S., Bettner, R. E. and Anderson, J. A. (1957) *Proc. 6th Congr. EBC*, Copenhagen, p. 92.
Merry, J. and Goddard, D. R. (1941) *Proc. Rochester Acad. Sci.*, **8**, 28.
Meyrath, A. (1966) *Process Biochem.*, **1**(4), 234.
Mikola, J. (1981) *Abhandl. Akad. Wiss. DDR, Abt. Math., Naturwiss. Techn.*, **5N**, 153.
Mikola, J. (1983) *Annu. Proc. Phytochem. Soc. Europe*, **20**, 35.
Mikola, J. and Enari, T.-M. (1970) *J. Inst. Brewing*, **76**, 182.
Mills, V. (1979) *Brew. Dist. Int.*, **9**(6), 33.
Milner, M., Christensen, C. M. and Geddes, W. F. (1947) *Cereal Chem.*, **24**, 182, 507.
Minch, M. J. (1960) *Proc. Annu. Meet. Am. Soc. Brew. Chem.*, p. 56.
Minch, M. J. G. (1976) *Brewers' Guard*, **105**(Jan), 31.
Minch, M. J. G. (1978) *MBAA Tech. Quart.*, **15**(1), 53.
Mirocha, C. J. and Christensen, C. M. (1982) in *Storage of Cereal Grains and Their Products*, 3rd edn (Christensen, C. M. ed.), p.241. Am. Assoc. Cereal Chem., St Paul, MN.
Moir, M. (1989) *Brewers' Guard.*, **118**(9), 64.
Moll, M. (1979a) *Brewing Science*, Vol. 1 (Pollock, J. R. A. ed.), p. 1. Academic Press, London.
Moll, M. (1979b) *Brewing Science*, Vol. 1 (Pollock, J. R. A. ed.), p. 539. Academic Press, London.
Moll, M. (1985) *Cerevisia*, **1**, 13.
Moll, M. and Duteurtre, B. (1976) *MBAA Tech. Quart.*, **13**, 26.
Moll, M., Flayeux, R., Mathieu, D. and Phan Tan Lun, R. (1982) *J. Inst. Brewing*, **88**, 139.
Moonen, J. H. E., Graveland, A. and Muts, G. C. J. (1987) *J. Inst. Brewing*, **93**, 125.
Morgan, A. G., Gill, A. A. and Smith, D. B. (1983) *J. Inst. Brewing*, **89**, 283.
Morgan, K. (1971) *J. Inst. Brewing*, **77**, 509.
Morrall, P. C. and Basson, A. B. K. (1989) *Proc. 2nd Conv. IoB*, (Central and S. African Sect.), Johannesburg, p. 56.
Morrall, P. C., Boyd, H. K., Taylor, J. R. N. and van der Walt, W. H. (1986) *J. Inst. Brewing*, **92**, 439.
Morrall, P. C. and Briggs, D. E. (1978) *Phytochemistry*, **17**, 1495.
Morrall, P. C., Lieberberg, N. van der W. and Glennie, C. W. (1982) *Scanning Electron Microscopy*, **III**, 571.
Morrison, W. R. (1978) *Advances in Cereal Science and Technology*, Vol. 2 (Pomeranz, Y. ed.), p. 221. Am. Assoc. Cereal Chem., St Paul, MN.
Morrison, W. R. (1988) *J. Cereal Sci.*, **8**, 1.

Morrison, W. R., Scott, D. C. and Karkalas, J. (1986) *Stärke*, **38**(11), 374.
Mosha, A. C. and Svanberg, U. (1983) *Food Nutr. Bull.*, **5**, 10.
Mulford, S. (1992) *Brew. Dist. Int.*, **22**(3), 10.
Muller, H. G. (1976) *Ghana J. Agric. Sci.*, **3**, 187.
Munck, L. (1993) *Ferment.*, **6**(2), 119.
Munck, L. and Lorenzen, K. (1977) *Proc. 16th Congr. EBC*, Amsterdam, p. 369.
Murphy, H. J. (1904) *J. Inst. Brewing*, **10**, 99.
Muspratt, S. (1855?) *Chemistry, Theoretical, Practical and Analytical as Applied and Relating to the Arts and Manufactures*, Vol. 1, p. 236. William Mackenzie, Glasgow.
Muts, G. C. J. and Pesman, L. (1986) *EBC Monograph-XI. EBC Symp. on Wort Production*, Maffliers, p. 25.
Mutter, P. A. (1981) *MBAA Tech. Quart.*, **18**(4), 180.
Narziss, L. (1966) *MBAA Tech. Quart.*, **3**(4), 237.
Narziss, L. (1969) *Proc. 12th Congr. EBC*, Interlaken, p. 77.
Narziss, L. (1975) *EBC Monograph-II. EBC Barley and Malting Symp.*, Zeist, p. 62.
Narziss, L. (1976) *Die Bierbrauerei*, Vol. 1, *Die Technologie der Malzbereitung*. Ferdinand Enke Verlag, Stuttgart.
Narziss, L. (1981a) *EBC Monograph-VI. EBC Symp, Relationship between Malt and Beer*, Helsinki, p. 99.
Narziss, L. (1981b) *Brauwelt*, **126**, 1406.
Narziss, L. (1987) *Proc. 21st Congr. EBC*, Madrid, p. 133.
Narziss, L. (1991a) *Brau. Int.*, **11**, 106.
Narziss, L. (1991b) *Brau. Int.*, **11**, 284.
Narziss, L. (1992) *Die Bierbrauerei*, Vol. 2, *Die Technologie der Würzebereitung*. Ferdinand Enke Verlag, Stuttgart.
Narziss, L. and Hunkel, L. (1968) *Brauwissenschaft*, **21**(3), 96.
Narziss, L. and Kieninger, H. (1967) *Brauwelt*, **107**, 1830.
Narziss, L., Kieninger, H. and Sailer, H. (1965) *Brauwissenschaft*, **18**(10), 390.
Narziss, L., Reicheneder, E., Dürr, P. and Eder, J. (1980) *Brauwissenschaft*, **33**(10), 253, **33**(11), 295.
Narziss, L., Reicheneder, E., Schwill-Miedener, A., Mück, E. and Freudenstein, L. (1988) *Monatss. Brau.*, **41**(4), 159.
Narziss, L., Reicheneder, E., Schwill-Miedener, A., Zinsburger, P. and Freudenstein, L. (1989) *Monatss. Brau*, **42**(8), 320.
Neve, R. A. (1991) *Hops*. Chapman & Hall, London.
NIAB (1966 or current) Detailed descriptions of varieties of wheat, barley, oats, rye and triticale. National Institute of Agricultural Botany, Cambridge, UK.
Nicholls, P. B. (1982) *Aust. J. Plant Physiol.*, **9**, 373.
Nielsen, E. B. (1971) *Proc. 13th Congr. EBC*, Estoril, p. 149.
Nielsen, N. (1937) *C. R. Trav. Lab. Carlsberg, Ser. Physiol.*, **22**, 49.
Niemeyer, H. M. (1988) *Phytochemistry*, **27**, 3349.
Nilan, R. A. and Ullrich, S. E. (1993) in *Barley: Chemistry and Technology* (MacGregor, A. W. and Bhatty, R. S. eds), p. 1. Am. Assoc. Cereal Chem., St Paul, MN.
Nordland, O. (1969) *Brewing and Beer Traditions in Norway*. Universitetsforlaget, Oslo.
Norris, K. and Lewis, M. J. (1985) *MBAA Tech. Quart.*, **22**, 154.
Northam, P. C. (1962) *Brewers' Guild J.*, **48**, 259, 296.
Northam, P. C. (1965a) *Brewers' Guard.*, May, 97.
Northam, P. C. (1965b) *Brewers' Guild J.*, **51**(May), 229.
Northam, P. C. (1971) *Brewers' Guard.*, **100**(Aug), 51.
Northam, P. C. (1985) *Proc. 20th Congr. EBC*, Helsinki, p. 635.
Northam, P. C. and Button, A. H. (1973) *Proc. 14th Congr. EBC*, Salzburg, p. 99.
Nout , M. J. R. (1980) *Chem. Mikrobiol. Technol. Lebensm.*, **6**, 137, 175.
Nout, M. J. R. and Davies, B. J. (1982) *J. Inst. Brewing*, **88**, 157.
Novellie, L. (1966) *Int. Brewer Dist.*, **1**(1), 27.
Novellie, L. (1968) *Wallerstein Labs. Commun.*, **31**, 17.
Novellie, L. (1977) in *Proc. Symp. Sorghum and Millets for Human Food*, Vienna, 1976 (Dendy, D. A. V. ed.), p. 73. Tropical Products Institute, London.
Novellie, L. (1981) in *Proc. Int. Symp. Sorghum Grain Quality* (Rooney, L. W., Murty, D. S. and Mertin, J. V. eds), pp. 113, 121. Icrisat Center, Patancheru, India.

Novellie, L. and de Schaepdrijver, P. (1986) in *Progress in Industrial Microbiology*, Vol. 23, *Microorganisms in the Production of Food* (Adams, M. R. ed.), p. 73. Elsevier, Amsterdam.
Nutbeam, A. R. and Briggs, D. E. (1982) *Phytochemistry*, **21**, 2217.
Nykänen, L. and Lehtonen, P. (eds) (1984) *Proc. Alko Symp. Flavour Res. Alcoholic Beverages*. Foundation for Biotechnical and Industrial Fermentation Research, Helsinki.
Nzelibe, H. C. and Nwasike, C. C. (1995) *J. Inst. Brewing*,**101**, 345.
Obretenov, T. D., Kuntcheva, M. J., Mautchev, S. C. and Valkova, G. D. (1991) *J. Food Chem.*, **15**, 279.
O'Farrell, D. D., Baxter, E. D. and Reeves, S. G. (1981) *J. Inst. Brewing*, **87**, 169.
Oh, S. Y. and Briggs, D. E. (1989) *J. Inst. Brewing*, **95**, 83.
Ohlmeyer, D. W. and Matz, S. A. (1970) in *Cereal Technology* (Matz, S. A. ed.), p. 173. Avi Publishing, Westport, ME.
Okafor, N. and Iwouno, J. (1990) *World J. Microbiol. Biotechnol.*, **6**, 187.
Okon, E. U. and Uwaifo, A. O. (1985) *Brewers' Dig.* Dec., 24.
Oksanen, J., Ahvenainen, J. and Home, S. (1985) *Proc. 20th Congr. EBC*, Helsinki, p. 419.
Oliver, W. B. (1994) *Brew. Dist. Int.*, **25**(10), 28.
Oliver, W. B. (1996) *Proc. 6th Int. Brewing Technol. Conf.*, Harrogate, p. 277. Brewing Technology Services Ltd, London.
Oloff, K. and Piendl, A. (1978) *Brewers' Dig.*, Mar., 39.
Opoku, A. R., Ohenhen, S. O. and Ejiofor, N. (1981) *J. Agric. Food Chem.*, **29**, 1247.
Østergård, K., Björk, I. and Vainionpää, J. (1989) *Food Chem.*, **34**, 215.
Osuntogun, B. A., Adewusi, S. R. A., Ogundiwin, J. and Nwasike, C. C. (1989) *Cereal Chem.*, **66**, 87.
Oxford, T. (1926) *J. Inst. Brewing*, **32**, 314.
Oxley, T. A. (1959) *Irish Maltsters' Conference – Selected Papers, 1952–1961*. p. 101. Arthur Guiness, Dublin.
Oxley, T. A. and Jones, J. D. (1944) *Nature* (Lond.), **154**, 826.
Pajunen, E. (1986) *EBC Monograph-XI. EBC Symp. Wort Production*, Maffliers, p.137.
Pal, A., Wagle, D. S. and Sheorain, V. S. (1976) *J. Food Sci. Technol.*, **13**, 75, 253.
Palamand, S. R., Hardwick, W. A. and Markl, K. S. (1969) *Proc. Annu. Meet. Am. Soc. Brew. Chem.*, p. 54.
Palmer, A. H. (1965) *Brewers' Guard.*, May, 17.
Palmer, G. H. (1969) *J. Inst. Brewing*, **75**, 536.
Palmer, G. H. (1974) *J. Inst. Brewing*, **80**, 13.
Palmer, G. H. (1975a) *J. Inst. Brewing*, **81**, 71.
Palmer, G. H. (1975b) *J. Inst. Brewing*, **81**, 408.
Palmer, G. H., Barrett, J. and Kirsop, B. H. (1972) *J. Inst. Brewing*, **78**, 81.
Palmer, G. H. and Bathgate, G. N. (1976) in *Advances in Cereal Science and Technology*, Vol. 1 (Pomeranz, Y. ed.), p. 237. Am. Assoc. Cereal Chem., St Paul, MN.
Palmer, G. H., Etokakpan, O. U. and Igyor, M. A. (1989) *MIRCEN J. Appl. Microb. Biotech.*, **5**, 265.
Palmer, T. J. (1987) *Proc. 21st Congr. EBC*, Madrid, p. 41.
Palmer, T. (1992) *Brewers' Guard.*, **121**(5), 26.
Papadopoulos, A. (1964) *Brewers' J.*, **100**, 617.
Parsons, N. E. and Marsh, J. B. (1981) *Proc. 18th Congr. EBC*, Copenhagen, p. 661.
Parsons, R., Wainwright, T. and White, F. H. (1977) *Proc. 16th Congr. EBC*, Amsterdam, p. 115.
Pathirana, R. A., Sivayogasundaram, K. and Jayatissa, P. M. (1983) *J. Food Sci. Technol.*, **20**(3), 108.
Paula, G. (1932) *Woch. Brau.*, **49**, 329, 339.
Patrick, A. (1996) *Indust. Archaeol. Rev.*, **18**(2), 180.
Pedersen, J. R. (1992) in *Storage of Cereal Grains and their Products*, 4th edn (Sauer, D. B. ed.), p. 435. Am. Assoc. Cereal Chem., St Paul, MN.
Percival, J. (1902) *Agricultural Botany*, 2nd edn. Duckworth, London.
Percival, J. (1921) *The Wheat Plant – a Monograph*. Duckworth, London.
Perez-Escamilla, R., Patino, H. and Lewis, M. J. (1988) *Brewers' Dig.*, Dec., 28.
Perissé, J., Adrian, J., Rerat, A. and le Berré, S. (1959) *Ann. Nutr. Aliment.*, **13**(1), 1.
Perry, R. H., Green, D. W. and Maloney, J. V. (eds) (1984) *Perry's Chemical Engineers Handbook*, 6th edn, Sections 12.3–12.6, 20.4–20.13. McGraw-Hill, New York.

Peterson, M. S. and Tressler, D. K. (1965) *Food Technology the World Over*, Vol. 2, *South America, Africa and the Middle East, Asia*, p. 130. Avi Publishing, Westport, ME.
Petit, P. (1899) *Brasserie et Malterie*. Gauthier-Villars, Paris.
Petters, H. I., Flanningan, B. and Austin, B. (1988) *J. Appl. Bact.*, **65**, 279.
Pfenninger, H. B., Schur, F. and Wieg, A. J. (1971) *Proc. 13th Congr. EBC*, Estoril, p. 171.
Philbrick, H. L. (1965) *Brewers' Guard.*, June, 39.
Piendl, A. (1971) *MBAA Tech. Quart.*, **8**(4), 196.
Piendl, A. (1975) *Brauwissenschaft*, **28**(6), 177.
Piendl, A. (1976) *MBAA Tech. Quart.*, **13**(2), 131.
Piendl, A. and Oloff, K. (1980) *Brauwissenschaft*, **33**(4), 85.
Piggott, J. R., Sharp, R. and Duncan, R. E. (eds) (1989) *The Science and Technology of Whiskies*. Longman, Harlow, UK.
Piglas, J., Coates, K. D. and Armitt, J. D. G. (1988) *Proc. 20th Conv. IoB* (Australia and New Zealand Sect.), Brisbane, p. 81.
Piratzky, W. (1936) *Woch. Brau.*, **53**, 105.
Piratzky, W. (1958) *Die Nahrung*, **2**, 239.
Piratzky, W. and Rehberg, R. (1935) *Woch. Brau.*, **52**, 89, 101.
Piratzky, W. and Wiecha, G. (1937) *Woch. Brau.*, **54**, 225.
Pitz, W. J. (1987) *J. Am. Soc. Brew. Chem.*, **45**, 53.
Pitz, W. J. (1990) *J. Am. Soc. Brew. Chem.*, **48**, 33.
Pitz, W. J. (1991) *J. Am. Soc. Brew. Chem.*, **49**, 119.
Pixton, S. W. (1982) *Trop. Stored Prod. Information*, **43**(1), 16.
Pixton, S. W. and Griffiths, H. J. (1971) *J. Stored Prod. Res.*, **7**, 133.
Pixton, S. W. and Henderson, S. (1981) *J. Stored Prod. Res.*, **17**(4), 191.
Pixton, S. W. and Warburton, S. (1968) *J. Stored Prod. Res.*, **4**, 261.
Pixton, S. W. and Warburton, S. (1971) *J. Stored Prod. Res.*, **6**, 283.
Platt, P. S. (1964) *Food Technol.*, **18**(May), 68.
Plevoets, Ch. (1974) *Fermentatio*, **70**(3), 103.
Plischke, R. and Beschorner, A. (1903) English Patent 28,264 (Dec. 23rd 1903).
Pollock, J. R. A. (1960) *J. Inst. Brewing*, **66**, 22.
Pollock, J. R. A. (1962a) in *Barley and Malt: Biology, Biochemistry and Technology* (Cook, A. H. ed.), p. 303. Academic Press, London.
Pollock, J. R. A. (1962b) in *Barley and Malt: Biology, Biochemistry and Technology* (Cook, A. H. ed.), p. 399. Academic Press, London.
Pollock, J. R. A. (1962c) *J. Inst. Brewing*, **68**, 308.
Pollock, J. R. A. (1981) *J. Inst. Brewing*, **87**, 356.
Pollock, J. R. A. (ed.) (1987, published in 1979, 1981 and 1987) *Brewing Science*, Vol. 1, p. 604; Vol. 2, p. 666; and Vol. 3, p. 611, respectively. Academic Press, London.
Pollock, J. R. A. (1989) *Proc. 2nd Conv., IoB* (Central and S. African Sect.), Johannesburg, p. 9.
Pollock, J. R. A., Essery, R. E. and Kirsop, B. H. (1955) *J. Inst. Brew.*, **61**, 295.
Pollock, J. R. A. and Pool, A. A. (1979) *J. Am. Soc. Brew. Chem.*, **37**(1), 38.
Pollock, J. R. A. and Pool, A. A. (1987) in *Cereals in a European Context* (Morton, I. D. ed.), p. 146. Ellis Horwood, Chichester.
Pollock, J. R. A. and Weir, M. J. (1976) *MBAA Tech. Quart.*, **13**, 22.
Pomeranz, Y. (1971) *Wallerstein Labs. Commun.*, **34**, 175.
Pomeranz, Y. (1974) in *Triticale: First Man-made Cereal* (Tsen, C. C. ed.), p. 262. Am. Assoc. Cereal Chem., St Paul, MN.
Pomeranz, Y. (1982) in *Storage of Cereal Grains and their Products*, 3rd edn (Christensen, C. M. ed.), p. 145. Am. Assoc. Cereal Chem., St Paul, MN.
Pomeranz, Y., Burkhart. B. A. and Moon, L. C. (1970) *Proc. Annu. Meet. Am. Soc. Brew. Chem.*, p. 40.
Pomeranz, Y. and Dikeman, E. (1976) *Brewers' Dig.*, July, 30.
Ponnet, J. (1969) *Bull. Assoc. Anc. Etud Louvain*, **65**(3), 133.
Ponton, I. D. and Briggs, D. E. (1969) *J. Inst. Brewing*, **75**, 383.
Pool, A. A. (1964a) *J. Inst. Brewing*, **70**, 221.
Pool, A. A. (1964b) *Proc. Irish Maltsters' Conference*, p. 71.
Pool, A. A. (1967a) *MBAA Tech. Quart.*, **4**(4), 217.

Pool, A. A. (1967b) *Brewers' Guild J.*, **57**(April), 188.
Pool, A. A. (1970) *Brewers' Dig.*, **45**(9), 84.
Pool, A. .A. (1978a) *J. Inst. Brewing*, **84**, 264.
Pool, A. A. (1978b) *Pauls and Whites Brewing Room Book, 1978–9*, 77th edn, p. 22. Pauls and Whites Ltd, Ipswich, UK.
Pool, A. A. and Pollock, J. R. A. (1962) *J. Inst. Brewing*, **68**, 201.
Pool, A. A. and O'Connor, T. N. (1963) *J. Inst. Brewing*, **69**, 382.
Pool, A. A. and Pollock, J. R. A. (1967) *Proc. 11th Congr. EBC*, Madrid, p. 241.
Porter, L. J., Hrstich, L. N. and Chan, B. G. (1986) *Phytochemistry*, **25**, 223.
Potter, R. T. (1953) *Chem. Ind.*, 787.
Pourmohseni, H., Ibenthal, W.-D., Machinek, R., Remberg, G. and Wray, V. (1993) *Phytochemistry*, **33**, 295.
Powell, G. E. and Barrell, G. W. (1974) *MBAA Tech. Quart.*, **11**, 244.
Préaux, G. and Lontie, R. (1975) in *The Chemistry and Biochemistry of Plant Proteins* (Harborne, J. B. and van Sumere, C. F. eds), p. 89, Academic Press, London.
Prentice, N., Dickson, A. D., Burkhart, B. A. and Standridge, N. N. (1963) *Cereal Chem.*, **40**, 208.
Prentice, N. and Sloey, W. (1960) *Proc. Annu. Meet. Am. Soc. Brew. Chem.*, p. 28.
Prescott, S. C. and Dunn, C. G. (1959) *Industrial Microbiology*, 3rd edn, p. 945. McGraw-Hill, New York.
Prior, G. (1996) *Brewer*, Mar., 117.
Quayle, J. P. (ed.) (1987) *Kempe's Engineers' Year-Book, 1987*, 92nd edn. Morgan-Grampion, London.
Radley, J. A. (ed.) (1968) *Starch and its Derivatives*. Chapman & Hall, London.
Ramakrishnan, C. V. (1979) *Baroda, J. Nutr.*, **6**, 1.
Ramsay, G. (1964) *Brewers' J.*, **100**, 786.
Ranki, H. (1990) *J. Inst. Brewing*, **96**, 307.
Rao, B. A. S. and Narasimham, V. V. L. (1976) *J. Food Sci. Technol.*, **13**, 119.
Rasmusson, D. C. (ed.) (1985) *Barley*. Am. Soc. Agron. and Crop Sci. Soc. Am., Madison, WI.
Ratnavathi, C. V. and Bala Ravi, S. (1991) *J. Cereal Sci.*, **14**, 287.
Raynes, J. G. and Briggs, D. E. (1985) *J. Cereal Sci.*, **3**, 55, 67.
Reed, R. J. R. (1990) *Ferment.*, **3**(6), 337.
Reed, R. M. (1965) *Brewers' Guard.*, June, 55.
Reeve, R. M. and Walker, H. G. (1969) *Cereal Chem.*, **46**, 227.
Reeves, S. G., O'Farrell, D. D. and Wainwright, T. (1980) *J. Inst. Brewing*, **86**, 226.
Reeves, S. G., Wainwright, T. and Bennett, H. O. (1979) *J. Inst. Brewing*, **85**, 31.
Reicheneder, E. and Narziss, L. (1989) *Brau. Int.*, **4**, 338, 345.
Reichert, R. D., Fleming, S. E. and Schwab, D. J. (1980) *J. Agric. Food Chem.*, **28**, 824.
Reiner, L. (1973) *Brauwissenschaft*, **26**, 4.
Reitzel, E. (1964) *Brewers' J.*, **100**(Mar), 146.
Renfrew, J. M. (1973) *Palaeoethnobotany*. Methuen, London.
Rennecke, D. and Sommer, W. (1979) *Lebensmittelindustrie*, **26**, 66.
Rennie, H. and Ball, K. (1979) *J. Inst. Brewing*, **85**, 247.
Reynolds, J. E. F. and Prasad, A. B. (eds) (1982) *Martindale, the Extra Pharmacopoeia*, 28th edn. Pharmaceutical Press, London.
Reynolds, T., Button, A. H. and MacWilliam, I. C. (1966) *J. Inst. Brewing*, **72**, 282.
Reynolds, T. and MacWilliam, I. C. (1966) *J. Inst. Brewing*, **72**, 166.
Reynolds, T., MacWilliam, I. C. and Griffiths, C. M. (1964) *Brewers' Guild J.*, **50**(Nov), 552.
Riffkin, H. L., Wilson, R. and Bringhurst, T. A. (1989) *J. Inst. Brewing*, **95**, 121.
Ringrose, D. W. (1979) *MBAA Tech. Quart.*, **16**(2), 68.
Riviere, M. V. B. (1961) *J. Inst. Brewing*, **67**, 55, 387.
Roberts, E. H. and Abdalla, F. H. (1968) *Ann. Bot.*, **32**, 97.
Roberts, R. H. and Rainbow, C. (1971) *MBAA Tech. Quart.*, **8**, 1.
Rockley, H. J. (1968) *Proc. Irish Maltsters' Tech. Meet.*, p. 69.
Rohrlich, M. and Brückner, G. (1966) *Das Getreide*, Vol. 1, *Das Getreide and Seine Verarbeitung*, 2nd edn. Paul Parey, Berlin.
Römheld, A. (1897) *Z. ges. Brau.*, **20**, 399.
Rosendal, I. and Madsen, N. B. (1982) *J. Inst. Brewing*, **88**, 23.

Ross-Mackenzie, J. (1927) *A Standard Manual of Brewing and Distilling and Laboratory Companion*, p. 4, 39. Crosby, Lockwood, London.
Ruf, H. W., Warrick, L. F. and Nichols, M. S. (1935) *Sewage Works J.*, **7**(3), 564.
Rumpf, H. and Kaiser, F. (1952) *Chem. Ingenieur Technik.*, **24**, 129.
Runkel, U.-D. (1975) *EBC Monograph-II. EBC Barley and Malting Symp.*, Zeist, p. 222.
Sacher, B., Ketterer, M. and Back, W. (1995) *Monatss. Brau.*, **48**(11/23), 370.
Sacher, B. and Narziss, L. (1991) *Brauwelt*, **131**, 2100.
Sacher, B. and Narziss, L. (1992) *Monatss. Brau.*, **45**(12), 404.
Samuel, D. (1996) *Science*, **273**, 488.
Samuel, D. and Bolt, P. (1995) *Brewers' Guard.*, **124**, 27.
Sandegren, E. and Beling, H. (1958) *Brau. Wiss. Beil.*, **12**, 231.
Sandegren, E. and Beling, H. (1959) *Proc. 7th Congr. EBC*, Rome, p. 278.
Sandstedt, R. M., Kneen, E. and Blish, M. J. (1939) *Cereal Chem.*, **16**, 712.
Sargeant, J. G. (1982) *Stärke*, **34**(3), 89.
Sauer, D. B., Meronuck, R. A. and Christensen, C. M. (1992) in *Storage of Cereal Grains and their Products*, 4th edn (Sauer, D. B. ed.), p. 313. Am. Assoc. Cereal Chem., St Paul, MN.
Saunders, J. (1905) *Brewers' J.*, pp. 398, 445, 525, 573, 631, 702, 766.
Saunders, J. (1906) *Brewers' J.*, pp. 47, 105, 166, 248, 312, 380, 444, 509, 568, 619.
Scamell, G. and Colyer, F. (1880) *Breweries and Maltings: their Arrangement, Construction, Machinery and Plant*, 2nd edn. E. and F. N. Spon, London.
Scherrer, A., Pfenninger, H. B., Kornicker, W. and Muller, P. W. (1973a) *Brauwissenschaft*, **26**(5), 129.
Scherrer, A., Pfenninger, H. B. and Wieg, A. J. (1973b) *Brauwissenschaft*, **26**(4), 101.
Schildbach, R. (1977) *Brewers' Dig.*, **52**(2), 42, 53.
Schildbach, R. (1977) *Monatss. Brau.*, **30**(9), 338.
Schildbach, R. and Burbridge, M. (1996) *Brau. Int.*, **14**(4), 368.
Schjerning, H. (1910) *C. R. Trav. Lab. Carlsberg*, **8**, 169.
Schlimme, G. (1972) *Brauwelt*, **112**, 1859.
Schlimme, G. S. (1985) *Brauindustrie*, **70**, 1835.
Schönfeld, F. (1932) *Handbuch der Brauerei und Mälzerei*, Vol. 2, *Das Mälzen*. Paul Parey, Berlin.
Schu, G. (1984) *Brau. Int.*, **1**, 48.
Schuster, I. (1985) *Brau. Int.*, **2**, 160.
Schuster, K. (1962) in *Barley and Malt: Biology, Biochemistry, Technology* (Cook, A. H. ed.), p. 271. Academic Press, London.
Schuster, K. and Grünewald, J. (1957) *Brauwelt*, **97**, 1446.
Schwarz, H. M. (1956) *J. Sci. Food Agric.*, **7**, 101.
Schwarz, P. B. and Horsley, R. D. (1995) *J. Am. Soc. Brew. Chem.*, **53**, 14.
Scott, L. (1951) *Antiquity*, **25**, 196.
Scott, R. W. (1972) *J. Inst. Brewing*, **78**, 411.
Scriban, R. (1963) *Brasserie* (No. 201), 178, 180.
Scriban, R. (1965) *Petit. J. Brass.*, **73**, 39, 393.
Scriban, R. (1970) *Ann. Nutr. Aliment.*, **24B**, 377.
Scriban, R. (1979) in *Brewing Science*, Vol. 1 (Pollock, J. R. A. ed.), p. 52. Academic Press, London.
Scriban, R. and Dancette, J.-P. (1979) *Bios*, **10**(4), 12.
Scriban, R. and Vazart, B. (1960) *Brasserie*, **15**, 344.
Scully, P. A. S. and Lloyd, M. J. (1965) *J. Inst. Brewing*, **71**, 156.
Scully, P. A. S. and Lloyd, M. J. (1967) *Brewers' Guild J.*, **53**, 29, 559.
Seidl, P. (1992) *Brau. Int.*, **3**, 247.
Serre, L. and Laurière, C. (1989) *Sci. Aliment.*, **9**, 645.
Sethi, V. B. and Bains, G. S. (1978) *J. Food Sci. Technol.*, **15**(2), 62.
Seward, B. J. (1986) in *Pauls Brewing Room Book, 1986–8*, 81st edn, p. 48. Pauls Malt Ltd, Ipswich, UK.
Seward, B. J. (1990) *Ferment.*, **3**(1), 50.
Seward, B. J. (1992) *Ferment.*, **5**(4), 275.
Sfat, M. R. (1966) *MBAA Tech. Quart.*, **3**, 22.
Shands, H. L., Dickson, A. D. and Dickson, J. G. (1942) *Cereal Chem.*, **19**, 471.

Shands, H. L., Dickson, A. D., Dickson, J. G. and Burkhart, B. A. (1941) *Cereal Chem.*, **18**, 370.
Shedd, C. K. (1953) *Agric. Eng.*, **34**, 616.
Sheneman, J. M. and Hollenbeck, C. M. (1960) *Proc. Annu. Meet. Am. Soc. Brew. Chem.*, p. 22.
Sheneman, J. M. and Hollenbeck, C. M. (1961) *Proc. Annu. Meet. Am. Soc. Brew. Chem.*, p. 93.
Sheorain, V. S. and Wagle, D. S. (1973) *J. Food Sci. Technol.*, **10**, 184.
Sheppard, G. (1986) in *Progress in Industrial Microbiology*, Vol. 23 (Adams, M. R. ed.), p. 237.
Shewry, P. R., Faulks, A. J., Parmar, S. and Miflin, B. J. (1980) *J. Inst. Brewing*, **86**, 138.
Shewry, P. R. and Miflin, B. J. (1985) in *Advances in Cereal Science and Technology*, Vol. 7 (Pomeranz, Y. ed.), p. 1. Am. Assoc. Cereal Chem., St. Paul, MN.
Shirk, H. G. (1942) *Am. J. Bot.*, **29**, 105.
Shirk, H. G. and Appleman, C. O. (1940) *Am. J. Bot.*, **27**, 613.
Siddappa, G. S. (1954) *J. Sci. Ind. Res.* (India), **13**, 33.
Silvey, V. (1986) *J. Nat. Inst. Agric. Bot.*, **17**, 155.
Simpson, A. C. (1966) *Process Biochem.*, **1**(7), 355, 365.
Simpson, J. R. (1970) *J. Water Pollution Control*, **1970**(part 2), 696.
Simpson, S. B. (1979) Br. Patent 1,557,549.
Sims, R. C. (1959) *J. Inst. Brewing*, **65**, 46.
Singer, C., Holmyard, E. J. and Hall, A. R. (eds) (1954) *A. History of Technology*, Vol. 1, p. 275. Clarendon Press, Oxford.
Singh, D. P. and Tauro, P. (1977) *J. Food Sci. Technol.(India)*, **14**, 255.
Singh, T. and Bains, G. S. (1984) *J. Agric. Food Chem.*, **32**, 346.
Singh, T., Harinder, K. and Bains, G. S. (1988) *J. Am. Soc. Brew. Chem.*, **46**, 1.
Singh, T., Maninder, K. and Bains, G. S. (1983) *J. Food Sci.*, **48**, 1135.
Singh, T. and Sosulski, F. W. (1985) *J. Food Sci.*, **50**, 342.
Sipi, M. I. and Briggs, D. E. (1968) *J. Inst. Brewing*, **74**, 444.
Sisler, W. W. (1987) *MBAA Tech. Quart.*, **24**, 55.
Slack, C. R., Hancock, D. A., Haslemore, R. M. and Tunnicliffe, C. G. (1986) *J. Inst. Brewing*, **92**, 262.
Slaughter, J. C. and Uvgard, A. R. A. (1971) *J. Inst. Brewing*, **77**, 446.
Sleeman, J. (1892) *Trans. Inst. Brewing*, **5**, 31.
Sleeman, J. (1908) *J. Inst. Brewing*, **14**, 566.
Sloey, W. and Prentice, N. (1962) *Proc. Annu. Meet. Am. Soc. Brew. Chem.*, p. 24.
Smart, J. G., Lukes, B. K., Tie, E. C. and Ford, A. T. (1993) *MBAA Tech. Quart.*, **30**(3), 80.
Smeaton, R. H. (1985) *Brewers' Guard.*, Aug., 20.
Smith, C. C. (1980) *MBAA Tech. Quart.*, **17**(1), 1.
Smith, C. C. and Maupoux, M. P. (1981) *MBAA Tech. Quart.*, **18**(4), 205.
Smith, D. B. and Lister, P. R. (1983) *J. Cereal Sci.*, **1**, 229.
Smith, J. B. (1977) *Brewer*, **63**(2), 50.
Smith, L. F., Linko, M., Enari, T.-M. and Dickson, A. D. (1964) *Proc. Annu. Meet. Am. Soc. Brew. Chem.*, p. 86.
Smith, L. J. (1986) *Water Sci. Tech.* (Singapore), **18**, 127.
Smith, M. T. and Briggs, D. E. (1979) *J. Inst. Brewing*, **85**, 160.
Smith, M. T. and Briggs, D. E. (1980a) *Phytochemistry*, **19**, 1025.
Smith, M. T. and Briggs, D. E. (1980b) *J. Inst. Brewing*, **86**, 161.
Smith, S. W. (1905) *Brewers' J.*, p. 759.
Snow, D., Crichton, M. H. G. and Wright, N. C. (1944) *Ann. Appl. Biol.*, **31**, 102, 111.
Sobral, J. (1989) *Proc. 16th Annu. Meet. Portuguese Brewers*, Vilamoura, p. 30.
Soderstrom, T. R., Hilu, K. W., Campbell, C. S. and Barkworth, M. E. (eds) (1988) *Grass Systematics and Evolution*. Smithsonian Institute Press, Washington, DC.
Søltoft, M. (1990) *Brygmesteren*, **47**(2), 30.
Sole, S. M. (1994) *J. Am. Soc. Brew. Chem.*, **52**, 76.
Sole, S. M., Stuart, D. J. and Scott, J. C. R. (1987) *Proc. 21st Congr. EBC*, Madrid, p. 243.
Sommer, G. (1975) *EBC Monograph-II. EBC Barley and Malting Symp.*, Zeist, p. 198.
Sommer, G. (1977a) *Monatss. Brau.*, **30**(8), 272.
Sommer, G. (1977b) *J. Am. Soc. Brew. Chem.*, **35**, 9.
Sommer, G. (1980) *Mitt. Versuchs station Gärungsgew.*, Vol. 7/8, p. 62.
Sommer, G. and Antelmann, H. (1967) *Brewers' Guild J.*, **53**(1), 7.

Sopanen, T., Takkinen, P., Mikola, J. and Enari, T.-M. (1980) *J. Inst. Brewing*, **86**, 211.
Sorensen, S. A. (1970) *Process Biochem.*, **5**(4), 60.
Sorensen, S. A. and Fullbrook, A. (1972) *MBAA Tech. Quart.*, **9**(4), 166.
South, J. B. (1991) *Proc. 23rd Congr. EBC*, Lisbon, p. 177.
South, J. B. (1992) *MBAA Tech. Quart.*, **29**, 20.
South, J. B. (1996) *J. Inst. Brewing*, **102**, 161.
Sparrow, D. H. B. (1964) *J. Inst. Brewing*, **70**, 514.
Sparrow, D. H. B. (1965) *J. Inst. Brewing*, **71**, 523.
Sparrow, D. H. B., Lance, R. C. M. and Cleary, J. G. (1986) *Barley Genet.*, **V**, 871.
Spillane, M. H. and Briggs, D. E. (1966) *J. Inst. Brewing*, **72**, 398.
Stars, A. C., South, J. B. and Smith, N. A. (1993) *Proc. 24th Congr. EBC*, Oslo, p. 103.
Stead, P. (1842) Br. Patent 9475.
Stentebjerg-Olesen, B. (1980) *Proc. 16th Conv. IoB* (Australia and New Zealand Sect.), Sydney, p. 127.
Stevens, B. (1993) *Brew. Dist. Int.*, **24**(9), 22.
Stevens, R. (1958) *J. Inst. Brewing*, **64**, 470.
Stewart, E. D. and Hahn, R. H. (1965) *Am. Brew.*, July, 21.
St Johnston, J. H. (1954) *J. Inst. Brewing*, **60**, 318.
Stoddart, W. E., Graesser, F. R. and Wesson, J. P. (1961) *Proc. 8th Congr. EBC*, Vienna, p. 105.
Stollberg, H., Gelhaart, R. A. and Baker, D. E. (1957) *Proc. Annu. Meet. Am. Soc. Brew. Chem.*, p. 98.
Stopes, H. (1885) *Malt and Malting*. F. W. Lyon, London.
Stowell, K. C. (1978) *Brewers' Guard*, May, 50.
Stowell, K. C. (1985) *Proc. 20th Congr. EBC*, Helsinki, p. 507.
Stuart, I. M., Loi, L. and Fincher, G. B. (1987) *J. Cereal Sci.*, **6**, 45.
Suhr, L. (1984) *Brau. Int.*, **1**, 45.
Sun, Z. and Henson, C. A. (1996) *Plant Physiol.*, **94**, 320.
Swain, E. F. (1976) *MBAA Tech. Quart.*, **13**, 108.
Swanston, J. S. (1990) *Adv. Malting Barley*, **5**, 31.
Sykes, W. J. (1897) *The Principles and Practice of Brewing*. Charles Griffin, London.
Tahara, S. (1969) US Patent 3,472,737.
Tahara, S. (1973) *Bull. Brew. Sci.*, **18**, 1.
Talve, I. (1960) *Bastu och Torkhus I Nordeuropa*, Vol. 53. Nordiska Museets Handlingar, Stockholm.
Taylor, A. G. (1979) *J. Inst. Brewing*, **85**, 168.
Taylor, B. R. and Blackett, G. A. (1982) *J. Sci. Food Agric.*, **33**, 133.
Taylor, J. R. N. (1989) *Proc. 2nd Sci. Tech. Conv. IoB* (Central and S. African Sect.), Johannesburg, p. 275.
Taylor, J. R. N. (1991) *Proc. 3rd Sci. Tech. Conv. IoB* (Central and S. African Sect.), Victoria Falls, p. 18.
Taylor, J. R. N. and Robbins, D. J. (1993) *J. Inst. Brewing*, **99**, 413.
Teng, J., Stubits, M. and Lin, E. (1983) *Proc. 19th Conv. EBC*, London, p. 47.
Teutonico, S. A. and Knorr, D. (1985) *Food Technol.*, Oct., 127.
Thausing, J. E., Brannt, W. T., Schwarz, A. and Bauer, A. H. (1882) *The Theory and Practice of the Preparation of Malt and the Fabrication of Beer*. Henry Carey Baird, Philadelphia, PA.
Thompson, R. G. and La Berge, D. E. (1977) *MBAA Tech. Quart.*, **14**(4), 238.
Thompson, R. M. (1950) *J. Inst. Brewing*, **61**, 305.
Thomson, T. (1849) *Brewing and Distillation*, with an addendum by Stewart, W. Adam and Charles Black, Edinburgh.
Thomson, T., Hope, T. C. and Coventry, A. (1806) *Report of the Experiments made, by direction of the Honorable Board of Excise in Scotland, to ascertain the relative qualities of Malt made from Barley and Scotch Bigg*. The House of Commons.
Thornton, J. (1990) *Food Sci. Technol. Today*, **4**(1), 9.
Tilton, E. W. and Brower, J. H. (1987) *Cereal Foods World*, **32**(4), 330.
Tizard, W. L. (1843) *The Theory and Practice of Brewing Illustrated: etc.* The Author, London.
Tombros, S. and Briggs, D. E. (1984) *J. Inst. Brewing*, **90**, 263.
Tottman, D. R., Makepeace, R. J. and Broad, H. (1979) *Ann. Appl. Biol.*, **93**, 221.
Tressl, R., Bahri, D. and Silwar, R. (1979) *Proc. 17th Congr. EBC*, Berlin (West), p. 27.

Tressl, R., Grünewald, K. G., Silwar, R. and Helak, B. (1981) *Proc. 18th Congr. EBC*, Copenhagen, p. 391.
Tressl, R., Kossa, T. and Renner, R. (1975) *Proc. 15th Congr. EBC*, Nice, p. 737.
Tschirner, M. (1986) *Brauindustrie*, **71**(23/24), 1660.
Tuerlinckx, G., Berckmans, D. and Goedseels, V. (1982) *Brauwissenschaft*, **35**(3), 70.
Tuerlinckx, G., Berckmans, D. and Goedseels, V. (1984) *Brewers' Guard*, June, 15.
Tuerlinckx, G. and Goedseels, V. (1979) *Proc. 17th Congr. EBC*, Berlin (West), p. 701.
Tuite, J. F. and Christensen, C. M. (1952) *Phytopathology*, **42**, 476.
Tuite, J. F. and Christensen, C. M. (1955) *Cereal Chem.*, **32**, 1.
Turcan, W. J. (1982) in *Pauls and Whites Brewing Room Book, 1982–3*, 79th edn, p. 53. Pauls and Whites Ltd, Ipswich.
Turner, B. K. (1986) *Food*, Sept., 63.
Turtle, E. E. and Freeman, J. A. (1967) in *Modern Cereal Chemistry*, 6th edn (Kent-Jones, D. W. and Amos, A. J. eds), p. 198. Food Trade Press, London.
Tusser, T. (1580, reprinted 1984) *Five Hundred Points of Good Husbandry*. University Press, Oxford.
Ullmann, F., Anderegg, P., Pfenninger, H. B. and Schur, E. (1987) *Proc. 21st Congr. EBC*, Madrid, p. 273.
Ulmer, R. L., Zytniak, R. and Hoskins, P. H. (1985) *J. Am. Soc. Brew. Chem.*, **43**, 10.
Ulsch, K. (1895) *Z. ges. Brau.*, **18**, 228.
Ure, A. (1860) *Ure's Dictionary of Arts, Manufactures and Mines*, 5th edn (Hunt, R. ed.), p. 18. Longmans, London.
Urion, E. and Barthel, Ch. (1961) *Brasserie*, **16**, 139.
Urion, E. and Chapon, L. (1955) *Proc. 5th Congr. EBC*, Baden-Baden, p. 172.
Valentine, G. (1920) *J. Inst. Brewing*, **26**, 573.
van den Berg, R., Muts, G. C. J., Drost, B. W. and Graveland, A. (1981) *Proc. 18th Congr. EBC*, Copenhagen, p. 461.
van den Beken, R., Devos, A. and Huygens, R. (1975) *Proc. 15th Congr. EBC*, Nice, p. 15.
van der Walt, J. P. (1956) *J. Sci. Food Agric.*, **7**, 105.
Van Eerde, P. (1983) *J. Inst. Brewing*, **89**, 195.
Van Heerden, I. V. (1989) *J. Inst. Brewing*, **95**, 17.
Van Heerden, I. V. and Glennie, C. W. (1987) *Nutr. Rep. Int.*, **35**(1), 147.
Van Oevelen, D. (1978) *Agricultura* (Heverlee), **26**, 353.
Van Oevelen, D., de l'Escaille, F. and Verachtert, H. (1976) *J. Inst. Brewing*, **82**, 322.
van Roey, G. and Hupe, J. (1955) *Proc. 5th Congr. EBC*, Baden-Baden, p. 158.
Vermeylen, J. (1962) *Traité de la Fabrication du malt et de la Bière*, Vol. I, II. Assoc. Roy. Anciens Elèves Inst. Sup. Fermentations, Gand, Belgium.
von Ugrimoff, A. (1933) *Die Mühle*, **70**, 603, 625 (in (1933) *J. Inst. Brew.*, **39**, 560).
von Wettstein, D., Jende-Strid, B., Ahrenst-Larsen, B. and Erdal, K. (1980) *MBAA Tech. Quart.*, **17**(1), 16.
von Wettstein, D., Nilan, R. A., Ahrenst-Larsen, B., Erdal, K., Ingversen, J., Jende-Strid, B., Kristiansen, K. N., Larsen, J., Outtrup, H. and Ullrich, S. E. (1985) *MBAA Tech. Quart.*, **22**, 41.
Vose, J. R. and Youngs, C. G. (1978) *Cereal Chem.*, **55**, 280.
Voss, H. and Piendl, A. (1978) *MBAA Tech. Quart.*, **15**, 215.
Vriens, L., van Soest, H. and Verachtert, H. (1990) *C.R.C. Crit. Rev. Biotechnol.*, **10**(1), 1.
Wackerbauer, K. and Toussaint, H.-J. (1984) *Monatss. Brau.*, **37**(8), 364.
Wade, A. and Reynolds, J. E. F. (eds) (1977) *Martindale. The Extra Pharmacopoeia*. Pharmaceutical Press, London.
Wahl, A. S. (1944) *Wahl Handybook of the American Brewing Industry*, Vol. 2, *Brewing Materials*. Wahl-Henius Institute, Chicago, IL.
Wahl, R. and Henius, M. (1908) *American Handy-Book of the Brewing, Malting and Auxiliary Trades*, 3rd edn, Vol. II, p. 795. Wahl-Henius Institute, Chicago, IL.
Wainwright, T. (1981) *J. Inst. Brewing*, **87**, 264.
Wainwright, T. (1995) *Brewers' Guard.*, **124**, 24.
Wainwright, T. and Buckee, G. K. (1977) *J. Inst. Brewing*, **83**, 325.
Wainwright, T., O'Farrell, D. D., Horgan, R. and Tempone, M. (1986) *J. Inst. Brewing*, **92**, 232.
Walker, E. W. (1986) *Proc. 2nd Aviemore Conf. Malting, Brewing and Distilling* (Campbell, I. and Priest, F. G. eds), p. 375. IoB, London.

Walker, E. W. (1990) *Proc. 3rd Aviemore Conf. Malting, Brewing and Distilling* (Campbell, I. ed.), p. 348. IoB, London.
Walker, J. F. (1981) in *Pauls and Whites Brewing Room Book, 1982–3*, 79th edn, p. 70. Pauls and Whites Ltd, Ipswich, UK.
Walker, R. B. (1978) in *Pauls and Whites Brewing Room Book, 1978–9*, 77th edn, p. 79. Pauls and Whites, Ltd, Ipswich, UK.
Wallace, W. and Lance, R. C. M. (1988) *J. Inst. Brewing*, **96**, 379.
Walton, C. (1995) *Brew. Dist. Int.*, **26**(3), 29.
Wang, Y.-Y. and Fields, M. L. (1978) *J. Food Sci.*, **43**, 1113.
Watson, B. (1987) *Food Manuf.*, Nov., 51.
Watson, S. A. (1987) in *Corn: Chemistry and Technology* (Watson, S. A. and Ramstad, P. E. eds), p. 53. Am. Assoc. Cereal Chem., St Paul, MN.
Watson, S. A. and Ramsted, P. E. (eds) (1987) *Corn Chemistry and Technology*, Am. Assoc. Cereal Chem., St Paul, MN.
Watson, T. G. and Novellie, L. (1976) *Agrochemophysica*, **8**, 61.
Weak, E. D., Miller, G. D., Farrell, E. P. and Watson, C. A. (1972) *Cereal Chem.*, **49**, 653.
Weast, R. C. (ed.) (1977) *CRC Handbook of Chemistry and Physics*, 58th edn. CRC Press, Cleveland, OH.
Webster, R. (1978) *Brewers' Guard.*, **107**(7), 51.
Weichherz, J. (1928) *Die Malzextract*. Julius Springer, Berlin.
Weith, L. (1960) *Brauwissenschaft*, **13**, 214, 262, 288.
Weith, L. (1967) *Proc. 11th Congr. EBC*, Madrid, p. 251.
Welch, R. W. (1995) *Oat Crop: Production and Utilization*. Chapman & Hall, London.
Wenzel, W. G. and Pretorius, A. J. (1994) *Angew. Bot.*, **68**, 143.
Weselake, R. J., MacGregor, A. W. and Hill, R. D. (1985) *J. Cereal Sci.*, **3**, 249.
West, C. J. and Sleight, G. B. (1994) *Proc. 4th Aviemore Conf. Malting, Brewing and Distilling* (Campbell, I. and Priest, F. G. eds), p. 209. IoB, London.
Whatling, A. J., Pasfield, J. and Briggs, D. E. (1968) *J. Inst. Brewing*, **74**, 525.
White, E. S. (1860) *The Maltsters' Guide; being a history of the Art of Malting from the Earliest Ages, etc.* W. R. Loftus, London.
White, F. H. (1977) *Brewers' Dig.*, **52**, 38.
White, J. (1966) *Process Biochem.*, **1**(3), 139, 155.
White, J. (1970) *Process Biochem.*, **5**(10), 54.
White, J. (1971) *Process Biochem.*, **6**(5), 21, 50.
White, W. (1844) *History, Gazetteer and Directory of Suffolk, 1844.* (reprinted 1970; David and Charles, Newton Abbot, p. 371) (Mr P. Stead's Maltings).
Whitmore, E. T. (1960) *J. Inst. Brewing*, **66**, 407.
Whitmore, E. T. (1963) *Barley Genetics I. Proc. 1st Int. Barley Genet. Symp.*, Wageningen, p. 332.
Whitmore, E. T. and Sparrow, D. H. B. (1957) *J. Inst. Brewing*, **63**, 397.
Wickens, D. (1973) *Brewers' Guard.*, **102**(11), 47.
Wiebe, G. A. and Reid, D. A. (1961) *US Dept. Agric. Tech. Bull.* No. 1224.
Wieg, A. J. (1987) in *Brewing Science*, Vol. 3 (Pollock, J. R. A. ed.), p. 533. Academic Press, London.
Wieg, A. J., Holló, J. and Varga, P. (1969) *Process Biochem.*, **4**(5), 33.
Wild, H. (1966) *Bull. Inst. Fran. Archaeol. Orientale*, **64**, 95.
Wild, J. (1910) *Z. ges. Brau.*, **32**, 321.
Wilderer, P. A. and Fedder, K. (1985) *Brau. Int.*, **1**, 100.
Wilkin, G. D. (1989) in *The Science and Technology of Whiskies* (Piggott, J. R., Sharp, R. and Duncan, R. E. B. eds), p. 64. Longman, Harlow, UK.
Williams, L. H. O. (1976) *MBAA Tech. Quart.*, **13**(4), 261.
Willmar, H. (1977) *Brauwissenschaft*, **30**(7), 209.
Willox, I. C., Rader, S. R., Riolo, J. M. and Stern, H. (1977) *MBAA Tech. Quart.*, **14**(2), 105.
Wilson, D. M. and Abramson, D. (1992) in *Storage of Cereal Grains and Their Products*, 4th edn (Sauer, D. B. ed.), p. 341. Am. Assoc. Cereal Chem., St Paul, MN.
Windisch, W. (1897) *Woch. Brau.*, **14**, 520.
Windisch, W. (1901) *Woch. Brau.*, **18**, 573.
Windisch, W. (1902) *Woch. Brau.*, **19**, 81, 93.

Windisch, W. and Kolbach, P. (1929) *Woch. Brau.*, **46**, 459.
Winthorp, Mr (1678) *Philos. Trans. R. Soc. London*, **12**, 1065.
Withey, J. S. and Briggs, D. E. (1966) *J. Inst. Brewing*, **72**, 474.
Witt, P. R. (1945) *Cereal Chem.*, **22**, 341.
Witt, P. R. (1959) in *The Chemistry and Technology of Cereals as Food and Feed* (Matz, S. A. ed.), p. 475, Avi, Westport, ME.
Witt, P. R. (1970a) in *Cereal Technology* (Matz, S. A. ed.), p. 129. Avi, Westport, ME.
Witt, P. R. (1970b) *Brewers' Dig.*, **45**(2), 48.
Witt, P. R. (1972) *MBAA Tech. Quart.*, **9**(2), 80.
Witt, P. R. and Admic, E. (1957) *Proc. Annu. Meet. Am. Soc. Brew. Chem.*, p. 37.
Woods, J. L., Favier, J. F. and Briggs, D. E. (1994) *J. Inst. Brewing*, **100**, 257.
Woodward, J. D. and Oliver, W. B. (1989) *Proc. 22nd Congr. EBC*, Zurich, p. 299.
Woodward, J. R., Fincher, G. B. and Stone, B. A. (1983) *Carbohydr. Polym.*, **3**, 207.
Wu, Y. Y. and Wall, J. S. (1980) *J. Agric. Food Chem.*, **28**, 455.
Wych, R. D. and Rasmusson, D. C. (1983) *Crop Sci.*, **23**, 1037.
Yamada, K. (1985) *Rep. Res. Lab. Kirin Brew. Co.*, **28**, 1.
Yamada, K. and Yoshida, T. (1976) *Rep. Res. Lab. Kirin Brew. Co.*, **19**, 25.
Yin, X. S. and MacGregor, A. W. (1988) *J. Inst. Brewing*, **94**, 327.
Yin, X. S. and MacGregor, A. W. (1989) *J. Inst. Brewing*, **95**, 105.
Yin, X. S., MacGregor, A. W. and Clear, R. M. (1989) *J. Inst. Brewing*, **95**, 195.
Yoshida, T. and Yamada, K. (1970) *J. Inst. Brewing*, **76**, 455.
Yoshida, T., Yamada, K., Fujino, S. and Koumegawa, J. (1979) *J. Am. Soc. Brew. Chem.*, **37**(2), 77.
Young, R. S. (1949) *J. Inst. Brewing*, **55**, 371.
Žila, V. V., Trkan, M. and Škvor, F. (1942) *Woch. Brau.*, **59**, 63.
Zimmerman, R. B. (1938) *Wallerstein Lab. Commun.*, **1**(4), 7.

Index